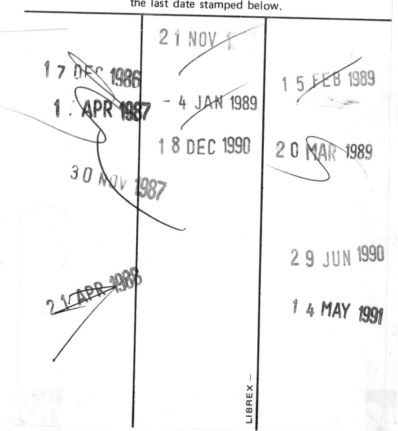

This book is to be returned on or before
the last date stamped below.

2 1 NOV

1 7 DEC 1986   1 5 FEB 1989

1   APR 1987   - 4 JAN 1989

1 8 DEC 1990   2 0 MAR 1989

3 0 NOV 1987

                                         2 9 JUN 1990

2 1 APR 1988   1 4 MAY 1991

LIBREX —

MINE VENTILATION

PROCEEDINGS OF THE 2ND U.S. MINE VENTILATION SYMPOSIUM/UNIVERSITY
OF NEVADA-RENO/23-25 SEPTEMBER 1985

Mine Ventilation

Edited by
PIERRE MOUSSET-JONES
Mackay School of Mines, University of Nevada-Reno, USA

*Sponsored by the Underground Ventilation Committee of SME-AIME, Joint Committee
of the Coal Division and Mining and Exploration Division & the Mackay School of Mines,
University of Nevada-Reno*

A.A.BALKEMA/ROTTERDAM/BOSTON/1985

COMMITTEES

Paper Committee

Madan M. Singh, President
Engineers International, Inc.
98 E. Naperville Road
Westmont, Illinois 60559

Y.J. Wang, Professor
College of Mineral Resources
West Virginia University
Morgantown, West Virginia 26506

Malcolm J. McPherson, Professor
Department of Mineral Engineering
University of California
Berkeley, California 94720

Don Trotter, Manager, Technical Services
Mines Accident Prevention Association
P.O. Box 1468
North Bay, Ontario, Canada P1B 8K6

Jack Barry, President
Peabody Fans
1622 Browning
Irvine, California 92714

James L. Banfield, Jr., Chief
Ventilation Mine Safety & Health Adm.
4800 Forbes Ave.
Pittsburgh, Pennsylvania 15213

Bruce Johnson
Magma Copper Company
P.O. Box 37
Superior, Arizona 85273

Wayne A. Sadik
Fluor Engineers, Inc.
D4-2, One Fluor Dr.
Sugarland, Texas 77487

Keith R. Notley, Professor
Queen's University
Mining Engineering Department
Kingston, Ontario, Canada K7L 3N6

J.W. Andrews
Mine Safety & Health Administration
P.O. Box 25367
Denver, Colorado 80225

Pierre Mousset-Jones, Associate Professor
Mackay School of Mines
University of Nevada-Reno
Reno, Nevada 89557

Jack Stevenson, General Manager
Jim Walter Resources, Inc.
P.O. Box C-79
Birmingham, Alabama 35283

Equipment Exhibit Committee

Michael D. Ashcraft, Supervisor
Hartzell Fan
P.O. Box 919
Pique, Ohio 45356

Richard Berg
Peabody Fans, Inc.
1622 Browning
Irvine, California 92714

Symposium Organizer

Pierre Mousset-Jones, Associate Professor
Mackay School of Mines
University of Nevada-Reno
Reno, Nevada 89557

*The texts of the various papers in this volume were set individually
by typists under the supervision of each of the authors concerned.*

ISBN 90 6191 611 9

Published by A.A.Balkema, P.O.Box 1675, 3000 BR Rotterdam, Netherlands

Distributed in USA & Canada by: A.A.Balkema Publishers, P.O.Box 230, Accord, MA 02018

Printed in the Netherlands

Foreword

The First Mine Ventilation Symposium was held in Tuscaloosa, Alabama and was organized by Dr. Howard Hartman. This was a most successful symposium with an excellent attendance, informative papers, equipment exhibits, and field trips to the Jim Walter Resources mines close to Tuscaloosa. The proceedings published by the Society of Mining Engineers have been well received by the Mine Ventilation professional community.

This success was a most satisfying experience for the organizing committee since it was the first such symposium organized by the newly formed Underground Ventilation Committee of the SME/AIME sponsored by the Coal Division and the Mining/Exploration Division. This committee was organized as a result of a need expressed by the SME members and others involved in Mine Ventilation, to bring some publicity to the Mine Ventilation Engineering profession in the U.S. and to promote the exchange of technical developments and research in this most important area of mining engineering. The committee is currently chaired by Mr. Bruce Johnson of the Magma Copper Company and next year the chairman will be Dr. Madan Singh of Engineers International, Inc. Those interested in the committee activities can contact these individuals, membership of SME/AIME is encouraged but not required.

This 2nd Mine Ventilation Symposium is the result of considerable effort on the part of the symposium committee all of whom are very active in the Mine Ventilation field. A total of 128 abstracts were received and much discussion took place before arriving at the final program for the symposium. Most of the sessions have familiar titles, however some new areas covered in this symposium are those of Recirculation, Nuclear Waste Site Ventilation and Microcomputers. The first topic is a result of the increasing use of Booster Fans in underground mines and the somewhat controversial application of controlled recirculation to maximize use of air and minimize ventilation costs. The second topic is due to the extensive program in the U.S. to find safe sites for the underground storage of nuclear waste. Adequate ventilation will be essential for these sites to operate and in particular to deal with the large amount of heat generated by the nuclear waste. The sites are indeed very similar to large mines, even though the "product" is being brought in rather than taken out of the mine. The final topic comes from the rapid development of microcomputers that can handle computer programs which in the past required the use of a large mainframe computer system. The result of this hardware development in the computer industry has meant that programs to solve many aspects of mine ventilation are now available for microcomputers with full graphics capability, and are thus much more accessible to the practicing mine ventilation engineer. The sixteen other sessions cover recent developments in the more typical areas of mine ventilation.

In addition to the main technical part of the symposium, an equipment exhibit, short courses and field trips to nearby Idaho and Utah are organized to give every opportunity for the participants to make the most of their time in the Reno area. Finally, a Teaching Workshop in Mine Ventilation will take place to enhance the importance and relevance of teaching this topic on the improved practice and recognition of mine ventilation engineering expertise in the mining industry. A variety of topics will be discussed during this workshop and it is hoped that the teaching profession will gain from this experience and thus ultimately the mining industry.

Mine Ventilation has come a long way since before the days of Agricola, which is the subject of the illustration on the symposium brochure. It is hoped that this symposium will further enhance this most important aspect of mining engineering, so that indeed the underground working environment will continue to be the best possible that technology, human ingenuity and expertise can provide.

Following on from the tradition set by the previous two symposiums, the 3rd U.S. Mine Ventilation Symposium will be held at the Pennsylvania State University. It is hoped that many of those attending this symposium will return to renew friendships and exchange technical information.

Pierre Mousset-Jones
Mackay School of Mines

Table of contents

4 Dust

5 Fans and shafts

6 Recirculation

7 Radon

11 Underground heat flow and climate

12 Microcomputers

13 Face ventilation I

14 Air cooling and refrigeration

18 *Face ventilation II*

19 *Monitoring*

1. Mine fires

Fire fighting expertise in French underground mines

C.E.FROGER
Centre d'Etudes et Recherches de Charbonnages de France (CERCHAR), Verneuil-en-Halatte, France

ABSTRACT : The use of nitrogen in the French underground mines has completely changed the methods used to fight open fires and to prevent or control spontaneous combustions. Some typical examples are given.
A permanent supply of nitrogen can prevent the development of spontaneous oxidation in seams liable to self heating and makes it possible to mine such seams.

With the large quantities of nitrogen now available on site it is possible to reduce the intensity of a fire and to extinguish it in a relatively short time. It is in general possible to come back to the workings in safe conditions a few hours or a few days after the beginning of the use of the inert gas, and restart working. These methods are most efficient when their use can be ensured rapidly after the start of the fire.

By the use of prevention techniques the number and seriousness of incidents are reduced and, if the control methods are well prepared in advance, fire fighting in mines can be carried out efficiently under safe conditions.

1 COAL MINING IN FRANCE

The production of coal in France comes from 3 coal basins whose deposits are very varied in many respects :
- coal quality which varies from anthracites to bituminous coals and lignites,
- steepness of coal seams which varies from 0 to 90°,
- seam thickness varies from 0.7 to more than 10 m and in some cases massive coal accumulations,
- depth varies from zero in the case of open pit mines to 1 250 m maximum,
- methods of mining, with or without stowing,
- and finally liability to self-heating, a characteristic which is nearly inexistent in some deposits but which can be very critical in others.

2 THE CONVENTIONAL APPROACH FOR CONTROLLING SPONTANEOUS COMBUSTION OF COAL

In French collieries, all conventional methods of prevention against self-heating of coal are applied :
- identification of the coal seams most liable to spontaneous combustion, often known through long experience,
- determination of the most favourable conditions likely to provoke coal self-heating, their elimination or at least some course of preventive action,
- manual or automatic monitoring of the most suspect mine areas, including the return airwuys of the working faces.

Methods of fighting against spontaneous combustions which occur despite the preventive measures applied, are also conventional :
- in some cases, direct extraction of the coal,
- in some cases, modification of the air flow in order to suppress or restrict the quantity of air feeding the fire,
- in most cases, obstruction of the leakages feeding air to the fire, by stowing or injecting either mineral or synthetic materials,
- finally, if all the above courses of action fail, construction of a stopping, as air-proof as possible, sealing off the area concerned which is often a working face, and possibly stowing of the area thus isolated.

Preventive measures are very cumbersome and the conventional way of fighting spontaneous combustion is generally long and difficult, sometimes dangerous and always very expensive.

3 CURRENT METHODS

The use of inert gas, generally large quantities of nitrogen, has considerably modified fire fighting methods as well as those for fighting and preventing spontaneous combustion.

We began treating gob self-heatings with nitrogen, and then fires but with greater flow rates and quantities of nitrogen. Amongst the numerous cases treated, we describe a selection of characteristic examples as follows :

3.1. Examples of gob heating

3.1.1 Systematic and continuous injection of nitrogen in the gob of an operating face. Rozelay Colliery

The first experiment made using this method was carried out at the Rozelay Colliery of Blanzy in the "Centre Midi" basin, on a retreating face with little methane emission and with a seam thickness of approximately 6 m and a slope of nearly 15° (fig 1). The face is 3 m high, caving the upper coal at the backside.

The face is 120 m long and its speed of advance is initially 1.30 m/day. The CO content in the return airway from the face, where the rate of air flow is approximately 9 m/s, is measured by a UNOR measuring unit installed in the Gardon gate (re- turn air-way) and is recorded on a monitoring screen (televigile) system at the surface.

Injection was carried out from the surface by a mobile installation comprising of a tank and an atmospheric evaporator both fixed on a truck.

The nitrogen gas was injected from the surface by means of the slurry flushing mains and it arrived at the main gate and tail gate of the face. Then it was injected to the gob by means of pipes which were abandoned in the mined area and which were displayed either in the gate

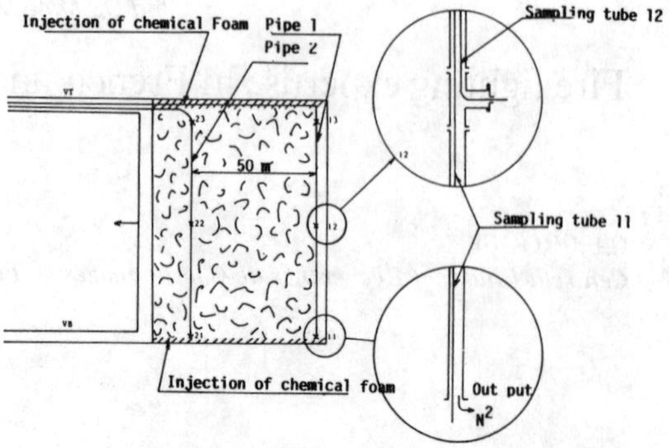

Fig 2. Principles of the prevention of fires in the gob of a retrading face caving coal at the backside

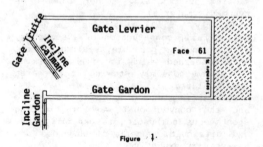

Figure 1.

or in the gob of the sublevel caving face (fig 2).

A system for controlling the nitrogen injection was put into service, based on the following principles :
- analysis of gas samples taken from the mined area by means of plastic tubes left inside the pipes used for injecting nitrogen gas (fig 3 and 4).
- results of temperature measurements obtained by thermal transducers in the gob of the face.

The face started on 15th june 1976, it had made an advance of 25 m when nitrogen injection was started on 13th june 1976 at an average flow rate of 100 m^3/hour. The control of the flow rate and selec-

Fig 3. Sampling tubes and thermocouple wires at the end of a steel pipe injecting nitrogen

Fig 4. Operator taking gas samples for analysis

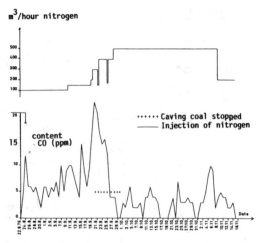

Fig 5. Rozelay colliery. Face S61 CO content in the return VS flow of nitrogen injected

tion of the mains effectively used for the injection depended on the readings supplied by the UNOR CO measuring unit in the return airway of the face and on the analysis of the atmosphere in the gob.

The graph in fig 5 indicates as a function of time, the CO content of the face return airway versus nitrogen flow rate used during the period betwen 23rd august 1976 and 16th november 1976.

In the first phase of the life of the face, that is from 13th july 1976 to 22nd august 1976, there was no cause for alarm and the nitrogen flow rate remained at 100 m^3/hour.

After 23rd august 1976, the following events were observed simultaneously :

1. An increased emission of CO in the return airway. This content reached 22 ppm on 20th september.

2. The non-return to zero of the CO contents of the face return airway on non-working days.

3. An 0^2 content which decreased only very slowly in the channel constituted by the abandoned gates and the initial face development heading and only slowly in the gob of the face.

The flow rate of nitrogen was then increased to the maximum value of 500 m^3/hour which was reached on 27th september. Moreover, caving coal at the backside of the face was discontinued from 20th september to 30th september corresponding to a distance of advance of 12.60 m of the face. This action was taken in order to create in the gob a barrier which would be less permeable than the one previously normally constituted by caved coal not drawn out from the gob and rocks fallen from the caved roof. Finally, after 20th september, in order to cope more efficiently with point 3 above, the blind ends of the main and tail gates were systematically sealed with synthetic foam injections.

The return to zero of the CO contents in the return airway was obtained after 29th september and this permitted the resuming of caving coal at the backside of the face whilst maintaining all the other measures i.e. nitrogen injection at 500 m^3/hour, injection of foam and the daily advance of the face was increased to 1.80 m/day.

The decrease of the 0^2 content in the gob became significant and stable only after the first week of november, at this moment the nitrogen injection was decreased to 200 m^3/hour. The rate of advance of the face was maintained at 1.80 m/day from then onwards and the sealing of gateends with foam was continued systematically.

In these conditions, the working of the face was able to be carried out to the

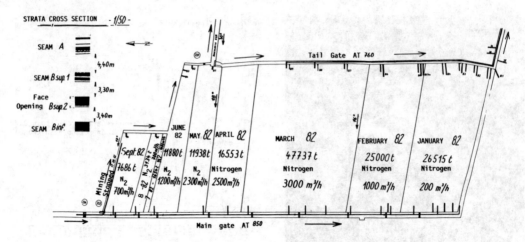

STRATA CROSS SECTION - 1/50 -

SEAM A

4,40m

SEAM B sup.1

3,30m

Face
Opening B sup.2

3,40m

SEAM B inf.

Tail Gate AT 760

Main gate AT 850

| JUNE 82 | MAY 82 | APRIL 82 | MARCH 82 | FEBRUARY 82 | JANUARY 82 |

Sept. 82 | 7686 t | N₂ 700m³/h | 8-82 N₂ 900m³/h | 7-82 N₂ 1200m³/h | 11880 t N₂ 1200m³/h | 11938 t N₂ 2300m³/h | 16553 t Nitrogen 2500m³/h | 47737 t Nitrogen 3000 m³/h | 25000 t Nitrogen 1000 m³/h | 26515 t Nitrogen 200 m³/h

Mining Stopped

Fig 6. The situation of face B2/2 - Ste Fontaine Colliery

end of the panel without any further spontaneous oxidization.

3.1.2 Control of CO emission in the gob of a working face (face B2/2). Sainte-Fontaine Colliery.

Situation : (fig 6).

The plan view presents the situation of the face : the initial layout was made up of the main-gate, the air intake and the face development heading (on the right of the plan). The face retreated on its main gate and advanced on its tail gate. Towards the end of the panel it was decreased in length due to faults.

At the roof of the seam, a bed which was part of the same seam was abandoned. Moreover there was another seam nearby in the roof.

During the entire mining of the panel, the pattern of ventilation was diagonal. The air flow rate was between 25 and 28 m³/s.

Control of CO :

A unit for monitoring the CO contents, "UNOR", monitored the return of face, whilst another controlled the air intake.

The tail gate, which was left open, was bordered by a gate end pack made with anhydrite. This zone of the face was traversed at regular intervals by :
- hydraulic stowing pipes of 150 mm diameter,
- measuring tubes for controlling CO, CH_4, O_2, and CO_2 in the atmosphere of the mined area. As a rule, the controls were made by taking samples which were analysed at the surface. One of the tubes was connected to

a "UNOR" unit and the signals produced were transmitted to the surface.

The main gate was isolated from the gob by an anydrite pack. It was sealed off at approximately every 30 m by a stopping made with anhydrite. 150 mm diameter pipes destined for the injection of nitrogen, penetrated into the gob area at regular intervals.

Control of CO by nitrogen injection :

A diagonal pattern of ventilation favours self-heating of the gob. This is why a control of CO was particularly necessary and appropriate measures for fire fighting, including nitrogen, were provided.

The CO content of the face return airway rapidly rose to 10-12 ppm. The CO content was much higher in the measurement tubes.

The level of nitrogen injection was adjusted to a flow such that the CO content remained stable at 10-12 ppm in the return airway of the face.

It can be noted that the nitrogen flow rate varied from 200 to 3000 m³/h during the entire duration of mining operations.

Due to the high cost of nitrogen, the mine operator always tries to limit the consumption of nitrogen to the minimum possible level, depending on the particular geographical conditions and on the rate of production of CO.

This panel was mined to completion without any development of spontaneous oxidization in the gob.

3.1.3 Conclusion

In coal seams liable to spontaneous combustion, the risk of fires development in the gob is very great.

Certain well-known dispositions, mining a retreating face for example, lessen this risk but are not unfailing.

Continuous injection of nitrogen in the gob considerably improves the prevention of spontaneous combustions, because this technique allows a reduction in the oxygen content sufficient to avoid the spontaneous oxidization of the residual coal in the worked seam, or of the coal in neighbouring seams not yet mined.

Monitoring of the oxygen content in the gob permits adjusting the flow of nitrogen injected to obtain the best results.

The point at which the nitrogen is injected is chosen so that the gas spreads throughout the whole volume being protected. The desired effect is obtained with limited amounts of nitrogen, and therefore at reduced cost, if the injection point (or points) is wisely chosen and if the precaution is taken to block all air leaks entering the gob as well as possible.

Safety of this method is guaranteed by the monitoring of the CO content, giving an alarm signal in the case where occurrence of new air leaks, bringing oxygen to certain gob areas, would provoke a beginning of spontaneous oxidization. In this case, the remedy consists of impro-ving the efficiency of the nitrogen injection by reviewing the choice of the injection points, increasing the rate of flow, and sealing of air leaks.

We have thus been able to avoid any loss at the mining face in coal formations prone to spontaneous combustion, that is to say, loss of :
- material,
- coal reserves,
- production potential.

3.2 Fire Examples

3.2.1 Inertization of a face in order to extinguish a seam fire. Seam F. Sainte-Fontaine Colliery

The panel of seam F/1 is situated between levels 850 and 930 (fig 7).

The face was retreating on its 2 gates.

Its tail gate is air-intake. The layout of this gate is irregular, in particular presenting a low point near the entry.

Geometrical characteristics
- length of the tail gate   : 350 m
- length of the main gate   : 402 m
- face length   : 245 m
- face height   :   3 m
- average slope of the face : 17°

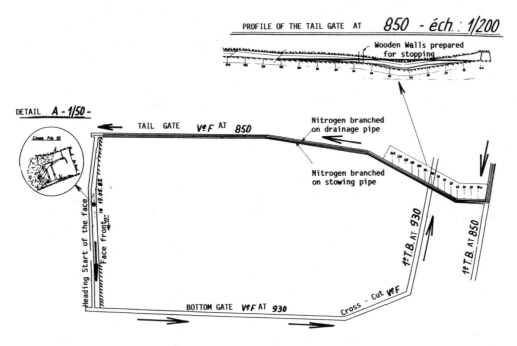

PROFILE OF THE TAIL GATE AT *850 - éch.: 1/200*

Fig 7. Face F 1 - Sainte Fontaine Colliery

7

Ventilation.

Ventilation was descending with a flow rate of 21 m^3/s. It must be noted that such a configuration is in principle favourable for decreasing the risk of self-heating because the gob area filled with methane is more stable and nearer to the face line.

The methane content in the intake airway was 0.2 % and 0.6 % in the return airway.

Control of CO was not yet ensured by a UNOR unit in the return gate because the face had only just started. However a UNOR was installed in a main roadway on the main return level and this permitted correct monitoring of CO in the face.

Situation of face operations on the day of occurrence of the incident (17th may 1982) :

The shearer had only just started as it had completed its 9th shear cut.

The face line was crossed by 3 fault zones, at 90 m, at 140 m and at the top of the face.

The section of the face between 90 and 140 m was disturbed by a buckling of the seam and consequently there were roof falls at front and above the canopies of the supports in this area. In the faulty zones, the presence of shaly sand-stone layers caused some difficulties for the shearer.

Brief description of the incident.

During the shearing of the 9th cut and in the faulty area at 140 m on the face line, the drum of the shearer was surronded by a yellowish light wich rapidly disappeared. Controls were immediately carried out but gave no evidence of any anomaly. It was 7.0 pm.

At 8.50 pm a support operator saw blue flames above the support canopies at about 90 m from the main entry of the face. The personnel was then mobilised in an attempt to extinguish these flames above the support canopies between 90 and 140 m on the face line. In particular, flames were burning at the front of the face in the fault zone situated at 140 m.

The extinction of the fire was only temporary and the flames started again.

Two sudden air blasts occurred, but there was no reversal of ventilation.

At 11.15 pm, CO contents were detected by the UNOR measuring units, in the 2 return airways.

At 0.10 am, the order of evacuation was given.

Fire fighting operations :

A light air barrier was built around 01.00 am in the tail gate of the face which was intake air-way.

Construction of a plaster stopping able to withstand an explosion was envisaged, but then abandoned in favour of a hydraulic seal at the low point of the tail gate. However to avoid the risk of an explosion at the moment of closing thereof, and to control the fire, inertization using nitrogen was adopted.

A pipe mains was therefore arranged for injecting nitrogen, this was the compressed air mains of the tail gate of the face, which was being used for this purpose and connected to the nitrogen injection network installed in a fixed position in the main cross measure roadway.

From then on, the sequence of events was the following :
- 02.18 am N_2 was injected at 3000 m^3/h
- 02.33 am N_2 was injected at 5000 m^3/h
- 02.43 am N_2 was injected at 7000 m^3/h
- 06.10 am N_2 was injected at 8000 m^3/h air flow to the face was 10 m^3/s
- 11.00 am N_2 was injected at 9000 m^3/s
- 12.00 am arrival of the equipment of the supplier of nitrogen of the neighbouring Sarrois coalfield with which there was a reciprocal assistance agreement.
- 2.00 pm N_2 was injected at 14500 m^3/h, the air intake to the face decreased to 7 m^3/h
- 3.00 pm N_2 at 17500 m^3/h
- 3.22 pm beginning of filling of the water seal in the tail gate
- 11.10 pm completion of the filling of the water seal. At this time there was only nitrogen in the area.

CO readings by the UNOR units dropped to 0 at 6.30 pm.

The filling up of the water seal had been conducted while the mine was evacuated. Nobody was exposed to risk.

On the following days several reconnaissance visits were made to the main gate of the face by miners equipped with oxygen breathing apparatus. These visits showed that :
- the face and its equipment had not suffered any damage,
- the blow of methane at 140 m of the face line had been the cause of the incident.

The flow rate of nitrogen was decreased to 7500 m^3/h on 19th may. On 23rd may, the water of the water seal was pumped out and mining operations were resumed on 24th may on the morning shift, that is one week after the fire.

None of the fire-fighting methods previously used would have made it possible to re-start work on the face after such a short delay.

3.2.2 Extinction of a conveyor fire in a metal mine.

In order to carry out extraction of a uranium mine an incline shaft was being sunk. A 700 m length had already been sunk with a slope of 25 %.

The incline was equipped with two conveyors :
- the first, had a reserve belt allowing the progression of the front to be followed,
- the second, in the upper part, had its driving head above ground. It had a length of about 650 m.

Ventilation was ensured by plastic and rigid ducts suspended above the conveyors.

The incident occurred at the beginning of a shift at around 5 o'clock in the morning. The weather was humid and a light drizzle was falling on the ground and on the unprotected conveyor driving head.

In the absence of the driving head foreman the motors were started by another employee while the rest of the personnel descended to take over from the nigth shift.

Soon afterwards a glow appeared at the driving head : this was the start of the fire. The driving drum rubbing against the belt had initiated inflammation of the rubber.

A by-stander stopped the motors and attempted to smother beginning of the fire using the powder extinguishers nearby, but without success.

Realizing his inability to stop the fire, the worker alerted the surface and underground mine personnel by telephone. It was around 5.15 am. At this point, the conveyor belt broke and the lower section, guided by its supports, fell downwards into to incline.

At around 5.30 am all the mine personnel had returned salefy to the surface. While passing near the flames activated by the mine ventilation, they attempted to extinguish them with the numerous extinguishers dispensed all along the inclined shaft. They also located the fire at about 70 m from the shaft opening. The belt had folded up and rested immobile on the lower framework of the conveyor, leaving a free passage way on the side.

Secondary ventilation was stopped (this could be the cause of an explosion if combustible gasses produced by the brumind material were ignited).

At this point, a thick black smoke was issuing in bursts from the mine shaft : heat was progressively increasing and convective currents were being set-up, with fresh air entering the incline at the lower part of its portal and hot smoke issuing from the upper part.

After reviewing the advantages and disadvantages of several possible methods of extinguishing the fire the following steps were taken :
- a barrier was constructed at the shaft entrance in order to reduce, if not eliminate, air introduction into the shaft : pipes and an air-duct were placed across the barrier,
- liquid CO_2 was introduced behind the barrier. This gas was chosen in place of nitrogen because of the characteristics of the location, in particular becaused of the steep slope. It was hoped that the high density CO_2 would rapidly replace the air in the shaft. Attempts were made to locate the quantity of CO_2 required to fill the shaft ($8000 \ m^3$). By 9.30 am a truck had been located with a CO_2 supply of 30 t ($10000 \ m^3$ under normal conditions). It was sent to the mine with qualified personnel.

Introduction of CO_2 began at 3.00 pm. During the operation, the air-duct across the barrier allowed the smoke to escape. The concentration of CO_2 in these gases only increased to over 2 % at around 5.30 pm at the end of the CO_2 introduction.

The following day a reconnaissance visit was made by a rescue team from the Cevennes Coal Basin who came specially equipped with breathing apparatus. They observed high concentrations of CO_2 and an absence of oxygen and smoke.

A second visit made it possible to observe the burnt piled-up belt which was still red-hot, and which had brought the surrounding metal parts also to a high temperature.

Water sprays were used to lower these temperatures and finish the extinguishing of the fire.

All that was left to do was to purge poisoned atmosphere, which was done with a double ventilation system. Thirty hours after the start of the fire the atmosphere, right to the bottom of the mine, had returned to normal.

The use of CO_2 to render inert the mine atmosphere thus allowed bringing the fire under control at low cost and without damage to equipment. It also made it possible to return the mine to normal conditions after only a brief delay.

Finally, and far from being the least important advantage of this method, all work carried out by workers was in safe conditions.

4 CONCLUSION

Fire-fighting has made considerable progress over the past few years, thanks to the use of inert gases, being mainly nitrogen but sometimes CO_2 :
- in coal deposits most prone to spontaneous self-heating the oxidization development can be limited without compromising mining by using a continuous low flow rate injection of nitrogen,
- in the event of a fire, the use of high nitrogen flow rates decreases explosion risk and allows bringing the fire rapidly under control ; damages are reduced, whilst the return to the mine face and the re-starting of work may take place after only a short lapse of time after the fire.

At the same time, safety has been considerably improved thanks to a better control of ventilation stability, and to materials and precautions adapted for individual protection.

In relation to this, we note :
- self contained self-rescuers,
- refuge chambers,
- calculation methods to evaluate ventilation stability and research for methods to prevent reversals,
- methods to evaluate the inflammability of underground atmospheres.

These different methods are the most efficient when they are put into effect in the shortest time possible after the discovery of the fire or self-heating.

For this reason, all our coal mines are equipped to enable the use of nitrogen after only a very short period of time. Everything necessary for the connecting and putting into action of a liquid nitrogen tank and of a portable evaporator transported by road convoy has been prepared and tested. Contracts guaranteeing delivery within fixed periods of time have been signed with liquid nitrogen suppliers. Moreover, the Collieries of the Lorraine Basin are connected to an industrial network of gaseous nitrogen distribution which is less expensive ($0.6 \ F/m^3$ instead of $1 \ F/m^3$).

If the number and seriousness of fires is reduced by a competent course of prevention, and if the necessary means for fighting against residual cases are prepared ahead of time, the fight against mine fires has the best chance of being carried out in good safety conditions and at low cost.

Acknowledgement.
The author wishes to thank his colleagues at the Charbonnages de France and the coal fields, for the information, help and advice gone towards the preparation of this paper, and especially :
Mr GARNIER, Chef du Service Securite de Charbonnages de France,
Mr CASADAMONT, Adjoint au Chef du Siege "La Houve" des Houilleres du Bassin de Lorraine,
Mr OLLIVIER, Chef du Service Securite du Bassin du Centre Midi,
Mr PETETIN, Chef du Siege "Vouters" des Houilleres du Bassin de Lorraine.

History of metal and nonmetal mine disasters and trends for a potential disaster

RALPH K.FOSTER
Mine Safety & Health Administration, Denver, CO, USA

ABSTRACT: The recommendations made by investigators following past metal and nonmetal disasters show much commonality. One recommendation which recurs is the need for ventilation control measures. Five metal and nonmetal mine fires and three explosion disasters are reviewed with the investigator's recommendations.

To avoid furture fire disasters, improvements have been made in eliminating much of the timber used in mines and particularly in the shafts. Research has produced more reliable systems for early fire detection and extinguishment.

The predominant ignition sources have changed over the past few years, with diesel-powered equipment components such as hot manifolds and brakes supplanting electrical equipment as the leading ignition source starting mine fires.

Training of people in evacuation, fire safety and ventilation control remains a critical factor in preventing disasters. Current ventilation programs are available to aid in preplanning and are extremely important in a rescue operation during a fire. Both on-site infrared and chromatograph instruments have proved very important in fire fighting and mine recovery operations during a fire or following an explosion.

In addition to the training and contests in which company and MSHA rescue crews participate, a mine emergency response program has been instituted by MSHA where a mine fire or explosion scenario is enacted with MSHA personnel playing the roles of various company, state and federal officals. This program has proved effective and has pointed out the need for organization and control by all parties so that cooperation can be achieved under disaster conditions.

1. HISTORY OF METAL & NONMETAL MINE DISASTERS

1.1 Introduction

Despite numerous underground fires in metal and nonmetal mines over the years, few have claimed lives; however, it is a misconception to believe large losses of life cannot occur. When a fire occurs at the right location under the right conditions, carbon monoxide concentrations can become lethal rapidly and a disaster claiming many lives can result. To illustrate the small amount of burning timber which can result in catastrophe, a fire that consumes 100 pounds of wood a minute (equivalent to a 12-in by 12-in by 3.7-ft piece of timber) will produce about 5,000 parts per million of carbon monoxide in a drift carrying 20,000 cfm of ventilation air. This 5,000 ppm of CO represents a concentration which can be lethal in 5 to 15 minutes.

Other disasters have been due to explosions, toxic strata gases, and hoist failures, but fire continues to be the principal threat.

1.2 Major Mine Fires

Records of metal and nonmetal mine fires are incomplete prior to 1978; however, there is information on disasters since the turn of the century. Because these disasters have much in common, they are worth reviewing because many of the recommendations are just as important today as they were at the time they were made.

On June 8, 1917, a fire in the Granite Mountain Shaft, Butte, Montana, caused the death of 163 men-the greatest number killed in any American metal mine disaster. The Granite Mountain Shaft served the North Butte mine and, with three other shafts, provided one of the best ventilation systems of the time.

Electric power was extensively used in the mine and was supplied at 2,300 volts to a substation on the 2600 level by a cable suspended in the Granite Mountain Shaft. A new lead-sheathed cable, 1,200 feet long, was being lowered when the cable slipped from its rigging and fell in the shaft. During the assessment of the damage from this accident, the oil-soaked hemp filler in the cable was ignited by a carbide light. This fire ignited the shaft timbers, causing smoke and combustion gases to spread rapidly throughout the mine and resulting in the heavy loss of life.

In 1922, a timbered shaft fire at the Argonaut Mine, Jackson, California, was responsible for 47 men being trapped and subsequently losing their lives. In this fire, the ignition source was thought to be electrical wiring.

Recommendations common to the Granite Mountain and Argonaut fires were:
1. Men should be warned immediately.
2. Mine fans should be readily reversible.
3. Doors should be hung and arranged to stay closed with air reversal.
4. Shaft and shaft stations should be fireproofed or fire protected.

In 1928 at the Hollinger Mine, Ontario, Canada, a fire claimed the lives of 39 men. The fire occurred in a stope where powder boxes and wrappers were thrown after frozen dynamite was thawed. The ignition source was not determined.

Recommendations made by the hearing judge included:
1. Men should be warned immediately in an understood language.
2. Combustible waste material should be removed from the mine.
3. Underground storage of oil and grease should be limited to one week's supply.
4. Fan reversal and doors for ventilation control should be available.

The worst fire disaster in this hemisphere occurred in 1945 at the Braden Mine in Chile, and claimed the lives of 355 people. At 7:45 a.m., the fire was discovered by a foreman when he looked into an underground blacksmith shop and saw a "raging furnace." The blacksmith shop was located along an intake adit leading to a shaft. At the time of the company's investigation, the ignition source was thought to have been sabotage; however, later information indicated a drum of oil had been placed on the forge and probably caught fire. The indications were that there was a rapid build-up of carbon monoxide concentration as some people perished within 35 to 50 minutes of the fire's discovery.

No recommendations were made in the com-

pany's investigation report; however, it is apparent that control of the amount of oil and combustibles along the intake and ventilation reversal would have helped control the rapid spread of the fire, carbon monoxide, and smoke.

In 1968, toxic fumes from a fire at the shaft station and in the shaft at the Belle Isle Mine in Louisiana claimed the lives of all 21 men underground. The exact ignition source was not determined, but it was thought to be an electric fault, an oxyacetylene torch or a frictional belt ignition. It is interesting to note that claims had been made that salt incrustation on timber prevents the spread of fire. Tests using a torch against the salt incrusted timber did appear to support this contention. However, during the fire, the incrustation did not prove effective against a very hot, long-lasting ignition source. The mine was ventilated using a single, plywood mid-walled shaft. This fire reemphasized the need for two escapeways from underground workings.

The mine disaster of which many here have intimate memories was the 1972 Sunshine Mine fire which claimed the lives of 91 miners. The fire occurred in a gob-filled stope, which, based on its location, probably contained trash along with timber supports. The ignition source was thought to be spontaneous combustion.

In the final investigation report, the reasons for the severity of the fire and for high casualties were:
1. Time elapsed in looking for the fire before ordering evacuation;
2. Burning through of a bulkhead separating the intake and return, causing recirculation from the pressure side of the two booster fans across the fire, back to the intakes which were the escape routes;
3. A single-channel telephone system leading to confusion on instructions;
4. The early loss of the chippy hoist and main hoist because of smoke and carbon monoxide; and
5. The lack of remote control to the booster fans.

There were 12 additional contributing factors, but these five factors probably represented the most important ones.

These disasters illustrate the need for good evacuation planning, good communications, early evacuation, preplanned ventilation control (particularly in the event of an intake fire), and control of combustible materials, particularly on intake air courses.

1.3 Explosions

Metal and nonmetal mines have been relatively free of explosions. Disasters include No. 1 Incline Mine, American Gilsonite Company, in Utah where 8 people were killed in a gilsonite explosion in 1953; the Cane Creek Mine explosion, also in Utah, where 21 men lost their lives in a methane explosion in 1963; and the Belle Isle Mine, where five men were killed in a methane explosion in 1979.

For those of you who are not familiar with gilsonite, it is a solidified hydrocarbon mined in Utah and its dust is extremely explosive. The dust explosion at the No. 1 Incline Mine, which killed eight miners, was initiated by a spark setting off a dust cloud at the collar of the inclined shaft with the explosion propagating down the shaft, starting a fire, and killing eight men. To prevent a recurrence of the ignition, no electrically powered equipment is used in the shafts or underground and better dust control is practiced. All blasting, when needed, is from the surface with no men underground.

The Cane Creek potash mine explosion occurred in a shop area and resulted from methane, liberated from a face blast near the shop, recirculating back into the shop area. The methane was ignited by an electric arc or spark, open torch flame, or hot metal surface. Twenty-five men were underground at the time; 18 died from flame, explosion forces, or asphyxiation. Barricading was done in two areas. Three men erected a barricade near a face and died behind it. At the other location, seven men erected a barricade in a drift; two of these men left the barricade and traveled to the shaft station where they were met by a rescue crew and brought to the surface about 19 hours after the explosion occurred. The other five men remained behind the barricade until a recovery crew contacted them and they reached the surface about 50 hours after the explosion.

Recommendations included (1) to operate the mine as a gassy mine, (2) to sink a second shaft for a secondary escapeway, and (3) to provide adequate ventilation to the face.

The Belle Isle explosion was the result of a high-pressure methane outburst following a face blast. The ignition source was thought to be an electric spark or burning insulation. The recommendations were (1) to operate the mine as a gassy mine, (2) to blast with all personnel on the surface, and (3) to locate the main fans on the surface.

2. RESULTANT ACTIVITY

As is true with most major disasters, the Sunshine disaster spurred interest in promoting safety in metal and nonmetal mines. Inspection funding was increased, safety standards committees appointed, and research money allocated to safety research. The research was carried out by the Bureau of Mines with MESA, later MSHA, overviewing the research projects. Immediately following the Sunshine disaster a great deal of concern was shown by the mining industry in upgrading their escape plans to insure a viable system for the escape of the underground work force if a fire occurred. With the passage of time, some of the urgency has waned. However, under the 1977 Act, the change of standards from advisory to mandatory has tended to keep the industry's attention on the danger of fire. The three fire standards involving mine ventilation which became mandatory required fire control doors at or near shaft stations and at underground shops, and required buildings within 100 feet of portals to be one-hour fire-resistant. These three standards have been rewritten and the new standards were effective in April 1985.

One of the recommendations from the Sunshine disaster was to find a reliable means of early fire detection. Extensive work on both sensor monitoring and tube bundles has been done, with both systems now far advanced from the state of the art 10 years ago. Sensors for early fire detection are currently installed at the Magma Superior mine and Sunshine mine.

For those who are not familiar with the tube bundles, the tube bundle technique has been used extensively in European coal mines and consists of a number of small, plastic tubes usually with about 1/4-in ID, pulling air samples from selected sites such as gob areas and returns. The collection tubes are bundled together and convey the gas samples to a surface or underground monitoring station. The samples from the individual tubes are analyzed for CH_4, CO, and CO_2, and O_2 by a chromatograph or infrared detectors. Particularly in larger mines, the tube bundle system installation and equipment involved is expensive so the application is limited to a long-term installation. As a result of a fire during December 1978 at the Bureau of Mines experimental oil shale mine at Horse Draw, Colorado, a tube bundle system was installed to collect data on normal methane emissions from the shale and emissions following the blast. A fire similar to the Horse Draw Mine fire occurred following a

Reported Underground Fire Incidents by Year

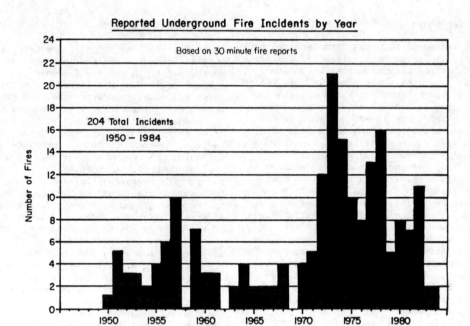

FIGURE 1.

Percent of Total Underground Fires
by Ignition Source

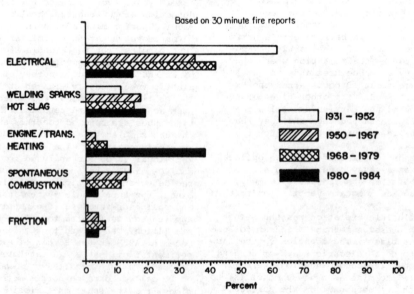

FIGURE 2.

14

pre-split blast in the oil shale strata at the White River Oil Shale mine near Vernal, Utah, December 5, 1983. Fortunately, the White River company personnel had discussed possible methane liberation and blasting problems prior to the fire, so they were aware of the problem and were able through air reversal to get to the fire for extinquishment with water. Monitoring was being considered in the long-range planning.

A fire in a timber-supported shaft serving as an intake air course or primary escapeway is one of the most life-threatening conditions. One of the first research projects in 1973 was to investigate the economics and advantages of a concrete and steel shaft vs. timber-supported shaft. Although many steel and concrete shafts have been sunk in the past 10 years, some mining companies have resisted the change to concrete-and-steel-supported shafts because of the fear of ground pressure. However, a concrete-and-steel-supported shaft has recently been completed in the Coeur d'Alene area in characteristically heavy ground. The sucess of the shaft should aid in the reduction of timber loading in future shafts.

To further protect miners against fires in timbered intake shafts, research work was done on better fire suppression by use of sprinkler systems. The water-ring deluge systems now used extensively have been shown to be limited in effectiveness because of poor water distribution. Fire door research to control the air and fire, particularly in large drifts, was done and has proved valuable in rewriting the standards, as well as providing operators with information for building fire doors on the job to meet their particular needs.

Decreased timber use in metal and nonmetal mines has been offset by the increased use of diesel equipment. Diesel fuel and oil storage, in relationship to ventilation and work areas, needs to be given a great deal of consideration.

Automatic fire suppression systems developed for surface installations for the early control of equipment fires have been adapted to underground diesel-powered units. Although the present systems are effective, additional development work may be needed to insure dependability in an adverse underground environment. Four years ago, research work was started on the possibilities of an oil-shale-dust explosion, but answers are still needed for blasting agents and procedures when methane is present in the oil shale.

Additional useful research projects have included:

1. Computer simulation software for escape

time and fire contaminant spread;
2. Large scale timber fire tests to establish fire spread in timbered drifts and the effect of air velocity on the fire spread;
3. Investigation into a better method of impregnating timber for fire retardancy; and
4. Prevention of spontaneous combustion of sulfide ores.

3. CHANGING IGNITION SOURCES

Early fires are tabulated in Bureau of Mines Information Circular 55. A Bureau of Mines research project to tabulate mine fires between 1950 and 1979 was completed by the Allen Corporation in 1980, and the report included both MSHA's and industry's information on these fires. MSHA's Health and Safety Analsis Center (HSAC) currently is recording data on mine fires reported by operators, collecting MSHA investigation reports covering mine fires, and providing up-to-date fire information.

Figure 1 (slide) shows the number of fires recorded and is based on the above sources of information. The reported incidents increased dramatically when P.L. 89-577 requiring the reporting of fires became effective in 1970. There have been a number of fires that have disrupted production while the fire was brought under control by sealing (Homestake, 1975, 13 days; Magma Superior Division, 1980, 5 days; Magma San Manuel, 1980, 16 days). Fortunately no fatalities occurred in any of these situations.

Figure 2 (slide) shows the change in ignition sources over the years. In 1923, Bureau of Mines Technical Paper 314 predicted that electrical-source ignition would replace carbide lights, cigarette, and candles as the principal ignition source. This 1923 prediction came true between 1931 and 1952 when 62 percent of the fires started from electrical ignition sources. The graph shows a steady decline of electrical ignitions which has to be the result of better electrical practices such as better separation of the electrical wiring from combustibles and better grounding protection. The percent of total ignitions by welding, sparks, and hot slag has stayed about the same over the years.

As would be expected with the increase of diesel-powered equipment use, fires ignited by hot manifolds, overheated brakes, and transmissions have increased dramatically. As the graph shows, the percentage of ignitions due to diesel-powered equipment increased from 7 percent in 1968 through 1979 to 38.5 percent in 1980 through 1984. In two instances where tires along with hydraulic oil become involved in equipment fires, the fire was allowed to burn itself out rather than endangering personnel in an attempt to extinguish the fire. When the oil shale mines become operational, this measure of letting the equipment burn itself out may not be possible because of the potential expansion of the fire into the oil shale. Consequently, special precautions need to be taken in preplanning for a fire in combustible ores. Ideally, extinguishment of the fire in the incipient stage is desirable. The research work done by the Bureau of Mines in semi-automatic dry chemical extinguishment on diesel equipment should prove extremely beneficial in acheiving this.

The 1973 fire which claimed the lives of two miners at the Lakeshore Mine, Casa Grande, Arizona, illustrates the rapid spread of fire and heat generated by an equipment fire. In that particular fire, a mud slide at the bottom of a borehole trapped two miners. An LHD was left running and subsequently caught fire, with the fire intensity being dramatically increased by a severed compressed air line blowing on the fire. The extreme heat caused the aluminum castings in the engine to melt. Following this incident, a great deal of work was done on automatic dry chemical fire suppression systems for LHDs and other diesel-powered equipment.

Many of the small shallow mines today are using inclines instead of shafts for ore haulage. Frequently, diesel-powered equipment is used. Some inclines are heavily timbered; consequently, these mines have to be given special attention, particularly since most inclined haulageways are on intake air.

A diesel equipment fire, particularly if timber is involved, could be disastrous unless immediate evacuation is instituted and ventilation control measures are quickly considered and acted upon. These considerations are essential in preplanning.

4. CONCLUSION

Because major fires are not a common occurrence, many mine managers and supervisory personnel have never experienced a mine fire, and the same applies to MSHA personnel. However, mining industry and MSHA personnel continue to receive training in how to cope with fire and explosion emergencies.

Preplanning for a fire or explosion emergency is the responsibility of management at each mine. They are at the scene where immediate, well-organized response to the emergency is essential. Trained rescue crews are required in mining areas, and many participate in contests which enhance their training and ability to perform under stressful conditions. Preplanning for emergency occurrences is equally important to MSHA in fulfilling its responsibilities, and the procedures are outlined in national and district emergency procedure manuals. An MSHA team of 12 persons, including the coordinator, J. D. Pitts, and trainer, Edmundo (Eddie) Archuleta, train in Denver four times a year. They can be flown to locations and assist in the underground recovery work along with setting up bench crews. Oxygen breathing apparatus is maintained for training purposes at locations within districts.

MSHA's Denver and Pittsburgh Technology Centers maintain portable chromatographs, infrared monitoring equipment, and trained personnel to aid inspection and mine personnel in interpreting the condition and control of the fire or explosive gases. The agency's Mine Emergency Operations (MEO) personnel have knowledge on measures and equipment available for rescue operations. These services are available for mine rescue or mine recovery endeavors. Once on the scene, MSHA's role is primarily that of advisor unless the mine management proves insensitive to the dangers involved in the recovery mission. It is essential that mine management and MSHA keep each other informed of their activities. Under emergency conditions, clear communication between management, rescue teams and MSHA is essential to the rescue operation.

To augment its trained rescue crew, MSHA has instituted a Mine Emergency Response program to train its supervisors, from district managers down through supervisory inspectors. Participants are assigned roles of company, state and federal personnel in a simulated emergency situation. A fire or explosion seminar also is presented to these participants. These training sessions, like the rescue crew contests, have proved invaluable in establishing the need for organization and cooperation between the company, union, state, and federal personnel on a fire or explosion scene. By having the command center and rescue crews in separate rooms the simulated emergency experience also points out how easy it is to lose or misunderstand communications between the rescue people and command center. The control of ventilation and interchange of correct information remains one of the most important facets of any recovery operation. A similar type of training exercise, with mining company management and staff participating, might be highly beneficial in preparing for a possible disaster at a mine.

5. REFERENCES

Harrington, D., Pickard, B., and Wolfin, H., Metal-Mine Fires, Bureau of Mines, Technical Paper 314, 1923, pp 9 and 10.

Harrington, D., Fire and Ventilation Doors, Bureau of Mines, Report of Investigation 2426, 1923, pp 5.

Harrington, D., Metal-Mine Fires and Ventilation, Bureau of Mines, IC 6678, 1933, pp 31.

Anonymous, An Annotated Bibliography of Metal and Nonmetal Mine Fire Reports, Allen Corporation of America, Final Report, Bureau of Mines Contract J 0295635, Dec. 1980, pp 381.

Anonymous, Fires, Gases and Ventilation in Metal and Nonmetal Mines, Bureau of Mines, Miners Circular 55, 1955, pp 120.

Jarret, S., et al., Final Report of Major Mine Disaster, Sunshine Mine, Bureau of Mines, 1970, pp 175.

Browne, H. F., et al., Final Report on Major Fire Disaster, Belle Isle Salt Mine, Bureau of Mines 1968, pp 47.

Anonymous, Hollinger Mine Fire, Canadian Mining Journal, June 8, 1928, pp 464-481.

Anonymous Braden Mine Fire, Bureau of Mines, Memorandum June 19, 1945, pp 4.

Plimpton, H. G., et al., Final Report of Mine Explosion Disaster, Belle Isle Mine, June 8, 1979, pp 135.

Westfield, J., et al., Final Report at Major Potash-Explosion Disaster, Cane Creek, Bureau of Mines, 1963.

Development of ventilation software on personal computers in France and the application to the simulation of mine fires

E.P.DÉLIAC & G.CHOROSZ
Centre Mines Infrastructures, Ecole Nationale Supérieure des Mines de Paris, Fontainebleau, France

N.D'ALBRAND
Centre d'Etudes et Recherches des Charbonnages (CERCHAR), Verneuil-en-Halatte, France

ABSTRACT : The use of professional micro-computers is expanding fast in the mining industry, it was therefore decided, between the CHARBONNAGES de FRANCE'S CERCHAR and the ECOLE des MINES de PARIS-CMI, to create some appropriate software - mainly based on the IBM Personal Computer to calculate ventilation networks and to simulate the effects of mine fires on their overall stability.

Because nothing was available in France on the IBM-PC, a first program, called P.C. VENT, was designed. In the same time another software, VENDIS, was created for DEC micro-computers with a special emphasis on interactive graphics and is now available on the IBM-PC. P.C. VENT (like VENDIS) is a mine ventilation analysis software, of the "user friendly" type, calculating pressure drops and air flows in a network consisting of up to 2 000 branches. Furthermore, it can accept fixed air flows (for leakages instead of resistances for instance), takes into account the effect of dry bulb temperatures and uses an accurate representation of fan characteristic curves. As it stands it is already a powerful tool for the ventilation engineer and has been used to calculate the overall network of a French colliery (consisting of 120 nodes and 200 branches) within one minute.

The P.C. FIRE program works as an additional module to P.C.VENT. It is based on the internationally well known equations derived by Simode and Vielledent to calculate the effects of a mine fire in a tunnel on the temperature, the mass-flow and the pressure variations. In situations whereby evaporation can be neglected and the fans are not stopped, it has been found to converge in most cases and the first results seem very promissive. However research is under way, partly to ensure systematic convergence, and P.C. FIRE should be considered as a tool for training or preventive purposes rather than for making decisions when fighting against a fire.

INTRODUCTION

Recent developments of professional micro-computers (like the IBM PC range) are currently generating a deep change in the production or design engineer's work. Among the reasons for this, some factors can be highlighted :

- machines becoming easy and friendly to use for non computer trained engineers
- power increasing fast (up to one million bytes central memory, several tens of megabytes storage memory)
- interactive graphics and even artificial intelligence becoming available

- strong effort within the universities to train the undergraduate students and develop software
- low costs (5000 to 10000 US dollars for the machines dealt with in this paper).

This is why the mining industry, like the other sectors, is making a fast increasing use of micro-computers and the needs of the French mining and quarrying profession have been clearly shown in a recent survey (Déliac, 1985). Among those needs, there appeared to be a strong interest for ventilation related problems solved by micro-computer software, the presently available one being run on expensive machines usually installed in head offices whereas the mining engineer would rather have a self sufficient computer on site, possibly connected onto the main frame.

As a result of this, the Charbonnages de France's research organization, CERCHAR, and the Ecole des Mines de Paris (mining department) have set a long term cooperation program aimed at building ventilation software which could be extended from network analysis to other problems such as mine fires. The machines chosen were mostly the IBM PC-XT, but also the DEC PDP/11 and the BULL MICRAL 30, because they were to become international standards. The former one has the advantage of the 8087 Intel arithmetic co-processor, very efficient in speeding up the calculations. All programs have been written in FORTRAN rather than BASIC because it is better structured and well adapted to lengthy calculations. As there was nothing available in France on the machines described previously, the first part of the work has been focused on building ventilation analysis software, and this will be the main scope of this paper, mostly concerned with the P.C. VENT program written for the IBM PC-XT (Chorosz, 1984). Because of

the interest of the results, in terms of calculation time particularly, and because of the well acepted modelling of the effect of mine fires on ventilation networks derived by Simode (1981) and Vielledent (1982) , it has been decided, as a second part of the research project, to integrate this model as an addition to P.C. VENT, called P.C. FIRE, which is detailed in the second section of the paper.

VENTILATION ANALYSIS SOFTWARE

General Comments
Ventilation software has been developed for approximately twenty years and the theory is now well documented (SIM-N3, 1976; Hartman et al., 1982). It will therefore not be dealt with in this paper. Suffice to say that all algorithms consist of two "modules" : a meshing one and an iterative network calculation one. The programs described here use the Sollin methodfor the mesh constitution and the Hardy Cross network calculation technique. The main differences between them consist in the data files management and the output of results (which can influence the calculation time). The general algorithm was developed by Mr. Gunther (consultant to CERCHAR) during the 70's, for main frame calculations.

P.C. VENT

Simplified Organization of the Program : it is illustrated on figure 1. After a pass-word, the program gives a choice of three unit systems (U.S., French, international). A main menu, shown on figure 2 is then displayed.

Input of data : to create a new file or to retrieve an existing file for possible modifications is possible through options 1 to 4

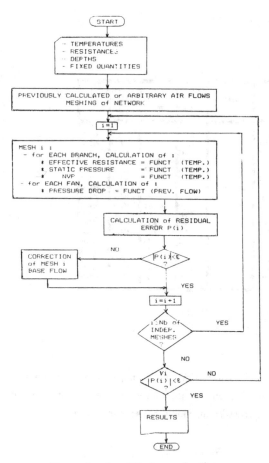

Figure 1 : simplified organization
of P.C. VENT

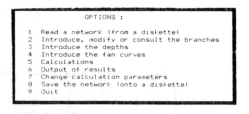

Figure 2 : main menu of P.C. VENT

of the main menu. At first the network is built, by entering the various branches (option 2). In its present version, The program can accept up to 2000 branches, and could easily be up-graded. The information required by the computer is basically the starting and ending nodes of the branch, the resistance and the mean dry bulb temperature. The latter is not compulsory but, if provided, P.C. VENT will account for it when computing the natural ventilation pressure (NVP). An original feature of the program is also the possibility of imposing fixed quantities of air flow; this can occur under two circumstances : ensure a fixed quantity to mine personnel in a working area or simulate a branch where the resistance is difficult to assess. In the latter case, ventilation doors for instance, it is sometimes impossible to determine the resistance, but the leakage of air is easy to measure; P.C. VENT will then accept the value of the airflow as a fixed quantity and any ordinary value can be given to the resistance : the actual high value will be determined by the program through the calculation. It must be pointed out, however, that the number of fixed quantities should not be too high (in practise, up to 7% of the total number of branches), in order to ensure the convergence of the calculation, and that it is not possible to fix the air flow through a fan. When the network is entered, it is necessary to input the depths of the nodes , which is done through option 3. Finally P.C. VENT needs the fans characteristics; it can accept up to ten different types of fans and allows for a fairly accurate representation of their curves since they are given by ten points chosen arbitrarily by the user except for the two ends (zero flow and zero pressure drop) and since the interpolation between two points is quadratic; the stall area, however, is not cared for as the curve must be continuously decreasing when the air quantity increases.

21

US Unit System

Ending nodes Numbers	Nominal Resistance 10⁻¹⁰ in.min²/ft⁶	Actual Factor	T° °F	Flow cf/min	Head Loss in.W.G.	Remarks	N V P in.W.G.	Power HP	Heads at ends in.W.G.	
1- 74	.31	.32	57	243828.	.0546		3.260	2.10	407.00	410.21
1-110	.31	.32	57	162223.	.0241		6.447	.61	407.00	413.43
1-112	.31	.32	57	22712.	.0005		6.447	.00	407.00	413.45
1-164	11167.27	12088.51	66	16836.	9.6981		.000	25.72	407.00	397.30
1- 13	321.98	335.45	62	16503.	.2586		16.234	.67	407.00	422.98
1-148	5.88	6.07	57	56436.	.0547		15.857	.49	407.00	422.80
1-167				-51914.	.8681	Fan N° : 3	-7.10		407.00	406.13
1-163	3.10	3.26	66	-42178.	-.0164		11.888	.11	407.00	418.91
1-164				-424445.	9.6980	Fan N° : 1		-648.41	407.00	397.30
10- 74	.31	.32	58	-243828.	-.0544		-2.637	2.09	412.79	410.21
10- 12	.93	.97	60	243828.	.1632		.000	6.27	412.79	412.63
12- 16	1.24	1.26	62	244054.	.2126		23.959	8.17	412.63	436.37
12- 13	14.24	14.79	64	-226.	.0000		10.352	.00	412.63	422.98
13- 38	71.21	73.04	64	16277.	.0548		.000	.14	422.98	422.92
15- 16	.31	.30	60	-166068.	-.0236		-8.860	.62	445.21	436.37
15- 17	.31	.30	58	48189.	.0020		1.217	.01	445.21	446.42
15- 18	1.24	1.19	58	117879.	.0469		.581	.87	445.21	445.74

Figure 3 : Example of printed results from a calculation with P.C. VENT

Calculations : the convergence of the calculation is obtained when two conditions are satisfied : firstly, the residual error must be less than the set value; secondly, the fixed air quantities must agree with the calculated values within set limits. As a matter of fact, the number of iterations, the maximum residual error and the accuracy on air quantities can be arbitrarily chosen (option 7 on fig. 2). In any case, the program displays a message when it has not converged.

Output of results : after the calculations, it is possible to visualize or print the results, through the option 6 of the main menu (fig.2). When displayed on the screen, they appear in a table which can be hard copied; they consist of branch identification, nominal (standard thermodynamic conditions) and actual resistances, temperature, air quantity, pressure drop, and NVP. A column for comments is also displayed (for fan branches or fixed quantities). When printed, they appear on a table containing the same information and, in adition, the power (mechanical) dissipated in the branch and the pressure at ending nodes (see figure 3).

Examples of application of P.C. VENT :

i)   test on an existing network : the program has been used to calculate the network of a French colliery consisting of approximately 120 nodes and 110 branches (see figure 4). The calculation required about ten seconds for meshing the network and half a second for each iteration and would usually converge within 60 iterations, thus needing less than one minute in total (Chorosz, 1984). The efficiency of the 8087 co-processor was clearly established.

ii)   a developing mine : P.C. VENT was used to determine the ventilation layout of the new uranium mine of Jouac (near Limoges, France), due to start production at the end of 1985; the grade of the ore

22

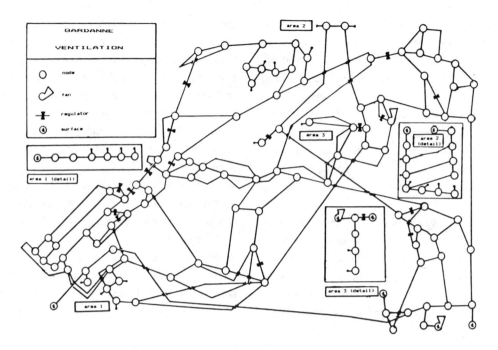

Figure 4 : sketch of the ventilation network of the Gardanne colliery
(belonging to Charbonnages de France)

being unusually high (locally well over 1%), a high attention is paid to the ventilation system because of a potential radon hazard. Within one month, the program has enabled the calculation of approximately thirty alternatives, year by year over a period of ten years, and has lead to the final choice as to the type and number of fans, the layout of up-cast shafts and the initial scheduling of production. It is currently used in order to investigate the possibility of a remote control of the ventilation network.

VENDIS

Introduction : This software has been developed by CERCHAR, with a special emphasis on graphical display of the results of a ventilation network calculation. Originally written for a DEC PDP/11 micro-computer, it is now available on IBM PC-XT, and requires a connection to an AFIGRAF 6080 graphics terminal (manufactured in France by CSEE). This terminal has a 21" refresh screen and a resolution of 1024 x 1024 (with a 4096 x 4096 image computation). A digitizing tablet is also connected, using a mouse to designate points on the screen. VENDIS is menu driven, with features similar to those of P.C. VENT, but also with an important use of interactive graphics. The calculation specifications are the same as for P.C.VENT.

Creation of a network : a network can be entered by digitizing a mine plan; the two nodes are designated to input a branch and their coordinates are entered with the keyboard. A branch can even be non-linear and be entered as a set of linear sections; intermediate "pseudo-nodes" are then entered with their coordinates, after the starting node. The program then asks for the same information as for P.C. VENT (resistance, temperature, fixed quantity). Fan curves can also be digitized; they are represented by ten points which appear on the screen and can be interactively modified.

Display of results : in addition to the standard table of results described previously (see fig.3), VENDIS offers a range of graphics features.

The network can be displayed as it was entered, i.e. an image of the mine plan appears (figure 5). Because the coordinates of each node

are known, the programme calculates the coordinates of the intersections of the different branches, projected on a plane which is transverse to the direction of viewing, in order to interrupt the plotting of those parts of the branches which are hidden by other ones. The banches where the air flow is reversed, as compared with a previous situation, start blinking.

One option of the program makes it possible to display nodes with their numbers and another one to display the branches with indication of the calculated flow.

When the number of branches of a network becomes too large, the use of the two above options can be difficult. Then it is possible to adopt the following procedure in order to

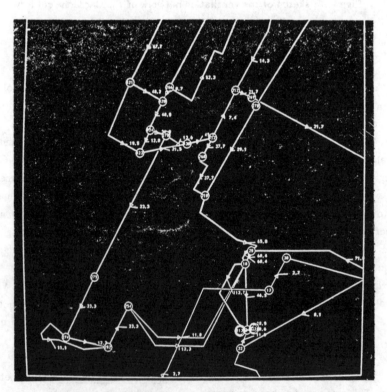

Figure 5 : view of a section of the ventilation network of the Gardanne Colliery using VENDIS

display the results of the calculations :
- designate successively each branch with the mouse. The flow rate is then displayed next to the designated position and the corresponding branch starts blinking. The whole set of data concerning this branch is at the same time displayed on the alphanumerical terminal.
- zoom the zone of interest by defining a window with the mouse or by using the keyboard; then use the option for the display of nodes and flow rates as mentioned above.
- list the complete set of results on a printer.

Another possibility of the software is to change the direction of viewing. Instead of projecting on a horizontal plane (vertical direction of viewing), it might be interesting to visualize vertical sections of the mine (shafts, steep workings, etc...) or even inclined sections. VENDIS allows to choose any direction of viewing and projects the results on the plane perpendicular to this direction.

SIMULATION OF MINE FIRES

It is felt by the authors that it is not possible to use micro-computer software for fire-fighting like the above programs for ventilation network analysis. When a fire occurs in a mine, decisions are to be made within a short time, with an often limited information, and this requires an experience which cannot be fed -as yet- into a computer. Furthermore, the network might not have been upgraded recently, resulting in erroneous data in the computer data files and dangerous mistakes in the calculation when human lives are threatened underground. In addition, as explained below, it is difficult to compute the effect of a fire under certain circumstances. Consequently one must be very careful when using a program simulating the effect of a mine fire, particularly when it comes to urgent decisions to make. However, such a program could prove quite useful; in fact its applications are mainly twofold : training of mining students or mine personnel, and preventive analysis of a ventilation network (detection of weaknesses, location of safety equipment, etc...). This is the main philosophy of the P.C.FIRE program, currently undergoing final development.

P.C. FIRE

General principle : this program has been written as an additional module to P.C.VENT and therefore runs on an IBM PC-XT. It is based on the Simode-Vielledent model which enables an easy computation of temperatures along a branch containing a fire, as a function of time, geometry of the branch, rock thermodynamic parameters, temperature of the air through the fire, distance from the fire and air velocity. Its main limitation is to neglect the effect of water vaporization, which could prove significant when the air velocity is low. The organization of P.C. FIRE is shown on figure 6.

The program starts with a standard P.C. VENT calculation, as though there was no fire; it then calculates pressure drop and temperatures along the fire branch; after this, it follows the circulation of the air from its ending node downstream, taking the temperature a unknown and calculating it as well as the new ventilation pressures (modified because of the fire and replacing NVP); at the end of this first stage, a new calculation is started, initialized with the new values of airflows, pressure drops, etc...; it is stopped if the network is stable, i.e. the quantities have kept the same value within

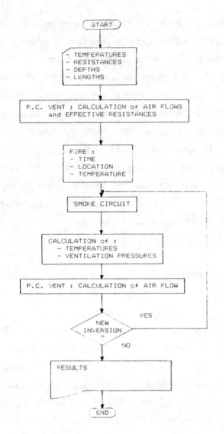

Figure 6 : general organization of P.C. FIRE

set limits of accuracy; if not the program iterates again, until stability is obtained.

Input and output of data : the information required by P.C. FIRE is that for P.C. VENT, plus the length of each branch, the location of the fire, the time since it broke out, and the temperature (dry bulb) of the air flowing through the fire. This last data is in fact difficult to assess, but it is not very sensitive in the calculation and the effect of a variation of 100°C or 200°C is quickly negligible. One difference from P.C. VENT is that P.C. FIRE cannot accept fixed quantities; it will adjust the resistances to previously calculated values and remove the fixed air flows. The results

displayed by the program, in addition to those from P.C. VENT, consist of the identification of branches affected by the fire (i.e. with potential smokes) as well as those which might soon have an reversed flow (low pressure drop).

Limitations of P.C. FIRE : as mentioned earlier, the program is still undergoing final development; as a matter of fact, it has two main downfalls in its present version; the first one is due to the theory on which it is based : it has been said that it makes no allowance for vaporization of water in the vicinity of the fire; this could lead to significant errors in some situations and it is due to the fact that the heat transfer equations are easily and explicitly solved in a dry atmosphere, whereas an analytical solution could not be found when mass transfers occur; in addition to this, the heat exchange from the air to the rockmass is calculated with the Rees' formula, established for a heat flow from the rock to the air. The second downfall of P.C.FIRE is due to the lack of systematic convergence; in fact, when there is no forced ventilation (fans stopped), some branches happen to be reversed at each iteration (particularly when they are in a structure similar to the electricians' Wheatstone bridge).

CONCLUSIONS

It is felt that the examples detailed here show the interest of professional micro-computers to solve problems which were previously dealt with using main frames or super mini-computers. In fact, the IBM PC-XT, with its arithmetic co-processor has proved surprisingly powerful in the ventilation calculations. In addition, the development of interactive graphics enables a new approach to the design of a ventilation layout.

As far as fire simulation is concerned, there is still some research to be done on P.C. FIRE, particularly to ensure a proper convergence, but it could already be used in certain conditions.

Reseach work currently undertaken between CERCHAR and the Paris School of Mines involves automatic consistency test of the experimental data used to feed ventilation software, fire simulation as explained above, and simple graphical output of results for P.C.VENT.

Finally it must be reminded that a simulation program is only valid when the treated data is accurate and in no case can it be substituted to the experienced engineer. Its basic aims are to perform tedious and lengthy calculations in a short time, help the decision making process and, possibly, support the training of newly appointed personnel.

ACKNOWLEDGMENTS

The authors acknowledge CERCHAR for permission to publish this paper; they also wish to thank managements of the Gardanne colliery and of the Jouac uranium mine for allowing them to quote the work done with their mines. The views expressed here are the authors' and do not necessarily reflect Charbonnages de France's position.

REFERENCES

Chorosz G. (1984) "Simulation sur micro-ordinateur de l'incendie dans un réseau d'aérage minier", final undergraduate year thesis, ENSMP, ed. by Centre Mines Infrastructures, Paris.

D'Albrand N., Froger C., Josien J-P. (1984) "Practical use of microcomputers for ventilation calculations", 3rd International Mine Ventilation Congress, Harrogate, ed. by I.M.M., London, pp. 27-32.

Déliac E.P. (1985) "L'outil informatique et l'exploitation des ressources minérales : un panorama de la situation française, des besoins, des perspectives", Revue de l'Industrie Minérale, paper accepted, to be published in vol. 67.

Document SIM-N3 (1976) "Aérage", special issue of Revue de l'Industrie Minérale, ed. by GEDIM, Saint-Etienne.

Hartman H.L., Mutmansky J.M., Wang .J. (1982) "Mine ventilation and air conditioning", 2nd edition, pub. by Wiley Interscience, New York.

Simode E. (1981) "Contribution à la maîtrise de l'aérage en cas d'incendie dans les travaux souterrains", Doctorate thesis, ENSMP, ed. by Centre Mines Infrastructures, Paris.

Vielledent L. (1982) "Stabilisation de l'aérage en cas d'incendie", Revue de l'Industrie Minérale, Les Techniques 7-82, pp. 317-348.

Mine stench fire warning computer model development and in-mine validation testing

STEVEN J.OUDERKIRK & WILLIAM H.POMROY
Twin Cities Research Center, Bureau of Mines, Minneapolis, MN, USA

JOHN C.EDWARDS
Pittsburgh Research Center, Bureau of Mines, PA, USA

JOHN MARKS
Homestake Mining Company, Lead, SD, USA

ABSTRACT: The stench system is commonly used in metal and nonmetal underground mines to warn miners of a fire or other emergency. Typically, a highly odoriferous chemical is released into the mine's ventilation and/or compressed air streams. The distinctive stench odor warns miners to evacuate the mine or to follow prescribed alternative emergency procedures. The layout of stench system components is seldom optimized, however, with release point locations chosen for convenience rather than performance. The result is uneven distribution of stench underground, exposure of some miners to excessive stench levels, long stench travel times to some workplacces, and some areas being missed altogether.

The Bureau of Mines has developed a mine stench fire warning computer model capable of determining stench levels at any location in a mine at any time after the release of odorant. The model enables mine ventilation personnel to develop an optimized stench system layout (release point locations, release rates, etc.) without resorting to costly trial and error fire drills in the mine.

This paper discusses the development of the computer model and the results of recent in-mine validation testing at the Homestake Mine, Lead, South Dakota.

The results of this work have shown that computer modeling of stench flow in underground mines is feasible with the use of most unmodified contaminant spread ventilation programs and if modeling is used in conjunction with remotely actuated stench injection, a significant reduction in mine fire warning times is possible. Modeling stench flow can also be an inexpensive alternative to the use of tracer gasses such as SF_6 in the gathering of ventilation data.

1 INTRODUCTION

Mine fires are a serious hazard to life and property. Between 1965 and 1980 there were over 115 reportable fires accounting for 119 fatalities in U.S. non-coal underground mines (Baker, 1980). Countless millions of dollars were spend on rescue and recovery efforts, equipment repair and replacement, and mine rehabilitation. In addition, mines shut down by fires were forced to forego hundreds of millions of tons of mineral production.

Prompt emergency evacuation, though essential to avoid or minimize loss of life, is difficult to accomplish. In a recent survey of 50 underground non-coal mines, emergency evacuation times averaged 27 minutes, ranging from 5 to 85 minutes with a strong correlation between the time required for evacuation and the maximum depth of the main shaft (Stevens, 1974). Compounding the problem of longer evacuation times for deeper mines, is the fact

that workforces in deeper mines are usually larger than workforces in shallower mines. The data show the average workforce in mines over 3,600-ft deep is more than double the average workforce in mines under 1,800-ft deep (Stevens, 1974). As mines become deeper with the depletion of shallower ore bodies, these problems of long evacuation times and more miner exposure will tend to get worse.

Once a fire has been detected, miners must be warned and they must respond accordingly by following the mine's emergency plan. Although mine fire detection has been the subject of significant research activity in recent years, mine fire warning has not. As a result, mine fire warning technology has not kept pace with the advances achieved in mine fire detection.

This paper describes research conducted by The Bureau of Mines, U.S. Department of Interior to upgrade the safety and reliability of the stench warning system. Although the authors acknowledge the inherent limitations of air-carried warning systems, the stench system was singled out for improvement because of its widespread use.

Release of a stench agent such as ethyl mercapton or tetrahydrothiophane is the primary method used to warn miners of fire in U.S. deep metal mines because it is the quickest and most reliable method available today. Unfortunately most deep metal mines have highly complex ventilation which typically results in some areas of the mine receiving concentrations of stench sufficient to cause nausea while miners in remote or poorly ventilated areas receive no warning or greatly delayed warning. The Bureau of Mines has approached this problem as a two phase project. Phase I was to develop, construct, and test an improved stench warning system. Phase II was to determine, through computer modeling, the number and location of stench injectors required to provide warning to all operating areas of a mine in some predetermined minimum time, and the stench release rates required to achieve the desire stench concentrations at each working area. The computer simulation was validation tested at the Homestake gold mine near Lead, South Dakota, by comparing computer predicted stench arrival times to actual stench arrival times reported by miners during a routine stench fire warning/evacuation drill.

2 STENCH SYSTEM DEVELOPMENT

Although the stench system has been used successfully for over 60 years, it suffers several serious shortcomings owing to certain chemical properties of ethyl mercaptan and to certain performace characteristics and limitations of present injection systems. The research program began by correcting these deficiencies.

Ethyl mercaptan mixed with Freon is the predominant stench agent used in U.S. underground metal and nonmetal mines. However, ethyl mercaptan's toxicity, corrosiveness, extreme order intensity at high concentrations, and tendency toward odor fade when transported in steel pipes limit the overall safety and reliability of this agent. Tetrahydrothiophene (also simply THT) was identified as a preferred alternative to ethyl mercaptan. THT is less toxic, less corrosive, and has a less intensive odor at high concentrations than ethyl mercaptan. Its vapors are nonreactive with iron oxide so the odor does not fade when transported in steel pipes.

Present injection systems release stench fluid in an uncontrolled fashion, resulting in unbearably high and potentially toxic levels of stench in some parts of the mine while other areas may receive no warning at all.

The improved injector features a pressure balance line and metering orifice to control agent concentrations and extend release time, thus protecting miners against overexposure to stench vapors while at the same time insuring better coverage of the mine (Pomroy, 1985).

Where underground workings are extensive, release of stench at a single, surface location (the collar of a main downcast shaft, for example) may result in unacceptably long stench travel times to the most remote workplaces. The improved stench system was designed to facilitate remote control operation of multiple injectors, including underground locations nearer to remote workings.

Prototype stench injection equipment was fabricated and bench tested in the laboratory. Field testing, complete with stench fire drills using THT stench agent, was conducted at Kerr-McGee's Church Rock No. 1 uranium mine near Gallup, New Mexico.

3 COMPUTER MODELING

Over the last twenty years computer modeling of mine ventilation has advanced from the research laboratory to common mine usage. The use of computers to solve ventilation networks has not only relieved the ventilation engineer of much tedium, but has forced a disciplined, thorough, and methodical approach to the gathering of ventilation data.

One of the areas of ventilation computer modeling that has received much attention in recent years is simulating the spread of contaminants in mine air such as methane, radionucleides, and products of combustion. In 1980, Dr. Rudolf E. Greuer of Michigan Technological University (MTU) was awarded a contract from the U.S. Bureau of Mines to develop a program to perform real-time precalculation of the distribution of combustion products and other contaminants in the ventilation systems of mines. The program was designed to predict ventilation in a complex, multilevel mine for normal and emergency conditions. The MTU code was designed to be fast and efficient, taking 1/3 the CPU time of other codes to solve network flows (Xintan, 1983).

To use the real-time capability of the program, the airway ventilation has to be established either as measured quantities that are entered as input data for the program, or calculated within the program from flow production factors such as fans and natural ventilation, as well as flow impedance associated with wall friction. The advantage of the latter approach is that as new drifts are opened, doorways closed, and fans added, a prediction can be made of the new mine ventilation. The ventilation calculations are made within the program by first representing the mine network as a collection of closed paths. Then the conservation of energy is applied to the air flow around each path, while frictional wall losses are used to establish pressure drops along airways. At each junction (airway intersection within the mine network) conservation of mass is applied to relate the airflow rates. Temperature variations within the mine are used to calculate natural ventilation head loss as part of the conservation of energy around each path. Fan head losses are determined from the fan characteristic curve and are entered into the energy balance.

In the case of a mine fire, the localized heat production rate of the fire is entered as part of the input data to the program. The heat addition alters the airflow, and its effect is evaluated by calculating the new temperature distribution and the associated natural ventilation head loss. The effect on ventilation can be calculated in a steady state approximation (Greuer, 1982), or in a dynamic mode as a short duration of steady state processes (Greuer, 1983). The real-time capability of the program is utilized to project the time dependent spread of contaminants, such as smoke from a fire or stench from a warning system, throughout the mine complex. This evaluation proceeds by associating control volumes with specific contaminant concentrations. Each control volume is transported with the airflow. At junctions where control volumes meet, perfect mixing is assumed and a new control volume (a new contaminent concentration) is formed. The contaminant sources are specified at fixed locations in the network, and for finite time durations. Greuer (1982) and Edwards (1982) describe in detail the input data procedures for the real-time capability of the program, as well as for the ventilation calculations.

4 DESCRIPTION OF TEST SITE

The Homestake gold mine near Lead, South Dakota, was chosen as a site suitable for modeling stench flow because of its large size and complex ventilation. Operations cover an area of approximately one mile wide by two miles long. There are 57 levels from the surface to the 8,000 level of which 47 are active. Approximately 1500 people are employed at the mine and 600 of those work underground dayshift. The ventilation in the mine is controled primarily by two surface fans, one underground fan on the 2150 level, and natural ventilation. Secondary ventilation is controled by the use of a small number of booster fans to produce fixed quantity airflows in several airways. About 860,000 cfm of air

flows through the underground workings. Primary intake shafts are the Ross, Yates, and No. 5; primary exhaust shafts are the Oro Hondo, Ellison, and No. 2. Wall rock temperature ranges from 10°C (50°F) to 56°C (133°F). Humidity approaches 100% throughout the mine and air temperature ranges from 10°C (50°F) to 29°C (85°F).

5 COMPUTER SIMULATION

The three to four thousand distinct airways of the mine are represented, through a combination of parallel airways into single airways, by 404 distinct network branches (see Figure 1). Based upon temperatures and airway resistances acquired through a ventilation survey and known fan characteristic curves, the ventilation throughout the mine was modeled. Several airway resistances were adjusted to produce better agreement between the measured and predicted airflows at the fans and within several mine airways.

The second stage in the prediction of stench spread requires the specification of injector location, injector duration, and total quantity of stench released. The stench consists of pure (99%) THT. Each injector is designed to release 850ml of stench within 5 minutes. This represents a release rate of .006 cfm of THT in the vapor state, which is subsequently diluted by the mine ventilation. Six injectors were used for the test and had observed release periods of 4.5 to 5.5 minutes. To simulate this process, the local injection rates were scaled according to the observed release times. One of the injectors was connected to an open-ended compressed air line that led to two airways. Simultaneous with the release of stench into this line, a solinoid valve was actuated, allowing compressed air to fill the line and blow the stench out the open ends. This case was represented, for simulation purposes, by two separate injectors located at the airways into which the compressed air line entered.

6 FIELD TEST

In-mines validation testing of the stench warning model was conducted at the Homestake gold mine during the week of April 14, 1985. For this test six injectors containing 850ml of Pure (99%) tetrahydrothiphene released stench into six different areas of the mine at rates ranging from 150ml/min to 180ml/min. The model was evaluated by comparing stench arrival times reported by miners to stench arrival times predicted by the computer. Previous discussions with the

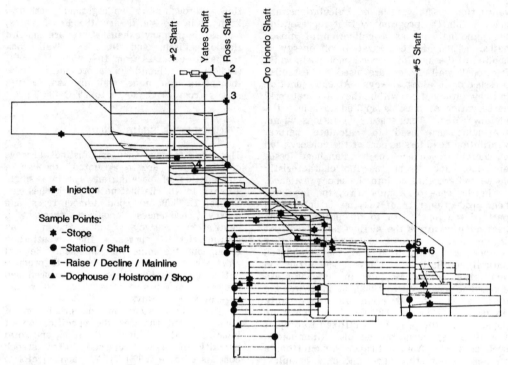

Fig. 1. Homestake Gold Mine Injector and Sample Point Locations

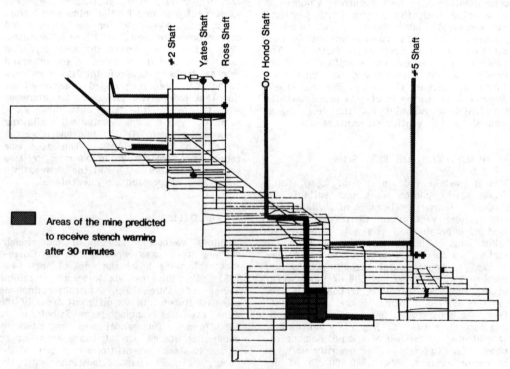

Fig. 2. Homestake Gold Mine Stench Distribution

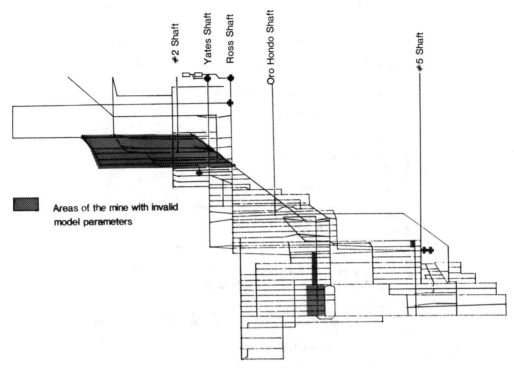

Fig. 3. Homestake Gold Mine Model

National Institute for Petroleum and Energy Research in Bartlesville, Oklahoma, had determined that there was no known way to collect samples of stench/air mixtures for later analysis since the stench tended to plate out onto the containers walls (Winisman, 1977). Therefore, stench arrival times at each workplace were determined from the shifters report to the mine manager. This report provided 40 arrival times for the stench at different locations within the mine as reported by miners to their shifters. The advantages of this procedure were the large number of data points and widespread mine coverage for evaluating the model. The disadvantage was that there was no guarantee that the time and location for the detection of stench would be accurate. It was decided to allow $\pm$ 5 minutes to account for arrival time fluctuations due to inaccuracies such as incorrect watches and imprecise reporting of locations. Another problem was that contract miners tend to ignore fire drills if they are nearly ready to blast, since missing a round could cost them bonus money. Although minor delays from predicted arrival times are to be expected in areas out of the main ventilation flow (such as stopes and doghouses) contract miners will

typically report greatly delayed arrival times in order to justify their tardy evacuation.

7 RESULTS

Initial comparison of measured and predicted arrival times for the stench provided poor correlation (about .155) so the field data was broken down into four subgroups for further analysis (see Figure 4). The subgroups were chosen to differentiate between areas of well defined ventilation (subgroups one and two) and areas where ventilation might be somewhat uncertain (subgroups three and four). Supgroup one consisted of 13 sample points in station and shaft areas; subgroup two consisted of 7 sample points located in raises, declines, and mainlines; subgroup three consisted of 12 sample points located in stopes; subgroup four consisted of 8 sample points located in doghouses, hoistrooms, and shops. Analysis of each of these subgroups yielded correlations of .470 for supgroup one, .435 for subgroup two, -.079 for subgroup three, and -.015 for subgroup four.

These results confirmed the expectation that sample locations out of the main

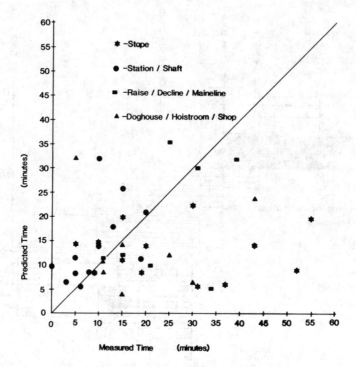

Fig. 4. Predicted vs. Measured Stench Arrival Times for all Sample Locations

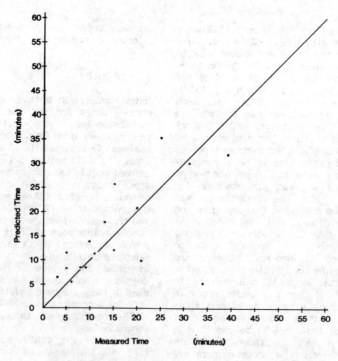

Fig. 5. Predicted vs. Measured Arrival Times for MainVentilation Sample Points

34

ventilation flow (subgroups three and four) would be poorly correlated. Furthermore, these results indicate that the model will accurately predict arrival times only for nodes or branches that correspond to single airways within the mine.

Of the twenty sample points in subgroups one and two, two were found to be in error. The first reported detection simultaneous with the release of stench even though the closest injector was over 4800 feet away. The second failed to report detection until 20 minutes after sample points further downstream had reported detection. Correlation for the remaining 18 points was .509 (see Figure 5). Analysis of outliers located three areas of the mine that were inaccurately modeled due to errors in the input data (see Figure 3).

8 CONCLUSIONS

This computer model is very useful in determining gross stench distribution to all working areas. The assumption that the stench would not absorb onto mine surfaces and thereby delay stench spread appears to be valid so stench flow should be suitable for modeling by normal, unmodified, contaminant spread programs. The use of stench distribution modeling is a potentially valuable method for refining ventilation flow data. Although less precise than using tracer gasses such as SF_6, it has the advantage of being very inexpensive since the data is a byproduct of routine mine evacuation drills. With care excellent results should be obtainable with this method.

REFERENCES

Baker, R.M. & Nagy, J., An Annotated Bibliography of Metal and Nonmetal Mine Fires, USBM/Allen Corp. Contract Final Report No. JO295035, 1980.

Edwards, J.C. & Greuer, R.E., Real-Time Calculation of Product of Combustion Spread in a Multilevel Mine, U.S. Bureau of Mines, Information Circular 8901, 1982, 117 p.

Edwards, J.C. & Ti, J.S., Computer Simulations of Ventilation in Multilevel Mines, Procedings Third International Mine Ventilation Congress, Harrogate, England, 1984, pp 47-51.

Greuer, R.E., Real-Time Precalculation of the Distribution of Combustion Products and Other Contaminants in the Ventilation Systems of Mines, NTIS PB82-183104, 1982, 236 p.

Greuer, R.E., A Study of Precalculation of the Effect of Fires on Ventilation Systems of Mines, NTIS PB84-159979, 1983, 293 p.

Pomroy, W.H. & Muldoon, T.L., Improved Stench Fire Warning For Underground Mines, U.S. Bureau of Mines, Information Circular 9016, 1985, 33 p.

Stevens, R.B., Mine Shaft Fire and Smoke Protection System - Volume I, Design and Demonstration, USBM/FMC Contract Final Report No. HO242016, 1974.

Whisman, Goetzinger, Cotton, Brinkman, and Thompson, A New Look at Odorization Levels for Propane Gas, BERC/RI-77/1, Sept. 1977, 91 p.

Xintan, C., Investigation on Ventilation Network Calculation Techniques, M.S. Thesis, Michigan Technological University, 1983, 133 p.

An investigation of the causes of underground fires and ignitions in South African coal mines

G.P.BADENHORST
The Government Mining Engineer of South Africa, Johannesburg, South Africa

R.MORRIS
Johannesburg Consolidated Investments Co., Ltd., South Africa

ABSTRACT: Over the past decade, publications have emerged from Britain, United States of America, Germany and Poland which indicate the major causes of underground coal mine fires. The authors have investigated the causes of all underground fires over a ten year period in South Africa and have subsequently subdivided these into the four major coal mining districts of Natal, Witbank, Ermelo and the Vaal Triangle.

The major causes are found to be those due to underground spontaneous combustion followed by heatings due to electrical causes. The authors conclude by indicating the approaches being made in South Africa to protect workmen in the event of an underground fire or an ignition.

1 INTRODUCTION

For the reader who has no knowledge of South African coal mining industry, Figure 1, adequately describes the areas and location of the Coalfields of South Africa. Throughout this paper fire incidents occuring in Natal, the area of Vereeniging, Sasolburg and South Rand known locally as the Vaal Triangle, and the areas of Ermelo and Witbank will be referred to continually.

These areas correspond to those of importance with respect to high production tonnages and are identified in Figure 1 as numbers 7, 9, 11, 12, 13, 14, 16, and 17. Figure 2 indicates the total tonnages produced in South Africa on an annual basis between the years 1885 - 1979, whilst Figure 3 shows the growth in the annual coal supply to the Electricity Supply Commission, (Escom), between the years 1950 - 1983 inc.

In South Africa, capital investments totalling well over R1 500 million have been committed to the expansion of the South African coal mining industry's production capacity in the past decade. This expansion has involved giant new mines, some of which have set both South African and world records for size, production or advanced technology. The industry has introduced multi-product, multi-seam, highly capital intensive coal mining and as a result has created thousands of new jobs.

Some of the major projects which have recently come into production are listed below:

Rietspruit Opencast Coal Mine: One of the new generation of modern, highly-mechanized export mines, and one of the industry's largest open-cast operations. Rietspruit came into production during September 1978 at a capital cost of R200 million, and is situated in the Witbank Coalfields, Refer Coalfield No 7, Figure 1.

Coal output at full production in 1983 was five million tonnes a year, all of which was exported. Three Bucyrus Erie 1570 W draglines, each with 53 cubic-metre buckets are used to strip overburden to expose the coal seams. Coal reserves total about 300 Million

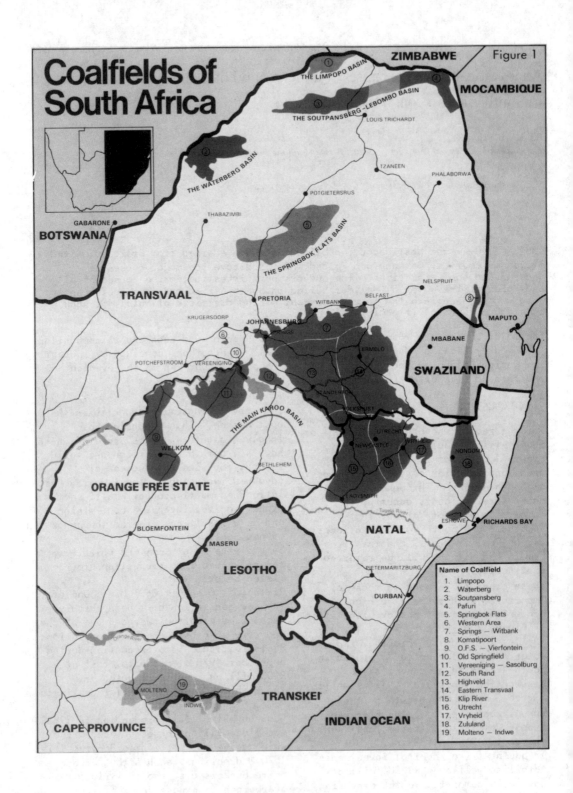

Coalfields of South Africa

Figure 1

Name of Coalfield

1. Limpopo
2. Waterberg
3. Soutpansberg
4. Pafuri
5. Springbok Flats
6. Western Area
7. Springs — Witbank
8. Komatipoort
9. O.F.S. — Vierfontein
10. Old Springfield
11. Vereeniging — Sasolburg
12. South Rand
13. Highveld
14. Eastern Transvaal
15. Klip River
16. Utrecht
17. Vryheid
18. Zululand
19. Molteno — Indwe

tonnes and an extraction rate of 90 percent or better is planned.

Duvha: This R200 million open-cast mine is being developed to supply coal to Escom's new Duvha power station. By 1986 the Duvha mine will supply 845 000 tonnes of coal a month to the 3 600 MW power station.

Coal is produced by strip mining, using three 53-cubic-metre draglines for overburden removal. Two 8,4-cubic-metre draglines are used for parting removal. To uncover the annual coal output of more than 10 million tonnes, some 40 million bank cubic metres of overburden will have to be removed.

Bosjesspruit Colliery: This mine is a direct result of South Africa's world leadership in the field of producing fuel from coal. The giant colliery that will feed coal to the new Sasol 2 and Sasol 3 plants will ultimately deliver 27 million tonnes of coal a year, easily the world's largest coal mine, and one of the most technologically advanced. It will be entirely an underground mine, employing modern, mechanized mining methods. Mining will be done by six large longwalls, conventional bord and pillars, and continuous mining with bord and pillar including pillar extraction.

Matla: This R188,8 million mine will supply 9,5 million tonnes of coal a year to an Escom 3 600 MW power station, working entirely by underground mining. Methods employed include conventional bord and pillar, continuous mining and longwalls. Development of the mine started in 1973.

Mining Survey, Number 1/2, 1981.

Grootegeluk: This mine, which opens up the Waterberg coalfields in the northern Transvaal, has been developed at a cost of R386 million, including the total infrastructure. Refer to Coalfield No 2, Figure 1. The planned output of phase one is 9,9 million tonnes a year which includes 1,8 million tonnes of blend coking coal and 8,1 million tonnes of steam coal. During 1980, it was announced that the steam coal middlings from Grootegeluk will be used to fuel the Matimba power station to be constructed by Escom in

the area. With the mine producing for diverse markets, the coal beneficiation plant is complex, and, at an average input capacity of 3 000 tonnes of run-of-mine coal an hour, is the largest of its type in the world.

Kriel Colliery: Tied to a 3 000 MW Escom power station, Kriel Colliery was one of the industry's first truly large producers. It started production in 1975 and has a capacity of 8,5 million tonnes per annum.

Kleinkopje; This venture, costing R109 million, was announced by Amcoal in 1977. The mine was planned to start operating in January 1979, and to reach its initial annual production rate of 2,7 million sales tonnes in June 1979.

Total sales output will comprise about 700 000 tonnes of low-ash metallurgical coal for Iscor, 1,6 million tonnes of coal for the TCOA trade and 2,5 million tonnes of coal to be exported. Iscor being the Iron and Steel Corporation of South Africa, and the TCOA being the Transvaal Coal Owner's Association which markets coal to the inland consumers.

Arnot: One of the earliest major opencast coal mines in South Africa, Arnot was brought to full production of 6,2 million tonnes of steam coal a year by mid-1976. The mine was developed to produce steam coal for the 2 100 MW power station, supplying 1,908 million tonnes from underground and 4,2 million tonnes by strip mining.

New Denmark: Amcoal disclosed in 1980 that the New Denmark Colliery would supply Escom's Tutuka power station which has a planned generating capacity of 3 600 MW. Coal output from New Denmark will be 10 million tonnes a year. Production commenced during 1984, ready for commissioning of the first of six 600 MW generating sets at Tutuka in the first half of 1985. Full production is scheduled for 1990. The mine will be the first to exploit the New Denmark coalfield in the Eastern Transvaal situated near Standerton. Refer Coalfield No 14 Figure 1. Initial plans are to mine

39

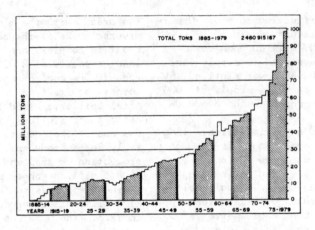

SOUTH AFRICA'S COAL SALES

FIGURE 2

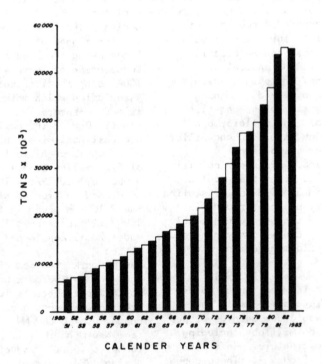

CALENDER YEARS

ANNUAL COAL SUPPLY
TO THE
ELECTRICITY SUPPLY COMMISSION

(ESCOM)

FIGURE 3

the 2m thick seam at depths of 200m
below the surface by an underground
operation using continuous miner and
longwall methods.

New Vaal: The New Vaal colliery will
fuel another planned Escom 1 800 MW
power station, supplying coal from
the Cornelia coalfield, some 15 km south
of Vereeniging, where a combined underground and openpit colliery is to be
established. Refer Coalfield No 11,
Figure 1. This colliery will commence
production in the latter part of the
1980s and, at full output, will supply
some 6,5 million sales tonnes to the
power station each year.

Ermelo Mines: Capital costs were about
R70 million when the mine came on
stream in 1978 to produce a total of
three million tons a year of export
coal. Mining is entirely from underground operations, the geology encountered being unsuitable for longwalling
techniques, however, mechanized
sections are used wherever possible.
The mine has extremely difficult
conditions and in fact Ermelo suffered
from two severe methane explosions
in 1983.

The authors have laboured over the
growth of the South African coal
industry over the past decade to
enable the reader some insight in the
vastness of the industry. Indeed
South Africa's remarkable rise in the
world's coal export markets, from
insignificance to a world leader is
remarkable and is indicative of the
soundness of the South African
economy and in the leadership controlling the South African coal industry.
But, what of the future? In South Africa
during the past decade, some extremely
large opencast mining operations have
occured. In the Witbank coalfield
of the Transvaal, these opencast
mines have mined thick and shallow
coal seams of extremely large coal
reserves. However, opencast mining
has a limited application in this
country because only 4 - 20 percent
of the country's reserves can be
classed as opencastable, Petrick et al.,
(1975). Thus the majority of the

large mines which will be opened in
the future in South Africa will be
underground mines.

2 THE EFFECT OF AN UNDERGROUND FIRE
ON COAL RESERVES

Underground fires result in a loss
of coal reserves, as follows:

1. The collapse of bord and pillar
workings due to the weakening effects
the fires have on pillars and roof.
Watson (1977), indicated that a fire
at Northfield Colliery resulted in a
loss of approximately 2 million
sales tonnes of high grade coking
coal.
Nakata (1980), stated that at the
Sumitomo - Akabira coal mine in
Hokkaido between 1938 - 1980, 104
incidences of spontaneous combustion
had occured. Out of 40 million
tonnes of total coal produced during
this time period more than 2 million
tonnes were abandoned due to sealing
- off these areas.
Further, Guney (1975), indicated that
in the Armutcuk Colliery in Turkey
over 60 fires in the period of 1966
- 1973 resulted in the sealing off of
2,5 million tonnes of coal.

2. Large scale roof falls can result
in severe problems in the case of
mining seams above an old fire area,
Morris, R. (1979).

3. In areas prone to spontaneous
combustion some mining methods become
excluded. Unfortunately, the high
extraction percentage methods, such
as pillar extraction and longwalling
which induces a collapse of the roof,
are particularly liable to heatings.
However, because the majority of
mining houses are looking to these
high extraction methods, spontaneous
combustion of coal can be expected
to become more prevalent in the
future and that fire precautions and
combating techniques must be well
understood by South African mining men.

4. Underground fires are life
endangering due to the following;
    a) The toxic effect of the atmosphere

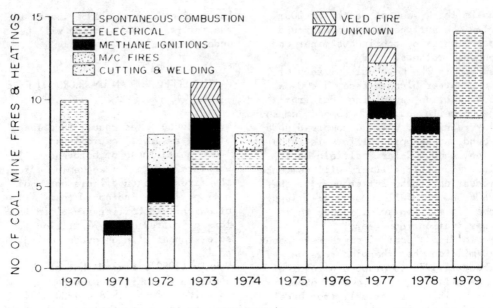

FIGURE 4

SOUTH AFRICAN COAL MINE FIRES AND HEATINGS 1970-1979
SOURCE : OFFICE OF THE GOVERNMENT MINING ENGINEER,
DEPARTMENT OF MINERAL AND ENERGY AFFAIRS

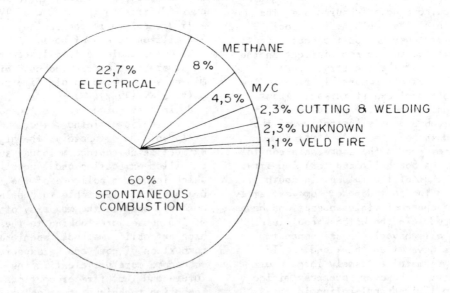

FIGURE 5

PIE CHART OF SOUTH AFRICAN COAL MINE FIRES &
HEATINGS 1970-1979 INDICATING THE CAUSES &
THE PERCENTAGES OF THE TOTAL

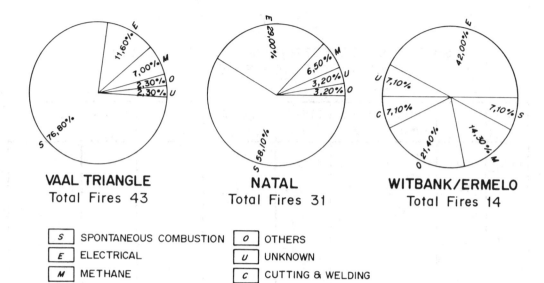

S	SPONTANEOUS COMBUSTION	O	OTHERS
E	ELECTRICAL	U	UNKNOWN
M	METHANE	C	CUTTING & WELDING

FIGURE 6

PIE CHARTS DEPICTING % INCIDENTS
ON AN AREA BASIS
1970 – 1979

b) The asphyxiation effect of the atmosphere.

c) The danger of the explosiveness of gas mixtures and coal dust.

5. When an underground fire has reached an advanced stage it can incur considerable high costs in fighting the fire and the disruptions in the production process can have far reaching effects on the mine.

3. SOUTH AFRICAN FIRE STATISTICS

From the office of the Government Mining Engineer the following statistics were obtained. Table 1 describes some 88 fires between the years 1970 – 79 inclusive on a 'type of fire' basis, whilst Table 2 describes the same fires over the same period on an area basis.

From these tables, Figure 4, 5 and 6 were derived which showed that during 1970 – 1979 inclusive, 60 percent of all underground fires in South Africa were due to spontaneous

combustion, which was further broken down on an area basis to 76,8 percent of fires due to spontaneous combustion in the Vaal Triangle and 58,10 percent in Natal. Whilst in the Witbank/ Ermelo area of 14 fires reported 42 percent were due to electrical causes. Fires are normally classified into two main sub-categories. Holding (1981), classified fires as;-

1. Fires caused by spontaneous combustion.

2. Fires due to other causes.

Whilst, Thorp (1981), classified fires as:-

1. Open Fires

2. Spontaneous combustion

However, Morris (1984), classified fires as:-

1. Natural fires, being those due to spontaneous combustion.

2. Man made fires, which in the order they are likely to occur include:

i) Electricity,

ii) Conveyors and other machinery,

TABLE 1

88 FIRES BETWEEN 1970 – 1979 INCLUSIVE ON A 'TYPE OF FIRE' BASIS

YEAR	TOTAL	SPONTANEOUS COMBUSTION	ELECT.	METHANE	CUTTING WELDING	OTHERS	UNKNOWN
1970	10	7	3	–	–	–	–
1971	3	2	–	1	–	–	–
1972	8	3	1	2	–	2	–
1973	11	6	1	2	–	1	1
1974	8	6	1	–	1	–	–
1975	8	6	1	–	–	1	–
1976	5	3	2	–	–	–	–
1977	13	7	2	1	1	1	1
1978	8	3	4	1	–	–	–
1979	14	9	5	–	–	–	–
	88	52	20	7	2	5	2

TABLE 2

88 FIRES BETWEEN 1970 – 1979 INCLUSIVE ON AN AREA BASIS

CATEGORY	WITBANK/ERMELO	NATAL	VAAL TRIANGLE	TOTAL
SPON COMB.	1	18	33	52
ELECTRICAL	6	9	5	20
METHANE	2	2	3	7
CUTTING & WELDING	1	–	1	2
OTHERS	3	1	1	5
UNKNOWN	1	1	–	2
TOTAL	14	31	43	88

iii) Burning and welding,
 iv) Dieseline and other oils,
  v) Explosives,
 vi) Friction, and
vii) Surface fires carried underground
It can be noted by the statistics
presented by the authors that natural
fires are most prevelent in the Vaal
Triangle and Natal whilst in the
Ermelo/Witbank area, electrical fires
are the most common and indeed are
an increasing danger.

The following sections of the paper
deal with the manner in which the Coal
Mining Research Controlling Council
and the Office of The Government
Mining Engineer, Department of Mineral
and Energy Affairs have combined their
theoretical and practical knowledge in
order to protect workmen in the event
of an underground fire or an ignition.

4.   THE COAL MINING RESEARCH
     CONTROLLING COUNCIL

After the Coalbrook disaster in 1960,
it was decided to appoint a Coal
Mining Research Controlling Council.
This Council appointed two Committees
namely, the Strata Control Advisory
Committee and the Explosions Hazards
Advisory Committee. The first author
of this paper, Mr. G. P. Badenhorst,
as the Government Mining Engineer of
South Africa is the Chairman of the
Coal Mining Research Controlling
Council. The second author, Mr. R.
Morris, is the Chairman of the
Explosion Hazards Advisory Committee,
and as such is also a committee
member of the Coal Mining Research
Controlling Council.

The Explosion Hazards Advisory
Committee appointed the following
Sub-Committees to assist with its
research program.
1.   Sub-Committee on the problems
resulting from the occurrence of
methane and the spontaneous combustion
of coal.
2.   The Sub-Committee on the use and

testing of explosive and ancillary
equipment.
3.   The Sub-Committee on the explosibility of Coal Dust.
4.   The Sub-Committee on Lightning and
Stray Currents.
5.   The Sub-Committee on the safe use
of Electricity in Coal Mines.
6.   The Sub-Committee on the Starting
and Stopping of fans in Collieries.
In this paper the authors only deal
briefly with the work of the first-
mentioned Sub-Committee. This Sub-
Committee is so constituted that its
members are representative of all the
interested parties namely, The Government, the Coal Mining Industry and
Scientists from the Fuel Research
Institute and the Universities.
It has also been common practice to
invite manufacturers of instruments or
agents to attend the meetings of the
Sub-Committee if and when such attendance is deemed necessary.
By involving all the parties the
necessary co-operation was created to
achieve substantial progress.
Particular mention must be made of the
invaluable assistance that was given by
successive managers and ventilation
officers of mines in the Vaal Triangle
by testing, under practical conditions,
the different instruments and by
regularly submitting reports to the
Sub-Committee. Badenhorst, Buchan
and Visser, (1976).

4.1 Work on methane detection.

The work initially carried out by the
Sub-Committee concentrated on problems
experienced with the detection of
flammable gas and the various instruments available for the testing of gas.
Amongst the instruments tested included:
1.   The Cambrian No G2 (Garforth type)
flame safety lamp.
2.   Four different types of methanometers.
3.   Continuous recording methanometers.

4.2 Spontaneous combustion of coal.

In the past reliance had been placed on
the non-instrumental detection of heat-

45

ings where dependence was placed on the human senses.

The detection of incipient heating by means of the human senses has never been satisfactory for the reason that the heating must have reached an advanced stage before even experienced operators could detect it.

This fact was driven home effectively when an open fire occurred at Coalbrook Collieries.
Thompson, (1962).

In 1972, the direct cost of fighting this fire was R43 300 (excluding the effect of loss of production). If it is taken into consideration that production had to be stopped for approximately one week on this mine which was producing 235,000 metric tons per month, it can be realised that the total cost of the fire was substantial.

Two very important points were re-emphasized by the fire at Coalbrook:-
(i) Firstly it brought home that it was of paramount importance to detect an incipient heating as early as possible so as to facilitate the controlling thereof.
(ii) Secondly it was realised that it was of even more importance to have a quick method of analysing gas samples taken in the area being sealed to provide for the safety of personnel employed on the sealing of the area.

As a result of the fire at Coalbrook Collieries the managers of the 4 coal mines in the Vaal Basin, the Chief Inspector of Mines, Heidelberg, and the Superintendent of the Rescue Training Station met to discuss these factors. Investigations resulting from this meeting showed conclusively that the monitoring of carbon monoxide was the most effective and efficient means of identifying a heating in its earliest stage. Joubert, Buchan and Beukes, (1976).

Over the years the Unor I and the Unor II carbon monoxide and gas indicators were tested at mines which were sus-ceptible to spontaneous heatings.
Today numerous mines in South Africa

have continuous carbon monoxide detec-tion systems which have had impeccable results in detecting heatings in their early stages.

The third instrument successfully tested and presently used throuthout South Africa was the Mobile Gas Analysis Laboratory or 'Mogul'. The work of the Sub-Committee showed that the main advantages of the Mogul are:-
(i) Results are available almost immediately. This is extremely important as immediate action can be taken should a dangerous situation develop during or after sealing operations.
(ii) Check analyses have proved that the accuracy of the analyses done in Mogul is in excess of the accuracy normally required for monitoring fires and therefore, completely adequate for the purpose for which it was acquired.
(iii) A large number of samples can be analysed and this can assist in building up information in regard to colliery fires which can lead to a better understanding.
(iv) The fact that the unit is on site during the fighting of fires, has a psychological advantage which should not be underestimated. Its use decreases the dependence upon the unknown, puts persons more at ease and leads to better efficiency of the sealing off operations. Joubert, (1975). Both the authors of this paper have been involved in numerous underground heatings where the use of the Unor gas detection system and the 'Mogul' have reduced the crisis of an underground mine fire to a problem of simple logistics due to the fact that the 'problem of the unknown' has been eliminated.

5. GUIDELINES AND RECOMMENDATIONS FOR THE PROTECTION OF WORKMEN IN THE AFTERMATH OF COLLIERY FIRES AND EXPLOSIONS.

In 1984, in the light of recent colliery

disasters, both in South Africa and overseas, attention was focused on the need for Guidelines and Recommendations to advise mine management on the best practicable means of protecting workmen underground, in the aftermath of colliery fires and explosions.

The Collieries Technical Committee, of the Chamber of Mines, in recognising this need, decided upon the formation of a Sub-Committee, members of which are experts in the fields of mine ventilation, mine fires and explosions and were drawn from the Groups, members of the Chamber, to advise on the preparation of such a document.

The Sub-Committee, in preparing the Guidelines and Recommendations, took cognizance of current technology and research documentation available in South Africa and overseas. These Guidelines and Recommendations are discussed under the following subheadings:
1. Mine design
2. Demarkation and maintenance of escape routes.
3. Communication of the locality, nature and extent of a disaster.
4. Self-rescuers.
5. Back up rescue systems.

5.1 Mine design.

It is of course impossible to dictate the best possible mine design to be employed in a colliery, as individual circumstances prescribe the best section and underground configuration. Amongst the items to be included in a mine design include, secondary intakes; two outlets to surface; the confinement of the extent of disaster and the usage of stonedust or water barriers.

5.1.1 Secondary intakes.
A secondary intake(s) can best be described as an independent fresh air source extending from the working face to as near surface as practicable. This intake can be used to provide an alternative escape route in the

aftermath of a fire or explosion, should the primary intake into the mine, or section of the mine, be affected. These secondary intakes are available in many mines in South Africa

5.1.2 Two outlets to surface
Regulation 6.1.1 of the Mines and Works Act (Act No 27 of 1956).

"In connection with every mine there shall be provided shafts or outlets to surface such that, except as is permitted in regulation 6.3.1, every person employed underground in such mine shall have available to him not less than two separate and independent shafts or outlets affording means of egress from underground to surface: Provided that it shall not be necessary for such shafts or outlets to be situated on the same mine".

The world wide design of longwall faces is that there is one intake and one return roadway. In South Africa where it is normal to **retreat** longwall there are invariably three intakes and three return airways to a longwall face. These are used as additional means of egress.

5.1.3 Confinement of the extent of a
             disaster.
Figure 7, illustrates how the reduction of the number of roadways into a section of the mine can be effected by keeping pillar entries to a minimum, thus helping to confine the extent of a fire or explosion.

5.1.4 Stonedust+water barriers
South African mine management are aware of the implementation of stone dust and/ or water barriers in conjunction with stonedusting. Regulation 10.14.9 deals with this as follows:

"Stone dust barriers for the purpose of suppressing a coal dust explosion shall be of a design and construction approved by the Inspector of Mines and located at such points as the manager, after consultation with the Inspector of Mines, may determine".

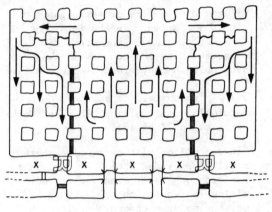

$\boxed{\text{X}}$ PILLAR USED TO REDUCE NUMBER OF ROADWAYS

FIGURE 7

5.1.5 Mine Ventilation

Methods of ventilating mines vary from one mining company to another, there are however accepted principles, which have proved efficacious in the past in limiting the extent of a disaster.

(i) Air coursing

It has sometimes been the practice in the past, to ventilate sections of a mine by coursing of used air (i.e. air that may contain dangerous or noxious gases) from another section of a mine. Coursing will increase the number of personnel affected by an incident and could increase the travelling time into fresh air.

Coursing of ventilation has been the subject of a Chamber of Mines investigation. Recommendations have been submitted to the Government Mining Engineer for consideration.

The Chamber of Mines recommendations are:-

1. Used air coursing should not be accepted as a ventilation policy on any coal mine. Each production unit should be ventilated with fresh air.
2. Where special circumstances make used air coursing a temporarily desir-

able procedure, this should only be permitted under controlled conditions as laid down by the Manager. These controlled conditions shall include specific reference to dust suppression, gas detection, stone dusting and any other items the Manager may deem necessary.

The Manager shall further indicate the time period during which used air coursing may be practised. Further he shall forward a copy of the controlled conditions to the Chief Inspector of Mines for his information and inform him of the relevant time period used air coursing will be practised.
3. In bord and pillar workings used air coursing should be permitted in a new section providing the line of faces has not advanced more than 100m or 3 lines of pillars (whichever is the greater in lateral extent from the air crossing that would be required to establish a fresh air.

(ii) Bleeding of goaf.

Where methane bleeding from goaf areas is felt to be a necessary part of the operation of a safe mining system, it should only be carried out under controlled conditions, with continuous gas monitoring in the area affected.

(iii) Methane emissions and explosions

Variations in barometric pressure have a direct effect on the rate of emission of methane from coal seams and old workings and goafs. Research has indicated that there are two important aspects to the relationship between methane emissions and barometric pressure. One is short term, over a period of say one hour, and the other longer term over a period of say, one day. Typical danger signs would be a drop in barometric pressure of more than 150 Pa in one hour or more than 500 Pa in twenty-four hours.

A continuous recording micro-barograph with integrated circuitry, is necessary to automatically monitor pressure changes of this nature and to give alarm that extra precautions are necessary.

Much research it still taking place in this field, both by individual Groups and by the Research Organisation of the Chamber of Mines of South Africa.

5.2 The demarkation and maintenance of escape routes.

Having taken into account the necessity for escape routes in the mining operation, the demarkation and maintenance of such escape routes is of prime importance. This is adequately covered under Regulations 6.2.2., 6.2.3, 6.2.4, 10.11.1., and 10.24.4 of The Mines and Works Act (Act No 27 of 1956).

In the aftermath of a mine fire and/or explosion, men will be disorientated, and visibility may be very poor or non-existent. Workers must therefore be aware of the escape routes to be used and this will be dependent on the locality, nature and extent of the disaster and what communication is possible to enable them to take the best route, with regard to information available.

5.3 Communication of the locality, nature and extent of a disaster.

The best course of action to be taken by management, in the aftermath of a mine fire or explosion, will be dependent on the speed and effectiveness of communication in determining where the disaster has taken place, who and how many people are involved, the hazards to which they have been or will be exposed and the area of influence of the hazards.

Once the situation has been analysed it is equally important to communicate instructions back to the survivors, i.e. where to travel, where not to travel, where to wait, etc.

Regulation 11 .8 states that:

"In the case of a fire occurring in any coal mine or of a fire due to or resulting in the ignition of inflamm-

able gas in any mine, all persons except those dealing with the fire or in services in connection therewith shall be withdrawn from the whole of the underground workings affected and shall only be allowed to return when safe conditions have been restored".

Whilst Regulation 11.10 states:

"The Manager shall not permit or direct any person to remain in or proceed to a ventilating district where there is a fire which cannot be brought under immediate control, unless and until he has satisfied himself that the safety of such person will not be endangered thereby: Provided that this prohibition shall not apply to such persons as are required to bring the fire under control, to conduct investigations or to do work incidental thereto".

All South African coal mines maintain an effective telephone network within the workings to provide contact with the surface and is dealt with under Regulations 6-3.2.8 and 24.16.

5.4 Self Rescuers.

The Government Mining Engineer has stated his intention to amend the Mines and Works Regulations, to make compulsory the carrying of an approved self-rescuer by every worker underground.

This will be mandatory on collieries with effect from the end of 1985.

The introduction of self-contained self-rescuers will therefore become an integral part of the whole strategy to protect workmen in the aftermath of fires and explosions.

The ideal industry specification takes into account the requirements of the Government Mining Engineer and is as follows:-

To meet the approval of the Government

Mining Engineer, the service life of
the apparatus must be 3o minutes at an
inspired rate of 30 1t/min. air equivalent - 100 minutes at rest (10 1t/min).
(A moderate work rate equivalent to
walking at 4.0 km/h at an inclination
of 5.7° approximates to an inspired
rate of 30 1t/min.)

The following guidelines should be
borne in mind:-
1.   The unit should be ergonomically
designed, to suit body contours and
must be comfortable to wear. The ideal
belt when this apparatus is worn, is a
continuous belt of suitable width.
2.   The carbon dioxide concentration
in the inspired air should not exceed
2,5% during the service life.
3.   The rate of oxygen supply should
be demand based.
4.   The oxygen supply should be available
within 10 seconds of removing the self-
rescuer from the carrying case.
5.   Both a chemically generated oxygen
and a compressed oxygen self-rescuer
will be considered. Attention should
be given to "shelf" life.
See (11).
6.   The inhalation temperature should
not exceed 40°C saturated air supply
and 85°C dry air supply and should be
as comfortable as possible.
7.   The breathing resistance should
not exceed 600 Pa.
8.   The self-rescuer, as worn on the
belt, should be as light and as small
as possible.
9.   The self-rescuer should have a
minimum life expectancy of 5 years in
the underground environment.
10. A means of checking for leakage
should be provided.
11. Provision should be made for the
re-charging of spent self-rescuers by
mine staff.

Presently in South Africa, the Chamber
of Mines Research Organisation are
evaluating a variety of self rescuers
in field trials in both low and thick
coal seams. The progress of the
Research Organisation is being monitored
on a regular basis by a Steering

Committee whose members represent the
Technical Advisory Committee of South
Africa, the Collieries Technical Committee of South Africa, the Collieries
Research Advisory Committee, the Committee of Environmental Engineers of South
Africa, the Industrial Hygiene Branch
of the Chamber of Mines, the Rescue
Training Service Department and the
Government Mining Engineers Department.
At the end of the trial period, a
standard procedure document will be
drawn up to advise and guide mine
management on the use and implementation
of self contained self rescuers in coal
mines.

5.5 Refuge bays.

The gold mining industry of South
Africa has successfully employed
refuge bays since 1972, when a team
leader led a body of men into a
development end during a mine fire
and by opening the compressed air
supply, created a positive pressure in
the development end, protecting the
workmen from the effects of the
noxious fumes from the fire.

Halasz, (1985) stated that the refuge
chamber concept defines that each
working section of the mine should
have access to a bay or chamber which
can provide a healthy, respirable
atmosphere, drinking water, telephonic
communication, first aid facilities
and a limited number of self rescuers.

In the event of a fire, which affects
intake air, such a chamber or bay will
provide each man in the workings with
a place of safety to wait in until the
rescue operation is completed.

Implicit in the concept is that the
chambers are in close proximity to
the workforce concentration and the
necessary training procedures to ensure
that each man knows where his place of
refuge is located.

The type of refuge bays are depicted
on the plates overleaf which illustrate

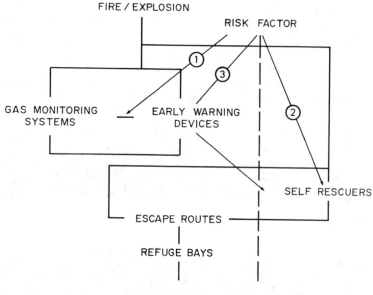

FIRE / EXPLOSION

RISK FACTOR

GAS MONITORING
SYSTEMS

EARLY WARNING
DEVICES

SELF RESCUERS

ESCAPE ROUTES

REFUGE BAYS

RESCUE TEAMS – LONG DURATION
SELF RESCUERS

FIGURE 8

an actual refuge bay at the Western
Areas Gold Mining Company Limited.

The procedural philosophy for the use
of refuge chambers depends upon making
each and every employee aware of the
existence and location of the facility
in his section via initial as well as
follow-up training. The idea is that
when threatened by a dangerous atmosphere people retreat to the chamber,
inform the necessary authorities of
their predicament and await the
successful completion of rescue
operations.

The use of refuge bays as a back up
to self rescuers are presently being
investigated for implementation into
the coal mines of South Africa.

Figure 8 is a schematic line diagram
which indicates how the workmen are
protected in the event of a fire or
an explosion.

6. CONCLUSIONS

The authors have attempted to give
both the reader and the audience
present today an insight into both
the causes of underground fires in
South African coal mines and the
means by which approaches have been
made over the past twenty five years
to protect workmen in the event of a
fire or an ignition.

7. REFERENCES

1. Watson, J.R. (1963)., The opening
    of an area which was sealed off
    after an explosion. Internal
    Report of the Coal Division of
    the General Mining and Finance
    Corporation Ltd., South Africa.

2. Guney, M. (1975)., Equilibrium
    humidities and water iso therms
    in relation to spontaneous

The entrance to the refuge bay must be clearly marked.

Necessary equipment inside a refuge bay includes a first aid unit, self rescuers and a telephone.

Sufficient seating should be provided to accommodate every man working in the section on the busiest shift.

After Halasz, (1985)

52

heatings of coal. Proceedings
of the International Mine
Ventilation Congress, Johannesburg, September.

3. Nakata, T. (1980)., Technologies
of deep mine safety under difficult conditions - spontaneous
combustion and gas outbursts at
Sumitomo - Akabira coal mine.
Fifth International Conference
on Coal Research. Sept 1 - 5.
Dusseldorf, Federal Republic of
Germany.

4. Badenhorst, G.P., Buchan, I and
Visser, C.P. (1977)., The Sub-
Committee on the problems relating
from the occurrence of methane
and spontaneous combustion.
Chamber of Mines of South Africa.

5. Morris, R, (1979)., Fire fighting
in South African Collieries.
Internal report, Amcoal.

6. Thompson, G.C. (1972)., The under-
ground fire at Coalbrook Collieries.
Colliery Managers' Association of
South Africa.

7. Joubert, F.E., Buchan, I.F. and
Beukes, J.D.R. (1976)., Journal
of the Mine Ventilation Society
of South Africa, April, P74 - 80.

8. Joubert, F.E. (1975)., Technical
memorandum No. 7 of 1975: Gas
analysis during control of fires
in collieries.

9. Morris, R. (1984)., Incompleted
Ph.D Thesis, Nottingham University
entitled "Spontaneous combustion
in coal mines and the interpretation of the state of a mine fire
behind the stoppings".

10 Holding, W. (1981)., Acute hazards
of explosive gases, their detection
and control. Journal of the Mine
Ventilation Society of South Africa
January.

11 Thorp, N. (1981)., A review of
ventilation practice on South
African Collieries. Presidental
Address to the Mine Ventilation
Society of South Africa. Vol. 1,
NOV., Vol. II, December.

12 Petrick, A. J. et al, (1975).,
Reports of the Commission of Inquiry
into the Coal Resources of the
Republic of South Africa. Pretoria.
April.

13 Halasz, L. (1985)., Establishment
and use of refuge bays at the
Western Areas Gold Mining Company
Limited. Mine Safety Division,
Loss Control Survey. A Chamber
of Mines of South Africa
Publication. Vol 4/ No 2, May.

14 The Guidelines and Recommendations
for the Protection of Workmen in
the Aftermath of Colliery Fires
and Explosions. Johannesburg
December, 1984.

2. Ventilation system analysis

Leakage between intake and return airways in bord and pillar workings

MICHAEL J.MARTINSON
University of the Witwatersrand, Johannesburg, South Africa

ABSTRACT: Leakage past stoppings separating intake and return airways in bord and pillar workings may have a detrimental effect on the ventilation of remote (distal) working places. Four ignitions of firedamp, in which stopping leakage may have been a contributory factor, are briefly reviewed to illustrate the possible consequences of inadequate ventilation. The leakage process is then examined in greater detail in separate discussions on the flow characteristics of airways and flow past stoppings. Finally, results from a computer network analysis of data relating to airways and stoppings in hypothetical bord and pillar workings are presented to illustrate the leakage phenomenon in quantitative terms.

1 INTRODUCTION

Mines are ventilated to abate a variety of atmospheric hazards, and notionally at least the risk of injury associated with each hazard can be maintained at predetermined levels by controlling ventilation flowrates throughout the system. For economic or technological reasons airflow may be supplemented by some form of air conditioning and other hazard abatement measures, but nevertheless controlled air circulation continues to provide primary protection for underground personnel.

The efficacy of a mine ventilation system in terms of hazard abatement depends ultimately on the quality of the data available for design and control purposes relating to:
- the prevalence of hazardous agents and conditions in the mine atmospheric environment;
- the dose-response relationship for each agent and condition;
- the flow characteristics for the entire ventilation system.

The present paper focuses on the flow characteristics of main intake and return airway systems in bord and pillar workings, where parallel intake and return roads are typically separated by numerous stoppings constructed in connecting crosscuts. Under operating conditions it may be difficult, if not practically impossible, to seal stoppings completely, and since ambient atmospheric pressure is normally lower in return airways than in intakes, air will leak from the intake to the return. The leakage rate past each stopping is a function of:
- the magnitude of the pressure difference across the stopping;
- the detailed geometry of the leakage flowpath(s);
- the nature of the flow regime in each flowpath.

The differential pressure across any discrete stopping depends in turn on where the stopping is located with respect to the rest of the ventilation system, the flow characteristics of the intake and return airway systems, and the respective flowrates.

Although leakage past a single stopping may appear insignificant, as the number of stoppings increases the aggregate leakage may eventually have a drastic effect on the ventilation of more remote working places. Unless appropriate steps are taken to offset the loss of air due to leakage the risk of injury attributable to all atmospheric hazards may increase accordingly. Historically the hazards associated with inadequate ventilation are dramatically illustrated by ignitions of firedamp in bord and pillar coal mines.

Based on views expressed in the literature and conclusions reached in the present study, it would seem that in the past less attention has been devoted to leakage in bord and pillar workings than might have been expected in relation to the risk of injury and the number of persons at risk.

2 IGNITIONS IN B & P WORKINGS IN COAL

In the last twenty years there have been several disasterous ignitions of firedamp in remote bord and pillar workings, of which four incidents are briefly reviewed below. In no case is there any evidence to suggest that stopping leakage (as opposed to direct short-circuiting) was solely or directly responsible for dangerous accumulations of firedamp, but nevertheless it seems likely that leakage was a contributory factor in all four cases. Apparently the subject was not canvassed in depth at official inquiries into the incidents, but there is a telling reference to leakage in the first incident considered below.

2.1 Bulli colliery, Australia

On 9 November 1965 an ignition of firedamp occured in a pillar extraction section of Bulli colliery in New South Wales, starting a fire that killed four mineworkers trapped in the section (Bulli 1966). The section was located just under 6,5 km from the inlet portal, and gas had been detected at the edge of the goaf on several occasions prior to the ignition. On the day in question an accumulation of firedamp in a shunt was apparently ignited by a piece of smouldering wood jammed in the brake assembly of a shuttle car. Ventilation arrangements were criticised by Judge Goran, who conducted the inquiry, and in his observations and recommendations the judge commented on the need in New South Wales to study 'the science of ventilation'; in this connection he refers (p 38) to 'evidence of the loss in certain ventilation splits of a substantial quantity of air and this loss could not be explained by anybody at the inquiry'.

2.2 Wankie No 2 colliery, Zimbabwe

On 6 June 1972 a massive explosion in Wankie No 2 colliery killed 427 mineworkers and resulted in the total destruction and closure of the mine. Before the accident the colliery produced about 2 Mt/a from several sections using conventional bord and pillar mining methods. Gas had been reported in the mine on several occasions prior to the explosion, and at least two localised ignitions had occured in the preceding decade.

On the limited evidence available the commission of inquiry appointed to investigate the disaster conjectured (Wankie 1973) that the explosion started as a methane explosion ignited by shotfiring in the Matura Main section located about 4,8 km from the Kamandama upcast shaft and 2,7 km from Bisa upcast shaft. The colliery ventilation system is criticised in the commission's report, but leakage is not specifically mentioned.

2.3 McClure No 1 mine, USA

On 23 June 1983 an explosion in the 2 Left entries of McClure No 1 mine in Virginia killed seven of the ten miners working in the continuous miner section (McClure 1983). Total methane emission rates of 91 000 m^3/d and 74 000 m^3/d for the whole mine under working conditions were measured by MSHA inspectors shortly before the explosion. At the time of the accident entries in sets of three and four were being driven from mains to establish blocks for longwall mining; the 2 Left entries were about 2,13 km from the intake shaft, and nine hours before the explosion a 2 Left crosscut holed into an adjoining set of entries, thereby affecting the ventilation of the 2 Left entries. In the words of the MSHA report 'The volume and velocity of air became inadequate to dilute, render harmless and to carry away flammable gases which were liberated in the area'. So-called area leakages are shown on ventilation schematics forming part of a post-disaster ventilation study (appendix L) but are not discussed in either the ventilation report or the main report.

2.4 Hlobane colliery, South Africa

On 12 September 1983 a gas and coal dust explosion in the neighbouring 5 and 10 sections of Hlobane colliery in Natal resulted in the death of 68 miners. An explosion in the same mine in 1944 killed 57 miners. Before the 1983 explosion eleven entries were advancing in a northerly direction using conventional methods in one-metre high workings. As entries approached an E-W dyke they were stopped and were being replaced by fifteen entries advancing in an easterly direction, but the new layout was incomplete when the explosion occurred. The two sections were at the time about 6,5 km from the intake portal, and difficulties had already been experienced in complying with the minumum ventilation standards prescribed in the regulations. The explosion occurred soon after the start of a Monday morning shift; during the preceding Saturday morning shift a crosscut in one of the sections holed into an adjoining sub-system and further reduced the amount of air entering the sections. The holing was not closed before work stopped an hour or two later for the weekend, and on Monday an extensive accumulation of methane was apparently ignited by a battery-operated scoop in a non-flameproof condition.

Within broad limits it may be supposed that at any given time there is, for each working place in a coal mine ventilation system, a relationship between the amount of air supplied to the place and the probability of the formation of a dangerous accumulation of firedamp. Weak and perhaps intermittent flows clearly increase the risk of a firedamp ignition, but may also have an indirect detrimental effect in that persons working in the area become so accustomed to indeterminate ventilation that they ignore or fail to notice the sort of flow aberration that presaged the Hlobane colliery explosion.

3 RESISTANCE OF INTAKE AND RETURN AIRWAYS

The aerodynamic resistance of intake and return airways is of fundamental importance in relation to all aspects of ventilation design, control and economics, and is also highly relevant in the present context since static pressure gradients in intake and return airways determine pressure differences across stoppings and thus affect leakage rates. Leakage also influences pressure gradients on both sides of stoppings, but for convenience is considered separately below.

Leakage apart, pressure gradients in airway systems in bord and pillar workings are primarily determined by the flowrate, the system length, and the overall resistance coefficient for the system. Discussion here is confined to resistance coefficients.

Various geometrical factors combine to fix the overall resistance coefficient of an airway system, among which the number of parallel roads included in the system often has a major impact. Other geometrical variables can be grouped into three main categories.

The first group consists of variables directly linked to the exploitation process itself, viz:
- the height and width of bords;
- the angle at which crosscuts intersect with roads;
- the extent to which corners of pillars are rounded-off by mining operations;

- pillar spacings;
- expansion/contraction characteristics associated with changes in cross-sectional area;
- the roughness of roof, floor and side-walls;
- changes in direction of the workings.

Secondly, changes in geometry following the extraction process may have long-term effects on aerodynamic resistance, such as:
- installation of permanent support in the airstream;
- the presence of fixed equipment such as conveyors and bunkers in roads;
- the location (and possibly mode of construction) of stoppings in crosscuts separating intakes and returns;
- the installation of overcasts, undercasts and regulators;
- spalling of roofs and sides;
- material and equipment dumped or abandoned in airways.

Thirdly, movements of conveyances and equipment in airways, repair operations and the like may effect transient changes in the resistance of airway systems.

The importance in relation to hazard abatement of being able to predict resistance coefficients with reasonable accuracy need hardly be stressed, but paradoxically there is a dearth of reliable design data relating to bord and pillar airways in the literature. Most of the available information has been ably summarised in a Pennsylvania State University (PSU) study undertaken for the US Bureau of Mines (Ramani et al 1977), where the authors comment (p 12) that 'There is relatively little published information in the United States in recent years on investigations of pressure losses due to friction in mine ventilation systems', and add that the pioneering studies of Greenwald and McElroy in the 'twenties are still the standard references on the subject. Later the authors stress the need for rigorous field studies - a need apparently shared on most mining fields where bord and pillar mining is practised.

The PSU study also examines the inconclusive evidence on the applicability of the square-law relationship to flow in mine airways and concludes that the relationship 'is not quite clear'.

4 LEAKAGE PAST STOPPINGS

Leakage in the context of mine ventilation can be defined as any unintended and undesired split or short-circuit of air along a flowpath or paths in parallel with the design flowpath or paths. In the circumstances presently under consideration, leakage reduces the amount of air available for ventilating active working areas - where the need for good ventilation is generally greatest - and may result in the under-utilization of airways or sections of airways due to air velocity limitations. Likewise the total leakage handled by the system fan(s) represents a waste of fan capacity and input power based on the prevailing fan total pressure and efficiency.

Leakage may occur in various circumstances in mine ventilation systems, but here discussion is limited to leakage past permanent stoppings constructed in crosscuts connecting intake and return airways.

Various methods of constructing stoppings are used, most of which consist basically of building a wall of hollow-core concrete blocks across the full width and height of the stoppings. Both dry- and wet-wall construction methods are used, while one face of the completed wall may be plastered and/or sealed. Careful site selection and preparation are essential prerequisites for effective stoppings.

The PSU study (Ramani et al 1977) contains a useful review of the literature on stopping construction and sealing; since then there have been further publications on the subject, including a US Bureau of Mines circular (Timko et al 1983) and two South African contributions (van der Bank 1983; Coetzer 1985).

Assuming that a pressure difference is applied across a completed wall, four types of potential flowpath can be visualised:
- diffusion paths through porous wall material;
- cracks in building blocks, between blocks, and between peripheral wall surfaces and surrounding coal or rock surfaces;
- diffusion paths through porous coal or rock;
- cracks in the surrounding coal and rock.

Flow in all four flowpaths simultaneously in response to a steady pressure difference across the stopping may be subject to different flow regimes, with the result that the overall pressure-leakage relationship may be more complex than is normally the case for airways. Furthermore, both leakage coefficients and flow exponents may change with time due to strata settlement, spalling and deterioration of the fabric of the wall.

As a first approximation it is assumed that the total leakage past a stopping is directly proportional to the cross-sectional area of the stopping.

Most of the published data on leakage coefficients and flow exponents are summarised in the PSU study, and some additional information is presented in the three more

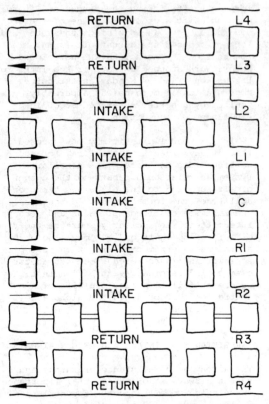

Fig. 1. Plan of intake and return roads separated by stoppings in hypothetical bord and pillar mine (bords 6,0 m wide, 2,5 m high; pillars 10,0 m by 10,0 m).

recent references mentioned above. However there is still considerable uncertainty surrounding stopping leakage and the subject warrents further rigorous study.

5 LEAKAGE IN HYPOTHETICAL B & P WORKINGS

To illustrate the detrimental effects of leakage past stoppings on air quantities delivered to distal workings, sets of data based on standardised hypothetical workings were used as input for a ventilation network analysis program developed for the Chamber of Mines of South Africa for use on the Witwatersrand University computer system (Harrison 1974). Characteristics of the basic layout are detailed below, where the pillar size is based on a factor of safety of 1,8 (Salamon 1967):

Depth of floor : 100 m below surface.
Inclination   : horizontal.
Working height :   2,5 m.

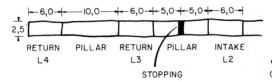

Fig. 2. Transverse section (part) through hypothetical bord and pillar mine shown in Fig. 1 (dimensions in metres).

Bord width   : 6,0 m.
Pillar size   : 10,0 m by 10,0 m.
Number of entries: 9.
Intake airway   : 5 centre roads.
Return airway   : 2 outside roads on
                      either side.

Permanent stoppings separate intake and return airways as shown in figures 1 and 2.

The hypothetical workings are not modelled on any specific prototype, but are thought to represent reasonable South African practice (Hardman et al 1971), and possibly compare with bord and pillar layouts found in other parts of the world.

The network used in the computer program is depicted schematically in figure 3; to simplify the network return roads R3 and R4 were combined with roads L3 and L4 to form a single return airway of four parallel roads, and each pair of 15 m^2 stoppings was replaced by a single 30 m^2 stopping.

Fixed values of resistance coefficients for 16-metre modules of intake and return airway were used in all the simulation exercises, and were based on preliminary results obtained from 1:100 scale model tests performed by the author using portions of the hypothetical layout illustrated in figure 1:
Intake airway : R_i = 0,000016 (5 roads).
Return airway : R_r = 0,000020 (4 roads).
R = (Pressure loss Pa)/(Flowrate m^3/s)^2.

Flow distributions were obtained for lengths of intake (and return) airway ranging from 25 pillars at 16-metre centres (0,40 km) to 400 pillars at the same spacing (6,40 km), with five intermediate lengths of 50, 75, 100, 150 and 200 pillars.

Three values of stopping leakage coefficients were used in the program; the upper and lower values correspond roughly with extreme values quoted in the PSU study (Ramani et al 1977), while the so-called moderate value is thought to approximate a median value in South African bord and pillar coal mines in recent times:
High leakage   : S_h =   50 (30 m^2).
Moderate leakage : S_m = 1000 (30 m^2).
Minimal leakage : S_s = 50000 (30 m^2).
S = (Pressure diff Pa)/(Leakage m^3/s)^2.

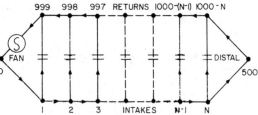

Fig. 3. Schematic ventilation network for hypothetical bord and pillar mine showing intake airways, return airways, and leakage paths past stoppings (N is the number - less than 500 - of pillars in the intake airway).

Note that the simulation program used in the present study applies a square-law relationship to flow in all branches (including leakage branches) in solving the network.

In each run the network included a single fan discharging notionally to atmosphere; for convenience in reporting results the fan in every case operated along a horizontal pressure-volume characteristic at fixed static pressures of 200 Pa, 400 Pa and 600 Pa. A flow of 375 m^3/s through the fan represents an air velocity of 5 m/s at the start of the intake airway system and 6,25 m/s in the return roads closest to the fan; larger flowrates may be impracticable but were used for illustrative purposes.

Results are summarised in table 1 and are seemingly self-explanatory. However a few points warrant comment or stress:
- the figures clearly illustrate the detrimental impact that leakage may have on the ventilation of remote working places;
- leakages that appear insignificant over short distances may become serious over longer distances and at higher pressures, eg compare moderate leakage over 50 pillars at 200 Pa (4 per cent) with 400 pillars at 600 Pa (55 per cent);
- taking the second case (55 per cent leakage) and ignoring velocity pressure, if the overall efficiency of fan and motor is 0,65 and the cost of power is R0,035/kWh, leakage wastes R47 000 out of annual power costs of R86 000;
- in practice many South African coal mines have longer airway systems and operate at higher fan pressures (van der Bank 1983);
- in real ventilation systems stopping leakage coefficients may differ markedly from one stopping to another but the variation can not be modelled at present due to lack of in-situ measurements;

Table 1. Fan, leakage and distal flowrates (m^3/s) in hypothetical bord and pillar mine for selected fan pressures (Pa), numbers of pillars at 16,0 m centres N, and leakage coefficients S (sub h = high leakage, m = moderate leakage, s = minimal leakage).

Number of pillars N	Flow elements	Fan static pressure								
		200 Pa			400 Pa			600 Pa		
		S_h	S_m	S_s	S_h	S_m	S_s	S_h	S_m	S_s
25	Fan	490,1	475,6	472,0	693,1	–	667,5	–	–	–
	Leakage	31,8	7,1	1,0	44,9	–	1,5	–	–	–
	Distal	458,3	468,5	471,0	648,2	–	666,0			
50	Fan	370,6	341,9	334,5	524,1	483,6	473,1	–	–	–
	Leakage	62,9	14,4	2,0	89,0	20,5	2,9	–	–	–
	Distal	307,7	327,5	332,5	435,1	463,1	470,2	–	–	–
75	Fan	326,4	285,1	274,0	461,7	403,2	387,5	565,5	493,8	474,1
	Leakage	91,6	21,7	3,1	129,7	30,7	4,4	158,9	37,6	5,5
	Distal	234,8	263,4	270,9	332,0	372,5	383,1	406,6	456,2	468,6
100	Fan	305,4	252,9	238,2	432,0	357,5	336,8	529,1	437,9	412,5
	Leakage	117,9	28,9	4,2	166,8	40,5	5,8	204,3	49,8	7,1
	Distal	187,5	224,0	234,0	265,2	317,0	331,0	324,8	388,1	405,4
150	Fan	288,2	217,7	196,2	407,7	307,8	277,4	499,3	377,0	339,7
	Leakage	161,9	42,4	6,2	229,0	59,9	8,7	280,4	63,4	10,7
	Distal	126,3	175,3	190,0	178,7	247,9	268,7	218,9	303,6	329,0
200	Fan	489,7	199,5	171,6	399,8	282,0	242,8	282,7	345,5	297,0
	Leakage	338,2	55,2	8,3	276,1	78,0	11,7	195,3	95,6	13,9
	Distal	151,5	144,3	163,3	123,7	204,0	231,1	87,4	249,9	283,1
400	Fan	277,9	175,5	127,6	393,8	248,2	180,4	482,8	303,8	221,0
	Leakage	257,2	97,3	16,3	363,4	137,6	23,0	446,2	168,4	28,3
	Distal	20,7	78,2	111,3	30,4	110,6	157,4	36,6	135,4	192,7

- the condition of minimal leakage can be achieved but entails good stopping design and siting, careful construction and sealing, and periodical inspection and maintenance.

6 CONCLUSIONS

A recent MSHA manual (MSHA Academy 1984) says that 'in some (bord and pillar coal) mines as much as 70 percent of the intake air may leak into the return before reaching the last open crosscut'. Although the statement refers to mines in the United States, similar situations probably obtain in other jurisdictions as well (van der Bank 1983).

High leakage rates can conceivably be accomodated in a rational ventilation control system to ensure that all working places are nevertheless adequately ventilated; however the available evidence suggests that this state of affairs is seldom found in practice, and that more often high leakage rates are synonymous with poor ventilation in the more remote workings.

Although leakage was not held to have been directly responsible for any of the four firedamp ignition incidents reviewed in the paper, inadequate ventilation, whether of a temporary or protracted nature, was apparently a major causative factor in all four incidents, and serves as a stark warning of the dangers that may be associated with excessive leakage.

Leakage can be controlled by paying careful attention to the design and maintenance of stoppings, and perhaps in appropriate circumstances by using chain or barrier pillars to reduce the number of stoppings separating intakes from returns. Leakage costs in economic terms may be high and may alone justify the cost of effective leakage control programmes.

7 ACKNOWLEDGEMENTS

The author gratefully acknowledges the assistance of Mr Frank von Glehn, Chamber of Mines of South Africa, in modifying the computer program used in the hypothetical mine exercises.

8 REFERENCES

Bulli 1966. Report of Judge A.J. Goran...
(on an) accident at Bulli colliery on 9th
November, 1965. Sydney, NSW Government
Printer.

Coetzer, W.J.J. 1985. Use of a sealant
paint to seal leakage through ventilation
stoppings constructed from precast concrete blocks. J.Mine Vent.Soc.S.Africa
38:9-11.

Hardman, D.R. & A. Richardson 1971. The
effects on productivity of some dimensional changes in bord and pillar mining.
Johannesburg, Chamber of Mines of South
Africa. (RR 12/71)

Harrison, M.W. 1974. A manual for users
of the ventilation network program on the
computer at the University of the Witwatersrand. Johannesburg, Chamber of
Mines of South Africa. (RR 43/74)

McClure 1983. Report of investigation:
Underground coal mine explosion, McClure
No 1 mine...McClure, Dickenson County.
Washington DC, Mine Safety and Health
Administration, US Department of Labor.

MSHA Academy 1984. Coal mine ventilation
awareness program. Washington DC, Mine
Safety and Health Administration, US
Department of Labor.

Ramani,R.V., R. Stefanko & G.W. Luxbacher
1977. Advancement of mine ventilation
network analysis from art to science. Vol
4: Sensitivity of leakage and friction
factors. Washington DC, Bureau of Mines,
US Department of the Interior. (OFR 123(4)
- 78)

Salamon, M.D.G. 1967. A method of designing
bord and pillar workings. J.S.African
Inst.Min.Met. 68:68-78.

Timko, R.J. & E.D. Thimons 1983. New techniques for reducing stopping leakage.
Washington DC, Bureau of Mines, US Department of the Interior. (IC 8949)

van der Bank, P.J. 1983. Moving air from
surface to the last through road in the
section. J.Mine Vent.Soc.S.Africa 36:53-58.

Wankie 1973. Report of the commission of
inquiry into the Wankie colliery disaster
and general safety in coal mines in
Rhodesia. Salisbury (Harare), Government
Printing and Stationery. (Cmd RR 4-1973).

Design of a ventilation system for the Ruttan deepening project

L.D.NEL
Sherritt Gordon Mines Ltd, Leaf Rapids, Manitoba, Canada

ABSTRACT: Ventilating multilense, multilevel mines has long been a ventilation engineer's dilema. This problem is compounded by a mining method employing longhole, open stoping and trackless diesel equipment. The Ruttan Operation of Sherritt Gordon Mines Limited is such a mine. In 1983 it was decided to expand the mine which involved deepening the underground operation to 430 metres below the existing mine. The paper deals with the design of the ventilation system for the operation emphasizing the application of CANMET's "Ventilation Network Analysis Program and Plot Program" to test the ventilation network and to predict ventilation changes in the existing mine during the development phase. The use of energy and massflow as opposed to pressure and volumetric flow is also discussed with reference to its application during the design process. A brief examination is done of the programs potential as an everyday tool and as an aid in trouble-shooting.

1 HISTORY

Sherritt Gordon's mining division commenced exploration along the then proposed 391 highway, north of Thompson, Manitoba. The Ruttan anomalies indicated by airborne surveys led to more intensive ground geophysical surveys. The first diamond drill intersected a wide mineralized zone containing copper and zinc in April 1969. This mineralized zone became known as the Ruttan ore body which presently supports a 6,000 tonne/day underground operation and the Town of Leaf Rapids. The mining operation started as an open pit which in time was replaced by an underground operation extending 430 metres below surface. In 1983 it was decided to deepen the operation to 800 metres below surface. The deepening is still in the late development phase and will be in full production by 1986.

2 INTRODUCTION

Sherritt Gordon Mines Limited is presently expanding its Ruttan underground operation. Successfully ventilating the existing mine is in itself an onerous task to say nothing of expanding the system into the lower mine development. Ruttan is a multilevel, multilense mine which employs an open stope mining method and trackless diesel equipment to extract the copper and zinc bearing ore.

The complexity of Ruttan's ventilation network and the problems experienced with controlling a descensional ventilation flow pattern in the existing mine greatly influenced the conceptual design of the ventilation system for the expanded mine. Simplicity, ease of control and the long term economic viability became the key factors in the decision to pursue an ascensional concept in the new system design. During the design process the long term ventilation plan was simulated on the CANMET Ventilation Network Program to determine its practicality. As the project progresses simulations are being done to assist in predicting changes in airflow and energy levels in the existing ventilation system of the upper mine.

3 LIMITATIONS ON VENTILATION SYSTEMS AT RUTTAN

At Ruttan, for any ventilation system to be effective, the design of such a system must overcome certain inherent difficulties.

The major factor is the large variation in natural ventilation energy levels throughout the mine. Natural ventilation energy is that energy created by the conversion of heat into mechanical energy which causes a natural draft similar to that created in smoke stacks and industrial cooling towers.

The causes of these wide variations are very obvious. Ruttan is presently mining under an open pit and in some cases has broken through into the pit leaving a number of large openings at elevations much lower than that of our main fans. This, together with the rapid temperature changes in Northern Manitoba, as can be seen in Figure 1, creates unstable energy levels within the mine itself. The end result is an ever changing energy conversion rate which actually reverses airflows in the system and has a de-stabilizing effect on the operating points of our main fans.

Another limitation to be contended with is the shape and economics of our ore body which dictates a scattered open stope mining method with delayed hydraulic backfill. This mining method, although practical and economical at Ruttan does not lend itself to a stable ventilation system.

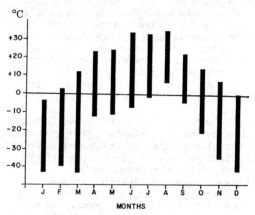

Figure 1. 1984 Temperature Range

4 THE EXISTING DESCENSIONAL VENTILATION SYSTEM (Surface to 430mL)

The primary ventilation system (Figure 2), consists of a 7.6 metre diameter intake raise in the footwall and two 4.9 metre diameter exhaust raises in the hangingwall. These raises, driven vertically from 260 metre level to surface individually form the main arterials for the intake and exhaust systems. From 260 level the intake system conveys air to each of the lower levels in a descensional pattern via three smaller diameter raises.

Both the east and west exhaust vent raises remove polluted air from all levels in a similar fashion. The isolation of the main vent raises on 260 level and a series of regulators and bulkheads on each level facilitate controlled splitting of ventilating air to the production areas.

The intake system is serviced by four 2.1m Sheldon axico fans with 186 Kw motors.

However, only three of the four fans operate at any one time, the fourth is kept as a backup unit in case of breakdown or an emergency that may require extra fan power. The east and west exhaust systems are serviced by three 1.8m diameter Sheldon axico fans and three 1.5m diameter Joy vane axial fans respectively. The number of fans operating at any one time on the exhaust depends upon the dynamics of the underground operation.

Finally, our ore passes are ventilated under a negative pressure by a raise driven between the ore passes down to the crusher chamber. Two 1.5m Joy fans exhaust air via this raise from the ore passes to surface.

5 THE NEW ASCENSIONAL SYSTEM (Surface to 860mL)

The ascensional system of ventilating Ruttan Mine stems from the problem of maintaining sufficiently high velocities in the workings without unduly increasing the overall ventilating quantity. The system is in essence an extension of the existing system (as shown in Figure 2) with one major difference however. Intake air will be ventilating the workings from the bottom of the mine as it ascends to surface rather than from the top down as previously described in this paper.

The bulk of intake air will flow via the existing intake system to 340 metre level from where a 3.66m diameter raise will convey it to 730 and 800 metre levels, the deepest production levels of the mine. These two levels set in parallel connect to the bottom of the east intake raise from where air is distributed to all the remaining working levels. As it ascends to surface the actual airflow on the levels is from east to west and from footwall to hangingwall in all cases. The flow constitutes a uni-directional pattern which is easier to control and allows for more accurate prediction of changes in airflow as the levels expand.

To establish the ascensional system requires very little modification to the existing raise system besides the installation of a few concrete bulkheads and minor slashing to enlarge some of the smaller raises.

The system will be split into two ventilation districts - an upper district and a lower district by a series of ventilation doors on 430 level. The purpose of dividing the network into two districts is to prevent polluted air created by maintenance activities and mining operations on 260 level and 370 level from entering the lower section of the mine via the vehicle decline.

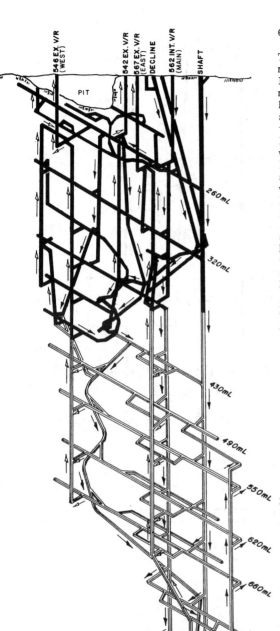

Figure 2. Ruttan Ventilation Raise
Network

LEGEND:
◄── INTAKE
◄── EXHAUST
▬▬ OLD RAISE NETWORK
═══ EXPANDED RAISE NETWORK

6 ASCENSIONAL vs DESCENSIONAL CONCEPTS

The discussion of these two concepts refers
mainly to the Ruttan Mine system as seen in
Figures and but is valid for many other
mines. The descensional airflow pattern
has merit in that it is easier to establish
an airflow direction opposite to the direction of these diesel units travelling from
loading point to the dumping point; in other
words, when diesels are operating at maximum
load. The reason being that as the intake
air reaches a level it splits both east and
west creating a bidirectional flow on the
level which facilitates bidirectional mucking.

Although this counterflow is desirable,
air usage is wasteful and expensive. At
Ruttan, because of the large drift sizes,
the resultant low air pressure differential
across tramming routes and the cost of
heating large volumes of air, relatively
low air velocities with very low ventilating
pressures prevail. It is not uncommon for
an LHD to cause airflow reversals in neighbouring drifts while tramming. It is therefore very difficult to establish a stable
and reliable ventilation system.

On the other hand the ascensional system
as previously described allows all the ventilating air for the level to travel in a
single direction from east to west.
Immediately higher pressure differentials
are realized and therefore higher air velocities along main tramming routes. The
advantage of higher air velocities in a
highly mechanized mine is self evident and
to further justify the ascensional concept
the whole system will require less air -
260.00 m³/s as opposed to 354.00 m³/s. A
further consideration is the relative ease
with which such a system can be controlled.

The economic advantage of the ascensional
system is based on the fact that this system
requires 94.00 m³/s less than the descensional system. The economic comparison below represents the projected total cost
savings over a seven year period.

Descending:-

Life of Mine	(7) Seven Years
Air Quantity	354.00 m³/s
Annual Propane Cost @ $0.25/Litre	$ 1,497,600
Total Operating Cost	$10,483,200
Capital Cost	$ 2,536,586
TOTAL COST:	$13,019,786

Ascending:-

Life of Mine	(7) Seven Years
Air Quantity	260.00 m³/s

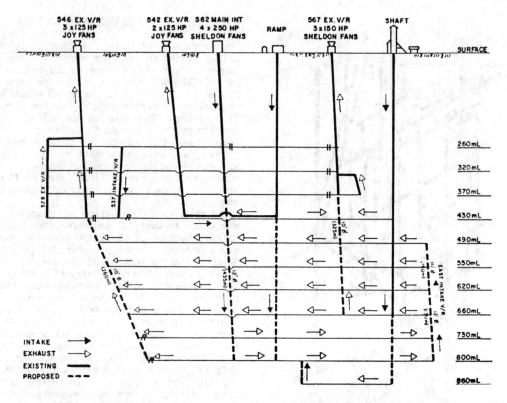

Figure 3. Ruttan Mine Ventilation Circuit for 260.00 m³/s Ascensional Pattern

Annual Propane Cost @ $0.25/Litre	$ 1,098,240
Total Operating Cost	$ 7,687,680
Capital Cost	$ 2,894,449
TOTAL COST:	$ 9,993,689

7 THE DESIGN PROCESS

The design procedure adopted for the ventilation system expansion involves four stages. The first stage being the conceptual design followed by a more specific economic and energy mass flow distribution study. The air flows were then simulated on computer using the CANMET Ventilation Network Analysis Program. The final stage being the commissioning of the system and the planning required to facilitate the interfacing of the new system with the old system.

8 THE CONCEPTUAL DESIGN

The conceptual design involved assessing the existing ventilation system and determining whether it would be feasible to either incorporate the old system into an

ascensional or descensional design for the lower mine, or isolate the lower mine forming a new ventilation district with an independent ventilation system. Economics immediately ruled out the later concept and based upon the results of further investigation the ascensional concept was adopted.

The two concepts were used in separate base case studies which revealed the following:- (i) the economic airway sizes in the descensional pattern called for raises of varying size. At virtually each level the raise diameters would have to be increased or decreased according to the air split on that level; (ii) existing fan arrangements would not be able to cope with the pressure losses created by these raise size changes; (iii) because the ascensional system required less air and fewer passes by raise borers the existing fan arrangements would be suitable; (iv) the airflow direction in the ascensional system features a major safety consideration in that the two main entrances to any level would be under a negative pressure ensuring clean air access routes in most emergencies.

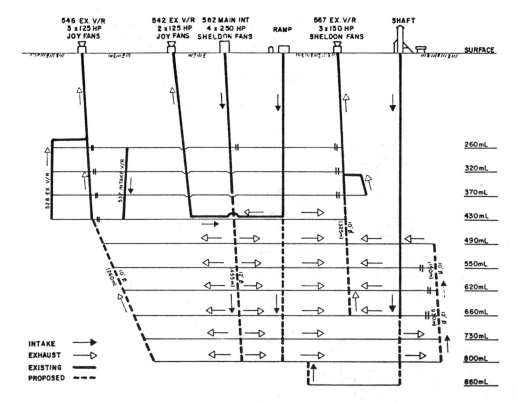

Figure 4. Ruttan Mine Ventilation Circuit for 354.00 m³/s Descensional Pattern

10 THE ENERGY MASS FLOW STUDY

The terms energy and mass flow are rarely used in a ventilation context in North America by the on site ventilation engineer, however, this powerful concept allows one to eliminate the frustration and confusion associated with converting volumes and pressures to representative values with common densities for comparison and calculation purposes. Both energy and mass flow have apparent density as one of its factors, the other being pressure and volume respectively.

The expressions energy and mass flow are best explained in the following equations.

10.1 Air Mass Flow (Equation 1)

$M = Q.w$ where M = Mass Flow Rate (kg/s)
Q = Volume (quantity) (m³/s)
w = Density (kg/m³)

10.2 Air Power (Equation 2)

$W = P.Q = E.M$ therefore $E = P/w$ where

E = Energy (Nm/kg), furthermore $P = RQ^2$
where R = Resistance (Ns²/m⁸) or $E = R'M^2$
therefore $R' = P/Q^2.w^3 = R/w^3$ (Equation 3)
where R' = Energy Basis Resistance
(Nms²/kg³).

Using the above equations the air distribution throughout the mine was calculated in accordance with the requirements of Kirchoff's Laws.

The whole system was designed with the hope of being able to establish an economically viable system using the existing fan power. To establish the initial raise network as a basis for the distribution calculation a modification of Atkinson's Equation was used.

Atkinson's Equation:- $R = \frac{k.C.L}{A^3} \times \frac{w}{1.2}$

Modified this reads:- $r = 5\sqrt{\frac{2k.C.L.w}{\pi^2.R.1.2}}$
(Equation 4), where
r = Radius (m), k = Friction Factor
(Ns²/m⁴), L = length (m), w = Density
(kg/m³) and A = Area (m²).

The resulting raise sizes were compared to the economic raise sizes calculated from

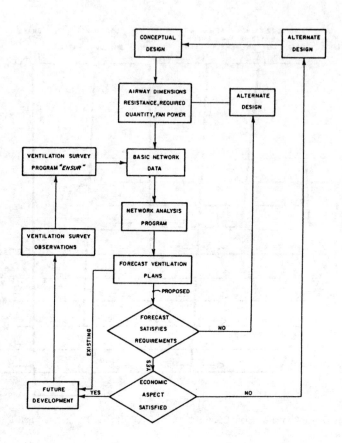

Figure 5. Flowchart: The Use of the Ventilation Network Program in the Ruttan Design Process and Development Phase.

the economic air velocity for the expansion project.

Economic Velocity $V = \dfrac{Q}{\frac{\Pi}{4} \cdot D^2}$ where

$D = \dfrac{dPV}{dD}$ Most Economic Diameter (m)

and PV = Present Value of Total Raise Cost including power costs ($).
Q = Quantity (m³/s)

11 THE CANMET VENTILATION NETWORK PROGRAM

The ascensional design was digitally simulated and analysed using this program in order to calibrate and "fine tune" the design. To my knowledge it is the first time that this program has successfully been used to simulate a multilevel ventilation system in North America. The results were very favourable. Not only did it verify the results of the energy mass flow study previously mentioned but it also proved its own capabilities to the extent that we now use the program to predict mass flow and energy changes created by changes in natural ventilating energies and future development. In the future it is planned to update the simulation on a more frequent basis from field data gathered daily by field technicians. Furthermore, CANMET has expressed interest in mass ventilation data transmittal from remote locations underground as part of an underground communications research project. This data would be processed by a program known as "Ensure" and used to update the simulation input. In effect this would allow the Network Analysis Program to be an everyday tool for the ventilation engineer. It is clear that the program played a major role in the design process and will be an invaluable asset in future decision making.

A brief outline of how the program fits

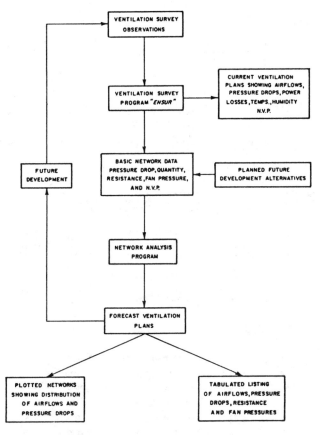

Figure 6. The Use of the Network Analysis Program as an Everyday Tool (after McPherson)

into the design process and the overall ventilation engineering of a mine is represented in the flow charts - Figures 5 and 6.

The CANMET Ventilation Network Program, like most other ventilation network programs, satisfies Kirchoff's Laws using iterative techniques modified from the Hardy Cross iterative technique for Waterflow Networks. The CANMET program differs from most other programs in that it is compressible flow model which permits the input of resistance using Atkinson's K factor and the physical parameters of the airway in addition to the traditional input formats used in other programs. The program also offers a plot option.

The plot program, in conjunction with a Calcomp plotter, creates a plot of the network with the energy levels at junctions and airflow quantities and pressure drops across branches. The plot depicts both the initial analysis results and the update

analysis results in different colours, viz: green and red respectively. Thus an unchanged condition would appear black because red could be printed over green.

12 THE RUTTAN VENTILATION NETWORK MODEL

The Ruttan model comprises approximately one hundred and seventeen (117) branches, seventy-six (76) junctions and eight (8) variable pitch fans. The iteration limit is set to 200 with a flow balance of +0.05 m^3/s. The results of the network analysis had, at the time of writing, correctly predicted two major changes in the underground ventilation system. The first being the changes associated with the completion of the east ventilation raise between 430 east level and 370 east level, and the enlargement of the existing east exhaust raise systems. Obtaining an accurate prediction of the overflow in the raise systems is important, however, it is not sufficient to

FANS
****

FAN NUMBER	BRANCH FROM TO	OPERATING PRESSURE (PA)	QUANTITY (M3/S)	DRY AIR MASS FLOW (KG/S)	STALLING VOLUME (M3/S)	STALL DISCREPANCY (%)	AIR POWER (KW)
1	3 -100	2149.	166.8	200.1	75.00	122.4	358.45
2	100 - 16	2580.	324.1	389.0	250.00	29.7	836.25
3	56 -100	1650.	185.5	222.6	100.00	85.5	306.07
4	100 - 58	426.	62.0	74.4	0.00	0.0	26.42

BRANCHES
****

---- BRANCH ----- NUMBER FROM TO	FRICTIONAL PRESSURE DROP (PA)	QUANTITY (M3/S)	RESISTANCE (NS2/M8)	DRY AIR MASS FLOW (KG/S)	OPERATING DENSITY (KG/M3)	AIR POWER (KW)	
1	2- 1	1	4.70	0.0824	5.64	1.2000	0.01	
2	1- 100	0	4.70	0.0	'5.64	1.2000	0.0	
3	8- 3	80	166.78	0.0029	200.14	1.2000	13.41	
4	5- 8	51	47.43	0.0233	56.91	1.2000	2.45	
5	6- 5	74	47.43	0.0336	56.91	1.2000	3.53	
6	7- 6	21	47.43	0.0100	56.91	1.2000	1.02	
7	10- 8	81	99.25	0.0084	119.11	1.2000	8.10	
8	11- 10	21	99.25	0.0023	119.11	1.2000	2.14	
9	7- 11	44	99.25	0.0046	119.11	1.2000	4.41	
10	12- 7	206	146.78	0.0096	176.14	1.2000	30.30	
74	73- 31	-11	4.70	-0.5927	5.64	1.2000	-0.06	FIXED QUANTITY - BOOSTER FAN REQUIRED
75	67- 28	2	4.70	0.1341	5.64	1.2000	0.01	FIXED QUANTITY - BOOSTER FAN REQUIRED
76	41- 74	745	4.70	33.7571	5.64	1.2000	3.50	FIXED QUANTITY - REGULATOR REQUIRED
77	42- 75	669	4.70	30.3630	5.64	1.2000	3.15	FIXED QUANTITY - REGULATOR REQUIRED
78	43- 76	595	4.70	26.9923	5.64	1.2000	2.80	FIXED QUANTITY - REGULATOR REQUIRED
79	44- 77	569	4.70	25.8532	5.64	1.2000	2.68	FIXED QUANTITY - REGULATOR REQUIRED

Figure 7. Example of the Analysis Program Output.

know only what to expect in the raise itself, probably more important would be all the associated changes in other areas of the mine indirectly and directly connected to the raise. Initial flow surveys subsequent to the predictions extracted from the simulation results indicated flows in 70% of the airways surveyed to be within 10% of the predicted airflows. These percentages are quite significant in that the accuracy of the input is limited by the accuracy of the underground thermodynamic surveys and the time lapse between actual measurement, simulation and remeasurement.

The second significant test of the program was one in which the program was required to analyse a network model using one of the proposed main intake raises between 660 level and 370 metre level as a temporary exhaust to facilitate development of the lower mine until a permanent exhaust could be established. From the simulation results

we were able to predict the airflows in six out of ten of the main sub system branches within 10% of the actual flows measured in the branches once they had actually been developed.

It must be remembered however that when simulating a network which has not yet been developed underground, variables such as size and friction factors must be given assumed values. These values may be totally inadequate once a branch is developed underground.

The program has proved its worth many times and if used judiciously can be an invaluable aid to the practising ventilation engineer.

13 THE COMMISSIONING AND INTERFACING OF THE NEW AND OLD SYSTEM

This is the fourth and final stage to realizing an ascensional system. Presently

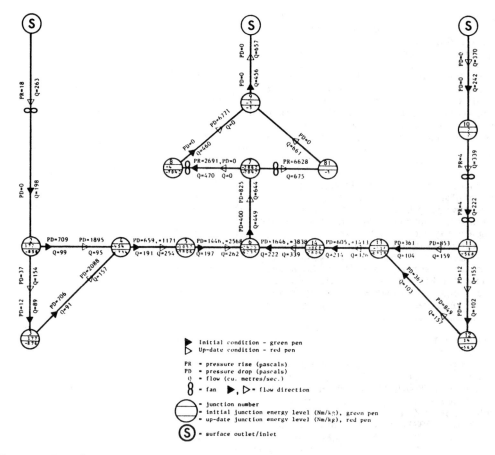

Figure 8. Example of the Plot Program Output. (after Hall, Stoaks and Gangal)

the ventilating of development to the lower mine raise locations is our primary concern. This development and all the ventilation raises will be completed in the latter part of this year.

The development phase is a very difficult period because airflow in development headings is solely dependent on auxiliary ventilation systems. At Ruttan air was forced down 812mm vent ducting to 660,800 and 860 levels. Two 56kw 965mm fans, in series, located on 320 shaft station together with booster fans on each level supplied the pressure necessary to move air through the system. Later a temporary exhaust was established on 660 metre level using the new intake raise. The fans on 320 level were moved to 620 shaft station cutout which eased the pressure requirements temporarily. Finally, by mid February, an exhaust was established from 800 level using the main ore pass to 660 level on a temp-

orary basis. This system will remain in effect until the permanent intakes and exhausts have been developed.

13 CONCLUSION

It is evident that as a mine expands it is not necessary to spend enormous sums of money on new fans and increased ventilating quantities because tradition dictates a certain ventilating concept.

The introduction of Ventilation Network Analysis Programs into the mining industry eliminates much of the uncertainty when investigating an untried ventilation concept for a particular mine. CANMET's Ventilation Network Program and Plot Program is designed to cope with the unique problems associated with deep Canadian mines. If used correctly with remote monitoring techniques, ventilation practice would improve to a point where trouble shooting and

73

emergency situations can be handled with
speed and accuracy never yet experienced
in the industry.

At Ruttan the CANMET program has been
translated to run on an IBM-PC which adds
versatility and mobility to the program.

I am convinced that without the program,
we, at Ruttan, would not have been able to
achieve the standard of ventilation practice
that we have and hope to see the program
used extensively throughout North America.

14 ACKNOWLEDGEMENTS

The author would like to thank the management of Sherritt Gordon Mines Limited for
permission to publish this paper, and to
Ruttan Mine management and staff for their
co-operation and encouragement.

Foremost the writer wishes to express
gratitude to Allan E. Hall, Professor of
Mining at University of British Columbia,
Canada, for his support and guidance.
Special mention is made of Edgar W. Wright
for his assistance with the computer work
and photography.

15 REFERENCES

Barenbrug, A.W.T. 1974, The thermodynamic
    approach to mine ventilation, the ven-
    tilation of South African gold mines. Mine
    Ventilation Society of South Africa.
    Johannesburg:Klem-Lloyd
Gangal, M.K. 1984, Private Communication
Hall, A.E., Stokes, M.A., Gangal, M.K. 1982,
    CANMET's thermodynamic ventilation program
    C.I.M. Bulletin, Volume 75, No. 848.
Hall, A.E. 1984, The determination of fans
    partially in series or parallel using
    energy and mass flow equations. Journal
    of the Mine Ventilation Society of South
    Africa, Volume 37, No. 2.
Hall, A.E. 1984, 1985. Private Communication.
McPherson, M.J. 1974, Ventilation network
    analysis. The Ventilation Society of South
    African gold mines. Mine Ventilation
    Society of South Africa. Johannesburg:
    Klem-Lloyd.
Stackulak, J. 1980, Computer network calcu-
    lation of Creighton Mine mass flow and
    natural ventilation. Presented at the
    Second International Mine Ventilation
    Congress, Reno.

Air leakage through longwall wastes in the Sydney Coalfield

A.W.STOKES
Cape Breton Coal Research Laboratory, CANMET, Sydney, Nova Scotia, Canada

ABSTRACT: The high losses of ventilation air through the caved wastes of the advancing longwall coal faces in the Sydney Coalfield are a continual problem to the mine operator. In excess of 80 percent of the air which enters some longwall sections may short circuit the face in this way. This can cause health and safety related problems and in addition limit production due to excessively high general body methane concentrations (greater than 1.25%).

This paper outlines the field studies which have been undertaken by the Cape Breton Coal Research Laboratory (CBCRL) to better understand this leakage with a view to controlling and reducing it. Although not a new phenomena in coal mining, this paper presents a novel approach to solving the problem.

Underground ventilation and ground control surveys were conducted to facilitate measurement of the resistance of the gob. This information enabled the roadway airflows and leakages to be accurately simulated using a digital computer. The gob was divided up into a series of 50 m wide corridors parallel to the face line. Air leakage through these was then measured together with the pressure drop across the ends. Using these measurements the resistances were then calculated.

A wide variation in the gob corridor resistances was noted. A review of the results revealed a strong agreement with the understood effect of other mining parameters such as pillar widths, abutments and rock consolidation and compaction.

Tracer gas was used in further studies to more accurately measure the air leakage and to trace the leakage paths through the gob area.

This technique, when refined, may eventually provide a method for determining caving characteristics of rock in inaccessible caved waste areas.

1 INTRODUCTION

The principal method of coal extraction used in the Sydney Coalfield is advancing longwall with full caving. Panels of up to 3.5 km in length are worked to an extracted height of 2 m using shearers and hydraulic powered supports. The gassiness of the coal and proximate strata, plus the high production rates, dictate that the face be supplied with large air quantities to dilute the liberated methane to within acceptable working limits (below 1.25%).

Each longwall face is serviced by two gateroads at the ends of the 210 m long faceline. One serves as the fresh air intake and coal conveyor road (maingate), the other as a return airway and alternate access road (tailgate). The gateroads are supported by steel arches and stone filled wooden crib packs built between the road-

way and the caved waste, see Figure 1. The relatively high porosity of the gateside packs allows leakage of the fresh intake air from the maingate through the caved waste (gob) into the tailgate. This short circuiting of the face increases as the panel is worked further away from the face startline. It is further exacerbated by the increase in ventilation pressure resulting from progressively longer roadways and larger volumes of air deliveries to the longwall section required to offset deteriorating panel ventilating efficiency.

To provide sufficient airflow at the face of some longwall panels, it has been necessary to implement leakage control. Three methods have been attempted; improvement of pack density, physical sealing of the pack and reduction of the differential pressure across the waste (cundy ventilation). To date the most satisfactory

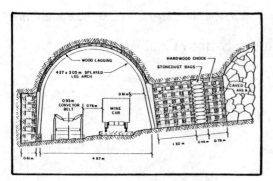

Fig. 1. Intake roadway arrangement.

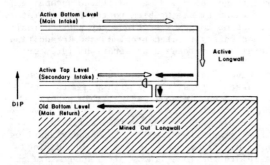

Fig. 2. Schematic of cundy system.

method has been cundy ventilation. Figure 2 shows the layout of this system.

Return air is diverted to an adjacent worked out longwall through a purpose driven roadway near the face end (cundy). The outbye portion of the return gateroad is converted into a bleeder intake. This reduces the differential pressure across the gob and hence gob leakage.

The current production schedule at Lingan Colliery requires full output from two adjacent coalfaces. This effectively rules out the use of cundy ventilation on the second panel. This has forced a reevaluation of gob leakage control methods to find an effective alternative to cundy ventilation.

The Cape Breton Coal Research Laboratory (CBCRL) agreed to evaluate the effectiveness of the alternative methods. This includes geotechnical investigations and the monitoring of improvements made to the ventilation. The results of the geotechnical work are reported elsewhere (Cain, 1985). This paper describes the techniques which have been developed to monitor the changes in the ventilation and the preliminary findings.

2 GOB LEAKAGE AT LINGAN COLLIERY

The field investigations were concentrated on three longwalls in Lingan Colliery, namely 4 West, 8 East and 9 East. A plan of Lingan in Figure 3 shows the relative positions of these longwalls.

The longwall panels are mined on strike extending out either side of a set of five parallel roadways (deeps) which are driven down dip. The severity of gob leakage differs on each side of the deeps. Leakage on the eastern longwalls is markedly worse than that on the west walls even though they use identical mining methods. Measurements made on the west side walls show that gob leakage accounts for approximately 55 percent of the total air volume entering districts that have reached their extremity (1300 m long) (Cain and Stokes, 1984). A leakage survey of 8 East wall just prior to its conversion to cundy ventilation (800 m long) measured more than 70 percent of the air passing through the gob.

The most probable explanation for this variation is the composition of the overlying strata. It is often observed that mudstones and similarly weak roof strata cave better and produce a compact waste. Conversely, massive competent strata like sandstone in the roof may inhibit caving and form a loosely compacted, porous gob. A substantial sandstone is known to overlie the coal seam mined in Lingan Colliery

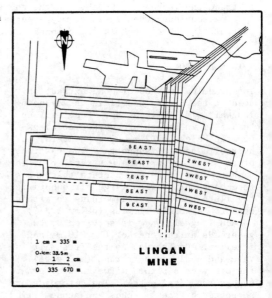

Fig. 3. Plan of Lingan Colliery.

and sandstone channels from this bed are occasionally encountered dipping down into the coal seam. There is evidence to support a variation in the composition of this sandstone on either side of the deeps (Forgeron, 1984).

3 PREVIOUS PREVENTIVE METHODS

The problem of gob leakage in the Sydney Coalfield has been recognised since rapid advance and high production rates became achievable standards in the early 1970's. Since then considerable effort has been expended in numerous methods of leakage control. Presenting an impermeable barrier at the intake pack using sealants has met with little success, even though eight different methods have been tried. These range from the application of spray sealants such as mandoseal, shotcrete and urea-formaldehyde foam, to the construction of an independant wall of stonedust bags between the intake gateroad packs (see Figure 1).

Two methods for improving the pack density have also been attempted: monolithic packing and pneumatic stowing. Both these trials were abandoned before significant lengths of pack had been constructed, preventing an assessment of their effectiveness at reducing gob leakage. In view of the success of monolithic packing at reducing air leakage elsewhere in the world (Collier, 1983) a more extensive trial was planned for 9 East longwall at Lingan Colliery. It is the effectiveness of this trial, together with the before and after effects of spraying shotcrete in the intake gateroad, that is the subject of this paper.

4 MEASUREMENT OF GOB RESISTANCE

Monolithic packing was initiated on 9 East wall in March 1985 when the wall had advanced 635 m from the face startline. The effectiveness of the monolithic pack as a gob leakage prevention technique required a comparison of the airflows that would have existed on 9 East had conventional packing been continued, with those obtained using the monolithic packing. To achieve this some form of airflow prediction was required. This could have been achieved in two ways:
- regular measurement of the airflows in a similar longwall that used the conventional wooden chock packing.
- computer simulation using a ventilation network analysis program.

Direct comparison with 8 East's airflows at the same advance would have worked, but 8 East was converted to cundy ventilation before it had advanced the required distance. This meant that the airflows had to be predicted using a computer. A data base of the gob leakages in existing conventionally ventilated longwalls in Lingan was compiled. This data was used to create computer models to simulate the longwall airflows.

Information for the data base was gathered through ventilation surveys. These results were used to calculate roadway and face resistances and the resistances of imaginary corridors running through the gob (leakage paths). Figure 4 shows the layout of these corridors.

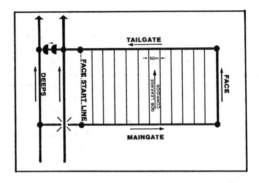

Fig. 4. The gob leakage corridors.

They are 50 m wide running from the maingate to the tailgate parallel to the face and extend from the face startline to the rear of the faceline. For ease of analysis and computer simulation all of the gob leakages were assumed to be parallel to the faceline and hence through the imaginary corridors. The justification of this assumption is discussed later.

Once the resistance of the gateroads, face and gob corridors of a longwall are found, these values can be entered into the computer and the district airflows simulated. Using typical resistance values for the gob corridors and gateroads, the computer model can then be extended to predict future airflows. This method also has the advantage that any improvement that the packing or sealant may make to the ventilation can be expressed either as a percentage improvement of the ventilation efficiency of the longwall or, more directly, as an increase in the gob corridor resistance. Expressing it as a resistance enables a direct comparison with other longwalls in the mine.

5 GATEROAD AIRFLOW MEASUREMENT

Four West was the first longwall to be surveyed. Beginning at the face startline a rope was used to chain off survey stations along both gateroads at 50 m intervals. Each station was photographed to allow an accurate determination of the cross-section area. On a non-production shift air velocity readings were made at each survey station using vane anemometer (type AM5000). Using the same measuring stations a gauge and tube survey of the district was done simultaneously. A total of 148 anemometer traverses were made at 54 measuring stations (27 in each gateroad and 3 along the face); the duration of the survey was almost five hours. Regular anemometer measurements were made at a control station near the return fireseal to monitor fluctuations in the airflow during the survey. Excessive variations or drifts may invalidate a survey.

The results clearly showed the gob leakage trend with 35 m^3/s entering 8 East and only 12 m^3/s measured at the face. Figure 5 shows the airflows measured in both gateroads plotted against the inbye distance from the face startline.

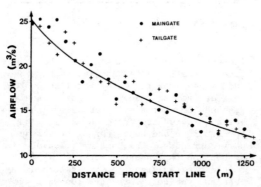

Fig. 5. Anemometer measurements made in 4 West gateroads.

A curve has been drawn through the points to represent the trend of the airflow profiles in the gateroads. The deviation of the measured values from the curve is as much as 25 percent, and is particularly noticeable for the readings taken in the intake gateroad. These deviations could be due to a combination of factors:
    (1) Obstructions in the roadway (e.g.

conveyors) which make it awkward to traverse with an anemometer.
    (2) Fluctuations in the ventilation caused by opening and closing of ventilation doors or the movement of machinery.
    (3) Instrument error.
    (4) Operator error (e.g. anemometer yaw or timing error).
Obviously a system for measuring the gateroad airflow which could reduce or elimate these errors would be advantageous.

Airflows in large mine openings not easily measured using conventional equipment have been measured by releasing a steady stream of tracer gas and analysing the diluted concentration in the airstream (Shuttleworth, 1967). This method offers the chance to reduce all four listed sources of error. The tracer gas sulphur hexafluoride (SF_6) can be measured in concentrations as low as a few parts per trillion (Gray, 1978) and is commonly used to solve mine ventilation problems. Using this fact, a system was developed where the airflows at each tailgate survey station could be measured with a single release of SF_6.

Four West longwall was surveyed again shortly after the first anemometry survey. This time SF_6 was used to measure the tailgate airflows. A continuous stream of SF_6 was released from a lecture bottle through a limiting critical orifice. This assured a steady flow of 33 ml/min which was dispersed into the face airstream 20 m from the tailgate. Sufficient time was allowed for the SF_6 release system to achieve steady state and for concentrations along the tailgate to stabilise. Samples of air were then collected at each measuring station. These were drawn through a 2 m long sampling wand by an MSA fixed flow pump and collected in 1 litre Tedlar gas sampling bags. The sampling wand was traversed across the roadway cross-section in a similar manner to an anemometer traverse. This ensured an integrated air sample was collected containing an average concentration of SF_6 from the roadway (not suprisingly spot air samples taken close to the gob side of the tailgate contained less than samples collected from the rib side).

The Tedlar sampling bags are fragile and easily punctured. Leaks in sample bags which are used in a survey become apparant when their contents are analysed as the concentration of tracer gas they hold is markedly lower than samples collected from adjacent stations. This diffusion of fresh air into the leaky bags is noticeable a few hours after sample collection.

Leaky bags are discarded and the analysis of their contents disregarded.

Table 1 contains the analysis of the sample bags from the 4 West survey.

Table 1. Four West tailgate SF$_6$ survey results

Inbye Distance (m)	Time Taken (hrs)	SF$_6$ Concentration (ppb)	Indicated Airflow (m^3/s)
0	0923	34.6	15.73
50	0925	38.0	14.30
100	0927	Leaky Bag	–
150	0929	Leaky Bag	–
200	0931	42.06	12.75
250	0932	Leaky Bag	–
300	0934	46.38	11.72
350	0937	47.64	11.41
400	0938	48.50	11.21
450	0940	50.22	10.82
500	0942	52.89	10.28
550	0947	54.69	9.94
600	0949	Leaky Bag	–
650	0951	54.45	9.98
700	0953	57.35	9.84
750	0955	57.51	9.45
800	0958	57.54	9.53
850	1000	68.09	7.98
900	1002	52.20	8.74
950	1004	61.58	8.83
1000	1009	63.91	8.50
1050	1011	61.58	8.82
1100	1013	66.87	8.13
1150	1015	67.96	7.99
1200	1017	69.98	7.76
1250	1019	Leaky Bag	–
1300	1022	72.48	7.50
1323	1024	Leaky Bag	–

The air quantities in column 4 are found from the tracer concentrations in column 3 using the following relationship.

$$Q = \frac{\dot{v}}{c}$$

where: Q = Airflow (m^3/s)
$\dot{v}$ = Release rate of SF$_6$ (m^3/s)
c = Measured Concentration of SF$_6$

Figure 6 is the airflow profile for 4 West obtained from this survey. Deviation of individual points from the profile line is much less than for the comparable anemometer readings made in the first survey. This allows the more subtle inflections of the profile to be seen. Four airflow

measurements were made with a vane anemometer during the survey; these values are marked on Figure 6 to give an indication of the relative accuracy of the two techniques.

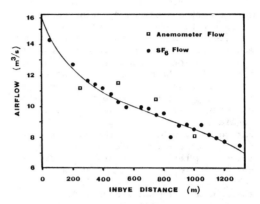

Fig. 6. Four West tailgate airflow profile.

Once the release system had stabilised the 27 sweep samples were collected in one hour. This is less than one-third of the time which was taken to make the same survey with an anemometer. By shortening the duration of the survey, errors caused by the longer-term fluctuations in the ventilation system are minimised.

6 SF$_6$ INJECTIONS THROUGH THE GOB

The method for measuring the tailgate airflows is relatively simple, but there is no such easy technique to measure the maingate flows. Anemometer traverses made in the maingate have been particularly inaccurate. As previously mentioned, this is caused by a number of factors, but is primarily due to the coal conveyor which precludes easy access to a large portion of the roadway cross-section. Dilution of a steady stream of tracer gas could be used. But as air leaks from along the entire length of the maingate it would require perfect mixing of the SF$_6$ into the airstream within a very short distance of the release point. In addition, the release system would need to be moved to each measuring station in turn to allow airflow measurement at that point. This process would be tedious and very time consuming.

The objective of measuring both the tailgate and maingate airflows was to deduce the main flow paths through the gob. This would hopefully validate the assumption that the main flow paths are parallel to the faceline, thus allowing the tailgate profile to be used to determine the leakage flows. Due to the difficulty in accurately measuring maingate airflows, a different approach was used. Flow paths in the gob were obtained by injecting tracer gas from the maingate through the gob and monitoring its emergence into the tailgate.

A 2 m injection lance was inserted into the gob from the maingate and a steady stream of SF_6 was then released for a period of 4 to 5 hours. Once the injection was started air samples were collected regularly at the tailgate fireseal. This was continued for the duration of the survey. After steady state was achieved (usually after 3 to 4 hours) samples were collected from the tailgate and adjacent gob. Sweep samples were collected from the roadway from three measuring stations inbye the comparable position of the injection point and at every measuring station out to the tailgate fireseal. Gob samples were taken at 25 m intervals for a distance of 100 m inbye the comparable position to the injection point to 150 m outbye that point, and then every 50 m to the startline. These were drawn from the gob through a probe inserted through the gateside pack and collected in Tedlar sampling bags.

Figure 7 shows the graphical plot of the SF_6 concentrations measured in 4 West versus the inbye distance from the face startline. This curve is for an injection which was made 458 m inbye the face startline of 4 West. Drawn on the same figure is a schematic of 4 West longwall with the flow of tracer gas through the gob and the known locations of sandstone channels. Both the gob and roadway samples indicate that most of the SF_6 entered the tailgate between 350 and 400 m inbye the startline. The gob samples show that the flow of SF_6 diffuses as it moves through the gob. But, if the main flow path is taken to be coincident with the line drawn between the injection point and the maximum gob concentration, and approximation of the angle of flow can be calculated. For this injection the maximum gob concentration was measured 380 m inbye the startline. This is a 78 m (458-380 m) displacement which equates to a flow path at an angle of 20 degrees to the face startline.

The gob injection illustrated in Figure 7 was interesting as analysis of the gob samples further outbye showed a second

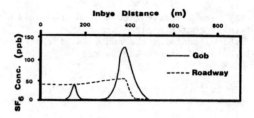

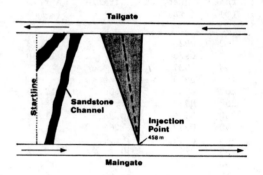

Fig. 7. Schematic of 4 West gob injection and measured tailgate and gob SF_6 profiles.

smaller peak 150 m inbye the startline.

At first it was thought that this was a second flow path through the gob. However, it was noticed that the maximum SF_6 concentration measured in this peak was the same as the air in the roadway. This suggested that it may have come from the tailgate. Further investigations with smoke tubes showed that air was moving back into the gob from the tailgate between 150 and 200 m inbye. This is where two major sandstone channels were found in the roof above the coal seam. It is probable that these structures have prevented the break-up and consolidation of rocks in the gob, maybe even creating large open areas supported by a sandstone cantilever. Whichever, the resistance of the gob at this point was low enough for air to flow back into the gob from the tailgate. This effect has been discovered at the intersection of sandstone channels in other longwalls since.

The 4 West gob injection illustrated in Figure 7 was the largest angular flow path measured. Flow paths on the east side walls tended to be towards the face at an angle of 2 to 3 degrees. While flow paths on 4 West varied between 20 and 5 degrees away from the face.

Figure 8 shows a typical build up of the concentration of SF_6 at the tailgate fire-seal versus the elapsed time since the start of a point source gob injection.   By

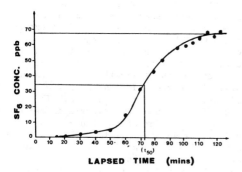

Fig. 8. SF_6 build up at 9 East return fire-seal.

making some assumptions the build up of the SF_6 concentration measured at the tail-gate fireseal (e.g. Figure 8) can be used to calculate the range of air velocities in the gob.   It has been shown (Stokes and Stewart, 1985) that for a continuous in-jection of tracer gas into a ventilation system, the time it takes to reach half the steady state concentrations (t_{50}) at a point downwind, equals the average time taken by air moving along the same route to reach that point.   This fact can be used to calculate the average velocity of air moving through the gob.   This value is needed to calculate the Reynold's number of the flow which in turn is used to deter-mine the type of flow (i.e. laminar or tur-bulent).   The appropriate fluid flow equa-tions can then be used to calculate the resistance of the leakage paths.

From Figure 8:

$$t_{50} = 72 \text{ minutes}$$

minus two minutes for the gas to travel the tailgate to the fireseal.

so, t_{50} = 70 minutes = 4200 seconds

The plan distance between the injection point in the maingate to its point of entry into the tailgate for this injection is 213 m.

So the average longitudinal velocity along this line is:

$$\text{Velocity} = \frac{213 \text{ m}}{4200 \text{ s}} \approx 0.05 \text{ m/s}$$

However, the flow path in the gob is not a straight line, but around broken rocks. Assume that this may increase the distance travelled by up to a factor of 5.   Hence, the average velocity may be anywhere in the range of 0.05 m/s to 0.25 m/s. Reynolds number is given by the equation:

$$Re = \frac{\rho \cdot u \cdot d}{\mu}$$

where ρ is the air density (kg.m^{-3})
       u is the air velocity (m.s^{-1})
and μ is the air viscosity (kg.m^{-1}s^{-1})

The component, d, is the hydraulic mean diameter of the flow path for which an average value must also be assumed.   The average distance between the broken rocks should lie in the range of 0.001 m to 0.2 m.   Using standard values for air den-sity, ρ, and viscosity, μ, the range of Reynold numbers found in the gob would be:

$$3 < Re < 3300$$

Clearly the flow could be either laminar, turbulent or more likely a mixed regime. This is supported by the curve in Figure 8 and similar curves from other gob injec-tions which have two distinct parts.   Fig-ure 8 shows that from twenty minutes after the injection (when SF_6 was first detected at the fireseal) until about fifty minutes after there was a slow gradual build up in the SF_6 concentration.   After fifty min-utes the SF_6 concentration began to in-crease rapidly towards its steady state value.   Curves with such pronounced changes in slope are indicative of the product of two or more competing flow regimes.

Given that the airflows in the gob are probably a mixture of turbulent and laminar flow, what value for n should be used in Atkinson's equation to calculate the gob corridor resistance and to predict the flows in the computer simulations?

After careful consideration it was decid-ed to treat all gob flows as turbulent. The justification for this is that the air-flows through most of the 50 m gob corri-dors lie in the range 0.7 to 1.8 m^3/s. And, as it is this value which is raised to the power, n, it closeness to unity will ensure that error from this treatment will be small.   This crude approach will be more clearly justified in the following section.

The t_{50} time can also be used to deter-mine the active volume through which air moves in the gob. The air loss through the 50 m wide corridor into which the injection in Figure 8 was made was 1.8 m^3/s. The average residence time for this air equals the t_{50} time, which was 4200 seconds. The active volume is the product of these two values which equals 7560 m^3, or 36 percent of the 21,000 m^3 of coal extracted from that corridor.

8 GOB CORRIDOR RESISTANCE

Tracer gas was used to measure tailgate airflows in 8 East and 4 West longwalls. From these surveys, values for the gob corridor resistance were calculated. Figure 9 is a histogram of the gob corridor resistances for both 8 East and 4 West.

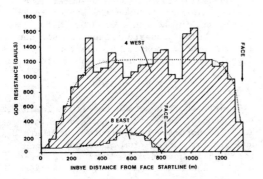

Fig. 9. 8 East and 4 West gob corridor resistances.

From this it is apparent that the resistance of the 4 West gob is much higher than 8 East. Resistance values for 4 West's gob corridors are predominantly in the range 1000 to 1400 gauls, whilst the values of 8 East gob corridors lie in the range of 50 to 250 gauls. This clearly shows the difference in the gobs of the two sides of the mine. This result is the justification for treating the gob flows as turbulent. Errors associated with the assumption will not obscure the large differences in gob corridor resistance that we are looking for.

A close examination of Figure 9 shows that both longwalls have a low resistance gob area close to the face startline. This gradually rises to a higher resistance value and then stabilises into a plateau. The low resistance close to the face startline is probably attributable to the "deeps" barrier pillar which causes incomplete caving of the gob near the face

startline. A similar low resistance area of gob is noticeable near to the coal face-lines. This is due to the rocks in the gob immediately behind the roof supports which have not had a chance to consolidate and the incomplete caving due to the pillar effect of the solid coal face.

When predicting airflows with computer models the average gob corridor resistance for a wall was taken from the mean value of the plateau. These were used to extend the longwall by adding leakage paths in the middle of the profile. In this way the two low resistance areas of gob at either end of the longwall were maintained.

A similar survey was done on 9 East longwall after 565 m of face advance. The values of the gob corridor resistances were the same order of magnitude as 8 East, varying from 45 to 75 gauls. This was considered sufficient justification to assume that the pattern of gob corridor resistance values would develop in a similar manner to 8 East's. The resistance values measured in 8 East were therefore used as the basis for the simulation of airflows in 9 East. From the prediction exercises the graph in Figure 10 was produced. This shows the predicted ventilation efficiency for 9 East longwall versus it's face advance. This curve is the control against which improvements in 9 East's gob leakage were measured.

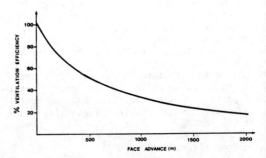

Fig. 10. Predicted ventilation efficiency.

9 NINE EAST GOB LEAKAGE

The first full ventilation survey of 9 East was completed on February 9, 1985. This comprised a pressure survey and measurement of the airflows in the tailgate using SF$_6$. The face had advanced 565 m in from the startline and the maingate packs were all of the conventional stone filled wooden chock type. The first 150 m of packs from the face startline in had been sprayed with a layer of shotcrete that week. This was to be extended to about 400 m inbye in the following weeks with the objective of reducing the air leakage near to the face startline. Measured airflows were 30.5 m³/s at the fireseal and 16.1 m³/s at the face. This equates to a ventilation efficiency of 54 percent, some 5 percent higher than the efficiency predicted using the computer model.

A second ventilation survey was repeated on 9 East on March 10. By this time the face had advanced another 70 m to an inbye distance of 635 m and the shotcrete spraying in the maingate had been completed to a total length of 375 m. In the survey 30.1 m³/s of air was measured at the fireseal and 15.8 m³/s on the face. This equals an efficiency of 52 percent which was a 7 percent improvement over the predicted value of 45 percent. Figure 11 shows the tailgate SF$_6$ generated airflow profiles from both surveys. The profiles

proved at least in part to be only temporary. In the four weeks between the surveys this section of shotcrete may have begun to crack through either drying or roadway closure. This would explain why an increase in leakage was measured through the 0 to 200 m section in the second survey. The theory is further supported by the reduced leakage measured through the 200 to 400 m section of gob in the second survey. This was the area of the recently sprayed shotcrete and also exhibited only temporary improvement as explained below.

A third survey was completed on May 25, 1985. By this time the face had advanced 880 m and monolithic packing had been introduced as the maingate packing system. The monolithic packing was started at 630 m inbye and ran 250 m up to the rear of the face. In this survey 27 m³/s was measured at the fireseal and 11.1 m³/s at the face; a ventilation efficiency of 41 percent which is some 6 percent higher than the predicted efficiency of 35 percent. Figure 12 shows the profiles measured on all three surveys. These are

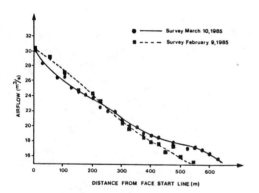

Fig. 11. SF$_6$ generated tailgate airflow profiles from first two 9 East surveys.

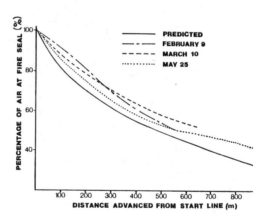

Fig. 12. 9 East ventilation efficiency profiles.

are of quite different shape. The first survey profile had an almost linear appearance for the first 200 m. This is unlike all the profiles which had been measured in the other longwalls which tended to look more like exponential decay curves. The profile from the second survey does have this exponential shape in the first 200 m but then flattens out in the 200 m to 300 m section. These abnormalities might be explained by the effects of the shotcrete which shrinks in curing and will crack with time dependant roadway closure.

The reduction in leakage effected by the shotcrete as measured in the first survey in the 0 to 200 m section of the tailgate

reduced to a percentage of the total air entering 9 East on each survey (i.e. ventilation efficiency). Also drawn is the predicted ventilation efficiency from the computer simulation. The distance between the predicted curve and the others give an approximation of the reduction to the leakage. From this it can be seen that the performance of the shotcreted section had deteriorated further since March. The profile of the May 25 survey is lower in the 0 to 400 m section than the earlier survey profiles and is returning more to the predicted exponential decay curve shape. It is still higher than the predicted curve suggesting that the shotcrete is still effecting a small reduction in the leakage. There is also a deviation of the May 25 profile from the predicted curve in final 300 m. This is probably attributable to the monolithic pack which is reducing the leakage close to the face. Of the 6 percent improvement in ventilation efficiency it is hard to judge what proportion of this is due to the monolithic pack and that due to the shotcrete. A good estimate would be in the range of 3 to 4 percent to the monolithic packing and 2 to 3 percent to the shotcreting.

Gob corridor resistance values were calculated from the information gained graphically in Figure 12. Values from the first survey lie in a range between 45 and 75 gauls. These are compariable with the same gob corridor values measured on 8 East. The resistance values from the second survey begin at about 50 gauls at the face startline and then gradually rise to between 300 and 550 gauls 400 m inbye. The higher values measured in the third survey are at least in part due to the monolithic packing and shotcreting. However, some of the increase should be attributed to the compaction of the rocks in the gob with time as the face moved further away. The values are higher than 8 East resistances, but not substantially so (values in the plateau area of 8 East's histogram were between 250 and 300 gauls). To reduce the leakage substantially the resistance of the corridors would need to be increased by far more. It is worth noting that there was only 250 m of monolithic packing in place at the time of the survey. This is not really enough to judge performance. In addition, it is still close to the working face - an area in which the gob flows are likely to be not always parallel to the face.

Use of monolithic pack close to the face startline may effect a greater improvement in leakage reduction than the shotcrete. This would also make it easier to measure it's effectiveness.

10 CONCLUSIONS

A steady release of SF_6 into the return airstream provides a quick and accurate method for measuring tailgate airflows. The results of the surveys performed using this method have enabled gob leakage to be measured more accurately than with anemometers. This in turn has enabled a better understanding of gob leakage in Lingan Colliery and provided a means to measure the effectiveness of remedial measures aimed at reducing the leakage.

Remedial measures used to date on 9 East longwall have met with some success. A small percentage improvement in the ventilation efficiency was measured after the application of shotcrete to the face of the maingate packs. Most of the reduction proved temporary and was lost within a matter of weeks. A similar improvement in efficiency has been achieved by the 250 m of monolithic pack now in place. This improvement should increase as more pack is added. An insufficient length of monolithic pack has been installed to allow an accurate assessment of it's contribution to the gob corridor resistance.

11 ACKNOWLEDGEMENT

The author wishes to thank the Cape Breton Development Corporation for their help and cooperation in performing the work for this paper.

12 REFERENCES

Cain, P., Stokes, A., Genter, D. 1984, Preliminary report on studies into gob leakage in the Sydney Coalfield. Division Report ERP/CRL 84-43(OP).

Cain, P. 1985, The effect of monolithic packing on gateroad performance at Lingan Colliery. Division Report ERP/CRL 85-66(TR).

Collier, L. 1983, Further developments of monolithic packing in the Warwickshire Coalfield. Presented to the Institution of Mining Engineers South Staffs. and South Midlands Branch, May 1983.

Forgeron, S. 1984, Private Communications.

Gray, D.C. 1978, SF what? Gas Engineering and Management, pp. 20-24, January 1978.

Shuttleworth, S. 1967, Measurement of mine fan drift and evasee air quantities by a tracer gas technique. Colliery Engineering, pp. 308-313, August 1967.

Stokes, A.W. and Stewart, D.B. 1985, Cutting head ventilation of a full face tunnel boring machine. For presentation at and publication in the proceedings of the 21st Inter. Conf. of Safety in Mines Research Institutes, Sydney, Australia, October 1985.

Simulation of unsteady-state of airflow and methane concentration processes in mine ventilation systems caused by disturbances in main fan operation

ANDREW M.WALA
University of Kentucky, Lexington, USA

BONG J.KIM
Chosum University, Kwang-Ju, South Korea

ABSTRACT: The results of a study of airflow and methane concentration in a mine ventilation system obtained by the numerical simulation method are presented. The main concerns of this study were the unscheduled shutdown and turn-on of the fan and their effects on methane concentration in atmosphere of the mine.

Two different situations for a summer and winter season were investigated. Because of the nearly vertical geometry of the mine, the natural ventilation pressure plays a substantial role, especially during periods when the fan is not operating.

The study was carried out using a computer program which implemented the mathematical models of dynamics of flow in the mine openings, and the dynamics of methane concentration in a zone where the methane emission occurred.

1 INTRODUCTION

Air flowing through mine openings is described by a system of nonlinear equations with initial and boundary conditions. Consequently, such processes have more than one state of equilibrium, with some being stable and others unstable. In certain circumstances, there may be no steady-state at all, in which case the process will oscillate.

Generally the airflow in mine ventilation systems tends towards a steady-state, but as was mentioned before, the actual steady-state established will depend on the initial and boundary conditions. Disturbances in flow, such as changes of resistance, fan characteristics, fires, etc., can cause the transition from one steady-state to another. In real mines the disturbances occur frequently, and they are mainly due to the technology, transport, and handling or mishandling of ventilation devices (doors, fans, regulators, etc.). This means that such a system could be permanently unstable, or alternately jumping from one state of equilibrium to another.

During recent years the idea of remotely controlling ventilation systems has attracted the attention of mining engineers, especially as related to methane concentration. Unsteady-state airflow and gas concentration are of particular interest when considering the automatic control for a mine ventilation system. It is impossible to predict the dynamic behaviour of such a system, therefore, either the actual process must be continuously monitored, or digital simulation can be used to compile the necessary information.

The main goal of this paper is to show the results obtained by the simulation method applied to study the unsteady-state of airflow and methane concentration in a hypothetical coal mine based on conditions in South Korea.

The objective of the study was to determine the time relations between the methane concentration changes in the working areas and disturbances in the main fan operation.

The following system situations were investigated:
- main fan shutdowns (summer and winter conditions);
- main fan turn-ons (summer and winter conditions).

Solutions were determined for the problems listed below:
- time required for methane concentration to return to a safe level after fan restart;
- after shutdown of the fan, the time in which underground personnel should be evacuated to an area furnished with fresh air;

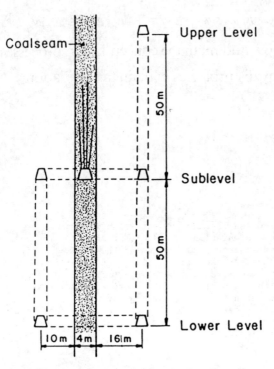

Fig. 1. The vertical cross-section through the mine.

- time in which fan should be restarted using electrical power generated by a diesel powered generator, before the methane concentration reaches explosive level;
- level of the methane concentration at which it will reach steady-state after the main fan has shutdown.

2 MINE DESCRIPTION

2.1 Mining Method

The study was carried out for a hypothetical, relatively small, gassy coal mine with nearly vertical geometry, located in a mountainous and geologically complicated area in South Korea.

The mine layout (see Fig. 1) follows a sub-level caving system, with upper and lower levels open from the side of the mountain by drifts, and sublevel. The upper level, developed in a hanging wall separated 16 meters from the coal seam, is used as a return airway. The lower level, developed in the foot wall separated 10 meters from the coal seam and 100 meters below the upper level, is used for intake air and for transportation. The drifts driven in the rock are parallel to the coal seam, but the levels are con-

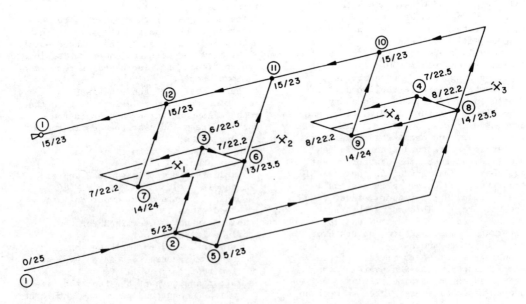

Fig. 2 Three-dimensional diagram of the ventilation system.

nected by the 60-degree inclined rock
raises.

Coal is excavated from four working
areas located in 50-meter long drifts
driven in both directions from each
crosscut, through the seam. When the
drifts reach the limit, then coal is mined
from the end of the drift by the longhole
blasting and retreating caving method.
Maximum production for each working section is 150 tons per six-hour working
shift.

Dimensions of certain airways in the
system are as follows:
- drifts in rock, 3 x 2.4 meters (area
  $7.2m^2$);
- crosscuts and rock raises, 2.5 x 2.2
  meters (area $5.5m^2$); and
- drifts in the seam, 2.4 x 2.2 meters
  (area $5.28m^2$).

A three-dimensional diagram of the
network created by the drifts, raises, and
crosscuts for this particular mine is
shown in Fig. 2.

2.2 Ventilation system

The exhaust ventilation system is used to
ventilate the entire mine with a main fan
installed at the portal of the upper level.
The diagram given in Fig. 2 can also serve
as a schematic of the ventilation network.
The nodes of the ventilation system are
marked by circled numbers, and directions
of flow, while the fan is operating, are
marked by arrows.

The blowing ventilation system with line
brattice is ventilating the working section, as shown by Fig. 3. The quantity of
air needed to provide a proper ventilation
in the working section has been calculated
based upon the dilution requirements.

According to the production and methane
content of the coal, the amount of methane
which is liberated in a particular working
area is as follows:
- section #1 (branch 7) and #4 (branch
  9), with a rate of 0.02 m^3/s; and
- section #2 (branch 6) and #3 (branch
  8), with a rate of 0.03 m^3/s.

To maintain the methane concentration
in the working section below 0.5%, each
section requires 7 m^3/s of air to flow
through the face. In addition to that,
5 m^3/s of air are needed to flow through
each chute (branches #10 and #11).

The total amount of air flowing through
the entire mine is 38 m^3/s.

The characteristic data of the ventilation system (node connection, lengths,
areas, perimeters, resistances of the
airway, air density, and NVP), are shown

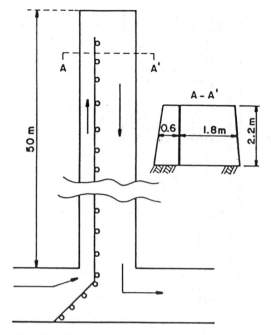

Fig. 3. Blowing ventilation system with
line brattice in the working section.

in Table 1. The system consists of 19
branches, 12 junctions, with 6 controlled
flows.

To obtain the initial conditions for
unsteady-state analysis, the steady-state
of the ventilation system was defined with
the help of a computer. Considering the
control of flow through certain areas of
the system, the locations and sizes for
the regulators were calculated by the
"fixed quantity branches" version of the
computer program. The axial fan cooper-
ating with network has the performance
characteristic as follows:

$$h(Q) = 1001 + 126.35 \ Q - 2.74 \ Q^2$$

Canonical diagrams of the ventilation
systems, with flow distribution controlled
by the regulators located in branches 6,
7, 8, 10, 11, for both summer and winter,
are shown in Fig. 4a and 4b, respectively.

2.3 Temperature Distribution and NVP

Due to the nearly vertical geometry of
this mine and the mining method being
used, the NVP plays a substantial role,
especially in case of the main fan
shutdown.

The values of NVP for each branch were

87

Table 1. Characteristic data of the ventilation system

Branch No.	From Node	To Node	Area A [m²]	Parameter P [m]	Length L [m]	Resistance for ρ=1.2 [N·s²·m⁻⁸]	Air Flow Q [m³·s⁻¹]	ΔZ Z₂-Z₁ [m]	Winter Avg. Density $\bar\rho_w$ [Kgm⁻³]	Winter Resistance R_w [N·s²·m⁻⁸]	Winter Acoustic Mass β_w [Kgm⁻⁴]	Winter N.V.P [Nm⁻²]	Summer Avg. Density $\bar\rho_w$ [Kgm⁻³]	Summer Resistance R_s [N·s²·m⁻⁸]	Summer Acoustic Mass β_s [Kgm⁻⁴]	Summer N.V.P [Nm⁻²]
1	12	1	7.2	10.8	800	0.4467	38[2]	-1.6	1.218	0.4534	135.33	-1.1611	1.182	0.4400	131.33	0.1883
2	1	2	7.2	10.8	1000	0.5560	38	2	1.279	0.5926	177.64	0.2550	1.176	0.5449	163.33	-0.2354
3	2	3	5.5	9.4	57.74	0.1512	14	50	1.264	0.1593	13.27	13.7293	1.1835	0.1491	12.42	-6.6195
4	2	4	7.2 / 5.5	10.8 / 9.4	200 / 57.74	0.2696	14	50.4	1.2615	0.2834	48.29	15.0748	1.1835	0.2659	45.30	-6.6724
5	2	5	5.5	9.4	30	0.0946	10	0.3	1.266	0.0998	6.91	0.0765	1.182	0.0932	6.45	-0.0353
6	3	6	5.5 / 1.32 / 3.96	9.4 / 5.6 / 8.0	30 / 50 / 50	6.7418[1] (5.5459)	7[2]	0.3	1.2445	6.9918	69.64	0.1397	1.182	6.6407	66.14	-0.0353
7	3	7	7.2 / 1.32 / 3.96	10.8 / 5.6 / 8.0	30 / 50 / 100	6.8904[1] (5.6320)	7[2]	0.5	1.24225	7.1330	86.77	0.2439	1.1805	6.7784	82.46	-0.0515
8	4	8	5.5 / 1.32 / 3.96	9.4 / 5.6 / 8.0	30 / 50 / 50	5.6092[1] (5.5459)	7[2]	0.3	1.23975	5.7950	69.38	0.1537	1.182	5.5251	66.14	-0.0353
9	4	9	7.2 / 5.5 / 1.32 / 3.96	9.4 / 5.6 / 8.0	30 / 50 / 50 / 57.74	5.6320	7[2]	0.5	1.23975	5.8186	86.60	0.2562	1.1805	5.5405	82.46	-0.0515
10	5	6	7.2 / 5.5	9.4 / 5.6 / 8.0	30 / 50 / 57.74	{13.9352[1] (0.1512)}	5	50	1.2465	14.4752	13.09	22.3101	1.1805	13.7088	12.39	-5.1485
11	5	8	7.2 / 5.5	10.8 / 9.4	200 / 57.74	{12.7292[1] (0.3084)}	5[2]	50.4	1.24425	13.1986	47.62	23.6007	1.1805	12.5224	45.18	-5.1897
12	6	7	5.5	10.8	100	0.0861	9.2	0.2	1.22475	0.0879	17.01	0.1319	1.1775	0.0845	16.35	-0.0147
13	6	11	5.5	9.4	57.74	0.0736	2.8	50	1.2225	0.0750	12.83	34.0781	1.1805	0.0724	12.39	-5.1485
14	7	12	5.5	9.4	57.74	0.0736	16.2	49.6	1.22025	0.0748	12.81	34.8999	1.179	0.0723	12.38	-4.4130
15	8	9	5.5	10.8	100	0.0861	3.6	0.2	1.2225	0.0877	16.98	0.1363	1.1775	0.0845	16.35	-0.0147
16	8	10	5.5 / 7.2	9.4 / 10.8	57.74 / 100	0.1328	8.4	49.8	1.22025	0.1350	29.76	35.0406	1.1805	0.1306	28.79	-5.1279
17	9	10	5.5	9.4	57.74	0.0736	10.6	49.6	1.22025	0.0748	12.81	34.8999	1.179	0.0723	12.38	-4.4130
18	10	11	7.2	10.8	100	0.0592	19	-0.2	1.218	0.0601	16.92	-0.1451	1.182	0.0583	16.42	0.0235
19	11	12	7.2	10.8	100	0.0592	21.8	-0.2	1.218	0.0601	16.92	-0.1451	1.182	0.0583	16.42	0.0235

1) Included regulator resistance.
2) Fixed flow.

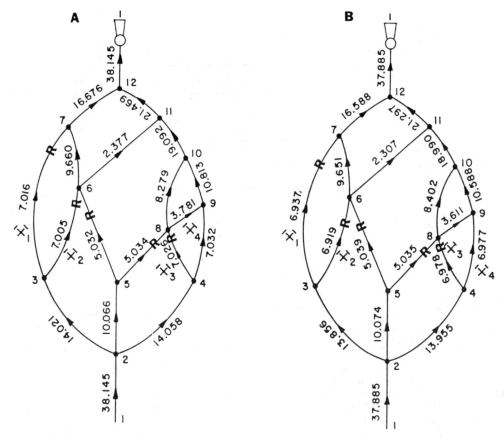

Fig. 4. Canonical diagrams of a ventilation system with flow distribution for summer season (a) and winter season (b), when fan is operating.

calculated for summer and winter, and are presented in Table 1 in corresponding columns. Determination of the NVP temperature distribution along the ventilation system was assumed for both winter and summer seasons. Figure 2 shows the average temperature distribution along the system respectively for winter and summer. The average density of outside air for winter and summer were assumed to be 1.29 kg/m³ and 1.17 kg/m³, respectively.

3 MATHEMATICAL MODEL

3.1 Dynamics of Flow in an Airway

Litwiniszyn (1,2), assuming one-dimensional flow through a mine airway, and modifying Euler's equation by adding the friction forces between the air and the walls, proposed the following equation:

$$\frac{\partial p}{\partial X} + RQ^2 + \frac{\rho}{S} \frac{\partial Q}{\partial t} + \frac{\rho}{S^2} Q \frac{\partial Q}{\partial X} = 0 \qquad (1)$$

Then simplifying the problem to the case in which the variation of air density in respect to time and space can be neglected, from the equation of conservation given below:

$$\frac{\partial \rho}{\partial t} + \frac{1}{S} \frac{\partial}{\partial X} (\rho \cdot v \cdot S) = 0 \qquad (2)$$

for ρ=const, the final conclusion is as follows:

$$Q(t) = v(x,t) * S(x) \qquad (3)$$

where:
Q = quantity of air;
v = velocity of air;
S = cross-sectional area; and
X = length coordinate.

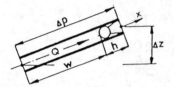

Fig. 5. A segment of airway.

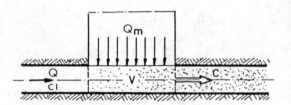

Fig. 6. Schematic of the zone where methane emissions occur directly from the wall and broken coal.

Taking into account equations (1) and (3), the mathematical model which describes the dynamical equilibrium of flow through the airways (see Fig. 5) is as follows:

$$B \frac{dQ}{dt} + RQ^2 = H \qquad (4)$$

or

$$B\frac{dQ(t)}{dt} + W(t) + \Delta p(t) + \rho\Delta Z = h[Q(t)] \qquad (5)$$

where
  H = total static pressures of all kind of pressure sources, $[N/m^2]$;
  $h[Q(t)]$ = fan pressure, $[N/m^2]$;
  $\rho\Delta Z$ = natural ventilation pressure, $[N/m^2]$;
  $\Delta p(t)$ = pressure difference between the ends of airway, $[N/m^2]$;
  $W(t)$ = $RQ^2(t)$ pressure losses, $[N/m^2]$;
  $Q(t)$ = quantity of air, $[m^3/S]$;
  B = acoustical mass ($B = \frac{L\rho}{S}$), $[kg/m^4]$;
  ρ = density of air, $[kg/m^3]$;
  X = length coordinate;
  L = length of airway, $[m]$; and
  S = cross-sectional area of airway, $[m^2]$.

3.2 Methane Concentration

Concentration of methane is defined as the ratio of the rate of methane and the rate of air flow:

$$C(x,t) = \frac{Qm(x,t)}{Q(t)}$$

Two different sources of methane emissions in coal mines can be distinguished. First, when methane emissions occur directly from the walls along the airway; and second, which is more complicated to simulate, when the airway is adjacent to gob areas (worked-out areas), and the amount of methane depends on a difference between the pressure in the airway and gob.
  For this particular study the first model of methane emission was adopted.
  For such a condition, the mathematical model of the dynamics of the methane

concentration in a zone where the methane emission takes place (see Fig. 6) is based upon the mass conservation law, and can be expressed by ordinary differential equation as follows:

$$V \frac{dC(t)}{dt} = Q(t) \times CI(t) + Q_m(t) - [Q_m(t) + Q(t)] C(t)$$

Where:
  V = volume of zone involved, $[m^3]$;
  $Q(t)$ = quantity of air flow through the zone, $[m^3/s]$;
  $Q_m(t)$ = quantity of methane emitted to the zone, $[m^3/s]$;
  $C(t)$ = average concentration in the zone; and
  $CI(t)$ = concentration of methane in the air flowing to the zone.

The mathematical model of unsteady-state airflow in a ventilation network (5) is easy to combine with the methane concentration process which is described by equation (6). As the results of such an arrangement, the methane concentration changes in the ventilation system can be investigated.

3.3 Numerical Simulation Method

The numerical simulation method is a means of solving equations which describe the unsteady-state of airflow in a mine ventilation network. Because nonlinear, ordinary differential equations are concerned, only approximation methods are available to solve numerous simultaneous equations that describe the flow processes in ventilation networks. A computer program being used for this study, written in Fortran, implements the numerical methods suggested by Trutwin (3-5).

4 RESULTS OF SIMULATION

The main concerns of this study were the

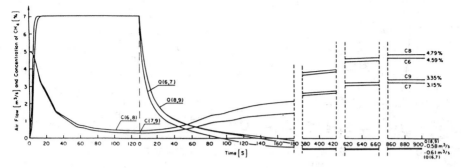

Fig. 7. The unsteady-state of flows and methane concentrations in working areas for the summer season.

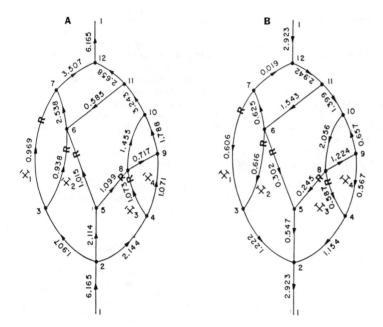

Fig. 8. Canonical diagrams of a ventilation system with flow distribution for summer season (a) and winter season (b), when the fan is shutdown.

unscheduled fan shutdowns caused by power failure; fan restarts, and their effect on methane concentration in the mine atmosphere. In other words, determination of the time relations between the changes in methane concentration in the working areas and disturbances in the main fan performance.

The ventilation system will react when changes of fan performance occur (e.g., stoppage, restart), generating unsteady-states of airflow, and consequently, ef-fecting changes in methane concentration. These were studied by means of numerical method utilizing computer programs called NETWORK, NONSTEADY, and METHANE.

The programs are based on the mathe-matical models of: unsteady air flow in mine ventilation networks (5), and methane concentration buildup in workings (6). They enable the simulation of transients of an airflow in the network and methane concentration in a zone where the methane emission takes place due to:

91

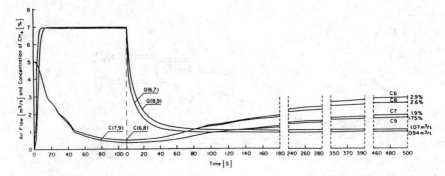

Fig. 9. The unsteady-state of flows and methane concentrations in working areas for winter season.

- initial conditions of flow and methane concentration;
- changes of parameters, i.e., fan characteristics, resistance, methane emission.

To illustrate the simulation data, four combinations caused by changes in main fan operation, for summer and winter conditions, were investigated. The results of computer simulation plotted in the form of graphs are shown in Fig. 7 and Fig. 9. Table 1 presents data of the network for each of the simulations being carried out. Note that the dynamics of the main fan with driving system are excluded from this study.

4.1 Fan Turn-on and Shutdown (Summer)

Figure 7 shows the changes of flow, "Q," and methane concentration, "C," in the working sections located in branches 6, 7, 8, and 9. The first part of the graph refers to changes after the main fan was turned on, and for initial conditions as follows:

$$Q_6(0) = Q_7(0) = Q_8(0) = Q_9(0) = 0$$

$$C_6(0) = C_7(0) = C_8(0) = C_9(0) = 5\%$$

The data show that after a fan is turned on, the flow in the workings stabilizes itself in less than 20 seconds, and methane concentration starts to drop immediately and reaches a steady-state (below 0.5%) in 120 seconds.

Finally, when flows and methane concentrations stabilize in the system, the effect of main fan being shutdown was simulated.

The second part of the graph in Fig. 7 shows the transients of flow and gas

concentrations in working areas as a response to those changes. As was expected, the airflow decreases and reaches zero at different times at different locations, in a range of 100 to 180 seconds. Then, because of the NVP existing in a mine in summer, the flow reverses in almost all branches. Figures 8a and b show the flow distribution in the mine due to the NVP for summer and winter.

After fan shutdown, the methane concentration starts to build up slowly and stabilized at different levels, depending on ventilation and methane emission conditions in a particular working section.

4.2 Fan turn-on and shutdown (winter)

Similar kinds of changes in main fan operation have been investigated for the winter season. The only major difference is in NVP, which affects the final quantity of air and direction of flow, through the system when the fan is shutdown. Figure 9 shows the changes of flow "Q," and methane concentration, "C," in the working sections.

5 CONCLUSIONS

The results obtained by unsteady-state analysis of the ventilation system can be used to determine:
- time in which the miners should be evacuated to an area having uncontaminated air;
- duration time that the fan can be shutdown before the methane concentration reaches a certain level, especially important when a diesel

powered generator is used to provide
electricity;
- time required for the methane to
  return to a safe level after the fan
  is turned back on;
- the duration of unsteady-state
  processes;
- all these values vary, not only for
  each mine, but also for the same mine
  due to the changes of the parameters,
  i.e., flow condition, gas emission,
  etc.;
- the developed computer program can be
  used to study transients in
  ventilation networks for different
  types of initial conditions and
  disturbing factors (controlled
  regulators, recirculation, etc.).

6 REFERENCES

1. J. Litwiniszyn, 1951. A problem of
    dynamincs of flow in conduit
    networks. Bull. Acad. Polon. Sci.
    Lett. Ser. Math., Vol. I, No. 3,
    pp. 325-339.
2. J. Litwiniszyn, 1959. Flow stability
    in pipe networks. Bull Acad. Polon.
    Sci., Ser. Techn., Vol. VII, No. 10,
    pp. 599-608.
3. W. Trutwin, 1970. Use of digital
    computers for the study of non-
    steady states and automatic control
    problems in mine ventilation
    networks. Int. J. Rock Mech. Min.
    Sci., Vol. 9, pp. 289-323.
4. W. Trutwin and D. Dziurzynski, 1978.
    Numerical method used to determine
    non-steady state of airflow in mine
    ventilation networks. Gornictwo,
    Vol. 2, No. 1, (in Polish).
5. W. Trutwin, 1979. On a simulation
    method of methane concentration
    control. Proceedings of the
    Second International Mine
    Ventilation Congress, Reno, Nevada,
    November, 1979.

Mine ventilation system design and optimization at a developing mine

KULDIP S.KHUNKHUN
Sundt International Inc., Tucson, AZ, USA

ABSTRACT: The main objective of this paper is to establish the ventilation design criteria and its application in the design of a mine ventilation system. Optimization of the primary and secondary airways, ventilation equipment and the total ventilation network are discussed based on site-specific development and production schedules. Capital costs for development of ventilation shafts, ventilation equipment costs and electric power cost for operating the designed ventilation plant are estimated.

1 INTRODUCTION

The deposit is located in the south-western United States and consists of complex multi-horizon copper, zinc and silver mineralization in calc-silicate (skarn) around the edge of quartz-monzonite stock. Ore-grade mineralization occurs in numerous isolated ore bodies localized at fourteen separate horizons representing approximately 300 m of stratigraphic thickness. Individual ore bodies are quite variable in size and vary from sheet-like to pod-like, to tabular masses varying from 1.8 m to 15 m thick. These ore horizons occur at elevations between 1,850 to 2,200 m above sea level. Room and pillar mining method is considered to exploit this ore deposit using diesel-powered, rubber-tired equipment.

2 VENTILATION DESIGN PARAMETERS

The ventilation system must meet the environmental needs of the designed daily mining schedule and the peak ventilation demand during the development and production periods. The various parameters which are considered important for the ventilation system design are listed below:
1. Working environment.
2. Mine resistance.
3. Mine production and equipment.
4. Criteria for air velocity.
5. Criteria for air volume.
6. Size of airway.
7. Location of airway.

3 VENTILATION SYSTEM DESIGN

Based on the ventilation design parameters, a volume of 350,000 cfm (165.2 m^3/sec) of fresh air is designed to ventilate the total mine. The ventilation schematic and mine configuration are shown in figure 1. The mine air flow and its distribution through the primary airways are shown in table 1.

Table 1. Mine air flow (Development and Production Phase).

Primary Intakes	Air Volume
Service Decline	200,000 cfm (94.4 m^3/sec)
Development Borehole	150,000 cfm (70.8 m^3/sec)
Total Mine Intake	350,000 cfm (165.2 m^3/sec)

Primary Returns	Air Volume
Exhaust Shaft No. 1	175,000 cfm (82.6 m^3/sec)
Exhaust Shaft No. 2	175,000 cfm (82.6 m^3/sec)
Total Mine Return	350,000 cfm (165.2 m^3/sec)

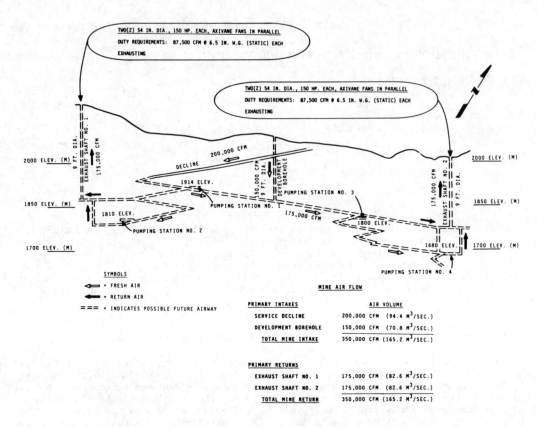

Figure 1. Mine ventilation schematic – development and production phase.

Table 2. Air volume design calculations (Development and Production Phase)

Description	Primary Development Headings	Secondary Development Headings	Production Headings	Total
No. of active headings projected	1	1	7	9
No. of spare headings assumed	1	–	2	3
Total no. of headings to be ventilated	2	1	9	12
Air volume designed per heading-cfm (m³/sec)	18,000 (8.5)	18,000 (8.5)	18,000 (8.5)	
Air volume required for total no. of headings to be ventilated-cfm (m³/sec)	36,000 (17.0)	18,000 (8.5)	162,000 (76.5)	216,000 (102.0)
Shops, pump stations, lunch room, etc. allow-cfm (m³/sec)				45,000 (21.2)
Total				261,000 (123.2)

Assuming Volumetric Efficiency of 75 percent, total air volume required = 348,000 cfm. Use 350,000 cfm (165.2 m³/sec).

The ventilation network shown in figure 1 is based on 1) a proposed mining rate of 1,200 metric tons per day of ore and waste and 2) the mining equipment fleet planned for the project. The recommended airflow, when properly distributed, will sufficiently meet the air volume requirements essential for the provision of a suitable working environment throughout the mine. The air volume design calculations for the total mine are shown in table 2.

The frictional resistance of the proposed mine ventilation network, after adjusting for altitude, is estimated at 10.08 inches (256.03 mm) of water gauge-static. The mine resistance calculations are based on nine foot (2.74 m) diameter (finished) ventilation boreholes and are shown in table 3. To assist in ventilation planning and extension or modification of the ventilation system, a mine characteristic curve has been developed and is shown in figure 2.

Table 3. Mine resistance calculations.

Airway	Friction Factor (x 10^-6)	Surf. Perimeter (p-ft)	Equiv. Length (l-ft)	Effective X-N Area (A-ft^2)	Resistance Factor (-R)	Air Volume (Q-cfm)	Air Velocity (V-ft/min)	Press. Drop - Static (in. W.G.)
Service decline	85	60	1,800	190	0.26	200,000	1,053	1.03
Development borehole	28	28	750	63	0.45	150,000	2,381	1.02
Decline access to west mining zone	85	60	2,800	190	0.40	175,000	921	1.23
Decline access to east mining zone	85	60	3,000	190	0.43	175,000	921	1.31
Mine workings - west	-	Allow 1.00 in. W.G.						1.00
Mine workings - east	-	Allow 1.00 in. W.G.						1.00
Exhaust shaft #1	28	28	1,100	63	0.65	175,000	2,778	2.03
Exhaust shaft #2	28	28	800	63	0.48	175,000	2,778	1.48
								10.10
Add for shock losses								2.50
Total static pressure drop								12.60
Static pressure drop when corrected for altitude of 6,500 feet above sea level =								10.08 in W.G. (256.03 mm)

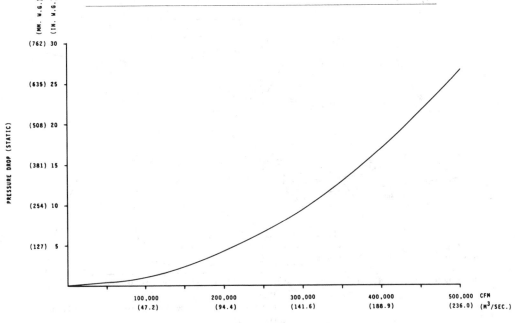

Figure 2. Mine characteristic curve.

The total power consumption to operate the ventilation plant is estimated at 1,500 horsepower. The power distribution between the primary and auxiliary ventilation systems is tabulated in table 4.

Table 4. Electric power requirement.

Primary Ventilation	Horsepower (estimated)
Exhaust shaft no. 1 mine fans	300
Exhaust shaft no. 2 mine fans	300
Auxiliary ventilation	900
Total electric power requirement estimated for mine ventilation	1,500

4 COSTS

The capital cost for the development and lining of the development borehole and the two ventilation shafts shown in figure 1 is estimated at $3,100,000.00.

4.1 Ventilation equipment costs

The costs of primary ventilation fans, auxiliary ventilation equipment and the air doors/fire doors required for ventilation control are estimated at $330,000.00 including installation.

4.2 Electric power cost

The electric power cost for operating the ventilation plant (primary ventilation system) designed for the project is calculated at $640.00 per day or $0.64/metric ton.

5 PROPOSED MINE VENTILATION SYSTEM DEVELOPMENT

5.1 Initial development phase

The development of an exploratory heading in the westerly-lying ore body and a heading connecting to the easterly-lying ore body will be accomplished using an overlap arrangement of the blowing and

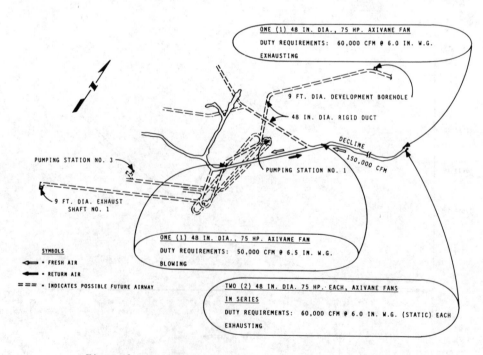

Figure 3. Initial mine ventilation development phase.

98

exhausting (auxiliary) ventilation systems. Such arrangement will establish the service decline as a fresh air base with approximately 120,000 cfm (56.7 m3/sec) available for development of the proposed headings. With proper installation, each development heading will have adequate ventilation for safe operation of one LHD 5 yard unit and one 25 ton haulage truck unit. The ventilation design for the initial development phase, indicating fan and duct arrangement, system operating mode, air volumes, and direction of airflow is shown in figure 3.

5.2 Interim development phase

With the completion of the proposed development borehole, located approximately half-way on the ramp access to the east-end ore body, an interim ventilation system will be developed. The system arrangement will eliminate the requirement of the exhaust fans at the service decline collar. The development of two headings, one connecting to the exhaust shaft no. 1 in the westerly ore body and the other connecting to the exhaust shaft no. 2 in the easterly ore body, will be carried out using auxiliary ventilation systems. The ventilation design for the interim development phase, indicating fan and duct arrangement, system operating mode, and air volumes and direction of airflow, is shown in figure 4.

5.3 Final development and operational phase

5.3.1 Ventilation system development

The mine ventilation system for the development and operational phase of the project will be developed using two primary intakes, centrally located between the two main ore bodies, and two primary exhaust shafts, located at or near the extremities of the ore bodies. The primary intakes consist of the service decline with an average cross-section of 4.5 meters by 4.5 meters, and the development borehole with an inside diameter of 2.74 meters. The exhaust shafts no. 1 and no. 2, will be bored to have a finished inside diameter of 2.74 meters each. Internal raises of smaller diameters and shorter lengths will be required to extend the proposed ventilation network into the currently planned mine workings as shown in figure 1.

5.3.2 Mine air flow

A volume of 350,000 cfm (165.2 m^3/sec) of fresh air will flow into the mine with approximately 200,000 cfm (94.4 m^3/sec) through the service decline, and approximately 150,000 cfm (70.8 m^3/sec) through the development borehole. Approximately 175,000 cfm (81.7 m^3/sec) of intake air is designed to ventilate mine workings in each of the two ore bodies. The total return air from the mine workings will

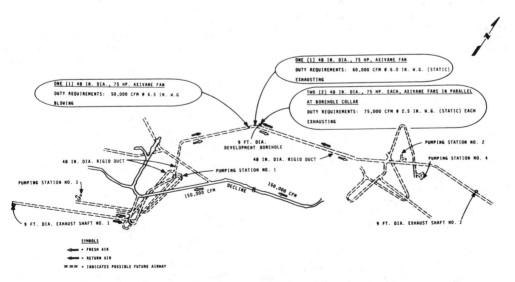

Figure 4. Interim mine ventilation development phase.

travel to the exhaust shafts no. 1 and no. 2 where it will be exhausted to the surface by the main mine exhaust fans.

The mine ventilation schematic indicating volumes and direction of airflow is shown in figure 1. Also indicated in the figure are the location and size of main mine fans, including fan-duty requirements and operating mode of the specified fan units. The mine ventilation controls such as air doors, fire doors and regulators, which will be required for the system operation, are not shown in the mine ventilation schematic.

5.3.3 Optimization of ventilation borehole and shafts

In the preliminary ventilation design of the project, the different sizes of ventilation boreholes have been considered with an average air volume requirement of 210,000 cfm (99.2 m^3/sec) through each

Table 5. Solution output for sizing the most economical circular airway lined with steel linear and concrete.

Inputs

Item	Value
Fan efficiency	75.00%
Capital return	15.00%
Friction factor	28.0 x 10^{-10}
Electricity cost	$0.065/KWHR
Quantity of air flow	210000. ft^3/min
Excavation and lining cost	$17.17/ft^3
Length of airway	875. ft
Mine life	10.0 yrs
Airway operating life	10.0 yrs
Insurance, taxes, etc.	0.00%

Solution

Item	Value
Diameter of circular airway	8.6 ft
Head loss (airway)	2.816 in W.G.
Horsepower (airway)	124.
Annual operating cost (airway)*	$184,591
Capital cost for excavation and lining of airway	$879,039

*Includes costs for electricity to operate the fan motor, capital return, insurance, taxes, etc.

Table 6. Solution output for sizing the most economic circular airway unlined.

Inputs

Item	Value
Fan efficiency	75.00%
Capital return	15.00%
Friction factor	28.0 x 10^{-10}
Electricity cost	$0.065/KWHR
Quantity of air flow	210000. ft^3/min
Excavation cost	$6.04/ft^3
Length of airway	875. ft
Mine life	10.0 yrs
Airway operating life	10.0 yrs
Insurance, taxes, etc.	0.00%

Solution

Item	Value
Diameter of circular airway	10.0 ft
Head loss (airway)	1.335 in W.G.
Horsepower (airway)	59.
Annual operating cost (airway)*	$87,521
Capital cost for excavation of airway	$416,786

*Includes costs for electricity to operate, capital return, insurance, taxes, etc.

borehole or shaft. A volume of 210,000 cfm (99.2 m^3/sec) through each of the boreholes will allow a 20 percent increase in the ventilation system design capacity. The borehole sizes used in the ventilation network shown in figure 1 have been optimized at nine feet (2.74 meters) diameter (finished). The optimization is based on fan-duty requirements, capital and operating costs of constructing and operating these ventilation boreholes for an average depth of 875 feet (266.8 meters). A computer printout, "Solution Output for Sizing the Most Economical Airway," showing a detailed account of the data input for a lined and unlined airway, is shown in tables 5 and 6.

ACKNOWLEDGEMENTS: The author wishes to thank the Project Management and Pincock, Allen and Holt Inc. for their permission to present this paper and to the staff of Sundt International, Inc. for their kind assistance and guidance in the preparation of this paper.

3. Methane drainage

Use of vertical wells for drainage of methane from longwall gobs

D.W.HAGOOD, J.E.JONES & K.R.PRICE
Jim Walter Resources Inc., Brookwood, AL, USA

ABSTRACT: The release of abnormally high amounts of methane in pillaring areas of coal mines is a well-known phenomenon in the mining industry. Methane emission rates are particularly high when coal seams exist within the strata above the seam being mined and become part of the broken, relaxed zone created after longwall extraction. Vertical wells drilled from the surface into the overlying strata of the longwall panels at the Jim Walter Resources, Inc. No. 4 Mine have proven to be an effective means of methane control in the longwall gob bleeders as well as an economically viable commercial venture. Methane flow rates in the longwall bleeders have been reduced as much as 80% as a result of the gob well program with savings in air volume currently estimated at 600 000 ft^3/min (283m^3/s). A total of 15 gob wells have been installed since April 1983 with a cumulative methane production of over four billion cubic feet (113 million cubic meters).

INTRODUCTION

No. 4 Mine, located approximately 50 miles southwest of Birmingham, Alabama, is one of four deep-shaft mines owned and operated by Jim Walter Resources, Inc. in the Warrior Coal Basin.

The mine produces a high-grade metallurgical coal at a current rate of 2.3 million tons per year with three longwalls and ten continuous mining units. The seams extracted, overlain by 2000 ft (610 m) of cover, include both the Blue Creek and Mary Lee with average thicknesses of 60 in. (1.5 m) and 16 in. (0.4 m), respectively. A rock parting separates the two seams and ranges in thickness from 16 - 54 in. (0.4 - 1.4 m). Approximately 80% of all development mining is accomplished by extracting both seams with the remaining 20% comprised of mining the Blue Creek only. Due to problems associated with the relatively thick and competent rock parting between the Mary Lee and Blue Creek seams, only the Blue Creek is extracted on the longwall sections.

The initial 100 ft (30.5 m) of overburden above the active mining horizon consist primarily of sandstone and includes the upper and lower New Castle seams as well as a band of marker coal. (Figure 2) Immediately above the sandstone lies a section of sandy shale which typically measures

200 ft (61 m) in thickness. The strata which exist 300 - 500 ft (91 - 152 m) above the Blue Creek is a mixture of shale, sandy shale and sandstone and includes the Gillespie and Curry seams located approximately

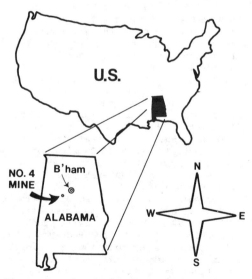

Figure 1. No. 4 Mine located approximately 50 miles southwest of Birmingham, near Brookwood, Alabama.

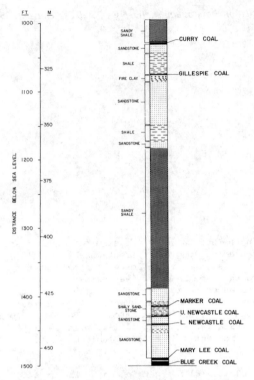

Figure 2. Typical lithology of immediate 500 ft (152 m) of overburden above the Blue Creek/Mary Lee coal measures.

and two return entries which allows a split system of section ventilation.

A total of 310 acres of gob have been generated during the 27-month period beginning in February 1983 when the first longwall began. This represents a long-wall retreat rate of approximately 800 ft (244 m) per month. Assuming that each ft of longwall retreat generates approximately 110 tons of coal, an estimated 2.38 million tons of coal have been longwall mined along with a cumulative methane production from the gob wells of some 4.3 billion ft[3] (121.8 million m[3]). Therefore, approximately 1800 ft[3] (52 m[3]) of methane has been produced and sold from the gob well system for every ton of longwall coal mined. In addition to the commercial value of the methane, an estimated 600 000 cfm (283 m[3]s) air has been diverted from the bleeder entries due to the effectiveness of the gob wells; an air volume sufficient to ventilate a longwall and three continuous miner sections or one-third of No. 4 Mine's productive force.

GOB WELL INSTALLATION

All gob wells, if possible are installed and ready for production prior to inter-ception underground by longwall retreat mining. Typical installation procedures (Figure 3) include: 1.) Drilling a 17½ in. (44.4 cm) hole to a depth of 160 ft (44.8 m)

400 ft (122 m) and 450 ft (137 m) above the Blue Creek, respectively. Roof joints are dominant in the N75°W direction with conjugate joint sets less prominent trending in the southwest - northeast direction.

Mine ventilation is currently supplied by three intake shafts and two return shafts each of which is equipped with two-3500 horsepower (2611 kw) TLT Babcock fans. With a total connected motor capacity of 14 000 horsepower (10 444 kw), the four fans exhaust 3.6 MMcfm (1700 m[3]/s) of air at a pressure of 18 in. water gauge (4478 N/m[2]) with a methane liberation of 20 MM cfd (566 400 m[3]/d).

A typical longwall panel at No. 4 Mine is 650 feet (198 m) wide and approximately 5000 ft (1524 m) long. Longwall produc-tion equipment includes Anderson-Mavor 500 double drum shearers with Thyssen shields and American Longwall armored face convey-ors. Panel development is accomplished with Joy 12 CM continuous miners driving a four-entry system, ie., a track entry (smoke-free intake) a belt entry (intake)

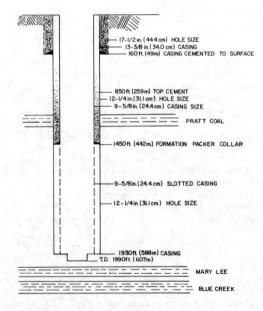

Figure 3. Detail of borehole completion.

to accomodate 13 3/8 in. (34.0 cm) surface casing, 2.) Setting the surface casing and cementing, 3.) Extending the borehole using 12¼ in. (31.1 cm) bit to within 20 ft (6.1 m) of the coal seam being extracted, 4.) Running 9 5/8 in. (24.4 cm) casing to the bottom of the hole with the lower 400 ft (122 m) of casing slotted for gas flow and 5.) Cementing the 9 5/8 in. (24.4 cm) casing by setting a Halliburton packer collar just above the slotted casing.

Plumbed immediately into the top of the well is a parallel filtering system which screens dust and fine particles from the methane flow and thereby prevents blockages from occurring in the flame arrestor or inlet screens on the gathering compressor. (Figure 4) The flame arrestor is installed just downstream from the filtering system and provides protection against any possible "back-burning". After exiting the flame arrestor, the methane enters a manifold which is designed to feed one or two gathering compressors for gas sales or vent to the atmosphere either by natural aspiration or under a vacuum by means of a North American centrifugal blower.

A variety of gathering compressors are currently in use (powered either electrically or by methane-fueled engines) with the selection of the unit depending upon the potential methane production from each individual well. Compressor units are often shifted between wells as dictated by changes in the production characteristics of the wells. Figure 5 is a typical well site which is equipped with two Ingersoll-Rand LR-165 compressors. The 165-HP (123 kw) units are methane-fueled and rated to move 1.3 MMcfd (36 817 m^3/d) methane at a 0 psig suction pressure and 50 psig (345 KPa) discharge pressure.

Each well is equipped with an oxygen analyzer which continuously monitors the oxygen content of the gas and allows adjustments to be made to prevent below-specification gas production. Each well site is also "hardwired" into a central monitoring station which allows continuous surveillance of compressor operation. In the event of compressor malfunction, the well is vented to atmosphere. If for any reason, the well must be "shut-in", the mine site is contacted immediately so that the affected underground bleeder area can be examined.

METHANE GENERATED FROM A LONGWALL PANEL AT NO. 4 MINE

Figure 6 is a graphical representation of methane flow rates for the second longwall panel at No. 4 Mine which began in August 1983 and finished in September 1984.

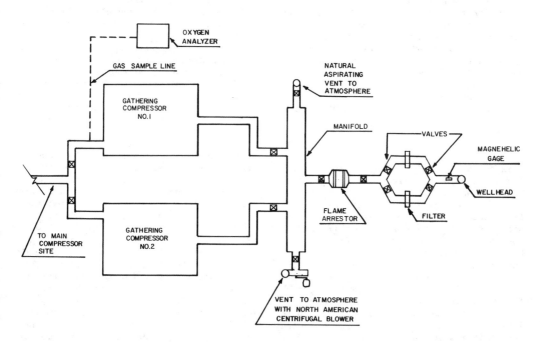

Figure 4. Plan view schematic of typical plumbing and compressor arrangement.

Figure 5. Standard gob well plumbing and compressor arrangement.

Underground bleeder air volumes and methane concentrations were determined on a weekly basis for the entire panel. Gob well flow rates were also recorded weekly in order to determine the methane capture ratio or rather that part of the total methane associated with a given longwall panel which is vented from the underground gob area to the surface via vertical gob wells.

The underground bleeder, gob well and total methane flow rates are each plotted relative to the amount of gob area which existed in the panel at the time of the measurements. The subject longwall panel is considered to be typical in size with a face width of 650 ft (198 m) and a panel length of 5000 ft (1524 m). As shown by the diagram in Figure 6, a total of four gob wells were installed in the panel; the first well located 845 ft (258 m) from the start-up line and the second, third and fourth wells located 1495 ft (456 m), 2260 ft (689 m) and 3700 ft (1128 m) from the start up line, respectively. Normally, only three gob wells would be planned for a 5000 ft (1524 m) panel. However, sufficient engineering data had not been collected at the time of panel start-up to layout and plan the optimum number of wells and their respective locations.

The methane flow rate in the bleeder when the panel started was approximately 450 cfm (0.2 m^3/s). The increase in the methane release rate as the longwall face retreated was relatively stable but accelerated rapidly after the gob area reached 6 acres and continued until the peak rate was measured at a gob area of 20 acres. Generally speaking, the methane rate increased about 185 cfm (0.09 m^3/s) per acre of gob for the initial 13 acres. From 13 acres to 20 acres the release rate increased approximately 380 cfm (0.18 m^3/s) per acre of gob generated. After the peak release rate of 5500 cfm (2.6 m^3/s) was attained at 20 acres, the total methane flow decreased and remained relatively constant at an average flow rate of 4000 cfm (1.9 m^3/s) throughout the remainder of the panel. The first gob well, located at the 13-acre point, began producing at a rate of 1000 cfm (0.47 m^3/s) or 1.44 MMcfd (40 780 m^3/d). This flow rate represented 35% of the total release rate with the remaining 65% being diluted in the underground bleeder. The second gob well began producing after approximately 23 acres of gob had been generated. The No. 1 and No. 2 Gob Wells were collectively producing 2450 cfm (1.2 m^3/s) or 3.5 MMcfd (99 915 m^3/d) methane which represented 59% of the total methane released from the longwall area. After the

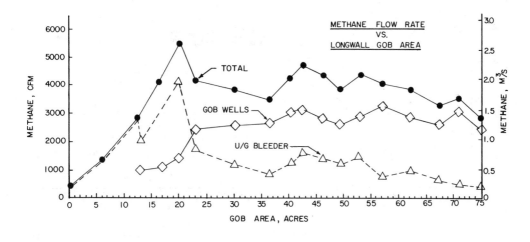

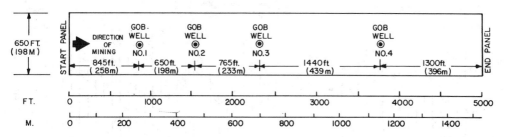

Figure 6.   Methane liberation history of a typical longwall panel at No. 4 Mine

third well was intercepted, methane production via the gob wells reached 3150 cfm (1.5 m³/s) or 4.5 MMcfd (128 462 m³/d) with a gob area of about 42.5 acres. With the total release rate at 4700 cfm (2.2 m³/s), this represented a capture ratio of 67%. Finally, Gob Well No. 4 was placed in service with about 1300 ft (396 m) of the panel remaining. The gob well capture ratio reached a maximum of 81% during this period with the four gob wells producing 3300 cfm (1.56 m³/s) or 4.75 MMcfd (134 579 m³/d) methane with the total flow rate from the longwall area being 4050 cfm (1.9 m³/s).

The graph indicates that an air volume exceeding 200 000 cfm (94.4 m³/s) was required to maintain bleeder methane concentrations less than 2.0% at the 20-acre level. However, after the No. 2 Gob Well was placed in service at a location 1495 ft (456 m) from the start-up line, the air requirements were reduced substantially to a level of about 100 000 cfm (47.2 m³/s).

Without the gob well program, air requirements for the subject panel would have approached 275,000 - 300 000 cfm (130-142 m³/s) in order to maintain methane concentrations in the underground bleeder at some point below 2.0% by volume.

GOB WELL SPACING

Ventilation data collected routinely in the bleeder entries coupled with gob well production data indicates that, in general, the methane liberation rate from the gob accelerates rapidly in the initial 25% of the panel with the peak methane flow rate developed with a gob area of 20 acres (Fig. No. 6). Total gob production appears to decrease somewhat after the peak rate is reached and remains relatively constant throughout the remainder of the panel.

Initial panels were equipped with gob wells located approximately 800 ft (244 m) from the longwall start-up line. However, monitoring of the bleeder during the early stages of panel extraction showed that substantial benefits could be gained by shifting the first well closer to the start-up line.

Methane flow rates in the bleeder can be as high as 2000 cfm (0.94 m³/s) after only 500 ft (152 m) of extraction. Of the 2000 cfm (0.94 m³/s) methane in the bleeder, at least 700 cfm (0.33 m³/s) can be captured with a gob well with 1300 cfm (0.61 m³/s) remaining in the bleeder. The 1300 cfm

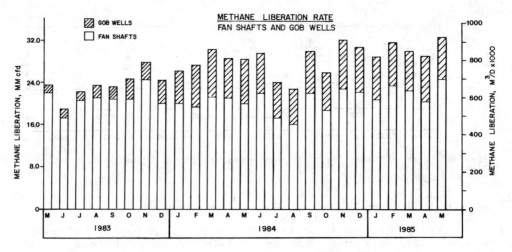

Figure 7. Total methane liberation rate - fan shafts and gob wells

(0.61 m³/s) represents the methane released from the active longwall face, the roof, floor and ribs of the ventilating entries and the "spillover" fraction of gas which is generated in the gob and released along the boundary of the gob into the adjacent ventilating entries. With peak methane production occurring in the gob at an areal extent of 20 acres, a second gob well is usually located at approximately 1200 ft (366 m) from the longwall start-up line. By doing so, two wells are provided for capturing the peak flow rates in the panel. A third well is tentatively located halfway between the second well and the end of the panel. If adequate communication is maintained between the initial two wells and the primary methane producing source (ie., newly-formed gob immediately behind the active faceline), a third well in the panel

may not be necessary. In most instances, unless a restriction develops in the bottom of the well, the 9 5/8 in. (24.4 cm) well casing allows adequate flow rates with compressor suctions from 0-10 inch H_2O (0-2488 N/m^2) vacuum. Gob wells have produced up to 3.5 MMcfd (99,122 m^3/d) at vacuums of 30 inches H_2O (7464 N/m^2). Therefore, from a capacity standpoint, two gob wells would service a single panel in almost all cases. The third well, if installed, is done so to improve the collecting efficiencies of the system by decreasing the distance from the primary methane source to the bottom of the well.

EFFECT UPON TOTAL MINE VENTILATION REQUIREMENT

The longwall gob well program has signifi-

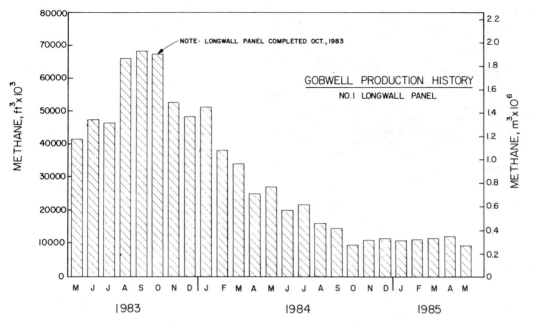

Figure 8. Gob well production history for initial longwall panel extraction
at No. 4 Mine.

cantly reduced the volume of methane in the
underground bleeder system and, hence, in
the overall mine ventilation network. Referring to the bar graph in Figure 7, the
average amount of methane released on a
daily basis through the mine fans has been
approximately 20 MMcfd (566 412 m^3/d) for
the past two years. In May 1983, the first
gob well was installed and placed in service
with a flow rate of 1.5 MMcfd (42 481 m^3/d).

As the longwall operation created additional gob, the gas production from the
vertical wells continued to increase for a
period of nine months. At that time, the
production from the wells appeared to
stabilize at a rate of 8.0 - 8.5 MMcfd
(226 565 - 240 225 m^3/d). The graph in the
upper part of Figure 7 shows that the gob
well system is handling an average 27.5%
of the total mine liberation rate.

GOB WELL PRODUCTION DECAY UPON COMPLETION
OF PANEL

No. 4 Mine currently incorporates a total
of 15 wells in the gob drainage program.
Nine of the 15 wells are located in panels
which have been totally extracted. The six
wells in the "active" panels are currently
producing 64% of the total methane production or approximately 5.3 MMcfd (150 099 $m/^3$d).
The balance of the daily make (36% of total

production) is produced from the "inactive"
wells which amounts to 3.8 MMcfd (84 962
m^3/d).

Figure 8 shows the collective methane production history of two gob wells located in
the first longwall panel extracted at No.
4 Mine. The panel was mined between February and October of 1983 and resulted in
a total gob area of 44 acres.

The methane production plot in Figure 8
shows that the wells exhibited an exponentially declining rate for a period of some
12 months following the completion of the
panel in October 1983. A steady-state rate
of production appears to have been established during the period from October 1984
through May 1985. The total monthly average
of 10 000 Mcf (238 000 m^3) is equivalent to
333 Mcfd (9440 m^3/d) methane which translates into approximately 7.5 Mcfd (212
m^3/d) methane per acre gob.

SUMMARY

Vertical holes drilled from the surface and
completed above the active mining level
prior to extraction can vent a substantial
amount of gob-generated methane, thereby
significantly reducing the ventilating air
requirements in the bleeder entries. No.
4 Mine currently produces approximately 8.5
MMcfd (240 725 m^3/d) of saleable methane

through a system of 15 gob wells which represents a savings in underground ventilation
capacity of an estimated 600 000 cfm (283
m^3/s) of air.

The effectiveness of the gob well program
in draining gob methane has had a significant impact on reducing ventilating air
requirements in the bleeders and, therefore,
has allowed No. 4 Mine to maximize the
production potential and remain on schedule
from a development standpoint. At the same
time, substantial economic benefit has
resulted from the commercial aspect of
methane sales.

A methane drainage project in Australia's Sydney Basin

WALTER L.RICHARDS
Methane Drainage Ventures, Placentia, CA, USA

ABSTRACT: A joint venture test project for methane drainage was conducted by Methane Drainage Ventures (MDV) at Australian Iron and Steel Pty. Ltd's (AI&S) Appin Colliery, Appin, New South Wales, Australia. The project for methane drainage in advance of longwall mining included five (5) horizontal boreholes into the Bulli Coalbed of the Illawarra Coal Measure in the south Sydney Basin, Australia.

Peak gas flowrates from a total of 1380 m (4500 ft) of horizontal boreholes reached 17,000 m^3 per day (600,000 standard cubic ft/d). The total horizontal drilling project was conducted in eight months of 1981. Produced gas of 97% methane was safely vented both during and after the test project due to the efforts of the colliery operators.

Test drilling was successfully conducted from two separate mining sections on opposite sides of the colliery. Data collected during and after the drilling campaign were used to define the properties of the coalbed. Computer simulation was used to predict commercial gas production.

The Appin project and a concurrent exploration corehole project were conducted to provide data to evaluate the Sydney Basin. Results of the evaluation indicate commercial feasibility of methane drainage for natural gas production. A subsequent design was completed for a natural gas production facility utilizing methane drainage technology.

INTRODUCTION

The coal fields of the south Sydney Basin are well known gassy mining regions. Collieries on the south coast have experienced gas problems in mining activity since mining began there in the 1800's. As is true with most coal seams, the Bulli coalbed's gas content increases with distance from the outcrop and overburden depth.

At the Appin Colliery, the distance to the outcrop is approximately 19 km (12 mi) and the overburden is greater than 300 m (985 ft). Appin Colliery is located at Appin, New South Wales (NSW), Australia shown in the map of Fig.1. The Bulli coalbed at Appin averages 3.0 m (10 ft) in thickness. Bulli coal at Appin is medium to low volatile bituminous coking coal used by AI&S for steel production. An analysis of Bulli coal is shown in Table 1.

Planning for the Appin Project and the associated methane drainage evaluation of the Sydney Basin began in 1980. A working agreement was reached between Occidental Petroleum Corporation (OPC) and the Australian Gas Light Co. (AGL) to jointly pursue a two-pronged evaluation program.

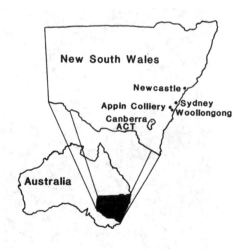

Figure 1

Location Map
Appin Colliery
New South Wales, Australia

Table 1
Analysis of Bulli Coal

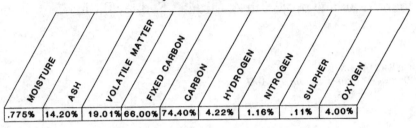

MOISTURE	ASH	VOLATILE MATTER	FIXED CARBON	CARBON	HYDROGEN	NITROGEN	SULPHER	OXYGEN
.775%	14.20%	19.01%	66.00%	74.40%	4.22%	1.16%	.11%	4.00%

This program included an in-mine evaluation project to assess gas production capability and an exploration core drilling project to assess the gas reservoir. A further agreement was reached with AI&S to coordinate an in-mine project at the Appin Colliery

The Occidental Research Corporation (ORC), MDV's predecessor, provided expertise and personnel for the Appin project. However, drillers for the drilling effort were recruited locally in the Sydney and Wollongong areas of NSW.

HORIZONTAL DRILLING

Some of the drilling equipment for the project was purchased in the Sydney area. The remainder of the equipment was brought from the United States and Great Britain. Since the drillers recruited for the job were unfamiliar with the equipment or the colliery, a training program was established. With the cooperation of the management and staff of the Appin Colliery, a short familiarization program was completed by all the project personnel. Continuing training of the drillers by ORC supervisors provided the necessary knowhow and direction for the project.

The equipment for the project was assembled and tested on the surface at Appin Colliery during March and April, 1981. Some of the equipment and personnel are shown in Fig. 2.

Figure 2

Appin Colliery operates a single longwall with four development working sections. Drilling for this project was conducted from two separate sections of the colliery on opposite sides. These are shown as Areas One and Two in Fig. 8. The first drilling was conducted from Area One in A heading of the southwest development.

After an initial shakedown and familiarization with the equipment, the first horizontal borehole was drilled to 315 m (1034 ft) in depth. The Bulli coalbed is strongly cleated in Appin with a middle section of very soft coal. The cleat at Appin is face and butt as shown in Fig. 3. The top and bottom sections of the seam in the vicinity of Appin No. 1 borehole had shaley inclusions with a 2 m soft friable coal in the middle of the seam.

Figure 3
1st Borehole Initiated From Left Return (A Heading) of the SW Dev.

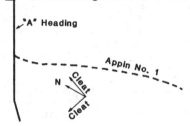

The Appin No. 1 borehole was drilled near the roof interface and nearly perpendicular to the number two cut-through of A heading southwest development as shown in Fig. 4. The borehole was terminated at 315 m when high gas flowrates caused drilling difficulties. The tools were pulled from the hole and a flow test was conducted on June 1, 1981. The initial flowrate was 14,442 m^3 per day (510,000 scfd). Early gas production history is shown in Fig. 5.

Figure 5
Early Production History Appin No. 1

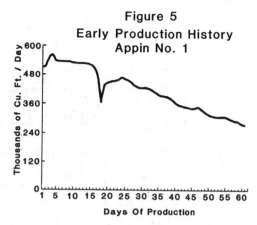

Due to the high rates of gas flow from Appin No. 1, a vent piping system to safely vent the gas was constructed by the colliery. This allowed for safe venting of the gas and prevented high local concentrations of methane in the return airways.

Other complications with the high initial rate are indicated in Fig. 5. Erosion of the borehole due to high gas velocity caused some sloughing of the friable coal within the hole near the standpipe. This caused some production separator plugging problems which are shown as a flowrate decline in the production data of Fig. 5.

Water production from Appin No. 1 was never accurately measured as it presented no problems to production. The gas produced was fully water saturated at 25°C (77°F) with the quality shown in Table No. 2. It is assumed that initial water rates were less than 50 liters/day (13.2 US gal/day). From these data we can see that both gas and water quality are good by comparison for coalbed methane drainage projects.

Once the safe venting network was established at Appin No. 1, the drill and equipment were moved 300 m "in-bye" to the selected location of Appin No. 2 from the same

Figure 4
Appin No. 1 Borehole Drilled Near Roof Interface

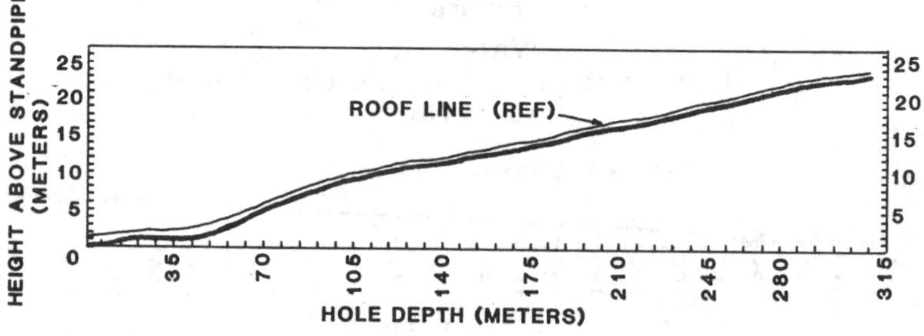

113

Table 2
Appin No. 1 – Gas Quality
Fully Water Saturated

| SAMPLE No. | DATE TAKEN | ANALYSIS, VOL. % | | | |
		METHANE	CARBON DIOXIDE	NITROGEN	TOTAL
1	6/11/81	97.1	2.3	0.6	100.0
2	6/11/81	97.0	2.3	0.7	100.0

heading and projected somewhat perpendicular to Appin No. 1 as shown in Fig. 6.

New drilling equipment and techniques were tried in drilling the initial part of the Appin No. 2 borehole. Due to the lack of stability and control, a serious vertical deflection of the borehole's course caused approximately 100 m of the floor rock to be intercepted near the mouth of the borehole. The remainder of the horizontal borehole was drilled in the bottom of the coal seam to 445 m (1460 ft) in total depth. The hole was

terminated due to excess stress on the drilling equipment on July 8, 1981.

As a result of the low section of the borehole near the mouth, a great deal of water tended to accumulate there. This resulted in a water trap as shown in Fig. 7. In fact, this trap functioned somewhat similar to its drainage namesake. The low spot near the mouth of the borehole caused slugging and surging flow of both gas and water from the borehole. This caused great difficulty in measurement of flowrates.

The third borehole, Appin No. 3, was drilled from a location some 200 m in-bye Appin No. 2 and relatively parallel as shown in Fig. 6. Appin No. 3 was drilled and completed to a total depth of 396 m (1300 ft) on August 24, 1981. The plan view of the borehole is shown in Fig. 6. It is obvious that the borehole's path was strongly influenced by the cleat structure of the Bulli coal. Appin No. 3 was terminated due to drilling stress and adjacent mining activity. Blasting activity 200 m in-bye created serious environmental problems in the return airway from which the borehole was drilled.

After initiating Appin No. 3 drilling, it was learned that a major igneous intrusion exists approximately 200 m in-bye the drilling location. As can be seen from Fig. 6, the course of the borehole actually came within approximately 50 m of the igneous "dyke". Fig. 6 also shows the dip angle of the coal seam at this point. It can be seen that the seam dips in the direction of the active face advance of southwest development (i.e., to the southwest).

Figure 6
Plan View of
Boreholes #1,2,& 3

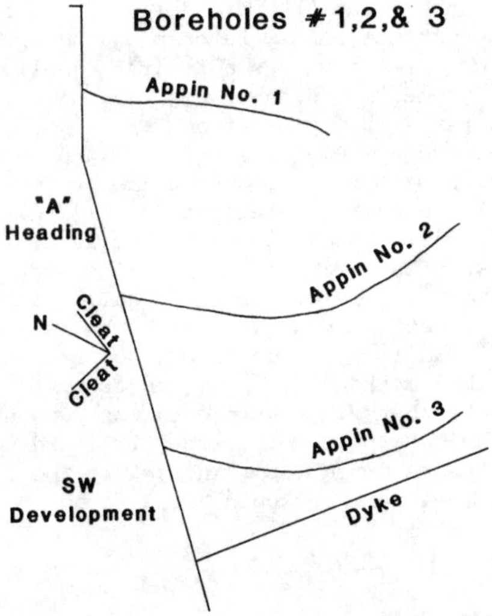

Figure 7
Water Trap
(Low Section of Borehole Near Mouth)
Appin No. 2

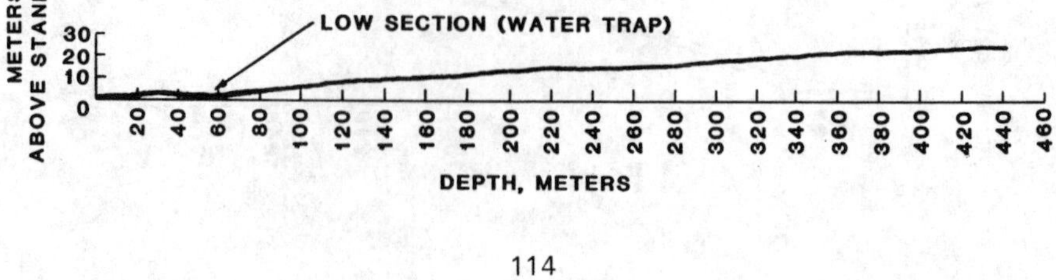

Figure 8
Equipment Transferred to Far Side of Workings in Red Panel (Area 2)

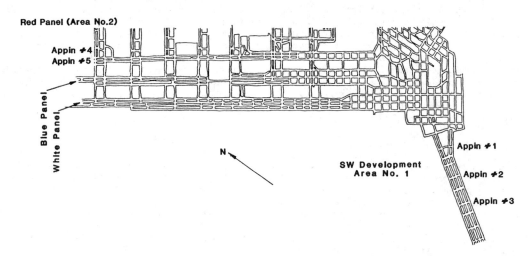

Red Panel (Area No.2)

Appin #4
Appin #5

Blue Panel
White Panel

N

SW Development
Area No. 1

Appin #1

Appin #2

Appin #3

The preliminary geological information available on the course of Appin No. 3 was in error. As a result, the borehole path was largely drilled near the floor rock interface. The last 20 m of the borehole were actually in the floor rock strata.

Water production from Appin No. 3 was initially very high at 570 liters/hour (150 US gal/hr). The initial gas production was limited to 3400 m^3/day (120,000 scfd). It was felt that the lower gas flowrates of Appin Nos. 2 and 3 were due to the water production influence. The unusual water production difference between the three holes could be correlated to geological phenomena and elevation within the seam.

Due to scheduling delays and short work stoppages, the drilling program was held up for approximately one month. Equipment was transferred to the far side of the active workings to Area Two in the Red Panel future developments. This area is shown in the map as Area No. 2 on Fig. 8. The area is at the retreat end of the longwall panel not yet outlined.

The choice of this area for drilling was for a two-fold purpose. First, it was thought that pre-drainage of this area would shield development advance of both Blue and Red Panels on either side of the planned borehole path. Second, the properties of the coal on this side of the active workings could be compared with those from the southwest development.

Another associated reason for choosing the site of Appin No. 4 was to evaluate the effects of drilling and production from a down dip direction with the true dip angle to the left.

Figure 9
Appin No. 4
Plan View of Borehole

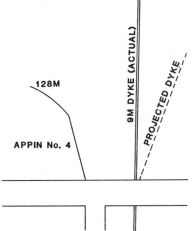

128M

APPIN No. 4

9M DYKE (ACTUAL)

PROJECTED DYKE

As shown in Fig. 9, the plan view of the borehole deviated dramatically, in the course of drilling, to the dip side. The borehole had to be prematurely terminated at 132 m (433 ft) because of the proximity to mining advance from a parallel blue panel development within 30 m distance to the left.

Initially the Appin No. 4 borehole showed some water problems. However, after a short dewatering period of 20 days, the gas flowrate increased to 1850 m^3/day (65,000 scfd).

Water flowrates were initially up to 2 liters/min but they declined to 48 liters/hr at 20 days.

Due to the premature termination of Appin No. 4, it was decided to drill Appin No. 5 nearby. The planned course of Appin No. 5 was to the right of the same intersection to avoid an igneous intrusion. However, the dyke was intercepted at 60 m (190 ft) and the borehole was terminated.

Since no further drilling was planned, all equipment was removed from the mine by the end of November, 1981.

PRODUCTION DATA

The long-term gas production history of the Appin No. 1 borehole is shown as Fig. 10. The decline curve is very predictable and can be easily fitted for computer simulation. Since no accurate water flow data were measured, the water production can only be approximated.

Even the initial water flowrates were estimated at less than 50 liters/day (13.20 US gal/day). This small water production diminished to virtually no water flowrate after a month of production.

The long term gas production from the Appin No. 2 borehole is shown in Fig. 11. As was mentioned in the text earlier, the gas production from this borehole was hindered by the surging water flow from the "trap" located near the mouth of the borehole. You can see that the flowrate decline is slower, but the high, initial flowrate of the first hole was not experienced. Accurate water production was not measured due to the surging nature of the flow.

A curious phenomena is illustrated in the gas production flowrate data for Appin No. 3 shown in Fig. 12. Initial gas flowrates were surpressed by high water flowrates. The water flowrate initially was as high as 600 1/day (160 US gal/day). Some water surging caused spurious readings as high as 1500 1/day (400 US gal/day). As the water decreased, the gas flowrate increased up to

7400 m^3/day (200,000 scfd) after 65 days of production. From this point, the gas flowrate declined similar to Appin No. 1 aside from the temporary decline due to separator plugging at 145 days.

Data on the Appin No. 4 borehole is limited due to its proximity to active mining. The active borehole terminated nearly 30 m from an active entry. As the entry advanced, the flowrate as shown in Fig. 13 decreased accordingly. A separator malfunction at 65 days created the decline in the measured flowrate.

EXPLORATION COREHOLES

As part of the joint venture program to evaluate the methane drainage potential of the Sydney Basin, a number of exploration coreholes were drilled. Two separate petroleum exploration licenses were obtained from the New South Wales Mineral Resources Department. These were designated PEL 255 and PEL 260. A map of the respective license areas and the coreholes described here is shown in Fig. 14.

The northern area which includes the city of Sydney itself is referred to as PEL 260 and covers an area of approximately 9900 sq km (3825 sq mi). The southern area outlined in the figure is referred to as PEL 255 and covers approximately 2500 sq km (965 sq mi) in the inland region of the southern coal fields of New South Wales, Australia. The area between PEL 255 and the coast of NSW is currently being intensively mined under coal licenses to various coal companies. Most of the current mining in the area is in either the Bulli coalbed which is the upper seam on the south coast or the lower seam which is the Wongawilli seam in this area. Major mining companies in the area are Australian Iron and Steel (AI&S), Kembla Coal and Coke (KCC), Clutha Development Corp. (CDC), West Ballambi and Austen and Butta.

The coreholes discussed in this paper are shown on the map of Fig. 14. The coreholes

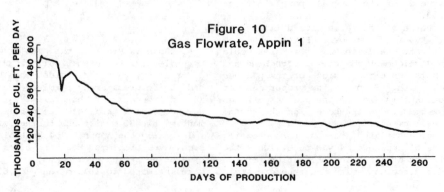

Figure 10
Gas Flowrate, Appin 1

THOUSANDS OF CU. FT. PER DAY

DAYS OF PRODUCTION

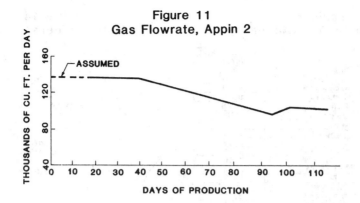

Figure 11
Gas Flowrate, Appin 2

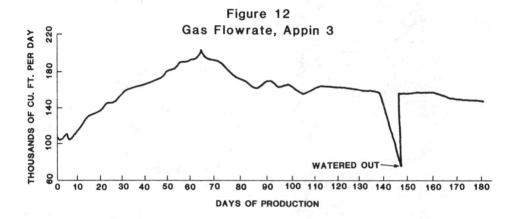

Figure 12
Gas Flowrate, Appin 3

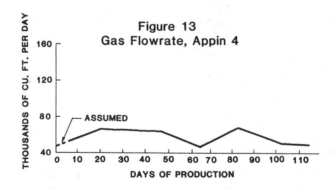

Figure 13
Gas Flowrate, Appin 4

in PEL 255 are referred to as the MS, Moon-shine Series, and the coreholes in PEL 260 are referred to as the BL, Bootleg Series. The holes are numbered in increasing numerical order according to the approximate order in which they were drilled or more accurately spudded. Some holes were completed out of sequence when using more than one rig or due to drilling and completion difficulties.

PEL 255

The northeast corner of PEL 255 was the first area of interest for exploration. This area was close to the horizontal drilling project located at the Appin Colliery about 10 km (7 mi) to the northeast. Since work in the upper seam of the Illawarra Coal Measure (i.e., Bulli Seam) near this location indicated high gas content, the core-

Table 3
Corehole Gas Content

COREHOLE	SEAM	GAS CONTENT (cc/g)
Moonshine #1	Bulli	12.12
	Balgownie	6.60
	Wongawilli	6.59
	Tongarra	9.39
Moonshine #2	Bulli	9.25
	Balgownie	N/S
	Wongawilli	7.79
	Tongarra	8.86
Moonshine #3	Bulli	8.32
	Balgownie	9.48
	Wongawilli	7.30
	Tongarra	9.25
Moonshine #5	Bulli	14.53
	Balgownie	10.07
	Wongawilli	8.70
	Tongarra	9.35
Bulli #1	Bulli	8.53
	Balgownie	6.70
	Cape Horn	N/S
	Wongawilli	1.67
Bootleg #1	Bulli	9.16
	Balgownie	12.59
	Woronora	8.52
	Wongawilli	6.12
	Upper American Creek	8.06
	Lower American Creek	10.94
	Tongarra	5.24
	Upper Woonona	9.70
	Lower Woonona	12.20
Bootleg #2A	1	16.53
	2	4.45
	3	14.07
	4	N/S
	5	15.27
	6	N/S
	7	15.51
	8	7.84
	9	N/S
	10	4.95
	11	N/S
	12	10.39
	13	6.52
	14	10.28
	15	9.51
	16	10.22
	17	N/S
	18	N/S
	19	N/S
	20	11.78
	21	10.41
	22	8.82
	23	10.32

N/S = SEAM NOT SAMPLED

Figure 14
Map of License Areas

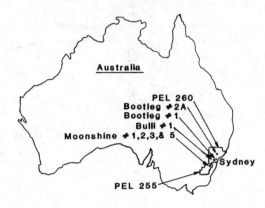

Australia

PEL 260
Bootleg #2A
Bootleg #1
Bulli #1
Moonshine #1,2,3,& 5

Sydney

PEL 255

Figure 15
Elevation Plot of Moonshine No. 1 Coalbeds

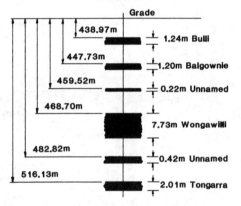

Grade

438.97m   1.24m Bulli
447.73m   1.20m Balgownie
459.52m   0.22m Unnamed
468.70m
  7.73m Wongawilli
482.82m   0.42m Unnamed
516.13m   2.01m Tongarra

hole project was initiated to define the extent of this seam and other lower seams as well as evaluate gas content and reservoir parameters.

Table No. 3 shows the data taken from the various exploration coreholes in PEL 255 and PEL 260. The coreholes in the southern part of PEL 260 have many of the same coal seams as found in PEL 255. In particular BU-1 and BL-1 indicate that Bulli, Balgownie and Wongawilli seams continue into the southern part of PEL 260.

Focusing on PEL 255, we notice in Fig. 15 that the coal seams observed in Moonshine No. 1 are the Bulli, Balgownie, Wongawilli and Tongarra. The thickness observed for the Bulli coalbed was less than expected at

1.24 m. Since this was one of the first coreholes in PEL 255, it was terminated below the Tongarra seam at 524 m.

The next two coreholes shown in both the table and in Figs. 16 and 17 are Moonshine No. 2 and Moonshine No. 3. Like the earlier Moonshine No. 1 corehole, these coreholes intercepted the same coal seams. Since they are in the same area of PEL 255 this was to be expected. However, the thickness of the coal seams varies according to the data shown in the figures. The Bulli coal seam as well as the Balgownie and Wongawilli seams vary considerably in this area. The most consistent coal seam in the area is the Tongarra seam at about 2 m in thickness.

As noted in the map of the area, Figure 14, notice that the Moonshine No. 5 corehole shown again in cross section of Fig. 18 is the most remote of the identified core-

118

Figure 16
Elevation Plot of Moonshine No. 2 Coalbeds

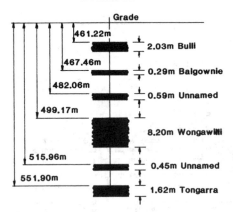

Grade

461.22m — 2.03m Bulli
467.46m — 0.29m Balgownie
482.06m — 0.59m Unnamed
499.17m
— 8.20m Wongawilli
515.96m — 0.45m Unnamed
551.90m
— 1.62m Tongarra

Figure 17
Elevation Plot of Moonshine No. 3 Coalbeds

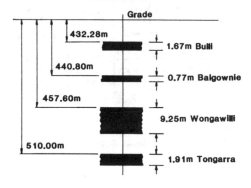

Grade

432.28m — 1.67m Bulli
440.80m — 0.77m Balgownie
457.60m
— 9.25m Wongawilli
510.00m — 1.91m Tongarra

Figure 18
Elevation Plot of Moonshine No. 5 Coalbeds

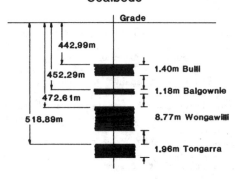

Grade

442.99m — 1.40m Bulli
452.29m — 1.18m Balgownie
472.61m
— 8.77m Wongawilli
518.89m
— 1.96m Tongarra

Figure 19
Elevation Plot of North Bulli 1 Coalbeds

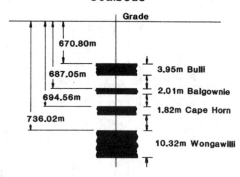

Grade

670.80m — 3.95m Bulli
687.05m — 2.01m Balgownie
694.56m — 1.82m Cape Horn
736.02m
— 10.32m Wongawilli

Figure 20
Elevation Plot of Bootleg No. 1 Coalbeds

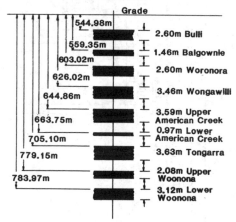

Grade

544.98m — 2.60m Bulli
559.35m — 1.46m Balgownie
603.02m — 2.60m Woronora
626.02m — 3.46m Wongawilli
644.86m
— 3.59m Upper American Creek
663.75m — 0.97m Lower American Creek
705.10m — 3.63m Tongarra
779.15m — 2.08m Upper Woonona
783.97m — 3.12m Lower Woonona

holes in PEL 255 as well as Moonshine No. 1. Since Moonshine No. 5 is further west we might expect different properties of the seam. You will note from the figure that the seams encountered and their thicknesses are higher than those in earlier coreholes to the east.

PEL 260

The geology of the Sydney Basin can best be described as an elliptical saucer with its narrow center near Sydney itself and out-lined by the Georges, Nepean and Hawksbury River systems.

This is best described by looking at the coreholes shown for PEL 260 in Fig. 14.

Figure 21
Elevation Plot of Bootleg 2A
Coalbeds

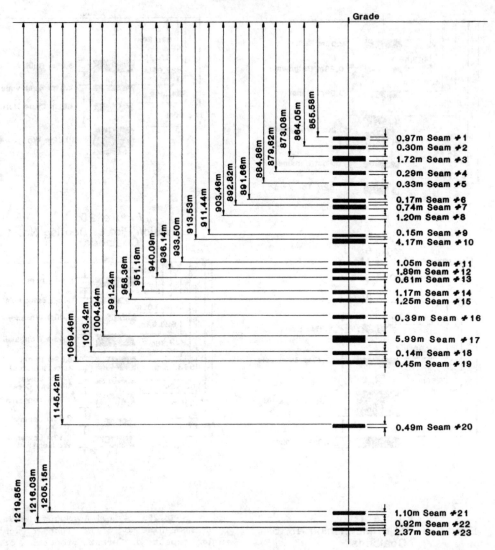

Bootleg No. 1 is shown on the southern coastal area of PEL 260 while Bulli No. 1 is shown on the southern interior part of PEL 260. The Bootleg 2A corehole is near the center of the basin just north of the city itself.

The first corehole completed in PEL 260 was the Bulli No. 1 shown in Fig. 19. Due to the inexperience in the area, this corehole was only completed through the Wongawilli seam. Information shown in Fig. 19 indicates that the Bulli and Balgownie seams

at this location are much thicker than those of the PEL 255 adjoining area to the south. Another curious phenomenon is the gas quality in this corehole. A significant level of ethane and propane was measured in the gas evolved from the core samples. Gas quantities however were slightly lower than those measured in the coreholes to the south in PEL 255. The Cape Horn seam was also identified at this location between the Balgownie and Wongawilli seams. Little data is available on this seam due to its limited areal extent. The same seam however has been identified to the south at the Appin Colliery. The Cape Horn has been located at Appin with a thickness of less than 1 m, but at other locations in PEL 255 it disappears entirely.

Bootleg No. 1 was drilled in the Darkes Forest area near the Royal National Park which is on the coast as shown in the map of Fig. 14. A number of new coal seams were exposed here both above and below the Tongarra seam which were not observed in the Moonshine Series coreholes of PEL 255. The corehole results are shown in Table 3 as well as Fig. 20. The additional seams included the Woronora, Upper and Lower American Creek and Upper and Lower Woonona. From the data in the figure and table it is obvious that the coal seams encountered in this corehole are thicker, more abundant, and higher in gas content than the same seams at the Bulli No. 1 corehole. Some of the coal seams (i.e., Balgownie) have significantly higher gas content than in the Moonshine Series coreholes of PEL 255.

Bootleg No. 2A was drilled just to the north of Sydney in the deepest part of the Sydney Basin. The location is shown on the map of Fig. 14. This corehole intercepted a total of 23 different coal seams as shown in Table 3 and Fig. 21. The top coal seam encountered was at 855 m depth. All but two of the seams were less than 4 m in thickness and only about six of the seams were of mineable thickness (i.e., greater than 1.2m). However, a number of the seams have high gas content and good gas quality. The lower four seams also have higher than normal CO_2 content in the gas. This corehole was completed to 1411 m total depth.

SUMMARY

Both the horizontal drilling program at Appin and the exploration corehole projects in PEL 255 and PEL 260 were considered successful. These programs have stimulated the participants to evaluate further exploration in the Sydney Basin.

Computer simulation was conducted to evaluate commercial production from methane drainage as a gas source for the Sydney area. Results of the simulation and concurrent economic analysis of subsequent costs were done to provide an economic evaluation of such a project.

Plans are currently being made to initiate a horizontal drilling project in the Sydney Basin to verify results of the analysis. This project will test the gas produceability of one of the lower seams in the area. Since no data exists on the gas produceability of the Wongawilli coal seam, it is important to verify the gas projections in this seam during the course of the new project.

Experience with cross-measure boreholes for gob gas control on retreating longwalls

J.CERVIK, F.GARCIA & T.W.GOODMAN
US Department of the Interior, Bureau of Mines, Pittsburgh, PA, USA

ABSTRACT: The surface gob hole is a common auxiliary method of controlling gob gas during longwall mining in the United States. Because surface gob holes cannot always be drilled, the Bureau of Mines is conducting studies to modify and adapt European cross-measure borehole technology to U.S. retreating longwalls. Tests conducted where overburden is 650 ft (198 m) or less indicate up to 71% of the methane produced by the mining operation can be captured by a properly designed cross-measure system. Hole spacing should be limited to 200 ft (61 m) except on about the first 600 ft (183 m) of the longwall, where spacing should be reduced to 100 ft (30 m). More holes are necessary near the start of the longwall to capture the large quantities of methane released when the first large roof fall occurs and to prevent it from overloading the return air system.

1 INTRODUCTION

Control of methane in gobs plays an important role in a successful and productive longwall mining operation. The primary method of controlling gob gas in the United States is by air dilution, which is extremely expensive. In the Warrior Basin of Alabama, power costs for the operation of high-pressure fans are expected to be about $4800 a day (Stevenson 1981). In many cases, sufficient air is not available to dilute large volumes of methane associated with the break-up of roof strata during longwall mining. In the 1970's the surface gob hole was introduced as an auxiliary method of methane control; today it is commonly used by the mining industry.

Typically, 2 to 3 gob holes are drilled in a 550 by 5000 ft (167 by 1524 m) longwall panel; in deeper coalbeds (2000 ft (610 m)), such as the Warrior Basin in Alabama, four gob holes are drilled per panel. The mining industry is experimenting with 800 ft (244 m) panel widths (Bourquin and Jaspal 1984), and the trend is to increase the panel width to 1000 ft (305 m) and length to 10 000 ft (3048 m). These large panels may require quadrupling of the number of gob holes per panel for methane control over the number used with the smaller panels. Unfortunately, gob holes cannot always be drilled because mining may be under populated areas, surface topography may be too severe, or access to private property may be denied. An alternative method of gob gas control is needed that is independent of the mine's ventilation system and surface right-of-way problems, and more effective in capturing larger quantities of gob gas than surface gob holes.

In Europe, small-diameter holes are drilled from underground locations into the roof strata above a longwall. Suction is applied to these holes to draw the methane from the strata being fractured by the longwall operation. These holes are connected to an underground pipeline that transports the methane to the surface. For the past several years, the Bureau in cooperation with the mining industry has been experimenting with this technique. This paper presents the results of two tests and compares the effectiveness of cross-measure boreholes and surface gob holes in controlling gob gas on retreating longwall operations.

2 STUDY AREA

The studies were conducted in the Lower Kittanning Coalbed, which is about

48 in (1.2 m) in thickness. The strata-
graphic column shows there are three
coal groups located about 15, 85, and 125
ft (5, 26, and 38 m) above the roof
of the Lower Kittanning Coalbed
(fig. 1). A total of 9 ft (2.7 m) of

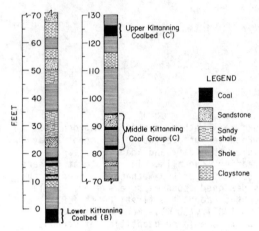

Figure 1. Stratigraphic column above
Lower Kittanning Coalbed.

coal lies within the first 130 ft (40 m)
above the roof. These coalbeds are known
sources of methane. Overburden is about
645 ft (197 m).

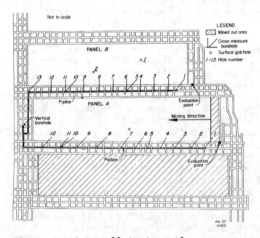

Figure 2. Longwall test panels.

Figure 2 shows the two test longwalls
(Panels A and B), which were mined in
succession. Cross-measure boreholes were
drilled over the test panels from
the center entry in the return side of
the longwall. They were drilled over
pillars which protected the collars of

the boreholes when the longwall face
passed. The center entry was supported
with cribs to protect the underground
pipeline and to prevent caving so that
access to the pipeline and cross-measure
boreholes could be maintained during the
life of the panels.

The design parameters for the cross-
measure boreholes on Panel A are similar
to the design used in earlier Bureau
studies by Schatzel et al. (1982) and
Campoli et al. (1983). These boreholes
were drilled 45° with respect to the
axis of the longwall, which is a common
practice in Europe. On Panel B, the
cross-measure boreholes were drilled
perpendicular to the axis of the longwall. Table 1 summarizes the design
parameters for the two panels. All

Borehole parameters	Panel A	Panel B
Vertical angle, °(rad)	28 (0.49)	35 (0.61)
Length, ft (m)	280 (85)	225 (69)
Terminal height, ft (m)	130 (40)	130 (40)
Horizontal angle, °(rad)	45 (0.79)	90 (1.6)
Spacing	200 ft (61m) on first half and 300 ft (91m) on second half of panel	

Table 1. Design parameters for cross-
measure boreholes.

boreholes terminated about 130 ft (40 m)
above the mined coalbed, and even though
the boreholes on Panel B are 55 ft (17 m)
shorter than those of Panel A, the
penetration depth of the boreholes into
the gob is about the same for the two
panels. The penetration depth is the
horizontal projection of the hole parallel
to the line of the face. All holes were
surveyed after completion of drilling to
determine the actual path of the bit
through the roof strata. In general, the
surveys show that the holes turned upward
more steeply and to the right because of
clockwise rotation of the bit.

The first 25 ft (7.6 m) of each borehole
was drilled with a 4 or 6 in (102 or 152
mm) diameter diamond bit core barrel and
the remainder with a 2 in (51 mm) bit. A
20 ft (6.1 m) plastic standpipe was then
grouted into the collar of each borehole.
Each borehole connection to the main
pipeline contained a valve to control the
negative pressure on the borehole and to
shut in the borehole when the methane
concentration in the gas flow (methane
plus air) fell below 25%. A venturi was
used to monitor gas flow from each
borehole.

A 6 in (15 cm) polyethylene pipeline was used to transport the gas to the bottom of a cased 8 in (20 cm) surface borehole. Two exhausters on the surface borehole were employed to produce a negative pressure in the underground pipeline and at each borehole. The capacity of the smaller exhauster ranged to 450 cfm (0.21 m³/s) with partial vacuums up to 8.7 in Hg (29 kPa). The larger exhauster requires a minimum flow of 350 cfm (0.17 m³/s). Maximum capacity is 1300 cfm (0.61 m³/s) and partial vacuums up to 12.6 in Hg (42 kPa). The smaller exhauster was used initially until the gas flow from the undermined cross-measure boreholes reached about 400 cfm (0.19 m³/s); thereafter, the larger capacity exhauster was used.

3 SURFACE GOB HOLES

Surface gob holes have been used at this mine since 1968 to assist in controlling methane in gobs. For Panel A, only one surface gob hole could be drilled about 1300 ft (396 m) from the start of the panel because of a populated area on the surface (fig. 2). Even this borehole had to be displaced about 75 ft (23 m) from the centerline towards the return side of the longwall to avoid surface structures. Panel length is 2700 ft (823 m). For Panel B, two surface gob holes were drilled 650 and 1350 ft (198 and 411 m) from the start of the panel. Both surface gob holes were displaced from the centerline of the panel because of surface right-of-way problems. Panel B is 2340 ft (713 m) long.

4 DATA ANALYSIS

4.1 Panel A

Panel A was worked 3 shifts/day but in a predominantly 3 day/week schedule because of the low demand for coal (fig. 3). The panel was completely mined in 69 working days, which were spread over a 240 day interval and included three idle periods of 73, 32, and 10 days.

Most boreholes did not produce gas until a partial vacuum was applied and the longwall face passed 75 to 100 ft (23 to 30 m) beyond the end of a borehole (but before the face reached its collar). Holes 5 and 10 were exceptions. A free flow of methane (60 cfm

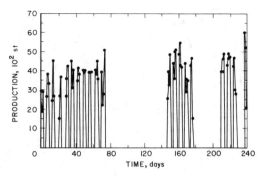

Figure 3. Daily longwall coal production (Panel A).

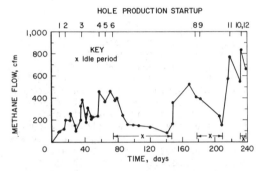

Figure 4. Methane flow from cross-measure system (Panel A).

(0.028 m³/s)) was measured from hole 5 before the face reached its collar. After the face passed beyond the collar, the free flow stopped and a partial vacuum was required to maintain flow. Gas production occurred from hole 10 only after the face passed well beyond its collar.

The methane flow from the cross-measure system is shown in figure 4. Also shown are the times when each borehole started methane production which is greatest when it first goes on production. That methane flow rates are dependent on whether coal is actively being mined is clearly demonstrated by the much lower flow rates during prolonged idle periods. During the first idle period (73 days), methane flow declined from 400 to 100 cfm (0.19 to 0.05 m³/s); when mining resumed, the flow increased to about 400 cfm (0.19 m³/s). These data indicate that about 75% of the methane in the gob comes from newly fractured roof strata near the face.

Figure 5 shows the variations in methane concentration in the gas flow from hole 2,

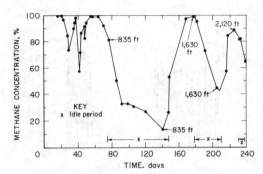

Figure 5. Methane concentrations in gas flow from hole 2 (Panel A).

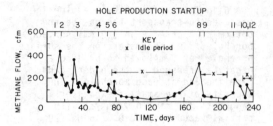

Figure 6. Methane flow in return air (Panel A).

which is typical of the methane concentration variations in gas flow from other cross-measure boreholes. The methane concentration averages about 90%, but drops to about 30% in 20 days during the first idle period. The distance between hole 2 and the longwall face is 835 ft (255 m) at this time. When mining resumed on day 140, the methane concentration in the gas flow began to increase and reached 100% on day 178 when the longwall face was 1630 ft (497 m) from hole 2. During the second idle period (day 180 to 210), the methane concentration again declined, but increased after mining resumed and reached 90%. At this time the distance between the longwall face and hole 2 was 2120 ft (646 m). These data clearly indicate the gob is quite permeable and that methane migrates easily through the gob for distances of at least 2120 ft (646 m).

The effectiveness of the cross-measure system in controlling methane flow in the return air from the longwall is shown in figure 6. The point in time when each borehole started to produce gas is shown along the top of the graph. Methane flows in the return air gradually decrease from 250 to 100 cfm (0.12 to 0.05 m³/s) as the number of boreholes producing gob gas increases. The effect of each borehole on methane flow in the return air is clearly demonstrated by the large decrease in flow in the return air immediately after each borehole starts gob gas production. These peak flows vary from about 200 to 450 cfm (0.09 to 0.21 m³/s) and suggest that the 200 ft (61 m) spacing should be reduced.

The first large roof fall after mining of the panel was started was accompanied by large flows of methane, and mining was suspended for short periods because sufficient air was not available to maintain methane concentrations in the returns at permissible levels. Generally only one borehole is on production at this time, and the quantity of gob gas that can be drawn through a 2 in (51 mm) diameter borehole is small in comparison to the tremendous quantities generated in the gob. Consequently, gob gas begins to spill into and overload the return air system. Additional boreholes drilled between holes 1 and 2 and between holes 2 and 3 would prevent large peak flows and would lower the general level of methane flow in the return. Therefore, boreholes should be spaced 100 ft (30 m) apart along the first 600 ft (183 m) of the longwall and thereafter 200 ft (61 m) apart.

The surface gob hole produced gob gas for about 6 days and was finally shut in because of low methane concentrations in the gas flow. The hole vented about 1 000 000 cu ft (28 300 cu m) of methane during its short productive life. Gas production started on day 156, and no indications exist that the gob hole affected methane flows in the return air, in spite of the fact that hole 7, which was undermined around day 160 and produced no gas, is in the same area as the surface gob hole. The large peak methane flow around day 180 (fig. 6) was caused by a

System	Methane, 10^6 cu ft	Capture ratio, %
Cross-measure boreholes	118	70.7
Return air	48	28.7
Surface gob hole	1	0.6
Total	167	100.0

Table 2. Panel A capture ratios.

partially water blocked pipeline; this may have masked the effect of the surface gob hole.

Table 2 shows that the cross-measure borehole system captured 70.7% of the methane produced by the mining operation. About 28.7% of the methane entered the mine ventilation system and only 0.6% was captured by the vertical gob hole.

4.2 Panel B

Panel B was mined 3 shifts/day in a 4 or 5 day/week schedule. The panel was mined in 61 working days, which were spread over a 121 day interval and included two idle periods of 11 and 26 days (fig. 7).

The concentration of methane in the gas flow (methane plus air) from the cross-measure system dropped from 100% to below 60% when Surface Gob Hole 1 (SGH-1) was intercepted around day 37 (fig. 8). The concentration remained around 60% for the duration of the study except during the second idle period (26 days) during which the concentration dropped to about 30% and then returned to about 60% when mining resumed. SGH-2 was intercepted around day 96; its effect on the cross-measure system was negligible. Figure 9 shows that the corresponding methane flow through the cross-measure system increases as the number of boreholes producing gob gas increases. Interception of SGH-1 arrests any further increase for a while. During the 26 day idle period the methane flow decreases by about 50% and then increases again when mining resumes.

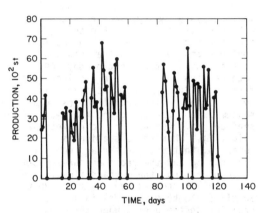

Figure 7. Daily longwall coal production (Panel B).

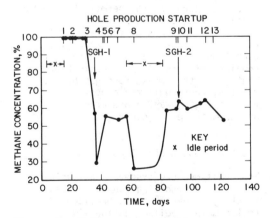

Figure 8. Methane concentrations in gas flows from cross-measure system (Panel B).

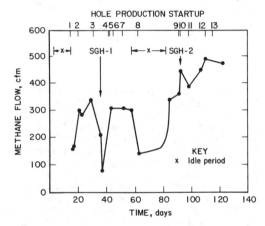

Figure 9. Methane flow from cross-measure system (Panel B).

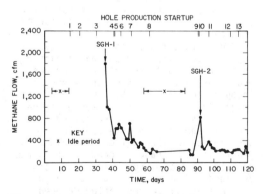

Figure 10. Methane flow through surface gob holes (Panel B).

127

System	Methane, 10^6 cu ft	Capture ratio, %
Cross-measure boreholes	8	28
Return air	2	7
Surface gob hole	19	65
Total	29	100

Table 3. Panel B capture ratios (day 37 to 59).

System	Methane, 10^6 cu ft	Capture ratio, %
Cross-measure boreholes	28	62
Return air	3	7
Surface gob hole	14	31
Total	45	100

Table 4. Panel B capture ratios (day 60 to 121).

Figure 10 shows the combined methane flow from SGH-1 and SGH-2. Methane flow from SGH-1 declined from 1800 to 200 cfm (0.85 to 0.09 m³/s) in about 23 days (day 37 to 59); during this time interval, it captured 65% of the methane generated by the mining operation (table 3). The cross-measure system captured 28%, and the other 7% entered the mine ventilation system. However, for the remainder of the panel (day 60 to 121), when the combined flow from the SGH-1 and SGH-2 averaged 200 cfm (0.09 m³/s), the cross-measure boreholes were the dominant system controlling methane. Table 4 shows that the crossmeasure system captured 62% of the methane compared with 31% for the two surface gob holes during this period.

The capture ratios for the two systems may have been much higher than indicated in tables 3 and 4 had only the surface gob holes or the crossmeasure system been in operation on Panel B. The surface gob holes did affect the performance of the cross-measure system and undoubtedly the cross-measure system affected the performance of the surface gob holes. On Panel A, the methane concentration in the gas flow from the cross-measure system averaged 80% or greater during mining. On Panel B, the concentration dropped from 100% to 60% after SGH-1 was intercepted (fig. 8). Tests conducted on

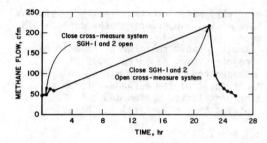

Figure 11. Methane flow in return air (Panel B).

Panel B during the second idle period (day 62) indicate that no change occurred in the methane flow in the return air over a 20-hr period when the cross-measure system was closed and the surface gob hole was operating. The test showed that the surface gob hole alone could control methane flow in the return air. However, on a latter test on day 109, methane flow in the return air increases from 50 to 220 cfm (0.02 to 0.10 m³/s) over a 22-hr period when the cross-measure system is not operating and the two surface gob holes are producing methane (fig. 11). When the situation is reversed, that is, the surface gob holes are shut in and the cross-measure system is in operation, the methane flow in the return air decreases rapidly from 220 to 50 cfm (0.10 to 0.02 m³/s) in about 3.5 hr. During the period when the cross-measure system is not in operation, over 50% of the methane produced by the mining operation enters the mine ventilation system. The test was terminated to avoid exceeding the permissible methane level in the return air.

Tables 3 and 4 show that about 93% of the methane generated by the longwall operation on Panel B is captured by the combined effects of the cross-measure system and the two surface gob holes. On Panel A, where only the cross-measure system was in operation, about 71% of the methane was captured.

5 SUMMARY AND CONCLUSIONS

The cross-measure system is a viable method of controlling gob gas on re-treating longwall faces, provided access to the main pipeline and individual boreholes can be maintained during the life of the panel. About 71% of the methane produced by the longwall operation can be captured by the crossmeasure system.

Methane flows in the return air indicate that borehole spacing should be limited to 200 ft (61 m) except on about the first 600 ft (183 m) of the longwall where spacing should be reduced to 100 ft (30 m). More cross-measure boreholes are necessary near the start of the longwall to capture the large quantities of methane that are released when the first large roof fall occurs.

About 75% of the methane in the gob emanates from fractured roof strata near the face. Prolonged idle periods show that low methane concentrations in the gob located over 2000 ft (610 m) from the face increase to above 90% after mining resumes. On Panel A where only the cross-measure system was operating, about 71% of the methane produced by the mining operation was captured. On Panel B where both the cross-measure system and surface gob holes were in operation, 93% of the methane was captured.

6 ACKNOWLEDGEMENT

The cooperation of the management of BethEnergy Mines Inc., Ebensburg, PA, is greatly appreciated.

7 REFERENCES

Bourquin, B. J.& J. S. Jaspal 1984. Mid-Continent has early success with the longest longwall face ever operated in the U.S. Min. Eng. 36:48-52

Campoli, A. A., J. Cervik & S. J. Schatzel 1983. Control of longwall gob gas with cross-measure boreholes (Upper Kittanning Coalbed). BuMines RI 8841.

Schatzel, S. J., G. L. Finfinger, and J. Cervik 1982. Underground gob gas drainage during longwall mining. BuMines RI 8644.

Stevenson, J. W. 1981. Methane control in the Warrior Coal Fields of Alabama. Min. Cong. J. 67:44-46.

Direct coal mine ventilation with dilute methane-fueled gas turbines

CARL R.PETERSON
Massachusetts Institute of Technology, Cambridge, USA

BYRON MILLER
Bell Laboratories, Holmdel, NJ, USA

ABSTRACT: Dilute methane in the ventilation discharge from gassy mines represents a valuable resource that is currently wasted, while ventilation costs are high. A system is described in which the dilute methane-air mixture is used directly as the working fluid and fuel in a modified industrial gas turbine, providing both ventilation and power with little auxiliary fuel. Gassy mines can become net power producers rather than large power consumers.

1 INTRODUCTION

Methane is a serious hazard in many coal mines. It is "controlled" by dilution to safe concentrations with sufficient air flow. The discharged methane represents considerable loss of valuable resource, while the power consumed in moving the ventilation air constitutes a substantial cost. The legal concentration limit is 1% (by volume) in work places and return mains and 2% in bleeders. However, the discharge air from a gassy mine usually contains no more than 1/2% methane because of leaking or "short-circuitry" of air from intake mains to return mains before it reaches the work place. That is, roughly half the air flow is simply wasted in the typical arrangement which circulates air from the intake to the work place deep within the mine and then back out again in a parallel path to a discharge in the vicinity of the intake.

The 2% methane permitted in bleeders happens to be very nearly the desired fuel-air ratio for a typical industrial gas turbine. If it could be vented directly to the surface (and indications are that it could – certainly it would be safe) and passed through such a turbine, both ventilation and power output could be provided. For example a 100,000 cfm gas turbine, a medium to large jet engine and a reasonable auxiliary ventilation "fan", would produce an output power of about 10,000 kW, more than the peak demand of a large coal mine.

All of this is possible only if one can devise a means to burn the lean methane-air mixture. Obviously, a 2% mixture will not burn, else it would not be permitted within a mine. Two masters theses, by Sarfatti (1) and Miller (2), have explored this potential. Miller describes a simple combustor concept which appears capable of burning such mixtures without the use of a catalyst. Sarfatti showed the system to be economically attractive, including the expense of additional ventilation shafts. The following summarizes these works.

2 METHANE SUPPLY

Concern with the wasted methane resource is not new. Studies (3,4) have considered a variety of proposals for utilization, but none has found an economic application for the dilute methane carried in the discharge air. Various studies (5,6) have indicated relatively long-lived methane flows in quantities of interest for significant power generation. Indeed, one might even wish to continue ventilation of a shut-down mine simply to exploit the lingering discharge from caved materials. Of course, not all coal mines discharge sufficient methane to support such a scheme. Some gassy mines emit more than 2000 cubic feet of methane per ton of coal mined while others emit negligible quantities. About 3% of U.S. coal mines producing 25% of the coal, primarily in the Appalachian region, yield almost 80% of the methane (7). The methane problem will increase in the future as deeper seams are **mined**.

Miller (2) derives a semi-empirical model to predict the methane flow from a mine in terms of mine geometry, history, present mining activity and formation properties. Methane flows from the active mining face(s), from coal on conveyors, from exposed ribs, from horizontal boreholes (i.e., methane drainage holes), and from adjacent strata are included and integrated in time to properly account for time-varying emission rates and mining history. An example case is presented. Such treatment will be necessary in assessing the applicability of the concept in any situation and in determining the best turbine location and associated ventilation system.

Some departure from conventional ventilation patterns is advantageous in fully exploiting the gas turbine's potential, not only in sustaining power generation but also in improving mine safety. In particular, a "one-way" pattern which discharges high concentration (2%) methane directly to the surface from the back of the mine, rather than carrying it(at lower concentration) all the way back past other work places to the vicinity of the intake, is indicated. One-way ventilation is in fact favored for safety but not often economically attractive. Sarfatti (1) showed that, with sale of excess power generated, the gas turbine scheme was economically attractive even when costs of extra discharge shafts are included (and such extra shafts are in themselves an added safety feature).

Gas turbines would not be used to handle the entire mine ventilation flow - there is not sufficient methane for such an arrangement. A total gas turbine flow of about 100,000 cfm, located to discharge safely and directly from problem methane areas, would, as noted, generate more than enough power to drive conventional fans for the bulk of the ventilation flow and all other power consuming devices.

It is not possible, nor desirable, that the turbine receive all of its fuel directly in the mine discharge air since that would not permit adequate control. A small, separate and controllable auxiliary fuel flow is necessary to act as the turbine "throttle". This might be high concentration (90%+) methane drained directly from the coal seam, low concentration (50%) methane in gob gas, or even liquid hydrocarbons.

3 GAS TURBINES

Gas turbines are most familiar as turbojet engines, but with a power turbine added to

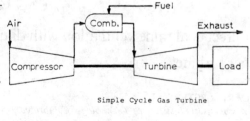

Simple Cycle Gas Turbine

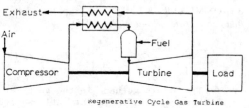

Regenerative Cycle Gas Turbine

Fig. 1. Gas turbine schematics

convert the power expended in the high velocity exhaust jet into shaft power, they are in common use as industrial power generators. Figure 1 illustrates schematically the two common types of gas turbine. In the simple cycle machine, air is first compressed, then heated by combustion of fuel added in the combustion chamber, and then expanded through a turbine where power is extracted to drive the compressor and the attached load. Typically, in excess of two-thirds of the turbine power goes to drive the compressor.

The regenerative cycle turbine is similar except that the still hot turbine exhaust flow is passed through a heat exchanger where it heats the compressor discharge flow prior to combustion, thereby requiring less fuel in the combustion chamber to reach the desired peak temperature (as limited by turbine material behavior).

Tables I and II present typical performance parameters for commercially available gas turbines (2) Simple cycles attain best performance at relatively high pressure ratios (16:1 or more), while regenerative machines utilize lower pressure ratios. Note that the "heat rate" of the latter, expressed as Btu of heat (fuel) input per horsepower hour of output, is significantly better. Offsetting this lower fuel consumption, the regenerative machines cost 30-40% more per horsepower. For sustained operation the fuel saving is often well worth the added initial cost. Note also that the full power output of the latter requires very little more than a 2% methane fuel content. In quite

similar service, remotely controlled
gas turbines are used to drive compressors
distributed along natural gas pipelines,
burning methane from the pipeline as fuel.
Turbine availabilities are consistently
99.5% or better (8).

For mine ventilation applications the
compressor inlet would be connected
directly to the mine discharge shaft
at the surface, perhaps with an intervening
filter if the discharge air carries
excessive dust. Auxilliary fuel of
whatever concentration, such as gob
gas, could be simply injected at the
compressor inlet (a low pressure region)
because the lean burning combustion
chamber will be capable of burning a premixed fuel.

4 HIGH TEMPERATURE COMBUSTOR

Complete combustion of a 2% methane-air
mixture gives a temperature rise of
about 920° F, adequate for our needs, but
this lean mixture simply will not burn
under ordinary circumstances. Conventional
gas turbine combustors use concentrated
fuel (pure methane for example), mixing
it wtih only a portion of the compressed
air to achieve a local stoichiometric
mixture (9.5% in the case of methane) which
burns readily, reaching a very high
temperature. This hot material is then
mixed with the remaining air to reach a
lower average temperature before entering
the turbine.

For stationary methane-air mixtures
with prolonged residence times, the
combustible-lean limit is a strong function
of temperature, as shown in Figure 2
(from reference 9). At 2400°F the lower
limit is essentially zero, while a 2%
concentration is flammable around
1470°F.*

Laboratory experiments have also been
conducted to determine combustion reaction
rates in turbulent flow conditions, with
methane-air mixtures from 0.5 to 5.0
percent (10). From these data the
reaction times for a 2% methane-air mixture
can be estimated as a function of initial
temperature and pressure as shown in Fig.
3 (2). A reaction time of 40 milliseconds

* By comparing Figure 2 with the compressor
  discharge temperatues in Tables I and
  II it can be seen that there is no
  danger of combustion within the compressor
  even in the event of a violent blade rub or
  other ignition source.

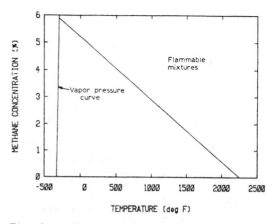

Fig. 2. Effect of Temperature on Lower
Limit of Flammability of Methane in Air
at Atmospheric Pressure (9).

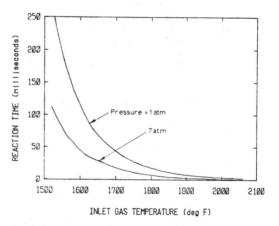

Fig. 3. Effect of Inlet Temperature on the
Combustion Reaction Time of a 2% Methane-
Air Mixture (2)

(1600°F, 7 atm) at a reference velocity
of 50 feet per second requires a combustor
reaction zone length of 2 feet.

Catalytic processes provide a means to
achieve reaction at less than normal
combustion temperatures (though not
so low as compressor discharge temperatures
in this case). However, this paper
presents a simple non-catalytic concept
that should be lower in cost and more
durable. Furthermore, since methane
is the most difficult to ignite
hydrocarbon, a combustor developed for
lean methane-air mixtures would find
applicaitons with a wide variety of

other low Btu mixtures, mixtures which
might poison catalytic combustors. Thus
a non-catalytic combustor will be of
wider interest.

The high-temperature combustor is
illustrated schematically in Figure 4.
As may be seen from Figures 2 and
3, it is only necessary to raise the initial
temperature of the lean methane-air
mixture to a level at which reaction
will occur at an acceptable rate, say
1600 ^{o}F for example. The illustrated
combustor achieves this in what might be
called a "thermal siphon": the fuel
is converted to its reacted state by
passing over an intermediate temperature
elevation using the energy of combustion
and an appropriate heat exchanger to
create the local high temperature. The
incoming (i.e., compressor discharge)
mixture at T_1 is heated in a counterflow heat exchanger to T_2, a temperature
selected to assure prompt reaction.
Reaction, perhaps triggered by a
continuous spark ignitor or small pilot
flame, then raises the temperature to
T_3, whereupon the combustion products
are passed through the other side of the
heat exchanger   where they are cooled
to T_4 while giving up heat to the
inflowing mixture. This "bootstrap"
action appears to violate some law of
thermodynamics, but it doesn't - the
general concept is used to incinerate low
Btu waste materials which also require
a high reaction temperature (2).

Actually the necessary heat exchanger can
not be constructed of stationary tubes
as in conventional low-temperature practice.
Metals cannot withstand the necessary
temperatures and ceramics have proven
inadequate in such configurations. Instead,
a rotating honeycomb structure "regenerator"
would be used, as illustrated in Figure 5
(2). As the honeycomb disc rotates it is
alternately heated by throughflow of hot
gasses and then cooled by opposite
throughflow of cold gasses. The
heat capacity of the solid ceramic material
is used to carry the heat from the hot
stream to the cold stream, but the net
effect is that of the stationary
counterflow heat exchanger illustrated
in Fig. 4. Figure 5 illustrates a high
temperature combustor on a simple cycle
gas turbine , with numbers corresponding
to an 11:1 compressor pressure ratio,
1600°F combustor inlet temperature for
reasonable reaction rate, 2.5% total methane
content (and complete combustion), and
an 1800°F turbine inlet temperature
(lower methane concentration would simply
provide a lower turbine inlet temperature

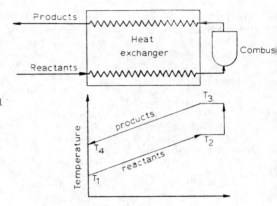

Fig. 4. High Temperature Combustion
Scheme (2)

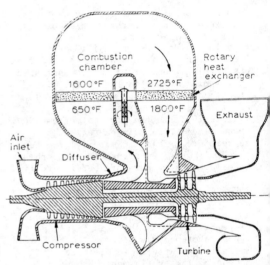

Fig. 5.   Cross Section of Simple Cycle
Gas Turbine with High Temperature Combustor(

and less power). The ceramic honeycomb
wheel would be 2.5 inches thick and
rotating at 23 rpm. For a 31500
cfm turbine the wheel would be 4.4 feet
in diameter. The power output would be
about 6000 horsepower.

The rotating heat exchange element is
similar to that in a conventional regenerato
as used for example on experimental automoti
turbines, but there is an important
difference. Note in Figure 5 that there mus
be a sliding seal against the bottom surface
of the rotating ceramic disc to prevent
direct leakage of compressor outlet

air to the turbine inlet passage since this leakage flow would bypass the heat exchanger and reactor. Actually, the seal need not be extremely tight since a modest leakage causes no harm in that leakage flow will still react (so long as it is not so large as to depress the reacted material temperature below reaction temperature) and it still passes through the turbine. Furthermore the pressure difference driving the leakage is only that pressure drop through the heat exchanger and combustor, perhaps a few psi at most. This is in sharp contrast to the seal problem in a regenerator where the cold flow is compressor discharge flow at high pressure and the hot flow is turbine discharge flow at essentially atmospheric pressure. For an 11:1 pressure ratio the driving pressure is thus about 160 psi. Furthermore, leakage in that case by passes the turbine and is lost entirely. Thus the suggested high temperature combustor is a much simpler design. New ceramic materials are available for the honeycomb element that are easily capable of withstanding the necessary temperature.

CONCLUSION

The suggested high temperature combustor seems, on the basis of available data, to provide a means to burn very dilute methane-air mixtures. Several coal mining companies have indicated a willingness to employ the concept for mine ventilation as soon as it can be developed. The predicted combustor performance should be verified by relatively simple experiments and data sufficient to design a full-scale combustor should then be gathered. A prototype should then be designed, fabricated, and field tested.

Table I.

SIMPLE CYCLE GAS TURBINES

	GE LM 500	I-R GT-22	Solar Centaur	Solar Mars
Horsepower	5000	4250	3830	10600
Pressure Ratio	14 to 1	8.5 to 1	9 to 1	16 to 1
Compressor Exit Temp.	725 F	540 F	590 F	760 F
Turbine Inlet Temp.	2120 F	1800 F	1450 F	1800 F
Turbine Exhaust Temp.	940 F	990 F	850 F	800 F
BTU/hp-hr	8305	9430	10000	7770
Air (lbm/sec) flowrate (cfm)	34.0 26770	33.8 26600	38.0 29920	80.0 62990
% CH4 in air * at rated power	2.87	2.79	2.37	2.42
CH4 (lbm/sec) flowrate (cfm)	0.54 768	0.52 742	0.50 709	1.07 1524

ISO conditions, Sea level, 59°F, no inlet or exhaust losses
*Assuming complete combustion, methane heating value=900 BTU/ft^3

Table II.

REGENERATIVE CYCLE GAS TURBINES

	GE M1502R	GE M3132R	Solar Centaur	Westinghouse CW182RMAA
Horsepower	5050	13750	3730	12800
Pressure Ratio	7.2 to 1	7.2 to 1	9 to 1	7.2 to 1
Compressor Exit Temp.	500 F	500 F	590 F	500 F
Combustor * Inlet Temp.	900 F	910 F	810 F	950 F
Turbine Inlet Temp.	1700 F	1700 F	1450 F	1850 F
Turbine Exhaust Temp.	970 F	980 F	850 F	1040 F
BTU/hp-hr	7970	7410	7780	7100
Air (lbm/sec) flowrate (cfm)	45.0 35430	115.0 90550	38.0 29920	102.0 80315
% CH4 in air ** at rated power	2.10	2.08	1.80	2.10
CH4 (lbm/sec) flowrate (cfm)	0.52 744	1.33 1883	0.38 539	1.19 1687

ISO conditions, Sea level, 59°F, no inlet or exhaust losses
*Assuming heat exchanger effectiveness = 85%
**Assuming complete combustion, methane heating value=900 BTU/ft^3

REFERENCES

Sarfatti, Michael, "Direct Coal Mine Ventilation Using Gas Turbines", Master of Science Thesis, Mechanical Engineering Department, MIT, November 1977.

Miller, Byron, "Direct Coal Mine Ventilation with Coalbed Methane Fueled Gas Turbine Powerplants", Master of Science Thesis, Mechanical Engineering Department, MIT, June, 1982.

"Economic Feasibility of Recovering and Utilizing Methane Emitted from Coal", Arthur D. Little, Inc., USBM PB-249-728, 1975.

"Methane Produced from Coalbeds", Vol. 1 & 2, TRW Energy Systems Group- Energy Systems Planning Division, ERDA Contract No. 3 (46-1) -8042, January, 1977.

Moore, T.D., Duel, M. and Kissel, F.N., "Longwall Gob Degasification with Surface Ventilation Boreholes Above the Lower Kittanning Coalbed", USBM RI 8195, 1976.

Krickovic, S. and Findlay, C., "Methane
    Emission Rate Studies in a Central
    Pennylvania Mine", USBM RI 7591, 1971.
Irani, M., Thimons, E., Bobick, T.,
    Deul, M., and Zabetakis, M., "Methane
    Emission from U.S. Coal Mines, A
    Survey", USBM IC 8558, 1972.
Sawyer, J.W. - editor, Sawyers's Gas
    Turbine Engineering Handbook -
    2nd ed. (Gas Turbine Publications,
    Inc.), Vol. 1-3, 1972.
Zabetakis, M., "Flammability Characteristics
    of Combustible Gases and Vapors",
    USBM Bulletin 627, 1965.
Dryer, F., and Glassman, I., "The High
    Temperature Oxidation of CO and CH_4"
    Fourteenth Symposium (International) on
    Combustion. The Combustion Institute
    1973, p. 987.

4. Dust

Generation and control of mine airborne dust

G.KNIGHT
Elliot Lake Laboratory, CANMET, Ontario, Canada

ABSTRACT: A study was undertaken to determine the major sources of airborne dust in mines and to investigate control methods. Experiments have identified three airborne dust types and their sources: mineral dust from all mining processes, oil mists from lubricating oil used in compressed air-powered machines, and particulates from diesel exhaust. Rock breaking and dust dispersion were studied and the results showed that the dust produced is directly dependent on the total breakage and that dispersion increases by the extent of the handling, the energy expended in creating new breakage, the kinetic energy of the rock itself, and the velocity and turbulence of the air.

Two main dust control methods are wetting and ventilation. The effectiveness of wetting depends on wetting time, rock surface properties, type of wetting agent, use of steam and on the extent of mechanical mixing; sprays are ineffective in removing fine dust that is already airborne. Controlling dust through ventilation is effected by adding air to dilute the dust concentration, exhausting or drawing dust-laden air (from the worker's zone), and by controlling airflow direction.

Gravimetric dust measurements from six mines are given for blasting, rock loading, transport and hauling, drilling, and ancillary operations. Blasting, crushing and vertical orepasses produced the most dust. In mining areas, ore handling rather than drilling was responsible for creating most of the dust.

Two sets of experiments were made to investigate the effect of water on dust control in LHD loading operations. Wet processes in milling operations produced substantial dust; enclosures and extraction could decrease general ventilation requirements.

Recommendations are given for dust control in hard rock mines.

1 INTRODUCTION

Dust has long been a major health hazard in the mining industry. There has even been evidence of silicosis in Stone Age flint workers.

This report forms part of a program to develop technology to solve problems related to underground mine workers' exposure to airborne, and thus, respirable, dust produced by various underground mining operations.

Studies were undertaken by the Mining Research Laboratories at Elliot Lake to determine the sources of fine, potentially airborne dust in mines and to develop control methods capable of reducing dust levels significantly below regulatory standards. Dust production measurements were taken with the CAMPEDS for gravimetric dust, respirable combustible dust, respirable mineral dust (ash), and respirable quartz dust (Knight 1978).

Rock breakage, dust dispersion and rock cutting processes are described as well as control methods by wetting and ventilation. Recommendations for improving dust control in hard rock mines are given.

The principles of dust generation and control are discussed in the following sections.

2 BREAKAGE

Dust is produced in all rock breaking processes except possibly when rock splits at a plane of weakness. In most processes examined the quantity of fine, potentially airborne dust was related to the quantity of coarser dust produced, that is, to the total amount of rock broken into finer sizes. Figure 1 shows the mean value and range of sizes of broken material produced from 1 cm lumps in a rotary crusher. This

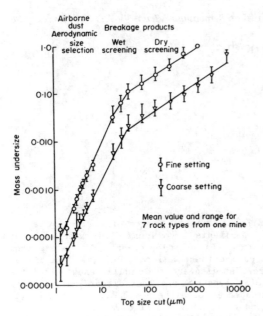

Fig. 1. Size distribution of breakage products and airborne dust in laboratory test.

information is important in evaluating the quartz content of airborne dust.

Discontinuities may occur in rocks with a marked grain size (Hamilton, 1957); These are most often found in sedimentary rocks with strong grains in a weak matrix. Despite this, the fact is that the amount of potentially airborne dust is directly dependent on the total breakage. Generally, breakage is directly proportional to the energy input and inversely to the rock strength. Table 1 shows the estimated quantity of potentially airborne dust in the respirable size range when rock is broken to various maximum sizes for transport assuming a uniform breakage procedure. Table 1 shows that a large increase occurs when the ore is broken into finer sizes.

3 DUST DISPERSION

Generally, only a proportion of the potentially airborne dust is dispersed. Therefore, it is expected that in breaking dry material, the dust generation and dispersion will resemble that shown in fig. 2. In a simple shatter test where 50 mm lumps were dropped 2 m onto a steel plate in a gentle airstream, less than 10% of the dust from a weak coal was dispersed compared with most of the dust from a strong coal (Hamilton 1957; Hamilton et

Table 1. Estimated potential airborne dust from breaking massive rock to various sizes.

Maximum size of broken rock (metres)	Potential airborne dust in respirable size range (1) (g/t)
1	19
0.75	23
0.5	31
0.25	50
0.1	105
0.05	290
0.01 (1/2 in.)	600

al. 1963). Presumably the stress relief from the stronger coal imparted more energy or speed to the broken products leading to dispersion of a higher proportion of the dust. Test results suggest that dust dispersion was directly proportional to the energy input and independent of coal strength. In contrast, potentially airborne dust increased with decreasing coal strength.

It can be surmised that dust dispersed in handling processes increases with:

1. the amount of fine dust created in the breaking process
2. the extent to which the material is stirred up during handling
3. the energy expended to create new breakage
4. the kinetic energy imparted to the

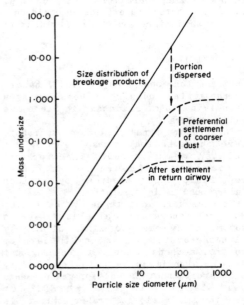

Fig. 2 Development of the size distribution of airborne dust.

freshly broken particles by the stress relief in the material broken

5. higher air velocities.

4 ROCK CUTTING PROCESSES

Rock cutting processes are used in coal mining machines, rotary drills, percussion drills, etc. Cutting is a misnomer; the characteristic of these machines is that they separate chips of rock from the main mass. The actual mechanics of raising a chip from the solid is complex, involving impact, fracture propagation, etc. Generally, two processes must be considered; cutting of the chip and its removal from the site.

For cutting the least amount of potentially airborne dust is produced when the chips are as large as possible. The design of cutting tools is complex, but minimum specific dust generation, i.e., dust generated per tonne, and minimum specific energy consumption are usually achieved together.

The problem of removing chips is most obvious in drilling where insufficient clearance or flushing agent may result in further breakage of the chips, binding and resultant loss of energy from the drill rod, decreased cutting speed, and increased specific dust generation and energy consumption.

5 DUST SUPPRESSION BY WETTING

Water is widely used in mines for dust control. Factors affecting dust suppression by wetting are:

1. Small potentially airborne dust particles can be attached to larger lumps and rendered non-dispersible by a thin film of liquid.

2. Water or other sprays are ineffective in removing respirable dust particles that are already airborne, unless very high accelerations are applied as in high pressure-drop wet scrubbers.

3. To be fully effective the liquid has to spread over the entire surface of the rock. The wettability of the rock depends on its surface properties and those of the liquid. Experiments have shown that most common rocks except sulphide minerals and coking coals are wettable.

4. It has been shown that it takes a long time for water to spread over the surface of a rock pile. Table 2 shows shatter test results for six coals after wetting and standing (Hamilton 1963).

5. Adding a wetting agent to water has been proposed. Laboratory experiments indicate improvement, which suggest that (Hamilton 1963):

a. wetting agents reduce spreading time

b. high concentration over 1% is required

c. the effect is most marked for difficult-to-wet rocks such as coking coals and sulphides

d. the use of wetting agents in sprays reduces the size of droplets which could also be achieved by increasing water pressure; no other measurable effect on the collection of airborne dust has been detected.

However, mine trials have failed to show any significant improvement in dust suppression. Wetting agents may only be of value in certain operations. It must also be noted that wetting agents may interfere with flotation processes used in ore preparation.

6. Obtaining a water layer by feeding steam into the rock pile is effective and decreases spreading time. Steam can be fed into moving rock in an enclosure at a transfer point on a conveyor belt.

7. Gentle mechanical mixing accelerates spreading of water over the rock surface. In laboratory tests two minutes of mixing in a low-speed tumble mill was sufficient for full control as shown in fig 3. Mechanical mixing may effectively improve dust suppression on conveyor belts and during drilling.

Table 2. Dust raised from various materials after treatment with water.

Material	Water Addition Level (mass %)	% Dust Dispersion			
		time of standing (min)			
		1	5	30	300
Coal coking (301)	10	93	87	79	70
Coal bituminous (902)	10	82	61	50	35
Coal bituminous (802/902)	10	–	50	32	20
Shale a	2.5	66	21	11	9
b	2.5	65	24	21	10
c	2.5	40	28	24	15

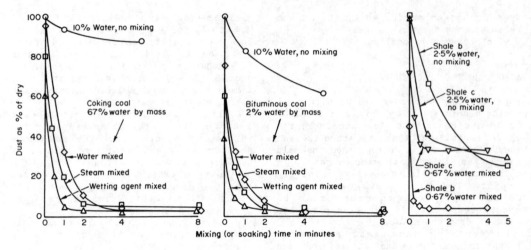

Fig. 3. The effect of mechanical mixing or soaking after the addition of water, steam and 2% wetting agent solution.

8. It has been shown in rock and coal cutting operations and drilling that water is most effective when applied as close to the cutting point as possible and that this can be achieved by feeding water through the tool bit and onto the cutting face.

9. Feeding water to machines has two major problems: 1) the human one of ensuring that the water supply is connected and turned on which can be avoided by interconnecting the water valves and the power supply and 2) clogging due to dirt and pipe scale for which it has been recommended that spray orifices have a diameter of not less than 1.5 mm and be protected by filter screens on or close to the machine.

10. Muck should be kept wet during handling and transport because of long wetting times required when dry.

11. Most liquids are effective dust suppressants. Oils and salt solutions have been used specifically to avoid drying or freezing. Drying of settled dust in underground roadways has been prevented by using hygroscopic salt as a binder.

12. Foam appears to show two main advantages, a barrier against airflows giving rise to dispersion and minimizing the quantity of liquid applied.

6 DUST CONTROL BY VENTILATION

Ventilation can control dust by: adding air to dilute the dispersed dust; exhausting or drawing dust-laden air away from the operation to prevent it from reaching workers, and controlling airflow direction.

It is usually assumed in dilution ventilation that an increase in airflow will lead to a corresponding reduction in dust concentration, i.e., doubling airflow will halve the dust concentration. However, this is not entirely true as the decrease in dust concentration is usually not as great as the increase in ventilation. It has been shown on English and German longwall coal faces that respirable dust concentration decreases with increased airflow at low velocities, reaches a minimum at 3 to 6 m/s and increases at higher velocities. Presumably increased air velocity tends to stir up the dust and disperse it, eventually off-setting the increased dilution.

Exhaust ventilation is effective when the dust source is enclosed and sufficient dust-laden air is drawn out to entrain the dust and prevent it from reaching workers. For this, air must have sufficient velocity to pick up and retain the dust. A velocity of 1 m/s at the entrance to a hood is usually considered suitable for design purposes. However, higher values may be necessary if the ambient velocities exceed 0.5 m/s or if material moves out of the enclosure as on a conveyor belt. The exhausted air can be either passed directly to a return airway or filtered and returned to the ventilation circuit.

The quantity of air required for dust control of an enclosure with exhaust ventilation is decreased by:

1. reducing openings to a minimum

2. proper maintenance
3. lowering general air velocities
4. decreasing air turbulence from moving parts or from material entering and leaving
5. avoiding pumping effects, i.e., rotating or moving components acting as fans can push air out.

Exhaust ventilation generally requires much smaller volumes of air than dilution ventilation to control dust concentrations.

Airflow direction is mentioned in most manuals as a means of controlling dust but these generally do not emphasize its possibilities. Basically, it can have an intermediate effect between that of dilution and exhaust ventilation. In underground mining where airflow is necessarily controlled, dust exposure may be greatly reduced by ensuring that air from dusty operations passes directly to return airways and that most work is done on the intake side. For instance, stope drilling and mucking should be planned so that drillers are not exposed to dust from mucking and the mucking machine should be operated from the air intake side.

A two-fold difference in the dust exposure of two men working in a development heading was found in one study over four sampling days. Presumably this was due to one man preferring to work in the stream of high velocity fresh air coming from the duct whereas the other did not.

Diffuser nozzles have been used on high speed ventilation ducts to supply a comfortable stream of clean low-velocity air to workers.

7 DUST MEASUREMENTS IN MINES

For discussion purposes, hard rock mining consists of:
1. separating ore from surrounding rock and breaking it into pieces small enough to transport
2. loading
3. transporting and handling
4. drilling
5. ancillary operations

7.1 Dust Measurement Technique

"Dust production" is the term most descriptive of dust generation and dispersion into the air at a mining operation. Dust was measured with the CAMPEDS gravimetric dust sampling system (Knight 1978). Samples were collected on silver membranes and were weighed three times to give tare mass, gross mass and mass after ashing. They were further analyzed by X-ray diffraction for mass of quartz. These measurements gave four assessments:
1. total respirable dust
2. respirable combustible dust (defined as loss in mass on ashing at $500^{\circ}C$),
3. respirable mineral dust (ash),
4. respirable quartz dust.

Possible minor damage to the filter when loading and unloading the sampler can lead to large errors in estimating the mass of total respirable dust. This occurred particularly on the light samples in most of these studies and frequently caused difficulties in estimating the respirable mineral dust, which should equal total minus combustible, and the percentage of quartz in the dust. Note however, that the quartz measurement is absolute and is an order of magnitude more sensitive and accurate than weighing.

To facilitate comparison between mines, estimates of respirable mineral dust production were made from the quartz measurements and from the average value of the quartz content in heavy ashed dust samples, as well as from the measured difference between tare and ashed masses.

Sampling stations were set up in the airways both upstream and downstream from each operation. The stations were chosen to give the most uniform mixture of dust and air possible. Replicate samplers were strategically placed to obtain best readings without obstructing passage for men and vehicles. The samplers were kept running after the operations stopped to allow time for the dust to pass the return station. Airflow measurements were made at the stations and a recording anemometer was used to observe changes in airflow and to assist in estimating the total air volume passing each station.

Total dust produced at each operation was determined by multiplying the dust concentration at each station by the total corresponding airflow and subtracting the resultant figures at the intakes from those at the returns. This figure was then divided by the unit of production - tonnes of ore, length of hole drilled, or other unit as applicable - to give specific values per unit of production.

7.2 DUST PRODUCTION MEASUREMENTS

The measurements taken in six mines are summarized below.

Blasting
Measurements were made on airborne dust

Table 3. Blasting.

Site	Airflow (m³/s)	Tonnes	Sampling Period (min)	Number of Tests	Respirable Dust Production (mg/t)			
					Total	Combustible	Mineral	Quartz
Total mine	450	2900	210	1	1200	540	450	300
Steep slope	2.5	180	330	1	2100	600	1400	700
Flat heading	20	270	360	1	1800	1400	300	150
Secondary (6 oversize rocks)	7	6	30	1	909	–	700	70
Mean					1500	900	750	–

reaching surface after production blasting between shifts and underground after blasting in a stope and heading, and after secondary blasting in a drawpoint (Table 3).

It is difficult to measure dust produced in a blast because of disturbance to the airflow. Future studies should determine the energy input from the explosives in at least a semi-quantitative form.

Clearly, blasting in hard rock mines produces large quantities of dust, emphasizing the importance of evacuating men and allowing adequate time for ventilation to remove dust and fumes.

Loading
The following equipment was used to load broken rock for transport (Table 4):

1. rail-mounted compressed air-powered mucker emptying into a rail car in development headings

2. electric and compressed air-powered slusher operating in drifts and stopes (2a,b,c)

3. diesel-powered load-haul-dump machines (LHD) of various sizes operating in headings and drawpoints (3a,b). In some cases LHD's loaded the rock into diesel-powered trucks. A separate measurement of dust produced in this transfer was not possible because the layout of the heading and the ventilation system prevented selecting sampling stations with adequate mixing of dust with air.

It can be seen that rock loading produces substantial mineral dust. The electric-powered slushers produced negligible combustible dust. The compressed air-powered equipment produced measurable quantities, presumably of lubricating oil mist. The diesel-powered equipment produced large quantities of combustible dust.

The quantities of mineral dust varied substantially from one site to another for which a number of factors were responsible:

1. Wetness of rock - although water was always used, apparently not all of the rock surfaces were wetted.

2. Roughness of floor - it was apparent that loading was more difficult and presumably more dust was produced on rough floors than on smooth.

3. Clean up and scaling - loading machines used for clean up at two sites produced much more dust per tonne of muck than normal loading operations.

4. Operator finesse - some drivers, especially on lower-powered machines, developed techniques for rapid loading, presumably with low energy input and low dust production.

Rock Transport and Handling
After loading, rock is transported horizontally and vertically for considerable distances to surface, usually via an underground crusher. A limited number of such operations were examined (Table 5):

1. Filling rail cars from a chute with a free fall of 1 to 2 m.

2. Dumping rock onto a grizzly from diesel-powered LHD's and trucks. This operation included breaking oversize pieces with a hydraulic pick. The measured dust was not the total produced but only that part escaping the local system below the grizzly and from the filter system.

3. Near vertical ore and waste passes.
a. This system had dump points at every 35 m vertically with interconnections between ore and waste passes at each level; an exhaust system drew dusty air from the lowest level.
b. The part of this system examined consisted of a 30 m section of orepass with fingers at top and bottom; the free fall of material induced airflow out through the lower finger.
c. The system consisted of a 600 m vertical orepass with dump points at the top and at 120 m intervals; dust leakage from the top was calculated on tonnage

144

from this level and at the second level on tonnage from both upper levels.

4. Underground crushers; although total dust was measured on systems a and b, only that leading from the crusher and air exhaust system with filter was measured on c and d; at two other crushers, dust leakage could not be estimated because of low ventilation rates.

5. Big rock handling; rocks too large for the crusher were handled at this mine by a crane and dropped into a side heading for secondary blasting.

6. Skip-loading; one skip-loading

Table 4. Loading of rock.

Site	Operation	Airflow (m³/m)	Tonnes	Sampling Period (min)	Number of Tests	Respirable Dust Production (mg/t)			
						Total	Combustible	Mineral	Quartz
1	Mucker in heading								
	Range	1.5-3	60-90	210-330	3	40-200	12-180	14-42	7-21
2a	Slusher in raise								
	Range		30-50	69-150	3	0-50	0-6	8-80	4-39
2b	Slusher in stopes								
	Range	0.6-4	50-200	220-280	4	11-150	0-10	40-170	20-84
	Range	.5-1.5	60-80	210-270	4	60-500	90-250	100-250	10-110
2c	Slusher in stope								
	Range	0.7-1.3	60-100	-	4	80-250	30-45	40-200	20-53
3a	LHD in heading								
	Range	0.6	500	-	1	136	43	100	23
	Range	10	200	100-180	9	225-980	240-450	100-400	29-315
3b	LHD in drawpoint								
	Range	3.5-8	250-500	90-345	4	200-450	170-280	50-200	14-39
	Range	1-17		100-350	5	260-3000	230-350	150-1300	15-130
Overall range of test means								26-540	
Overall mean								140	

Table 5. Ore handling.

Site	Operation	Airflow (m³/s)	Tonnes	Sampling Period (min)	Number of Tests	Fall (m)	Respirable Dust Production (mg/t)			
							Total	Combustible	Mineral	Quartz
1	Loading mine cars	3-73	140-420	30-114	2	3	35	15	70	8
2	Dumping on grizzly	70	800	300	1	1-3	270	-	54	27
3a	Orepasses	40	2000	320	1	300	700*	low	800*	400
b	Orepasses	1	200	120	1	30	105	7	70	16
c	Orepasses									
	top level, mean	40	600	200	3	300	270	90	180	18
	182 m down, mean	25	1600	140	7	300	250	80	160	16
4a	Underground crusher	3	2400	325	2	3	600	20	800	400
b	Underground crusher	3	800	200	1		200		250	130
c	Underground crusher	2.5	200	120	1	3	100		60	15
d	Underground crusher	3	750	390	1	3	150		90	22
5	Big rock	3	20	390	1	3	110		35	10
6	Skip loading	5	514	210	1	6	115	15	30	16

*Lower total than mineral assessment is an indication of the weighing errors.

Table 6. Drilling.

Site	Operation	Airflow (m³/s)	Metres	Sampling Period (min)	Number of Tests	Respirable Dust Production (mg/m)			
						Total	Combustible	Mineral	Quartz
1	Jackleg mean	2	105	224		9	8	1.2	0.6
	Mean	6	85	200		2.4	0.6	0.6	0.3
2	Bar-arm mean	1.5	180	153	2	20	15	3	0.6
3	Jumbo mean	1.7	300	300	2	4.5	3	1.5	0.3
	Mean	10	180	225		20	18	1.8	1
4	Mini borer				2	85	6	3.5	1
						(8)*	(0.6)*	(0.35)*	(0.1)*
5	Down the hole range		46-120		4	120-1000	60-700	60-350	16-110
						(90)*	(70)*	(20)*	(10)*

*Values in brackets are those for equivalent length, by rock volume removed, in 50-mm diameter hole.

facility fed from an orepass was examined.

Most ore handling operations produce large quantities of dust and many achieve partial or complete dust control using exhaust air systems.

Drilling

Drilling is a major operation in most hard rock mines and uses a substantial proportion of the total manhours. Before the advent of wet drilling it was considered the most hazardous occupation leading to silicosis.

Five drilling systems were studied as shown in Table 6. Dust production is given as mg/m of drillhole.

1. Jackleg compressed air-powered rotary percussive drills; these data could not be separated from drilling for roof bolts using a stoper; dust production is given in terms of total length of hole; bit diameter was about 40 mm.

2. Bar and arm compressed air-powered rotary percussive drills; these drilled long holes of about 50 mm in diameter.

3. Jumbo-mounted compressed air-powered rotary percussive drills; bit diameter was about 50 mm.

4. Mini-borer; rotary drill, drilling down-holes of about 150 mm in diameter.

5. Down-the-hole drills; four compressed air-powered rotary percussive drills were used in the same stope for which the total production was measured on four separate shifts; bit diameter was 110 mm.

Table 6 indicates that all drills except the down-the-hole were characterized by low dust production. This made it difficult to measure the increase in dust concentration between intake and return air except at low rates of airflow.

The down-the-hole drills produced copious quantities of airborne dust because compressed air was used as the flushing agent. Although some water was used it was apparent that dust control was less effective than with any other drill even allowing for the greater quantity of rock broken per metre of hole.

Ancillary Operations

A few ancillary operations were examined for which production results are given in Table 7:

Operations 4 to 8 in Table 7 were in a surface mill processing iron ore. Even though all the processes were wet, substantial quantities of dust were produced. Enclosures should be considered.

7.3 SUMMARY OF DUST PRODUCTION

In most mines examined gravimetric

assessments show that rock handling and specifically loading are a a much greater hazard than the historically hazardous operation of drilling. Although partly due to the parameter change from number to mass for the finer dust in drilling compared with other operations, the main factor is the effective dust control achieved by feeding water to the bit. In most of the mines examined the major mineral dust exposure was due to loading. Secondary blasting in the mining area or substantial contamination of the intake air by the ore transport system was equally important in some mines.

Mine transport systems can produce large quantities of mineral dust but effective control has been achieved by entraining dust from non-mobile operations in an airstream and filtering or directing it to a return airway.

Table 8 shows the estimated dust production for the mine where most of the experimental work was undertaken. This mine operated at depths between 240 and 600 m using horizontal track drifts and stopes on the 2 to 6 m thick ore horizon with jackleg drills and electric slushers. The ore output was 2700 t/shift and production blasting as carried out between shifts.

It can be seen that most dust was produced in the ore transport outbye of the active mining area, where in this particular mine the orepasses are near the shaft and away from the extraction area. The main dust sources were well controlled by exhaust ventilation and only a small proportion of the transport dust leaked into the working areas and travelways. The production blast was the next major source and exposure was avoided by blasting between shifts.

The mineral dust production in the active mining areas is only about 5% of the mine total. Most of this arises from handling with only little from drilling.

An LHD operator's exposure was analyzed in a second mine which worked a near vertical orebody. The orepasses were placed close to the intake allowing dust contamination of most of the mining zone. The operator's exposure comprised:

a. 15% in the orepass heading;

b. 12% leakage from orepass into intake air; this also formed a major part of dust exposure for other miners on the level.

c. 10% loading muck;

d. 30% directly from engine exhaust.

7.4 DUST CONTROL EXPERIMENTS

Loading is one of the dustiest

operations to which miners are exposed. To date two sets of experiments were made to investigate the effect of water control on dust in LHD loading operations. In the first set carried out in a drawpoint in high sulphide ore, loading was carried out dry or with 1-, 2-, or 4-bar pattern sprays. In the second set, muck piles in headings six or more metres wide were examined using various wetting times with simple jets from hand-held or blocked-in-position hoses and nipples.

Results from the first set, based on differences between return and intake as well as between drawpoint and intake air samples, are given in Table 9.

It was clear that adding water decreased mineral dust production but not fully as dry muck was always visible immediately after loading.

Results from the second set of tests on highly siliceous rock muck piles are shown in Table 10. Sixteen tests were made on muck piles in various headings and represented normal mine operation. The last two tests were made on a similar muck pile at a site having through-ventilation. It was obvious that water additions could reduce dust levels to less than 20%. It was also evident during the first few buckets at the start of the shift that dry muck was frequently visible. The extra wetting tests in which loading of a muck pile was continued after the lunch break led to a further substantial reduction in mineral dust.

The two tests showed somewhat erratic results on combustible dust production. Although adding water could affect ease of loading and power requirements or even modify the soot production by the engine, it is believed that the erratic results are probably due to the errors in assessing dust, or to variation between engines.

7.5 COMPOSITION OF AIRBORNE DUST IN MINES

The experiments on dust production have identified three airborne particulate types and their sources:
1. mineral dust from all mining processes.
2. oil mists from lubricating oil used in compressed air-powered machines.
3. diesel exhaust particulates.

Mineral dusts are not all equally hazardous to health. Free silica minerals and mineral fibres (asbestos) are of an order of magnitude more hazardous than most other common minerals. Of the free silica minerals, quartz is the most common, and in this study was the only

"more hazardous" mineral measured separately. In most mines tested the respirable quartz formed a constant proportion, within the limits of experimental error, of the respirable mineral dust. This proportion was always less than that in the rock mined. In one mine with a massive sulphide ore there was some evidence that the quartz content was higher in the airborne dust in the mining zone than in the crusher room. Presumably, by the time the rock reached the crusher the silicate components were well wetted and their dust was better controlled by the various water additions en route.

In jackleg drilling it was found that 90% of the dust collected on the filter was combustible. It was presumed but not proven that this was lubricating oil atomized by the high energy of the exhaust air. A simple calculation showed that the measured quantity of respirable combustible dust was about 0.33% of the amount of lubricating oil sent underground for use in drills.

The diesel exhaust particulates are a complex mixture of soot, unburnt hydrocarbons, and sulphuric acid. The sulphuric acid mist is produced by oxidation of some of the sulphur dioxide in the catalytic purifier, and may be absorbed onto the soot particles.

In this report only the total exhaust particulates as indicated by the respirable combustible dust, i.e., loss in mass on ashing, are considered.

Table 11 shows the composition of the dust produced at various operations and its probable source. In view of the limited number of operations examined, the extent of variation in composition is probably too low and the figures should be used only as an indication of the possible hazards.

8 RECOMMENDATIONS FOR DUST CONTROL IN HARD ROCK MINES

8.1 Blasting

Because of both dust and fumes, blasting should be carried out so that men are not exposed to contaminated air either in the area itself or to the air leaving it. Blasting techniques should aim at producing a minimum of fines and oversize lumps, and producing the smoothest floor possible (Hagan, 1979).

8.2 Rock handling

Rock breaking produces large quantities of

Table 7. Ancillary mining operations.

Site	Operation	Airflow (m³/s)	Tonnes	Sampling Period (min)	Number of Tests	Respirable Dust Production (mg/t)			
						Total	Combustible	Mineral	Quartz
1	Drill prep. (ST 5)	10		30	1	1800	1000	800	400
2	Hyd. backfilling	2.5	800	180	1	270	-	-	-
3	Conveyor belt³	4	4000	360	12	23	1	14	5
4	Autogenous mill	2	2000	360	2	65		27	9
5	Pebble mill	2.2	500	360	2	5		11	4
6	Magnetic separator	0.2	320	360	2	1		1	3
7	Flotation	0.2	320	360	2	3		2	7
8	Balling drum	2.7	400	360	1	13		13	-

[1]Toe scraping ST 5.   [3]Dry muck conveyor excluding feed point.

Table 8. Estimates of total dust produced for one mine shift.

Operation	Tonnes	Respirable Dust Production g/shift				Mineral Dust as % of	
		Total	Combustible	Mineral	Quartz	Entire Mine	Mining Area
Between shift main blast	2700	5600	2800	2600	1300	34	
Mining area							
Secondary blasting	100*	100	20	80	40	1	15
Slushing	2400	350	20	330	165	4	63
Mucking	400	40	30	10	5	0.1	2
Car loading	2700	115	60	56	27	0.7	10
Drilling	2700 (2200 m)	22	20	3	1.5	0.04	0.6
Scaling etc.**		10	-	10	5	0.1	2
Total		637	150	487	243	5	100
Outbye							
Rail transport	2700	65	60	5	2	0.1	
Dumping into orepasses**	2700	60	10	50	25	7	
Orepasses	2700	2700	60	2600	1300	34	
Underground crusher	2700	1800	-	1800	900	23	
Skip loading	2700	160	40	100	50	1.3	
Total		6800	170	4600	2300	60	
Entire underground mine		11000	3100	7700	3850	100	

*Estimates based on a guess as to variable secondary blasting carried out.   **Estimate.

Table 9. Effect of water on LHD loading high sulphide ore.

Sprays		Respirable Dust production (mg/t)				Relative Respirable Dust Production (ng on filter/bucket)				Dust as % of Dry	
No.	Cond.	Total	Combustible	Mineral	Quartz	Total	Combustible	Mineral	Quartz	Combustible	Quartz
Mean	Dry	280	330	21	59	58	30	21	7	100	100
1	Sprays	18	50.15	47	20	30	20.67	8	3.5	60	42
2	Sprays	9	36.11	-11	16	9	4.13	5	0.9	12	20
4	Sprays	200	130.40	110	13	3	0	1.4	0.5	0	15

Table 10. Effect of water on LHD loading high silica ore.

Description	Runs	Tonne	Respirable Dust Production (mg/t)				Dust as % of Dry	
			Total	Combustible	Mineral	Quartz	Combustible	Quartz
Normal mine practice								
Range	9	152-424	225-1300	240-1100	59-630	29-315		
Mean		1800	700	525	150	75	105	25
Early start mean	2	160	825	675	185	92	135	31
Extra wetting range	5	32-200	600-1500	600-1500	28-96	14-48		
Mean	4	600	900	825	62	31	165	10
Dry test	1	120	900	500	600	300	100	100
Same pile after 1½ hours wetting	1	88	800	700	110	55	140	18

Table 11. Composition of airborne dust produced by various operations.

	Ash %	Mineral Dust Total %	Combustible Dust %
Blasting	88	50-70	12
Loading			
C.A. mucker	25	25	75*
C.A. scraper	70	70	30*
Electric slusher	95	95	5
Diesel LHD	20-40	20-40	60-80**
Drilling	10	10	90*
Orepasses	100	100	0
Crushing	100	100	0

*Lubricating oil mist.
**Diesel exhaust particulate.

airborne dust, and although the broken ore needs reducing further, additional breakage should be minimized at all sites where men may be exposed and confined to such areas as crushers where enclosures and filtration can capture most of the dust. In handling rock, attention should be given to:

1. Minimizing free fall as the amount of dust produced is proportional to the distance dropped.

2. Using slides as sliding a given vertical distance reduces dust production compared with falling freely.

3. Making floors as smooth as possible to minimize energy required to pick up or drag rock.

4. Locating and designing sites for secondary breakage of oversize rock such as grizzlies or dumps, so that the dust dispersion can be entrained and either directed to a return airway or collected by filtration; in particular, they should not be located in main airways where high air velocity increases dust dispersion.

5. Vertical orepasses. These are a major source of dust because of the piston effect of falling rock pumping air out of the orepass into the work place. Pressures can be developed high enough to lift 1 cm thick steel plate doors.

Although pressure can be reduced by interconnecting ore and waste passes at each level, there will still be some leakage of dusty air and elaborate precautions may be necessary to divert this away from the work areas. Siting orepasses close to the return airways may be well worthwhile.

The use of off-vertical orepasses could possibly decrease dust production by transforming free falling to a sliding motion, decreasing fresh breakage. A controlled feeder might prevent the formation of plugs and greatly decrease the pumping of dusty air.

6. Wetting broken rock. This can prevent fine dust from becoming airborne by binding it to large pieces of muck. To be effective a thin liquid layer must cover all free surfaces in the rock pile. For the liquid to spread over the rock its surface must be wettable. Silica rocks are usually readily wettable whereas some sulphide minerals are not. Even for silica rocks, wetting times of more than two hours are required to completely wet a muck pile and minimize dust dispersion during loading. For sulphide rocks wetting times may be much longer.

The quantity of water required however is not great - a few tens of litres per tonne and the use of mist sprays applying water at a low rate evenly over the top of the muck pile for hours is probably most effective. A hand-held hose jet is usually not satisfactory and encourages inadequate wetting.

The mechanical mixing involved in rock movement in orepasses and crusher operations spreads the water much faster and quickly traps any dust created by new breakage.

7. Some ventilation is essential to dilute the dust concentration. Whenever possible, airflow direction should be such that the air flows away from workers towards the dust source.

8. Enclosure of dust handling operations is effective in decreasing air velocity over moving rock and reduces the dispersion of dust.

9. Where breakage is unavoidable, such as in an orepass and crusher operations, enclosures and air extraction are almost always required for dust control.

8.3 Drilling

Normal wet drilling is not a substantial source of dust as the application of water close to the cutting edge and its virtual immersion in water in the hole prevents dust escape. However, large quantities of potentially respirable dust are formed so that dust control must be considered in any change in drilling technique, such as when going to down-the-hole drilling (Table 6).

LHD's, and probably other machines, when preparing faces for drilling, apply high forces and expend much energy, thereby causing fresh breakage and creating more

potentially airborne dust. The LHD bucket
is a poor tool for cleaning faces and
alternatives are required.

11. REFERENCES

Hagan, T.N. 1979. The control of fines
    through improved blast design. Proc.
    Aust. Inst. Mech. & Metal. p.9.
Hamilton, R.J. & G.Knight 1957. Some
    studies of dust size distribution and
    the relationship between dust formation
    and coal strength. In: Mechanical
    Properties of Non-Metallic Brittle
    Materials. W.H.Walton (ed), p 365.
    London: Butterworth.
Hamilton, R.J. & G.Knight 1963. Laboratory
    studies of the suppression of dust from
    broken coal and shale. Int J.Rock Mech.
    Min. Sci. 1:105.
Knight, G. 1978. Mine dust sampling system
    - CAMPEDS. CANMET Report 78-7. CANMET,
    Energy Mines and Resources Canada.

An update of the reduced standards problem and overview
of improved control technology for respirable coal mine dust

ROBERT A.JANKOWSKI & FRED N.KISSELL
Pittsburgh Research Center, Bureau of Mines, PA, USA

ROBERT E.NESBIT
Mine Safety and Health Administration, Arlington, VA, USA

ABSTRACT: Over 10 pct of active coal mine sections are on reduced dust standards due to quartz, with an average applicable standard for the underground designated occupation at 1.2 mg/m^3, while the applicable standard for the surface designated work position averages 0.8 mg/m^3. It is estimated that by the end of FY85 over 15 pct of all coal mine workers will be affected by overexposure to respirable quartz and that a significant percentage of mining sections will be placed under more stringent dust standards.

Studies have shown that the primary source of quartz dust on surface mining operations comes from dust escaping the collector system of the highwall drill. On underground operations, the primary source of quartz is the continuous mining machine, cutting roof, floor and rock partings in the seam.

With the cooperation of industry and the Mine Safety and Health Administration, the Bureau of Mines has identified new or improved controls to reduce the amount of airborne dust, controls to reduce dust rollback, and procedures to correct the deficiencies of the roof bolter dust collector system. The Bureau has also initiated extensive research to control the quartz dust exposure of the continuous miner operator, control quartz dust levels downwind of the continuous mining machine, and identify novel concepts for reducing quartz dust in underground coal mines.

1 INTRODUCTION

1.1 Identification of the problem

Until a few years ago, reduced respirable dust standards represented only a fraction of entities at all underground and surface mining operations (fig. 1). The present increase has been brought about in part by a change in the coal mine quartz dust sampling strategy enforced by MSHA, but even more by improvements in analytical techniques. In the past, because of the amount of sample required for analysis (approx. 5 mg) and the technique of combining up to five samples from different occupations in order to get sufficient mass for analysis, few of these composite samples showed greater than 5 pct quartz. With the introduction of the present low temperature ash furnace and infrared method of analysis, a much smaller size sample (approx. 0.5 mg) can now be analyzed and composite samples are no longer necessary. As a result, MSHA has been able to identify many quartz dust problems that were only suspected a few years ago previously.

1.2 Current status of reduced standards

Reduced standards resulting from exposure to coal mine dust containing more than 5 pct quartz are currently one of the most serious health compliance problems facing the underground and surface coal mining industries. Awareness of quartz problems in the mine environment during the past three years has increased dramatically. Today, approximately 2,400 sites have been identified where the allowable level of 5 pct quartz is exceeded. These sites constitute the work environment of over 20,000 coal mine workers, who represent over 10 pct of today's coal mining work-force. Over 50 pct of these standards are

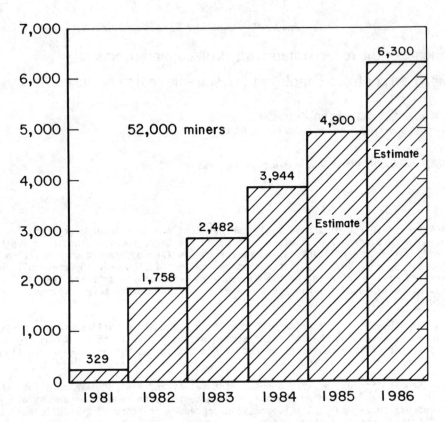

FIGURE 1. Number of Entities Placed on Reduced Dust Standards, 1981-1986.

reduced to a level where no known technology can insure the health of the mine worker. Some respirable dust samples contain 50 to 60 pct quartz, corresponding to a 0.2 mg/m^3 standard. Existing technology will not work, or works with only minimal success, in achieving these very stringent standards.

The average applicable reduced standard for the 874 underground designated occupations (continuous miner operator) on reduced standards is 1.2 mg/m^3. For designated work positions at surface mines (highwall drill operator), the average standard is 0.8 mg/m^3, with 546 positions affected. Figure 2 shows the number of entities in each of the various reduced standards levels. Average reduced respirable dust levels for the 1,239 non-designated entities is approximately 1.6 mg/m^3. This category includes such occupations as roof bolter and bulldozer operator. Some

understanding of the complexity and scope of the problem may be obtained from the most recent projections of the Mine Safety and Health Administration (MSHA), which predict that by the end of FY86 (September 30, 1986) approximately 6,300 entities representing well over one-third of today's coal mining workforce will be on reduced respirable dust standards because of quartz, unless immediate corrective actions are taken.

1.3 IMPROVED QUARTZ MONITORING STRATEGY

The significant increase in the number of reduced dust standards in effect in coal mines today and projected in the near future has prompted MSHA to develop a more effective monitoring policy for these mines. Beginning October 1, 1985, when more than 5 pct quartz is found in coal mine dust the new policy will allow a combination of up to three mine operator

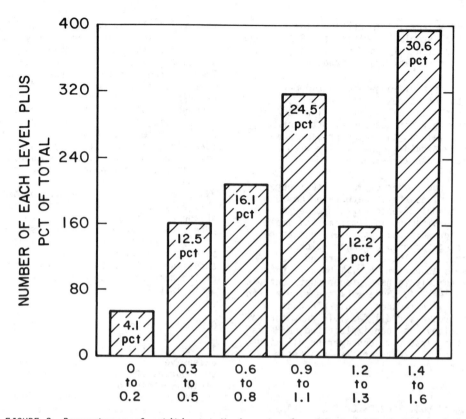

FIGURE 2. Percentages of entities at Various Levels of Reduced Dust Standards.

and MSHA samples to be used in setting the standard for this mine. For the first time, MSHA will establish "designated areas" on mechanized mining units whenever more than 5 pct quartz is found around the roof-bolting machine operator. Coal mine operators will then be required to collect a respirable dust sample from the affected environment every two months, as with other designated areas. At six-month intervals, MSHA will select and analyze an operator bi-monthly sample having sufficient weight gain from each designated area on a reduced standard. If the quartz content is within two percent of the value used to set the standard, MSHA will either average the sample with previous MSHA and mine operator samples to establish a new reduced standard, or leave the existing standard in effect. For example, a sample from a roof-bolting operation on a reduced standard is identified by the respirable dust computer system as the sample to be analyzed for the required 6-month reevaluation and is found to contain 16 pct quartz.

Since this is not within plus or minus 2 pct of the values of 11 pct used to set the standard currently in effect, it cannot be averaged with this pre-established value (11 pct) and the dust standard cannot be adjusted based on the sample analyzed. The operator is given the option via a computer generated message to collect a dust sample for quartz analysis within seven calendar days after receiving MSHA notification. This sample is received within the 10 calendar days of its collection, and is found to contain 12 pct quartz. The average percentage of quartz of the two samples (16 pct + 12 pct / 2 = 14 pct) is then averaged with the pre-established value of 11 pct (14 pct + 11 pct / 2 = 12.5 pct or 12 pct). The 12 pct value is used to adjust the dust standard from 0.9 to 0.8 mg/m^3. If the operator fails to submit a sample when given the option, or there is insufficient dust for quartz analysis, or samples are not collected in accordance with the regulations, the pre-established standard of

0.9 mg/m^3 remains in effect. These optional samples submitted at MSHA request will be used only for quartz analysis to adjust respirable dust standards.

The concept of long-term continuous monitoring is new in this country. The consensus is that it will allow MSHA to concentrate enforcement efforts more effectively on the major problem areas, thereby providing coal miners with a healthier work environment.

At a large number of work sites where the amount of quartz is between 6 and 10 pct, the new policy encourages greater management involvement in sampling and implementing additional engineering measures to bring the problem under quicker control. At locations where quartz exceeds 10 pct, MSHA will be able to move more quickly to have proper dust controls installed. At those work areas unable to comply with reduced respirable dust standards after all feasible dust control technology has been implemented, MSHA is for the first time considering allowing the use of respirators as an interim control.

2 QUARTZ DUST SOURCES ON SURFACE AND UNDERGROUND COAL MINE OPERATIONS

2.1 Surface

Based on MSHA data, the designated work positions with the highest exposure to respirable quartz are, in order of severity: Highwall driller; drill helper; truck driver; and bulldozer operator. Compliance sampling has shown that many samples contain more than 60 pct quartz, and thus pose a serious and immediate health threat to surface coal mine workers. The highwall driller and helper receive the highest respirable quartz exposure, since their primary function is overburden removal. Drilling through various rock formations naturally presents a strong potential for quartz dust generation, as this host material often contains between 60 to 80 pct quartz.

The objective of a recent Bureau of Mines study was to identify and investigate the respirable quartz dust sources and their magnitudes during the removal of overburden material at surface coal mine operations. Results of the survey indicate

that the drill rig has three major dust generating sources during the dry drilling operation: the collector dump, drill shroud leakage, and drill stem leakage. These sources can contribute up to 89 pct of the total respirable quartz dust generated. The collector dump cycle can account for 41 pct of the respirable quartz dust, shroud leakage 32 pct, and drill stem leakage 16 pct. Upwind sources (dragline, second drill rig) contribute only 11 pct of the respirable quartz dust generation. These dust contributions do not necessarily reflect the actual exposure of the drill operator or helper; they only estimate the relative magnitude at the dust source and, therefore, potential exposure. The study showed also that short-term dust concentrations of more than 12 mg/m^3 can be expected from shoveling drill cuttings, and maintenance of the dust collector can result in approximately 8 mg/m^3.

3.2 Underground

The Bureau of Mines has also completed extensive contract research to identify the sources of quartz dust for underground mining sections on reduced dust standards. The objectives of this were to determine the various sources of respirable quartz and determine the percentage of quartz in the coal seam and surrounding strata. It was found that outby sources contributed 2 pct of the respirable quartz dust and roof bolting 25 pct. The major source, 73 pct, was the continuous mining machine, cutting roof, floor, or rock bands within the seam (table 1).

Table 1. Sources of respirable quartz dust on continuous mining sections

Intake conc.	
mg/m^3.....................	0.3
(pct quartz)...............	(2.7)
Percentage contribution...	2.0 pct
Bolter, return -	
outby miner	
mg/m^3.....................	1.9
(pct quartz)...............	(7.0)
Percentage contribution...	25.0 pct
Miner return -	
Inby bolter	
mg/m^3.....................	5.7
(pct quartz)...............	(8.7)
Percentage contribution...	73.0 pct

Table 2 shows the range in percentages of quartz in the coal seam and surrounding strata. The significant sources of quartz dust are drilling the roof, and mining the roof, floor, and rock parting within the seam.

Table 2. Bulk Sample Analysis, Coal Seam and Surrounding Strata

Material	pct quartz,(average)
Roof...............	22-55 pct (33)
Rock partings......	38-47 pct (42)
Coal...............	1- 4 pct (3)
Floor..............	18-82 pct (44)
Collection box, bolting machine...	36-66 pct (43)

Results of these surveys indicated four steps that must be taken to resolve the problem of reduced standards due to quartz:

o   Identify new or improved controls to reduce dust rollback to the continuous miner operator's controls. This should significantly reduce the exposure of the miner operator and helper.

o   Identify new or improved controls to reduce the amount of airborne dust generated by the continuous mining machine. This will not only lower the exposure of the continuous miner operator and helper, but also the exposure of the roof bolter operator when working inby the continuous mining machine.

o   Identify and correct deficiencies of the roof bolter dust collector. This should significantly reduce the quartz dust exposure of the roof bolter operator.

o   Identify new or improved controls for the highwall drill on surface mining operations, including the drill collar and shroud design, and collector dump cycle.

3 Recent advances in dust control technology for surface coal mine operations

As previously noted, 89 pct of the respirable quartz on surface coal mine

operations can be attributed to the highwall drill. Dust sources include the collector dump cycle, leakage around the shroud, and leakage around the drill stem. There are several ways of reducing the dust generated during the collector dump cycle. Instead of discharging the dust onto the ground, at least one commercially available collector has provisions where the dust is discharged into a collection bag, for temporary containment and subsequent disposal at the end of the shift. The dust may also be stored in a hopper beneath the collector, where it is held for periodic dumping at a less frequent rate away from the active work site. Finally, a simple and inexpensive method consists of attaching a large diameter tube to the collector, to act as a dump chute. This last method prevents the dispersion of dust which normally occurs when the dust falls 2 to 3 ft, as in most systems.

Additionally, the tube can act as a receiving chamber and as the drill rig moves to the next hole, the dust trickles out at ground level.

There are several possible means to control the dust from drill shroud leakage. Increasing the collector airflow can decrease the dust which escapes from around the shroud. As a minimum, the collector airflow should be equal to the flushing airflow, to prevent or at least minimize dust leakage around the shroud. All seams around the shroud should be tightly sealed, thereby reducing the number of points where dust can escape. Increasing the size of the shroud has two advantages: more volume is provided for the cuttings, reducing the frequency with which the shroud needs to be raised, and the escape velocity is decreased, allowing for more effective dust capture. A "double-shroud" arrangement can be used, creating a shroud-within-a-shroud. Such an arrangement would allow the collector airflow to be confined to the outer shell; providing a significantly higher dust capture rate within the system, and thus, less leakage. Typical operations frequently raise the shroud above the ground or cutting pile to observe when the coal is reached, or to prevent the cuttings from falling back into the hole. The use of a telescoping shroud arrangement would provide more volume for cuttings and enclose the hole until near the end of the cut.

155

Two primary methods can be used to control the dust which escapes from the drill stem. An air ring seal may be attached to the drill collar forming an air curtain around the stem, thus preventing the dust from escaping around the drill steel. A rubber seal can also be used around the drill stem, however, maintaining the tightness and integrity of the rubber seal is difficult due to abrasion.

4 Results of "Cleanest Continuous Miner" survey

The Bureau of Mines has recently completed a survey of 12 continuous miner sections which were at or below 0.5 mg/m^3 during the past 1-1/2 years. Four features were noted which were employed in all or most of the sections: good ventilation, good water supply and spray system maintenance, water application on the flight conveyor, and a modified cutting cycle. These features helped to minimize dust rollback to the miner operator's cab, as well as to control the amount of respirable dust generated by the mining machine.

4.1 Good ventilation

At all mines surveyed, the quality and quantity of face ventilation appeared to be the most important factor in controlling dust exposure. The mean entry air velocity ranged from 63 to 335 fpm (0.32 to 1.70 m/s), and averaged 122 fpm (0.62 m/s). In all cases, the brattice/tubing setback distance was 15 ft (4.6 m) or less, and all the sections employed exhaust ventilation. Eight of the mines used exhaust tubing with an auxiliary fan while at the other mines, the exhaust brattice was very well maintained and leakage was minimized by sealing the floor/brattice interface.

The high face ventilation minimized dust rollback. Dust generated by coal extraction was usually confined to the face area, and any operator exposure was usually from intake sources such as shuttle car loading and haulage.

4.2 Good water supply and spray system maintenance

Water flowrates ranged from 22 to 43 gpm (100 to 200 l/min), with an average of 29 gpm (130 l/min). Most sections used higher than normal flowrates. There was a great variety of nozzle locations, types, and numbers. However, all spray systems were well maintained and completely functional.

These high water flowrates helped to control dust in several ways. The added flowrate at the miner cutting head helped to reduce the amount of airborne dust generated, thus reducing rollback as well as dust levels in the return. The high water flows also helped to increase the total moisture content of the run of mine product, thus reducing dust levels during loading and transport.

4.3 Water applied to the conveyor

In 11 of the 12 mines surveyed, sprays were mounted on the mining machine flight conveyor. Flowrates to the sprays averaged 5 gpm (23 l/min). These sprays served to add water to the cut material prior to discharge onto the shuttle car, thus minimizing operator's exposure to this intake dust source.

Normally, intake dust levels are not considered a significant contributor to worker dust exposures. However, loading the shuttle car can increase the mining machine operator's dust exposure since any dust which becomes airborne during coal transfer is carried over the miner operator. Redistributing a small portion of the water from the miner head to the flight conveyor may be necessary to ensure that the loaded coal is sufficiently wet to minimize dust re-entrainment. This additional moisture will also serve to minimize the shuttle car operator's exposure when discharging coal at the section dump point.

4.4 Modified cutting cycle

Of the 12 sections surveyed, eight used a modified cutting cycle. The usual cutting pattern is to sump into the coal at the roof and then shear down to the floor. Previous Bureau studies have shown that the sump cut creates at least 70 pct of the dust generated, while producing less than half of the coal. With the modified cutting cycle (fig. 3), the machine sumps into the coal face 1 to 2 ft (0.3 to 0.6 0.6 m) from the roof and then shears down to the floor. This is continued for at

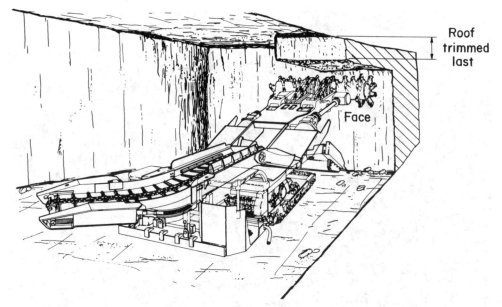

FIGURE 3. Modified Sump Cycle to Reduce Respirable Dust Liberation.

least two sump/shear sequences. The miner then backs up and trims the remaining coal from the roof. This modified cutting cycle has several advantages: it leaves the roof rock in place until it can be cut out to a free face which generates less dust; and it places the sump cut nearer to the floor, confining the dust cloud under the boom where it can be effectively suppressed instead of being released into the main airstream.

With regard to production, this modified cycle provides better machine control preventing the machine from "climbing into the roof" when sumping high. Mining machine type and/or roof conditions will determine how far this modified sump cycle can be advanced before trimming the roof. However, the cycle should offer particular control advantages when conditions require that roof rock be removed.

5 Recent advances in dust control technology for continuous mining sections

5.1 High Pressure Underboom Sprays

In most conventional water spray systems of continuous mining machines, the spray nozzles are located along the top of the

miner boom. Laboratory studies have shown that the dust knockdown capability of a given water spray type is directly related to the operating water spray pressure. However, other studies have shown that nozzles located across the top of the boom and operated at higher pressure [>150 psi (7.2 kPa)] actually increase dust rollback to the operator.

The Bureau has recently completed laboratory tests using two nozzles located under the boom, operated at 2,500 psi (120 kPa), for a flowrate of 2.8 gpm (12.8 l/min). This system was compared with 20 nozzles, located across the top of the boom and operated at 100 psi (4.8 kPa), for a flowrate of 18.8 gpm (85.7 l/min). It is apparent from table 3 that both the conventional and the high pressure spray systems are equally effective at suppressing airborne respirable dust concentrations in the return. A more significant result is that when the two are combined, the efficiency for airborne dust suppression is the sum of the efficiencies for the systems when operated independently. Preliminary results indicate that the two spray systems act on two size fractions of respirable dust, with the finer spray droplets produced by the high pressure sprays suppressing dust in the lower size range. This is particularly interesting

FIGURE 4. Half-Curtain for Increasing Air Velocity Over Mining Machine Operator.

from the standpoint of quartz dust control since preliminary studies suggest that it is the smaller size quartz dust particles that deposit in the lungs causing the highest potential for silicosis.

Table 3. Efficiencies of conventional and high-pressure spray systems

Spray system	Water flow, gpm, (l/min)	Pressure, psi, (kPa)	Return effi- ciency
Conven- tional.	18.8 (85.7)	100 (4.8)	30 pct
High pressure.	2.8 (12.8)	2,500 (120)	30 pct
Combina- tion.	21.6 (98.5)	100/2,500 (4.8/120)	59 pct

5.2 Half-curtain face ventilation system

Federal regulations require a minimum mean entry air velocity of 60 fpm (0.3 m/s), and Bureau studies have shown that this is the minimum velocity required to effectively confine the dust cloud to the face and

and carry the dust into the return ventilation system. Many mines have difficulty meeting this requirement due to mine layout or entry cross-section.

The Bureau of Mines has recently evaluated a half-curtain face ventilation technique, first identified at the Colorado-Westmoreland Mining Company. The system consists of exhaust tubing or brattice and a piece of brattice cloth supported by two pogo sticks. The curtain is placed perpendicularly to the rib at or inby the operator's position and extends from roof to floor (fig. 4). This significantly reduces the cross-sectional area of the entry, thus increasing the air velocity over the operator. Results of a laboratory study indicate that the half-curtain performance depends largely on placement. The greatest improvement (86 pct) was achieved when the curtain was outby the brattice mouth and just inby the operator. Underground tests indicate that with the half-curtain in place, the respirable dust exposure of the operator was reduced by 50 pct. However, lab studies also indicate that the curtain may significantly increase recirculation on the off-curtain

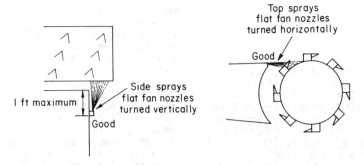

Excess air stirring is reduced when spray droplets move only a foot or so before impact. Good spray coverage involves covering only the cutter head.

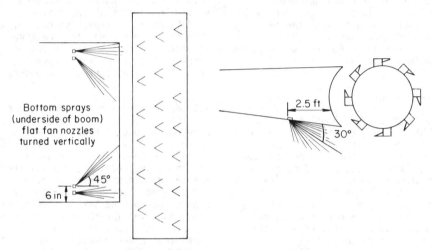

Underboom sprays located 2.5 feet back sustain much less damage than those
located at the front edge. Access is facilitated by keeping them close to the
sides.

FIGURE 5. Anti-Rollback External Water Spray System for Continuous Mining Machines.

side of the mining machine, inby the
curtain. Thus in gassy mines, caution must
be used to assure that hazardous
accumulations of methane do not build up
behind the curtain during the box cut.

5.3 Anti-rollback water spray system

Most spray systems consist of multiple
nozzles (15 to 30) located across the top
and along the side of the miner boom.
Research by the Bureau has shown that many
water spray systems produce enough air
turbulence to overwhelm the primary
airflow, causing dust rollback. Factors

that promote rollback are:

o High spray pressure [over 100 psi
   (4.8 kPa)], which increases air
   turbulence at the face more than it
   suppresses dust.

o Top and side sprays with wide-angle
   cones that overspray the cutter head
   or are set too far from the head.
   The longer the spray path, the more
   air is set in motion. Air movement
   may be desirable in a gassy mine, but
   it often stirs up more dust.

o Low primary airflow. Increasing

159

forward air velocity at the operator's position reduces rollback.

Tests have shown that a moderate spray pressure of 100 psi (4.8 kPa), measured at the nozzle, is a practical maximum. However, water flows should be as high as possible. A typical miner spray does most of its work in the first 12 in, thus top and side nozzles should be arranged for "low" reach and no overspray (fig. 5). Flat fan sprays are best suited for this application since the entire flow from the nozzle can be directed onto the cutter head. An improved spray system for continuous mining machines has been identified and evaluated by the Bureau. In recent underground trials, the new system reduced dust levels at the operator's position by 40 pct, as compared to conventional sprays.

5.4 Remote control

One of the most effective solutions to reduce dust exposure of the miner operator is to avoid dusty areas. If the operator can control his position with respect to the dust generating source and remain in uncontaminated air, his dust exposure will be minimal. One of the most effective ways to accomplish this is by use of remote control. The Bureau has conducted several underground studies to evaluate the impact of remote control on both exhaust and blowing face ventilation systems.

With exhaust ventilation, operator exposure comes from either dust rollback from the face, or intake (loading) sources. Although adequate mean entry air velocities and proper spray utilization can minimize rollback, the use of remote control can minimize the miner operator's dust exposure. During one underground evaluation, dust levels at the remote control location were 90 pct less than at the operator's cab. During a second underground study, tests results indicated a 97 pct reduction in the 8-hour respirable dust concentration, when comparing the usual cab position to the remote (radio controlled) position. The use of remote control could also increase productivity, by allowing greater advance.

With blowing ventilation, the machine operator is placed in the immediate return from the face. Although sections using blowing face ventilation usually have machine-mounted scrubbers, the operator can still be exposed to all the respirable dust escaping the dust collector system. Judicious selection of the remote operator position can result in significant reductions in the dust exposure. Underground studies were conducted to determine the dust reduction benefits associated with the remote operator location. Figure 6 shows the "return" and "intake" remote locations, with an average dust concentration of 3.1 mg/m^3 at the return location, and a 0.2 mg/m^3 level at the intake location. These results verify a 94 pct reduction in operator exposure was achieved by relocating the operator relative to the curtain discharge. It should be noted that local mining conditions with respect to gas emissions, MSHA guidelines regarding brattice setback, and roof control plans may preclude the most advantageous location from the viewpoint of dust exposure.

5.5 Improved roof bolter dust controls

A recent Bureau of Mines study has shown that while most of the roof bolter operator's dust exposure comes from upwind sources (C.M. machine), approximately one-third of the bolting machines underground allow a significant amount of dust to escape the dust collector system, thus contaminating the work environment. As previously mentioned, 25 pct of the continuous miner operator's quartz dust exposure can be attributed to dust from the bolting operation. One of the most efficient techniques to prevent worker exposure to dust from upwind sources is double-split face ventilation, but it has several problems from an operational and cost standpoint and is not applicable on a large number of underground sections.

5.6 Collector maintenance and bit selection

Dust in the blower exhaust is the most common and serious problem encountered on roof bolting machines. Common causes for this are damaged or improperly seated filters, and disconnected lines to the pressure check gage. Cloth bag (sock) type filters are often poorer dust collectors than pleated paper cartridge filters allowing dust to bleed through the system and out the exhaust. Poor quality bag-type filters may allow concentrations ranging from 5 to 7 mg/m^3 to continuously bleed through the exhaust, whereas, the

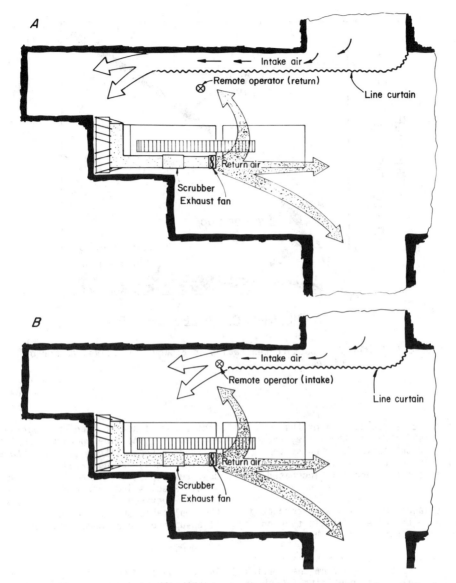

FIGURE 6. Potential Operator Locations Used with Remote Control and Blowing
Face Ventilation.

dust collection efficiency of the pleated paper filters increases as the cartridge is loaded with dust (fig. 7). Many of the roof bolter dust collectors showed accumulations of dust between the filters and blower, resulting from past or current filter leaks. With the filters removed and the access door open, this dust can be removed by backflushing the system with compressed air or by running the blower for several minutes.

Inadequate airflow to the chuck or bit can be detected as a visible plume from the collar of the drill hole. Air leaks into the system occur primarily at loose hose connections, through the pressure relief valve, and through poorly fitting dust

161

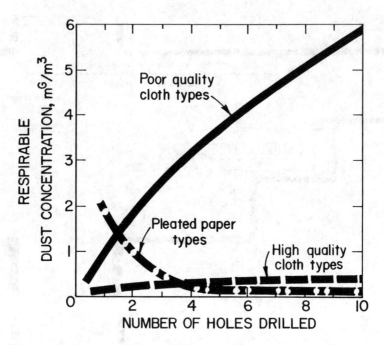

FIGURE 7. Filter Test Dust Concentration in Blower Exhaust vs. Number of Holes Drilled.

collector access doors. All of the systems examined during the survey showed more than 50 pct leakage. When system leaks were corrected and filters changed, airflow at the bit generally increased by 30 pct.

Extensive laboratory tests were conducted to determine the effects of bit type on dust escaping from the drill hole.

Shank-type bits allowed from three to ten times more dust to escape from the drill hole collar than "dust hog" bits. Most of this dust escaped during the first few inches of bit penetration. Typically, the dust hog bits generate one-fifth of the dust generated by the shank bits in the initial 12 in, and one-third of the dust over the full length of the hole. In relation to bit life expectancy, tests showed that the dust hog bits typically had an average maximum temperature of 210°F (99°C) as compared to 270°F (132°C) for the shank bits.

MSHA conducted an extensive cooperative program with mine management to evaluate the effectiveness of these recommendations from a compliance standpoint. During the MSHA survey, the mine operators replaced all duct hoses, filters, and the blower muffler, repaired the vacuum system and dust box seals, and cleaned the blower unit. Results indicate that 90 pct of the bolters surveyed were in compliance with the applicable standard after these clean-up procedures were instituted.

5.7 Upwind dust sources

The Bureau of Mines recently had the opportunity to evaluate a novel face ventilation system designed to minimize roof bolter operator exposure to dust generated by the continuous mining machine, operating upwind. This system consisted of an auxiliary fan and tubing to ventilate the active face, while a section of collapsible tubing on the discharge of the fan bypasses the roof bolter, going directly into the return (fig. 8). Results of underground studies indicate that the roof bolter operator's exposure with the collapsible tubing disconnected (fig. 8a) was 7.0 mg/m^3, whereas with the tubing in place (fig. 8b) it was only 0.7 mg/m^3 (90 pct reduction).

162

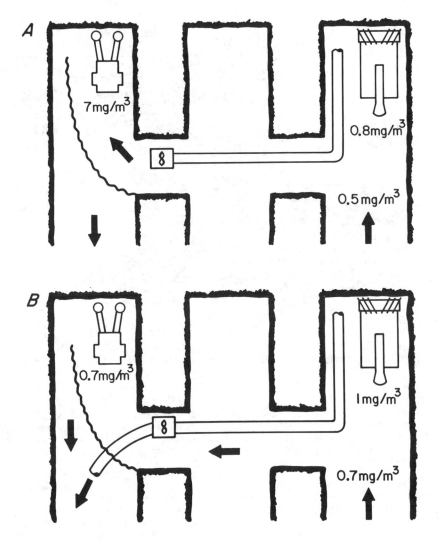

FIGURE 8. Auxiliary Fan-Tubing System Used to Direct Dusty Air Past Roof
Bolter Operator.

This system should be a viable and practical alternative when the roof bolter operator works downwind of the continuous mining machine, and is required only once during the right to left mining cycle.

6 CURRENT RESEARCH

The Bureau of Mines has initiated major contract research during both FY84 and FY85 to identify and evaluate concepts to control the quartz dust exposure of the continuous miner operator and downwind personnel. Four major techniques have been identified for controlling the dust exposure of the continuous miner operator. The location and efficiency of the underboom water spray system (fig. 9) must be improved to reduce dust rollback to the operator and to more efficiently suppress the secondary dust generated by the gathering and loading process. An evaluation is to be conducted to assess the efficiency and practicality of increasing the water flowrate to the

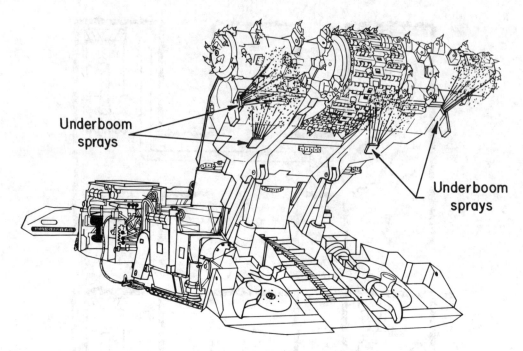

Underboom sprays

Underboom sprays

FIGURE 9. Potential Locations for Improved Underboom Water Spray System.

mining machine, in an effort to reduce the amount of dust which becomes airborne from the cutting and loading processes. This effort will also address dust generation from secondary sources, such as loading and transport, following up on the work using flight-conveyor sprays. Finally, improvements in mining practices, such as cutting roof and rock bands, will be identified and evaluated.

Addressing the problems of downwind dust levels, the Bureau will investigate a dust collector system located on the roof bolting machine (fig. 10) in an effort to provide localized clean air to the roof bolter operator. To reduce the amount of dust released into the return, the dust suppression potential of small water-powered dust scrubbers located on the continuous mining machine will be evaluated. Novel ventilation systems, such as the auxiliary by-pass system previously described and a tubing system such as that shown in figure 11, will also be evaluated.

To promote broad-based research oriented towards establishing new technological means for achieving systematic compliance

with more stringent dust standards, the Bureau recently completed a multiple contract award to research new concepts for reducing silica dust. This involves:

o Determining the size characteristics of silica dust particles in coal worker's lungs.

o Determining the relationship between the fraction of silica in the material being cut and in the respirable dust.

o Identifying and testing a novel air atomizing water spray system.

o Identifying and evaluating an optimum water spray/water powered scrubber system for the control of silica dust generation.

With this effort now underway, and through the cooperation of industry, MSHA and the Bureau, solutions should be identified which will allow compliance with applicable standards, thus protecting the health of the coal mining workforce, and ensuring a prosperous coal mining industry.

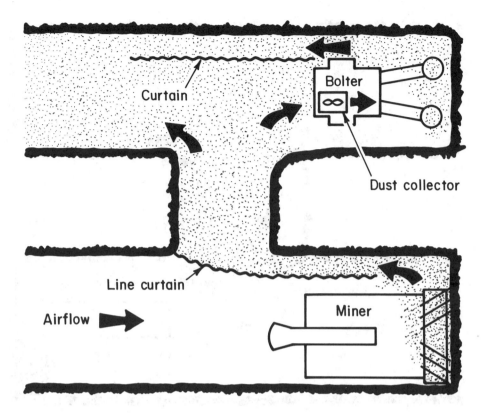

FIGURE 10. Dust Collector Mounted on Roof Bolting Machine, Designed to Provide Operator with Localized Clean Airflow.

7 SUMMARY

Reduced standards resulting from exposure to coal mine dust containing more than 5 pct quartz are currently one of the most serious health compliance problems facing the coal mine industry. Over 10 pct of active coal mine sections are on reduced standards due to quartz, with an average applicable standard for the underground designated occupation at 1.2 mg/m^3, while applicable standards for the surface designated work position average 0.8 mg/m^3. It is estimated that by the end of FY85 over 15 pct of all coal mine workers will be affected by overexposure to respirable quartz and that a significant percentage of mining sections will be placed under more stringent dust standards.

The primary source of quartz dust on surface mining operations comes from the drilling of overburden material and dust escaping the dust collector system of the highwall drill. On underground coal operations, the primary source of quartz is the continuous mining machine, cutting roof and floor, or rock partings in the seam. Unless properly maintained, the roof bolting machine also becomes a significant source of respirable quartz dust.

The Bureau of Mines has been actively pursuing improved technical controls to limit the quartz dust exposure of all underground coal mine workers. New or improved methods to reduce the amount of airborne dust generated by the continuous mining machine include: Improved water supply and spray system maintenance, modified coal cutting cycle, and high pressure underboom sprays. New or improved controls to reduce dust rollback include: Half-curtain face ventilation, anti-rollback water spray system, and remote control. Procedures to identify

165

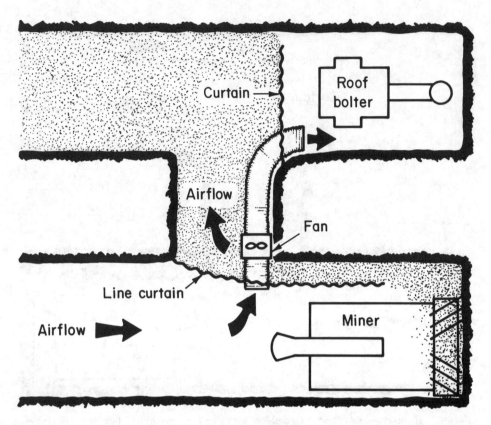

FIGURE 11. Auxiliary Fan-Tubing System, Designed to Provide Intake-Air Split for Roof Bolter Operator.

and correct the deficiencies of the roof bolter dust collector have been developed.

The Bureau of Mines has initiated an extensive contract research effort over the past two years to: 1) Evaluate concepts to control the quartz dust exposure of the continuous miner operator and helper; 2) identify strategies to control quartz dust levels downwind of the continuous mining machine; and 3) research new concepts for reducing silica dust in underground coal mines.

With the progress to date and with the effort currently underway, solutions to these problems should be forthcoming. However, only through the cooperation and dedication of industry, MSHA and the Bureau can solutions be formed that will allow compliance with applicable

standards, thus ensuring a healthy and productive coal mining industry.

8 REFERENCES

Bureau of Mines, Standard Method P-7, Determination of Free Silica (Quartz) in Ashed Respirable Coal Mine Dust by Infrared Spectroscopy.

Nesbit, Robert E., George E. Niewiadomski and Wayland M. Jessee, MSHS's Improved Strategy for Monitoring and Controlling Quartz Dust Exposures in Coal Mines, Presented at the American Industrial Hygiene Conference, May 19-24, 1985, Las Vegas, NV.

Maksimovic, S. D., and S. J. Page, Quartz Dust Sources During Overburden Drilling at Surface Coal Mines, presented at the

International Conference on Health of Miners, Pittsburgh, PA, June 2-7, 1985.

Control of Respirable Quartz on Continuous Mining Sections, BuMines Contract J0338077, Conoco Research, (contact J. Organiscak, PRC); Contract J0338033, BCR National Laboratories, (contact E. Divers, PRC); Contract J0338078, (contact S. Page, PRC).

How Twelve Miner Sections Keep Their Dust Levels at 0.5 mg/m^3 or Less, BuMines Technology News No. 220, July 1985.

Matta, Joseph E., Effects of Location and Type of Water Sprays for Respirable Dust Suppression on a Continuous Mining Machine, BuMines TPR 96, May 1976, 11 pp.

Dust Knockdown Performance of Water Spray Nozzles, BuMines Technology News No. 150, July 1982.

Jayaraman, N. I., F. N. Kissell, and W. E Schroeder, Modify Spray Heads to Reduce Dust Rollback on Miners, Coal Age, v. 89, no. 6, June 1984, pp. 56-57.

Code of Federal Regulations, Part 75.301-4, Velocity of Air, Minimun Requirements.

Kingery, D. S., et al, Studies on the Control of Respirable Coal Mine Dust by Ventilation, BuMines TPR 19, Oct. 1969, 13 pp.

Concepts to Control the Quartz Dust Exposure of the Continuous Miner Operator and Helper, BuMines Contract H0348031, Foster-Miller, Inc., (contact R. Jankowski, PRC).

Underground Tests of a New and Novel Water Spray System for Continuous Mining Machines, BuMines Contract H0199070, Foster-Miller, Inc., (contact N. Jayaraman, PRC).

Divers, E. F., N. I. Jayaraman, and J. L. Custer, Evaluation of a Combined Face Ventilation System Used with a Remotely Operated Mining Machine, BuMines IC 8899, 1982, 7 pp.

Better Roof Bolter Dust Collector Maintenance Reduces Silica Dust Levels, BuMines Technology News No. 198, March 1984.

Control of Respirable Quartz Dust on Continuous Mining Sections, BuMines Contract J0338033, BCR National Laboratories, (contact E. Divers, PRC).

Thaxton, R. A., Maintenance of a Roof Bolter Dust Collector as a Means to Control Quartz, proceedings of the Coal Mine Dust Conference, Morgantown, WV, Oct. 8-10, 1984, pp. 137-143.

Identification and Control Strategies to Control Quartz Dust and Dust Created During Auger Extraction of Coal, BuMines Contract H0348030, Foster-Miller, Inc., (contact N. Jayaraman, PRC).

Research New Concepts for Reducing Silica Dust in Underground Coal Mines, BuMines Contract H0358011, Foster-Miller, Inc., NIOSH, (contact R. Jankowski, PRC); Penn State Univ., (contract S. Page).

Evaluation of portable hand-held respirable dust monitors

RAYMOND R.GADOMSKI
Mine Safety & Health Administration, Pittsburgh, PA, USA

A program to evaluate portable hand held respirable dust monitors was conducted
by the Dust Division, Pittsburgh Health Technology Center, Mine Safety and Health
Administration. The instruments evaluated which utilize light-scattering techniques
included the Miniram (GCA Corporation) and the HAM (PPM Incorporated). Instruments
were evaluated by comparing results obtained between the two instruments and results
obtained from gravimetric sampling techniques. The evaluation was made in various
underground coal mines located in different coalbeds. The evaluation includes an
analysis of the effect of particle size distribution, material and dust
concentration on the comparison to gravimetric readings, as well as, an evaluation
comparing readings obtained from the two instruments. Results of this study provide
guidelines for using portable instantaneous instruments to evaluate dust generating
sources as well as the effect of various dust control techniques.

Portable coal dust/stone dust analyzer

P.A.KRZYSTOLIK, K.LEBECKI, P.BARTOŃ, R.DWOROK & J.ŚLIŻ
Central Mining Institute, Katowice, Poland

ABSTRACT: Description of a portable instrument with digital readout destined for quick measurements of solid incombustible content serving as protection against coal dust explosions. The discussed instrument meets the basic requirements of CFR 30, Part 29.

1. INTRODUCTION

The US Code of Federal Regulations, Title 30, determines, in paragraph 29.60, the basic requirements put to analyzers of solid incombustible content in coal dust.

According to these requirements a portable analyzer should be a self-contained unit suitable for service in underground coal mines equipped with a quantitative indicating device that is capable of indicating the incombustible content over the range of 50-100% incombustible and having batteries capable of failure-free operation.

The need for an instrument which quickly determines the incombustible content is very great where stone dusting is the basic method for safeguarding against coal dust explosions. This is true for the US coalmining industry and in a very large degree also for the Polish coalmining industry in which stone dusting, together with anti-explosion barriers, constitute the main protection against the propagation of coal dust explosions.

In result of long-year research work done in the Central Mining Institute an instrument was designed and its purposefulness tested in mines.

At present the Polish mining regulations require that in the Polish coal-mining industry 50-90% of solid incombustible matter be maintained, the percentage depending on the coal-dust explosion hazard degree.

The total length of stone-dusted zones is given in Table 1.

Table 1.

% of solid incombustible matter	Total length of stone-dusted zones km
90	35
80	416
70	131
50	600

The maintaining of the zones is becoming more and more difficult since growing mechanisation and concentration of output cause a more and more intensive deposition of new dust layers. The deposition rate is estimated at 2-100 $g/m^2/$ day in return air ways and at 2-80 $g/m^2/$day in galleries driven by heading machines.

To maintain the proper quantity of solid incombustible matter in stone dusted zones frequent checks must be done. The regulations do not state precisely how often the checks are to be carried out, they demand only that the required amount of incombustible matter be maintained continuously.

The quick shifting of the zones poses additional requirements to the stone dusting and the settling of new dust layers calls for smaller intervals between the checks. Due to this situation the determination of the content of solid incombustible matter is becoming a more and more difficult problem for colliery laboratories doing check analyses. These analyses depend on combusting dust samples taken underground during three hours at a temperature of 480°C and on determining the loss of mass. In many collieries more than 1000 samples are to be analysed each month.

The need has arisen to develop a method for quick determination of solid incombustible content without the necessity of combusting, weighing and so on.

In connection with the progress of isotope technique the idea emerged to take advantage of the radiometric methods - based on interaction of beta and gamma radiation with material - for determination purposes.

2 PHYSICAL BASIS OF THE METHOD

The following variants were taken into consideration:
- absorption of gamma radiation from the sources ^{241}Am ^{137}Cs
- back-scattering of gamma radiation from the sources
  ^{241}Am ^{60}Co
- back-scattering of beta radiation from the sources ^{201}Ta ^{90}Sr.

The intensity of penetrating radiation or back-scattered radiation depends on the effective atomic number of the material interacting with radiation. The basis for introducing the radiometric method is the predominance of one constituent of the stone dust used for dusting.

According to the requirements of the Standard, stone dust should contain over 90% of calcium carbonate. A typical analysis of the stone dust composition and the effective atomic numbers of the constituents are shown in Table 2.

Table 2

Stone dust constituents	Content %	Effective atomic number Z_{ef}
$CaCO_3$	90	12.56
$MgCO_3$	1.2	14.30
Fe_2O_3	0.1 - 1	20.32
Al_2O_3	0.2	10.60
SiO_2	5	10.80

The atomic number of stone dust taken as a whole can be estimated at 12, i.e. twice as much as that for coal C, which, at a square or stronger dependence of the intensity of penetrating or back-scattered radiation on Z_{ef}, ensures good sensitivity and accuracy of the method.

From among the constituents of stone dust the ferrous oxide is characterized by the greatest atomic number. By the way ferrous oxide occurs not only in stone dust but also as a constituent of natural ash. Thus ferrous oxide content over 5% is an interfering factor, chiefly for absorption and back-scattering of low-energy gamma quantums for which interaction

with material is proportional to
Zef 4. Therefore discrepancies between results obtained by the radiometric method and by combustion
may reach 6%.

To reduce the influence of heavy
additives it was decided to use
the method of back-scattered beta
radiation with a maximum energy of
about 1 MeV. This method secures
good rectilinear dependence betwe-
en the intensity of back-scattered
radiation I and the content of solid incombustible matter /n/.

From the available radioactive
sources ^{90}Sr - ^{90}Y has been chosen
having a sufficiently long half-
life period /33 years/. Taking into consideration radiologic protection only sources having an activity of 5 mCi /185 mBg/ or 2 mCi
/74 MBq/ were used although the
inclination of straight line J /n/
increases with the activity of the
source. It was found that the best
results are obtained at a disper-
sion angle of $\varphi \approx \alpha \approx 20°$.

Measurements geometry is shown
in Fig.1.

Basing on thorough investigations of the phenomenon of beta
radiation back-scattering on stone dust an instrument called
INFLABAR PC has been designed
which will be discussed below.

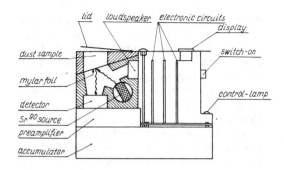

Fig.2. Internal view of INFLABAR PC

3 DESIGN OF THE INSTRUMENT

The internal view of INFLABAR PC is
shown in Fig.2.

The electronic system based on
integrated circuits ensures correct
operation of the instrument and a
direct digital readout of solid incombustible matter content.

The switch on the back wall of
the instrument has two positions.

Beta radioactive stront 90 is
the radiation source. The radiation of stront falls on the dust
sample in the container and after
back-scattering falls on the semi-
conductive radiation detector
which is specially made to the requirements of this instrument in
the Institute for Nuclear Problems.

The radiation source turns round
together with the cover of the container with the sample. Thanks to
this during the change of the
sample the operator is not exposed
to radiation.

The radiation source, detector
and sample container are housed in
one block made of organic glass
which ensures steadfast-ness of
mutual position of these elements,
small dimensions and small distances between the elements.

On the back wall of the instrument there is a switch having two
positions. Position "Calibration"
serves for regulating zero and the
accuracy of the instrument. In

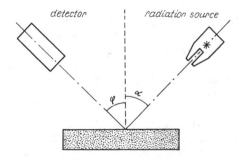

Fig.1. Measurements geometry.

this position 9 subsequent measurements are done at a shortened time and at only two significant digits. The tenth measurement has full time and accuracy. In position "Measurement" one measurement without calibration is done. Position "Calibration" enables quick calibration of the instrument. After finished measurement the readout switches-off automatically.

Technical data of the instrument are as follows:

- Measurement range
  50-100% solid incombustible matter content
- Measuring time   about 30 s
- Ambient temperature range
  278-308K
- Dust sample mass   abt.25 g
- Supply
  Cd-Ni battery, 7.2 V, 3.5 Ah
- Supply autonomy
  minimum 250 measurements
- Radiation source
  stront 90, activity 5 mCi
- Maximum radiation intensity
  not greater than 2.5 mR/h at a distance of 5 cm from the instrument
- Mass of instrument   3 kg
- Dimensions
  215x80x175 mm
- Electric system
  intrinsical safety II BI /category i_a according to IEC and CENELEC/

4 MEASURING PROPERTIES OF INFLABAR PC

Due to the accepted principle of operation the indications of the instrument can undergo interference, this means that they can differ from values found by the combustion method.

Interference can be caused by:

- Moisture of dusts over 5% adherent moisture,

- Occurrence of higher contents of ferrous oxides or of rust in dust samples.

Moisture exerts two kinds of influence. Firstly, water lowers the mean atomic number of the sample material and therefore the indications of the instrument are much lower. It is relatively easy to remove the influence of moisture simply by shaking the dust sample together with admixed silica gel grains for about 3 minutes. The second interference, resulting from high moisture content of the sample, is lumping of the grains, formation of cavities on the surface of the foil separating the sample from the detector.

To remove this factor a vibrator has been installed operating in the acoustic band but its activity is of little effectiveness at moistures above 6-7%.

In Polish collieries the problem of natural moisture is not severe. In seams occuring at shallow depths /up to 400 m/ and providing 30% of the mined coal, the minimum moisture content of dusts is about 3.3%, the maximum 24.4% and the medium content is 16.0%. In deeper lying seams total moisture of dust ranges from 1.4 to 15% and the medium one is 5%.

Stone dusting is done most often in seams having natural moisture of up to 5%.

Below a much detailed description of laboratory investigations of the metrologic properties of INFLABAR PC and of the influence which interfering factors exert on the measurement is given.

4.1 Reproducibility of measurement results

Two dust samples in air-dry condition were used, 85% of the samples consisting of grains in the size below 75 μm.

Sample A: incombustible content
/determined by the com-
bustion method/
n_s = 54.5%
total moisture
w_c = 3.5%

Sample B: n_s = 99.9% w_c = 0.7%

The reproducibility of measurement results was tested by means
of two procedures:
x - placing the sample once and
measuring 20 times
y - each time placing a new sample
of the same dust /two persons
in turn/

The results are shown in Table 3
where n_i denotes indications of
INFLABAR PC.

In case of placing the dust sample once, the maximum deviations
of individual results do not surpass 1.3% of solid incombustible
matter content and the mean second
power error does not surpass 1% of
the mean value.

In case of placing each time a
new sample, the corresponding parameters are 2.2% and 1.2% respectively.

The applied Student's t-test
shows that the hypothesis of the
equality of INFLABAR PC indications and results obtained by the
combustion method is better proved
in case of high incombustible
matter content. The reason is that
in case of higher incombustible
content the influence of interfering factors /moisture, ferrous
oxides/ is smaller. The differences between indications obtained
with one-time and each-time placing of the sample result from
small differences in the bulk
density of the sample.

4.2 Influence of dust granulation
on INFLABAR PC indications

Investigations were done on six
dust samples constituting two series with growing content of solid

incombustible matter, differing in
grain sizes. All dust samples were
air-dried.

Results are shown in Table 4.

Increasing the grain sizes of
coal dust causes higher indications if the instrument has been
calibrated on samples containing
dust of smaller grain sizes. Therefore calibration should be done
on samples of dust having grain
sizes typical for coal-mining conditions.

Deviations of INFLABAR PC indications from values obtained by
the combustion method do not surpass +3% of solid incombustible
content.

4.3 Influence of moisture on
INFLABAR PC indications

As mentioned above this influence
is very great. It is shown in
Fig.3.

Good results are obtained with
drying the samples, even the very
wet ones, by the help of silica
gel. The addition to the dust
sample of silica gel grains with
over 1 mm diameter in a mass proportion of 1:1 and subsequent
shaking of the sample for 3 minutes brings about drying of the

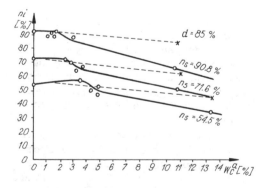

Fig.3. Influence of moisture on
INFLABAR PC indications

175

Table 3. Reproducibility of measurement results obtained by using INFLABAR PC

No	Dust A /n_s = 54.5%/		Dust B /n_s = 99.9%/	
	n_i X	n_i Y	n_i X	n_i Y
1.	54.9	55.2	98.6	100.8
2.	56.0	56.1	99.3	98.9
3.	55.2	55.0	100.6	98.6
4.	56.1	54.8	98.6	98.7
5.	55.4	55.9	100.1	98.2
6.	55.7	56.8	99.6	97.6
7.	55.9	54.9	100.3	98.6
8.	56.2	56.2	101.0	97.2
9.	56.3	55.3	99.6	100.4
10.	56.5	56.3	98.8	97.9
11.	56.1	56.2	99.8	99.8
12.	55.9	55.1	100.0	100.1
13.	55.2	54.5	100.2	100.6
14.	54.8	55.5	99.8	100.7
15.	55.2	56.0	99.8	100.6
16.	56.4	56.8	100.3	100.2
17.	54.9	55.2	100.7	98.8
18.	55.0	56.0	100.8	99.7
19.	56.2	55.4	99.2	100.1
20.	55.9	55.1	101.0	99.4
	n_i X = 55.6	n_i Y = 55.6	n_i X = 99.9	n_i Y = 99.4
	x = 0.14	y = 0.15	x = 0.17	y = 0.2
	/t/ = 8.29	/t/ = 7.74	/t/ = 0.06	/t/ = 1.92
	0.001	0.001	0.1	0.073

dust without lumping which is characteristic for drying by means of a drier.

The necessary screening-off of the silica gel substantially improves the structure of the dust. One dust sample can be used even twice without the need of regeneration.

In order to check the effectiveness of drying dusts by means of silica gel, samples of "d85" dust having a 54.5% solid incombustible matter content and samples of "d25" dust with 72.2% solid incom-

bustible content were wetted to have moisture contents of 14.4 and 12.2%. About 40 g of wet dust was put into a 150 cm3 container, adding 1-5mm-diameter silica gel grains at a voluminal ratio of 1:1. The container was shut, the contents shaken and subsequently sieved through a screen with 1 mm apertures.

After 3 minutes of shaking the moisture content in dust samples was 3.3 and 2.5%. After 1.5 minutes of shaking it was 3.2 and 4.8% respectively. If to analogous sam-

Table 4. Indications of INFLABAR PC for dusts with grain sizes "d25" /25% below 75 μm/ and "d85" /85% below 75 μm/

No	"d25"			"d85"		
	$n_s=54.5\%$ $w_c = 3.6\%$	$n_s=71.6\%$ $w_c = 2.4\%$	$n_s=90.8\%$ $w_c = 1.2\%$	$n_s=54.5\%$ $w_c = 3.5\%$	$n_s=71.8\%$ $w_c = 2.4\%$	$n_s=90.6\%$ $w_c = 1.2\%$
1.	54.8	71.9	92.3	54.9	69.8	88.2
2.	55.5	73.0	93.1	56.0	70.6	89.9
3.	56.2	73.0	92.6	55.2	70.5	88.9
4.	56.1	72.9	92.1	56.1	72.0	88.9
5.	55.8	73.2	94.1	55.4	71.4	89.7
6.	55.5	72.9	93.4	55.7	70.6	89.7
7.	55.4	73.5	93.4	55.9	70.9	90.5
8.	54.3	72.0	93.2	56.2	71.2	90.7
9.	54.8	72.7	92.5	56.3	70.2	90.3
10.	56.1	72.7	93.1	56.5	69.9	91.0
	55.5±0.56%	72.8±0.49%	92.9±0.61%	55.8±0.53%	70.7±0.69%	89.8±0.95%

Table 5. INFLABAR indications for samples of wet dust dried by the help of silica gel during 1.5 minutes

	n_2	w_c	n_i
d85	54.5	13.3	50.1
	71.6	10.8	68.2
	90.8	10.6	86.4
d25	54.5	11.8	55.7
	71.6	10.8	70.3
	90.8	9.2	88.0

ples a two-times greater volume of silica gel was added than after 1.5 min - shaking moistures of 2.8 and 2.4% were obtained.

In table 5 INFLABAR's indications for dusts of various solid incombustible matter contents and various moisture contents are given.

In conditions of Polish mines the drying of samples is necessary only in particular cases. The majority of seams are namely dry enough so that the moisture of dust remains in balance with the moisture of air.

The humidity of dusts containing 70-80% of solid incombustible matter does not surpass 5%.

If the INFLABAR has been calibrated with taking into account medium humidity of dusts then the maximum deviation in determining solid incombustible matter content does not surpass ±3% of the values obtained by combustion.

4.4 Environmental and strength testing of the INFLABAR instrument

In accordance with Polish Standards the following tests have been done to state:

- resistance to cold /two hours at a temperature of 278+3K, eight hours at a temperature of 263 +3K and bringing to room temperature/,

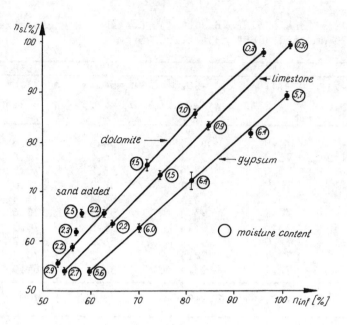

Fig.4. INFLABAR PC indications for mixtures of coal
dust with dolomite, limestone and gypsum

- resistance to dry heat /two hours
  at a temperature of 313K/,
- resistance to wet heat /ten days
  at a temperature of 313K and
  moisture of 93%/
- impact strength /free falling
  from a height of 0.25 m onto
  smooth concrete/,
- vibration resistance /amplitude
  of 0.15 mm, 50 Hz frequency,
  accelerates 2 g during 30 minu-
  tes/.

During none of those tests has
the instrument shown unfavourable
changes of operation parameters.

Tests were also done with
INFLABAR PC indications concerning
mixtures of coal dust with dolomite and gypsum /in dry condition/.
Results are shown in Fig.4.

5 FIELD INVESTIGATIONS OF INFLA-
   BAR PC

The field investigations were done
at the 1 Maja colliery, in which
- due to high gas and dust hazards-
regulations demand stone-dusting
of all workings; out of those 40
km of workings are stone-dusted to
a 80% content, the remaining, over
100 km, are stone-dusted to a 50%
content. The dust is dry and the
number of samples analysed monthly
in the laboratory by combustion
amounts to about 1000.

The samples were taken by the
"stripe" method which is valid in
Poland. This method consists in
taking dust samples in 3-5 places
of the checked stone-dusted zone,
the samples being taken in 20 cm
wide stripes from the whole circumference of the working. The
distance between the stripes is
15-30 cm.

At the spot the dust is sieved
through a screen with 3 mm apertures and before the laboratory
test through a screen with 1 mm
apertures. When measuring the solid incombustible matter content

178

by means of the INFLABAR the screening is done directly at the place of measurement. Samples from all the stripes are mixed carefully and form one representative sample.

During the first stage 102 samples taken from various workings were compared, their incombustible content being determined in the laboratory by combustion and by the INFLABAR PC instrument. Those samples had an incombustible content ranging from 64.5 to 100%. By the help of the χ^2 test the veracity of the hypothesis of the equality of results obtained by both methods was proved.

The critical χ^2 value was 8.09 which ensures the veracity of the hypothesis at a confidence level of 0.1 χ^2 = 10.64%. Then measurements were done below ground. They were performed in taking samples according to regulations and from the INFLABAR $/n_i/$. The conformity proved to be very good.

Thus at present the usefulness of the INFLABAR as a measuring instrument is unquestioned because the instrument gives indications very quickly, without the need of long lasting preparations of the samples /drying in the laboratory/.

5 CONCLUSIONS

1. The INFLABAR PC instrument meets the basic requirements of CFR 30/29 and of Polish regulations and is suitable for use in mines.

2. The instrument can be used for checking the state of stone dusting according to the so far valid methodology of taking samples. It also provides the possibility of quick and easy detection of places in which stone dusting is not adequate.

The INFLABAR is very useful for checking during the operation of pouring out the stone dust itself.

The possibilities of the instrument will become evident during its operation on a large scale.

REFERENCES

Krzystolik P.A., Lebecki K., and Śliż J., 1979
Radiometric method for quick analysis of incombustible content in coal dust deposited in mine workings – 18th International Conference on Scientific Research in the Field of Safety at Work in Mining Industry – Cavtat /Dubrovnik/ 1979, Paper B-5.

5. Fans and shafts

An analysis of mine fan irregularities relative to underground conditions, ventilation, and potential fan defects

STEPHEN R.HARRISON
Consolidation Coal Company, Pittsburgh, PA, USA

VIC KUTAY
Consolidation Coal Company, Moundsville, WV, USA

Irregularities in mine fan performance have long been an enigma for mine management. Unusual fan operating performance (such as fluctuations in fan pressure) can be a symptom of a serious mechanical malfunction, normal changes in natural ventilation pressure, or a cycling condition in the mine.

Generally, the only continuous fan operating variable immediately available for evaluation is the weekly pressure recording chart. Other critical fan monitoring is generally controlled by limit switches that terminate fan operation when mechanical failure is imminent. (This can result in expensive production downtime.) These limit switches (and the attached gauges) are often checked daily, but this is a limited diagnostic tool. Cyclic fan aberrations, particularly with a low frequency occurrence, can easily be missed during periodic checks.

In this paper, a systematic approach for isolating the cause and severity of irregular mine fan operation is addressed. Several case studies are presented where "trial and error" analysis was used to diagnose and resolve potential mine fan problems before they become fan failures.

A cost-basis justification for streamlining service-equipped ventilation shafts

JERRY L.FULLER
Mine Safety & Health Administration, Denver, CO, USA

ABSTRACT: This paper contains a cost-basis justification for streamlining ventilation shafts equipped with internal service structures. The particular shaft design used in this analysis was presented by Harrison Western Corporation. A plan drawing is included in the appendix as figure A-1. The method of analysis was presented by V.A.L. Chasteau and D. Gillard, both of the National Mechanical Engineering Research Institute of the Council for Scientific and Industrial Research at Pretoria, South Africa. Use is made of their method, as applied to the Harrison Western 18-foot-2-inch-diameter shaft design, to show the considerable ventilation cost savings obtainable through the streamlining of internal shaft structures.

1. INTRODUCTION

In the majority of cases, shafts must, out of necessity, have a dual role in the mining process. Not only must they provide the necessary services for the mining operation but they must serve as a major air-course as well. From a ventilation viewpoint, such a dual role is extremely costly.

The major reason for increased ventilation costs is a greater resistance to airflow caused by electrical cables, water and compressed air lines, skip and cage guides and supports, and other equipment in the air-carrying shaft. In order to overcome this increased resistance, additional fan motor horsepower is required and, consequently, a greater quantity of electrical power is used. This higher resistance and subsequent cost increase can be reduced by streamlining the included equipment.

A large amount of experimental and practical application work on streamlining of shaft structures has been done in South Africa. Shafts in the South African gold mines are, in general, greater than 20 feet in diameter. Since ventilation costs are high, the subject of streamlining has received careful study. The method of analsis proposed by Chasteau and Gillard (2) is claimed, by the authors, to be the most accurate having an error of $\pm$ 15 percent in the final resistance value. Their method will be used for this analysis.

2. BASIC FORMULAS

The following list contains an explanation of the abbreviations used in the formulas:

A_b = frontal area of form drag components (these structures whose drag is due to their frontal area, i.e, I-beams, pipe flanges, etc.), sq ft,

A_{bt} = total blockage area of all shaft equipment = $A_b + A_s$, sq ft,

A_n = nominal cross-sectional area of the shaft = $\frac{\pi}{4}D^2$, sq ft,

A_s = cross-sectional area of pipes and guides, sq ft,

A_t = effective true shaft area = $A_n - A_s$, sq ft,

B = empirical constant,

C_d = drag coefficient of shaft equipment,

D = nominal diameter of a circular shaft, ft,

d = I-beam frontal width, ft,

h = web depth of I-beam, ft,

k = Atkinsons's friction factor, $lb\text{-}min^2/ft^4$,

L = set spacing, ft,

m = total number of form drag components in a set,

N_i = a factor allowing for items not recurring at exactly one set spacing,

P_n = nominal shaft perimeter = πD for a circular shaft, ft,

P_t = perimeter contributing skin friction, P_t = P_n plus individual perimeters of pipes and guides, ft,

S = empirical constant,

W = set spacing coefficient in wall friction calculations (subscripted "b" for unstreamlined shafts and subscripted "s" for streamlined shafts),

λ = resistance coefficient, dimensionless (2),

λw = resistance coefficient for walls in an equipped shaft,

λwo = resistance coefficient for walls in an unequipped shaft.

The general formula for the resistance coefficient is:

$$\lambda_{total} = \frac{SD \sum_{i=1}^{m} N_i\, C_{d_i}\, A_{b_i}}{LA_t} \left(\frac{A_n}{A_t}\right)^2 \left(1 + \frac{2.5}{A_t} \sum_{i=1}^{m} N_i\, C_{d_i}\, A_{b_i}\right)$$

$$+ \lambda_{wo}\, \frac{P_t}{P_n} \left[1 + BW \left(\frac{A_{b_t}}{A_n}\right)^2\right] \quad (1)$$

where the subscript "i" is for individual component values. A listing of the values used for the Harrison Western shaft is included in table A-1 in the appendix. Additional values shown as summations in table A-1 or calculated for the formula are:

A_b = 29.82 ft^2,

A_{bt} = A_b + A_s = 33.27 ft^2,

A_n = $\frac{\pi}{4}$ (18.17)^2 = 259.20 ft^2,

A_s = 3.45 ft^2,

A_t = A_n - A_s = 259.20-3.45 = 255.75 ft^2,

D = 18 ft 2 in. (18.17 ft),

L = 10 ft,

P_n = 18.17 x π = 57.07 ft.

The remaining values used in the calculations are those suggested by Chasteau and Gillard or estimated. The values for the empirical constants, S and B, recommended were 1 and 176.4, respectively, for streamlined and 1.18 and 253.3, respectively, for unstreamlined. The value of the set spacing coefficient (W) was taken from the authors' graph reproduced as figure A-2 in the appendix. The values of W_b and W_s were chosen from the graph as 2.8 and 1.5,

respectively, using an average weighted L/d value of 20.11 (see table A-1). The drag coefficient (C_d) values were also found from graphs supplied by Chasteau and Gillard reproduced as figure A-3 and figure A-4 in the appendix.

The value of Atkinson's friction factor (k) was assumed to be 15 lb-min^2/ft^4 for a smooth "clean" concrete-lined shaft. The resistance coefficient (λ) and the friction factor (k) are related in British usage by:

λ = 12.3K, and, in U.S. usage, by

λ = 1.23 x 10^{-3} k (2) (the decimal point discrepancy is due to British practice of expressing Q in units of 1,000 cfm; U.S. practice specifies Q in units of 100,000 cfm. The air pressure varies as the square of the quantity, therefore, British resistances equal U.S. resistances times 10^{-4}).

Values for λ suggested by Chasteau and Gillard varied from 0.010 to 0.022 which correspond to values of k of 8 to 18. A friction factor of 15 corresponds to a λ of 0.01845 and this value was used as λ_{wo} in the calculation. The value of P_t, the perimeter contributing skin friction, was figured assuming an enclosed ladderway and a brattice between the skip and cage, both of which would contribute skin friction. Both sides of the brattice were counted as contributing skin friction. To this number was added the perimeter of the pipes and guides with a resultant value of P_t of 108.35 ft. The total effective perimeter of two guides mounted opposite one another was taken as the perimeter of the exposed guide surfaces. For a single guide, the entire perimeter was multiplied by 0.8 to obtain an effective perimeter.

The punched metal or lattice grating used for ladderway landings was treated as a compressed solid member and as being streamlined.

These values were then used to solve the Chasteau–Gillard equation.

Unstreamlined

$$\lambda_t = \frac{SD \sum_{i=1}^{m} N_i\, C_{d_i}\, A_{b_i}}{LA_t} \left(\frac{A_n}{A_t}\right)^2 \left(1 + \frac{2.5}{A_t} \sum_{i=1}^{m} N_i\, C_{d_i}\, A_{b_i}\right) \quad (3)$$

$$+ \lambda_{wo}\, \frac{P_t}{P_n} \left[1 + BW_b \left(\frac{A_{b_t}}{A_n}\right)^2\right]$$

$$\lambda_t = \frac{1.18 \times 18.17 \times 22.10}{10 \times 255.75} \left(\frac{259.2}{255.75}\right)^2 \left(1 + \frac{259.2}{255.75} \times 22.10\right)$$

$$+ 0.01845 \times \frac{108.35}{57.07} \left[1 + (235.3)\,(2.8) \left(\frac{33.27}{259.20}\right)^2\right]$$

$\lambda_t = 0.19 \, (1.03) \, (1.22) + 0.04 \, (11.85)$

$\lambda_t = 0.71.$

Since, from equation 2,

$\lambda = 1.23 \times 10^{-3} \, k,$ therefore,

$k = 579 \, \text{lb-min}^2/\text{ft}^4.$

Streamlined

$$\lambda_t = \frac{SD \sum\limits_{i=1}^{m} N_i \, C_{d_i} \, A_{b_i}}{LA_t} \left(\frac{A_n}{A_t}\right)^2 \left(1 + \frac{2.5}{A_t} \sum\limits_{i=1}^{m} N_i \, C_{d_i} \, A_{b_i}\right)$$
$$+ \lambda_{wo} \frac{P_t}{P_n} \left[1 + BW_b \left(\frac{A_{b_t}}{A_n}\right)^2\right] \tag{4}$$

$$\lambda_t = \frac{1 \times 18.17 \times 11.71}{10 \times 255.75} \left(\frac{259.20}{255.75}\right)^2 \left(1 + \frac{2.5}{255.75} \times 11.71\right)$$

$$+ \, 0.01845 \times \frac{108.35}{57.07} \left[1 + (176.4)\,(1.5)\left(\frac{33.27}{259.20}\right)^2\right]$$

$\lambda_t = 0.08 \times 1.03 \, (1.11) + (0.04) \, (5.36)$

$\lambda_t = 0.31.$

Again from equation 2,

$\lambda_t = 1.23 \times 10^{-3} k,$ therefore,

$k = 249 \, \text{lb-min}^2/\text{ft}^4.$

3. Cost Comparison

As a basis for figuring costs, the air input horsepower will be used with the accepted value of $65.35 per horsepower per year at 1¢ per kilowatt-hour. The additional cost of a larger fan and related equipment to overcome larger resistances will not be considered.

For a reference dollar amount, the cost of ventilating a clean shaft will be calculated. Using the friction factor (k) of 15, the shaft resistance is found by (4):

$$R = \frac{kLO}{5.2A^3}$$

where

R = resistance, Kingerys,
k = friction factor,
   lb-min^2/ft,
L = length, ft,
O = perimeter, ft,

and

A = area, sq ft. (5)

Using the values for the proposed Harrison Western shaft with a length of 3,000 ft:

$$R = \frac{(15)\,(3000)\,(57.07)}{5.2 \, (259.20)^3}$$
$$R = 0.03 \text{ Kingerys.}$$

Assuming a quantity (q) of 400,000 cfm is required, the resultant fan pressure (at 0.075 lb/ft^3 air density) is found from:

$$H = RQ^2,$$

where

H = pressure differential,
   in. w.g.,
R = resistance, Kingerys,

and

Q = air quantity, 100,000 cfm. (6)

Small "q" is used for an air quantity expressed in cfm units whereas Q represents an air quantity expressed in 100,000 cfm units. Substituting the values obtained for R and Q:

$$H = 0.03\,(4)^2$$
$$= 0.48 \text{ inches of water gage (in. w.g.).}$$

Because the air horsepower (A.H.) is independent of fan motor configuration, considerations for motor efficiencies, fan efficiencies, and fan motor drives can be eliminated. The air horsepower is found from:

$$\text{A.H.} = \frac{H \times q}{6350}. \tag{7}$$

where

A.H. = air horsepower, hp,
H. = pressure differential,
   in. w.g.,
q = air quantity, cfm,
6350 = conversion factor.

For a q of 400,000 cfm, the resultant air horsepower is

$$\text{A.H.} = \frac{0.48 \times 400 \ldots 00}{6350}$$
$$\text{A.H.} = 30.24 \text{ horsepower (HP).}$$

At $65.35 per horsepower per year at 1¢ per kilowatt hour, this ventilating capacity would cost $1,976 per year.

A similar cost analysis can be made for the streamlined and unstreamlined equipped shafts using the same procedure. For the unstreamlined equipped shaft, the calculated k factor was 579 lb-min^2/ft^4 and the resultant resistance is found by substituting into equation 5 as follows:

$$R = \frac{(579)(3000)(57.07)}{5.2\ (259.20)^3}$$

$$R = 1.09 \text{ Kingerys.}$$

The required fan pressure for 400,000 cfm against this resistance is:

$$H = 1.09\ (4)^2$$

$$H = 17.52 \text{ in. w.g.}$$

The air horsepower required would be:

$$\text{A.H.} = \frac{17.52 \times 400,000}{6350}$$

$$\text{A.H.} = 1,103 \text{ hp.}$$

The cost of providing this quantity of air on an annual basis would be:

$$1,103 \text{ hp} \times \$65.35/\text{hp-yr}$$
$$= \$72,100.$$

For streamlined equipment, the k factor derived was 249 lb/min^2/ft^4 and the resultant resistance is:

$$R = \frac{(249)(3000)(57.07)}{5.2\ (259.2)^3}$$

$$R = 0.47 \text{ Kingerys.}$$

The required fan pressure for 400,000 cfm against this resistance is:

$$H = 0.47\ (4)^2$$

$$H = 7.53 \text{ in. w.g.}$$

The air horsepower required is:

$$\text{A.H.} = \frac{7.53 \times 400,000}{6350}$$

$$\text{A.H.} = 475 \text{ hp.}$$

The annual cost of this value of horsepower is :

$$475 \text{ hp} \times \$65.35/\text{hp-yr}$$
$$= \$31,000/\text{yr.}$$

4. DISCUSSION

At this point, some precautions should be given. The calculations presented here are not intended to be, in any way, related to a fan sizing exercise. The effects of motor and fan efficiencies and motor drive configurations were not considered. However, in attempting to choose a fan for a particular application, these effects must be taken into account as they do increase the horsepower requirements. Large fans are available which will produce the required quantity of 400,000 cfm with a pressure capability of over 15 in. w.g. Such fans generally have horsepower requirements in excess of 1,300 hp and efficiencies around 80 percent.

The effect of the additional resistance to airflow caused by the mine circuit itself was not considered. The mine circuit resistance will also affect the fan pressure and horsepower requirements. Generally, the resistance of shafts is considered to be from 15 to 30 percent of the total mine airflow resistance (3).

The effect of skips and cages moving in the shaft was not considered. The process of choosing an adequate fan for a specific shaft application requires that the fan and its intake airway be sized and configured so that the fan will not stall when skips and cages are in the most critical positions. Surface mounted fans with connections to the shaft may stall when a shaft conveyance partially obstructs the fan's intake or exhaust shaft airway. Another critical situation develops when all conveyances are at the same location inside the shaft, resulting in maximum blockage. Benefits obtainable by streamlining skips and cages and some equipment were also not considered.

All the above-mentioned effects and omissions would serve to increase the potential cost saving value above the calculated $40,000 per year amount.

5. CONCLUSION

Considerable increases in resistance and ventilating costs can be expected when air shafts must also serve as service shafts. Typically, a 30-fold increase in the friction factor caused by the installed services is in agreement with values measured in South African mines. In those mines, the value of λ, the resistance coefficient, for a "clean" shaft has been measured at 0.010 and at 0.303 for a shaft equipped in a similar fashion to the Harrison Western shaft. Streamlining only the I-beam members can represent a considerable saving in ventilation cost (5). In other applications, the Russians reuse discarded conveyor belting as an inexpensive aerodynamic cover for the I-beam (1). Experiences in South Africa and the Soviet Union find payback periods for streamlining costs of less than 4 years to be commonplace.

6. ACKNOWLEDGMENT

The permission granted by Harrison Western Corporation for use of their shaft design and the permission granted by the Mine Ventilation Society of South Africa secretaries for use of the copyrighted material is gratefully acknowledged.

7. REFERENCES

1. Abramov, F. A., Shinkovskii, V. A., and Dolinskii, V. A. Shiszenie Aerodinamiczeskogo Soprotiv lenia Stvolov put k Ekonomii Energeticzeskikh Resursov. (Reducing Aerodynamic Resistance of Shafts-A Way of Energy Resource Conservation). Ugol (Coal), v.2, Feb. 1981, pp. 27-28.

2. Chasteau, V.A.L. and Gillard, D. The Prediction of the Resistance to Air Flow of Mine Shafts Equipped with Internal Structures (Part I). The Journal of the Mine Ventilation Society of South Africa, v. 18, No. 10, Oct. 1965, pp. 133-146.

3. _____. The Prediction of the Resistance to Air Flow of Mine Shafts Equipped with Internal Structures (Part II-Empirical Approach and Comparison of Methods). The Journal of the Mine Ventilation Society of South Africa, v. 18, No. 11, Nov. 1965, pp.149-158.

4. Kingery, D. S. Introduction to Mine Ventilating Principles and Practices. BuMines Bull. 589, 1960, pp.54.

5. Martin, H. Modelluntersuchungen zur Verringerung des aerodynamischen Widerstandes Zweier Schachte (Model Tests Made to Reduce the Aerodynamic Drag of Two Shafts). Neue Bergbautechnik, v. 8, No. 3, Mar. 1978, pp. 152-156.

Appendix

TABLE A-I-Calculated set component values

Component description	Dimensions used	Frontal area, A_{bi}, sq. ft.	Set spacing recurrence factor, NI	Bunton depth: width ratio, h/d, ft	Downstream spacing ratio, L/d, ft	Unstreamlined drag coefficient C_{di}	Streamlined drag coefficient C_{di} [3]	(NI C_{di} A_{bi})$_u$ sq ft	(NI C_{di} A_{bi})$_s$ sq. ft.
I-Beam W 12 x 35 [1] ..	14'11" x 6 5/8"	15.61 [2]	1.00	1.85	18.11	0.72	0.375	11.24	5.85
I-Beam W 8 x 24 ...	6'9" x 6 1/2"	3.32 [2]	1.00	1.21	18.46	0.84	0.375	2.79	1.25
Channel C8 x 11.5 ..	9'8 1/2" x 2 1/4"	1.82	1.00	3.56	53.33	0.93	0.375	1.69	0.68
Angle L 3 x 3 x 3/8 ..	11'10 1/2" x 3"	2.97	1.00	1.00	40.00	1.20	0.375	3.56	1.11
Rectangular Grating ..	2' x 6" x 1/2"	1.00	0.17	4.00	20.00	0.70 [4]	0.70	0.12	0.12
Trapezoidal Grating ..	2' x 8" x 1/2"	1.33	0.17	4.00	15.00	0.62	0.62	0.14	0.14
Electrical Bracket ...	See Footnote 5	1.16	1.00	—	—	1.00	1.00	1.16	1.16
Pipe Bracket	3' 7 1/2" x 3 3/4"	1.13	1.00	—	—	1.00	1.00	1.13	1.13
Ladder	See Footnote 6	0.13	0.30	—	—	1.00	1.00	0.04	0.04
Large Pipe Flange ...	12" I.D., 15" O.D.	0.64 [7]	0.30	—	—	1.00	1.00	0.02	0.02
Medium Pipe Flange ..	8" I.D., 10 1/2" O.D.	0.41	0.30	—	—	1.00	1.00	0.12	0.12
Small Pipe Flange [1] ..	4" I.D., 6" O.D.	0.30	0.30	—	—	1.00	1.00	0.09	0.09
Summation, $\sum$		29.82			20.11 [8]			22.10	11.71

[1] There are two of these items per set. The frontal area value, A_{bi}, represents the total area of both.

[2] These values include allowances for the non-form drag areas of the guides. See reference 3, pp. 157.

[3] The streamlined values used here are those recommended by Chasteau and Gillard.

[4] As no unstreamlined values for these items were suggested, these components were treated as streamlined to obtain a minimal unstreamlined-streamlined resistance differential. This and the remaining column values are streamlined values.

[5] For the electrical cable brackets, the dimensions of the pipe bracket was used plus 6" x 3/4".

[6] The ladder was treated as discontinuous components with dimensions of 3/4" x 1'6" plus 1/2" x 10". The L3x3x3/8 angle members were retained every 10 ft for extra ladder support.

[7] This area value contains an allowance for the coupling "ears". The two remaining values in the column also contain those areas.

[8] The sum of the L/d component was found by a weighted average using A_{bi}. See reference 3, pp. 154.

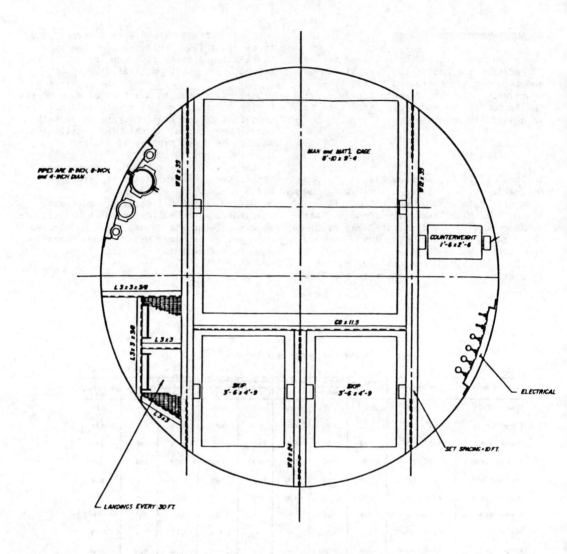

FIGURE A-I.- 18'2" diameter shaft plan (courtesy
of Harrison Western Corporation).

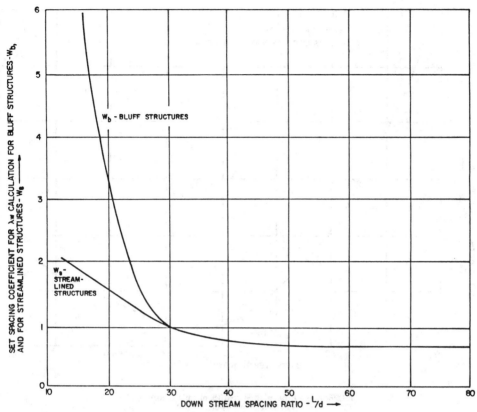

FIGURE A-2.—Set spacing coefficient (courtesy of The Mine Ventilation Society of South Africa).

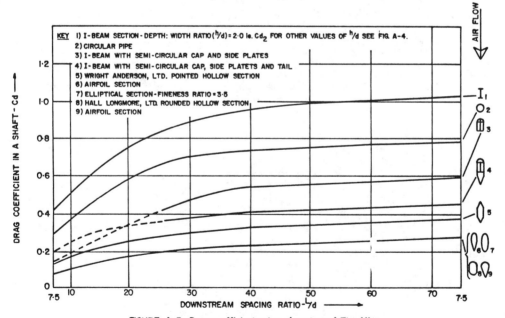

FIGURE A-3.- Drag coefficient values (courtesy of The Mine Ventilation Society of South Africa).

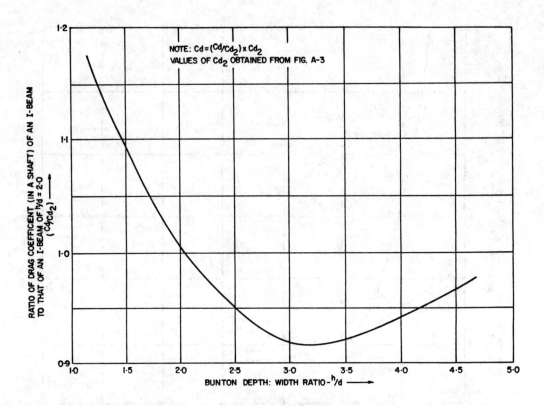

FIGURE A-4.- I-beam drag coefficient ratio (courtesy of The Mine
Ventilation Society of South Africa).

Reduction of miner noise exposure using reactive silencers

C.J.OPATRNY
VSM Corporation, Cleveland, OH, USA

D.L.JOHNSTON
Joy Manufacturing Company, New Philadelphia, OH, USA

ABSTRACT: A major source of noise that miners are subject to, emanates from portable exhaust fans. These fans are used to ventilate the mine face, thereby reducing dust concentrations and methane levels at the miner and bolter opertor stations. Conventional absorptive silencers used in this application have failed to perform due to the adverse airstream conditions.
A study was undertaken to design and test an underground fan silencer that is compact, anti-fouling, low in pressure drop, and yields a 12-14 dB reduction in noise level at the mine face. The silencer is of the reactive design and is constructed using a series of resonant cavities which are tuned to the blade passage frequency of the fan.

INTRODUCTION

The roof of a mine is prevented from collapsing by the installation of roof bolts at regular intervals. The bolts are installed using an electrically powered roof bolting machine which has two operators. The operators perform four bolting operations per day, each one lasting approximately 51 minutes. During each operation, the miners are subjected to noise from four different sources:

1. Drilling bolt holes.
2. Inserting and torquing of bolts.
3. Drive motors of bolting machine during tramming.
4. Fan noise at duct intake.

Figure 1 shows the operators exposure time and noise level during a typical shift.

OVERALL EXPOSURE LEVEL

Using Equations (1), (2) & the data from Figure 1, we can calculate the daily exposure level for the operator.

CUTS/ DAY	TIME IN HRS (C)	NOISE LEVEL dBA
21.7 min.drill/cut 4	1.45	99
24.0 min.bolt/cut 4	1.6	98
5.0 min.tram time 4	0.33	95

FIG. 1 OPERATOR NOISE EXPOSURE WITHOUT SILENCERS

$$T = \text{Antilog } (6.322 - 0.602 \, NL) \quad (1)$$

Actual Exposure =

$$\frac{C1}{T1} + \frac{C2}{T2} + \frac{C3}{T3} \quad (2)$$

Where T = allowable time in hours at a specific noise level (NL) in dBA, and where C = actual time at the same noise level.

@99 dBA T_1=Antilog[6.322-0.0602(99)]

$$T_1 = 2.30 \text{ hrs.}$$

@98 dBA T_2=Antilog[6.322-0.0602(98)]

$$T_2 = 2.64 \text{ hrs.}$$

@95 dBA T_3=Antilog[6.322-0.0602(95)]

$$T_3 = 4 \text{ hrs.}$$

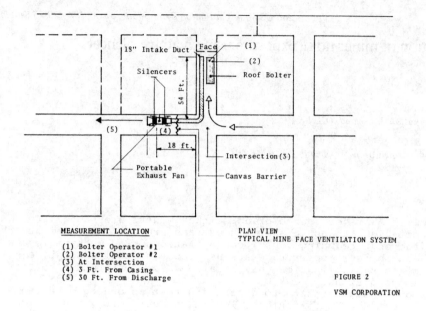

MEASUREMENT LOCATION

(1) Bolter Operator #1
(2) Bolter Operator #2
(3) At Intersection
(4) 3 Ft. From Casing
(5) 30 Ft. From Discharge

PLAN VIEW
TYPICAL MINE FACE VENTILATION SYSTEM

FIGURE 2

VSM CORPORATION

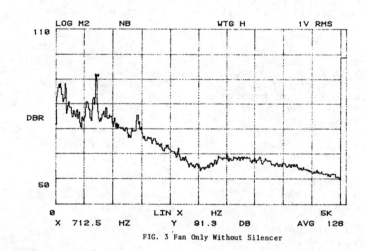

FIG. 3 Fan Only Without Silencer

Actual Exposure =

$$\frac{C1}{T1} + \frac{C2}{T2} + \frac{C3}{T3}$$

1.45/2.3 + 1.6/2.64 + 0.33/4

Actual Exposure = 1.31

Exposure levels in excess of 1.0 require feasible administrative or engineering controls, or personal hearing protection if the controls fail to produce level of 1 or less.

FIELD TESTS

Figure 2 shows the test setup in our operating coal mine. Noise level readings were taken at five locations under two operating conditions with and without fan silencers:

1. Fan on rock drill off.
2. Fan on rock drill on.

Figure 3 and 4 shows a spectum

194

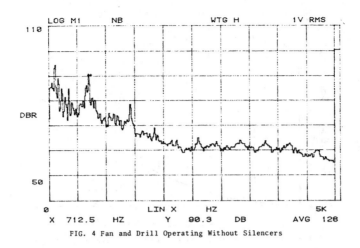

LOG M1   NB   WTG H   1V RMS

FIG. 4 Fan and Drill Operating Without Silencers

analysis of the above conditions for unsilenced fans. The fan blade passage frequency Eq. (3) is approximately 20 dB above average spectrum level at that frequency.

$$BPF = \frac{(No.\ Blades)(RPM)}{60}$$

$$BPF = \frac{12 - 3550}{60} = 710\ hz \quad (3)$$

For the purpose of this paper it was determined that the reduction in fan noise at the intake duct was the only feasible engineering approach to lowering the exposure level.

FAN NOISE ABATEMENT

The conventional approach to reducing fan noise has been to install absorptive type silencers on the inlet and discharge. Silencers of this type have been used with limited success due to the severe dust and moisture conditions prevalent in the mine environment. The mine fan under normal operating conditions handles large quantities of water, rock and coal dust. In addition, fans are often equipped with rock dusters, an integrally mounted hopper and feeder which automatically adds limestone dust to the exhaust air in order to render the discharge air non-combustible. Absorptive

silencers with a perforated metal retainer which has 3/32" dia. holes on 3/16" staggered centers become clogged and loose their acoustical effectiveness, when they are operated under these conditions.

Size and weight of the silencer is a major consideration because they are attached to a portable mine fan that is 8'-6" long; weighs 1,500 lbs. and must be maneuvered manually over uneven ground.

A cylindrical resonant-cavity silencer shown in Figure 5 consist of a series of resonant cavities tuned to the fan blade passage frequency. The cavities are constructed of solid metal vanes

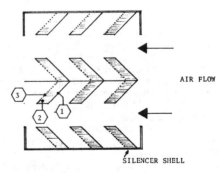

AIR FLOW

SILENCER SHELL

1. Aerodynamic Vane Construction

2. Acoustical Damper

3. Mesh Retainer

FIGURE 5   TUNED RESONANT CAVITY SILENCER

195

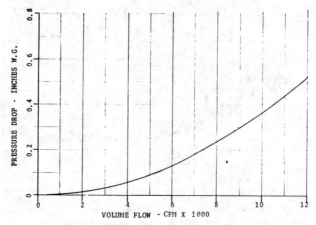

FIGURE 6 Pressure Drop Through Cylinder Reactive Silencer .075 #/Cu.Ft. Density

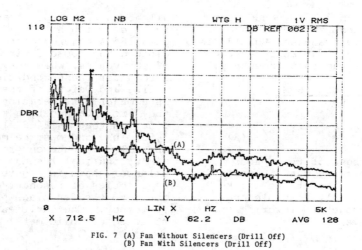

FIG. 7 (A) Fan Without Silencers (Drill Off)
(B) Fan With Silencers (Drill Off)

sloped in the direction of air flow to minimize turbulence and each cavity is lined with a thin layer of inert absorptive material to maximize the broad band attenuation of the silencer. Laboratory flow tests indicate that particulate entrained in the air stream passes over the cavity with minimal dropout due to the aerodynamic shape of the cavity. The center hub of the silencer is designed to match the fan rotor or motor diameter, thereby reducing pressure losses due the change in velocity head at thesilencer entrance. Laboratory air flow tests Figure 6,

show 0.42 in w.g. pressure drop when the silencer is operating at 11,000 CFM. Assuming a fan efficiency of 80% the increase in motor brake horsepower due to the silencer can be calculated Eq. (4) as follows:

$$BHP = \frac{CFM}{6356} \times \frac{TP}{Fan\ total\ efficiency}$$

$$BHP = \frac{11,000}{6,356} \times \frac{0.42}{0.08} = 0.91 \quad (4)$$

The final silencer design for an 11,000 CFM exhaust system consisted of inlet and discharge silenc-

ers that were 31-1/2 in. in diameter and 24 inches long.

FIELD TESTS WITH SILENCERS INSTALLED

Noise level measurements taken
before and after silencer installation in Figure 7 indicate an overall reduction in fan noise level at
the operators station from 95 to 83
dBA with a 28 dB reduction at the
blade passage frequency.

Figure 8 shows the operators
exposure time and noise level with
the fan silencers in place.

	CUTS/ DAY	TIME IN HRS (C)	NOISE LEVEL dBA
21.7 min.drill/cut	1.45		92
24.0 min.bolt/cut	1.6		90
5.0 min.tram time	0.33		90

FIGURE 8 OPERATORS NOISE EXPOSURE
        WITH SILENCERS

Exposure values were then calculated Eqs. (1) and (2) and indicate a
reduction in exposure from 1.31 to
0.43 using the same exposure time
as the previous unsilenced fan
test.

@92 dBA T_1=Antilog[6.322-0.0602(92)]

    T_1 = 6 hrs.

@90 dBA T_2=Antilog[6.322-0.602(90)]

    T_2 = 8 hrs.

@90 dBA T_3 = 8 hrs.

Actual Exposure =

1.45/6 + 1.6/8 + 0.33/8

0.48, permissible 1.0

SUMMARY

Field tests indicate that resonant cavity type silencers will
reduce fan generated noise levels
at the mine face by 12 dBA, which
brings the roof bolter operators

into MSHA compliance. The silencer is anti-fouling in design and
will withstand the high dust loading conditions in typical coal
mining operations.

The operating point and selection of the fan in mine areas
of high altitude

ZHAO ZI-CHENG
Kunming Institute of Technology, People's Republic of China

ABSTRACT: The atmosphere pressure, temperature and air density decrease with the increase of the altitude.But the air quantity in the mines and the air quantity supplied by the fan remain constant in spite of the change of the altitude.
   As the altitude increases,the decrease of air density causes the pressure losses of the mine and the air pressure of the fan to fall down proportionally; and for the same reason the output power of the motor is reduced, too. Thus the selection procedure of the fan is simplified. It isn't necessary to carry out complicated altitude correction in the ventilation design. Both the fan and the motor can be selected with normal methods. The only difference is that the practical operating point of the fan is lower than the one which is expected according to the design. The air pressure of the operating point of the fan is the designed air pressure of the operating point multiplied by the altitude correct coefficient. The air quantity of both operating points of the fan are equal.
   The paper expounds in theory the mine ventilation in areas of high altitude. The theory has been proved by a lot of data from tests. The paper serves as a theoretical reference for the design of mine ventilation in areas of high altitude and is of value to the work in practice.

When carrying out the design of selection of the fan, we take the air quantity, air pressure, power, efficiency etc. of the operating point (i.e. The point of intersection of the characteristic curve of the mine ventilation network and the characteristic curve of the fan) as indexes, and to meet the requirements of mine ventilation, we select the fan which is efficient and cheap as the power source of mine ventilation.
   The characteristic curve of the fan provided by the manufacturer is drawn according to the data of the model test and to the condition under normal meteorology, but the fan in mines in areas of high altitude is mounted on the ventilation system of the mine, therefore, owing to the reduction of air density, the practical characteristic curve of the fan varies to some extent. This variance makes the operating point change somewhat. So the problem how to select the fan correctly is raised.

1 INFLUENCE OF THE ELEVATION ON MINE VENTILATION

1.1 The change of the air density in mines

in areas of high altitude

With the increase of the altitude, the atmosphere pressure decrease and the air becomes thinner and the decrease of air density can be calculated with the following formula:

$$r' = r_o \left(1 - \frac{Z}{44300}\right)^{4.256} \qquad (1)$$

Here, r'=air density at the elevation kg/m^3
   r_o= air density at sea level kg/m^3

   (when we calculate the mine ventilation $r_o = 1.2 \ kg/m^3$)
   Z=The height of elevation,m

   In order to make the calculation simpler, let the ratio of air density in mines in areas of high altitude to that in sea level be the correct coefficient of altitude

$$K_r = \frac{r'}{r_o} < 1 \qquad (2)$$

1.2 Determination of the air quantity required by mines in areas of high altitude

At present, the air quantity required in mines is calculated both according to displacement of smoke and dust. In general, the air quantity which is calculated according to displacement of dust is greater, and it is independent of air density, so it needn't be corrected for its altitude.

1.3 The change of pressure loss in mines in areas of high altitude

Since the friction coefficient in mine is directly proportional to air density, and the pressure loss in mine decrease with the increase of the altitude above the sea level, the air resistance and pressure loss in mines in areas of high altitude should be corrected for altitude.

$$\frac{R'}{R_o} = \frac{r'}{r_o} \quad \text{or} \quad R' = K_r R_o \qquad (3)$$

$$\frac{h'}{h_o} = \frac{r'}{r_o} \quad \text{or} \quad h' = K_r h_o \qquad (4)$$

Here, R' and R_o represent air resistance in mines in areas of high altitude and that in mines at sea level respectively N.sec/m⁸. h' and h_o represent the pressure loss in mines in areas of high altitude and that in mines at sea level respectively, Pa.

2 INFLUENCES OF THE ELEVATION ON THE PROPERTIES OF THE FAN

2.1 Air quantity of fan

The air quantity of the fan in mines is measured in volume. It is independent of air density. So the change of the height above sea level has no influence on it.

2.2 Air pressure of fan

The air pressure of the fan is expressed as follows:

$$P = r.U.C_u \qquad (5)$$

Here: r=air density Kg/m^3
U=circular velocity of working wheel, m/sec.

C_u=twisting velocity, m/sec.

From the above formula, we can see that when the blade angle setting of the fan and the running per minute are constant, the only variate in the formula is air density and hence the air pressure exerted by the fan is proportional to air density, that is,

it decreases with the increase of the height above sea level.

2.3 Efficiency of fan

Since the air becomes thinner in areas of high altitude, the pressure loss and the axle power of the fan falls down correspondingly. When the condition of heat radiation of motor is bad, the capacity of isolation is poor, and the output power of the motor is also reduced. Hence, the efficiency of the fan is unchanged.

2.4 Characteristic of fan

The relations between the air quantity, air pressure and the power of the fan running in areas of high altitude and those of the fan running in areas of sea level are as follow:

$$Q' = Q_o \qquad (6)$$

$$P' = K_r P_o \qquad (7)$$

$$W' = K_r W_o \qquad (8)$$

Here, Q_o and Q' are the air quantity of the fan in mines in areas of sea level and that in mines in areas of high altitude respectively, m/sec.
P and P' are the air pressure of the fan in mines in areas of sea level and that in mines in areas of high altitude respectively, Pa.
W_o and W' are the power of the fan in mines in areas of sea level and that in mines in areas of high altitude respectively, KW.

According to formula (6) in the P-Q figure, we can draw a series of straight lines of the equal air quantity, as shown in fig 1. Under the condition of air density to be r_o, the characteristic curve of the fan is I with intersects a series of lines of the equal air quantity at points 1,2,3, and 4. When the fan is running at another height above sea level, and its air density is r' according to formula (7), we can find out points 1', 2', 3'and 4' of the corresponding value P' on series of lines of the equal air quantity.
The connecting line II of these points is a running characteristic curve of the fan. r'. The equi-ratio change of the air resistance in the mines is caused by the change of air density.
Hence, points 1',2',3'and 4' on curve II are the corresponding operating point of points 1,2,3,and 4 on curve I.
Thus, according to the standard charac-

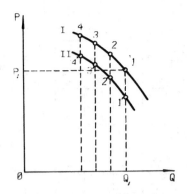

Fig 1. The characteristic curves of the fan running at different heights above the sea level

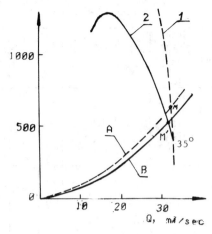

1—the standard characteristic curve of the fan
2—the practical running characteristic curve of the fan
A—the characteristic curve of the air resistance, when the mine is at the sea level
B—the characteristic curve of the practical air rasistance

Fig 2. The characteristic curve of the fan of Huang-Mao-San Mine

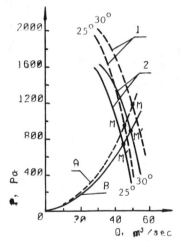

1—the standard characteristic curve of the fan
2—the practical running characteristic curve of the fan
A—the characteristic curve of the air resistance, when the mine is at the sea level
B—the characteristic curve of the practical air rasistance

Fig 3. The characteristic curve of the fan of Da-Yao Mine

teristic curve of the fan, we can draw the corresponding characteristic curve of the fan running on the arbitrary height above the sea level, so as to provide conveniences for analysing the operating point of the fan in mines in areas of high altitude.

3. PROPERTY TEST ON THE FAN RUNNING AT DIFFERENT HEIGHTS ABOVE THE SEA LEVEL

3.1 Property test on the main fan

For years, in order to find out a rule for the fan property varying with the heights above the sea level, we have made experiments one after the other on the main fan $70B_2$-11.No.12 of Huang-Mao-San Mine(2340m above sea level), the main fan $70B_2$-11 No. 18 of Da-Yao Mine(1980m above the sea level) and the main fan BY-18 of Yin-Ming Mine (2472m above the sea level) and have obtained a lot of data, by means of which we can analyse the running state of the fan in mines in areas of high altitude. The parameters of air pressure, air quantity etc. determined are in table 1 and Fig 2-4.

From Table 1 and Fig2-4, we can see that compared with standard characteristic curve of the products, the practical characteristic curves apparently fall down. The causes are as follows: (1) The motors mounted on the main fan are asynchronous, which require fewer rated revolutions than the products of the fan. (2) The air density decreases with the increase of the height above sea level, so that the air pressure of the fan decreases and its characteristic curve falls down.

3.2 Test on the auxiliry fan at different altitudes

The property test on the main fan in mines has shown that characteristic curve of the

Table 1. The characteristic and the determined data of the fan of the partial mine

Name of mine	Fan type	Height	Air density	Angle	Item	point of determination									
						1	2	3	4	5	6	7	8	9	10
Huang Mao-San Mine	$70B_2$-11No.12.	2340	0.938	35°	Q	31.3	31	30.8	30.2	29	25.9	24.6	22.5	17.4	12.3
					P	609	624	633	657	744	884	996	1110	1314	1170
Da-Yao Mine	$70B_2$-11No.18.	1989	0.97	25°	Q	49.8	44.4	44.7	47.2	47.6	43	44	30.7	29	
					P	284	379	435	490	644	773	1105	1410	1600	
				30°	Q	53	53.4	48.5	49.4	45.3	46.6	35.4	32.4		
					P	413	463	580	645	1110	1005	1206	1620		
Yin-Ming Mine	BY-18	2472	0.939	30°	Q	21.5	33.4	38.7	45	46.6	49.9	51	51.9	53.4	54.8
					P	1190	1140	1420	1200	1080	880	780	680	610	550
				35°	Q	21.7	40	44.6	49.9	53.6	56.4	60.2	63.5	66.5	67.9
					P	1110	1610	1550	1450	1330	1210	1030	880	760	680

Table 2. The characteristic curve of the auxiliary fan and its determined data

Name of place	altitude (m)	air density (Kg/m³)	Item	point of determination									
				1	2	3	4	5	6	7	8	9	10
Standard curve	0	1.2	Q	100	122	145	175	193	210	225			
			P	2740	2600	2400	2000	1500	1000	500			
Jia-jie-san	1360	1.04	Q	18	30	46	60	82	105	117	128	143	153
			P	2300	2270	2400	2420	2410	2280	2180	2040	1870	1680
Han-gu-di	1720	0.99	Q	19.5	30	48	64	88.5	102	120	129	142	155
			P	2260	2200	2320	2340	2280	2170	1920	1770	1560	1340
Xian-jin-keng	1920	0.97	Q	31	52	70	87	102	116	127	135	148	
			P	2160	2200	2180	2075	1940	1760	1590	1440	1210	
Da-lian-Hau-san	2880	0.87	Q	14	23.5	44	68	81	108	117	126	135	139
			P	2120	2030	2125	2020	1945	1740	1580	1410	1230	1130

fan apparently falls down with the increase of the height above the sea level, but since the type of fan is different from each other and their quality of manufacture and installation condition are influencial factors, the fan properties may thus be influenced. in order to further prove that the fan characteristic curve varies with different heights above the sea level, by means of the same experiment pipe we selected auxiliary fan (type JPT-52) and made experiments on the fan characteristic at Jia-Jie-San (1,360m above sea level) Han-Gu-Di (1,720m above sea level), Xian-Jin-Keng (1,920m above sea level) and Da- Lian-Hau-San (2,880m above sea level) respectively, within one day and the data of air quantity and air pressure shown in Table 2 and Fig 5, were obtained from the determination.

From the result of the experiments, we see that the fan characteristic curve falls down with the rise of the height above the sea level. This result corresponds to the conclusion from the former theoretical analysis, saying that the falling down of the

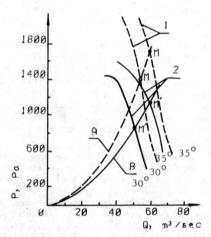

1—the standard characteristic curve of the fan
2—the practical running characteristic curve of the fan
A—the characteristic curve of the air resistance, when the mine is at the sea level
B—the characteristic curve of the practical air rasistance

Fig 4. The characteristic curve of the fan of Yin-ming Mine

202

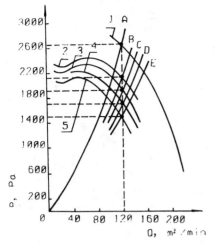

1--the standard characteristic curve of the fan
2,3,4,5--the characteristic curve of the fan on Fia-
jie-san,Hau-gu-di,Xian-jin-keng and Da-lian-
san respectively
A--the characteristic curve of the assumed mine
at sea level
B,C,D,E--the characteristic curve of the practical
experiments which were made at Jia-jie-san,
Hau-gu-di,Xian-jin-keng and Da-lian-hau-san,
respectively

Fig 5. The practical characteristic curve
of the auxiliary fan type JBT-52-2 at different altitudes

fan characteristic curve is caused by the
decrease of the air density.

4. THE OPERATING POINT OF THE FAN IN THE
MINE SYSTEM ON DIFFERENT HEIGHTS ABOVE
THE SEA LEVEL

As said above, the fan characteristic curve
falls down with the rise of the height
above the sea level, and at the same, the
pressure loss in the mine also decreases,
so as to cause the operating point of the
fan to change correspondingly. Now, take
Fig.2 and 5 as an example for analysis. Let
curve 1 in Fig. 2 be the standard characteristic curve of the fan product and curve
A be the air resistance curve ($r_o=1.2kg/m^3$)
when we assume the mine at the sea level.
The point of intersection M of the two curves is the operating point of the fan under
this condition, and we name it "the standard
operating point". Here, the air quantity is
$30m^3$/sec, the air pressure is 800 Pa. Curve
2 is the characteristic curve of the practical running of the fan at 2340 meters
above the sea level. Curve B is the characteristic curve of the air resistance of the
practical ventilation network in the mine.

The point of intersection of the two curves
is operating point M' of the practical
running of the fan. Its air quantity is
$30m^3$/sec, and its air pressure is 600 Pa.
from the change of the operating point in
Fig.2, we know that the ratio of the air
pressure is equal to that of the air density. Hence, the relation between the operating point as the fan on different heights
above sea level and the standard operating
point may be shown as follow:

$$Q'=Q_o \qquad (9)$$

$$P'=K_r P_o \qquad (10)$$

That is, the operating point of the fan
in mines in areas of high altitude varies
with the change of the height above the sea
level, and its air pressure varies along
the line of the equal air quantity. The air
pressure of the operating point of the fan
is that of the standard product multiplied
by the correct coefficient K_r for the altitude.

According to some reports, we know, with
the rise of the height above the sea level,
the air becomes thinner, the isolation
strength of electrical material falls down
and the output power of the motor decreases. If in the place of 1000 meters
above the sea level, every 100 meter is increased, the motor power must accordingly
decreases by 5%.

5. THE SELECTION OF THE FAN IN MINES IN
ARAES OF HIGH ALTITUDE

According to the above analysis, the problem of selecting the fan can be solved
more easily. In general, the characteristic
curve the fan in the design is provided by
a fan manufacturer and is obtained by a
model test with the air density being 1.2
kg/m^3. The fundamental parameter for calculating the air pressure of mine ventilation network—the coefficient of friction is
also obtained with r_o being 1.2 kg/m^3.
As the altitude increase, the decrease
of air density causes the pressure losses
in the mine and the air pressure of the fan
to fall down proportionally; but the air
quantity required in mine and air quantity
of the fan remain constant. For the same
reason the output power of the motor is
reduced, too. Thus the selection procedure
of the fan is simplified. It is not necessary
to carry out complicated altitude correction in the ventilation design. The only
difference is that the practical operating
point of the fan is lower than the one which

203

is expected according to the design. The
air pressure of the operating point of the
fan is the designed air pressure of the operating point mutiplied by the altitude
correct coeffient. The air quantity of both
operating points of the fan are equal.

REFERENCE

1. Wang Yin-ming "Mine Ventilation and
   Safety" 1978 (Chinese Edition)

2. Zhao Wei-yi "Air Density and Ventilation" "Matullary Safety" 1977. No.2
   (Chinese Edition)

3. Zhao Zi-cheng, He Wei "Influences of
   the Elevation on Mine Ventilation"
   "The Technology on Chemical Industrial
   Mine" 1978. No. 2 (Chinese Edition)

4. "Mine Main Fan" compile group of
   Ventilation and Dust Control (chinese
   Edition)

6. Recirculation

The use of controlled recirculation ventilation to conserve energy

A.E.HALL
University of British Columbia, Vancouver, Canada

ABSTRACT: Canadian mines are extensive users of air heating systems in winter. The use of ventilation recirculation on a whole mine basis to reduce heating costs is examined and found to have a good potential for application to mines using large amounts of diesel power underground.

1 INTRODUCTION

Canadian mines are often situated in climatic areas with winter temperatures of -30 to -40°C. This results in a need for intake mine air heating. The mines use gas, electric and waste heat recovery systems for this heating, the most common method being the use of propane burners on surface.

The cost of air heating may exceed $1M per annum for individual mines and this contributes significantly to operating costs. The possibility of reclaiming heat from mine exhaust air has been studied and some installations are in service but the temperature of the exhaust air makes it a low grade heat source. This low temperature coupled with the distance between intake and exhaust airways and the seasonal heating requirement have proved major obstacles to the development of economic recovery systems.

Controlled recirculation ventilation provides a method of reclaiming waste heat for reuse in a mine at economic cost but a practical system must be safe and capable of maintaining good ventilation conditions. Canadian mines are extensive users of diesel equipment which provides a challenge to implementing recirculation. Uncontrolled recirculation has been considered a hazard because it can create excessive dust and gas levels. Many legislations specifically prohibit air recirculation and prescribe precautions which mines must take to prevent it.

A controlled recirculation system must prevent dust and gas accumulations and there must be no chance of noxious combustion products from a mine fire being carried back into the intake air. Recent British and South African experiments have shown that the use of on-line monitoring systems permits safe controlled air recirculation which provides improved ventilation conditions.

This paper examines the viability of using controlled recirculation for energy conservation and outlines the requirements of a practical system for Canadian mines.

2 CURRENT PRACTICE

A 1980 study of Canadian underground mines investigated the potential for recovering heat from the exhaust air (Freyman, 1980). Ninety seven mines were examined and they were responsible for a daily production of 321,000 tonnes using 15,000 m^3/s of ventilating air. The air volume is dependent on the tonnage mined but many mines base their planning on the total installed underground diesel power, using values of 0.06 to 0.1 m^3/s of air per kiloWatt. Most mines use propane heating but the dramatic increase in price from 2 cents per litre in 1972 to the present 25 cents has resulted in some mines changing to electric or natural gas heating (Summers, 1984).

Table 1 sets out the operating statistics for 10 large mines using air heating. Considerable variations in energy costs were reported by individual mines. Electric costs ranged from 0.5 to 3.0 cents per kiloWatt hour and propane costs from 15 to 30 cents per litre. Costs were standardised on the basis of 2 cents per kW hour for electricity and 25 cents per litre for propane, which are representative of present average costs. Table 2

Table 1.   Heating   and   Ventilation   in   Canadian   Mines

Mine	Air Quantity M^3/S	Pressure kPa	Fan Power kW	Air Temp. ^{o}C	Heating Days	Propane Used l/day	Heating Systems
1	165	2.5	750	-20	100	8300	Propane
2	210	1.8	540	-10	160	10100	Propane
3	990	3.1	4300	-26	120	50000	Propane
4	310	-	3500	-4	160	*	Natural Gas ($\bar{}$*14800 1/day)
5	250	0.5	180	-9	140	13300	Propane
6	440	1.5	1200	-11	160	30500	Propane
7	2280	2.5	8900	-14	120	115000	Propane
8	225	-	1130	-9	120	6300	Propane & Elec (270 kWelec.)
9	220	-	800	-14	140	3600	Propane & Oil (4800 1/day)
10	415	-	1340	-30	140	12800	Propane & Elec (4300 kWelec.)

Table 2.   Annual Heating and Ventilation Costs.

Mine	Air Quantity M^3/S	Ventilation Cost $	Heating Cost $	Total Cost $	Annual Cost per M^3/S Ventilation	Heating	Total
1	165	131,400	259,400	390,800	796	1572	2368
2	210	94,600	505,000	599,600	450	2405	2855
3	990	753,400	1,875,000	2,628,400	761	1894	2655
4	310	613,200	300,000	913,200	1978	968	2946
5	250	31,500	581,900	613,400	126	2328	2454
6	440	210,200	1,525,000	1,735,200	478	3466	3944
7	2280	1,559,300	4,312,500	5,871,800	684	1891	2575
8	225	198,000	255,000	453,000	880	1133	2013
9	220	140,200	567,000	707,200	637	2577	3214
10	415	234,800	1,185,000	1,419,800	566	2855	3421
Average	550	396,700	1,136,600	1,533,300	721	2067	2788

All above costs are given in Canadian dollars.
(One $Canadian = $0.72 U.S.)

shows the standardised costs of the heating and ventilation systems. It is clear that the heating cost is considerably higher than the ventilation cost and that these mines are spending an average of $1M each per year for heating at an average cost of $2067 per cubic metre per second circulated.

The 1980 survey indicated that some 3200 m^3/s(i.e. 21%) of the total exhaust air had good potential for heat recovery because of its temperature and proximity to the intake shafts. Only 6 mines reused the exhaust air and only one company used exhaust air to heat intake air. This installation at the Strathcona mine of Falconbridge Nickel Mines used a "run-around" glycol system for heat recovery (McCallum,1969). The system is shown in Figure 1. The intake and return raises are within 120 metres of each other and the system comprises copper finned intake and exhaust coils at each end with 56% by mass glycol solution as the circulating medium. The shell and tube heat exchanger recovers heat from the mine air compressors and transfers it to the glycol. The system is designed to cool 283 m^3/s of exhaust air from 10°C to 3.3^{o}C and to heat the same volume of intake air by 13.8^{o}C. The air heat recovery is 4700 kWatt and the compressor recovery is a further 2900 kWatt. This total heat is not adequate at the low winter temperatures, which reach -34°C,

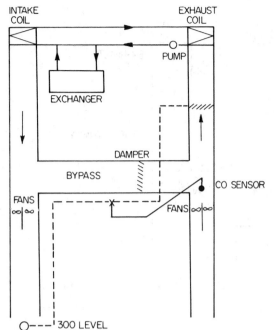

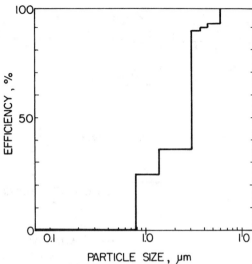

Figure 2. Dust collection efficiency of Lorraine Mine water spray chamber (after Burton).

Figure 1. Exhaust air bypass system-
Strathcona Mine, Falconbridge
Nickel (after McCallum)

because the exhaust air cannot be cooled below freezing without icing the exhaust coil.

A controlled by-pass of exhaust air to the mine air intake is used under these conditions. A thermostat is located in the intake on 300 level. When more heat is required a controller responds to the thermostat signal and opens a set of dampers in the by-pass duct. The dampers close again as the temperature reaches the required level. The by-pass is monitored by a continuous carbon monoxide detector in the exhaust air stream. Detection of a CO increase results in an alarm sounding and the by-pass dampers being closed regardless of intake air temperature. The system does not reduce the air volume circulating in the mine but some portion consists of recirculated air.

3 RECIRCULATION PRACTICE

Initial British experiments were designed to improve dust control in the advance headings of longwall coal mine faces. Pickering and Aldred, 1977 gave a full description of these tests. The use of controlled recirculation was found to be inherently safe and methane concentrations were found to depend only on the rate of gas emission and the rate of fresh air flow through the section. Pickering, 1984 reported that in 1982 there were 63 controlled recirculation systems in use in advanced headings and driveages. This paper also described a series of trials of district controlled recirculation at Wearmouth Colliery. These tests were performed to investigate the feasibility of using recirculation to extend the limit of conventional ventilation at undersea mines in North East England. The results indicated that recirculation was feasible and the fans should be installed close to the face to reduce power requirement.

The South African experiments at Lorraine Gold Mine were described by Burton et al, 1984 and by December 1984 five more schemes were operational with a further 15 planned. These systems were planned for district rather than whole mine recirculation to reduce ventilation power costs. The South African experiments have indicated that the water spray chambers provided to cool the recirculated air have been extremely successful in filtering dust out of the air. The performance of the Lorraine chamber is shown in Figure 2. Several tests were made to determine the blast contaminant decay rates with varied percentages of recirculated air. It was

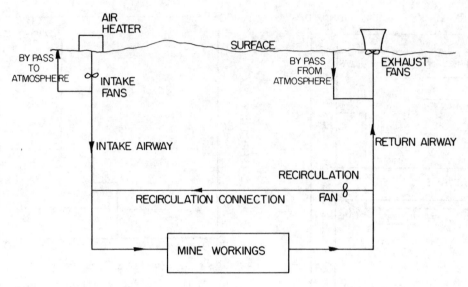

Figure 3. Push-pull ventilation system with atmospheric bypasses for recirculation.

found that blasting fumes were cleared more quickly with an increased intake air quantity. Both these findings are extremely important in designing an energy conservation recirculation system.

4 RECIRCULATION FOR ENERGY CONSERVATION

The cost figures in Table 2 indicate that for Canadian mines using air heating, the heating costs far outweigh the ventilation circulation costs. An optimum energy conservation system should therefore recirculate as much air as possible and reduce the intake volume of a mine until the maximum desirable contaminant levels prevail.

This indicates that for an energy conservation system it is preferable to use whole mine recirculation rather than a district or local recirculation scheme. Intake airways and shafts require heating so the recirculation should take place as close to the mine intake as possible. Optimum locations would be at points where intake and exhaust airways are in close proximity whether underground or on surface.

Controlled recirculation can either operate as an on/off system or as a variable quantity recirculation scheme. The majority of the British and South African systems have been designed as on/off systems with a constant volume of air recirculated. Recirculation takes place until emergency

conditions such as a fire are detected by the system, in which case recirculation is stopped by shutting down the fan. Recirculation operates throughout the blasting period because the decay of blast contaminants is only dependent on the rate of fresh air circulated through the mine. District recirculation does not affect the fresh air intake and hence does not affect the contaminant decay, also because the recirculation area is limited to the workings, the main intake airways are not contaminated by recirculation.

In a whole mine energy conservation system it is preferable to switch off the recirculation during the blasting period as this will increase the fresh air intake of the mine and hence optimise the blast contaminant decay time. Stopping recirculation during blasting will also prevent contamination of the intake airways.

Variable quantity recirculation systems are much more complex than fixed quantity systems and the fan types and positions are critical if the system is to be practical. The typical push-pull mine ventilation layout shown in Figure 3 can pose severe problems for a variable recirculation system. The intake and exhaust fans can stall when the quantity handled by them is reduced as the recirculation quantity increases. Usually the air heater precedes the intake fan so that both intake and exhaust fans operate on warm air. Providing by-pass dampers to

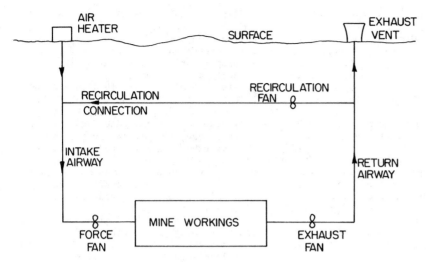

Figure 4. Recirculation ventilation system with underground main fans.

atmosphere to prevent fan stall is only partially satisfactory because the exhaust fan would then have to handle cold ambient air. The intake fan would also have to handle ambient air with the propane burners shut off or else hot air would be exhausted direct to atmosphere, which is unacceptable. Placing the fan upstream of the air heater may not be feasible because of cold temperature lubrication problems. The preferred fan position for an energy recovery system is downstream from the recirculation connection so that the quantity handled by the main fans is constant whether the recirculation system is operating or not. Figure 4 shows a layout of this type using underground main fans.

An alternative for a new mine is to use variable flow fans such as variable pitch axial fans as both main and recirculation fans. The fan selection must be such that they will not stall when the maximum recirculation occurs and must also be capable of handling the maximum volume and pressure required at zero recirculation.

A possible solution for existing mines where the intake and exhaust fans are in close proximity is to connect the exhaust fan outlet to the intake fan inlet by the recirculation duct. This is the method used at Falconbridge's Strathcona mine, shown in Figure 1. Unfortunately for many installations this will not be possible because of the distance between the two airways.

5 CONTAMINANT CONTROL

In a district recirculation system the objective is to increase the airflow through a small part of the mine whilst maintaining the overall flow through the whole mine at the same quantity. The effect of this is to prevent any excessive build-up of gas and dust beyond the levels existing prior to the introduction of recirculation.

Whole mine recirculation systems with variable quantities do not necessarily prevent build-ups of gas and dust. They, therefore, are required to be very strictly monitored and controlled. One index that can be applied to dieselised mines is the revised Air Quality Index (AQI) proposed by French, 1985. Indices are defined for both gaseous and particulate contaminants:

$$AQI(gas) = \frac{CO}{TLV-CO} + \frac{NO}{TLV-NO} + \frac{NO2}{TLV-NO2}$$

$$AQI(part) = \frac{RCD}{TLV-RCD} + \left[\frac{SO2}{TLV-SO2} + \frac{RCD}{TLV-RCD}\right] + \left[\frac{NO2}{TLV-NO2} + \frac{RCD}{TLV-RCD}\right]$$

Where
AQI = Air quality index
CO = CO concentration, ppm
NO = NO concentration, ppm
NO2 = NO2 concentration, ppm
SO2 = SO2 concentration, ppm
RCD = Respirable combustible dust, mg/m3
TLV-X = Threshold limit value for pollutant X

The AQI(gas) should not exceed 1, and no individual component should exceed its TLV. It is recommended that AQI (particulate) values should not exceed 2.0 and no single component should exceed its TLV, as dictated by current ACGIH values. If the level of SO2 or NO2 are 25% or less of their TLV, they are omitted from the equation.

Recent surveys of Canadian dieselised mines have indicated that the major pollutants are respirable combustible dust, nitrous oxide and sulphur dioxide in terms of their percentage contribution to the air quality index. All these pollutants are related to the number and quality of diesel powered units in operation. Diesel emissions are highly dependent on the type of engine used, the exhaust treatment system and the fuel consumed in quality and amount.

Johnson and Carlson, 1984 carried out extensive tests on diesel equipment in a salt mine. The tests showed that the mine carbon dioxide concentration level was an excellent indicator of the levels of NO2,CO,SO4,SO2 and diesel particulates. The mine air diesel produced CO2 concentration was also found to be a measure of the amount of ventilation per unit of fuel consumed or per unit of output power. Thus the diesel produced CO2 concentration is directly related to the air supplied per actual output kilowatt. This means that by monitoring the CO2 levels in a mine, unless there are major changes in the characteristics of the diesel equipment used, the actual diesel power and fuel consumption can be determined at any instant in time.

Usually there are large fluctuations in output diesel power during the working shifts and the pollutant levels vary significantly. Ventilation systems have therefore, been designed to cater for the "worst case" contaminant levels as determined by experience. It is clear that when diesel equipment is fully utilised at or close to peak loads large quantites of ventilating air are required to dilute the exhaust. When the loading drops, however, the same amount of air is supplied despite the reduction in pollutant levels. Given that the Air Quality Indices are satisfactory for full load conditions, then the air circulated through the mine could be reduced as the diesel loading is reduced without increasing the AQI. This is not done at present by operating mines because of the lack of monitoring facilities and also fluctuating airflows through a mine are undesirable for control purposes.

A viable alternative is to monitor the quality of the return air from the mine and the intake air and to recirculate the maximum quantity of return air which will not increase the mine AQI. This quantity is proportional to the unused diesel capacity at that time, i.e. if a mine circulates 400 m3/s of air to dilute 4,000 kilowatts of diesel power, then if the diesel output drops to 3,000 kW, it is only necessary to introduce 300 m3/s of fresh intake air into the system and 100 m3/s of exhaust air can be recirculated without raising the maximum mine AQI. Use of a variable flow system of this type could result in significant cost savings for a mine using air heating in severe winter climates.

Mineral dust levels in mine air also fluctuate over the working shift, dependent on the mining operations. Excessive dust levels must be prevented so it is necessary to monitor respirable dust levels in exhaust and intake to determine the allowable recirculation. The provision of a water spray chamber in the recirculation system can be expected to reduce the dust level considerably. Figure 2 shows that the Lorraine spray chamber removed 90% of the plus 2μm dust and 40% of the dust between 1 and 2 μm. It was unable to remove the minus 0.8μm dust and further research is required to improve the capture of fine dust in recirculation systems. Closing down the recirculation system during blasting will assist in reducing the recirculating mineral dust load.

Gaseous contaminants produced by mining such as methane and radon can pose problems in recirculation systems dependent on the amounts produced and in the case of radon the residence time of the gas in the system. These gases must be considered in the feasibility study of a recirculation system and the design must ensure that safe conditions are always maintained.

Whole mine recirculation systems are capable of causing a lowering of the oxygen level due to oxygen depletion in the recirculated return air. Research is required to investigate the oxygen levels in return air to determine whether it can be related to CO2 concentration. Until this work is done it is recommended that the oxygen level in the exhaust and recirculated air be monitored to prevent oxygen deficiency.

Fire product detection is required to close down the recirculation system in the event of a mine fire. This system is extremely important and it is recommended that parallel sampling be used to

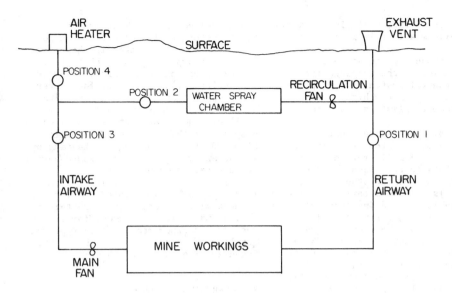

Sensors.

Position 1 CO_2, O_2, dust, airflow,   fire detection. Position 2 CO_2, dust, airflow

Position 3 CO_2, O_2, dust, airflow,   fire detection. Position 4 CO_2, dust, airflow

Figure 5. Recirculation control system showing sensor positions.

prevent a failure of the emergency system if the sensors fail.

6 MONITORING SYSTEMS

The discovery by Johnson that the CO2 concentration is an excellent indicator of diesel pollutant levels is extremely important because there are several commercially available CO2 monitors which are well suited to mining use. One monitor of this type which has been well proven in severe mining environments is the SPANAIR detector developed by Anglo American Corporation and described by van der Walt, 1975. On-line monitoring of CO2 is therefore feasible at reasonable cost.

On-line dust monitoring poses problems but recent work by CANMET and Metrex Instruments has resulted in the development of a nephalometer for the differential sensing of diesel and mineral dusts. The instrument has been described by Hardcastle, 1985 and in laboratory tests was found to be an excellent mineral sensor. It is expected that this instrument should prove to be a relatively inexpensive sensor for rapid data acquisition of respirable dust and possibly diesel particulates.

Oxygen monitors are readily available

as are combustion product detectors for fire detection. Flow monitors are also available with adequate accuracy for incorporation in a recirculation control system.

The instrumentation therefore exists for the implementation of a whole mine recirculation system and it is possible to provide safe, reliable on-line monitoring of air quality during recirculation. Telemetering and microprocessor systems are available for the overall control of such installations with adequate response time at reasonable cost.

Figure 5 shows a monitoring layout for a whole mine recirculation system. Sampling position 1 gives the return air quality, position 2 will provide data on the effect of the spray chamber in cleaning the recirculated air and position 3 gives the quality of the mixed air entering the mine workings. Sampling position 4 is optional in a basic control system but can be used to provide data balances to check that the other 3 sampling points are giving accurate readings. The fire product detection sampler at position 1 will be used to stop the recirculation fan when combustion products are detected. All the CO2 detector signals can be used to act as back-up fire detectors as the

CO2 levels during a fire will significantly exceed the maximum allowable recirculation values.

Shutting down the recirculation system during blasting can either be done manually or the control system will stop the recirculation when the high levels of gas and dust due to blasting reach the sensors.

The control system settings must be determined from a mine survey of normal operating conditions. A full profile of the relationship between the diesel produced CO_2 concentrations and the other diesel pollutant levels must be determined. The maximum allowable CO_2 and dust concentrations for recirculation can then be determined from a review of the mine air quality indices for gas and particulates. These values are then used in the control system to cut off recirculation. Johnson and Carlson, 1984 have described the calculation method in detail. In this way the mine AQI's can be controlled so that they do not exceed their current full load values. The values should be set such that the worst of the two AQI values, whether gas or particulate, is always used as the recirculation control parameter.

7 ESTABLISHING A SYSTEM

The first information required is the determination of the nature and pattern of the mine diesel equipment usage over the working shifts. This information can then be combined with the routine mine ventilation measurements to determine the recirculation potential from the existing Air Quality Indices.

The potential economic benefits of implementing alternative recirculation schemes should then be calculated to select the optimum scheme from both economic and practical viewpoints.

Discussions should take place with Mines Inspectorates during the outline design of the control system to ensure that there is full agreement on the feasibility and safety of the installation.

Prior to implementation of the system all underground employees should be informed of the basis of the system and assured of its safety and that there will be no extra health risks incurred by them.

Detailed records of the recirculation system's operations should be kept to permit a full analysis of its performance. Routine ventilation surveys should be integrated with the recirculation control system to ensure that the design pollutant profile is still valid. Any change in the

mining method or equipment must be evaluated to determine its effect on the pollutant profile. Any changes in the profile must be incorporated into the control system to prevent unsatisfactory AQI values.

A recirculation system of fan, water spray chamber and control system with sensors can be constructed from commercially available equipment. It is estimated that a system of this type could be built for $100,000 and by reducing the heating by 10% on an average size mine would have a payback period of one year.

There is a need for a trial system of this type to be implemented and researched to determine the operational factors involved and to act as a model for future systems. An experiment of this type would quantify the potential savings for the Canadian mines using air heating.

8 SUMMARY AND CONCLUSIONS

The majority of Canadian mines using air heating employ propane burner systems. The tenfold increase in the price of propane from 1972 to the present has resulted in these mines incurring annual heating costs of over $2,000 per cubic metre of air circulated, which contribute significantly to overall mining costs.

Conventional energy recovery systems for mine exhaust air have been investigated by mining companies but have achieved very limited acceptance because of the low grade heat source, the distances between intake and return airways and the seasonal heating requirement.

The ventilation air quantities for dieselised mines are usually based on worst case conditions with the majority of equipment considered to be working at or close to full load. Variations in the actual diesel output power due to the fluctuations of ore production will permit controlled variable flow recirculation of ventilation air on a whole mine basis. Such systems can be controlled by monitoring CO_2 and dust concentrations. The amount of recirculation will normally be proportional to the amount of design diesel output power not in use at that time.

The provision of a water spray chamber in the recirculation system will improve the quality of the recirculated air by reducing its respirable dust content and removing soluble pollutant gases.

Fan selection and siting are very important in implementing recirculation systems and the fans and control system must be capable of handling the full range of recirculation flows to be used.

214

Instrumentation and control equipment are commercially available to establish a safe monitoring system capable of controlling a whole mine variable flow energy conservation system, without lowering air quality in the mine workings. A system of this type could be installed for $100,000 and would have a payback period of only one year if a 10% saving in air heating costs could be achieved on an average size mine. Savings will depend on the operating conditions on individual mines but it is expected that savings would normally exceed 10%.

The financial potential and technology to implement a system of this type exist. There is a need now to establish a test installation to obtain hands-on experience and to serve as a model for future systems.

REFERENCES

Burton,R.C. et al. 1984. Recirculation of air in the ventilation and cooling of deep gold mines. Third International Mine Ventilation Congress. Harrogate.

French,I. and associates. 1985. Diesel exhaust and the miner's health. Energy, Mines and Resources Canada Symposium. Toronto & Vancouver. Full report to be published in 1985.

Freyman,A.J. 1980. Reclaiming heat from ventilation exhaust air of underground mines: technology and application in Canada. Energy, Mines & Resources Canada, Conservation & Renewable Energy Branch, Industry Series Publication #8.

Hardcastle,S.G. 1985. Comparison of prototype diesel/mineral monitors against a Simslin II, Energy, Mines & Resources Canada, Minerals Research Program, Mining Research Laboratories. Draft report to be published.

Johnson,J.H. & Carlson,D.H. 1984. The application of advance measurements and control technology to diesel powered vehicles in an underground salt mine. Canadian Institute of Mining & Metallurgy Annual General Meeting. Ottawa, paper 102.

McCallum,V.I. 1969. Design of mine air heating plant at Strathcona Mine. Falconbridge Nickel Mines Ltd. Canadian Mining J. October: 62-65.

Pickering,A.J. & Aldred,R. 1977. Controlled recirculation of ventilation - a means of dust control in face advance headings. The Mining Engineer. London. March:329-345.

Pickering,A.J. & Robinson,F. 1984. Application of controlled air recirculation to auxiliary ventilation systems and mine district ventilation circuits. Third International Mine Ventilation Congress. Harrogate.

Summers,E. 1984. Electric mine air heating. Canadian Institute of Mining & Metallurgy Annual General Meeting. Ottawa:paper 11.

Van der Walt,N.T.,Bout,B.J.,Anderson,Q.S.& Newington,T.J. 1975. A fire detection system for South African Gold Mines. 1st Int'l Mine Ventilation Congress. Johbg.

Firedamp accumulations and their dispersal with special reference to the use of controlled recirculation systems

I.O.JONES
Western Australian School of Mines, Kalgoorlie, Australia

ABSTRACT: Firedamp released as the result of coal extraction is still a major hazard to the welfare of personnel in underground coal mines. It is therefore essential that those responsible for the management of mines have a clear understanding of the problems associated with the safe dispersal and dilution of this dangerous mixture of gases. The hazard is of course created by methane which is the major constituent of firedamp since it establishes an inflammable mixture when mixed with air in the proportion of 5 to 15 percent. Due to its buoyancy the firedamp tends to create roof layers and having analysed this problem emphasis is placed on the use of controlled recirculation ventilation systems as a means of improving environmental conditions in longwall panels, advance headings and drivages.

1 RELEASE OF IN-SEAM GASES

Firedamp is a generic term used to describe a mixture of gases released as a result of coal extraction; its main constituent being methane. Other gases present in small quantities include nitrogen, carbon dioxide, hydrogen and helium. These gases are predominantly contained within the cellular structure of the coal in the adsorbed state. The ability of coal seams to adsorb gases increases dramatically with coal rank and seam pressures. It is not surprising therefore that some coal seams at depth contain large volumes of gas within their micells.

The excavation process disturbs the virgin stress field. The stresses in the seam at the exposed surface (ie coal face or rib side) can be assumed to be zero, whilst a short distance into the solid coal the stress will rise to a maximum value (abutment pressure) and thereafter diminish to the pre-mining value.

These changes influence the behaviour of the in-seam gases. The initial increase in strata pressures caused by the approach of a coal face or development heading will give rise to fracturing of the strata and the compression of the seam.

The release of large volumes of adsorbed gas due to compression will give rise to a rapid increase in gas pressure. However, the permeability of the coal seam may be temporarily reduced in this region as

shown in Figure 1, due to the partial closure of the in-seam drainage paths (Curl 1978)

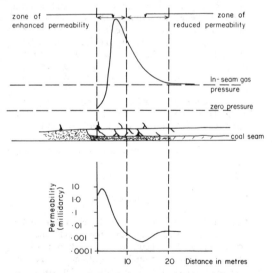

Figure 1: Model of inter-relationship between in-seam gas pressure and permeability

This situation may describe the root cause of some gas outbursts experienced in coal mines. Be that as it may, the rapid fall off in strata pressure which occurs from the abutment pressure towards the free face of the coal is associated with the opening up of slip planes and cleats and the fracturing of the coal mass. Consequently the permeability of the coal in this region is greatly enhanced, which assists in the migration and **controlled** release of the in-seam gases.

The released in-seam gases may also migrate through fractures, joints and bedding planes in the adjacent roof and floor strata; thus finding their way to the mine ventilation circuit through breaks in the strata which intersect the mine airways.

2 FIREDAMP CONTROL AND DISPERSAL

2.1 Dispersal of firedamp

The major constituent of firedamp, methane, is a hazard because when it is mixed with air in the range 5-15% it forms an inflammable mixture. Consequently one of the essential features of a ventilation system in a coal mine is its capacity to disperse and dilute the released firedamp to safe levels. However, it is difficult to achieve this objective at all locations and at all times. Indeed, those of us with any experience of working in gassy mines will know the reality of this problem. At its worst the evidence is there in the recorded ignitions of methane-air mixtures.

Of course, one means of minimising this problem is to pre-drain the coal seam and associated strata prior to coal extraction. However, in-seam gas drainage cannot be successfully applied in all cases and in many coal mines the only means of controlling the firedamp problem is by good ventilation.

One major problem associated with the safe dispersal of firedamp is its buoyancy compared with air. Its major constituent methane, has a specific gravity of 0.55 and firedamp accumulations at or near the roof of mine airways can be difficult to disperse.

2.2 The Richardson Number

Let us look briefly at this problem and use Prandtl's mixing length theory (Prandtl 1952) to analyse the situation when the buoyancy of the methane is sufficient to overcome the mechanism of turbulent diffusion.

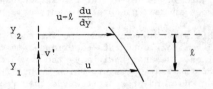

Assuming the flow is in the x-direction we have that $v = w = 0$ (v and w are the velocities in the y and z directions, respectively).

The mean turbulent shear stress between the two layers, a distance ℓ apart is given by:

$$\tau = \rho \ (\ell \frac{\overline{du}}{dy})^2$$

where: ℓ = mixing length

In the presence of a stable density gradient such as that caused by the presence of a firedamp roof layer the kinetic energy of the vertical fluctuations will decrease due to the additional work which has to be done in overcoming the buoyancy of the lighter gas.

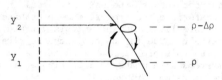

The work done in raising a ball of fluid a distance 'y' is given by the downward force at the upper level, so that:

Work done = $F = -g \ \Delta\rho \ V$

where: F = downward force
$\Delta\rho$ = density difference
V = volume of fluid

Thus over the distance $y_1 + \ell$, we have that:

$$\text{Work done} = \int_{y_1}^{y_1 + \ell} F \ dy = -g \frac{d\rho}{dy} V \frac{\ell^2}{2} \quad (2.1)$$

The kinetic energy of the vertical velocity fluctuations is given by:

$$\frac{1}{2} \rho \ V \ \overline{v'}^2 = \frac{1}{2} \rho \ V \ \ell^2 \ (\frac{\overline{du}}{dy})^2 \quad (2.2)$$

Equating the kinetic energy to the work done gives:

$$-g \frac{d\rho}{dy} = \rho \ (\frac{\overline{du}}{dy})^2$$

The ratio of these quantities is known as the Richardson Number (Ri), such that:

$$Ri = \frac{-\frac{g}{\rho} \frac{d\rho}{dy}}{(\frac{\overline{du}}{dy})^2} \quad (2.3)$$

In reality turbulent diffusion will cease if any particular disturbance is so enfeebled that the next one is weaker and it was suggested by Prandtl that the limit of stability for free flow can be taken as Ri = 0.25.

It is also interesting to note that the Richardson Number is synonomous with the so-called densimetric Froude Number, F_Δ, such that:

$$F_\Delta = \frac{\bar{u}^2}{g \frac{\Delta\rho}{\rho} \ell} = Ri^{-1} \qquad (2.4)$$

The Richardson Number as defined in equation (2.3) relates to the situation at a point in the flow regime. However, it is possible to obtain a so-called 'pipe' Richardson Number by making the following assumptions (Bakke and Leach 1960).

$$\frac{d\rho}{dy} = \frac{\rho_a - \rho_\ell}{h} = \frac{\Delta\rho}{h}$$

$$\frac{d\bar{u}}{dy} = \frac{\bar{u}_a - \bar{u}_\ell}{h}$$

$$h = \frac{q}{w\bar{u}_\ell}$$

where h = layer depth
q = inflow of firedamp
w = width of rectangular airway

and the suffixes a and ℓ refer to the air and layer flow respectively.

Thus the pipe Richardson Number, Ri_p, is given by:

$$Ri_p = \frac{g \frac{\Delta\rho}{\rho} \frac{q}{w}}{\bar{u}_\ell (\bar{u}_\ell - \bar{u}_a)^2} \cos\alpha \qquad (2.5)$$

where: α = angle of inclination of airway

It appears that the critical value of Ri_p for stable density layers in turbulent pipe flow is 0.8. At this value turbulent diffusion ceases across the layer interface. Below this value turbulent diffusion is operative and when Ri_p is zero, the mixing is that which would prevail in the absence of a density layer.

2.3 The layering number

It is difficult to calculate values for the pipe Richardson Number and it is generally accepted that the so-called Layering Number T, can be used to better effect in identi-

fying areas where firedamp layering may be a problem.

Dimensional analysis can be used to examine this phenomenon and the functional equation defining the variables influencing the layering of firedamp in mine airways is given by (Jones 1969):

$$f_1 (\ell, L, \rho, \Delta\rho, \bar{u}, q, \nu, g, \varepsilon, \alpha, \ell_s) = 0 \quad (2.6)$$

where:
ℓ = characteristic dimension of airway
L = layer length
ρ = density of air
$\Delta\rho$ = density difference
$\bar{u}$ = mean velocity of airflow
q = inflow of firedamp
ν = kinematic viscosity of air
g = gravitational constant
ε = mean height of roughnesses
α = inclination of airway
ℓ_s = characteristic dimension of gas entry

This equation can be reduced by dimensional analysis to give:

$$f_2 (\frac{L}{\ell}, \frac{\Delta\rho}{\rho}, \frac{\nu}{\bar{u}\ell}, \frac{g\ell}{\bar{u}^2}, \frac{q}{\bar{u}\ell^2}, \frac{\varepsilon}{\ell}, \alpha, \frac{\ell_s}{\ell}) = 0$$

$$\qquad (2.7)$$

Assuming the same fluids are involved in geometrically similar situations then:

$$\frac{\Delta\rho}{\rho}, \frac{\varepsilon}{\ell}, \alpha, \frac{\ell_s}{\ell} \text{ are no longer variables}$$

and the following simplified functional equation is obtained for the dependent variable $\frac{L}{\ell}$.

$$\frac{L}{\ell} = f_3 (Re, F_\Delta, \frac{q}{Q}) \qquad (2.8)$$

where: $\frac{L}{\ell}$ = dimensionless layer length
Re = Reynolds Number of airflow
F_Δ = densimetric Froude Number

It is permissible to carry out any transformation on this equation providing that the dimensional relationships are not disturbed and following is the one such rearrangement:

$$\frac{L}{\ell} = f_4 (Re, F_\Delta \text{ or } \frac{q}{Q}, \overline{\frac{q}{Q} F_\Delta}) \qquad (2.9)$$

and one definition of the $\overline{\frac{q}{Q} F_\Delta}$ term is the Layering Number, T, since

$$T = F_\Delta^{2/3} (\frac{q}{Q})^{-1/3}$$

$$\dot{T} = \frac{\bar{u}}{\sqrt[3]{g\,\frac{\Delta\rho}{\rho}\,\frac{q}{w}}} \qquad (2.10)$$

2.4 Experimental results (Jones 1969)

Experimental work carried out in horizontal rectangular ducts of various sizes both unlined and lined with model props and bars produced the results depicted in Figure 2.

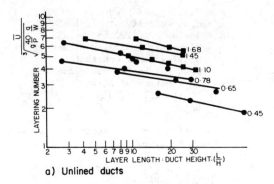

a) Unlined ducts

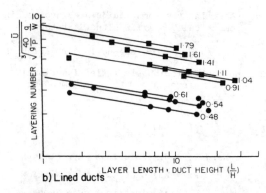

b) Lined ducts

Figure 2: Relationship between layering number, layer length and mean air velocity

This indicated a problem with the use of the Layering Number since the relationship between the layer length $\frac{L}{\ell}$, and the Layering Number, $\dot{T}$, appears to be velocity dependent. However, it is widely used and a recent reference to its use states that 'if the Layering Number is <5 then the likelihood of layering occuring is great whereas if it is >8 then the possibility is slight. Between values of five and eight the slope of the airway, roughness of the surfaces and other physical factors will determine whether layering occurs'(Burrows et al 1982). Used in this way the Layering Number can be a useful indicator of those

areas within a mine ventilation circuit where firedamp layering could be a problem.

3 CONTROLLED RECIRCULATION IN COAL MINES

3.1 General application of controlled recirculation systems in coal mines

In coal mining there are two circumstances which may warrant the use of controlled recirculation. Firstly there is the problem of insufficient fresh air being available to ventilate the more remote longwall panels. This problem could be overcome by any of a combination of the following methods:

a) Increasing the volumetric efficiency of the ventilation system.
b) Increasing the capacity of the main fan.
c) Introducing a controlled recirculation system in the panel.

Secondly there is the special problem associated with environmental control in mechanised headings and drivages (Allan 1982). When an auxiliary ventilation system is being designed for these situations the following environmental factors have to be taken into account:

* Firedamp dilution and removal.
* Dust control.
* Heat and humidity.
* Dispersal of shotfiring fumes.
* Dilution and dispersal of diesel exhaust fumes.
* Noise pollution.
* Visibility for machine operators.
* Congestion of equipment.

Controlled recirculation systems offer better control over many of these variables and many such systems have been and are being used in longwall advance headings and drivages.

3.2 Controlled recirculation on longwall faces

For many years the concept of controlled recirculation in mine ventilation was frowned upon due primarily to the mistaken belief that it would result in the build up of pollutants to hazardous levels. Indeed, for this reason the legislation relating to the ventilation of coal mines in the UK and Australia prohibits the use of controlled recirculation. The pollutants of major concern are:

* firedamp,
* dust,
* heat and humidity, and
* blasting fumes.

The general-body concentration of any pollutant in a ventilation circuit is governed by the rate at which the pollutant enters the circuit divided by the flow of fresh air in the circuit and does not in any way depend on any recirculation that may be taking place (Leach and Slack 1969). In order to verify this statement let us look at the situation in a coal producing section where recirculation has been established by the use of a booster fan in a cross-cut driven between the main intake and return roadways as shown in Figure 3 (Jones 1983, Burton et al 1984).

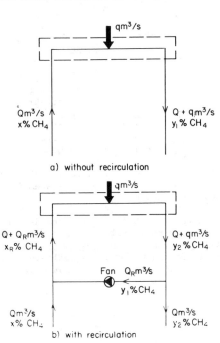

a) without recirculation

b) with recirculation

Figure 3: Ventilation of longwall face

Without recirculation

The maximum concentration of methane in the return airway is given by:

$$y_1 = \frac{Q \frac{x}{100} + q}{Q + q} \times 100\% \qquad (3.1)$$

With recirculation

The maximum level of methane in the return airway is given by:

$$y_2 = \frac{Q \frac{x}{100} + Q_R \frac{y_1}{100} + q}{Q + q} \times 100\% \qquad (3.2)$$

The concentration of methane in the intake airway following the mixing of fresh air and recirculated air is given by:

$$x_R = \frac{Q \frac{x}{100} + Q_R \frac{y_1}{100}}{Q + Q_R} \times 100\% \qquad (3.3)$$

Introducing the recirculation factor, R_F, defined as:

$$R_F = \frac{Q_R}{Q + Q_R} \qquad (3.4)$$

we have that:

$$x_R = \left[(1-R_F) \frac{x}{100} + R_F \frac{y_1}{100} \right] \times 100\% \qquad (3.5)$$

Let us now introduce some realistic figures to define a working situation. Assuming an inflow of fresh air into a longwall panel of 15 m³/s with a methane concentration of 0.05%, a recirculation factor of $\frac{1}{3}$ and a make of methane of 25 ℓ/s, the maximum methane concentration in the return roadway with and without recirculation would be as follows:

Without recirculation

$$y_1 = \frac{15 \times \frac{.05}{100} + .025}{15} \times 100\%$$

$$= 0.22\%$$

With recirculation

$$y_2 = \frac{15 \times \frac{.05}{100} + 7.5 \frac{.22}{100} + .025}{22.5} \times 100\%$$

$$= 0.22\%$$

It is therefore clear that the maximum level of methane concentration in the return airway with recirculation does not exceed that which existed prior to the introduction of recirculation.

The maximum methane concentration in the mixed intake airflow will increase to the level given by:

$$x_R = \frac{15 \frac{.05}{100} + 7.5 \frac{.22}{100}}{22.5} \times 100\%$$

$$= .107\%$$

It follows therefore that whilst the maximum methane concentration in the return airway does not increase, the average level of methane concentration in the circuit will do so due to the increase in methane concentration in the mixed intake airflow. Similar results would have been obtained if the pollutant were dust particles or heat. Indeed, in this regard reference must be made to the full scale experiment carried out at the Loraine Gold Mine in South Africa (Fleetwood et al 1984) where 35 m³/s of air was recirculated in a stoping section to increase the airflow rate to a total of 50 m³/s. This controlled recirculation system has worked successfully for over two years and other installations will be commissioned in other South African gold mines in the near future.

3.3 Controlled recirculation in advance headings

A conventional ventilation system installed in an intake advance heading is shown in Figure 4. The inflow of firedamp (methane) is assumed to be 10 ℓ/s and the system is designed in such a way that recirculation cannot possibly occur.

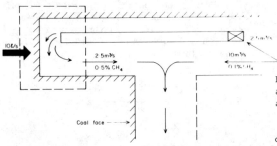

Figure 4: Conventional auxiliary ventilation system in intake advance heading

Using a controlled recirculation system with an exhausting fan as shown in Figure 5, the methane concentration at various positions for three different recirculation airflows is clearly shown.

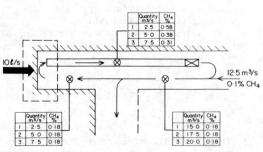

Figure 5: Controlled recirculation system in intake advance heading

One of the advantages of using a recirculation system is that the air velocity in the advance heading can be increased thereby giving much better conditions for the dispersal of firedamp accumulations.
It would also allow for better control of

the dust problem with an in-duct filtration system.

Acceptable conditions can also be maintained with a relatively low inflow of fresh air and a high recirculation quantity. Analysing such a problem when the recirculation flow is 1.5 times that of the fresh air inflow leads to the situation defined schematically in Figure 6.

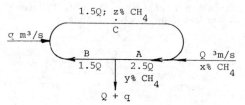

Q = inflow of fresh air, m³/s
$x\%$ = methane concentration in fresh air supply
y & $z\%$ = methane concentrations at A and B, respectively
q = inflow of methane, m³/s

Figure 6: Diagrammatic representation of a controlled recirculation system in an advanced heading

Assuming that fresh air extends throughout the circuit initially and that the inflow of methane commences when the fan is switched on, we have the following build up of methane percentages per cycle.

Cycle 1:
$$z_1 = \frac{1.5Q \times \frac{x}{100} + q}{1.5Q} \times 100\%$$

$$y_1 = \frac{Q \times \frac{x}{100} + 1.5(Q+q)\frac{z_1}{100}}{2.5Q} \times 100\%$$

Cycle 2:
$$z_2 = \frac{1.5Q \times \frac{y_1}{100} + q}{1.5Q} \times 100\%$$

$$y_2 = \frac{Q \times \frac{x}{100} + 1.5(Q+q)\frac{z_2}{100}}{2.5Q} \times 100\%$$

Cycle n:
$$z_n = \frac{1.5Q \frac{y_{n-1}}{100} + q}{1.5Q} \times 100\%$$

$$y_n = \frac{Q \times \frac{x}{100} + 1.5(Q+q)\frac{z_n}{100}}{2.5Q} \times 100\%$$

Given that the inflow of fresh air is 10 m³/s with a methane concentration of .1% and that the intake of methane in the heading is 10 ℓ/s, the system settles down fairly rapidly to give the following conditions at points A, B and C.

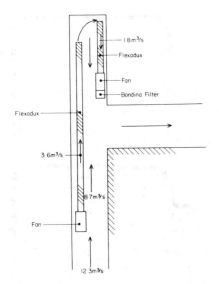

Figure 7: Intake gate advance heading
with conventional ventilation system

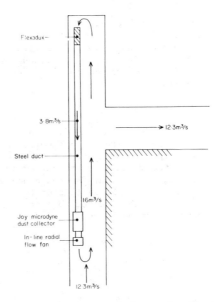

Figure 8: Intake gate advance heading
with recirculation system

Point	Methane Percent
A	.27
B	.20
C	.20

3.3.1 Practical examples

A conventional non-recirculating system in
an intake advance heading is shown in Figure 7. It involved the use of a forcing
fan to provide suitable velocities at the
face of the heading to disperse the firedamp and a short overlap exhaust system
with a dust extractor to control the dust
produced by a Dosco Mark 2A heading machine.
The system was not successful in controlling
the dust problem and the recirculation
system shown in Figure 8 (Mining Environmental Bulletin) was introduced. It was
much more successful with the concentration
of respirable dust in the heading being
reduced from 8.1 mg/m^3 to 2.16 mg/m^3.

The controlled recirculation system also
had the advantage of reducing the noise
level and congestion of equipment by doing
away with the 'overlap' exhaust system.

The use of large heading machines in
development drivages has also produced a
dust control and/or a firedamp dispersal
problem. Once again the inadequacy of a
conventional system as shown in Figure 9a
(Allan 1982) can be overcome by the introduction of the controlled recirculation
system shown in Figure 9b. Not only does
it allow for better air cleaning, but it
also gives higher air velocities which
ensures better mixing of firedamp.

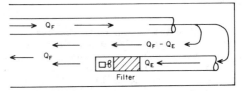

a) Conventional - Main forcing / overlap exhaust
($Q_F < Q_E$)

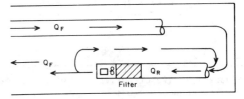

b) Recirculation - Main forcing / overlap exhaust
($Q_R > Q_F$)

Figure 9: Conventional and recirculation
systems in development drivages

3.4 Localised improvements in mixing using fans and airmovers

In longwall mining despite all precautions
firedamp accumulations often occur at

ripping lips, stable holes and the buttock of the cut being taken by a shearer loader. Enhanced mixing and dispersal of firedamp can be achieved in all such locations by the use of fans or airmovers (Leach and Slack 1969).

3.4.1 Ripping lip

If it is intended to use an airmover or jet fan to disperse a firedamp accumulation at a ripping lip then care must be taken to site it in the ventilation airstream as shown in Figure 10 otherwise the enhanced mixing will tend to be restricted to the eddy flow where the methane concentration will continue to build up.

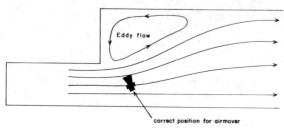

Figure 10: Airflow pattern at return ripping lip showing correct position for airmover

An auxiliary fan could also be used to improve the dispersal and mixing of firedamp at the ripping lip as shown in Figure 11. This system makes use of recirculatory flow to increase the airflow rate within the active zone where the enhanced mixing is required.

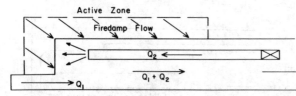

Figure 11: Use of auxiliary fan to increase airflow in active zone

3.4.2 Stable holes

It is difficult to adequately ventilate the stable holes on longwall advancing faces because the natural tendency of the airstream is to move directly from the intake roadway into the coal face leaving poorly ventilated regions near the ribside and coal face. Once again an airmover could be used to good effect if correctly positioned as shown in Figure 12.

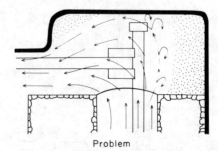

Problem

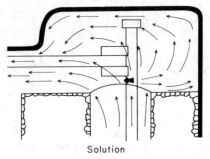

Solution

Figure 12: Airmover used to improve ventilation in stable hole

Of course brattice cloth hurdles or barriers can be used to re-direct the airflow in both these situations. However, such passive devices need to be well constructed and maintained if they are to be effective and this is often not the case.

3.4.3 Shearer cuts

Frictional ignition caused by coal cutting is an ever present problem with such machines as shearer loaders. Whilst the dust problem is alleviated with water sprays directed to the front of the picks, the problem of frictional heating can only be tackled by fan sprays directed to the back of the picks. Needless to say, it would also be helpful if the ventilation of the cut could be improved and this could once again by effected by the careful use of airmovers or small fans.

4 CONCLUSIONS

4.1 Precautions

In assessing the problems associated with controlled recirculation systems it is as well to consider the basic assumptions under-pinning the theoretical analyses. These are that:
* steady state conditions prevail,

* the operation of the recirculation fan does not adversely affect the inflow of fresh air,
* the mixing of the firedamp and/or other pollutants is immediate and effective.

Steady state conditions may not prevail at all times and a variable input of pollutants may occur. It is therefore essential that a controlled recirculation system in longwall panels is well monitored and controlled so that the system is discontinued if a concentrated pulse of hazardous pollutant is introduced into the return airway whilst personnel are in the section.

In this regard it is as well to remind ourselves of the full scale experiment carried out at the Loraine Gold Mines where over a period of three years a carefully designed monitoring and safety system provided adequate safeguards

In the case of controlled recirculation in coal headings or drivages in coal measure strata care must be taken when starting up the auxiliary ventilation system since it may cause a pulse of rich firedamp air mixture to move into the fresh air stream. If a firedamp accumulation exists at the face of the advance heading the system should be brought into being gradually with constant monitoring of the methane concentration in the mixed intake flow. Once again, careful management of the system is important.

In regard to the use of airmovers or jet fans to disperse localised accumulations of firedamp, care is needed to position the devices correctly so that they draw air from the main air stream and cause as little recirculation as possible within the eddy stream. Once again there is no substitute for regular monitoring of system performance.

4.2 Benefits

Following are the benefits of using a controlled recirculation ventilation system in a longwall panel.
a) The use of a controlled recirculation system results in an increase in the rate of airflow through the face. This means that the inertial energy of the air flow is increased thereby enhancing mixing. The increase in the mass flow of air also gives rise to increased values of cooling power.
b) A controlled recirculation ventilation system presents management with a cost effective means of overcoming the problem of providing increased air flows at the longwall faces.

The use of controlled recirculation systems in mechanised headings or drivages have the following benefits.
a) It ensures increased rates of airflow at and near the face of the heading/drivage thereby giving rise to enhanced mixing.
b) The systems also allow for a much better control of the dust problem created by the use of heading machines.
c) Applying controlled recirculation systems to advance headings can also result in a reduction of noise levels and less congestion of equipment.

In tackling localised problems such as firedamp accumulations at ripping lips or face cuts using jet fans or airmovers it is important that the devices be properly located and managed. With appropriate care and attention the system should work well and give rise to better conditions than would otherwise prevail.

5 REFERENCES

Curl, S J. 1978, Methane prediction in coal mines. Report Number ICTIS/TR04 IEA Coal Research 1978.

Prandtl, L. 1952, Essentials of fluid mechanics. Blackie & Sons Ltd.

Bakke, P & Leach, SJ. 1960, Methane roof layers. SMRE Research Report No. 195.

Jones, I O. 1969, A report of model and full scale experiments on firedamp roof layers. The Mining Engineer.

Burrows, J. (editor) 1982, Environmental engineering in South African mines. The Mine Ventilation Society of South Africa.

Allan, J A. 1982, A review of controlled recirculation in UK collieries. H.Q. Ventilation Branch, NCB.

Leach, S J & Slack, A. 1969, Recirculation of mine ventilation systems. The Mining Engineer.

Jones, I O. 1983, The benefits of recirculation in mines. Proceedings of seminar on Environmental Quality Control in Mines. WA School of Mines.

Burton, R C. Plenderleith, W. Stewart, J M. Pretorius, B C B. Holding, W. 1984, Recirculation of air in the ventilation and cooling of deep gold mines. Proceedings of Third International Mine Ventilation Congress, UK. 1984

Fleetwood, B R. Burton, R C. Pretorius, B. Holding, W, 1984. Controlled recirculation of ventilation air at Loraine Gold Mines Ltd. Proceedings of Association of Mine Managers of South Africa.

Mining Environmental Bulletin No. 1, 1977. Controlled recirculation of air. NCB (UK) Mining Department Publication.

The feasibility of controlled air recirculation around operating longwall coal faces

I.LONGSON, R.D.LEE & I.S.LOWNDES
University of Nottingham, UK

ABSTRACT: This paper discusses the difficulties being experienced by some UK coal mines in supplying adequate airflows, using their existing ventilation systems, to working faces which are long distances from the surface connections. Conventional solutions to this problem are provided and the contribution large scale recirculation might make recognised.

1 INTRODUCTION

There have been relatively few new shafts constructed since the nationalisation of the United Kingdom coal mining industry in 1947. The major part of present production comes from collieries which were in existence prior to this date and at a large proportion of these more than one seam is being or has been worked. Current coal production areas are up to 10 km from the surface connections and in some cases are also under the sea. Intensive face mechanisation has resulted in very large increases in coal production rates and consequently the rates at which the major pollutants of methane, dust and heat enter the ventilating airstream. Air flow rates on the faces have increased by between three to six fold over the past 30 years (Morris & Walker 1982) and this has been principally achieved through the extensive use of underground booster fans. Many of the shafts and trunk airways in use were not designed to carry the present flow rates often resulting in mine ventilation networks (MVN's) which have high frictional losses, high ventilating pressures, large leakages, low volumetric efficiencies and high ventilation costs. In one coalfield ventilation is responsible for 40% of the total power consumed and the current estimate of the NCB total ventilation power costs is £55m/annum (Hornsby 1985).

In the immediate and even medium term future the developments described above are likely to continue as it is believed that the existing methods of working are capable of providing further increases in productivity. Thus additional loads will be imposed on the ventilation systems which will be servicing an increased number of faces at even greater distances from the surface connections. This will result in some ventilation systems not being able to cope and many will develop excessive operating costs.

The problem to be resolved, therefore, is the maintenance or possible increase of face air flow rates subject to the physical constraints described above.

2 POSSIBLE SOLUTIONS

These include the enlargement or duplication of airways but often this is not feasible on the grounds of both the time required to implement and the cost involved.

Satellite ventilation shafts can provide the most efficient ventilation solution but in urban areas difficulties relating to site acquisition, the time required to obtain permission for and the carrying out of the sinking, together with the major capital sum needed often removes this alternative.

There has been an increase in the use of underground booster fans in the UK in recent years. This solution is attractive as it can normally be effected quickly and, by the careful siting of the installation, a medium term improvement in the efficiency of the network can be achieved. However, a number of mines which have been continuously increasing the number and size of their booster installations are fast approaching a safe upper limit.

The recirculation of ventilation air within a MVN is illegal in the UK although some carefully designed auxiliary ventilation arrangements for headings, which feature controlled recirculation, have been granted exemptions (Robinson 1972, Pickering and Aldred 1977, Pickering and Robinson 1984). These have functioned very satisfactorily affording significant improvements in the environmental control of heading conditions. This is the background leading to the study at the University of Nottingham, into the feasibility of controlled recirculation circuits within MVN's.

3 POLLUTANT DISTRIBUTION AND
  CONCENTRATIONS IN SIMPLE RECIRCULATION
  CIRCUITS

The principal role of ventilation particularly in the production areas is to rapidly dilute pollutants to a safe level and then to remove them. The recognised main forms of pollutant are gases, dust and heat which will be present to varying extents in different situations. The local velocity of the airstream has the most important effect on the rate of dilution and the volumetric flow rate on the level of dilution. Filters and air-conditioning units are available and are effective for the general body removal of dust and heat but there is no such device available for gases.

3.1 Gases

The most common fear or doubt about the concept of recirculating a controlled proportion of the return air around a district is that it will result in a build up of gaseous pollutants within the recirculation zone. It has been shown in earlier papers (Leach and Slack 1969) that, theoretically, the maximum concentration level in a simple recirculation circuit under steady state conditions is given by the rate at which concentrated pollutant (say methane) enters the zone divided by the rate at which fresh air enters the zone. If these rates are identical to those experienced on a conventionally ventilated district then the maximum general body concentration experienced in both circuits will be the same. This maximum methane concentration will not be altered by the proportion of return air that is recirculated, provided that the fresh air flow rate is maintained and, therefore, there will be no accumulation of pollutants within the zone.

Figure 1 summarises the layout and relat-

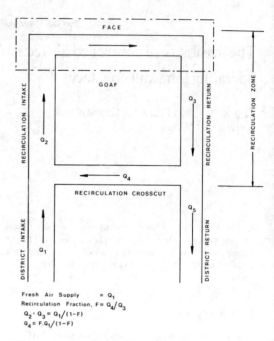

Fresh Air Supply $= Q_1$
Recirculation Fraction, $F = Q_4/Q_3$
$Q_2 \cdot Q_3 = Q_1/(1-F)$
$Q_4 = F.Q_1/(1-F)$

Figure 1. Recirculation district layout and basic relationships.

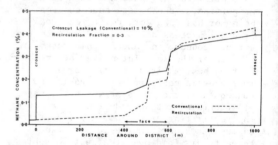

Figure 2. Effect of recirculation on district methane concentrations.

ionships associated with steady state flow around a simple recirculation circuit.

A change to recirculation, however, alters the methane concentration distribution around the district when compared with that present with conventional ventilation and this is shown in Figure 2. As would be expected the general body methane concentration in the intake is increased resulting from the introduction of the polluted recirculated fraction. It is very important, however, to recognise that this is methane which has already been diluted to well

228

below the explosive range and so does not
present any risk of causing an ignition.

3.2 Dust

Recent work has included the development of
a computer prediction program to study the
effect of increased airflows on the concentrations of both respirable and coarse dust
in the district airstream. This has shown
that recirculation should not pose any
additional health or discomfort problem
with regard to dust other than might equally
be caused by the increased air velocities
produced by similar increases in face airflow by the supply of additional fresh air.
    However, since this is the topic of
another paper at this Symposium by Dr. S.G.
Hardcastle, who was a member of the original
study group at Nottingham, the subject is
not dealt with in any more detail in this
paper.

3.3 Heat and Humidity

The complexity of the heat and moisture
exchanges with the ventilation airstream
experienced through a highly mechanised
coal face makes exact computer simulation
difficult. In addition, as there are no
working faces practising controlled recirculation, no data for correlation purposes
are available. However, existing simulation
programs produced for climatic prediction
with conventional ventilation have been
modified to enable recirculation circuits
to be modelled. Up to the present time the
study has only looked at situations with
VRT's below 30 deg C and where no air
conditioning is being carried out. For this
case it is found that the recirculation
results in the elevation of both dry bulb
and wet bulb temperatures in the intake
compared with those present in the return
(see Figure 3). The big improvement, when
compared with a conventional ventilation
circuit in which the air flow rate is equal
to the fresh air component of the recirculation case, is the climatic benefit derived
from the increased air velocity, as reflected in reduced Effective Temperature or
increased Air Cooling Powers. This is
particularly marked where the initial face
velocity is under 2 m/s. The magnitude of
the effect on the recirculation intake air
is directly related to the proportions and
properties of the combining airstreams.
    The recirculation crosscut is a convenient
position to carry out air conditioning or
dust filtration if this is desirable.
However, in a real system this crosscut may
be relatively remote from the face and, if

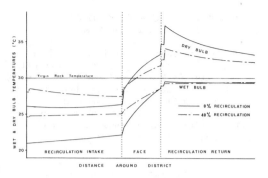

Figure 3. Effect of recirculation on air
temperatures.

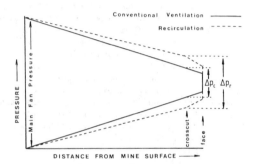

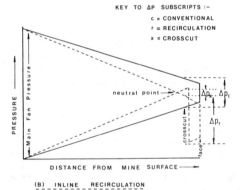

Figure 4. Generalised pressure distribution
in the mine.

this was the case, only a small proportion
of the air cooling introduced would reach
the desired location, which is often the
return end of the face, i.e. the refrigeration system would have poor positional
efficiency.

229

4 PRACTISING CONTROLLED RECIRCULATION WITHIN A VENTILATION NETWORK

In order to create an efficient and flexible controlled recirculation circuit within an existing MVN it will be necessary to reposition or add one or more booster fans. These fans will alter the pressure distribution within the network and two essentially different situations have been investigated, namely;

4.1 Siting the fan in the recirculation crosscut (x-cut recirculation)

Practically this is a convenient position and gives the fan the very specific function of moving the desired proportion of the return air back into the intake. The fan has to handle only the recirculated flow rate and the typical modifications to the network pressures and pressure gradients in changing from conventional ventilation are illustrated in Figure 4(a). However, with the fan in this position it is in opposition to other fans in the network and if no additional provision is made this will result in a reduced flow rate of fresh air into the recirculation zone when compared with the previous conventional ventilation.

4.2 Siting the fan in the intake or return inbye of the crosscut (in-line recirculation)

Under steady flow conditions, the air flow rate distribution in the circuit is the same whether an intake or return inline site for the recirculation booster fan is chosen and the choice will be determined principally by the relative safety/convenience of the two positions. However, pressure values around the network will be different and some important implications of this are discussed later in the paper. The inline position of the fan results in both the fresh air and the recirculation proportions of the airstream passing through it, thus requiring a larger capacity fan. This siting results in the recirculation fan aiding the main system fans and contributing a boosting effect which increases the fresh air flow rate into the recirculation zone when compared with the same circuit practising conventional ventilation. Typical changes produced in the network pressures and pressure gradients with this arrangement are shown in Figure 4(b).

5 CASE STUDY AT AN ACTUAL MINE

In order to confirm the applicability of recirculation theory to a practical situation, a series of computer studies was made on a real MVN. The mine selected lies on the outskirts of Nottingham and was a 'prima facie' case in which recirculation might be beneficial. The workings were all in the same seam, substantially level at a depth of about 400 m. Neither methane nor heat were a problem at this mine. As the plan (Figure 5) shows, there were two main working areas A and B, both homotropally ventilated, each with its own booster fan (approx. 3 km from the shafts) in addition to a main surface fan. The more remote Eastern faces (A1 – A3) lay at a distance of approximately 9 km from the shafts although there was another face A4, nearing the end of its life, only about 2 km inbye of booster fan A. The two Northern faces, represented as a single face (B5), lay some 3 km inbye of booster fan B.

Pressure, airflow and temperature surveys were carried out at this mine and it was found that each of the B faces and also A4 had airflows of about 18 m^3/s whilst the A1 – A3 faces had flows of only 5.8 – 7.0 m^3/s. It was decided, therefore, to simulate recirculation in the Eastern end of the mine with the ventilation objective of increasing airflows on these three faces by firstly 50% and then by 100% of the conventional values. The choice of recirculation crosscut, for practical reasons, lay between the crosscuts denoted V, W, X and Y on Figure 5 although the possibility of uprating booster fan A and driving a new recirculation crosscut (Z) was also considered.

It has already been shown, in the theoretical studies, that the effect of inline recirculation would be to increase the fresh air, Q_1, (see Figure 1) to the recirculation zone at the cost of air to the other faces in the mine, whilst x-cut recirculation was expected to have the opposite effect – namely a reduction of fresh air to the recirculation zone. The objective of this study, using a normal steady state ventilation network analysis program, was to show the extent to which this would apply in a real mine situation.

The effect of progressively increasing the pressure developed by a recirculation booster fan suitably situated in the intake airway or in the crosscut is shown in Figure 6 which shows the relationships between the fresh air supply (Q_1) reaching the recirculation zone and the total airflow (Q_2) inbye of the crosscut. At this stage, no changes were made to the existing main and booster fans. The figure can

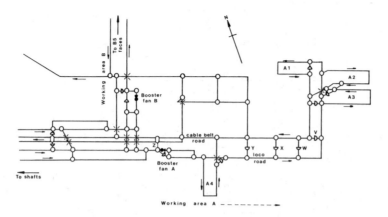

Figure 5. Schematic plan of case study mine.

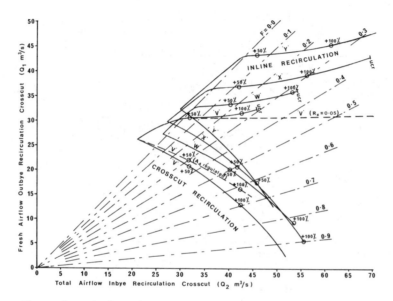

Figure 6. Relationships between Q_1 and Q_2 with recirculation.

be used to determine the Recirculation Fraction (F) necessary to give 50% and 100% increases in the face airflows and it also shows the limiting inbye airflows beyond which uncontrolled recirculation (UCR) through other crosscuts takes place - a feature which is unacceptable from a safety point of view.

The graph confirms that inline recirculation tends to increase the fresh airflow Q_1 and shows that the value of this increase is, to some extent, dependent on the x-cut resistance both before recirculation starts and also on its resistance (Rx) when air is being recirculated. In the case of cross-cut recirculation there is the expected decrease of fresh air supply, independent of crosscut resistance, and the rate of decrease is progressively greater as recirculation crosscuts further outbye are selected. The graph shows clearly that this fresh air decrease may severely limit the increase of face airflow which may be achieved by this system if corrective

231

Table 1. Effect of recirculation on fresh airflows to faces

Recirc. System	Recirc. Crosscut	Fresh Airflow Change on Face					
		A1-3	A4	B5	A1-3	A4	B5
Inline	$V(R_x = 1)$	+46%	- 7%	-0.4%	+52%	- 8%	-0.5%
	$V(R_x = 0.05)$	+46%	- 8%	-0.4%	+46%	- 7%	-0.4%
	W	+27%	-10%	-0.6%	+39%	-15%	-1.0%
	X	+35%	-15%	-0.9%	+43%	-20%	-1.2%
	Y	+44%	-28%	-1.7%	+52%	-35%	-2.2%
Crosscut	V	- 1%	+10%	+1.0%	-39%	+24%	+2.1%
	W	-23%	+14%	+1.3%	-65%	+34%	+2.7%
	X	-24%	+21%	+1.8%	-80%	+48%	+4.1%
	Y	-42%	+34%	+2.7%	Not achievable		
Total Face Airflow		50% increase			100% increase		

Table 2. Effect of regulated recirculation on fresh airflows to faces

Recirc. System	Recirc. Crosscut	Fresh Airflow Change on Face					
		A1-3	A4	B5	A1-3	A4	B5
Inline	V	+25%	+0.5%	+0.2%	+25%	+0.5%	+0.2%
(Q_1 regulated)	W	+ 4%	+0.5%	+0.2%	+ 4%	+0.5%	+0.2%
	X	+10%	+0.5%	+0.2%	+10%	+0.5%	+0.2%
	Y	+ 8%	+0.5%	+0.2%	+ 8%	+0.5%	+0.2%
Crosscut	V	+ 5%	0	+1.1%	-23%	0	+2.4%
(A4 face	W	-16%	0	+1.5%	-47%	0	+2.8%
regulated)	X	-13%	0	+2.0%	-53%	0	+4.5%
	Y	-24%	0	+2.8%	-	-	-
Total Face Airflow		50% increase			100% increase		

Table 3. Total mine ventilation costs relative to conventional ventilation

Recirc. System	Recirc. Crosscut	Regulation Applied	Recirc. Fraction	Cost Change	Recirc. Fraction	Cost Change
Using existing main and booster fans						
Crosscut	V	No	35%	+2%	(70%)	+11%
	V	A4	31%	+2%	(62%)	+10%
Inline	V	No	5%	+4%	26%	+11%
	V	Q_1	18%	+5%	38%	+12%
	W	No	17%	+7%	33%	+20%
	W	Q_1	32%	+8%	49%	+22%
Using derated booster fan A and $R_x = 0.05$*						
Inline	V	No	15%	-7%	37%	- 1%
	W	No	28%	-4%	45%	+ 6%
Total Face Airflow			50% Increase		100% Increase	

* Note:- It was necessary to use the lower recirculation crosscut resistance in order to eliminate the risk of uncontrolled recirculation at higher face airflow increases.

additional steps are not taken.

The effect of recirculation on the fresh airflow reaching the various faces in the mine, relative to those in the case of conventional ventilation, is shown in Table 1. This table relates to a recirculation crosscut resistance $Rx = 1 \ Ns^2/m^8$ except in one case for the use of crosscut V. In this the resistance Rx was reduced to $0.05 \ Ns^2/m^8$; this had only a small effect on the fresh airflow changes, even at a 100% increase in the total airflow on the Eastern faces, but it reduced the tendency to uncontrolled recirculation.

Table 1 shows that, whilst the introduction of recirculation can cause very significant changes in Q_1 reaching the recirculation zone itself, the effect on the other faces in the mine decreases with increasing distance from the recirculation zone. In particular, the airflow changes in the B faces was normally less than 2-3%.

In order to minimise the airflow loss in the A4 face with inline recirculation it is possible to regulate Q_1 to the recirculation zone, but at the cost of slightly higher recirculation fan pressures and, hence, higher operating power costs.

A more important use of regulation is when using the x-cut recirculation system. Table 1 clearly shows that, when using this system, the recirculation crosscut should be kept close to the face in order to minimise the fresh air loss to the system. This loss can be further minimised by regulating the airflow to the A4 face so that it is maintained at the value for conventional ventilation; this also has the effect of reducing the recirculation fan pressure. One example of this (for crosscut V) is shown by the broken line in Figure 2.

Table 2 shows that, by using regulated inline recirculation, the airflows to faces A4 and B5 are virtually unaltered and that the increase in Q_1 merely reflects the recovery of the conventional ventilation leakage through the recirculation crosscut selected; it is advantageous to select the crosscut which has most leakage - in this case crosscut V. The table also shows that x-cut recirculation, at least through crosscut V, is more feasible since larger total air flows on the Eastern faces could be achieved without excessive losses of fresh air supply, and hence increases in the maximum methane concentrations in the general body of the air. It is this latter consideration, in any mine considering recirculation, which may well decide which system should be employed.

It was found that obtaining recirculation by uprating booster fan A to a pressure of

7 kPa (the maximum considered desirable) and by driving the new recirculation crosscut (Z), with a resistance of $1 \ Ns^2/m^8$, it was not possible to achieve the desired increase of airflow on the Eastern faces. The maximum face airflow increase was rather less than 38% and this was accompanied by airflow increases of 41% on A4 face and of 2.5% on B5. Regulation of A4 airflow, as before, allowed an increase in the airflow on faces A1 - A3 of 45%, and on B5 of 1.8%. The operating power costs for this layout were, however, significantly higher than any of the others (44% higher costs than with the existing conventional ventilation).

Whilst safety must be the prime consideration, it is very important that the choice of recirculation system and layout should also take economics into account and the operating costs for each of the options considered were calculated (Table 3). The optimum solution for this mine from a purely economic standpoint, assuming no changes to the existing mine fans, was to use x-cut recirculation through crosscut V with regulation of the airflow in face A4 but, particularly if a 100% increase of the airflow in faces A1 - A3 was necessary, the recirculation fraction became rather too high (62%) - indicating an appreciable loss of fresh air. The total mine ventilating costs for 50% and 100% increases of A1 - A3 airflows were respectively 2% and 10% greater than for conventional ventilation; these small increases show that recirculation is a very economic method for increasing airflows where safety considerations allow.

Other layouts, however, could also be applied with little further additional cost increase as Table 3 shows. The table also shows that, if using inline recirculation, it would be possible to increase the Eastern face airflows at a lower ventilation cost than the conventional case by reducing the number of stages of booster fan A, and hence reducing its pressure and power costs. This option would, however, reduce the airflow on face A4 by 18 - 21% but this may be acceptable since the face is well ventilated and soon to close.

Taking both economics and changes of general body methane concentrations into account, therefore, recirculation at this mine would probably best be achieved by the use of inline recirculation through crosscut V or W either including the existing main and booster fans and regulating Q_1 or by reducing the pressure of booster fan A and accepting some reduction of the airflow on A4 face.

If only a modest increase of the Eastern

face airflows were required, however, x-cut recirculation through crosscut V could be used at limited additional cost by maintaining the existing main and booster fans.

6 SAFETY

Many of the established practices and safeguards associated with conventional ventilation are equally applicable to controlled recirculation. It is essential in a recirculation circuit, as in a conventional one, that a desired flow rate of fresh air is maintained in order to provide adequate and continuous dilution of all pollutants. Monitoring, followed by the sensible interpretation of any variation in general body concentrations of pollutants is very desirable.

The catastrophic potential of methane makes it very important to monitor and include an alarm provision for the return general body concentration. The higher velocities and increased rates of flow present in the recirculation zone improve the rate of pollutant dilution and result in less effect from any changes in the rate of pollutant production.

In the UK there is a legal requirement that in an intake airway to a working face the methane concentration must be kept below 0.25% and if this is to apply to recirculation systems it will in many situations limit the recirculation fraction that can be used.

Products of combustion from fires and other sources should be monitored and it is important that predictive computer simulation studies are carried out in advance of any serious incident occurring so that the correct interpretation and approved remedial procedure can be initiated by an authorised person without delay. It is considered that practical use of recirculation systems should be restricted initially to mines that do not suffer from particularly high methane emission or spontaneous combustion.

There are now available reliable continuous monitors for most pollutants and the signals from these can be transmitted to a central computer which will provide not only a continuous record of circuit conditions but also be used as input for suitable interrogative software which will automatically interpret and display temporal variations.

Many factors can result in the ventilation flow rates suffering major deviations from the analysed steady state and some of these are considered in the next section.

7 TRANSIENT PHENOMENA

The work presented so far has assumed the existence of steady-state operating conditions. If, however, a major disturbance is introduced to a MVN the resulting unsteady pressures and airflows may cause a potentially dangerous situation. For example, if a fan stops or an obstruction occurs in the vicinity of a coal face the redistribution of pressure across the waste might cause an abnormal emission of firedamp. The subsequent transport around the recirculation zone and beyond of any such pollutant emission will also be time dependent.

There is therefore a great need, from standpoints of safety and the design of suitable control devices and procedures, to be able to simulate the causes and to determine the effects of transient behaviour.

The transient methane phenomena associated with a district recirculation system can be modelled by a set of coupled mathematical equations which describes the emission of firedamp into, and its subsequent mixing and transport around, the recirculation circuit. The unsteady, incompressible and turbulent airflow experienced in mine airways may be described by a first order non-linear ordinary differential equation first proposed by Litwiniszyn (1956). This equation can be generalised using Kirchoff's Circuit Theorems to form a set of equations the solution of which describe the unsteady airflow and static pressure distribution within any given MVN. In the absence of reliable data relating to the natural reservoir gas pressures and available voidage in the waste this initial study considered that the firedamp emission from the goaf was laminar and governed by Darcy's Law. The theoretical model employed to describe the mixing and transport of firedamp around a single working face practising district recirculation is a generalisation of the well known Dilution Equation first communicated by Trutwin (1982).

A preliminary investigation into the unsteady air and methane flow phenomena associated with district recirculation circuits has been conducted at Nottingham University. This study has been concerned with the simulation of a wide range of system disturbances which may initiate major deviations from normal steady state airflow and methane emission patterns. A comparison of the different transients which may occur due to the siting of the recirculation booster fan and recirculation

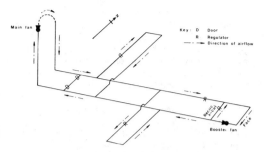

Figure 7. Mine layout used for transient predictions.

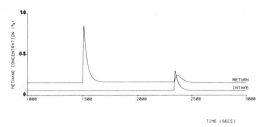

Figure 8. Transient effect due to a brief step input of methane.

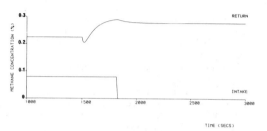

Figure 9. Transient effect due to stoppage of recirculation fan (crosscut doors kept open).

crosscut has also been made. The above system of equations is solved by replacing them by suitable finite difference approximations which are incorporated within a suite of flexible computer programs developed by Lowndes (1984). These programs made use of a range of mathematical disturbances, such as step functions, in order to predict the response of a MVN to a range of abnormal occurrences such as the tripping of booster fans, major changes in circuit resistance and step increases in pollutant emissions.

In order to demonstrate the transient phenomena predicted by the mathematical model, consider the MVN given in Figure 7. This model mine with level workings is practising single face recirculation around its Eastern face, which is 10 km from the surface connections. The recirculation crosscut is situated 800 m outbye of the return end of the face. In the following examples the recirculation booster fan is situated inline in the recirculation return; the other possible sitings of the fan are discussed later.

Figure 8 demonstrates the transient response caused by the introduction of a step input of methane to the recirculation circuit whilst the airflow remains steady. The increase, which lasts for 20 seconds, results in an initial peak in the general body concentration level experienced in the return airway. This peak is subsequently transported around the circuit until a smaller peak, whose size is directly scaled by F and Q_1, is registered at the intake end of the face. A series of diminishing peaks will continue to be recorded at regular intervals until the additional source of methane has been fully removed and the initial steady state conditions are restored. The period between successive peaks corresponds to the effective lap time of the circuit.

The transient phenomena produced on the tripping of the booster fan are shown in Figure 9. The fan stoppage is accompanied by an increase in the mid-face static pressure which inhibits the flow of firedamp from the waste. This reduction of emission is reflected in an initial drop in the return airway general body concentration before it begins to rise due to the reduction in the fresh air quantity delivered to the face. This increase continues until the polluted air-gas mixture remaining in the recirculation intake has been flushed out, at which point the concentration drops slightly to its new steadystate value, i.e. a return to conventional ventilation, although with the crosscut doors assumed to be still open.

The transients introduced on the tripping of a return recirculation fan and the simultaneous closure of the crosscut doors are portrayed in Figure 10. The decrease in methane flow from the waste caused by the rise in the mid-face static pressure is reflected by a depression in the return general body concentration. As the fresh air supply to the face drops, this causes the return concentration to rise gently before decreasing to a new lower steadystate value once the polluted gas remaining

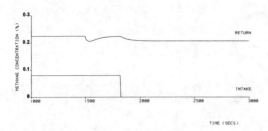

Figure 10. Transient effect of fan
stoppage (and closure of crosscut doors).

in the recirculation intake has been
flushed out. The closure of the crosscut
doors limits the amount of fresh air leakage experienced.

The above examples confirm that conditions
within the circuit are very dependent on
both booster fan and crosscut resistance
and corrective measures following deviations
from the operating steady state may require
their combined control. For example, to
enable the simple single face recirculation
system to provide the maximum fresh air
supply to the face requires the crosscut
doors to close on the failure of the
booster fan. (This could practically be
achieved by a direct monitoring of the fan
performance with a signal link to automatically alarm and to initiate closure of
the doors). An example of the operation of
such a mechanism is illustrated in Figure
10.

From the simulations conducted as part
of this initial study the following conclusions emerge:

There are significant differences between
crosscut sited and exhausting or forcing
inline recirculation fans regarding the
transient methane phenomena recorded at the
intake and return ends of the face during
the start-up or stoppage of the booster fan.
This is a result of the different pressure
distributions discussed earlier in the
paper.

Any transient methane phenomena effected
in a recirculation zone will be transmitted
around the circuit until the system
achieves new steady-state conditions.

The lap time for any given circuit depends
on the airflow quantities and the circuit
volume.

Methane transients are effected by
changes of the recirculation flow as well
as of the flow of fresh air.

The methane transients experienced at the
intake of the face are scaled functions of
the recirculation fraction as well as the
fresh air supply.

Simulations concerning the combined
operation of the booster fan and the crosscut doors illustrate the benefits that
would result from the continuous monitoring
and possibly linked automatic control of
the components in a district recirculation
system when handling major deviations from
steady state.

Finally, experiments conducted on a
laboratory rig were able to provide valuable qualitative support for the predictions of the simple mathematical models
used.

CONCLUSIONS

Up to the present time the recirculation of
ventilating air has been viewed with
suspicion and it has generally been
considered bad, if not illegal, ventilation
practice. The particular fear has been in
connection with the accumulation of methane,
and to a lesser extent other dangerous
gases, passing undetected and leading to
serious incidents resulting in injury and
possibly loss of lives.

Two important changes have taken place
which merit the above situation being
reviewed.

Firstly, a detailed study has been undertaken which has established the relatively
simple principles and theory which hold for
controlled recirculation circuits and these
have been confirmed by a series of practical
experiments including trials with a laboratory model.

Secondly, only in recent years has it
become possible to monitor concentrations
of all pollutants reliably and where
required transmit, continuously store,
process and present the information using
computer systems. These welcome developments have also enabled ventilation system
performance to be confirmed and consequent-
ially a review of ventilation practice can
be considered from a better informed
position.

It is within this current context that
the technique of controlled recirculation
must be judged together with its potential
for improving on the safety and economics
of ventilation practice.

The number of environmental pollutants
present and their individual production
rates vary widely between mines. Controlled
recirculation is not, therefore, being
proposed as the universal solution to
delivering a desired face airflow but
rather that it merits serious consideration,
particularly in situations where gaseous
pollutant concentrations are relatively
low and stable.

In addition to being a means of improving both the safety and environmental conditions, controlled recirculation can contribute very significant power savings. For example in the ventilation of an extensive MVN which has workings remote from adjacent surface connections it can be used to improve both the network efficiency and the use of air. It also opens up opportunities for undersea working which would not be possible using conventional ventilation systems.

The selection of the optimum layout and size of the recirculation circuit is currently being investigated but this is likely to be determined on an individual network basis using established network planning techniques. Single face to virtually whole mine recirculation circuits have been examined and shown to have merit in certain situations.

REFERENCES

Anon. 1984. 'Final Report ECSC Project 7220-AC/810'. Dept. of Mining Engineering, University of Nottingham.
Hardcastle, S.G. 1983. Analysis of airflow patterns and pollutant concentrations in mine ventilation systems. Ph.D. Thesis, University of Nottingham.
Hardcastle, S.G., Kolada, R.J. & Stokes, A.W. 1984. Studies into the wider application of controlled recirculation in mine ventilation. The Mining Engineer.
Hornsby, C.D. 1985. Underground mining environment engineering. IME Mining 85 Conference, Birmingham.
Kolada, R.J. 1983. Methane concentrations in recirculating mine ventilation systems. Ph.D. Thesis, University of Nottingham.
Leach, S.J. & Slack, A. 1969. Recirculation of mine ventilation systems. The Mining Engineer, No. 100.
Lee, R. & Longson, I. 1984. A feasibility study of recirculation in main ventilation systems of mines. University of Nottingham, Mining Dept. Magazine, vol. 36 p. 41.
Litwiniszyn, J. 1951. A problem of the dynamics of flow in conduit networks. Bull. de l'Academie Polon. des Sciences et des lettres Serie A: Science Mathematiques, Vol. 1, pp 326-339.
Lowndes, I.S. 1984. The computer simulation of transients associated with the controlled recirculation of air around working longwall faces. Ph.D. Thesis, University of Nottingham.
Morris, I.H. & Walker, G. 1982. Changes in the approach to ventilation in recent years. The Mining Engineer, pp 401-412.
Pickering, A.J. & Aldred, R. 1977. Controlled recirculation of ventilation: A means of dust control in stone drivages. The Mining Engineer.
Pickering, A.J. & Robinson, R. 1984. Application of controlled air recirculation to auxiliary ventilation systems and mine district ventilation circuits. Proceedings 3rd Int. Mine Ventilation Congress, Harrogate IMM/IME.
Robinson, R. 1972. Trials with a controlled recirculation system in an advance heading. The Mining Engineer.
Stokes, A.W. 1983. The effects of recirculating flows on mine climate. Ph.D. Thesis, University of Nottingham.
Trutwin, W. 1982. On gas concentration behaviour in mine workings with recirculating air. University of Nottingham, Mining Dept. Magazine, Vol. 34, p. 33.

ACKNOWLEDGEMENTS

The authors wish to thank the National Coal Board, the European Coal and Steel Community and the Science and Engineering Research Council for providing the financial support for this project. In particular, the authors wish to thank Professor T. Atkinson for his encouragement throughout the life of this project, Professor M.J. McPherson for his initiation of the project and Drs. S.G. Hardcastle, R.J. Kolada and A.W. Stokes, original members of the Nottingham study group, for their contributions. In addition the authors wish to record their thanks to Dr. I.H. Morris of HQ Ventilation Staff and Mr. A.J. Pickering and Mr. R. Robinson the Area Ventilation Engineers for the South Nottinghamshire and North East Areas respectively for their valuable assistance and advice. Lastly, the authors wish to thank the Area Director of the South Nottinghamshire Area of the National Coal Board for his permission to make underground measurements.

The views expressed in this paper are those of the authors and are not necessarily those of the sponsors.

Computer predicted airborne respirable dust concentrations in mine air recirculation systems

S.G.HARDCASTLE
CANMET, Mining Research Laboratories, Elliot Lake, Ontario, Canada

ABSTRACT: A history is given of investigations into, and field trials of, large-scale controlled recirculation in mine ventilation systems. A review is also presented of selected transportation and filtration studies of airborne respirable mine dust that may be relevant in a recirculation circuit.

Using a generalized theory for pollutants in a recirculation system, the effects of dust sedimentation and size differential dust filtration are added; this theory has been converted into a computer simulation model executable on most micro-computers.

The computer program requires little input of data. Resultant values are obtained for wide ranges of recirculation percentage and rated filter efficiency. The program predicts: changes of airborne dust mass flowrates and concentrations both in return air and in the air being re-introduced into the intake of the ventilation system.

Data from the computer model are then presented in graphical formats. These demonstrate the possible benefits of recirculation towards airborne respirable dust.

1 INTRODUCTION

Ventilation, as a prerequisite of underground mining, is necessary to dilute and remove the pollutants that are introduced into the mine atmosphere. These not only include the common gases, solids and vapours but also extreme heat or cold.

As a means of enhancing ventilation recirculation has been suggested and has since been the subject of both extensive research and field trials. In both of these, consideration of gas contaminants and heat aspects has been predominant.

1.1 The definition of recirculation as applied to mine ventilation

For the mining environment, it is only being proposed and considered that truly controlled recirculation be employed.

The definition given by Jones (1983) for recirculation is "the movement of the same body of air more than once past a given point in a ventilation system." He also emphasized that this method is widely used in domestic and industrial heating, ventilation and air conditioning (HVAC). An ammendment to this should be 'the movement of part of an original body of air more than once ....'.

Recirculation is induced by 'booster' fans. Morris and Walker (1982) stated a further condition to the definition "A booster fan controlled recirculation system is stipulated as a fan installation causing the recirculation of a pre-arranged, continuously monitored, proportion of return air ."

1.2 The concept of recirculation

The basic concept of recirculation is shown in fig. 1 (Hardcastle, Kolada & Stokes,1984). A mine ventilation system has inherent leaks which tend to reduce the amount of air reaching the working place, which leads to inadequate ventilation and the associated difficulties outlined in fig. 1a.

Depending on how and where recirculation is introduced the volume flow through a working place may be increased, firstly, by virtue of a natural leakage path being removed and, secondly, by the recirculated amount, (fig. 1b.). Where a new airway is used no leakage is regained.

The benefits of recirculation can include; improved fresh and total air quantities, increased air velocities, improved pollutant mixing and increased air cooling powers.

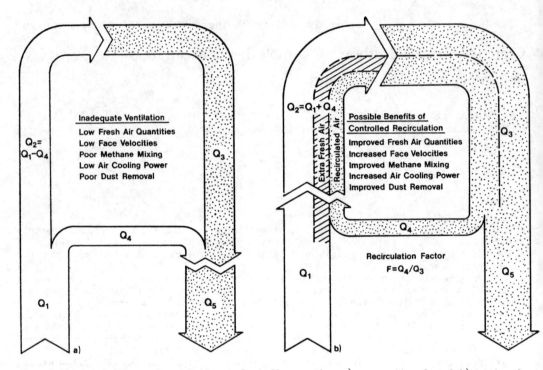

Fig. 1. Diagrammatic representations of airflows under a) conventional and b) recirculatory ventilation (Hardcastle. Kolada & Stokes, 1984).

As a point of reference it is necessary to define recirculation quantitatively, and a standard measure common to most studies is the recirculated fraction or factor. This theoretically has the range 0 to 1.0, and it is often converted to a recirculation percentage.

The rationale adopted here, with reference to fig. 1b, is :

(i) Recirculation factor, F

$$F = Q_4/Q_3 \qquad (1)$$

in which Q_4 = the recirculated quantity
Q_3 = the quantity flowing in the return directly inbye of the recirculation path

(ii) Recirculation percentage

$$= F \times 100 = Q_4/Q_3 \times 100$$

2 HISTORY OF RECIRCULATION

Studies and trials only pertaining to recirculation on a large-scale are discussed, Hardcastle (1983) demonstrating that heading recirculation regimes were generally dissimilar. Heading recirculation has already been widely employed and is the

subject of numerous publications.

2.1 Coal-mine applications

Recirculation as proposed here was first suggested for coal-faces (Leach & Slack, 1969). Generally any ventilation system has difficulties in maintaining adequate airflows at the work-place as mines continually develop. In coal mines conventional ventilation circuits are becoming so long and extensive that only limited improvements are possible even if they were economically viable. Such improvements include airway enlargement, airway duplication and satellite shafts.

A preliminary feasibility study into the wider application of controlled recirculation in coal-mines was performed between 1979-82, (Stokes,1982), (Kolada,1982), (Hardcastle,1983), (European Communities Commission,1983), (Lee & Longson,1984).

The first coal-mine trial of controlled recirculation was performed at Wearmouth colliery, U.K., July 1983, (Pickering & Robinson,1984). This is an undersea mine with working areas presently 7.5 km from shaft bottom and expected to exceed 11 km in the future.

The field trial and the feasibility study proved the following effects (summarized) :

A)There should be sufficient fresh air to dilute gaseous pollutants produced both inside and outbye of the recirculation area to acceptable levels, regardless of the recirculated amount.

B)Providing A) exists then the gaseous pollutant concentrations in the general body of the air in the recirculation area will reach a maximum value above which they will not increase.

C)It is predominantly the mixed-intake containing the recirculated air whose gaseous pollutant concentration will increase in relation to the recirculation factor.

D)The airflow required, apart from gaseous pollutant considerations, is also determined by optimum air velocities. These are important for dust control, dispersion of methane layers, and reducing effective temperatures.

E)The air velocity required for D) can be obtained using recirculation without increasing returnside gaseous pollutant concentrations to undesirable levels.

2.2 Metal-mine applications

Van Der Walt (1978) suggested recirculation for the hot, deep metal-mines of S. Africa. Here the objectives were; to reduce the amount of tertiary and secondary cooling; to fully utilize airways; and to provide better control over the distribution of cooling.

Conversely, in cold climates recirculation could be used to conserve heating. This could form a major cost saving in such places as Alaska and Canada.

A major field trial at Loraine Gold Mine again demonstrated that recirculation was a practical, effective and safe aid to normal ventilation practice (Burton, Plenderleith et al,1984). Here attention was given to combining the benefits of controlled recirculation with refrigeration and dust control. The recirculation percentage was normally 68% and a spray chamber was used to remove airborne dust from the recirculated fraction. This field trial of over 1 year duration has proved effective for temperature and dust control with no major new procedures required for its use. This trial also demonstrated again that recirculation per se will not cause a general increase of contaminant concentrations in the return air.

Blast contaminant decay proved itself to be an exponential function related to the fresh air flow. This is similar to steady state produced pollutants but with a time integration included.

The Loraine trial was successful in using recirculation to control dust and distribute cooling. Here only gas control remains to dictate the fresh air ventilation levels.

3 AIRBORNE RESPIRABLE DUST AND CONTROLLED RECIRCULATION

The theory for gaseous pollutants in a recirculation circuit for steady state conditions, equations 1-6, are already well documented. The theory with respect to respirable dust is not totally analogous to a gaseous contaminant.

Apart from the inclusion of filtration (Burton et al,1984), (Jones,1983) in the steady state gas equations little other consideration has be given to dust excepting Hardcastle (1983).

Recirculation will increase the air velocity; this in turn controls dust mass flows. Velocity can increase the air's dust dispersion efficiency from a source (Hall,1950), (Ford,1975); then once airborne, in association with the dust's size and concentration, it can also cause human discomfort, (Vincent & Gibson, 1980), (Gibson & Vincent, 1980), (Ford, 1975). Numerous other parameters also control dust availability and generation prior to entering the airstream (National Coal Board,1978), (European Communities Commission,1978)

Unlike gasses airborne dust can be deposited. The mechanisms causing this are dependent on air velocity and particle size.

Deposition and filtration are the two mechanisms which control concentrations once the dust is airborne and will be very significant in a recirculation circuit, thus the theory for dust is more complex than for gases.

Using a simplified model the desired preliminary estimate of the likely effects of controlled recirculation on airborne dust is obtained.

3.1 Relevant field studies to model dust

Theoretically values for deposition can be calculated for the mechanisms of gravitational and inertial settling, and diffusion (Knight & Hardcastle,1985), however, when compared, these are much less than the field values of Reinhardt (1972) and Ford (1976). The field studies, figs. 2 & 3, detail dust transportation through coal-faces and along return mine

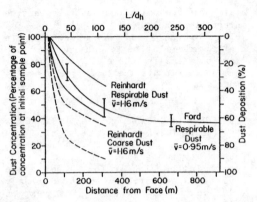

Fig. 2. The reduction of airborne dust concentrations along mine roadways (Ford, 1976; Reinhardt, 1972)

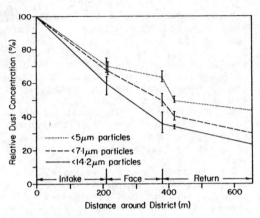

Fig. 3. Deposition of sized fractions of stone dust through a district (Ford, 1976).

airways. In fig. 2 the agreement between the two researchers is good, the curves show rapid deposition of relatively new dust over 300 m dependent on size. With increasing age of the dust cloud the rate of deposition exponentially decays. For respirable dust up to 60% of that introduced at the face will be lost within 600 m of the return gateroad.

Figure 3 shows the deposition around a district of tracer dust (stonedust) introduced at the intake. This again shows the size dependency of dust deposition and its exponential decay with age of the dust. In this instance the deposition is much less in the return airway. For <7.1 μm uds stonedust the system deposition was 70%.

These two field studies show deposition as a major airborne dust removal mechanism. As such this could be extremely significant where dust laden air is being re-used.

It is generally envisaged that a controlled recirculation system should include dust filtration of the recycled air. In the Loraine test a spray chamber was employed.

Filtration, as with deposition, is a size selective process with the larger sizes being predominantely removed. James (1979) performed tests with two identical MRDE high capacity filters in-line, and showed that the efficiency of the second filter was 50% of the rated value. This resulted from the second filter cleaning an airborne dust cloud of much finer composition.

The decay of a filter's efficiency on a second, third or fourth pass could be extremely important in a recirculation circuit where a proportion of the airborne dust will be filtered more than once.

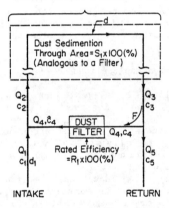

Q_1–Q_5 Air Flows (m³/s)
d & d_1 Respirable Dust Flow & Make (mg/s)
c_1–c_5 Respirable Dust Concentration (mg/m³)
c_4 Concentration After Filtration (mg/m³)

Fig. 4. Dust theory notation in a simple recirculation circuit.

3.2 Dust modelling with recirculation

Figure 4 shows the basic terminology used for dust modelling in a recirculation area. For simplification all dust producing sources in the recirculation area have been summed to a single input, d.

Dust sedimentation in the recirculation area has been assumed to act the same as a mechanical filter and treated as such.

General recirculation, (x = pollutant contribution):

$$F = \frac{Q_3}{Q_4} \quad (1) \qquad\qquad Q_2 = \frac{Q_1}{(1 - F)} \quad (2)$$

$$Q_3 = Q_2 = \frac{Q_1}{(1 - F)} \quad (3) \qquad\qquad Q_4 = FQ_3 = \frac{FQ_1}{(1 - F)} \quad (4)$$

$$C_2 = \frac{(C_1 Q_1 + Fx)}{Q_1} \quad (5) \qquad\qquad C_3 = \frac{(C_1 Q_1 + x)}{Q_1} = C_4 = C_5 \quad (6)$$

Standard geometric composition, (Subscripts, T and i, denote total and initial values):

$$Q_{3T} = Q_1 + FQ_1 + F^2 Q_1 + \ldots\ldots + F^n Q_1 = Q_1/(1 - F) \quad (7)$$

$$d_{3T} = d_{3i} + Fd_{3i} + F^2 d_{3i} + \ldots\ldots + F^n d_{3i} \quad (8)$$

$$d_{3i} = d_1 + d$$

Including filtration, R:

$$d_{3i} \text{ is unaffected} = d_{3i} f^0 (1 - R)$$

$$Fd_{3i} \longrightarrow (1 - R_1)Fd_{3i} = d_{3i} f^1 (1 - R)$$

$$F^2 d_{3i} \longrightarrow (1 - R_1)(1 - R_2)F^2 d_{3i} = d_{3i} f^2 (1 - R)$$

Substituting in (8)

$$d_{3T} = d_{3i} \left[1 + F.f(1 - R) + F^2.f^2(1 - R) + F^3.f^3(1 - R) \ldots\ldots + F^n.f^n(1 - R) \right] \quad (9)$$

$$d_{3T} = d_{3i} \left[f\{F , f(R)\} \right] = d_{3i} \left[f(F , R) \right] \quad (10)$$

Including sedimentation, S:

$$d_{3i} \longrightarrow (1 - S_1)(d_1 + d) = f^1(1 - S).(d_1 + d)$$

$$Fd_{3i} \longrightarrow F(1 - S_1)(1 - S_2)(d_1 + d).f(1 - R) = f(1 - R).f^2(1 - S).F(d_1 + d)$$

$$F^2 d_{3i} \longrightarrow F^2(1 - S_1)(1 - S_2)(1 - S_3)(d_1 + d).f^2(1 - R) = f^2(1 - R).f^3(1 - S).F^2(d_1 + d)$$

Substituting in (8)

$$d_{3T} = f(1 - S).(d_1 + d) + f(1 - R).f^2(1 - S).F(d_1 + d) + f^2(1 - R).f^3(1 - S).F^2(d_1 + d)$$

$$+ f^3(1 - R).f^4(1 - S).F^3(d_1 + d) \ldots\ldots + f^n(1 - R).f^{(n+1)}(1 - S).F^n(d_1 + 1) \quad (11)$$

Removing $(1 - S_1)(d_1 + d)$ or d_{3i} from each term

$$d_{3T} = d_{3i}[1 + F.f(1 - R).f^*(1 - S) + F^2.f^2(1 - R).f^{2*}(1 - S) + F^3.f^3(1 - R).f^{3*}(1 - S)$$

$$\ldots\ldots + F^n.f^n(1 - R).f^{n*}(1 - S) \quad (12)$$

* Sedimentation, $(1 - S_1)$, already removed in d_{3i}, equations start with $(1 - S_2)$.

$$d_{3T} = d_{3i}[f\{F , f(R , S)\}] = d_{3i}[f(F , R , S)] \quad (13)$$

$$c_{3T} = d_{3i}\frac{[f(F , R , S)]}{Q_3} = d_{3i}\frac{[f(F , R , S)].(1 - F)}{Q_1} \quad (14)$$

$$c_{4T} = c_{3i}[f'(F , R , S)] \quad (15) \qquad\qquad c_{2T} = (1 - F)c_1 + Fc_{4T} \quad (16)$$

243

Basic two element composition for dust and airflow in the recirculation area. Initial assumption of no losses reduces both to a summated geometric series.

Enlargement of the recirculation section depicts theoretical series composition for dust and air assuming no losses.

After one complete circuit of the recirculation area there is a new input of dust and air, and old elements have all been reduced by the recirculation factor, F, e.g. Q_i becomes $F.Q_i$ etc.

Pass number around the circuit.

Including a dust filter in recirculation path of the circuit reduces the amount of recirculated dust. Dust no longer reduces as a geometric series.

Pass number through the dust filter.

Enlarged dust section shows the effects of filter removal on each element. The filtration efficiency decays for each pass through the filter.

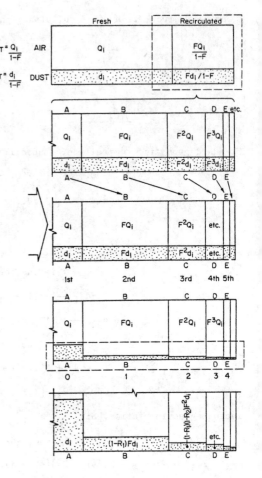

Fig. 5. Block representation of the theoretical air and dust composition under recirculation as a series of elements and the selective effects of filtration on each element.

Theoretically the air flowing in the recirculation circuit can be shown to be made up of elemental parts that are on their first, second, third, etc. pass through the system, fig. 5. Each of these elements is increasingly smaller as it is repeatedly cut by the recirculation factor. At steady state these form the components of a geometric series, equation 7.

The airborne dust recirculation theory is based on the same premise, equation 8. Element parts have been recirculated an increasing number of times and each recirculation reduces the size of the element. Initially dust production is assumed constant and not affected by the changes of air velocity induced with recirculation.

The theoretical composition of air and dust flows described in equations 7 & 8 and depicted initially in fig. 5 do not account for filtration or deposition in

the circuit.

Introducing a filter to the system, as in fig. 4, will reduce the amount of dust being introduced from the return to the mixed-intake. For the elemental analysis given here, in association with each element being progressively reduced by the recirculation factor, the dust in those elements is also increasingly removed by filtration; fig. 5 and equation 9 show the theoretical effects. In equation 9 the filtration operator is maintained as a function to include filter efficiency decay for each pass. Generally the dust flow in the return during recirculation is the product of the dust being produced and entering the area and a function of the recirculation factor and filter efficiency, equation 10.

Treating sedimentation as a filter allows it to be similarly included,

244

operating an increasing number of times for each pass around the circuit, equations 11 to 13. In equation 12 the sedimentation function starts at the value for the second pass as the original is already accounted for in the dust mass. This last form is used to determine the final return concentration, equation 14.

Another concentration of prime interest is that of the mixed-intake, c_2, fig. 4; this can not be as easily determined as it is controlled by, and a combination of, the availability and quality of fresh air and the recirculated fraction. By a similar treatment to that for the return-side the quantity of dust escaping through the filter and its concentration in the air, $\hat{c}_4$, may be found, equation 15.

Equation 16 gives the simplest form for calculating the mixed-intake concentration, this assumes that the fresh airflow does not increase with recirculation.

4 COMPUTER PREDICTION OF DUST CONCENTRATIONS UNDER RECIRCULATION

To fully evaluate the inter-relationship of recirculation factor, filter efficiency, sedimentation rates and possible decay of the latter two on dust concentrations necessitates extensive calculation. This is best suited to a computer.

A computer program has been written for the IBM PC. The two primary relationships used are; one, equation 12, for return dust mass flow and the other for dust mass in the cleaned air being mixed with the intake air. The program also calculates the dust in air concentrations for these masses and any change of rated filtration efficiency.

Input data required by the program are; the sedimentation factor, the decay rate of filter efficiency on each pass through the filter and the decay rate of sedimentation on each pass through the system. All three of these inputs have a 0 to 1.0 range.

For any set of input parameters, the output from the program is normally in a 10 x 11 tabular format, relating, for example, how the return dust concentration varies for the full range of filter efficiencies (0 to 1.0, i.e. 100%), and a practical range of recirculation factors (0 to 0.9, i.e. 90%). Both are incremented in 0.1 steps.

Contoured graphical plots of the tabled data may be produced using standard contouring packages. These can require a smaller increment, (i.e 0.025 or 37 x 41 table), between points to interpolate between and provide smooth curves.

The program presently assumes that the decay rates are constant, for example extrapolating from filter tests (James, 1979), where the second filtration efficiency was half that of the first; third would be half the second, the fourth half the third and so on, that is for a 0.5 decay rate.

Similar to mechanical filtration, sedimentation may be expected to reduce as the age of the airborne dust increases and size decreases; this is demonstrated in figs. 2 & 3.

As both dust removal mechanisms are size selective, both can be expected to interact and cause a general reduction in efficiency even before one recirculation is complete.

4.1 Computed example of the effects of recirculation on dust concentrations

In the following analysis the input parameters were;
a) Sedimentation was conservatively entered as 0.4(40%), this is lower than that found in field tests but in those studies there was no interaction with a filter.
b) The decay rate of filter efficiency was taken as 0.5(50%) throughout the analysis; this agrees with filter studies for the second pass. The decay rate probably does vary in later passes, assuming it remains constant will only produce small errors as the actual elements at this stage are also small.
c) The decay rate of sedimentation was also taken as 0.5(50%) throughout the analysis. As with to the mechanical filter it is assumed to stay constant.

The above assumptions are not expected to cause any greater inaccuracies in the theoretical equations than already exist due to the limited information available.

Figures 6a to 6e are contour plots of the computer predicted data for the above parameters; these all assume dust production is constant and not affected by the increased air velocity. Figures 6a & 6c are the primary dust mass plots from the central two series equations of the program, both are relative to the original dust flow in the return; for fig. 6c the original dust flow in the recirculation path is of no consequence. The mass flows are converted to concentrations for figs. 6b & 6d; these have been related to the initial return concentration c_5 directly outbye of the recirculation path and assumes that the fresh air flow Q_1 has not increased. If no recirculation path previously existed as a leakage path then the concentration changes relate to the

245

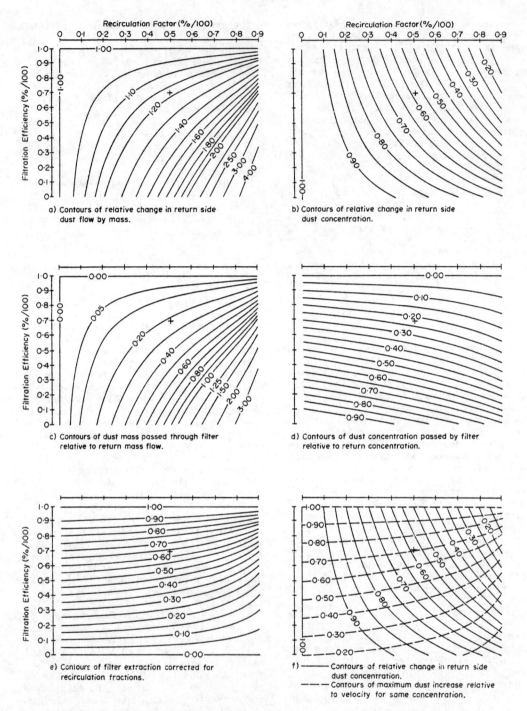

Fig. 6. Computer predicted contours showing the relative effects of recirculation with filtration, over conventional ventilation, on dust mass flows and concentrations; a) to e) assume no increase in dust make, f) indicates the make increase limit related to air velocity improvement for concentrations not to increase. (Input parameters - sedimentation factor = 0.40, filtration decay per circuit = 0.50, sedimentation decay per circuit = 0.50).

initial return concentration c_3.

From the dust mass flows it is also possible to determine how much dust is being removed by the filter and thus its efficiency. This will demonstrate how filtering previously filtered dust with recirculation causes the overall efficiency to be lower, fig. 6e.

The last contour plot, fig. 6f, is a composite; superimposed over the return concentration plot, as in fig. 6b, are contours of the proportional increase of dust mass flow, relative to the proportional increase of air velocity necessary to maintain the same concentration in the return.

4.2 Analysis of contour plots

From fig. 6 the following may be obtained;

1) Recirculation increases the mass of dust flowing in the return airway. This is irrespective of filtration and sedimentation efficiencies.

2) Due to filtration and sedimentation dust does not remain in the general air body as with a gas; this will produce a disproportionate increase of air flow in relation to dust and results in a general reduction of return air dust concentrations. This is providing dust makes do not increase.

3) Similar to the return the relative increase of dust flow entering the fresh air stream even after the filtration can be significant.

4) The airflow increase in the recirculation path is also disproportionate to the dust increase, thus after the filter the air being recycled is relatively clean; in certain instances the filtered air can be much cleaner than the normal intake air and so provide better dust conditions in the mixed-intake.

5) The rated filter efficiency gradually decays with increasing recirculation.

6) Figure 6f introduces the option of dust mass flow rates increasing with velocity; this demonstrates that if dust make increases faster than velocity then the return concentration will deteriorate, however depending on filter efficiency and recirculation factor a limited increase may occur without an increased concentration.

4.3 Theoretical example

Using the plots of fig. 6 the following example has been worked. Initial dust and airflow conditions were:

1) Intake concentration, c_1 = 0.5 mg/m^3

2) Return concentration, c_3 = 1.0 mg/m^3

3) Fresh air (no leakage), $Q_2 = Q_1$

4) Air velocity = 1.5 m/s

Recirculation conditions were:

5) Recirculation factor, F = 0.50(50%)

6) Filter efficiency, R_1 > 0.70(70%)

7) Fresh air unchanged

The air velocity increase was given by F/(1 - F) = 1, i.e. 100% or doubled. The following was obtained from each part of fig. 6 in turn, initially assuming no dust make increase:

a) The returnside dust mass flow will increase by up to 18%, (1.18 - 1).100.

b) The actual return concentration will reduce to at most 0.58 of its former value, 0.58 mg/m^3.

c) Of the dust flowing in the return only 0.22(22%) re-enters the intake as compared to 50% for the air.

d) The relative concentration of dust in the cleaned air will be up to 23% of the return, that is, 0.23 mg/m^3.

e) The minimum in situ filter efficiency with recirculation has reduced from 70 to 64%.

f) Using equation 16 provides the mixed-intake concentration as 0.365 mg/m^3 for no increase of dust make.

g) With a 100% increase of velocity, a 71%(0.71x1.0x100) increase in dust make would cause the return concentration to remain unchanged at 1.0 mg/m^3.

h) Recalculating d) & f) for an increased dust make of 71%; the cleaned air concentration will be 0.39 mg/m^3 and the mixed-intake concentration 0.44 mg/m^3.

5 DISCUSSIONS

The above treatment deals with the respirable dust cloud as a whole and accounts for size dependent filtration and deposition through the use of decay factors. An alternative method would have been to use actual filtration or deposition efficiencies on specific size ranges and apply them at each recirculation and change towards a finer size distribution. The accuracy or merits of either method will only be proven with more field studies and trials. The second requires greater detail and longer computation.

This analysis, although relatively simple, indicates how recirculation will affect dust concentrations. Generally it may be seen that good filtration is necessary, this being especially important when the dust make increases. Dust make should be controlled to as little as possible by avoiding sources. For a system with a good filter, dust make and air velocity can virtually increase at the same rate without causing higher concentrations in the system than occurred prior to recirculation. In the Loraine test (Burton et al), with 68% recirculation, the ore production in the area trebled while an acceptable concentration was maintained in the return.

For air velocities under 4 m/s, (a human comfort level from dust impaction) it can be generally assumed that dust make does not greatly increase with velocity. Therefore recirculation with a filter should always provide better return respirable dust conditions. Improved intake dust conditions are also possible when the filtered air is cleaner than the natural intake.

The quantitative benefits of a specific recirculation regime should always be checked through theoretical analyses to indicate how respirable dust concentrations are affected. The above analysis predicts the general concentrations either side of the recirculation path and not necessarily at the working place. With judgement it can be applied to work places but care is necessary when anti- and homotropal ventilation are being evaluated. However, this analysis has adequately demonstrated that dust concentrations can be controlled by a recirculation system.

In recirculation systems it remains the gaseous pollutants that control the acceptability of the system. Dust, heat, temperature and refrigeration, although important, are secondary.

REFERENCES

BURTON R.C., PLENDERLEITH, W., STEWART, J.M., PRETORIUS, B.C.B. & HOLDING, W. 1984, Recirculation of air in the ventilation and cooling of deep gold mines, Mine Vent, Third Int Congr, Harrogate, England, (Ed Howes, M.J. & Jones, M.J.), Inst Min Metall, London.

EUROPEAN COMMUNITIES COMMISSION 1978, Industrial health and safety - Health in mines, Synthesis Report on Research in the Third Programme 1971 - 1976, EUR 5931, Luxembourg.

EUROPEAN COMMUNITIES COMMISSION 1983, Controlled recirculation of air in mine workings, Final Report on ECSC Research Project 7220-AC/810, M.R.D.E., U.K.

FORD, V.H.W. 1975, The effect on respirable dust concentrations of varying the ventilation air quantities around a district, N.C.B.-M.R.D.E Report 62, U.K.

FORD, V.H.W. 1976, Investigations into deposition of airborne respirable dust along underground airways, N.C.B. - M.R.D.E. Report 65, U.K.

GIBSON, H. & VINCENT, J.H. 1980, On impaction of airborne coarse dust into the eyes of human subjects, Ann Occ Hyg, 23.

HALL, D.A. 1950, Factors affecting airborne dust concentrations with special reference to the effect of ventilation, Trans Inst Min Eng (London), 115, U.K.

HARDCASTLE, S.G. 1983, Analysis of airflow patterns and pollutant concentrations in mine ventilation recirculation systems, PhD Thesis, University of Nottingham, U.K.

HARDCASTLE, S.G., KOLADA, R.J. & STOKES, A.W. 1984, Studies into the wider application of controlled recirculation, Min Eng (London), 273, U.K.

JAMES, G.C. 1979, Controlled recirculation and dust control in drivages, N.C.B. - M.R.D.E. Dust Physics Group, U.K.

JONES, I.O. 1983, The benefits of recirculation in mines, Mine Quarry Mech,:133-39.

KNIGHT, G. & HARDCASTLE, S.G. 1985, Deposition of dust in mine galleries, Division report MRP/MRL 85-29(TR), CANMET, Energy, Mines and Resources Canada.

KOLADA, R.J. 1982, Methane concentrations in recirculating mine ventilation systems, PhD Thesis, University of Nottingham, U.K.

LEACH, S.J. & SLACK, A. 1969, Recirculation of mine ventilation systems, Min Eng (London), 100, U.K.

LEE, R.D. & LONGSON I. 1984, A feasibility study of recirculation in main ventilation systems of mines, Min Mag University of Nottingham, 36, U.K.

MORRIS, I.H. & WALKER, G. 1982, Changes in the approach to ventilation in recent years, Min Eng (London), 244, U.K.

NATIONAL COAL BOARD 1978, Dust control in coal mining, M.R.D.E. Handbook 15, U.K.

PICKERING, A.J. & ROBINSON, R. 1984, Application of controlled recirculation to auxiliary ventilation systems and mine district ventilation circuits, Mine Vent, Third Int Congr, Harrogate, England, (Ed Howes, M.J. & Jones, M.J.), Inst Min Metall, London

REINHARDT, M. 1972, Studies in mine roadways on the behavior of dust in air currents, Gluckauf-Forschungsh, Y33, 1.

STOKES, A.W. 1982, The effects of recirculating airflows on mine climate, PhD Thesis, University of Nottingham, U.K.

VAN DER WALT, J. 1978, Cooling mines having high rock temperatures - the case for recirculation of the ventilation air, Environ Eng Laboratory Project Report, Chamber Mines, Johannesburg.

VINCENT, J.H. & GIBSON, H. 1980, The response of human subjects to the facial impaction of airborne coarse dust, Atmos Environ, 14.

7. Radon

Membrane barriers for radon gas flow restriction

JAMES F.ARCHIBALD & H.JAMES HACKWOOD
Queen's University, Kingston, Ontario, Canada

ABSTRACT: Research was performed to assess the feasibility of barrier membrane substances for use within mining or associated high risk environments, in restricting the
diffusion transport of radon gas. Specific tests were conducted, both within laboratory and in-situ emanating environments, to determine radon permeability parameters of
a variety of membrane materials. Several sealant materials have been demonstrated to
be effective in restricting the emanation of radon gas into initially radon free
environments.

1 INTRODUCTION

Tests were conducted both in laboratory
and underground environments where concentrations and diffusion flows of radon
gas were known to exist. Radon gas concentrations were permitted to develop and
were monitored in sampling chambers, initially radon-free, which were emplaced
adjacent to high strength gas sources.
The source and sampling chamber volumes
were separated using specified membranes
substances, mounted on a special filter
assembly. Several membrane barriers were
demonstrated to result in significant reduced emanation concentrations of radon
gas within sampling chamber atmospheres.
Minimum gas concentrations were evidenced
to occur where the barrier materials were
shown to exhibit lowest radon permeability characteristics. The prime objective
of the research program was to assess the
sealant capabilities of selected membrane
materials for use in environments subject
to radon hazard potential

2 METHODOLOGY

2.1 Laboratory Phase

A series of permeability determinations
were conducted within the Mine Environment
Laboratory of the Mining Engineering Department, Queen's University at Kingston,
Ontario, Canada. Techniques have been
reported in the literature which develop
basic sampling procedures for permeability assessment (Jha, Raghavayya &
Padmanabhan 1982)(Culot, Schiager & Olsen

1976)(Pohl-Ruling, Steinhausler & Pohl
1980)(Franklin, Meyer & Bates 1977).

The technique utilized, as developed by
Jha, Raghavayya and Padmanabhan (1982),
involves measurement of radon concentration ratios, across membrane substances of
interest, under steady state conditions.
The general configuration of the laboratory
test apparatus is illustrated in the
schematic diagram of Figure 1.

A high strength radon gas source was
established within a sealed steel vessel
having a wall thickness of 0.64 cm. A
continuous flow of radon gas, of known
concentration, was pumped into this vessel
from a Pylon Electric Development Company
radon gas source, using a Manostat peri-
staltic pump. The in-line source of radon
gas is produced from a radium salt concentration of 1.83×10^6 pCi strength. Radon
quantities were circulated through the
source chamber continuously at the rate
of 100 pCi/minute, during each test. Air
volumes, at the rate of one litre per minute, were circulated between the Pylon
source and the source chamber.

A second steel vessel, identical to the
source chamber and exhibiting physical
dimensions as illustrated in the schematic
of Figure 1, was utilized as a detector
vessel or sampling chamber. Between the
source and detector chambers, a perforated
aluminum disc was inserted to act as a
carrier stage upon which the various membranes were mounted. Membranes, emplaced
upon the carrier stage, were clamped in
place using '0' rings, as illustrated.
The '0' ring seals prevented radon flow

outward to the atmosphere, from either chamber, and simultaneously restricted radon gas flow, between the two chambers, solely to diffusion through the barrier membrane of interest.

Detection chamber air was continuously circulated through an alpha scintillation assembly using a Gilian Instrument Corporation portable sampling pump, at a flow rate of one litre per minute. Alpha activity within this circulating airstream was continuously monitored using calibrated scintillation cells mounted in contact with a photo-multiplier/counter assembly.

Air flows, both prior to entering into the source chamber from the radon source and into the scintillation cell from the sampling chamber, were filtered to remove radon daughters. No differential pressure effects were observed to occur across the membrane substances during all phases of research.

Prior to each test, contaminant radon gas concentrations were removed from the source chamber, detector chamber, and filter stage assemblies by continuous circulation of radon-free air. The source and detector chambers were subsequently coupled with the filter stage and appropriate filter media was placed between. Circulating air flows within the source and detector vessels were initiated upon complete installation. Alpha activity within the detector vessel volume was monitored on a routine basis until equilibrium alpha activity response was observed. In most cases, sampling periods up to twenty days were necessary to demonstrate equilibrium conditions. Under such conditions, equilibrium radon gas concentrations were estimated to exist within the detection vessel solely as a result of diffusion transport of radon through the filter membrane barriers.

2.2 Description of Membrane Materials

A series of six commercially available membrane materials were selected for testing on the basis of availability and on the basis of potential for use in underground mining or construction environments, where radon barriers may be required.

Samples of polyurethane (PU), aluminized mylar (PT), latex rubber (LR) and polyethylene (PE) were obtained, in sheet form, from standard commercial sources. Such samples were available in thicknesses of 0.635 cm, 0.0089 cm, 0.0203 cm and 0.010 cm, respectively.

Samples of Rock Coat 82-3 (polyvinyl chloride copolymer, PVC) and polysulphide copolymer (PS) were specifically manufactured for use in this test program, and were made available in liquid form only.

Rock Coat 82-3, manufactured by Halltech Incorporated, is a 30% by weight solution of polyvinyl chloride copolymer in an acetone solvent base. This material was formulated for use within the mining industry as a spray type wall coating capable of providing both lighting and underground wall support benefits. Under certain conditions, the use of such a coating agent would be beneficial in uranium mine environments, should it prove to be a suitable radon gas flow retardant. For such reasons, Rock Coat 82-3 was selected for investigation.

The polysulphide copolymer product, manufactured by the Thiokol Corporation, is an alkyl polysulphide copolymer in a water dispersion. On the basis of manufacturer's product data, polysulphide compounds exhibit significant resistance to gamma irradiation with little physical degradation; the possible utility of such compounds for radon flow restriction was therefore anticipated and addressed.

2.3 In-Situ Investigation

A series of in-situ emanation rate studies were performed underground at a uranium mining facility in the Elliot Lake region of Ontario, during a four month period in 1984. Such tests were effected to both verify the results of controlled laboratory experiments as well as to determine the efficiency of installing various membrane materials on emanating excavation surfaces.

Analysis of in-situ emanation rate behaviour of rock and bulkhead surfaces is accomplished by standard characterization techniques which have been utilized often in the Canadian uranium mining industry (Thompkins & Cheng 1969)(Archibald & Nantel 1979). For the purposes of this research project, surface area emanation rate parameters (Js) and radon gas concentrations, within sealed chambers, placed on a wooden bulkhead, were measured to prove the viability of membrane barrier use.

A bulkhead site was selected which provided separation between a sealed stope, exhibiting no air flow conditions, and an exhaust airway, exhibiting significant air flow. All tests and barrier material placements were performed upon the exhaust airway side of the bulkhead which exhibited low level radon concentrations with respect to the sealed stope side. Induced diffusion flow of radon from the sealed stope, high concentration zone to the low concentration airway side was presupposed to occur. Figure 2 shows the typical sampling chamber arrangement upon the bulkhead surface.

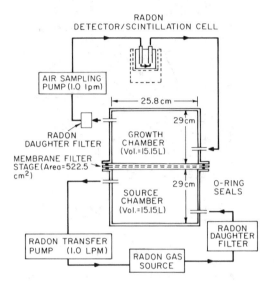

Figure 1: Schematic View of Membrane Barrier Test Apparatus

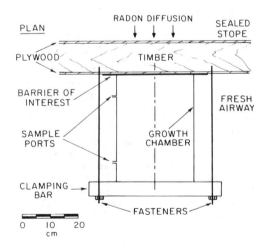

Figure 2: In-Situ Test Set-Up

The surface chamber technique which is utilized to perform emanation rate tests involved the following general procedure: Cylindrical, airtight, stainless steel sampling chambers were mounted upon the bulkhead, as shown. For test purposes, one chamber was placed directly upon the bulkhead surface to monitor existing conditions and the radon flow retardation effect caused by the bulkhead materials. The bulkhead surface was comprised of one untreated 0.95 cm thick plywood sheet, nailed directly to a timber frame. Each side of the bulkhead was sheathed by a similar thickness of plywood.

A second chamber was similarly placed; however, a separating layer of aluminized mylar was used to seal the bulkhead surface and act as a surface membrane layer. Silicone caulk was used to bond the mylar sheet upon the bulkhead surface.

A third chamber was placed upon a layer coating of polysulphide copolymer, which was brushed onto the surface of the bulkhead. The edges of each chamber, in contact with either the membrane layers or bulkhead surface alone, were sealed using silicone caulk to preclude any leakage of radon gas, in or out of the chambers, except by diffusion through the bulkhead.

Two valves were affixed to each chamber to permit sampling and initial voidance of the chambers with radon-free compressed air. Sampling was achieved using evacuated scintillation flasks, in line with millipore filters, to pre-screen radon daughters. Radon concentrations on both sides of the bulkhead were also monitored using similar calibrated flasks.

The principle of emanation rate calculation is based upon measurement of incremental radon concentrations within each chamber over short time intervals; therefore initial time of chamber voidance was recorded; subsequently, radon which may diffuse through the bulkhead and appropriate membrane layers will collect within the chamber.

The activity of the contained radon gas was measured to determine radon gas concentrations. From these results, the radon flux, or emanation rate (Js) parameters, for diffusion flow from the treated bulkheads surfaces were calculated according to the relationship:

$$Js = K \times C \times \frac{V}{A}$$

where, Js = emanation rate;
$(Ci/cm^2/s)$

253

K = depletion factor, equal to

$$\frac{\lambda}{1-e^{-t}}$$

λ = decay constant for radon equal to 2.1×10^{-6} disintegrations per second

C = radon concentrations at time t; (pCi/L)

V = volume of air within the chamber; (L)

A = exposed emanating surface area (cm^2)

t = time interval between chamber voidance and air sample collection; (hours)

During each interval period of site measurements, one sample of air was recovered from each chamber. For each case, radon gas concentration measurements and emanation rate determinations were performed between each set of sampling tests.

3 DESCRIPTION OF EXPERIMENTAL RESULTS

3.1 Laboratory Phase Test Results

On the basis of previously published research (Jha, Raghavayya & Padmanabhan 1982), it has been shown that the use of open cell polyurethane foam (PU) barrier membranes yields unrestricted flow of radon gas from source to detection chambers, while retarding the transport of radon daughter products. During each membrane test phase, scintillation cell readings of circulated detection vessel air were recorded until steady state count rates were achieved. Upon measurement of steady state radon concentration conditions within the detection stage, air samples were collected from the source vessel, and absolute radon concentrations were determined. In such a manner, uniform source chamber radon concentrations were able to be observed and checked.

During the initial test, using polyurethane foam as the sole membrane medium, the observed steady state count rate (Co) was measured to be 1.6226×10^4 cpm. The radon concentration within the source chamber, as determined by scintillation cell tests, averaged 6.64×10^4 pCi/L. For all membrane barrier tests, a value of Co equal to 1.6226×10^4 cpm was assumed to be the maximum possible detection chamber activity rate.

The experiment was thereafter repeated without removal of the polyurethane bed on the aluminum support stage. Other membrane materials were subsequently placed

directly upon the polyurethane material, and tests were repeated until steady state activity was recorded.

Curves illustrating the detection chamber activity, as a percentage of maximum activity, versus growth time are shown in Figure 3.

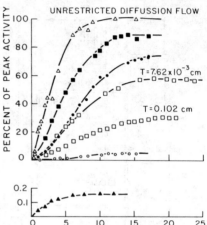

Figure 3: Radon Growth Through Membrane Barriers

The concentration of radon gas in the detection vessel is identical to that within the source vessel for the case in which polyurethane, alone, was used as the membrane medium. Where both concentrations were equivalent it was indicated that polyurethane exhibits complete permeability to radon diffusion flow.

Where a detection vessel concentration, less than that which exists in the source vessel, occurs, a measure of filter impermeability to radon diffusion is inferred. The concentration of radon in the detection chamber will be proportional to the peak alpha count rate (Co) which is measured. The ratio of detection vessel count rates versus peak rate, Cd/(C),is expressed as a measure of filter permeability to radon, and is designated as a ratio parameter, R (Jha, Raghavayya & Padmanabhan 1982). This parameter may be used to derive the membrane permeability constant, k, values based upon the expression:

$$R = [1 + \frac{\lambda T V_1 V_2}{kA(V_1 + V_2)}]^{-1}$$

Table 1: Filter Membrane Physical Property Summary

FILTER MATERIAL	THICKNESS (cm)	R (Cd/Co)	PERMEABILITY CONSTANT, k (cm.sec^{-1})
POLYURETHANE (PU)	6.35×10^{-1}	1	-
ALUMINIZED MYLAR (PT)	8.90×10^{-3}	1.633×10^{-3}	4.426×10^{-10}
ROCK COAT 82-3 (PVC)	1.27×10^{-1}	7.365×10^{-1}	1.079×10^{-5}
POLYSULPHIDE COPOLYMER (PC)	1.02×10^{-1}	4.659×10^{-2}	1.515×10^{-7}
LATEX RUBBER (LR)	2.03×10^{-2}	8.859×10^{-1}	4.794×10^{-6}
POLYETHYLENE (PE)	1.02×10^{-2}	3.012×10^{-1}	1.331×10^{-7}

where, R = count ratio Cd/Co

Cd = detection chamber steady state activity rate

Co = maximum or peak steady state activity rate

λ = radon decay constant; (disintegrations per sec)

T = membrane barrier thickness; (cm)

V_1 = source chamber volume; (cm^3)

V_2 = detection chamber volume; (cm^3)

k = membrane material permeability constant; (cm^2/sec)

A = membrane area through which radon diffusion flow occurs; (cm^2)

A summary of filter physical property parameters is listed in Table 1.

3.2 In-Situ Phase Test Results

The results of radon concentration and emanation rate behaviour determinations performed upon the three bulkhead sites are listed in Table 2 and illustrated in the curves of Figure 4. No estimates of permeability ratio (R) values or permeability constant (k) parameters were inferred for direct comparison with laboratory derived data due to the additional or masking effects of bulkhead constituent materials. However, the in-situ concentration and emanation rate parameters did serve to illustrate the relative benefits achieved through the application of each membrane substance material.

An approximate ratio parameter value, designated R^1, was inferred by comparing the emanation chamber peak radon concentration versus the sealed airway radon concentration values. Such data is listed in Table 2.

4 DISCUSSION OF RESULTS

On the basis of laboratory tests on various membrane materials, two substances were demonstrated to exhibit superior performance for reduction of radon emanation flows.

In general, the effectiveness of radon barrier materials will be demonstrated by the radon count rate ratio parameter, R. Variation in the R value, as evidenced in the results of Table 1, is dependent only upon the membrane thickness and perme-

Table 2: In-Situ Growth Activity and Emanation Rate Summary

DATE	Δ T (MINUTES)	RADON CONCENTRATION (pCi/L)					EMANATION RATE, Js (pCi/cm^2/s x 10^{-18})		
		UNTREATED BULKHEAD	POLY-SULPHIDE COPOLYMER SURFACE	ALUMINIZED MYLAR SURFACE	EXHAUST AIRWAY	SEALED AIRWAY	UNTREATED BULKHEAD	POLY-SULPHIDE COPOLYMER SURFACE	ALUMINIZED MYLAR SURFACE
23/5/84	0	-	-	-	60.7	147.8	-	-	-
28/5/84	7155	-	-	-	47.5	128.7	-	-	-
30/5/84	10275	-	-	-	43.4	122.1	-	-	-
1/6/84	11720	3.7	-	Too low to measure	-	-	33.3	-	-
1/6/84	11810	-	2.9		-	-	-	8.5	-
5/6/84	17710	21.2	14.5	0.8	31.2	129.4	2.0	1.1	0.075
6/6/84	18950	22.7	14.3	1.2	-	-	0.1	-	0.20
13/6/84	30270	-	-	2.5	34.6	98.2	-	-	0.068
22/6/84	43160	$C_d = 23.7$	$C_d = 17.5$	$C_d = 2.6$	-	-	0.024	0.059	0.046
29/6/84	53240	-	-	-	$C_d = 38.0$	$C_d = 100.7$	-	-	-
$R^1 =$		0.235	0.174	0.026					

255

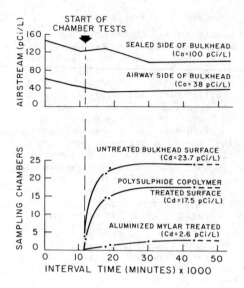

Figure 4: In-Situ Chamber Growth Activity and Ambient Radon Concentrations

ability constant (k) parameters. Due to the nature of fabrication, no uniform membrane thickness was possible to be maintained for the purpose of this study. Samples of Rock Coat (82-3) and the poly-sulphide copolymer were applied in the form of liquid coatings which were painted onto the membrane stage elements. All other materials were commercially supplied in variable thickness sheet form. The ratio parameter, R, may therefore only be used as a reference indicator of the effectiveness of the various membrane materials in restricting radon diffusion flows. On this basis alone, the results shown in Table 1 indicate that the poly-sulphide copolymer and aluminized mylar membranes offer the greatest restriction potential. The barrier effectiveness against radon flow, however, is directly proportional to the permeability constant values which have been derived; the same approximate order of effect between the two membrane materials is exhibited by the derived permeability constant values:

Of the six membrane materials tested under laboratory conditions, only two were capable of maintaining radon concentrations within initially radon-free environments at levels less than 5% of the source concentration levels. The equilibrium diffusion concentration ratio, R, measured for the polysulphide copolymer material approximated 4.7%; the aluminized mylar membrane effected substantially better performance, approximating a

ratio value of 0.16%. The remaining material suppression ratios varied between 30.0 and 88.6%, and would not be deemed to be suitable for use as radon barrier materials.

On the basis of laboratory investigations, the application of membrane coatings has been shown to be effective in preventing radon gas migration into designated environments under ideal conditions.

A complementary series of in-situ tests was conducted to verify the effectiveness of placed membrane materials under dynamic working conditions. Limited field study trials precluded continuous monitoring capabilities.

The two most effective barrier materials as evidenced by laboratory tests were installed upon the low concentration side of the bulkhead and were henceforth examined. The results listed in Table 2 indicated that the equilibrium radon concentrations resulting from emanation flows, approximating 17.4 and 2.6 percent, respectively, of initially present levels, were achieved through use of bulkheads coated by polysulphide copolymer and aluminized mylar barrier materials. The untreated plywood surface of the bulkhead yielded a sampled radon concentration of approximately 23.5 percent compared to the source radon concentration level. The plywood, comprising a mixed cellulose fibre, restricted approximately 76.5 percent of the radon source concentration which was available to emanate through the bulkhead from the sealed stope volume.

Measured emanation rate parameters are similarly listed in the data of Table 2. The flux rate trend indicates that a rapid initial diffusion flow of radon occurred through the untreated and the polysulphide copolymer-treated plywood surfaces, as expected from Fick's Law. Significantly lower emanation rate behaviour, at levels between 4 to 7 percent of the measured levels for the untreated and polysulphide coated barriers, respectively, was monitored for the aluminized mylar barrier.

A constant radon concentration is impossible to achieve in-situ, as evidenced by the curves of Figure 4, due to changes in absolute pressure and other fluctuating environmental conditions. Non-uniform airflow conditions or production activity in the vicinity of the bulkhead site may also have affected the concentration levels on each side of the bulkhead. Despite these factors the relative merits of the materials were verified.

5 CONCLUSIONS

The use of wall sealant materials has been demonstrated to be effective in restricting the emanation of radon gas into initially radon-free mine environments.

The application of such sealant materials would be suitable for particular environments, notably in uranium mining areas where residential and industrial exposure requires that radon contamination be minimized.

Optimum utilization would be gained where a particular membrane barrier material would yield maximum resistance to diffusion flow of radon. On the basis of test results, two membrane materials have been demonstrated to effectively restrict emanation of radon into radon-free environments and to thereby satisfy the above constraint. The substances, notably aluminized mylar, in sheet form, and a polysulphide copolymer compound, in liquid form, offer potential benefits for radon control applications, Each is capable of reducing radon emission concentrations by better than 95 percent; the aluminized mylar, in particular, is capable of reducing atmospheric concentrations by more than 99.8%. However, application of such barrier membrane materials within residential or industrial sites, on a large scale, may prove to be impractical. The application of aluminized mylar, for example, in sheet form, would be physically restrictive where irregular surfaces must be treated. Additional factors, such as cost and availability of supply, must also bear investigation before the use of effective membrane materials may be implemented.

REFERENCES

Jha, G., Raghavayya, M. and Padmanabhan,N. 1982. Radon permeability of some membranes. Health Physics 42 5:723-725

Culot, M., Schiager, K. and Olson, H. 1976. Prediction of increased gamma fields after application of a radon barrier on concrete surfaces. Health Physics 30:471-478

Culot, M., Schiager, K. and Olson, H., 1978. Development of a radon barrier. Health Physics 35 2:375-380

Pohl-Ruling, J., Steinhausler, F. and Pohl, E. 1980. Investigation on the suitability of various materials as Rn^{222} diffusion barriers. Health Physics 39:299-301

Franklin, J., Meyer, T. and Bates, R., 1977. Barriers for radon in Uranium mines. Washington:US Dept. of the Interior. Bureau of Mines

Thompkins, R. and Cheng, K. 1969. The measurement of radon emanation rates in a Canadian Uranium mine. C.I.M. Bulletin 62 692

Archibald, J. and Nantel, J. 1979. Determination of radiation levels to be encountered in underground and open pit mines. Reno Second International Mine Ventilation Congress.

Archibald, J. and Nantel, J. 1981. Assessment of radiation properties of hydraulic backfills for underground uranium mines. Golden, Co.: Conference on Radiation Hazards in Mining

A controlled study of the evolution of radon gas and decay products in radioactive mine environments

F.A.CALIZAYA & R.H.KING
Colorado School of Mines, Golden, USA

J.C.FRANKLIN
Spokane Research Center, Bureau of Mines, WA, USA

ABSTRACT: This paper discusses three aspects related to radon emissions and control of radon daughters in mine environments: (1) study of the effects of environmental parameters on radiation levels, (2) prediction of their concentration in the mine air, and (3) control of radon daughters by means of mechanical ventilation. To study these aspects a full scale experimental model was devised at the Colorado School of Mines' Edgar Mine. The model consisted of the Edgar Mine ventilation facilities, a bulkhead isolated mine drift containing low grade uranium ore, and the U.S. Bureau of Mines radiation monitoring system. The system continuously collects information such as radon concentration, working level, air temperature, humidity, barometric pressure and air velocity.

Time series analysis was used to evaluate the data and formulate models for predicting concentrations. These models may be useful in estimating ventilation requirements to control the health hazard from radon decay products.

An important finding of this study was that concentrations of radon and radon daughters in the mine model were inversely related to changes in barometric pressure.

1 INTRODUCTION

This research was made possible by a joint effort of the U.S. Bureau of Mines – Spokane Research Center and the Colorado School of Mines. The main purpose of the study was to gain experience in controlling radon and its decay products in mine environments. Obviously, as with any other air contaminant, a good control practice requires a valid and reliable means for both data collection and analysis.

The fact that radon concentration changes with changes in barometric pressure was first notices by Brandes in 1905 while radon emissions from soil were measured (Edwards, 1980). Since then, various experiments were conducted. In 1976, the research personnel of the U.S. Bureau of Mines of the Spokane Research Center, replicated Brandes' findings at the Twilight Mine (Franklin, et. al., 1976). In 1982, the U.S. Nuclear Regulatory Commission reported for two underground mines (D and T) that "increases in radon concentration tends to occur when the barometric pressure was fallen". Most of the measurements however, were conducted for short periods of time (less than 10 days) and the results of data analysis showed poor correlations thus, there is a need for further experiments.

In an attempt to understand the process, a full scale field model was devised at the Colorado School of Mines' Experimental Mine, near Idaho Springs, Colorado. The model consists of the Edgar Mine ventilation facilities, a bulkhead isolated mine drift containing four tons of low grade uranium ore (less than 0.05% U3 08) and a computer based radiation monitoring system owned and developed by the U.S. Bureau of Mines. The system continuously collects information such as radon concentration, working level, air temperature, relative humidity, barometric pressure and air velocity from monitors located throughout the mine. Although radiation level together with mine parameters affecting these levels were monitored for a period of one year, this study is based on one month information only.

2 MINE MODEL AND INSTRUMENTATION

The model is basically represented by a 200 ft long mine drift which has been isolated from the rest of the mine

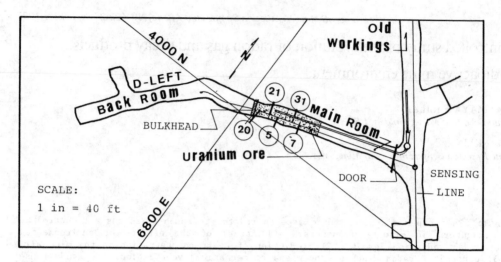

Figure 1. C.S.M.' Experimental Mine Model

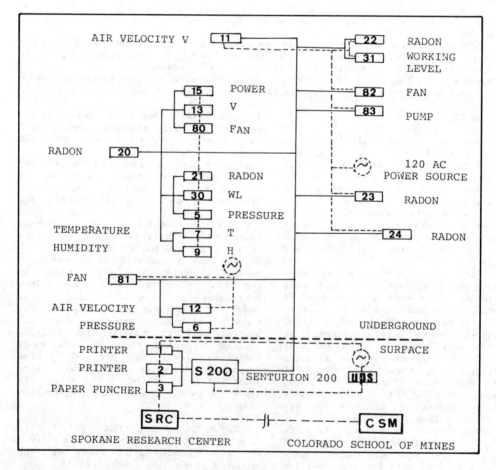

Figure 2. Computer Based Data Collection System

openings by means of an airtight steel door. The drift was divided into two rooms by means of a concrete block bulkhead: the innermost room (Back) used to monitor both natural and man-induced radiation behind the bulkhead and the outer-most room (Main) to monitor emissions of radon and its decay products from the low grade uranium ore placed in the room. The latter room is equipped with an exhaust ventilation system, and all the facilities to operate the monitors (Figure 1).

The radiation monitoring system used with the model consisted of a central data processing unit, a data gathering network, and various sensing units (Figure 2). A Senturion 200 microcomputer was used to collect data and monitor all the sensing and control devices. The data gathering network consists of point multiplexers (accessors), four wire communication cables branched to allow for accessors, and a trunk-processor connection box. The system was operated with fourteen transducers: five to monitor radon concentrations, two each for working level and barometric pressure, one each for temperature and relative humidity, and three for air velocity. In addition to the transducers, binary accessors were used to control three underground fans from the surface.

The system was operated with two sets of programs: operating and control programs (System) and application programs (Status). These programs enable the system to monitor field devices and activate trend log and event printers (Sheeran and Franklin, 1984). Samples were collected for periods of 10 minutes and printed every hour.

3 SYSTEM OUTPUT

The system provides the user with two types of output: trend logs and event sequences. Trend logs are used to generate monitor outputs in engineering units; this information was printed and stored on tapes. The event sequences are digital information that can either represent an analog reading outside a preset normal operating region or ON/OFF positions of binary points. All events were printed with their time of occurence. In this study only the output of five monitors were considered: two for radon and one each for working level, barometric pressure and temperature. All samples were taken in the D-Left drift (Edgar Mine) between February 22 and March 22. Figures 3a-3b and 4a depict the output of monitors as graphs of time

series. Figure 4b is a summary of commandable fans and power conditions in the mine.

Measurements of radon were performed in the two rooms separated by the concrete bulkhead. The radon laden air from the back room was sampled by means of 45 ft long tubing, passed through Rn Monitor No. 20 and returned to the room. This unit measures radon from the predominantly massive granitic-gneiss country rock and man induced radiation that resulted from the introduction of uranium ore in the main room. Monitor No. 21 measures concentrations of radon in the main room. Concentrations in this room are more affected by natural and mechanical ventilation than the back room.

Concentrations of radon decay products in the main room were determined by WL Monitor No. 30. In Figure 3b, part of the peak concentrations recorded on day 8, were caused by diesel smoke from a mine locomotive brought to the test section for 15 minutes to pull out a mine car that delivered one ton of low grade ore.

The barometric pressure in the testing room was measured with a Staham PA 822-25 transducer (Monitor No. 5), with an output range from 0 to 1 Vdc for 0 to 1000 mm Hg. Because of the position of the transducer in the mine (D-left Drift main room), the instrument output is not affected substantially by changes in main fan pressure (Figure 4a).

The fifth transducer mentioned in the study consisted of an HMP-14ut probe and an HT-100 IR indicator (Monitor No. 7). The units were used to measure the air temperature and relative humidity in the main room.

All the mining activities (such as drilling, blasting, mucking), except the eighth day, took place outside the testing room and were limited to one-8 hour shift per week (Fridays). During this particular day however, the mine was kept well ventilated.

4 GRAPHICAL COMPARISONS

The data points of Figures 3 and 4 were collected simultaneously to form discrete time series of 667 observations. Each series represents the historical behavior of a process that was considered to bo a crucial in determining ventilation requirements to control the health hazards in radioactive mine environments. Because of the very nature of the radiation process (growth and decay), observations within a series are neither independent nor stationary; a causal inspection of these graphs for instance, may anticipate

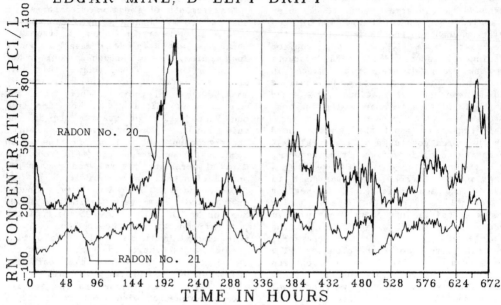

Figure 3a. Radon Concentration Series

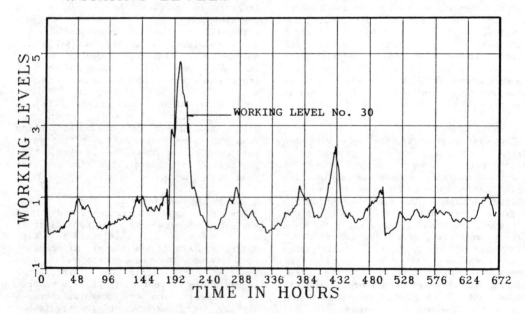

Figure 3b. Working Level Series

Figure 3. Concentrations of Radon and Radon Decay Products

a strong negative relationship between atmospheric pressure and concentrations of either radon or radon daughters. However, in order to apply statistical tools such as Box-Jenkins Forecasting Method, each series can be regarded as a stochastic process where observations are sample realizations from an infinite population.

The two time series of Figure 3a basically show the same pattern, the only difference is that concentrations of radon in the back room are greater than the ones found in the main room. A major portion of this difference is attributed to contaminant isolation caused by the bulkhead. On the other hand, the concentration of radon in the test room is more sensitive to local ventilation conditions. In any case, from this graph one can observe that for a given amount of radioactive material, radon concentration is inversely related to barometric pressure. Since mathematical models to reflect particular conditions are adjusted by intrinsic local factors, this study assumes that in developing a prediction model no information is lost by using either time series. The authors' choice was the series generated by the radon monitor behind the bulkhead.

Comparing Figures 3a and 3b one finds that the measured concentrtions of radon and its decay products show basically the same pattern. As in the case of radon concentration, a substantial fraction of changes in working levels was explained by variations in barometric pressure and a small fraction was attributed to other factors such as general ventilation, amount of condensation nuclei (smoke infusion) and plate-out of unattached ions on the filter. From a visual inspection of Figures 3b and 4b, one observes that radon daughter concentrations in the room declined sharply to almost zero level when the mine was mechanically ventilated and the test room door open. However, when the fans were switched off and the door sealed, it took a longer period for the decay products to come back to its normal level showing that the removal rate of radon products was greater than the growth rate, a common pattern found with efficient ventilation.

The calculated equilibrium factors (1.0 WL* 100/ radon concentration in pCi/1), for the WL monitor output as compared to radon concentrations measured by monitors 2 (Back) and 1 (Main) were 0.14 and 0.36 respectively.

Figure 4a shows changes in barometric pressure and the measured temperatures in the testing room. The mine temperature remained fairly constant at 58 DF; therefore, the effect of temperature change on concentrations of either radon or radon daughters could not be evaluated.

5 TIME SERIES METHOD

The statistical approach used in the analysis of the radiation data was the Box Jenkins Method. This method is a self-projecting time series forecasting technique, which allows the user to systematically select the appropriate model for his data. The method consists of seven building blocks: Autoregressive Model (AR), Moving Average (MA), Regular Difference (RD), Seasonal AR, Seasonal MA, Seasonal Difference and Trend. The modeling process consists of the following steps: model identification, parameter estimation and forecasting. All these steps are performed by means of computer programs. Box Jenkins programs are currently accessed through almost all commercial and non-commercial computer services.

Model identification is the process of selecting one or a combination of any of the above seven blocks for a time series; this process is accomplished by analyzing simultaneously the Autocorrelation (ACF) and Partial Autocorrelation (PACF) functions of the data, usually displayed by the computer as a graph of correlations called correlogram. For a stationary data, the ACF's of an AR(p) model declines exponentially with the number of lags between observations whereas the PACF's will cut off after p lags, where p is the order of the autoregressive model. The ACF's and PACF's of an MA(q) model of order q, produce patterns that are exactly the reverse of those produced by an AR model. Similar graphs of ACF's and PACF's are developed with seasonal models, except that the non-zero autocorrelations occur at lags that are multiples of the number of periods per season (Hoff, 1983). For non-stationary data, the ACF's of the series tend to remain large for many successive lags beginning at lag one. This type of correlograms indicate that a regular or seasonal differencing is required (Box Jenkins, 1976).

Once a model is identified and its parameters selected, the next step is model validation. To check the validity of the model, two types of tests are performed: residual diagnostics (Chi-square test and Residual standard error) and parameter diagnostics (Parameter confidence limits and correlations between parameters). Most of the Box Jenkins programs will provide these statistics

BAROMETRIC PRESSURE
EDGAR MINE, D-LEFT DRIFT

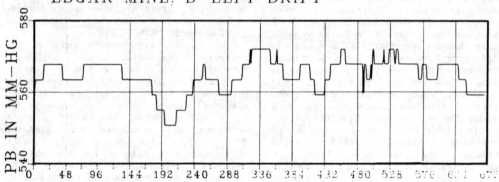

AIR TEMPERATURE

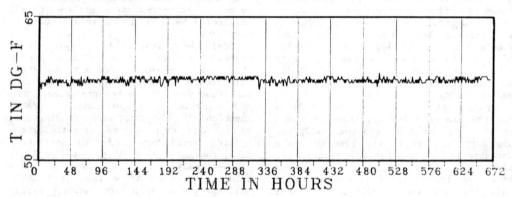

Figure 4a. Pressure and Temperature Series

	FEBRUARY						MARCH																					
DAY	23	24	25	26	27	28	⧄	2	3	4	5	6	7	⧄8	9	10	11	12	13	14	⧄15	16	17	18	19	20	21	⧄22
Main Fan	OFF	OFF	OFF	OFF	OFF	ON 13-16	ON:9 -20	OFF	OFF	OFF	ON:9-16	OFF	OFF	ON:7-16	OFF	OFF	OFF	OFF	OFF	OFF	ON:13-20	OFF	OFF	OFF	OFF	OFF	OFF	ON:14-20
D-L Fan	OFF	OFF	OFF	OFF	OFF	OFF	ON:16-20	ON:10-12	OFF	OFF	OFF	OFF	ON:10-12	OFF	OFF	OFF	OFF	OFF	OFF	OFF	ON:17-19	OFF	OFF	OFF	OFF	OFF	OFF	ON:16-20
Power	ON	ON	ON	ON	ON	ON	ON-ALARM	ON	ON	ON	ON	ON	ON	ON	ON	ON	ON	OFF:1 ON	ON	ON	ON	ON	ON	ON	ON	ON	ON	ON

Figure 4b. ON/OFF Positions of Binary Points

Fifure 4. Parameters Affecting Radiation Levels in a Mine

264

when parameters are calculated. It is left to users' judgement to interpret the data output.

The last step of the process, forecasting, is performed by substituting past observations, residuals, and previous forecasts in the formula just generated. The random error component of the formula is assumed to be zero. Again, all Box Jenkins programs will calculate forecasts automatically. The user must assess the accuracy of the forecasts and compare them with new data points.

6 TIME SERIES RESULTS

In this paper, Box Jenkins methodology is applied to radon concentration data of the Back room. Figure 5a illustrates the first 14 ACF's of the original series. A visual inspection of this graph shows that autocorrelations are decreasing at almost constant rate, indicating that the series requires at least a regular differeincing of order one. The ACF's for the differenced series is shown in Figure 5b, this correlogram contains one large spike at lag one indicating that the model includes one MA(1) parameter. Figure 5c shows the rapidly decaying PACF's for the differenced series which confirms an MA(1) parameter in the model. When this model and the once-differenced series are entered into the Box Jenkins programs, the estimation results of Figure 6a were obtained. The estimated MA(1) parameter was 0.274, which is statistically significant, and its calculated Chi-Square was 47.74, which is greater than 35.17 (test value at 95% confidence limit). Furthermore, the residual ACF's for this model, shown significant spikes at lags 3 and 4, which suggests the presence of another MA parameter in the model.

In subsequent trials, new parameters were introduced into the model, MA(3) and MA(4) were examined as possible alternatives. Diagnostic tests indicated that a Moving Average model of order 3 was more adequate than a moving average of order 1. The final model for the series is therefore represented by (Figure 6b):

$$Z_t = Z_{t-1} + a_t - 0.30006\, a_{t-1} + 0.10433\, a_{t-3}$$

where, both of the model parameters are statistically significant (95% confidence limits exclude zero), the new calculated Chi-square is 36.70, which is not substantially greater than its test value (33.40). Finally, the model parameters were tested to insure the invertibility condition of the process by solving the following equation:

$$\propto(B) = 1 - 0.30006B + 0.10433B^3 = 0$$

the roots of this equation lie outside the unit circle, indicating that the process is invertible therefore, the model is correct.

Based on the above formula, the following forecasts were developed:

Period (hrs)	Forecast (PCi/1)	95% limits (increments	Actual (PCi/1)
668	581.48	95.40	547.53
669	584.06	113.60	654.09
670	573.31	127.76	616.21

Similar MA models can be developed for the main room radon concentration series and the working level series however, since these series had the same pattern as the one shown by monitor 20, the model parameters were not identified. In any case these prediction models together with equilibrium factors determined for similar conditions can be used to predict concentrations of radon decay products in the room. Based on the above model the following concentrations were predicted for the testing room: 0.81, 0.82 and 0.80 WL, the measured working levels for these periods were: 0.65, 0.66 and 0.59 respectively.

The predicted working levels, now can be used to calculate air quantity requirements using simple formulae such as Harris and Bales equation (Harris and Bales, 1964) or the contaminant delution equation.

Example: The actual exhaust ventilation system of the model circulates 300 CFM of fresh air through the mine drift where the predicted concentration of radon daughters is 0.81 WL. Calculate the amount of fresh air needed to lower such a concentration to its threshold limit value of 0.30 WL.

Solution: $Q(1) = 300$ CFM, $WL(1) = 0.81$ and $WL(2) = 0.30$

1. Applying Harris and Bales equation:
$$Q(2) = Q(1)\left(\frac{WL(1)}{WL(2)}\right)^{0.56}$$
$$= 477 \text{ CFM}$$

2. Applying the simple dilution equation:
$$Q(2) = Q(1)\left(\frac{WL(2) - WL(1)}{WL(1)}\right)$$
$$= 510 \text{ CFM}$$

Which in this case can be achieved by changing the size of the regulator (opening the blast gate) of the vent system.

265

```
-.1000E+01                              .0000E+00                              .1000E+01    VALUES
.+++++++++.+++++++++.+++++++++.+++++++++.+++++++++.+++++++++.+++++++++.+++++++++.++++++++.
                                        XXX XX XX XXXXXXXXXXXXXXXXXXXXXXXXXXXXXXXXXXX XXX XXXX XXXXX    .96303E+00
                                        X XX XX XX XXXXXXXXXXXXXXXXXXXXXXXXXXXXXXXXX XXX XXX XXXX     .94878E+00
                                        X XX XX XXXXXXXXXXXXXXXXXXXXXXXXXXXXXXXXXXXX XX XXX XXX     .93572E+00
                                        X XX XX XXXXXXXXXXXXXXXXXXXXXXXXXXXXXXXXXXXX XXX XXX     .92123E+00
                                        X XX XX XXXXXXXXXXXXXXXXXXXXXXXXXXXXXXXXXXX XXX XXX YX     .90177E+00
                                        X XXXXXXXXXXXXXXXXXXXXXXXXXXXXXXXXXXXXXX XXX XXX X     .88511E+00
                                        X XX XX XXXXXXXXXXXXXXXXXXXXXXXXXXXXXXXX XXX XXX XXX     .86639E+00
                                        X XX XX XXXXXXXXXXXXXXXXXXXXXXXXXXXXXXXX XXX XX XX     .84482E+00
                                        X XXXXXXXXXXXXXXXXXXXXXXXXXXXXXXXXXXXX XX X     .82635E+00
                                        X XX XX XXXXXXXXXXXXXXXXXXXXXXXXXXXXXXX XXX     .80691E+00
                                        X XX XX XXXXXXXXXXXXXXXXXXXXXXXXXXXXXX X     .78651E+00
                                        X XX XX XXXXXXXXXXXXXXXXXXXXXXXXXXXXXX     .75810E+00
                                        X XX XX XXXXXXXXXXXXXXXXXXXXXXXXXXXX     .73435E+00
                                        X XX XX XXXXXXXXXXXXXXXXXXXXXXXXXXXX     .70610E+00
```

Figure 5a. ACF Graph of Observed Series

```
-.1000E+01                              .0000E+00                              .1000E+01    VALUES
.+++++++++.+++++++++.+++++++++.+++++++++.+++++++++.+++++++++.+++++++++.+++++++++.++++++++.
                          XXXXXXXXXXXXXXXXX                                            -.33133E+00
                                      XX                                              -.13137E-01
                                      XX                                               .27267E-01
                                      XXXX                                             .58868E-01
                                      XX                                              -.24261E-01
                                      XXX                                              .32514E-01
                                      XXX                                              .33513E-01
                                    XXX                                               -.46007E-01
                                      X                                                .96416E-02
                                      X                                                .15283E-01
                                      XXXXXXX                                          .10046E+00
                                  XXXX                                                -.64372E-01
                                      XXXX                                             .65692E-01
                                      XX                                               .19602E-01
```

Figure 5b. ACF Graph of Differenced Series

```
-.1000E+01                              .0000E+00                              .1000E+01    VALUES
.+++++++++.+++++++++.+++++++++.+++++++++.+++++++++.+++++++++.+++++++++.+++++++++.++++++++.
                          XXXXXXXXXXXXXXXXX                                            -.33133E+00
                                XXXXXXX                                               -.13807E+00
                                    XX                                                -.26834E-01
                                      XXXX                                             .66940E-01
                                      XX                                               .27605E-01
                                      XXX                                              .47817E-01
                                      XXX                                              .65249E-01
                                    XX                                                -.14289E-01
                                    XX                                                -.11325E-01
                                      X                                                .12978E-02
                                      XXXXXXX                                          .11837E+00
                                      XX                                               .21541E-01
                                      XXXXX                                            .70021E-01
                                      XXXX                                             .62985E-01
```

Figure 5c. PACF Graph of Differenced Series

Figure 5.   ACF and PACF Correlograms for Radon Concentration Series

DATA -   Z = RADIATION DATA - EDGAR MINE   667 OBSERVATIONS
DIFFERENCING ON   Z - 1) 1 OF ORDER 1
****
UNIVARIATE MODEL PARAMETERS
****

PARAMETER NUMBER	PARAMETER TYPE	PARAMETER ORDER	ESTIMATED VALUE	95 PER CENT	
				LOWER LIMIT	UPPER LIMIT
****
| 1 | MOVING AVERAGE 1 | 1 | .27412E+00 | .19930E+00 | .34895E+00 |
****
OTHER INFORMATION AND RESULTS
****

RESIDUAL SUM OF SQUARES   .94704E+06   665 D.F.   RESIDUAL MEAN SQUARE   .14241E+04
NUMBER OF RESIDUALS   666   RESIDUAL STANDARD ERROR   .37737E+02
BACKFORECASTING WAS SUPPRESSED IN PARAMETER ESTIMATION

THE ESTIMATED RESIDUALS - MODEL 1
GRAPH OF OBSERVED SERIES ACF
GRAPH INTERVAL IS   .2000E-01
-.1000E+01   .0000E+00   .1000E+01   VALUES
.+++++++++.+.+++++++++.++++++++++.++++++++++.++++++++++.++++++++++.++++++++++.++++++++++.+++++++++.
                                   X   -.13863E-01
                                  XX   .14820E-01
                                  XX   .89657E-01
                                   X   .89657E-01
                                XXXXX   .97064E-01
                                XXXXXX   .72671E-02
                                   X   .95019E-01
                                XXXXXX   .54614E-01
                                XXXX   -.28790E-01
                                  XX   .37929E-01
                                  XAX   .11001E+02
                                XXXXXXX

Figure 6a. Estimation Results for One MA(1) Model

DATA -   Z = RADIATION DATA - EDGAR MINE   667 OBSERVATIONS
DIFFERENCING ON   Z - 1) 1 OF ORDER 1
****
UNIVARIATE MODEL PARAMETERS
****

PARAMETER NUMBER	PARAMETER TYPE	PARAMETER ORDER	ESTIMATED VALUE	95 PER CENT	
				LOWER LIMIT	UPPER LIMIT
****
| 1 | MOVING AVERAGE 1 | 1 | .30006E+00 | .22572E+00 | .37441E+00 |
| 2 | MOVING AVERAGE 1 | 3 | -.10433E+00 | -.17878E+00 | -.29875E-01 |
****
OTHER INFORMATION AND RESULTS
****

RESIDUAL SUM OF SQUARES   .93565E+06   664 D.F.   RESIDUAL MEAN SQUARE   .14091E+04
NUMBER OF RESIDUALS   666   RESIDUAL STANDARD ERROR   .37538E+02
BACKFORECASTING WAS SUPPRESSED IN PARAMETER ESTIMATION

AUTOCORRELATION FUNCTION
DATA - THE ESTIMATED RESIDUALS - MODEL 1   666 OBSERVATIONS

ORIGINAL SERIES
MEAN OF THE SERIES = .29746E+00
ST. DEV. OF SERIES = .37460E+02
NUMBER OF OBSERVATIONS =   666

1- 12   -0.00   0.02 -0.02   0.07   0.00   0.09   0.04 -0.04   0.03   0.10   0.07   0.01
ST.E.   0.04   0.04   0.04   0.04   0.04   0.04   0.04   0.04   0.04   0.04   0.04   0.04

13- 24   0.09 -0.01 -0.04   0.00 -0.04 -0.01 -0.04 -0.02 -0.06   0.02 -0.02   0.07
ST.E.   0.04   0.04   0.04   0.04   0.04   0.04   0.04   0.04   0.04   0.04   0.04   0.04

MEAN DIVIDED BY ST. ERROR = 0.20483E+00

TO TEST WHETHER THIS SERIES IS WHITE NOISE, THE VALUE 0.36733E+02
SHOULD BE COMPARED WITH A CHI-SQUARE VARIABLE WITH 22 DEGREES OF FREEDOM

Figure 6b. Estimation Results for One MA(3) Model

Figure 6. Model Estimation Results for Radon Concentration Series

267

7 CONCLUSIONS

1. The time series of Figure 3a show that concentrations of radon in the mine model fluctuate around a fixed level (mean), taking trips away from it but always returning to that neighborhood. A major portion of these fluctuations were caused by changes in barometric pressure. Comparing the graphs of Figures 3a and 4a one can conclude that there is a strong negative correlation between these two variables.

2. For a given amount of radioactive material, the concentration of radon in the model is represented by an Integrated Moving Average Process IMA(1,3), which indicates that any value of the differenced series is only related to random perturbances that occurred in the past three time periods.

3. The concentration of radon decay products basically follows the same pattern developed by their parent products. In addition to barometric pressure, fluctuations in working levels in the mine were caused by dust and smoke particles generated in the test room.

8 REFERENCES

1. Box, E.G., and Jenkins M.G., 1976, Time Series Analysis Forecasting and Control, Holden Day, Inc., Oakland, California, 575 p.
2. Edwards, C.J., and Bates, C.R., 1980, Theoretical Evaluation of Radon Emanation under a Variety of Conditions, Health Physics, Pergamon Press Ltd., Vol. 39 (August), pp. 263-274.
3. Franklin, C.J., et.al., 1976, Data Acquisition System for Radon Monitoring, BuMines RI 8100, 19 p.
4. Harris, R.L., and Bales R.E., 1964, Uranium Mine Ventilation for Control of Radon and its Decay Products, Radiological Health and Safety in Mining and Milling, IAEA, Vienna, volume 2, pp 49-60.
5. Hoff, C.J., 1983, A Practical Guide to Box Jenkins Forecasting, Lifetime Learning Publications, Belmont, California, 316 p.
6. Jackson, P.O., et.al., 1981, An Investigation of Radon 222 Emmissions from Underground Uranium Mines, Pacific Northwest Laboratory, Richland, Washington, 33 p.
7. Sheeran, C.T., and Franklin, J.C., 1984, Microcomputer Based Monitoring and Control System with Uranium Mining Application, U.S. Bureau of Mines, IC 8981, 28 pp.

Characterization of radioactive dust in Canadian underground uranium mines

J.BIGU & M.G.GRENIER
CANMET, Mining Research Laboratories, Elliot Lake, Ontario, Canada

ABSTRACT: Measurements have been carried out in several Canadian underground uranium mines in order to determine a) the long-lived radioactivity (Ra-224, Ra-226, Th-232, Th-228, U-238) and the short-lived radioactivity (radon daughters and thoron daughters) associated with aerosols in the submicron range (0.05-1 μm), dust in the respirable range (1-10 μm), and dust beyond the respirable range (10-25 μm); b) the concentration and size distribution of aerosol and dust in the 0.05-25 μm size range; c) the electrical characteristics of radon and thoron daughters associated with submicron aerosols. Radioisotope identification was done by α-spectroscopy using silicon-barrier detectors, and by γ-spectroscopy using an NaI(Tl) detector and a high purity germanium detector. Dust in the respirable range, silica dust, combustible dust and total dust, were determined by dust samplers designed in-house. Aerosol size distribution was measured in the 0.05-25 μm range by means of a 10-stage cascade impactor. Radioactive size distributions corresponding to radon daughters, thoron daughters and long-lived radioactive dust (LLRD) were determined from radioactivity measurements carried out on the 10 stages of the cascade impactor.

1 INTRODUCTION

Extended inhalation of airborne radioisotopes poses a potential health problem to occupational workers. Because of this, monitoring of radiation and dust levels for dose exposure calculation purposes is a subject of considerable practical interest.

CANMET (Energy, Mines and Resources Canada) has long since been engaged in a comprehensive monitoring program aimed at determining dust concentrations in the respirable size range, and in the radioactivity associated with these particulates in underground uranium mines, particularly the short-lived decay products of radon and thoron, and some long-lived decay products of the U-238 and Th-232 natural radioactive chains.

Early measurement of radon and thoron daughter activity as a function of size in a horizontal elutriator indicated that the specific α-activity, i.e., α-activity per unit mass, remains nearly constant over the entire respirable dust size range (Bigu et al., 1980, 1981). The above result indicates that little activity is associated with respirable dust and hence most activity is associated with particles in the submicron range. These data were confirmed using Nuclepore filters of different pore sizes from 0.05 μm to 5 μm (Bigu and Kirk, 1982).

Until recently, the short-lived decay products of radon were the only airborne radioisotopes of concern from the occupational health viewpoint in uranium mines and other uranium-related industries. However, increasing experimental evidence shows that attention should also be paid to the short-lived decay products of thoron and the long-lived radioisotopes (U-235, U-238 and Th-232 and some of their decay products) associated with aerosols in the submicron range and dust in the respirable size range. Some recent concern has been expressed particularly as to the inhalation of respirable dust (1-10 μm) containing long-lived radioisotopes, as once inhaled and lodged in tissue they will remain active for long periods of time unless eliminated by natural biological processes.

Although size distribution measurements have been conducted to determine the submicron aerosol size distribution associated with the short-lived decay products of radon and thoron (Blanc et al.

1967; Busigin et al. 1978; Jacobi 1963; Mercer and Stowe 1971), more studies are necessary. Furthermore, very little information is available regarding the long-term effects of continuous exposure to long-lived radioactive dust (LLRD). Sparse data are also available on LLRD size distribution in uranium mines and mills. Thus, it is important to identify the major constituents of LLRD, their relative and absolute amounts in air and their size distribution as the latter determines the attachment characteristics of LLRD in the respiratory system (Bigu and Grenier 1984; Duport and Edwardson 1985).

This paper presents data collected in several Canadian underground uranium mines on:

a. the long-lived radioactivity (Ra-224, Ra-226, Th-232, Th-228, U-238) and the short-lived radioactivity (radon progeny and thoron progeny) associated, respectively, with dust in the 1-25 μm size range, and with aerosols in the submicron range (0.05-1 μm). It should be noted that short-lived radioactivity is also associated with dust in the respirable range and beyond;

b. the concentration and size distribution of aerosol and dust in the 0.05-25 μm range;

c. the electrical characteristics of the radon and thoron progeny associated with submicron particulates.

2. EXPERIMENTAL APPARATUS AND METHODS

Size distribution analyses of radioactive dust, radioactive aerosol and dust were conducted by means of a 10-stage cascade impactor, in conjunction with a cyclone preseparator, Model 210 manufactured by Sierra Instruments Inc. (U.S.A.). The cascade impactor was operated with either 10 stages at a nominal flowrate of about 2.6 L/min or with 8 stages with a flow-rate of approximately 14 L/min. In the latter case, the last two ultrafine impactor stages were eliminated at the expense of losing some size distribution information but with the obvious benefit of substantially increasing the amount of dust collected on the remaining 8 impactor stages. Glass fiber filters (47 mm diameter) were used as substrates to collect the samples. The cascade impactor was operated for 24 h at a time.

Total dust and LLRD samples were also collected on open-face filters (glass fiber and Millipore 0-8 μm filters, 25 mm and 100 mm diameter) at flow-rates ranging from 0.05 L/min to 700 L/min. Furthermore, one-stage impactor dust samplers were used to collect samples on silver membrane filters for X-ray diffraction analysis at a flow rate of 2 L/min (Knight 1978).

The sampling time used to collect samples ranged from 2 h to 30 days. The total volume of air sampled was in the range 2 to 84 m^3.

Filters placed behind the stages of the cascade impactor enabled determination of the size distribution (mass median aerodynamic diameter, MMAD, and geometric standard deviation) of dust by determining the weight of the filters before and after the sampling period. The filters were dried before and after sampling to eliminate moisture. Ambient temperature and pressure were carefully noted during sampling and results were corrected according to standard operating procedures. Total dust was also estimated from cascade impactor data and from samples collected with open-face filters. Total, combustible and silica dust was measured (the latter using X-ray diffraction analysis) using the one-stage impactor samplers indicated above. In all cases, the weight of the filters before and after sampling was carefully recorded for dust measurement purposes.

Radioactivity measurements were done as follows. Filters were measured by α-spectroscopy (Si-barrier detector) and γ-spectroscopy (high purity Ge-detector, i.e., HPGe) shortly after the end of the sampling period to identify the short-lived decay products of radon and thoron. Gross α-count was also measured. LLRD measurements were done by gross α-count, and α-spectroscopy and γ-spectroscopy of the same samples after a convenient time to ensure complete decay of the airborne radon and thoron progeny collected on the samples. For practical reasons, four weeks elapsed between the end of sampling and the time at which LLRD analyses were carried out.

Because a substantial number of samples were collected in underground uranium mines, which also contained appreciable amounts of thorium, the mine atmospheres consisted of a mixture of the short-lived decay products of radon and thoron and the long-lived decay products of U-235, U-238 and Th-232, including these radioisotopes. The short-lived radioactivity from the radon and thoron progeny was determined as follows. Activity measurements were taken 15 min and 60 min after the end of sampling for 3 min or 5 min. Activity measurements were again taken about 6 h after the end of sampling for 5 min or 9 min. The latter measurements were intended to determine the thoron progeny contribution as the radon progeny decays completely 3.5-4 h after the end of sampling. Thoron progeny activity measurements were extrapolated to the time the radon progeny activity measurements were conducted in order to subtract the contribution to the total from the former and hence obtain the 'net' radon progeny activity. All activity measurements were then extrapolated to the same time, usually end of sampling, for direct comparison purposes.

The procedure outlined above required knowledge of the radon and thoron progeny disequilibrium ratios, i.e., [Pb-214]/ [Po-218] and [Bi-214]/[Po-218] ratios, and the [Bi-212]/[Pb-212] ratio, respectively, where the square brackets are used to denote atmospheric activity concentration. Data for the disequilibrium ratios for the radon progeny were obtained by the Thomas-Tsivoglou method (Thomas 1972; Tsivoglou et al. 1953), and by means of a grab-sampler manufactured by Pylon Electronics, Model WL1000C. The grab-sampler was also used to determine the thoron progeny disequilibrium ratio. The average values obtained, and adopted in this work, were as follows: [Pb-214]/[Po-218] = 0.6; [Bi-214]/[Po-218] = 0.4; [Bi-212]/[Pb-212] = 0.45. It should be noted that the activity from the long-lived decay products was also taken into consideration. However, this activity was in most cases very small compared with the short-lived activity and, hence, could safely be ignored.

The procedure used for the determination of dust, activity, and size distribution from the cascade impactor data was as follows:

1. Activity (dpm, i.e., disintegrations per minute) and dust mass collected on each impactor stage were carefully noted.

2. Total activity and total dust mass from all the stages of the impactor, including the backfilter (BF), were estimated.

3. Percentage (%) activity and % dust mass for each impactor stage were calculated.

4. Cumulative % of dust mass and cumulative % of activity, less than $D_{p,50}$ (see below), were estimated as follows. Dust mass (or activity) % on the BF was used as cumulative % for the last ultrafine stage, i.e., stage 8 or stage 10. The cumulative % for the next stage was obtained by adding the % of dust mass (or activity) to the cumulative % dust mass (or activity) corresponding to the previous stage, and so on.

5. Cumulative % dust mass (or activity), less than $D_{p,50}$, versus EAD was plotted.

The variable $D_{p,50}$ is defined as the particle size cut-off at 50% collection efficiency for spherical particles. The magnitude EAD is the Equivalent Aerodynamic Diameter defined as the size of a spherical particle of density $1 g/cm^3$ which has the same terminal settling velocity as the sampled particle.

Dust collected in filters was also analyzed for total uranium and thorium content. Uranium content was determined using fluorescence of the sample excited under exposure to light from a N_2-laser. Thorium content was estimated by spectrophotometric means. Ra-226 α-activity was also estimated after precipitation from a solution of the sample with $CaSO_4$.

Measurements were carried out in several Canadian underground uranium mines of widely different ore grades, mining practices (e.g., conventional and trackless mining) and during the course of a number of mining operations such as slushing, mucking, drilling, rock crushing and backfilling. However, the data presented here do not specifically refer to any particular mining operation or mine.

3. EXPERIMENTAL RESULTS AND DISCUSSION

Some of the experimental data collected in several underground uranium mines are shown in Figures 1 to 9 and in Table 1.

Figure 1 shows the % cumulative total respirable dust and submicron aerosol collected of size less than $D_{p,50}$ versus EAD. The data were obtained with the cascade impactor and represent an average value obtained during a series of independent measurements carried out at different times and mine locations. The average MMAD derived from these data is about 2.2 µm.

Figure 2 shows the % cumulative radon progeny (RnD) or thoron progeny (TnD) α-activity, corresponding to dust of size less than $D_{p,50}$, versus EAD. In this graph, the α-activity is that corresponding to direct collection of the radon and thoron progeny (attached to airborne dust) during the sampling period. This activity should clearly be differentiated from the radon and thoron progeny activity derived from the decay of long-lived radioisotopes such as Th-232, Ra-226, Th-228 and Ra-224, respectively, contained in the collected dust. In the first case, the radon progeny activity decrease to a negligible amount 3-4 h after the end of sampling whereas the thoron progeny activity reduces to

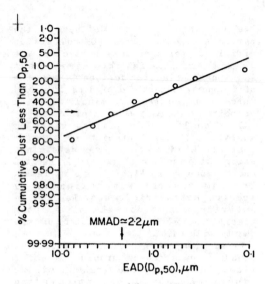

Fig. 1. Percentage (%) cumulative dust and submicron aerosol of size less than $D_{p,50}$ versus EAD.

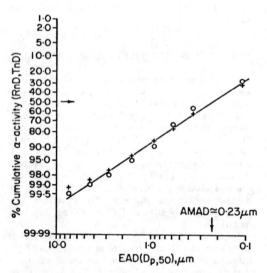

Fig. 2. Percentage (%) RnD (+) or TnD (o) α-activity versus EAD.

insignificant levels about 4 days after the sampling period. However, the radon and thoron progeny activity derived from the decay of long-lived radioactive products will be in effect for as long as the long-lived decay products remain active. Figure 2 permits determination of the radon progeny (crosses) and thoron progeny (encircled dots) AMAD (Activity Median Aerodynamic Diameter). The value

obtained is ∿0.23 μm. On the average no significant difference in the AMAD was found between the radon progeny and the thoron progeny when data were plotted in the range 0.1 to 7 μm. However, when data were plotted in the 0.1 to 0.6 μm range the AMAD for the thoron progeny seemed to lie somewhat higher than the AMAD for the radon progeny. This result is not conclusive and requires further verification. The data also show that attachment of the above short-lived radioisotopes to airborne particulates occurs preferentially at a size less than about 0.2 μm. These data are consistent with other data published elsewhere (Bigu and Grenier 1984) and with data published by other authors (see for instance Busigin et al. 1978 and refs. therein). The data presented in fig. 2 were obtained with the cascade impactor and with the same samples as those corresponding to fig. 1. Individual data points represent average values from several samples.

Figure 3 shows the % cumulative LLRD α-activity corresponding to dust of size less than $D_{p,50}$, versus EAD for several independent experiments. As before, the data were obtained with the cascade impactor and with the samples used to determine the MMAD (dust) and the AMAD (radon and thoron progeny). The scatter in the results shown in fig. 3 is mainly due to the relatively low α-activity of the LLRD samples. The line drawn through the data points represents the average, or 'best fitted' line. From the latter, a value of about 2.8 μm for the AMAD is obtained. It should be noted that although the MMAD and AMAD for the LLRD do not differ by much, AMAD data in the low size range lies significantly higher than the corresponding MMAD for the carrier particulate (dust). Furthermore, one may conclude that most LLRD is associated with the size range where a large percentage of dust, by mass, is found (see also fig. 4).

Figure 4 shows figures 1, 2 and 3 combined. The graph shows that the dust and LLRD lines cross-over at about 3.5 μm. It is also shown that the LLRD and radon and thoron progeny lines have the same slope. The significantly different slopes corresponding to dust and LLRD is not clearly understood, but it should be borne in mind that under our experimental conditions the errors associated with estimating α-activity from LLRD are substantially higher than those corresponding to the determination of dust.

Figure 5 shows the thoron progeny α-spectrum, taken with a Si-barrier detector, from a radioactive dust sample

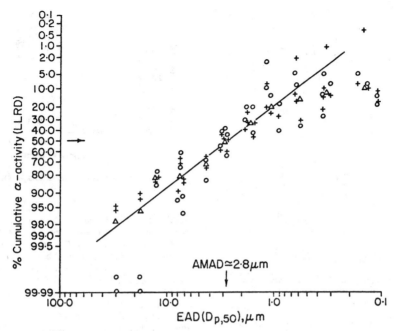

Fig 3. Percentage (%) cumulative LLRD α-activity versus EAD.

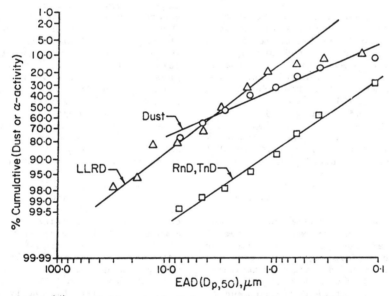

Fig. 4. Percentage (%) cumulative particulate or α-activity versus EAD.

collected in an underground uranium mine. Because of practical considerations, the sample could not be measured in the laboratory before about 4 h after the end of the sampling period. Hence, no radon progeny is present. Also shown in the graph is the α-spectrum from a radon progeny/thoron progeny mixture on a metal disc, used as a reference source. A comparison of both spectra shows the α-spectrum from the radioactive dust sample to be significantly broadened and energy-shifted towards the lower α-energy range of the spectrum, relative to the

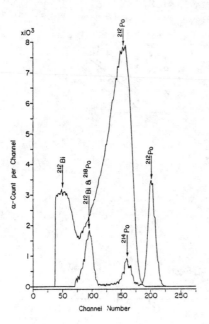

Fig. 5. TnD α-spectrum from a radioactive dust sample. Also shown is the α-spectrum from a RnD + TnD reference source.

reference source. This is so because of absorption of α-particles in the dust i.e., self-absorption. It should be noted that the dust collected during the sampling period may amount to quite an appreciable amount, i.e., thickness. The effect observed is, therefore, not surprising. Failure to take self-absorption and energy shifts into consideration may lead to significant errors in α-count, which are compounded by the inherently low α-activity from LLRD samples taken in uranium mines of low and/or medium ore grade.

Figure 6 shows a typical α-spectrum from a LLRD sample taken in an underground mine. The spectrum was taken using a Si-barrier detector, as above, under vacuum conditions. The spectrum was taken several months after the end of the sampling period. The radioisotopes measured have tentatively been identified as indicated by the vertical arrows. No conclusive data for the Th-232 series could be seen by α-spectroscopy. Some long-lived radio-isotopes are indicated. The short-lived radioisotopes originate, as previously indicated, from decay of long-lived radioactive members of the U-238 chain.

Figure 7 shows a typical γ-spectrum from a LLRD sample taken in an underground uranium mine. The spectrum was taken with a HPGe detector, as indicated above.

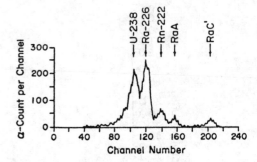

Fig. 6 LLRD α-spectrum from a radioactive dust sample.

Clearly distinguishable in the spectrum are the several members of the Th-232 and U-238 decay chains. The spectrum was taken several months after the sampling period.

Figures 8 and 9 show histograms of the LLRD α-activity concentration in mine air, within a given concentration range, versus the number of samples (normalized to unity) measured within this activity range. The data shown in figures 8 and 9 correspond to different underground uranium mines and mining operations. The data of fig. 8 correspond to the same general area collected over a period of several months. The data of fig. 9 consist of samples from many different locations and several mining operations taken over a period of about 2 years. The data presented show a wide range of LLRD concentrations from about 5 mBq/m^3 to over 1.0 x 10^5 mBq/m^3. Average values were in the range 50-80 mBq/m^3.

Table 1 shows dust data collected with open face filters and special dust samplers, namely: radioactive dust concentration (total uranium and total thorium) and mineral dust (total respirable dust (RD) in the 1-10 μm range, silica dust (SD) and combustible dust (CD)). Table 1 also shows airborne Ra-226 concentration in terms of α-activity. Samples in Table 1 were taken side by side with the cascade impactor in one of the several working locations in an underground uranium mine.

On the average, the radon progeny specific activity, i.e., activity per unit mass of dust collected in the stage, was about 40% higher than that corresponding to the thoron progeny. For a typical cascade impactor run, the combined radon progeny specific activity ranged from 5.2 x 10^5 mBq/mg to 2.0 x 10^6 mBq/mg. Hence, the combined specific activity per unit volume of air sampled (∿3.7 m^3) is 1.4 x 10^5 mBq/mg.m^3 to 0.53 x 10^6 mBq/mg.m^3. As

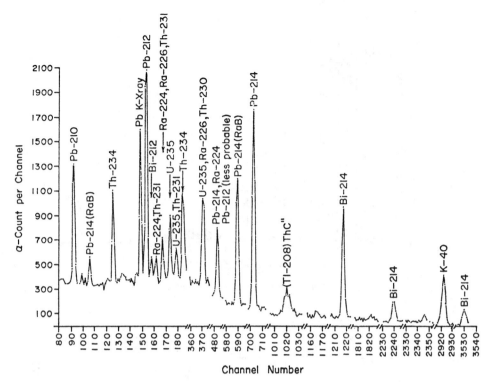

Fig. 7. LLRD γ-spectrum, taken with a HPGe, from a radioactive dust sample.

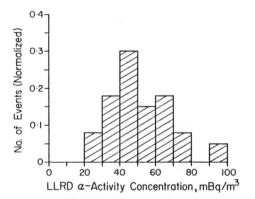

Fig. 8. LLRD α-activity concentration in mine air.

indicated above, data for the thoron progeny were significantly lower. The average value for the specific activity per stage can easily be obtained by dividing the above figures by the number of stages. As shown in figures 1 to 4, the maximum radon and thoron progeny activity was found in the lower size range whereas maximum dust mass was measured at higher size ranges.

4. ELECTRICAL CHARACTERISTICS OF SUBMICRON RADIOACTIVE AEROSOLS IN MINE AIR

The radon and thoron progeny are initially formed in an atomic, positively charged, state which rapidly combine with submicron mine aerosols. As mine aerosols are found in a positively charged, negatively charged and neutrally charged state, the resulting mine atmosphere consists of a complex mixture of charged and neutral particles of size covering a wide range. Because of their charge, radon and thoron progeny submicron aerosols can be influenced by external electric fields.

The electrical characteristics of submicron radioactive aerosols were investigated by sampling mine air through a glass fiber filter located at the end of a long aluminum tube equipped with a thin coaxial conductor. A DC voltage in the range 0-3000 V could be applied between the coaxial conductor and the tube. Measurement of the activity deposited on the filter and the inner wall of the tube, as a function of the electric field applied and the air sampling flowrate, permitted the determination of the radon and thoron progeny charged fraction in the submicron aerosol range.

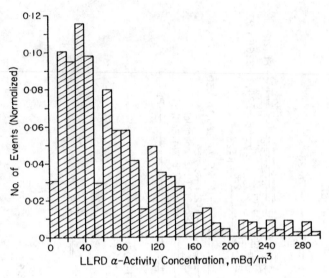

Fig. 9. LLRD α-activity concentration in mine air.

Table 1. Dust and radioactive dust data taken in several underground uranium mines.

Date	[U] μg/m³	[Th] μg/m³	RD mg/m³	SD mg/m³	CD mg/m³	[Ra-226]* mBq/m³
Dec 1/83	3.11±0.3	<0.43	–	–	–	10.12±1.47
Dec 15/83	–	–	1.05±0.3	0.12±0.02	0.93±0.3	–
Dec 21/83	–	–	1.69±0.4	0.04±0.04	1.65±0.4	–
Jan 4/84	–	–	1.94±0.5	0.10±0.03	1.74±0.5	–
Nov 1/84	0.72±0.01	0.32±0.03	–	–	–	6.71±0.71
Nov 2/84	0.81±0.01	0.18±0.02	–	–	–	4.79±0.48
Average:	0.77**	0.25**	1.56	0.07	1.41	5.75**

Remark: The square brackets stand for concentration. RD, SD and CD stand, respectively, for total respirable dust, silica dust and combustible dust. The values
given are averages from a substantial number of samples.
*Ra-226 is given in α-activity concentration.
**Average from Nov. 84 data taken at the same location.

The result of numerous measurements in an underground uranium mine shows that the thoron progeny charged fraction is about 65%. About 35% of the thoron progeny associated with mine aerosols in the submicron range was found to be neutrally charged. These data are consistent with radon progeny data collected in a radon/thoron test facility (Bigu 1982). Because of their electrical charge, radon and thoron progeny aerosols in the submicron range can be influenced by electrostatic means, thereby providing a means to reduce and control radiation levels in mine working areas.

5. CONCLUSIONS AND RECOMMENDATIONS

The data presented in this paper can be summarized as follows. The AMAD for the LLRD was about 2.8 μm, a value for the size somewhat larger than that corresponding to the MMAD of the carrier particulate, i.e., 2.2 μm. The AMAD for the radon and thoron progeny was approximately 0.23 μm. The AMAD for the thoron progeny seemed to be slightly higher than the AMAD for the radon progeny, but these data require further verification. The average α-activity corresponding to the LLRD was in the range

50-80 mBq/m^3. The total uranium dust and total thorium dust were in the ranges 0.7-3 μg/m^3 and 0.18-0.43 μg/m^3, respectively. As shown, the uranium content of the LLRD was substantially higher than its thorium content. Furthermore, about 65% of the short-lived decay products were electrically charged, whereas 35% were neutral.

The values given in this paper represent typical values for the locations, mining operations and underground uranium mines where measurements were taken. These data may not be truly representative of general typical conditions in other underground mines. Hence, in order to apply radioactive dust data to dose exposure estimation and use these data for epidemiological studies, more measurements over extended periods of time are necessary to better estimate the range and average values for the MMAD and AMAD.

6. ACKNOWLEDGEMENTS

The authors would like to express their appreciation to the Atomic Energy Control Board (AECB) of Canada for analyzing some of the radioactive dust samples for total uranium and total thorium content, and to Dr. S. Hardcastle for conducting some dust measurements in one underground area of interest. The authors would also like to thank the mining companies for giving permission to conduct experimental work on their premises.

REFERENCES

Bigu, J., Gangal, M.G., Knight, G.,Regan, R. & W. Stefanich 1980/81. Radiation, ventilation and dust studies in several mines. Division Reports MRP/MRL 80-111 (TR), MRP/MRL 80-114(TR) and MRP/MRL 81-41(TR), CANMET, Energy, Mines and Resources Canada.

Bigu, J. & B. Kirk 1982. Experimental determination of the unattached radon daughter fraction and dust size distribution in some Canadian uranium mines. Can. Mining J. 103:39-45.

Bigu, J. & M.G. Grenier 1984. Studies of radioactive dust in Canadian uranium mines. CIM Bull. 77:62-68.

Bigu, J. 1982. Electrical characteristics of radon daughters (unpublished data).

Blanc, C., Fontan, J., Chapuis, A., Billard, F., Madelaine G. & J. Pradel 1967. Dosage du radon et de ses descendants dans une mine d'uranium - repartition granulometrique des aerosols radioactifs. In Assessment of Airborne Radioactivity, IAIA (Vienna), 229-237.

Busigin, A., Van der Vooren, A. & C.R. Phillips 1978. Attached and unattached radon daughters: measurements and measurement techniques in uranium mines and in the laboratory. AECB Report.

Duport, P.J. & E. Edwardson 1985. Characterization of radioactive long-lived dust present in uranium mines and mills atmospheres. In Proc. Occup. Radiation Safety in Mining 1:189-195, Canadian Nuclear Association (Toronto). H. Stocker (ed).

Jacobi, W. 1963. Biophysik 1:175.

Knight, G. 1978. Mine dust sampling - CAMPEDS. Division Report MRP/MRL 78-7 (TR), CANMET, Energy, Mines and Resources Canada.

Mercer, T.T. & W.A. Stowe 1971. Radio-active aerosols produced by radon in room air. In Inhaled Particles III, v.2. Unwin Brothers Ltd. W.H. Walton (ed.).

Thomas, J.W. 1972. Measurement of radon daughters in air. Health Physics 23:783-789.

Tsivoglou, E.C., Ayer, H.E. & D.A. Holaday 1953. Occurrence of non-equilibrium atmospheric mixtures of radon and its daughters. Nucleonics 11:40-45.

An approach to calculate amount of air for eliminating
the radon daughters in uranium mines

WU GANG
Design & Research Institute of Uranium Mining and Metallurgy, Shijiazhuang, People's Republic of China

ABSTRACT: This article has a brief introduction of an approach to calculate amount of air for eliminating Rn and its daughters in the design of uranium mines of China, an inguiry into the hyperbola regression equation being as an accumulative equation or Rn daughters' α potential, a calculating formula is derived for eliminating radon daughters in uranium mines.

1 PREFACE

It is well known that the ventilation design calculation in mines consists of two parts:ventilating air volume calculation and ventilation resistance calculation. The method of ventilating resistance calculation in uranium mines is the same as the other mines adopt, and there is a character in calculating amount of air owing to the harm by Rn and its daughters.

Among the endangerness of Rn and its daughters, the later has more danger to human bodies, as a result of RaA, RaB and RaC releasing a large number of α particles from RaC' decay to RaD, after being breathed in, inhaled α particles would adhere to bronchus wall and boombard its epithelial cell for a long time, leading to come into being lung cancer.

Radon and its daughters in the atmosphere of mine could be diluted and eliminated by ventilation so as to reduce and avoid their harms to health. It is an important task for mining engineers to determine more economical and more resonable ventilating air volume in the design of uranium mines.

The present method of calculating ventilation in uranium mines of China has been laid down and adopted in the late fifties and in the early part of sixties, the techniques in ventilation of uranium mines and of radiation protection have made a great progress. Therefore, a new method should be engaged in research.

The article summarises radon emanation rates in uranium mines of China and an approach to calculate ventilating air volume for eliminating Rn and its daughters, and has an inquiry into the hyperbola regression of Rn daughters' α potential, on the bassis of the protective standard recommended by ICRP in 1981, a formula of calculating amount of air is derived for eliminating radon daughters.

2 THE PROTECTIVE STANDARD FROM RADON AND ITS DAUGHTERS IN URANIUM MINES

Uranium prospecting and mining began in the late fifties in China. The protective standard from radon was equal to $3.7 \times 10^3 \text{Bq/m}^3 (1 \times 10^{-10} \text{Ci/1})$ for maximum allowable concentration at that time in the atomosphere of mine. Control and measurement for Rn daughters began in 1963.

In the calculation of ventilation design, the standard for controlling the total return air flow presents $2 \times 10^{-10}\text{Ci/1}$ for Rn concentration, and Rn daughters' concentration is taken to be analogous to 1.85×10^3 $\text{Bq/m}^3 (0.5 \times 10^{-10}\text{Ci/1})$ of Rn concentration in equilibrium, i.e. the concentration of radon daughters is $10.4 \times 10^{-6} \text{J/m}^3 = 0.5\text{WL}$ in underground uranium mine of China.

The national standard " The Regulation of Radiation Protection" (GBJ8-74) promulgated in 1974 continues to have the standard of Rn concentration as $3.7 \times 10^3 \text{Bq/m}^3$, has an increase of the standard of Rn daughters' concentration that α potential value of Rn daughters could not exceed $6.4 \times 10^{-6} \text{J/m}^3 (4 \times 10^4 \text{Mev/L}$ or 0.30 WL).

A new protective standard from Rn daughters recommended by ICRP in 1981 as follows:

Allowable irradiation limit is equal to 0.02J of α potential of Rn daughters a man per year; as a result, Allowable irradiation energy limit is equal to 0.017Jh/m^3, i.e.

4.8WLM and the concentration is limited to $8.3 \times 10^{-6} \text{J/m}^3 (0.40\text{WL})$ for Rn daughters in the atmosphere of mine.

As stated above, the protective standard is more rigorous in China, being in accord with the standard recommended by ICRP in 1959.

3 CALCULATION METHOD OF RADON GAS EMANATION QUANTITY

A calculation of Rn gas emanation quantity is more important in computing ventilating air volume for eliminating Rn and its daughters. It consists of three parts: emanation from the wall of mine, emanation from ore piles and emission from the water of mine in the design of ventilation in uranium mines.

3.1 Randon emanation from surfaces of mine walls

Its calculation formula as follows:

$$D_1 = \delta S_d, \quad \text{Bq/s} \qquad (1)$$
$$S_d = (S_1 a_1 + S_2 a_2 + \dots S_n a_n) K_p, \quad \text{m}^2 \% \qquad 2)$$

where D_1 = amount of radon gas emanation from surfaces of mine wall, Bq/s;

δ = radon equivalent emanation rates, $\text{Bq/(S.m}^2 \%)$;

S_d = areas of equivalent emanating, $\text{m}^2 \%$;

$a_1 a_2 \dots a_n$ = U grade of ore and rock, %;

$S_1 S_2 \dots S_n$ = areas of mine wall of different ore grade, m^2;

K_p = coefficient in equilibrium of uranium and radium.

3.2 Radon gas emanation quantity in ore piles

Radioactive activity of uranium-238 per tonne is equal to 1.258×10^{10} Bq, decay constant(λ) of radon-222 is equal to $2.1 \times 10^{-6} s^{-1}$ and uranium grade(a) of ore puts percent as a unit so that calculation of Rn emanation quantity in ore piles can be expressed as:

$$D_2 = 0.264 \times 10^3 Pa\eta Kp, \text{Bq/s} \qquad (3)$$

Where D_2=Rn emanation quantity in ore piles, Bq/s;

P=amount of ore piles, tonne;

a= U grade of ores, %;

η =emanating coefficient of rock.

3.3 Radon gas emanation quantity in water of mine

Rodon gas emanation quantity in water of mine can be calculated with the following formula:

$$D_3 = \frac{q}{3600}(C_1 - C_2), \text{ Bq/s} \qquad (4)$$

Where D_3=Rn emanation quantity in water of mine, Bq/s;

$C_1 \& C_2$=seperately, radon concentrations of water pouring out from the underground and discharging on the surfaces, Bq/m^3;

q=amount of water pouring out in a mine, m^3/h.

3.4 Emanation quantity of radon in the active mines

It is calculated as the following formula:

$$D = (C_2 - C_1)Q, \text{ Bq/m}^3 \qquad (5)$$

Where $C_1 \& C_2$=concentration of radon in inlet air flow and in return air flow, respectively, Bq/m^3;

Q=amount of air, m^3/s.

4 APPROACH TO CALCULATE AMOUNT OF AIR FOR ELIMINATING RADON AND ITS DAUGHTERS IN URANIUM MINES

4.1 Calculating method of total air volume in a mine

It consists of two kinds:

One is called "the total calculating method", with that the emanation quantity of Rn and its daughters would be calculated with leakage coefficient(K=1.30 to 1.40, corresponding to 77-69% of effective ventilating air rates) on the basis of amount of the total emanation of radon in a mine and the volume of ventilation areas.

The other is called "respectively computing method", with that the calculation is on the basis of respective ventilating air volume level and a sum of ventilating air areas in stopes, tunnelling workings, chambers, etc. The method has been being adopted in the ventilation design of small and middle uraniun mines in China, calculating with the following formula:

$$Q=K(K_1 \Sigma q_1 n_1 + \Sigma q_2 n_2 + \Sigma q_3 n_3), m^3/s (6)$$

Where Q=the total ventilating air volume, m^3/s;

$q_1 \& n_1$ =ventilating air volume level and number of stopes with certain mining method;

$q_2 \& n_2$=ventilating air volume level and sum of certain tunnelling workings;

$q_3 \& n_3$=ventilating air volume level and sum of some chambers;

K_1=coefficient of amount of air for alternate ventilation in stopes:

K=leakage coefficient in a mine

In this calculation, amount of air in a stope is conducted on the basis of requirement for eliminating radon and its daughters, choosing maximum value as a parameter: amount of air in tunnelling workings is calculated by means of requirements for removing blasting mist, the calculation of amount of air in chamber is conducted on the basis of its usefullness and requirement.

4.2 A formula of calculating ventilating air volume for eliminating radon gas

Amount of air for eliminating radon gas can be calculated by means of radon concentration that it cann't be diluted to exceed maximum allowed concentration, it is expressed:

$$Q= \frac{D}{C-C_0}\ m^3/s \qquad (7)$$

When ventilating with the unpolluted air ($C_0=0$), the above formula becomes:

$$Q=\frac{D}{C},m^3/s \qquad (8)$$

Where D=amount of Rn emanation, 10^3Bq/s;

C=radon concentration in return air flow, in calculating of total amount of air in a mine getting 7.4×10^3Bq/s; in a stope, 3.7×10^3Bq/s^3;

C_0=radon concentration in inlet air flow, in calculation of total amount of air in a mine, adopting $C_0=0$.

4.3 Calculating method of amount of air for eliminating radon daughters

Amount of air for eliminating Rn daughters can be calculated on the basis of Rn concentration which is diluted to 1.85×10^3Bq/m^3(0.5×10^{-7} Ci/m^2) in equilibrium, with notice of radioactive equilibrium factor Z between radon and its daughters in the condition of ventilation, assuming $C_0=0$. the formula will be:

$$Q=\frac{D}{C}Z,m^3/s \qquad (9)$$

The coefficient z is the function of the exchanged air time by ventilation, representing as $Z=\varphi(t)$,showed as in Fig.1.

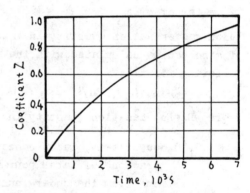

Fig.1 The relation of $\varphi(t)$ value and time t

The funtion is too complicated and too difficult to express clearly.The relational expression, $Z=CQ/D$, $Z= (t)$ & $t=V/Q$, can be calculated with graphic method acording to the emanation rates(D) of radon gas and the volume(V) of ventilation, showed in Fig.2.

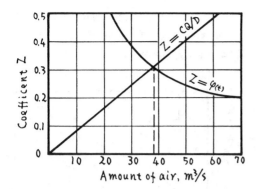

Fig.2 The example of calculating
amount of air with graphic
method for eliminating Rn daughters (assuming D=230x10³Bq/s,
V=50x10³ m³)

The expression of coefficient
$Z=\varphi(t)$ would be simplified with regression analysis method, the calculation
of Z value builds up according to
the following formula in which a
relative error would not exceed 4%
when the time t of air exchanging is
equal to 0.06-3.00(10³s):

$$Z=(t)=0.267t^{0.76} \qquad (10)$$

Where t=time of air exchanging,10³s

$$t=V/Q$$

Where V=ventilation volume,10³m³.

By substituting for equation(10)
into equation(8), the following
formula can be brought about:

$$Q=0.33(DV^{0.76})^{0.57},m^3/s \qquad (11)$$

5 ACCUMULATIVE EQUATION OF RN DAUG-
HTERS' α POTENTIAL

5.1 Accumulative equation of radon
daughters' concentration

Accumulative equation of Rn daughters'concentration has been put forward by Tivoglou,E.C.(U.S.)and Ayer,
H.E.(U.S.) in which it is assumed
that the area neccesary for ventilation would be held in the shape of a
chamber, when ventilating with unpolluted clear air, it is expressed as:

$$C_i=\{\frac{\lambda_2\lambda_3\cdots\lambda_i}{(\lambda_2+Q/V)(\lambda_3+Q/V)\cdots(\lambda_i+Q/V)}\}C_1 \qquad (12)$$

Where C_i=concentration of the i-th
element in series of radon
decay (i=2,3,...);
λ_i =decay constant of the i-th
element in series of radon
decay (i=2,3,...);
C_1=radon concentration;
Q/V=frequency of exchanging
air per unit time(ventilation strengh).

According to the author's study
and practice, it is more reasonable
to take the time of air exchanging
instead of the frequency of exchanged
air in the above equation because of
time for exchanged air being equal
to time for decay accumulation of
Rn and its daughters and the concentration of Rn daughters is directly
portional to time of air exchanging.
Owing to t=V/Q, the accumulative
equations of concentration of RaA,
RaB and RaC are expressed as following on the basis of the above formula:

$$C_A=(\frac{\lambda_A t}{\lambda_A t+1})C_{Rn} \qquad (13)$$

$$C_B=(\frac{\lambda_A t}{\lambda_A t+1})(\frac{\lambda_E t}{\lambda_B t+1})C_{Rn} \qquad (14)$$

$$C_C=(\frac{\lambda_A t}{\lambda_A t+1})(\frac{\lambda_B t}{\lambda_B t+1})(\frac{\lambda_C t}{\lambda_C t+1})C_{Rn} \qquad (15)$$

Where $C_A C_B \& C_C$=concentration of RaA,
RaB & RaC, separately,
10³Bq/m³;
$\lambda_A \lambda_B \& \lambda_C$=decay constant of RaA,
RaB & RaC,respecti-
vely,λ_A=3.79(10³)⁻¹,

$\lambda_B=0.431(10^3 s)^{-1}, \lambda_C=0.586(10^3)^{-1}$;

C_{Rn}=concentration of radon,10^3

Bq/m^3;

$$C_{Rn}=\frac{D}{Q} \qquad (16)$$

5.2 Accumulative equation of Rn doughters'α potential

Radon daughters'α potential (an other method to represent concentration) can be calculated as follows:

$$E=E_A+E_B+E_C=f_A C_A+f_B C_B+f_C C_C$$
$$=\frac{\lambda_A t}{\lambda_A t+1}[f_A+\frac{\lambda_B t}{\lambda_B t+1}(f_B+f_C\frac{\lambda_C t}{\lambda_C t+1})]C_{Rn} \qquad (17)$$

where   E=radon daughters'α potential, 10^{-6}J/m^3;

E_A, E_B&E_C=αpotential of RaA, RaB & RaC, separately, 10^{-6}J/m^3;

f_A, f_B&f_C=coefficient of α potential of RaA, RaB & RaC, respectively, when $C_{Rn}=1\times10^3$Bq/m^3, then $f_A=0.579\times10^{-6}$ J/m^3, $f_B=2.86\times10^{-6}$ J/m^3, $f_C=2.10\times10^{-6}$ J/m^3.

Konwing that as above, Rn daughters'α potential is a function of time of exchanged air. And supposed that,

$$\frac{\lambda_A t}{\lambda_A t+1}\{f_A+\frac{\lambda_B t}{\lambda_B t+1}(f_B+f_C\frac{\lambda_C t}{\lambda_C t+1})\}=f(t)(18)$$

therefore, equation(17) becomes:

E=f(t)C_{Rn} (19)

In fact, f(t) is α potential of Rn daughters from that radon, in which its concentration is equal to 1×10^3 Bq/m^3, decays in the condition of ventilation.

Value of f(t) is calculated on the basis of equation (17), showed as in table 1.

Figure 3 represents interrelated curve between f(t) value and time t, plotted by means of table 1.

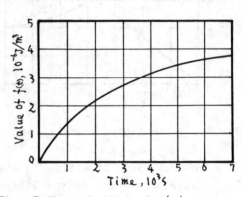

Fig. 3 The relation of f(t) value and time t

From Figure 3, f(t) value of Rn daughters'α potential increases in curve with that the time of exchanged air is prolonged, and the extent cf increase gets small and small.

Equation (17) should be simplified by regression analysis method because of its being complicated so as hardly to derive a formula for calculating amount of air.

The credential curve of f(t) value and time t, by analyzing, is close to logarithmic curve, power function curve or hyperbola. It is stated as above that the formula of amount of air would be derived hardly by logarithmic equation, and when there is no contamination in inlet air flow(i.e. $C_0\neq 0$ & $E_0\neq 0$), formula can be derived by power function equation for calculating amount of air, but on the contrary, it cann't. The author inquired into, therefore, accumulative equations of Rn daughters'α potential, simplified by hyperbola

Table 1 Value of f(t) when $C_{Rn}=1\times10^3$ Bq/m^3

Time 10^3 s	Concentration of radon daughters,10^3 Bq/m^3			Radon daughters'α potential, 10^{-6} J/m^3			Value of f(t), 10^{-6} J/m^3
	C_A	C_B	C_C	E_A	E_B	E_C	
0.06	0.1853	0.0048	0.0002	0.1703	0.0137	0.0004	0.121
0.12	0.3126	0.0154	0.0010	0.1810	0.0440	0.0021	0.227
0.18	0.4055	0.0292	0.0028	0.2348	0.0835	0.0059	0.324
0.24	0.4763	0.0446	0.0055	0.2758	0.1276	0.0116	0.415
0.30	0.5321	0.0609	0.0091	0.3081	0.1472	0.0191	0.501
0.60	0.6946	0.1427	0.0371	0.4022	0.4081	0.0779	0.888
0.90	0.7733	0.2161	0.0746	0.4477	0.6180	0.1567	1.222
1.20	0.8198	0.2795	0.1154	0.4747	0.7994	0.2423	1.516
1.50	0.8504	0.3340	0.1562	0.4924	0.9552	0.3280	1.776
1.80	0.8722	0.3811	0.1956	0.5050	1.0899	0.4108	2.006
2.10	0.8844	0.4221	0.2329	0.5121	1.2072	0.4891	2.208
2.40	0.9010	0.4582	0.2677	0.5217	1.3105	0.5622	2.394
2.70	0.9110	0.4899	0.3002	0.5275	1.4011	0.6304	2.559
3.00	0.9192	0.5183	0.3304	0.5322	1.4823	0.6938	2.708
3.60	0.9317	0.5666	0.3844	0.5395	1.6205	0.8072	2.967
4.20	0.9409	0.6061	0.4310	0.5448	1.7334	0.9051	3.183
4.80	0.9479	0.6390	0.4714	0.5488	1.8275	0.9899	3.366
5.40	0.9534	0.6669	0.5068	0.5520	1.9073	1.0643	3.524
6.00	0.9579	0.6907	0.5378	0.5546	1.9754	1.1294	3.659

regression equation, so as to derive a formula of amount of air.

Hyperbola regression equation can be expressed as:

$$\frac{1}{Y}=a+\frac{b}{X} \tag{20}$$

Where Y=dependent variable, Y=f(t);
X=independent variable,X=t;
a=regression constant;
b=regression coefficient.

After a comparison has been made between some schemes of regression equations, a hyperbola regression equation is adopted in the condition of air exchanging time that is 0.60 to 6.00(10^3s), interval, 0.60(10^3s), expressed as:

$$\frac{1}{f(t)}=0.180+\frac{0.570}{t} \tag{21}$$

The interrelated coefficient r of $\frac{1}{f(t)}$ and $\frac{1}{t}$ is equal to 0.9999, it has a good interelation. Equation(21) becomes:

$$f(t)=\frac{1.75t}{0.316t+1} \tag{22}$$

1.75(numerator) of above equation might be resolved as 5.54x0.316, and assuming 5.54=E_e, 0.316=λ_e, so that equation(22) becomes:

$$f(t)=Ke\frac{\lambda_e t}{\lambda_e t+1} \tag{23}$$

In equation(23), E_e is corresponded to the coefficient of Rn daughters' α potential,λ_e is commensurate to the equivalent decay constant of radon daughters. It is of a merit that the numerator(1.75) of the formula can be just resolved into 5.54x0.316, in which this is a result of selection from the regression schemes.

As a result of calculation, equation (23) is compared with equation (17), it is found that the relative error repressents 0-8%(average:1.2%) when the time of air exchanging expresses the extent of 0.18 to 6.00 $(10^3 s)$.

By substituting for equation(23) into equation(19), an accumulative equation of Rn daughters'α potential can be obtained in which there is no contamination of inlet air flow, expressed as:

$$E=Ke\frac{\lambda_e t}{\lambda_e t+1}C_{Rn} \qquad (24)$$

When inlet air flow would be polluted, the equation can be expressed as:

$$E=E_0+K_e(C_0+C_{Rn})\frac{\lambda_e t}{\lambda_e t+1} \qquad (25)$$

Where E_0=concentration of Rn daughters in inlet air flow, $10^{-6} J/m^3$;

C_0=radon concentration in inlet air flow, $10^3 Bq/m^3$.

The other symbols represent the meaning and the unit as above.

6 FORMULAS OF CAICULATING AMOUNT OF AIR FOR ELIMINATING RN DAUGHTERS

6.1 Principle of formula derivation of ventilating air volume

It is ussually considered that the highest concentration of Rn daughters is present in the ends of ventila-

ting air route or return air flow. The formula derivation of amount of air must be thought over on the basis of concentration of Rn daughters in the ends of air route and in return air flow being diluted to max. allowable level so as to ensure that concentration of Rn daughters cann't exceed protective standard in stopes, tunnelling workings, etc.

According to the protective standard recommended by ICRP in 1981 for removing radon daughters in uranium mines, ALI_p is equal to 0.02 J, which results in that the concentration of Rn daughters in the atmosphere of mine is limited to $8.3x10^{-6} J/m^3$ =0.40 WL. As a result, max. allowable concentration is adopted as $8x10^{-6} J/m^3 (5x10^7 Mev/m^3)$ in the formula derivation of amount of air in this article.

6.2 Formula of calcutating amout of air for elimilating Rn daughters

If there is no contamination in inlet air flow, formulas of $C_{Rn}=D/Q$ and $t=V/Q$ would be intergrated into equation(24) and then resolved, it becomes:

$$Q=0.5\sqrt{\frac{4K_e}{E\lambda_e}DV+V^2}-V \quad ,m^3/s \qquad (26)$$

If V^2 & V of above formula would be gotten rid of, and a revised coefficient 0.82 would be put in it, above formula can be simplified as follows:

$$Q=0.82\sqrt{\frac{K_e}{E\lambda_e}DV}, m^3/s \qquad (27)$$

By substituting for $E=8(10^{-6} J/m^3)$ and the value of $K_e\lambda_e$ into above formula, it can be expressed as follows:

$$Q=0.38\sqrt{DV} \ , \ m^3/s \qquad (28)$$

When there is contamination in inlet air flow, by substituting for $C_{Rn}=D/Q$ and $t=V/D$ into equation(25), resolved and expressed as:

$$Q =\frac{0.5\lambda_e}{E-E_o}\left[\sqrt{\frac{4K_e}{\lambda_e}(E-E_o)DV+(E-E_o-K_eC_o)^2V^2} \right. \\ \left. -(E-E_o-K_eC_o)V\right], m^3/s \qquad (29)$$

If contamination in inlet air flow wouldn't exceed the protective standard over 0.30 times, the second and the third item of above formula may be omitted, then it becomes:

$$Q= \frac{\lambda_e}{E-E_o}\sqrt{\frac{K_e}{\lambda_e}(E-E_o)DV}, m^3/s \qquad (30)$$

By substituting for $E=8(10^{-5}J/m^3)$ and the value of K_e, λ into the above formula, it can be expressed:

$$Q=\frac{1.32}{8-E_o}\sqrt{(8-E_o)DV} \ , \ m^3/s \qquad (31)$$

6.3 Selection and application of formulas to calculate amount of air

According to the author's study, there is, generally speeching, a little contamination in all of inlet air flow in mines, it may be omitted and the total amount of air in a mine can be calculated on the basis of equation(28) with "the total calculation method". It is noticed that contamination is higher in inlet air fiow of stopes and tunnelling workings, amount of air must be calculated by means of equation(31).

In calculation of ventilation design, the value E_o & C_o of inlet air contamination can be adopted to exceed the protective standard over 0.1 to 0.3 times.

Formulas of amount of air described as above has been tested and verified in uranium mines of China, proved to be practical.

7 CONCLUSIONS

An approach to calculate amount of air is outmoded for which it removes Rn and its daughters adopted at the present in uranium mines of China, a new method is being studied and improved, in which it is established on the basis of reseaching the decay law of radon and its daughters.

There is a small relative error in accumulative equation of Rn daughters' α potential expressed by hyperbola regression equation in this article and the formula is simple.

Konwing accumulative equation of Rn daughters' α potential described as above, the derived formula of calculating amount of air can be applied to the calculation of ventilation design in uranium mines on the basis of the protective standard from Rn daughters recommended by ICRP in 1981.

8 REFERENCES

ICRP, Pergamon Press, 1981, Limits for Inhalation of Radon Daughters by Worders, ICRP Publication 32.

The provisional regulations for eliminating radon and its decay producs in underground mines, 1959.

R.L.Harris, R.E.Bales, 1962 Control of radon and its daughters in uranium mines by ventilation.

The safe design of a uranium mine

R.W.THOMPKINS
Queen's University, Kingston, Ontario, Canada

Abstract: The paper will briefly review the physics of Radon Gas emission by diffusion as detailed in the author's manual "Radiation in Uranium Mines" published in September, October and Novemner, 1982, in The Canadian Institute of Mining Bulletin.

The paper will concentrate on design techniques used to control radiation in uranium mines. It will also provide sample calculations showing how the volumes of air required may be estimated as well as the design of the air routing through the mine.

1.0 Introduction

During the development stage of a uranium mine, knowledge of the basic emanating characteristics of the uranium ore and waste rock is essential in order to design an economical ventilating system.

The author's manual "Radiation in Uranium Mines", describes a method of maximizing radon gas production and estimating the volume of air required to ventilate that mine so that a miner's exposure will not exceed 0.33 W.L..

2.0 Discussion of the Diffusion Process

There are two phases to radon gas release from ore rocks. The first is the release from the mineral grain to the pores of the rock mass. At this point the release is dependent on two factors: (1) the rate of escape from the crystal (2) the pore filling liquid.

Because the diffusion coefficient of uranium minerals is very small (10^{-21} to 10^{-65} cm) the probability of radon escaping from the mineral crystal is very small. The release percent (emanating coefficient) at this stage is very low.

The second stage of radon gas release is from the rock pores to an open air interface in the mine. At this point the stress microfracturing of the mine walls confuses the rock porosity factor and multiplies the effect of radon gas release from the mineral grains. The rate of release then is dependent on four factors: (1) the pore concentration, (2) the pore filling liquid, (3) the rock porosity including microfractures, (4) the transfer force (flowing water, flowing air or diffusion.

The physics of all of these phenomena is well documented so that by altering any of these variables the engineer can control radon gas emanations into a mine. Some methods are more practical than others.

All of these factors are measurable but because they change so rapidly in a working mine any attempt to include all of them in a preproduction model would be an exercise in futility. For example, micro fracturing will change with every blast depending on the inherent rock stresses and their direction. In addition, water used to control dust results in changes to the pore liquid.

Therefore if it is assumed that radon gas is released to an essentially dry ore and if it is assumed that all the sequential events of mining such as drilling, blasting, shovelling and conveying are all done simultaneously the radon gas emissions can be maximized.

3.0 Factors of Diffusion

Section 8.0 of the manual deals with these factors in detail. The first important relationship is that the flux rate of radon is dependent upon the pore concentration, the distance from the air interface and the diffusion coefficient. It is represented by the equation:

$$J = -D\frac{dc}{dx}$$

where β = emanating power
α = emanating coefficient
A = radium activity

The value of A can be calculated from the ore grade (see 16-1 manual). β can be measured from the core sized samples.

7.0 The Diffusion Length

The diffusion length is defined as the distance over which radon will diffuse during its mean half life. It is described by the equation.

$$L_m = \sqrt{\frac{D}{\lambda S}}$$

where

L_m = diffusion length

D = diffusion coefficient

λ = decay constant

S = porosity

TABLE I. Some indices of diffusion length in some media

Medium	Diffusion Length (cm)
Air	218
Water	2.18
Coarse- and fine-grained sand - 40% porosity	148 - 183 (165)
Loose deposits - 20% porosity	102 - 129 (115)
Alluvium granite diorite	50
Granite diorite	6.8
Dense rock	4.9

For air, $L_m = \sqrt{\dfrac{1.0 \times 10^{-1}}{2.1 \times 10^{-6} \times 1}}$

$\qquad = 218$ cm

The diffusion length becomes a very useful tool because it can be related to the surface emanating rate and the emanating power.

$$J(ci/cm^2/sec = \beta(ci/cm^3/sec \times L_m(cm))$$

Thus radon gas emanating from ore piles and tailings fill can be estimated if the pore medium is known and the emanating power of the ore has been measured.

The physics also tells us that after 6 or 7 half lifes most of the radon gas will decay. Therefore an ore pile must be about 40 ft (13m) in radius before a maximum concentration in an ore pile will be achieved. It also tells us that it will take about 26 days to reach a maximum concentration.

Thus the size and length of time ore piles remain in a stope is a very large factor in the radon gas produced.

8.0 The Maximum Concentration

The maximum concentration is related to the emanating power, the decay constant and the porosity. It is described by the equation:

$$C_{max} = \frac{\beta}{\lambda S}$$

where C = pore concentration
β = emanating power
λ = decay constant
S = porosity

This is a useful tool for estimating concentrations likely to develop in dead ends or behind bulkheads.

9.0 Basic measurements required for Estimating Purposes

In addition to measuring all the emanating characteristics from core sized samples, wall emanating rates in the mine itself should be measured. Because of the deep microfracturing patterns in ore faces the emanating rates measured in core samples cannot be applied to a wall emanating rate. In addition some of the walls might be in low grade ore or waste.

The emanating characteristics of the core sized samples can be used to estimate the radon produced from broken ore piles in the mine.

10.0 The Tunnel Concept Expanded

Section 14 of the manual gives the details for expanding the tunnel theory of Evans and Schroeder so that it can be applied to any shaped mine opening.

The radon gas emanating from the walls, floor, backfill, broken ore piles not removed, ore in transit, etc., can be individually calculated. Assuming that the radon gas is all released at the same moment a sum in pCi/min can be estimated. This sum is then used in the following equation to estimate the time it would take to grow to 0.33 W.L. if no air was displaced.

$$T_{min} = \sqrt{2261 \frac{V(1)}{E_T(pCi/min)}}^{1.85}$$

where V is in litres
E_T is the sum of radon produced in pCi/min
T is the time to grow 0.33 W.L. of radon daughters.

If the air in the working place is replaced in the same time frame the working

290

where:

J = the average flux in a cross section-
    al area perpendicular to the x axis
D = the diffusion coefficient
c = radon pore concentration
x = distance from an air interface

Figure one is a pore concentration graph
for a particular ore.

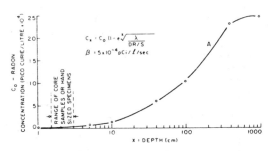

**FIGURE 1. Radon interstitial pore concentration predicted by
diffusion theory.**

Note that when the distance from the face
is about 2 cm to 3 cm the concentration
gradient is practically constant.

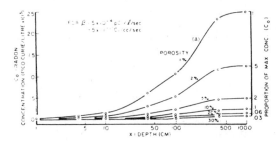

FIGURE 2 Scaling factor curves for rocks at different porosity.

Figure 2 shows that at even high porosi-
ties when the distance from the face is
about 2 cm to 3 cm the concentration gradient is practically constant.

For this reason, it is believed that
measuring the radon emanating characteristics of core sized samples will be a
good practical estimate of the emanating
coefficients.

4.0 The Diffusion Coefficient

The distance over which diffusion can take
place is limited by the time (1/2 life)
that radon exists as a gas. Therefore the
diffusion coefficient varies for many conditions of the pore filling liquid.

TABLE 2. Diffusion coefficients of radon in various
porous media

Medium	Condition	Temperature	Diffusion Coefficient (cm^2/s)
Air	Continuous	18°C to 20°C	$1.0\text{-}1.2\times10^{-1}$
Water	Continuous	"	1.13×10^{-5}
Soil	Alluvial-detrital	"	$3.6\text{-}4.5\times10^{-2}$
Concrete	Set-varying cement ratios	"	$1.69\text{-}3.08\times10^{-5}$
Mud	85% H_2O	"	2.2×10^{-6}
Polyethylene	Sheets	"	4.0×10^{-7}
Mylar	Sheets	"	2.0×10^{-9}
Rocks	6.2% porosity (1)	"	2.0×10^{-3}
	7.4% porosity (2)	"	2.7×10^{-3}
	12.5% porosity (3)	"	5.0×10^{-3}
	25% porosity (4)	"	3.0×10^{-2}
Dense rock			0.5×10^{-3}

Note: If calculated from $D = 0.66\ D_{air} \times S$ (porosity) Rock Diffusion Coefficients would be:
(1) 4.5×10^{-3}
(2) 5.3×10^{-3}
(3) 9.0×10^{-3}
(4) 1.8×10^{-2}

5.0 The Emanating Coefficient

Theoretically the emanating coefficient is
the fraction of the radon atoms escaping
from the mineral grains to the rock pores.
Tests show that when cores are ground the
emanating coefficient increased from 1.9%
(core) to 15.6% (fine grind). Other tests
have shown there is a limit to the coefficient with fine grinding, because fine
grinding can fracture the mineral crystals
thereby producing an error of too high a
coefficient.

Tests conducted on core samples have
measured coefficients varying from 1% to
over 25%.

Tests on tailings to be used for backfill showed that the classified portions
had a lower emanating coefficient than the
unclassified portions. As 60% of the contained radium remained in the fines this
is understandable. However, when cement
was added to the classified portion the
emanating coefficient increased sometimes
by as much as a factor of 3. (See Table
9 manual)

All of this indicates that the introduction of pore filling liquid other than air
creates many unpredictable changes in the
emanating characteristics of radon gas and
complicates the theoretical approach to a
mine design.

6.0 The Emanating Power

This is the measure of the total amount of
radon escaping from the mineral grain per
unit of volume or unit of mass. It is described by the formula:

$$\beta = \alpha\ A^{226}$$

place will never exceed 0.33 W.L.
Knowing the air volume, the size of airways, length, etc., can be economically
designed.

11.0 Typical Ore Analysis

Measured Emanating Rate from ore face

$$J(ci/cm^2/sec) = \begin{array}{l} 1.86 \times 10^{-15} \\ 4.40 \times 10^{-15} \end{array}$$

$$Av. \; 3.13 \times 10^{-15}$$

Note: J of ore and J ore face differ by a factor of 10 approx.

12.0 Comparison of Shrinkage and Cut and Fill Stope

Based on 100 ft of stope length.

A. Shrinkage Stope based on worst conditions (last cut)

	Estimated Rn Production. pCi/min
Exposed ore faces	1.96×10^6
Stope walls	38.46×10^6
Broken ore in stope	$4,803.0 \times 10^6$
Total	$4,843.42 \times 10^6$

Volume of open space in stope
$$= 2.1 \times 10^6 \text{ litres}$$
Time to grow to 0.33 W.L.
$$= 1.012 \text{ min}$$
Ventilation required = 74,151 cfm

B. Cut and Fill Stope (same dimensions) using classified Mill Tailings as fill

	Estimated Rn Production. pCi/min
Fill floor	16.3×10^3
Exposed ore faces	1.96×10^6
Stope Walls	38.46×10^6
Broken ore in stope	47.60×10^6
Total	88.04×10^6

Volume open space in stope
$$= 2.1 \times 10^6 \text{ litres}$$
Time to grow to 0.33 W.L.
$$= 8.84 \text{ min}$$
Volume of ventilating air required
$$= 8,388 \text{ cfm}$$

TABLE 3 - Radiation parameter values from typical core specimens.

ORE ANALYSIS

No.	J Emanating Rate ci/cm²/sec	β Emanating Power ci/cm³/sec	α Emanating Coefficient %
1 Core	3.48×10^{-16}	8.42×10^{-16}	25.2
2 Core	2.91×10^{-16}	7.09×10^{-16}	21.0
3 Core	2.14×10^{-16}	5.47×10^{-16}	16.9
4 Core	1.69×10^{-16}	2.57×10^{-16}	77.1
5 Core	2.04×10^{-16}	2.62×10^{-16}	87.4

Not only are the volumes required in a cut and fill stope smaller but it is easier to control the air flows into and out of the working face.

13.0 Large Tonnage Base Metal Mine Containing Uranium

As economically conceived it was planned to mine 25000 Tons/day using a blast hole caving system of mining. Based on the Elliot Lake criteria of 350-400 cfm/Ton mine management were faced with supplying between 9 and 10 million cfm to ventilate the mine. This was an impossible situation.

After analyzing core sized ore samples a ventilation system was designed to extract Radon gas at its source basically following a system as shown in the following Figure.

Based on core sample measurements the following criteria was used to estimate ventilation requirements.

β(ore) $= 32 \times 10^{-18} \; Ci/cm^3/sec$

β(low grade) $= 7.1 \times 10^{-18} Ci/cm^3/sec$

β(waste) $= 2.0 \times 10^{-18} Ci/cm^3/sec$

J(core) $= 16.0 \times 10^{-16} Ci/cm^2/sec$

J(low grade) $= 3.5 \times 10^{-16} Ci/cm^2/ sec$

J(waste) $= 1.0 \times 10^{-16} Ci/cm^2/sec$

Estimated Air Required:
During Waste Development
6,780 cfm per face
During Ore Development
47,500 cfm per face
Total Development of a single stope
80,500 cfm
Stoping Block with Development Complete   55,296 cfm
Open Stope Without Ore Stocks
31,200 cfm
Open Stope With Broken Ore Stocks
495,752 cfm
Broken Ore in Conveyor System
25,000 cfm
Total Air Required at 70% Usage Factor
2,750,000 cfm

Note:
Positive control of the air flows is absolutely essential. Therefore the air supply drifts and exhaust collecting drifts had to be properly located in the footwall and hanging wall. The capital required to develop the mine had to be increased but this was offset by the lowered operating costs.

Essentially the diesel requirements were the governing factor in the air required.

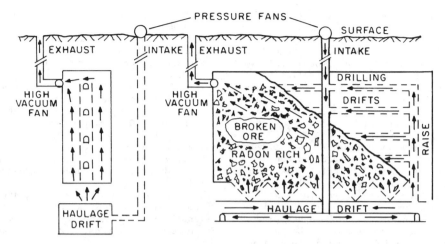

FIGURE 3 – SCHEMATIC PLAN OF STOPE VENTILATION

14.0 Verification of the Method

(1) At Denison mine where the method is used extensively for planning future air requirements and particularly for developing a tailings backfill method of mining, routine sampling of some areas under controlled conditions indicated estimates to be high by about 20%.

(2) A study of another project where the ore body had been developed by an adit. Measurements were made in the adit and the laboratory.

In the development adit

(1) Rn daughter levels in working environment
(2) Ventilating air flow measurements
(3) Rn concentrations in working environment
(4) Rn emanation characteristics from ore faces.

Laboratory

(1) Rn emanation characteristics from selected ore samples

Gamma radiation levels were also measured but were found to be of no consequence in a low grade ore. A record of the ore grade and rock density of all sample locations was measured also. The emanating coefficient from the in-situ measurements varied from 10% to 24%. The emanating coefficient from core sized samples varied from 1.4% to 17.4%. The basic figures developed for estimate calculations were:

β(ore) = 2.7 x 10^{-15} Ci/cm^3/sec
Waste = 4.9 x 10^{-17} Ci/cm^3/sec
J(ore walls) = 4.2 x 10^{-16} Ci/cm^2/sec
(Waste walls)= 2.1 x 10^{-17} Ci/cm^2/sec

Theoretically, the adit produced 83.17 x 10^6 pCi/min of Radon gas with 19,000 cfm ventilating the adit. The air residence time was 8 min. Therefore the exhaust air should contain no more than 0.01 W.L. of radon daughters. Actual background measurements taken during the tests did not exceed 0.01 W.L..

Using mostly a cut and fill mining method with a few shrinkage stopes it was estimated that the mine could be ventilated with 90,000 cfm without exceeding 0.33 WL. The diesel requirements were the governing factor and because the mine would be cold and damp excessive use of air had to be avoided.

(3) The Atomic Energy Control Board of Canada commissioned an independent study of mining conditions that existed prior to 1968 in the Elliot Lake Area. A mathematical model of a raise as it was driven during that early period indicated that 4.78 x 10^5 pCi/min of Radon would be produced. With one drill operating the air ventilating the raise would be 150 cfm. Under these conditions 0.84 W.L. of Radon daughters would be produced in 2 hours and with 2 drills operating (300 cfm) 0.25 would be created.

Under controlled conditions the AECB studies were below 0.5 W.L.. The maximum 3.84 W.L. was obtained when the raise was unventilated for several hours.

A comparison of drift headings was more difficult because the ventilating conditions before 1968 could not be duplicated.

In one test with broken ore left in the heading concentrations of 80 pCi/l of Radon was obtained. Theoretically it could

293

have been 112 pCi/l. An overestimate of 40%.

Another test showed normal concentrations of 20-30 pCi/l of Radon which compared to 21 pCi/l expected from a mathematical simulation. In a partially controlled test in two raises concentrations of Radon gas of 130 pCi/l and 1000 pCi/l were obtained. Theoretically both could have contained concentrations of 2000 pCi/l

Verification of the method is difficult to establish because a mine is not a research laboratory and controlled conditions are costly to duplicate. However, the few instances cited above plus the writer's observations of routine sampling in a mine after using the method to plan mine expansion as well as some new mine operations seems to establish the fact that it is indeed an overestimate of true conditions but the degree of overestimation is not certain.

When designing a mine, especially one as complicated as a uranium mine, with many environmental restrictions, the engineer must be certain that his design not only will work but that it represents the best chance of meeting the bureaucratic restrictions of the future easily and economically.

The preparation of preproduction comparisons of mining methods not only achieves this goal but it provides a basis for the staff environmental engineering department to study the system and to modify it so that it operates at maximum efficiency.

References

Andrews, J.N., and Wood, D.F., 1972, Mechanism of Radon Release in Rock Matrices and Entry into Groundwater. Trans. I.M.M., Section B, Vol. 81, November, 1972.

Bernhardt, D.E., Johns, F.B. and Kaufmann, R.F., Radon Exhalation from Uranium Mill Tailings Piles, U.S. EPA, Technical Note ORP/LV/-75-7.

Havlik, B., Grafova, J., and Nyceva, B., Radium 226 Liberation from Uranium Ore Processing Mill Waste Solids and Uranium Rocks into Surface Streams-I, Health Physics, Vol. 14, 1968.

Holaday, D.A., et al, 1957. Control of Radon and Daughters in Uranium Mines and Calculations on Biological Effects, U.S. Dept. of Health Education and Welfare, Public Health Service, Publication 494, U.S. Government Printers.

Kraner, H.W., et al. Measurement of the Effects of Atmospheric Variables on Radon-222 Flux and Soil-Gas Concentrations, reprint from The Natural Radiation Environment, RIce University Publications.

Muzard, P.P.N., 1975. Characteristics of Radon Emanations from a Conglomerate Uraniferous Ore, Masters thesis, Queen's University.

Newby, J.N., 1973. A Practical Investigation of the Parameters Controlling Radon Emanations from An Impermeable Uraniferous Ore, Masters thesis, Queen's University.

Ramsay, W., 1976. Radon for Uranium Tailing: A Source of Significant Radiation Hazard. Environmental Management, Vol. 1, No. 2.

Schroeder, G.L., and Evans, R.D., 1969. Some Basic Concepts in Uranium Mine Ventilation, AIME, September 1969.

Schroeder, G.L., Evans, R.D., and Kraner, H.W., 1966. Effect of Applied Pressure on the Radon Characteristics of an Underground Mine Environment, AIME, March 1966.

Tanner, A.B., 1964. In The Natural Radiation Environment, eds. J.A.S. Adams and M.M. Lowder, University of Chicago Press, pp. 161-184.

Thompkins, R.W., 1972. Radiation Controls in North American Mines and their Effects on Mining Costs. World Mining Congress, September 1972.

Thompkins, R.W., and Cheng, K.C., 1969. The Measurement of Radon Emanation Rates in a Canadian Uranium Mine. CIM Bulletin, Vol. 62, No. 692, December 1969 .

Thompkins, R.W.. The Emanation Characteristics of Radon Gas and their Effects on the Concentrations of Air-Borne Radon Daughters in Mine Atmospheres. Staub-Reinhalt, P. 117.

Thompkins, R.W., 1974. Characteristics of Radon Gas Concentrations in Underground Mines. World Mining Congress, Lima, Peru, November 1974.

Thompkins, R.W., 1982. Radiation in Uranium Mines. CIM Bulletin, Vol. 75, No.'s. 845, 846, 847, September, October, November, 1982

8. Nuclear waste site ventilation

Critical aspects of ventilating a civilian nuclear waste repository in salt

F.DJAHANGUIRI
Battelle Memorial Institute, OH, USA

J.GOZON
Ohio State University, Columbus, OH, USA

ABSTRACT: The current design indicates that safe disposal of high-level civilian nuclear waste in a salt medium requires excavation of entries (7 m wide X 8 m high) at a depth of 850 to 900 m with ambient temperatures ranging from 34°C for bedded salt to 62°C for domal salt. To date, ventilation engineers have been accustomed to extracting natural heat from deep mines to improve the safety and the working conditions of miners. Also, reduced heat stress on miners results in higher productivity as evidenced by the results of ventilation research conducted primarily in deep South African gold mines. Disposal of nuclear waste in salt presents a challenge to subsurface technology-oriented mining engineers, because canisters emplaced in salt could raise the host rock (salt) temperature to 250°C in the canisters' immediate neighborhood. This would result in a steady rise in the temperature of the salt horizon for some time before leveling off. In any event, these phenomena need to be evaluated as part of the development of waste disposal technology. This paper identifies and examines ventilation-related issues that are critical to successful repository excavation, waste emplacement, backfilling, and possible waste retrieval, namely regulatory, geotechnical, thermomechanical, and ventilation issues. Some conclusions of this study are that (1) salt as a host rock for the repository is different from other geologic media due to salt's creep characteristics and unique thermal properties; (2) if emplacement rooms are back-filled, they need not be ventilated, thereby requiring a smaller ventilation system (the effects of early backfilling on retrieval requirements, however, need to be further assessed); and (3) sub-surface ventilation should employ two separate ventilation systems, one for the development area and one for the emplacement area, because such a system would minimize the potential for exposure of underground staff to radiation during transport and emplacement of waste.

The concept of designing two independent ventilation systems
for the nuclear waste repository in basalt

PAUL C.MICLEA
Raymond Kaiser Engineers Inc., Oakland, CA, USA

ABSTRACT: The feasibility of the storing commercial nuclear wastes in deep geological formations depends on the successful resolution of several technical problems. The appropriate ventilation and conditioning of air to ensure safe and suitable working conditions underground is one of these problems.

The life of a repository will span several decades and the ventilation must satisfy the requirements imposed by its various stages: mine development, waste emplacement, care-taking and monitoring, and complete backfilling.

Information obtained from several deep boreholes drilled at Hanford, Washington through the basalt formations indicates that the repository can be developed at a depth close to 1 km, where the virgin rock temperature is 52° C.

Two independent ventilation and air cooling systems are required to separate mining development from waste handling activity. This paper outlines the criteria and the baseline for designing these ventilation systems for the Nuclear Waste Repository in Basalt.

1.0 INTRODUCTION

In 1982, the U.S. Congress passed the Nuclear Waste Policy Act establishing that the first national repository for commercial nuclear wastes must be built by 1998.

At the end of 1984, the U.S. Department of Energy (DOE) published the Draft Environmental Assessment document, designating the Hanford Site in the State of Washington as one of the three recommended sites for the first nuclear waste repository.

In the framework of the Basalt Waste Isolation Project (BWIP), the joint venture of Raymond Kaiser Engineers Inc. and Parsons Brinckerhoff Quade & Douglas, Inc. (RKE/PB) completed a conceptual design study for the construction of an underground repository in the basalt formations at the DOE Hanford Reservation near Richland, Washington. Based on the information at hand, and as concluded by recent engineering studies, the repository is proposed to be located in the Cohassett basalt flow, at a depth of between 960 and 990 meters.

There are four distinct phases in the life of the repository, each of them requiring different ventilation considerations.

During the initial development phase, several shafts need to be sunk and interconnected through a large shaft pillar. Main access drifts are to be developed delineating the waste emplacement panels and rooms. Based on the latest studies and schedule of construction, this phase requires more than 9 km of shafts, 13.5 km of drifts, or the equivalent of 1.25 million tons of excavation to be completed in five to eight years.

Only one ventilation system is required during this phase.

Second is the operational phase when, in parallel with developing new panels, the emplacement of waste containers takes place in previously developed and isolated panels; for this, a second and separate and confined ventilation system is required to support all activities related to waste handling operations.

Both ventilation systems for mining development and for confinement activities require a combination of air conditioning equipment on the surface and underground.

The caretaker phase occurs when the repository has been completely developed and filled to capacity, but further observations and instrumentations are required. Waste retrieval might be required during this phase. No mining activity takes place and, consequently, a single ventilation system is required as the whole repository is in confinement. The main ventilation concern during this phase is for cooling down the rooms for access when required for inspection and maintenance or for retrieval of any container.

The last phase of the repository is backfilling. Each room will be backfilled with engineered material. A single ventilation system is required, with major changes in the airflow network and air conditioning capacities.

This paper covers the ventilation requirements for the operational phase only, when both mining development and waste handling activities will be conducted simultaneously.

Schedule and cost constraints dictate that all the shafts be blind drilled. Nine shafts will be required, of which six are exclusively for ventilation.

2.0 DESIGN CRITERIA FOR THE REPOSITORY -- REQUIREMENTS AND CONSTRAINTS ON VENTILATION

In order to accommodate annual waste receipts, the repository is modular in design. The basic module unit is a panel. Following various trade-off studies, it has been concluded that a four-quadrant repository around a central rectangular pillar offers the best layout (Figure 1).

The waste will be packed in metallic containers with a heat output of 2,200 watts per container.

Horizontal placement in short holes accommodating one container per hole is considered the most acceptable solution of emplacement, both technically and economically. The storage rooms have been designed as a horseshoe shape, with a width of 7 m and a maximum height of 3.3 m. A minimum space of 6.7 m has been established between placement holes. The placement rooms of a panel will be connected by crosscuts spaced at a maximum of 300 m.

The rooms of a panel are provided with one or more panel entries and exit drifts, connecting the emplacement area with the main access and return corridors, respec-

tively. These corridors provide the connection between the shaft pillar area and the emplacement area.

Due to the nature and uniqueness of the underground nuclear waste repository, the ventilation planning and design process cannot follow the "classical" pattern of mine ventilation.

The following is a brief presentation of the principal requirements for ventilation under the imposed conditions.

2.1 Virgin Rock Temperature and Climatic Conditions Required Underground

Current information from the boreholes indicates a virgin rock temperature of $52°$ C at the repository level. The criteria for maximum allowable wet bulb globe temperature (WBGT) at the work places is $26.7°$ C for continuous work. For this, the heat released by the rock must be removed continuously which means that air conditioning is required throughout the underground facility, where people are present.

After the waste is emplaced, it releases heat that increases the temperature of the surrounding rock. The maximum allowable temperature of the surrounding rock is established by the design criteria at $300°$ C. Heat transfer analysis through basalt shows that the wall temperature of an emplaced room can reach $152°C$.

Once a panel is filled to capacity, no other activity will be conducted in the panel and, consequently, no ventilation and cooling are required, except for a small bleed to allow for monitoring. However, when access is required for either inspection and maintenance or waste retrieval, the rooms need to be cooled down to the acceptable $26.7°$ C WBGT. Studies conducted by RKE/PB indicate that to achieve this temperature requires up to 90 days, with the maximum airflow allowable per room of 46 m^3/s.

2.2 Groundwater

Information from boreholes indicates a possible water inflow of 2.84×10^{-7} m^3/min/m^2, at a hydrostatic pressure of 9.7 MPa and a temperature of $52°C$.

The ventilation also takes into account other sources of water in the underground facility (drilling, dust control, condensation).

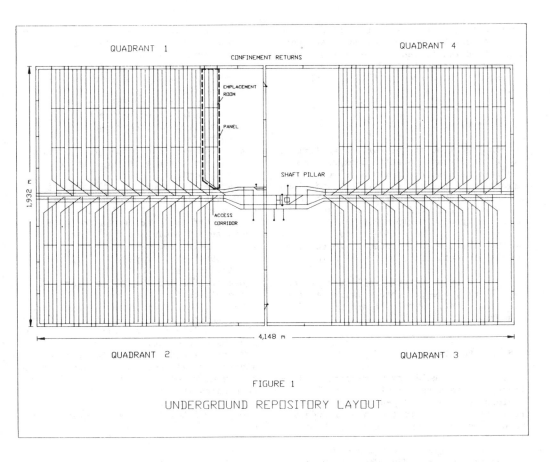

QUADRANT 1

CONFINEMENT RETURNS

QUADRANT 4

EMPLACEMENT
ROOM

PANEL

SHAFT PILLAR

1,932 m

ACCESS
CORRIDOR

4,148 m

QUADRANT 2

QUADRANT 3

FIGURE 1

UNDERGROUND REPOSITORY LAYOUT

2.3 Heat Stress and Air Cooling Requirements

In addition to the heat released by the virgin rock and groundwater, there are other sources of heat that impact the mine climate. Mining and drilling equipment, as well as the waste transporter and emplacement equipment, are also sources of heat. In addition, an average of 250 watts of heat is released by each person during moderate activity underground.

2.4 Presence of Methane Gas

Current information does not indicate any probability of free methane in the basalt formation. However, 0.62 g/L of dissolved methane has been detected in the groundwater. Assuming the extreme case that all of this dissolved gas will be liberated and accounted for in the water inflow when the repository is fully developed, the maximum methane inflow to the entire

repository would be 0.028 m^3/s, requiring 11.3 m^3/s of fresh air to dilute the methane below 0.25%, as mandated by MSHA.

2.5 Shaft Construction and Drifting Methods

Based on several engineering studies for the selection of shaft sinking, as well as for the drifting methods, it is concluded that the shafts should be blind drilled and all the drifts and rooms excavated by conventional drilling and blasting methods. The shafts will be steel lined, while the drifts, crosscuts and placement rooms will be supported by rock bolts and wire mesh, to the degree necessary.

2.6 Airways Size

Shafts, access and return drifts, as well as placement rooms and crosscuts between adjacent rooms, are the main airways. Most are designed for criteria other than ventilation, such as equipment size, required clearances, operational use, and regulation requirements. For the shafts, the maximum size limit (3.7 m I.D.) is imposed by the current drilling capabilities.

2.7 Air Velocities

In consideration of the various activities along the airways, practical experience and regulations, the maximum velocity allowable in the air shafts is 20 m/s, and in hoisting shafts 10 m/s. Along the main entry and return drifts, as well as in the placement rooms, maximum 7.6 m/s is allowed. In the development and hole boring areas, where personnel are continuously present, air velocity is held to a maximum of 2 m/s.

The minimum velocity at all work places is 0.25 m/s, except where drilling and/or blasting occur, where 0.31 m/s is the minimum.

2.8 NRC Regulations

U.S. Nuclear Regulatory Commission regulations for the disposal of high-level radioactive wastes in geologic repositories are contained in the Code of Federal Regulations -- Energy, Title 10, Chapter 1, Part 60.

This code requires compliance with mining regulations (MSHA Chapter 60.131 (b) (g)) for those structures, systems, and components important to safety. Specifically for underground facility ventilation, it mandates to "separate the ventilation of excavation and waste emplacement areas".

3.0 PROPOSED VENTILATION SYSTEM AND DESIGN APPROACH

3.1 Overall Concept

The repository ventilation systems must satisfy the imposed functional and design requirements during all four life phases of the repository: initial development, operation, caretaker, and backfill.

Specific requirements for ventilation during development and operations are:
o Provide a safe and comfortable working environment for all underground personnel
o Promptly remove blasting fumes and control airborne dust

o Prevent flow of contaminated air from waste handling areas into mining development areas
o Control and minimize radiation exposures to public and to repository personnel
o Protect the accessible environment from releases of radioactive contamination from repository operations
o Permit periodic inspection, testing, and maintenance of the ventilation equipment
o Preserve the option of waste retrieval.

The ventilation system design must also comply with all applicable regulations and established design criteria for this facility. The minimum fresh air requirements are:
o 0.0944 m^3/s per person (California Administrative Code)
o 6.33 m^3/s per 100 kW for diesel equipment.

The California mining and tunnel safety codes under which the design is performed also stipulate that the direction of airflow be reversible.

Trade-off studies conducted for the access and exit corridors to and from the placement rooms, respectively, led to the conclusion that five parallel drifts, 5.1 x 3.9 m each, are required for access and two 4.5 x 3.3 m drifts are required for exit.

3.2 Data Acquisition

The overall repository layout with the required number of storage panels and an appropriate shaft pillar was analyzed to arrive at the most suitable ventilation design. Given the various operational activities that must take place simultaneously, at least three shafts are required prior to considering the ventilation:
o One shaft dedicated to waste handling; not available for ventilation
o One shaft for basalt hoisting; can be used for ventilation, as intake or as exhaust from mine development only
o One service shaft for hoisting personnel, equipment, and materials; can be used for air intake only, due to the presence of personnel most of the time.

Given their functions and air velocities allowed, maximum shaft airflows after deducting the area occupied by guides, buntons, pipes, and cables, is 97.2 m^3/s for the basalt hoist shaft and 99.1 m^3/s for the service shaft.

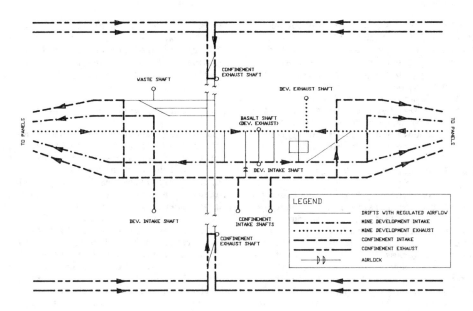

FIGURE 2

AIRFLOWS IN THE SHAFT PILLAR AREA

The actual schedule and nature of operations taking place simultaneously is used to arrive at the total airflow required during various phases of the repository life.

Minimum air per person and per kW of equipment, as well a minimum air velocity at work places, are calculated for each area to be ventilated. The airflow so calculated is checked for cooling the room down to the temperature required for safe working conditions. Resistance calculations are made for all the airways, as input for network analysis.

3.3 Ventilation Systems Planning

During the operational phase of the repository, mining development of new panels will take place simultaneously with the activity of waste handling and emplacement.

Due to the nature of nuclear waste handling, it is required that no air flows from any waste areas can be present in areas of mining development or support facilities. Consequently, the waste handling activity must be "confined". Restrictions on the shaft size require that more than one shaft be used for air intake. Similarly, more than one access drift to the emplacement panels is required. Finally, because the return airflows cannot be mixed, more than one

shaft is necessary for air exhaust.

Adding to the foregoing the specific requirements for monitoring continuously the nuclear waste handling activity, it appears logical to define and design two different and completely separate ventilation systems during the development and operations phase.

The design of the two ventilation systems is treated as completely separated and independent networks although some interference and connections between them is unavoidable. The systems must be kept as separate as possible, and any leakage between the two must be from the mining development system into the confinement system. The design incorporates several features to ensure this.

The two systems utilize completely separate fans, shafts, and intake and return drifts. In the shaft pillar area, there is only one connection between the two systems to accommodate occasional access. Here, the two systems are separated by an airlock (Figure 2).

At other points of contact, the two systems are separated by solid bulkheads.

It is not possible to completely prevent leakage through the bulkheads and airlocks, but the direction and quantity of leakage can be controlled by imposed pressure differential. The mining development system utilizes forcing fans, which results in a high-pressure airflow throughout the system. The confinement system utilizes a combination of forcing and exhausting fans. By adjusting the pressures on these fans, adequate airflow can be ensured while also maintaining the confinement system at a lower pressure than the mining development system. The quantity of leakage from the mining development system is kept to an acceptable level by adjusting the pressure differential between the two systems.

3.3.1 Flow Pattern and Design

Once the required airflows for various areas of the repository is calculated, a "worst-case" scenario is worked out. Then the flow pattern through the repository are modeled as a ventilation network, and schematic diagrams are created. Most airflows are shown individually on the diagrams, while some airflow paths are combined into parallel paths.

As for the panel rooms and drifts, the airflow is based on minimum air velocity required for personnel and equipment. For instance, a 7.0 x 3.3 m room requires 7.2 m^3/s to satisfy minimum air velocity during development. This air would be sufficient for 75 people. An economic analysis demonstrates that minimum airflow with more cooling is most desirable, because increased airflows require larger shafts and drifts, with attendant increased capital costs.

To allow for mine development and borehole drilling as scheduled, 201.3 m^3/sec of fresh air is necessary, and 308.7 m^3/s is required to support the waste handling activities in a confined network. Given the shaft size constraint, a pair of shafts is required for each system for air intake, and another pair for exhaust. The main fans will be located on surface, with the design parameters shown in Table 1.

3.3.2 Mining Development System

The mining development ventilation system provides air for the development of placement rooms and access drifts, drilling of placement holes, as well as shop and office activities in the shaft pillar area. The system is sized to allow room development at a rate equal to the maximum room emplacement rate.

The largest portion of the ventilation air travels down a development intake drift to the room development and hole boring areas. The air passes directly from the intake drift into the work areas, with the quantity in each area controlled by an auxiliary ventilation system. After ventilating a work area, the air is ducted across the development intake drifts to the development return air drift. Thus, the air used to ventilate one work area does not pass through any other work area.

In most cases, two drifts will be used for the mining development return air. This permits a low air velocity in the central drift, which is used for haulage of mined basalt. Upon returning to the shaft area, the air is exhausted up shafts R3 (basalt hoisting shaft) and R4 (Figure 2). No fans are located at the top of these two shafts.

The mining development intake air must be prevented from leaking directly into

Table 1. Maximum Shaft Airflows and Fan Pressures During the Operational Phase

Ventilation System	Shaft Designation	Shaft Size Airflow Diameter m	Capacity m^3/s	Designed Parameters Fan Airflow m^3/s	Pressure kPa
Mine Development	R1 (Intake)	3.7	99.1	95.6	2.70
	R2 (Intake)	3.1	141.6	105.7	4.27
	Subtotal	– –	240.7	201.3	– –
	R3 (Exhaust)	3.7	97.2	92.3	– –
	R4 (Exhaust)	3.1	141.6	72.3	– –
	Subtotal	– –	238.8	164.6	– –
Waste Handling	R8 (Intake)	3.7	183.1	154.4	3.69
	R9 (Intake)	3.7	206.7	154.4	3.69
	Subtotal	– –	389.8	308.8	– –
	R6 (Exhaust)	3.7	203.9	174.3	2.17
	R7 (Exhaust)	3.7	203.9	174.3	2.17
	Subtotal	– –	407.8	348.6	– –

returning development air. This is accomplished by placing masonry stoppings in the crosscuts connecting the intake and return drifts. In crosscuts through which the haulage locomotive must pass, a brattice formed of hanging, overlapping conveyor belting is used to reduce leakage without restricting access.

The mining development ventilation system is provided with instrumentation and alarms to detect and record flow rates, pressure differentials, dry and wet bulb temperatures, and all air quality related parameters.

3.3.2.1 Panel Ventilation During Mining Development. The sequence of panel development is planned to ensure that sufficient panels are opened up in time to allow hole drilling and waste emplacement at the prescribed annual rate.

At the advancing face of a drift or room, auxiliary ventilation with an overlap forcing system will deliver fresh air to a point some 10 meters back from the exhaust duct positioned as close to the face as possible. In addition to the panel rooms, the five main access drifts are kept just ahead of the panel to allow for air circuitry and in preparation for the next panel development.

Airlocks and stoppings will eliminate any short circuits between fresh and exhaust air in various airways. When all four rooms of a panel are advanced approximately 300 meters, a crosscut is made between them. A continuous circuit of fresh air from the intake main entries is then established through the first two rooms and used as a base flow, from which air is routed to the advancing face of each room. Exhaust air is ducted from the face and delivered beyond the crosscut into rooms three and four. Temporary curtains will be installed to separate the flows. A positive pressure will be maintained in the intake air, to prevent return air leaking into fresh air. The curtains need to be flexible to allow for trains to pass with a minimum of air escaping. The curtains also allow the passage of rubber-tired equipment and personnel carriers. All personnel and materials transportation will be carried out exclusively in the development intake. However, locomotives will travel in the return airway. To minimize the exposure to risk, all the locomotives will be equipped with smoke detection alarms which will provide early warning and allow enough time for crews to safely evacuate to an intake airway.

Figure 3 shows panel ventilation during various steps of mine development.

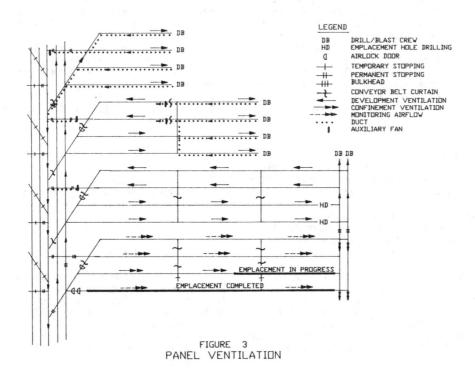

FIGURE 3
PANEL VENTILATION

3.3.3 Confinement Ventilation System

When a panel unit is completely developed, including placement hole drilling, it will be converted to the confinement ventilation system and made available for waste handling and emplacement activities. To accomplish this, physical barriers will be built to prevent both personnel and equipment from entering one system from the other, as well as isolating one ventilation system from the other. Such physical separation exists in the shaft pillar as well.

The confinement ventilation system provides air for all activities relating to waste handling and emplacement. It provides air for ventilating a room during emplacement, several rooms ready for emplacement, and the waste handling area at the base of the waste handling shaft. It also allows for the cooling of a room where emplacement is complete, for maintenance or retrieval activities.

The main fans are arranged as a "push-pull" system; fans are located at the top of both the intake and exhaust shafts. This arrangement allows for excellent control of the pressure differential and minimizes the leakage between the mining development and confinement air systems.

The intake air travels down the two intake shafts and then along the confinement intake drifts to all four quadrants of the repository. A regulated quantity of air passes into the waste handling area. Part of this air is allowed to exhaust up through the waste handling shaft, the rest travels through the waste handling area and joins returning confinement system air. Two confinement return drifts are provided in order to circumvent any hazardous situation that would be created by the blockage of a single return airway.

Prior to waste emplacement, stoppings are installed in all crosscuts connecting the placement rooms within a panel. After waste emplacement, two sets of doors are shut and sealed at the entrance and exit to each room. This procedure allows access to emplaced rooms without difficulty and also allows a single room to be cooled for maintenance or retrieval operations.

Although the placement room doors are assumed to be well sealed, a small quantity of air will leak through an emplaced room. This leakage air will be used for monitoring the conditions in the emplaced room. If there is no activity in any of the placement rooms in a given quadrant the entire airflow to that quadrant is kept to a minimum.

Figure 3 shows the airflow for the confinement ventilation system. The maximum airflow occurs near the end of the operations phase, when emplacement has been completed in almost all the rooms. The shafts, drifts, and fans are sized for this maximum airflow.

The confinement ventilation will use a multiple fan and air filtering system to provide fail-safe operation. A radiation monitoring system will extend throughout the subsurface confinement facilities and in the exhaust shaft to permit detection of unacceptable levels of airborne radioactive or toxic materials in sufficient time to switch the airflow to HEPA filters in the confinement exhaust building. Under normal operating conditions, confinement exhaust air is unfiltered. HEPA filtration will only be activated if the instrumentation system detects radioactivity in the ventilation circuit.

4.0 VENTILATION NETWORK SIMULATION

The Ventilation Network Analysis Program (VNET) developed at the Mining Research Laboratories, University of Nottingham (U.K.) and recently updated by Mine Ventilation Services, Inc. of Lafayette, California, is being used to simulate the "worst-case" network scenarios for the two systems. A total number of 272 branches and 161 junctions were identified and analyzed. For most of the branches, the calculated airway resistance is specified as input, while for a few, the minimum air quantity is also specified.

A complete list of branch airflows is computer generated, giving the frictional pressure drops, branch resistance values, airpower and annual individual branch ventilation cost.

Branches where the airflow needs regulation are identified.

Figure 4 shows the ventilation network for the two systems during the last stage of mining development in the fourth quadrant.

5.0 CLIMATIC SIMULATION

The Climatic Simulation (CLIMSIM) computer program developed by Mine Ventilation Services, Inc. is employed in conjunction with the Ventilation Network Analysis (VNET) program.

The program provides a practical and quick method to simulate the environment

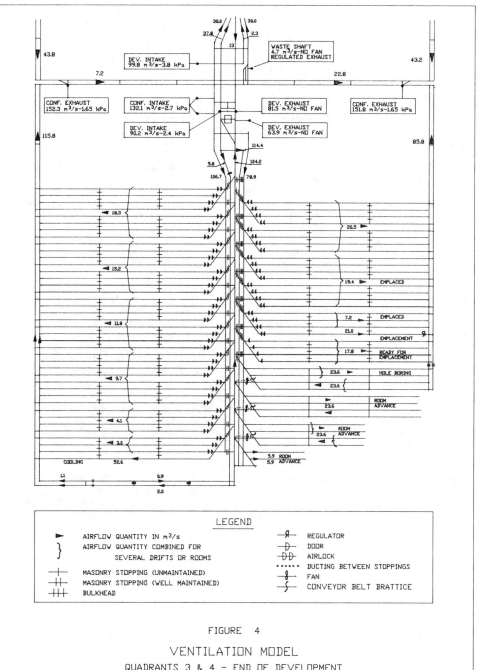

FIGURE 4

VENTILATION MODEL

QUADRANTS 3 & 4 – END OF DEVELOPMENT

in the repository, based on physically
changing factors, such as heat transfer at
the rock/air contact, autocompression
effects down the shafts, various sources
of heat underground (operating equipment,
personnel, fissure water, etc.), psychrometric changes in the airflow, wetness of
the airway perimeter.

The program also handles variations in
air pressure due to frictional flow, as
well as changes in elevation. Part of the
program input can be generated by the VNET
program (airflow, pressure, etc.), while
other information must be provided by the
program operator, such as: air temperature at inlet, virgin rock temperature and
geothermal step, rock conductivity and
diffusivity (or specific heat), airway
friction coefficient, wetness and age.
Artificial sources of heat must be identified and specified in relation to the
branch intake end, either as a stationary
(spot) or as a linear source.

The program output gives the quality of
air at various locations along the airway
(as requested) and can also produce a
graph showing dry bulb and wet bulb temperature variations along the airway.

6.0 SYSTEMS EVALUATION AND FURTHER PROBLEMS TO BE SOLVED

The design of the nuclear waste repository
in basalt is still in the conceptual phase
and more work is needed before the
advanced conceptual design will be prepared. Furthermore, two exploratory
shafts and an underground testing facility
in the repository horizon are scheduled to
be constructed in 1986-87. As more information is gathered by these studies and
tests, the repository design may be
altered or improved. As a consequence,
the ventilation will need reconsideration
as well. In any event, the concept of two
separate and independent ventilation
systems must be maintained.

Problems that require further examination and solutions, with impact on ventilation, are:
o Shaft Sizes and Method of Construction. Larger diameter shafts would
  allow a reduction in the number of
  shafts and main fans with a simplification of the network inside the shaft
  pillar.
o Location of Main Fans. There are
  advantages to locating some of the
  main fans underground.
o Airflow Reversibility. This is
  required by California Administra-

tive Code, and poses serious problems
due to the fact that a permanent
pressure differential must be
maintained between the two systems so
that no leakage is allowed from waste
handling circuit into mining
development circuit.
o HEPA Filters. The efficiency and
  effectiveness of HEPA filters as a
  means to totally control radionuclide
  migration outside of the confined
  system is questionable and needs more
  consideration.
o Computer Programs. Improvements in
  mining ventilation and air conditioning computer programs.
o Air cooling. Position and duties of
  cooling plant, 'coolth' distribution
  and heat exchangers.

7.0 CONCLUSIONS

To ventilate the underground repository in
basalt is a complex task with implications
and ramifications that often go beyond
ordinary mining ventilation.

Because the facility is unique in its
concept, nature, and size, no previous
data and site-specific information for
comparison are available.

The design of the facility, including
the ventilation, is a continuing effort to
be addressed over the next 10 - 15 years,
with many interesting aspects that cannot
be expressed in a single paper.

ACKNOWLEDGMENTS: The author wishes to
thank the Department of Energy for permission to present and publish this paper.

Thanks are also due to my colleagues in
the Department of Mining and Minerals
Projects of Raymond Kaiser Engineers for
their valuable suggestions and assistance.

Grateful acknowledgment is made to
Professor Malcom J. McPherson - President
of Mine Ventilation Services, for the
helpful advice in the preparation of this
paper.

While the concepts and results presented
are mostly derived from various studies
prepared for the Nuclear Waste Repository
in Basalt, the opinions presented do not
reflect the policies of DOE, Rockwell, or
any other authority, but rather they are
those of the author. Any errors are the
sole responsibility of the author.

REFERENCES

D.O.E. 1984: Generic Requirements for a
Mined Geologic Disposal System.
RKE/PB 1983: Conceptual System Design
Description, Nuclear Waste Repository in
Basalt, Project B-301.
RKE/PB 1984: Basalt Waste Isolation Proj-
ect, Task V, Engineering Study No. 7,
Waste Emplacement Optimization.
RKE/PB 1985: Basalt Waste Isolation Proj-
ect, Task V, Engineering Study No. 9,
Underground Repository Layout.

Effects of retrieval on ventilation and cooling requirements for a nuclear waste repository

DOUGLAS F.HAMBLEY
Argonne National Laboratory, IL, USA

ABSTRACT: The Nuclear Waste Policy Act of 1982 (Public Law 97-425) and related federal regulations require that it be possible to retrieve high-level nuclear waste from an underground repository for a period of 50 years following the initiation of waste storage. Two scales of retrieval can be envisaged -- local retrieval and full retrieval -- as well as three room conditions -- open and ventilated, bulkheaded but unbackfilled, and bulkheaded and backfilled. This paper discusses the implications for ventilation and cooling requirements of the various retrieval options. An example of a retrieval scenario in a hypothetical repository in salt is presented.

1 INTRODUCTION

Much of the electricity used in certain parts of the United States is generated by nuclear power plants. The heat generated as a result of fission reactions is used to convert water into steam, which drives the turbines that generate the electricity. Thus, nuclear power plants are similar to fossil fuel power plants, the only difference being the source of the heat.

Nuclear fission occurs when certain radioisotopes (uranium-235, uranium-233, and plutonium-239) subdivide, as a result of neutron bombardment, into radioactive isotopes of lighter elements. In the process, neutrons are liberated, which bombard the remaining fissionable radionuclides, thus prolonging the reaction. Fission continues until insufficient neutrons are present to sustain the reaction. The reaction is called subcritical if the rate of neutron production is less than the rate of neutron loss, primarily through absorption in the reactor walls.

In a reactor, the fissionable radionuclides are in the form of individual pellets contained within Zircalloy-cladded fuel assemblies. When the fuel within a fuel assembly becomes subcritical, or "spent," the assembly must be replaced. In light-water reactors in use in the United States, fuel assemblies have an average useful life of three years. The spent fuel assemblies are not inert and contain highly radioactive fission products as well as nonfissionable, but still radioactive, uranium and transuranic elements.

As a result of almost 30 years of generating power in nuclear power plants, large numbers of spent fuel assemblies have been stored in water pools at reactor sites or at "away-from-reactor" storage sites such as the facility at Morris, Illinois. As the limited capacity of such temporary storage facilities is approached, it has become clear that a more permanent solution is required.

Several options for permanent disposal of radioactive wastes have been proposed; however, the only option that is currently deemed feasible given today's technology is to isolate them underground in mined space in a suitable geologic environment. The Nuclear Waste Policy Act of 1982 includes a schedule for construction of two repositories for high-level nuclear waste.

2 REGULATORY FRAMEWORK

Three U.S. government agencies have regulatory responsibilities related to repositories for high-level nuclear waste, namely:

- U.S. Department of Energy (DOE), which is responsible for site selection and characterization, and for repository design, construction, operation, closure, and decommissioning.

- U.S. Nuclear Regulatory Commission (NRC), which is responsible for licensing all nuclear facilities, including geologic repositories.

- U.S. Environmental Protection Agency (EPA), which is responsible for setting acceptable limits for toxic and radioactive substances.

Through its draft regulations embodied in Title 40, Part 191 (40CFR191) of the Code of Federal Regulations, EPA (1984) has proposed limits for the concentrations of radionuclides and levels of radiation that can be allowed to reach the so-called "accessible environment," that is, the environment to which the public has access.

To facilitate the licensing process and to ensure that 40CFR191 requirements are met, NRC (1984) has promulgated licensing and technical requirements for geologic repositories for high-level radioactive wastes in Title 10, Part 60 (10CFR60) of the Code of Federal Regulations. Among the performance objectives for a repository is the requirement [10CFR60.111(b)] that repositories be designed to allow retrieval of any or all of the waste within a time period similar to that for repository construction and emplacement of the waste, at any time up to 50 years from the commencement of waste storage or until the end of the performance confirmation period.

It is apparently not NRC's intent that retrievability directly influence repository design; however, retrievability must be shown to be possible for a given repository design.

3 HOST GEOLOGIC FORMATION

Upon completion of draft environmental assessments (EAs) (U.S. Department of Energy 1984a–i), three sites (in basalt, welded tuff, and bedded salt) were selected as worthy of further site characterization. Two other sites, one each in bedded and domed salt, are backups should any of the three recommended sites be found to be unacceptable prior to publication of the final EAs (expected in late 1985).

The recommended sites are:

- The Cohassett basalt flow in the Pasco Basin (Hanford Reservation in Washington).

- A welded-tuff horizon within the Topopah Springs Tuff (Nevada Test Site).

- A salt bed in the Palo Duro Basin (Deaf Smith County, Texas).

The backup sites are:

- A salt bed in the Paradox Basin (Davis Canyon, Utah).

- Richton salt dome (Perry County, Mississippi).

4 REPOSITORY CONCEPTUAL DESIGN

Two stages have been identified for a repository: preclosure and postclosure. During the preclosure phase, which comprises construction, emplacement, and retrieval, a repository consists of:

- Surface facilities, including waste-handling and other service facilities.

- Underground storage rooms, accesses, and ventilation drifts.

- Shafts or ramps connecting the surface to the underground facilities.

From now on, however, the discussion shall be confined to the underground facilities, and more specifically to the storage areas. The layout and sizing of the storage rooms depend on the following considerations:

- Storage position, that is, either horizontally in the walls of the rooms or vertically in the floors of the rooms. A third option (in-room, or in-vault, storage) may come under consideration in U.S. designs).

- Areal thermal load, which depends on the heat generated by a waste container, the spacing (pitch) of the waste containers within a storage room, and the spacing of the storage rooms.

- Dimensions of the transporter used to bring the waste containers (contained within transfer casks) from the waste-handling shaft to the storage rooms.

With few exceptions, notably the preconceptual design for the Hanford site (Rockwell Hanford Operations 1980), early repository designs assumed that individual waste containers would be placed in vertical holes in the floor of the storage rooms. Later designers opted for storage of multiple containers in long horizontal holes between storage rooms (U.S. Department of Energy 1982). This second option has been generally, but not universally, discarded. The extreme difficulty of retrieving containers from the end or middle of these long holes remains a major problem. Further, the practicability of excavating 30-in. diameter or larger holes horizontally over lengths in excess of 20 ft remains unproven.

Early repository designs were based on thermal loadings on the order of 150 kW/acre. This criterion was based on the assumption that, by the time a repository

were to be built, the number of nuclear power plants (and hence the volume of waste) would have increased significantly. However, because of the recent recession, conservation efforts, and the aftermath of the Three Mile Island accident, nuclear power plants have not been completed at the rate predicted earlier. This situation does not alter the need for a repository, but it does reduce the required storage capacity. Current repository designs have thermal loadings of about 60 kW/acre.

5 RETRIEVAL OF WASTE CONTAINERS

There are two basic scenarios for container (canister) retrieval: (1) full retrieval, whereby all the containers are removed from the repository, and (2) partial retrieval, whereby containers are retrieved from a portion of the repository, whether on a single-container, single-room, single-panel, or several-panel basis. The nature of the host rock and the repository design will determine whether the rooms from which containers must be retrieved are:

- Open and ventilated.

- Bulkheaded but unbackfilled (ventilated by leakage into and out of the panel).

- Bulkheaded and backfilled.

In the first case, ventilation air removes the heat conducted from the canisters through the rock to the perimeter of the storage room. Hence, no special ventilation requirements are needed for retrieval, assuming that the air quantity is sufficient, both to allow operations in the room and to cool the rock.

In the second case, retrieval requires prior breaching of the bulkheads followed by precooling. The cooling time depends on the temperature of the rock and the airflow. (In the case of salt, slashing may be necessary because of the reduction in opening dimensions as a result of creep.)

In the third case, a pilot heading must be excavated under adverse conditions -- high temperature -- to establish flow-through ventilation for precooling. Alternatively, pipes could have been laid in the backfill to allow precooling by a fluid prior to remining. However, such piping might considerably complicate the remining scheme. In any case, this last scenario is clearly the most problematic. Retrieval would still be possible under such conditions, but it would be difficult and would require a well-conceived plan.

Because 10CFR60.111(b) requires that retrieval be possible from the beginning of emplacement, it could be required while storage operations are still in progress. In the case of local retrieval, it would be desirable that retrieval not interfere with storage operations. Hence, air volumes required for retrieval must be in addition to those required for storage operations. If full retrieval is required, storage operations would cease.

6 EXAMPLE OF A RETRIEVAL SCENARIO IN SALT

The ventilation requirements for retrieval in a repository in salt are presented as an example. (This presentation does not necessarily reflect current designs, which are continually evolving.) It is assumed that mining development operations have ceased so that it is no longer necessary to have two separate ventilation systems as required by 10CFR60. Furthermore, it is assumed that the rooms from which waste is to be retrieved have been bulkheaded and backfilled.

Retrieval will therefore require remining. Three steps will be necessary before retrieval can be accomplished:

- Mining of a pilot heading with a remotely controlled roadheader or continuous miner, with air provided to the back of the machine via a ventilation duct.

- Precooling of the salt in the room by creating a ventilation circuit through the pilot heading.

- Excavating a bench to bring the remined room to the required height. As the temperature at the floor of the enlarged opening may be too hot, additional precooling may be required prior to retrieval.

Table 1 summarizes assumed properties and dimensions.

For the first stage, whereby the pilot is mined by remote control, it will not be possible to provide an environment suitable for human presence. Although one might argue that personnel could be provided with heat-resistant suits with air-conditioned interiors (i.e., space suits), such suits can not be assumed to provide radiation protection (Post 1982). Hence, humans can not be present during the remining stage.

However, once a flow-through ventilation circuit has been reestablished and the rock has been sufficiently cooled, humans can be present. For the assumed temperatures in Table 1, the maximum temperature at the exhaust end of the room would be 104°F. For a three-month precooling period, the required airflow in a room with a cross section of 8 ft by 15 ft (first pass) would be about 13,500 cfm. This airflow was determined using tables

Table 1. Dimensions and rock properties of hypothetical storage room.

Room dimensions	
First pass	8 ft h x 15 ft w x 540 ft long
Second pass	20 ft h x 15 ft w x 540 ft long
Average room perimeter wall-rock temperature	192°F (assumed)
Desired room perimeter wall-rock temperature after cooling	120°F
Intake air temperature	65°F (maximum)
Salt thermal conductivity, k	2.45 Btu/hr-ft-°F
Salt specific heat, c	0.215 Btu/lb-°F
Salt density, p	137 pcf

developed by Starfield (1966) and the method described in Hartman et al. (1982). (To perform this cooling, 47 tons of refrigeration would be required.)

If retrieval is to be completed in a panel within one year -- a time period equivalent to the emplacement rate -- precooling or retrieval operations must occur in approximately one-third of the rooms in the panel at any given time. Thus, if a panel contains 74 rooms, 25 must be ventilated at any given time. Therefore, the total airflow required would be 337,500 cfm for full retrieval. Because the airflow for local retrieval depends on the number of affected rooms, it would be a multiple of the 13,500 cfm determined above. (Note that these airflows are small by mining standards.)

As discussed previously, local retrieval could be required while storage operations are still in progress. These operations, which comprise storage-hole drilling, waste emplacement, and backfilling, would be carried on in several rooms, all of which would require ventilating air. The air requirements in these rooms would depend on diesel- and electric-horsepower requirements, based on 100 cfm/HP and 20 cfm/HP, respectively. (At the depths currently assumed for repositories in the Palo Duro and Paradox basins, the virgin rock temperature is low enough (about 88°F) that precooling or refrigeration is unnecessary from a temperature standpoint. However, in locales having significant humidity, conditioning or stilling chambers are required on the intake side of the ventilation system, even for temperatures in the range of 70°F to 80°F (Jacoby 1985).

Assuming a 250-HP diesel transporter, a 78-HP diesel storage-hole drill, and 800-HP electric pneumatic backfilling equipment, the airflow needed for each room during storage is dictated by the flow required by the transporter. (Requirements are not cumulative.) The number of rooms to be ventilated would then dictate the total flow for storage requirements. Presuming that four rooms must be ventilated at a time as a minimum,

the required airflow for storage is 100,000 cfm. Therefore, if storage and local retrieval occur concurrently, the 13,500 cfm per room for retrieval must be provided in addition to that amount.

The governing airflow is, therefore, the 337,500 cfm required for full retrieval. This airflow assumes a literal interpretation of the retrieval-time requirement; slightly longer times could reasonably be allowed. Thus, the number of rooms to be treated at a time could be halved without major adverse effect. This alternative would reduce the airflow requirement to 169,000 cfm, which could likely be provided by combining the capacities of the development and emplacement ventilation systems.

CONCLUSION

Airflow requirements for retrieval (if necessary) will depend on whether local or total retrieval is needed. In the case of local retrieval, the additional airflow required may affect operations if emplacement operations are still in progress. Otherwise, the combined development and containment ventilation systems should, in most cases, have sufficient total flow to allow local retrieval.

In the case of full retrieval, other operations are assumed to cease. Therefore, available airflows should be sufficient. However, from a licensing standpoint, it is necessary to prove sufficiency. Retrieval and concomitant ventilation requirements must therefore be part of any repository operations analysis.

ACKNOWLEDGMENTS

The author thanks Wyman Harrison, Associate Director for Geoscience and Engineering, Energy and Environmental Systems Division, Argonne National Laboratory, for encouraging the preparation of this paper, and Mary W. Tisue of the same division for editing the manuscript. He also thanks Francis S.

Kendorski of Terraform Engineers, Inc., for reviewing the manuscript and offering useful comments and suggestions.

REFERENCES

Hartman, H.C., J.M. Mutmansky, & Y.J. Wang (eds.) 1982. Mine ventilation and air conditioning, 2nd Ed. New York: Wiley.

Jacoby, C.H. 1985. Jacoby and Associates, Waxhaw, N.C., personal communication.

Kaiser Engineers 1978. Special study no. 3, Retrieval from backfilled regions, Report No. 78-56-R.

Post, R.G. 1982. University of Arizona, Department of Nuclear Engineering, personal communication.

Rockwell Hanford Operations 1980. Nuclear waste repository in basalt, Project B-301, Preconceptual design report, Report No. RHO-BWI-CD-35.

Starfield, A.M. 1966. Tables for the flow of heat into a rock tunnel with different surface heat transfer coefficients. J. South Afr. Inst. Min. Met., 66:692-694.

U.S. Department of Energy 1982. Site characterization report for the basalt waste isolation project, Report No. DOE/RL 82-3, 3 Vols.

U.S. Department of Energy 1984a. Nuclear Waste Policy Act, draft environmental assessment, Lavender Canyon site, Utah, Report No. DOE/RW-0009.

U.S. Department of Energy 1984b. Nuclear Waste Policy Act, draft environmental assessment, Davis Canyon site, Utah, Report No. DOE/RW-0010.

U.S. Department of Energy 1984c. Nuclear Waste Policy Act, draft environmental assessment, Cypress Creek Dome site, Mississippi, Report No. DOE/RW-0011.

U.S. Department of Energy 1984d. Nuclear Waste Policy Act, draft environmental assessment, Yucca Mountain site, Nevada Research and Development Area, Nevada, Report No. DOE/RW-0012.

U.S. Department of Energy 1984e. Nuclear Waste Policy Act, draft environmental assessment, Richton Dome site, Mississippi, Report No. DOE/RW-0013.

U.S. Department of Energy 1984f. Nuclear Waste Policy Act, draft environmental assessment, Deaf Smith County site, Texas, Report No. DOE/RW-0014.

U.S. Department of Energy 1984g. Nuclear Waste Policy Act, draft environmental assessment, Swisher County site, Texas, Report No. DOE/RW-0015.

U.S. Department of Energy 1984h. Nuclear Waste Policy Act, draft environmental assessment, Vacherie Dome site, Louisiana, Report No. DOE/RW-0016.

U.S. Department of Energy 1984i. Nuclear Waste Policy Act, draft environmental assessment, reference repository location, Hanford site, Washington, Report No. DOE/RW-0017.

U.S. Environmental Protection Agency 1984. Report on the proposed environmental standards for the management and disposal of spent nuclear fuel, high-level and transuranic radioactive wastes, Code of Federal Regulations, 40 CFR Part 91.

U.S. Nuclear Regulatory Commission 1984. Disposal of high-level radioactive wastes in geologic repositories, Code of Federal Regulations, 10 CFR Part 60.

An experimental determination of mixed and forced convection heat transfer coefficients in a modeled nuclear waste repository

R.L.OSBORNE & R.N.CHRISTENSEN
Ohio State University, Columbus, USA

ABSTRACT: An experimental model was developed for a nuclear waste repository storage room. Data were taken over a Reynolds number range of 6,000 to 180,000, covering both the forced and mixed (combined natural and forced) regimes of convection. Data are presented for several circumferential boundary conditions. Results indicate that the natural convection component is significant.

1 INTRODUCTION

The nuclear waste disposal plan for the United States calls for high level waste and/or spent fuel to be stored underground in deep geological formations. One of the design constraints on such a repository is that these storage areas must be ventilated. Ventilation is required for for both cooling and air supply for worker safety and comfort. Ventilation may also be required to limit temperatures in both the repository and the waste package.

In order that the ventilation and cooling requirements can be determined in the repository, accurate values of the convective heat transfer coefficients on the interior of the disposal rooms are needed. General flow and heat transfer conditions suggest that the rooms will be operating in the fully turbulent, mixed convection regime (Byrne, et. al. 1979:H1-H32). This fact is illustrated in fig. 1, showing the different flow regimes for horizontal pipe flow. Thus, standard forced convection analyses need to be expanded to include the natural convection component of the heat transfer. Also shown in fig. 1 is the range of data from the present work.

Several studies, although mostly analytical, have addressed the ventilation in a repository. Svalstad and Brandshaug (1983) considered "blast cooling" to cool a repository drift rapidly prior to waste retrieval. The study was analytical and used a finite element code to solve the governing equations. A constant velocity of 1 m/sec was used, which corresponds to a Reynolds number of 280,000 (see section 5 for definitions). Flows of this type are purely forced convection. Boyd (1982) also considered the ventilation required for retrieval. In this study, also analytical, mixed convection was investigated explicitly, using convection coefficients from various workers. Boyd's results indicate mixed convection reduces the total heat transferred, but this effect appears to come from the overall reduction of the convection coefficient, h, not the presence of mixed convection. Danko and Cifka (1984) determined the convection heat transfer coefficients of an underground copper mine in situ. Although their data were quite scattered, these workers determined that the effects of the natural convection component were significant.

The boundary conditions in a repository drift differ significantly from most standard mine ventilation problems. Although many mines, particulary extremely deep mines, experience high heat fluxes from the rock surfaces, the distribution of the fluxes appears to be unique to the nuclear waste repository. The heat sources located in the floor create an extremely skewed heat flux profile that is present for several years after waste emplacement.

The aim of this work was to determine the heat transfer characteristics in an experimental system which models the repository storage rooms, both thermally and hydrodynamically. The results must be

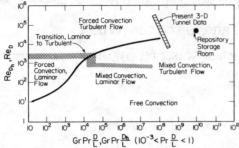

Fig. 1. Flow regimes in a horizontal tube (Metais and Eckert 1964) and estimated waste repository drift conditions.

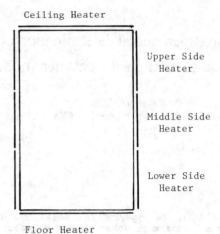

Fig. 2. Heater pad arrangement.

non-dimentionalized so that they apply in general to any similarly designed repository.

This work was supported by the United States Department of Energy, Office of Nuclear Waste Isolation under Contract No. ONWI/E512-03900 with The Ohio State University Research Foundation.

2 EXPERIMENTAL APPARATUS

The experimental apparatus was a flow channel of rectangular cross section with heated walls. The flow channel was constructed of galvanized sheet steel and had a cross section of 0.305 m wide by 0.457 m high. The total available heated length was 4.42 m. These dimensions were chosen so that the aspect ratio of the model was consistent with preliminary designs of repository rooms.

The heat sources were silicon rubber resistance heaters. The heaters provided a continuous heat flux axially down the tunnel, thus in effect smearing the discrete heat sources that actually exist in the repository. The work of Sisson (1978) and Christensen (1981) show that this assumption is reasonable, depending on the waste canister spacing. The heaters were arranged as shown in fig. 2, with the opposing side wall heater pads connected in parallel to give the same heat output. This arrangement gave a total of five separately controllable sets of heaters around the circumference of the tunnel. Each of these sets of heaters was powered by a continuously variable power supply. Thus a wide variety of circumferential heat flux profiles could be imposed on the tunnel walls.

The current and voltage inputs to the heaters were measured to determine the electrical power produced by the heaters. Thermocouples were placed at each heater to determine the wall surface temperatures. Thermocouples were also placed on the outside of the tunnel walls to determine the heat lost through the walls to the laboratory. All temeperature measurements were made with an electrically compensated datalogger. The datalogger signals were then sent to a VAX 11/750 computer for data reduction and heat transfer coefficient calculation. Thermal radiation was calculated by using a network scheme (Chaung 1982). The h values were found from the following equation

$$q"net = h(Twall - Tbulk)$$

where:

$$q"net = q"electrical - q"loss - q"rad$$

From these h values, the Nusselt numbers (Nu) were calculated.

Velocity measurements were made using a pitot tube array on a flow nozzle. Figure 3 shows a schematic of the complete setup. Only one heater set is shown for simplicity.

The air flows were in the fully turbulent regime as described by the Reynolds number (Re). The experimental Re ranged from 6,400 to 180,000. Preliminary designs have Reynolds numbers of approximately 170,000 (Sterns-Roger Services, Inc 1983:5-18). Thus the apparatus' range covers the expected repository Reynolds number.

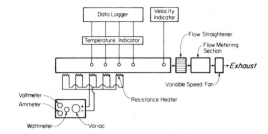

Fig. 3. Schematic of experimental setup.

3 RESULTS

Five different heat flux profiles were tested. Figures 4 through 9 show the fully developed Nusselt number for each of the five surfaces. The line labelled "Dittus-Boelter Equation" is the classical forced convection correlation developed for turbulent flow in long tubes with constant circumferential heat flux (Dittus and Boelter 1930):

$$Nu = 0.023Re^{0.8} Pr^{1/3}$$

The runs in fig. 4 had these boundary conditions, i.e., all heaters had the same heat flux. As would be expected, the data match the Dittus-Boelter correlation quite well at Re greater than 50,000. Below Re = 50,000, the data vary widely. For example, the Nusselt number of the floor at Re = 12,000 is approximately 110, three times the Nu given by Dittus-Boelter. From these results, it is apparent that natural convection is aiding the forced convection component of the heat transfer. Also apparent from fig. 4 is the fact that on the ceilng the natural convection is unaiding, or opposing, the forced convection component. This again is expected, as the buoyancy-driven flows would create a stagnation layer on the ceiling, thus reducing the heat transfer as predicted by forced convection analysis. Figure 5 has similar results for the constant circumferential wall temperature, except the ceiling Nu did not decrease as much as in fig. 4. The reason for this was, in order to maintain a constant circumferential temperature, the ceiling heat flux had to be reduced to counteract the buoyancy driven upward flow of hot air. This upward moving hot air increased the ceiling temperature. The reduction in the heat flux in turn reduced the stagnation layer, thus giving Nusselt numbers closer to the Dittus-Boelter correlation. A phenomenon occurring in

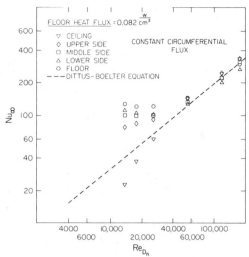

Fig. 4. Nusselt number versus Reynolds number for the constant heat flux case.

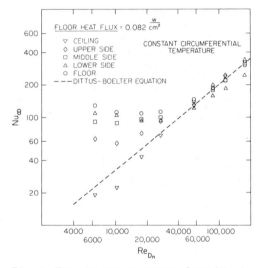

Fig. 5. Nusselt number versus Reynolds number for the constant temperature case.

both figs. 4 and 5 is the upward curvature of the data as the Re is decreased. This shows that as the flow is decreased, the increased natural convection component more than overcomes the drop in the forced convection component.

The trends in fig. 6 appear much like those in fig. 5 because, as previously described, the heat flux decreased from the floor to the ceiling to keep a constant wall temperature. The resulting heat flux profile of fig. 5 is close to that profile shown in the inset on fig. 6.

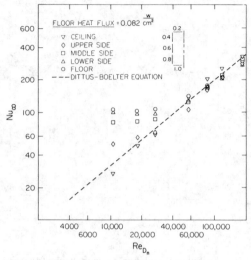

Fig. 6. Nusselt number versus Reynolds number for the indicated heat flux.

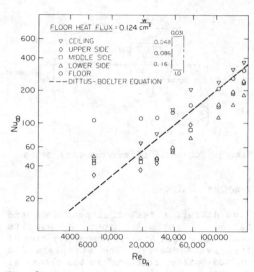

Fig. 8. Nusselt number versus Reynolds number for the indicated heat flux.

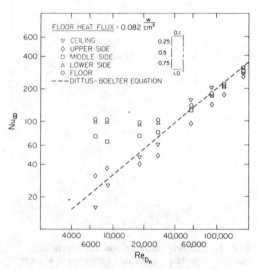

Fig. 7. Nusselt number versus Reynolds number for the indicated heat flux.

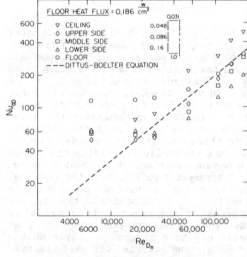

Fig. 9. Nusselt number versus Reynolds number for the indicated heat flux.

The numbers on this sketch indicate that, for example, the ceiling had a flux of 20% of the floor flux (0.082 watts/sq cm), or 0.016 watts/sq cm. Figure 7 shows a slightly different flux profile, with similar results to those in figs. 5 and 6. The curve shapes and the Nusselt number values of figs. 5, 6, and 7 agree closely.

Figures 8 and 9 show data from a heat flux profile as expected in a nuclear waste repository. The profile was

calculated from estimated temperature profiles (Science Applications, Inc. 1976:fig 3-2). The two figures have the same flux profile but different absolute fluxes. In the forced convection region (Re greater than 50,000), the Nusselt numbers for the five surfaces differed significantly from each other. This was due to the extremely skewed input flux profile. However the slopes of the curves (the exponent of the Reynolds number) are very close to each other and to the slope of the Dittus-Boelter correlation. This similarity in slope demonstrates that although the magnitudes differ, the dependence of the Nusselt numbers on the Reynolds number is consistent in the purely forced convection region. As the Reynolds number decreases, the Nusselt numbers appear to reach a horizontal asymptote around Re = 30,000. Below this Reynolds number, the Nusselt number curves are essentially flat, indicating that as the flow decreases, the forced convection component decreases, with, however, a corresponding increase in the natural convection component. Thus in the range of Re = 6,000 to 30,000, the mixed convection Nusselt number is independent of the flow rate. Although their data were scattered, the figures by Danko and Cifka (1984) show the same general trend.

This last result could be very significant if large flow rates are not required in the repository rooms, i.e. during the period that the rooms are open for emplacement activities or when the rooms are not undergoing blast cooling for retrieval. Using these results, the same heat transfer coefficient can be achieved at reduced flow rates.

4 SUMMARY

This work presents convective heat transfer coefficients for an experimental apparatus that models a nuclear waste repository storage room. Data are shown for several circumferential heat flux ratios, including a profile expected in a room. These data indicate that for Reynolds numbers below 30,000, the convection coefficient remains essentially constant for the range down to Re = 6,000. If ventilation flow rates are in this range, significant savings can be achieved by choosing the low end of the range.

5 DEFINITIONS

c_p = air specific heat at constant pressure

D_h = hydraulic diameter,

$$= \frac{4(\text{flow area})}{\text{flow perimeter}}$$

h = convective heat transfer coefficient

k = air thermal conductivity

q''_{net} = wall heat flux (in units of energy per unit area)

$q''_{electrical}$ = electrical power input to heaters per unit area

q''_{loss} = energy lost through tunnel's exterior walls, per unit area

q''_{rad} = energy loss from one heater set to another by thermal radiation, per unit area

$$Re_{D_h} = \rho V D_h / \mu$$

$$Nu_{D_h} = h D_h / k$$

$$Pr = c_p \mu / k$$

6 REFERENCES

Boyd, R.D. 1982. Forced convection cooling of a nuclear waste repository mine drift - a scoping analysis. Nuclear Engineering and Design 73:405-410.

Byrne, R.J., M.S. Giuffre, C.M. Koplik, L.R. Meyer, S.G. Osten, and D.L. Penty 1979. Information Base for Waste Repository Design, Vol. 4, NUREG/CR-0495. Reading, MA: Analytical Sciences Corp.

Chuang, F. 1982. The influence of (thermal) radiation on heat transfer in a rectangular duct - application to nuclear waste repositories, Master's Thesis. Columbus, OH: The Ohio State University.

Christensen, R., D.W. Gear, F.A. Kulacki 1981. Mixed convection in a rectangular channel flow with smooth surfaces and one heated wall. 20th National Heat Transfer Conf., Milwaukee, WI, Paper #81-HT-6.

Danko, Gy. and I. Cifka 1984. Measurement of the convective heat-transfer coefficient on naturally rough tunnel surfaces. Third Intl. Mine Vent. Congress, p.375-380. London: Institution of Mining and Metallurgy.

Dittus F.W. and L.M.K. Boelter 1930. University of California (Berkeley) Pub. Eng., vol. 2, p. 443. Berkeley: Univ. of Cal. (Berkeley).

Metais, B. and E.R.G. Eckert 1964. Forced, mixed, and free convection regimes. Transact. ASME, J. Heat Transfer 86:296-296.

Scientific Applications, Inc. 1976. Thermal operating conditions in a national waste terminal storage facility, Y/OWI/SUB-76/47950. Oak Ridge: Science Applications, Inc.

Sisson,   C.E. 1978. Predicted temperatures
in a bedded-salt depository resulting
from burial of DOE high-level nuclear
waste canisters, SAND78-0924.
Albuquerque, NM: Sandia National
Laboratory.
Stearns-Roger Services, Inc. 1983. Deaf
Smith/Swisher Counties, Texas :
Repository Design Concepts and Costs,
ONWI Report No. E512-06500/6, p. 5-18.
Denver: Stearns-Roger Services, Inc.
Svalstad, D.K. and T. Brandshaug 1983.
Forced Ventilation Analysis of a
Commercial High-Level Nuclear Waste
Repository in Tuff, SAND81-7206.
Albuquerque, NM: Sandia National
Laboratories.

The exploratory shaft and the design ventilation system

B.K.SCHROEDER
Rockwell Hanford Operations, Richland, WA, USA

ABSTRACT: This paper presents a summary description of the nuclear waste storage program and the role of Rockwell Hanford Operations, a description of the Exploratory Shaft facility at the Hanford Site, a description and analysis of the ventilation system design, and a loss-of-ventilation accident scenario. Underground ventilation is provided by air compressors and chillers on the surface.

1 ACKNOWLEDGEMENT

The author wishes to acknowledge the valuable contributions to this paper provided by those individuals listed below. The specific contribution made by each individual is identified by the section number in parentheses. In some cases their original work has been modified, as appropriate, to reflect current ventilation system design configuration or criteria.

S.R. Coleman, Environmental Health Sciences, Hanford Environmental Health Foundation, Richland, Washington (5.3).

M.J. McPherson, Consultant to Raymond Kaiser Engineers Inc./Parsons Brinckerhoff Quade & Douglas, Inc., Mine Ventilation Services, Lafayette, California (4.2).

J.V. Mohatt, Health, Safety and Environment representative, Rockwell Hanford Operations, Richland, Washington (5.2).

2 THE NUCLEAR WASTE STORAGE PROGRAM

In 1975, the U.S. Energy Research and Development Administration (now the U.S. Department of Energy) established the National Waste Terminal Storage (NWTS) Program to investigate a number of geologic formations to determine their suitability for long-term disposal of commercial radioactive waste. The NWTS Program has since been reorganized and retitled the Office of Civilian Radioactive Waste Management (OCRWM).

The Nuclear Waste Policy Act of 1982, (NWPA 1983) requires that the U.S. Department of Energy (DOE) recommend three sites for further investigation (site characterization). The presently proposed sites are (1) the Hanford Site in southeastern Washington State, (2) the Nevada Test Site, north of Las Vegas, and (3) Deaf Smith County in the Panhandle of Texas. Once the President of the United States approves the sites, then the detailed site characterization process may begin.

The Basalt Waste Isolation Project (BWIP), an organizational element of Rockwell Hanford Operations (Rockwell), has been chartered by the DOE to investigate the Columbia River basalts on the Hanford Site. Within the BWIP is the Exploratory Shaft (ES) project that will provide the facility for in situ testing at the candidate reference repository horizon provided the Hanford Site is selected. The site characterization program will provide the necessary information to determine the suitability of the Hanford Site for a nuclear waste repository.

3 EXPLORATORY SHAFT FACILITY

3.1 Background

Conceptual design for Phase I of the ES began in the fall of 1981. This phase consisted of a 1.83-m (6-ft) I.D. shaft that was 1,189 m (3,900 ft) deep, with a

15.2-m (50-ft) drift at the breakout depth of 1,159 m (3,802 ft) below the surface (the Umtanum flow). Phase II was to consist of approximately 305 m (1,000 ft) of drift development at the breakout depth.

The ventilation system design effort during Phase I was intended to accommodate both Phase I and Phase II. During the review of the final design for Phase I in 1983, heat stress resulting from a loss of ventilation posed the most serious unresolved safety question, and thereby required the most attention. A loss-of-ventilation study ensued (section 5).

In 1984, two major changes occurred that impacted the design of the ES project. First, the preferred breakout horizon was changed from the 1,159-m (3,802-ft) depth in the Umtanum flow to the Cohassett flow where breakout was identified at 965 m (3,164 ft). The change in depth reduced the virgin rock temperature from 58 °C (138 °F) to 52 °C (125 °F) (Rockwell 1983). Second, an alternate method of egress was required for compliance with DOE Order 5480.1A, "Environmental Protection, Safety, and Health Protection Program for DOE Operations"; hence, a second shaft was identified in the second phase of the project. This second shaft not only enhanced the schedule for meeting test objectives, but also provided a alternate method (a flowthrough exhaust system) for ventilation of the underground facility. The Phase I compressed air system will be used until the flow-through system becomes available, approximately 6 mo after breakout of the first shaft.

3.2 Surface, Shaft, and Subsurface Description

The location of the ES facility is shown in figure 1. Utilities to the site include (1) a 20-cm (8-in.) potable water line, (2) a 13.8-kV overhead powerline with a voltage stepdown transformer to 2,400 V, and (3) a 2.4-km (1.5-mi) access road from State Highway 240. The surface facility will occupy up to 1.6 x 10^5 m^2 (40 acres) enclosed by a security fence and will include standard services for electrical power, water, sewer, telephone, and fire protecion. Each 1.83-m (6-ft) shaft will have a headframe and a main hoist. The Phase I shaft will have an additional standby hoist. Warehouse and office requirements will be met by

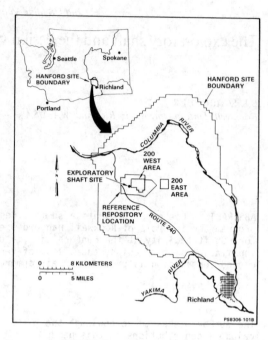

FIGURE 1. Site Location.

trailers and/or temporary buildings. Most of the surface area will be devoted to mud pits, excavated-material storage, cement storage and mixing, and laydown areas for the shaft casing.

The shaft arrangement is illustrated in figure 2. Total shaft depth will be approximately 1,037 m (3,400 ft) with breakout in the Cohassett flow at 965 m (3,164 ft). The shaft casing is a welded steel liner with a hemispherical head at the bottom for floating the casing into place. The liner varies in thickness from 1.3 cm (0.5 in.) at the top to 4.4 cm (1.75 in.) at the bottom and is reinforced with stiffener rings with various spacings and sizes for material optimization. Utility lines (compressed air and dewatering) and grout lines are mounted external to the casing before installation. The casing is grouted in place and the shaft outfitted with guides (Karri-wood) and utility lines (ventilation, water, electrical, and communications).

The conceptual subsurface layout is illustrated in figure 3. Underground utilities, including electrical power, water, equipment compressed air, and communications, will be provided from

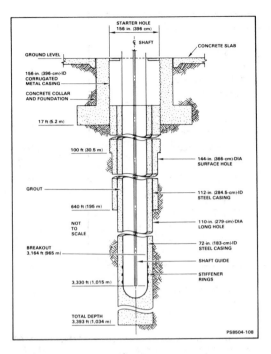

FIGURE 2. Shaft Arrangement.

the surface. The subsurface facility is designed to accomodate fifteen tests that include twelve geomechanic tests and three hydrology tests.

4 VENTILATION SYSTEM DESIGN

4.1 System Description (RKE/PB 1984)

The underground ventilation system supplies compressed air from the surface at a flow rate of 5.66 m^3/s (12,000 stdft^3/min), a temperature of 10 $^\circ$C (50 $^\circ$F), and pressure of 354 KPa (50 lbf/in^2 (gage)). The normal operating system consists of three air compressors, two chiller/dryers, two cooling water towers, and a distribution system to the underground facility. The equipment is located on the surface (fig. 4).

The ventilation air compressors are cenrifugal type, each rated at 2.83 m^3/s (6,000 stdft^3/min), and are complete with intake air filters and all necessary controls and instrumentation. Two compressors are required for the normal 5.66 m^3/s (12,000 stdft^3/min) operation. A third compressor is available as a spare for use during compressor maintenance.

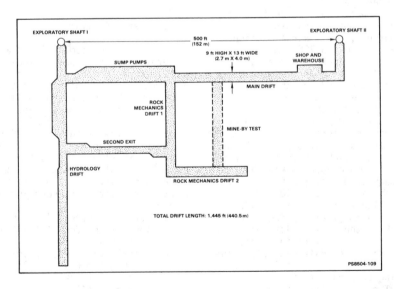

FIGURE 3. Underground Layout.

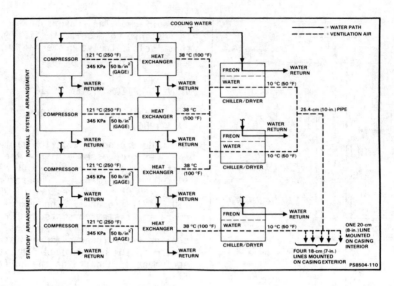

FIGURE 4. Ventilation Schematic.

Each chiller/dryer is a package unit
with refrigeration and water cycles.
The refrigeration cools the water that
in turn provides the final cooling of
the ventilation air from 38 °C (100 °F)
to 10 °C (50 °F). Each unit is sized to
cool 5.66 m³/s (12,000 stdft³/min) of
ventilation air. The second unit provides 100 percent redundancy for use
during maintenance.

The cooling water system consists of
pumps, cooling towers, and associated
piping, controls, and instrumentation.
This system provides 3,785 L/min
(1,000 gal/min) of cooling water at
27 °C (80 °F) to the compressors and
the chiller/dryer, and precools the ventilation air from the air compressors
from 121 °C (250 °F) to 38 °C (100 °F).
Again, there is 100 percent redundancy
for use during maintenance.

The compressors are arranged so that
the three units discharge ventilation
air into a common header to the chiller/
dryer through a 30.5-cm (12-in.) line.
After the air is cooled to 10 °C
(50 °F), it leaves the chiller-dryer
through a 25.4-cm (10-in.) line to a
ring header at the shaft collar where it
joins with four 17.8-cm (7-in.) lines
mounted outside the shaft casing and one
20.3-cm (8-in.) line mounted inside the
shaft casing that runs to the breakout
horizon. The supply lines into the shaft
terminate at the shaft station, where air
silencers limit the noise level to 90 dB.

The standby ventilation system, powered
by either the normal electric distribution or a diesel generator, provides
2.83 m³/s (6,000 stdft³/min) at 10 °C
(50 °F), and is independent of the normal
ventilation system. This system ties
into the normal distribution system at
the shaft collar for distribution
underground.

4.2 System Analysis

The complete system analysis was an
iterative process and included a variety
of assumptions and configurations. Only
a summary of the portion that is applicable to the present ES configuration
will be presented here. Parameters provided as input to the analysis included:
- Rock temperature - 51 °C (124 °F)
- Drift length - 305 m (1,000 ft)
- Drift size - 2.7 by 4 m (9 by 13 ft)
- Depth of breakout - 945 m (3,100 ft)
- Water inflow - 7.6 L/min (2 gal/min)

The energy transfer mechanisms considered
were:
- Vertical descent (ΔT = 9.5 °C(17 °F))
- Ventilation duct (ΔT = 5.4 °C
  (9.6 °F))
- Water inflow (2.8 L/min/
  (0.75 gal/min))
- Personnel (3 kW)
- Muck pile (36 kW)
- Electrical loads (30 kW)
- Rock surfaces (35 kW)

The vertical descent calculations assumed an adiabatic pipe wall and the thermodynamic properties calculated by the steady-flow energy equation that included turbulent friction. Because the air expands as it descends the pipe (a transient condition), 20 steady-state segments were used to approximate the actual condition. Also, 65 percent of the heat absorbed by the air in the ventilation duct originates from the drift air. The predicted air temperatures as the air exits the drift are 37 °C (98.5 °F) dry bulb and 23.5 °C (74.3 °F) wet bulb. This corresponds to a relative humidity of 32 percent.

4.3 Environmental Requirement

The DOE prescribed standard (DOE 1984) states that fully clothed acclimatized workers with adequate water and salt intake should be able to function effectively under working conditions without exceeding a deep-body temperature of 38 °C (100.4 °F). Since measurement of deep body temperature is impractical for monitoring purposes, the Wet-Bulb Globe Temperature Index (WBGT) is a practical method for determining the suitability of the working environment.

The ES ventilation design was based on not exceeding the WBGT threshold limit of 26.7 °C (80 °F) for continuous but moderate work load (ACGIH 1985). The WBGT is determined by the relationship:

$$WBGT = 0.7\ T_{wb} + 0.3\ T_g$$

where:
  T_{wb} = natural wet bulb temperature
  T_g = globe temperature

The globe temperature is a measured quantity and would be difficult to calculate. The globe thermometer consists of a temperature sensor located in the center of a 15-cm (6-in.) hollow copper sphere painted matte black on the exterior surface. The purpose of the globe thermometer is to determine the thermal radiation effects on the worker. In reality this temperature would be somewhere between the dry-bulb temperature and the rock surface temperature. Also, the natural (at rest) wet-bulb temperatures will be greater than the thermodynamic wet-bulb temperatures. As a

first approximation, assume that these temperatures are the dry-bulb and wet-bulb temperatures. Then, the WBGT for the ventilation air as it exits the drift would be 27.5 °C (81.5 °F). The actual WBGT would be higher than this and would require that a work/rest regimen be initiated at this location (shaft station).

4.4 Water Inflow

The water inflow of 2.8 L/min (0.75 gal/min) is twice the value predicted by computer modeling studies and permeability testing at the Hanford Site. The predicted temperatures in section 4.2 are based on 100 percent evaporation of this water into the air. If an inflow of 7.6 L/min (2 gal/min) is assumed, then the predicted air temperatures are 38.5 °C (101.3 °F) dry bulb and 30.2 °C (86.3 °F) wet bulb, corresponding to a relative humidity of 55 percent. This condition would exceed the WBGT working environment established for the ES project. The conclusion is that the working environment will be very sensitive to water inflow and water control.

4.5 Mitigating Considerations

a. Most work being performed will be at the working face where the air exiting the ventilation duct has predicted temperatures of 24.8 °C (76.7 °F) dry bulb and 14.6 °C (58.2 °F) wet bulb.
b. The third compressor could be used to supply an additional 30.5 m³/min (6,000 stdft³/min) of ventilation air through the 20.3-cm (8-in.) line located inside the shaft in parallel with the four 17.8-cm (7-in.) lines outside the casing.
c. Administrative measures for water control could be implemented, particularly during construction. For instance, all drilling water should be contained.
d. For construction, most of the electrical requirements will be replaced with compressed-air requirements.
e. An air turbine generator can be installed underground so that the kinetic energy of the air produces work and not a thermal rise in the ventilation air temperature.

5 LOSS OF VENTILATION

A study was made to determine if there would be a heat stress risk involved during escape if the ventilation system were to fail. This study involved determining (1) the changing environmental condition underground, (2) a person's metabolic rate during escape, (3) the allowable "stay-time" underground, and (4) the time required to recover personnel. All calculations were based on environmental conditions at the shaft station.

5.1 Environmental Conditions

If the ventilation system fails, the wet-bulb temperature increases due to continued water inflow, and the dry-bulb temperature approaches the rock surface temperature. The rock surface temperature was determined using the relationship

$$Nu = 0.023 \ Re^{0.8} \ Pr^{0.4}$$

where:

  Nu = Nusselt number
  Re = Reynolds number
  Pr = Prandtl number
  (Schenck 1959)

and the air temperature was determined from the relationship

$$dQ_{air} = Q_{rock} \ (dt)$$

where:

  dQ = incremental heat adsorbed
   Q = heat loss rate
  dt = incremental time

The calculated rock temperature was 4.7 °C (8.4 °F) higher than the air dry bulb temperature of 41.6 °C (106.9 °F). After 20 minutes the wet-bulb temperature was 27 °C (81 °F) and the dry-bulb temperature was 41 °C (106 °F).

5.2 Metabolic Rate

A comparative analysis was conducted of the projected ES environmental conditions against two published studies. Abstracts of the two comparative studies follow.

Six men, with mean age of 45 (with assumed poor-to-good fitness), were asked to escape a coal mine under timed conditions, in poor lighting, with full clothing and rescue equipment. They exerted moderate-to-heavy activity levels. It was assumed that the existing mine ventilation system was unchanged for this test. Heart rates and oxygen utilization were recorded (Kamon et al. 1983).

Nine fit young men were subjected to light-to-moderate exercise under increasingly hot and humid conditions. The men wore one layer of comfortable loose-fitting clothing and excercised in a familiar, well-lighted environment. Heart rates, oxygen utilization, rectal temperature, and mean skin temperatures were taken for 45-min exercise periods (Pulket 1980).

Using empirical adjustments, best estimates, and reasonable assumptions, the comparison of study variables (that contributed to variation in the metabolic rates) were adjusted from each of the above study situations in order to approximate the ES conditions. When the final adjusted metabolic rates of the two studies were made to simulate the projected ES conditions, a metabolic rate of 586 W (2,000 Btu/h) was chosen.

5.3 "Stay-Time"

The National Institute for Occupational Safety and Health (NIOSH) states that under transient conditions the deep-body temperature of 39 °C (102.2 °F) may be permitted, but only briefly (NIOSH 1973). Based on this, the underground "stay-time" was based on an increase in body temperature of 1 °C or 33 Wh absorbed (2 °F or 250 Btu adsorbed).

The methodology used was the heat stress index, stated algebraically

$$M + R + C = E_{max}$$

where:

    M = metabolic rate
    R = radiant heat exchange
    C = convective heat exchange
    Emax = maximum evaporative heat loss

Assuming a person walks for 152 m (500 ft) to the shaft station at a rate of 38 m/min (125 ft/min), and using the relationships from NIOSH (1973), the M+R+C term is 282 W (964 Btu/h) greater than the evaporative cooling term (Emax). This equates to an increase in body temperature of 0.30 °C (0.5 °F) or 18.8 W (64.3 Btu) in a 4-min period. At the shaft station the metabolic rate reduces to 140 W (476 Btu/h) at rest, and Emax = C = 0 because the air velocity is zero. Metabolic rate plus radiant heat exchange then equals 200 W (684 Btu/h) and the "stay-time" is 16 min, or a total allowable "stay-time" of 20 min.

5.4 Recovery Time

The recovery time is based on the following parameters:
- hoist speed - 229 m/min (750 ft/min)
- cage travel time - 5 min (one way)
- cage capacity - 6 persons
- persons underground - 12 (maximum).

If it is assumed that loss of ventilation occurs with a loaded skip, then the last persons to be hauled to the surface could take 20 min. If the loss of ventilation was due to a power failure, then an additional 3 min would be required for hoist operations via the standby electrical system (diesel generator).

5.5 Conclusions

The risk associated with heat stress during personnel evacuation is probably negligible. The assumptions made were conservative in that (1) the longest distance from the shaft station was selected, (2) the actual metabolic rate would probably be less than the value used in the calculations, and (3) the recovery time was a worst case condition because of the skip's location. The methodology used in the loss-of-ventilation study can only be used as an indicator because each individual might respond differently to the same or similar conditions.

6 REFERENCES

ACGIH 1985. Threshold Limit Values for Chemical Substances and Physical Agents in the Work Environment and Biological Exposure Indices with Changes for 1984-85, American Conference of Government Industrial Hygienists, Cincinnati, Ohio.

DOE 1984. Environmental Protection, Safety, and Health Protection Standards, DOE Order 5480.4., U.S. Department of Energy, Washington, D.C.

Kamon, E., D.Doyle & J.Kovac 1983. The Oxygen Cost of an Escape from an Underground Coal Mine, American Industrial Hygiene Association Journal, 44:552.

NIOSH 1973. The Industrial Environment - Its Evaluation and Control, 017 001-00396-4, Chapter 31, National Institute for Occupational Safety and Health, Washington, D.C.

NWPA 1983. Nuclear Waste Policy Act of 1982, Public Law 97-425.

Pulket, C., A.Henschel, W.R.Burg, & B.E.Saltzman 1980. A Comparison of Heat Stress Indices in a Hot-Humid Environment, American Industrial Hygiene Association Journal, Vol. 41:442.

RKE/PB 1984. Exploratory Shaft Phase I Title II Design Report System Design Description, SD-BWI-DR-001, Rockwell Hanford Operations, Richland, Washington.

Rockwell 1983. Repository Horizon Identification Report, SD-BWI-TY-001, Rockwell Hanford Operations, Richland, Washington.

Schenck, H.Jr. 1959. Heat Transfer Engineering, p. 89, Englewood Cliffs, New Jersey, Prentice-Hall, Inc.

9. Methane control

Methane occurrence in US metal/nonmetal mines:
Emanation mechanisms and gas volumes

WILLIAM E.BRUCE
Mine Safety & Health Administration, Denver, CO, USA

WILLIAM H.DONLEY
Mine Safety and Health Administration, Rapid City, SD, USA

ABSTRACT: This paper discusses gassy metal and nonmetal mines in the United States. A tabulation is presented which enumerates those United States metal and nonmetal mines classified as gassy which are currently operating or which have operated recently. The rationale used to classify the mines gassy is briefly described and is included in the tabulation. Twenty-four gassy metal and nonmetal mines in the United States are grouped and discussed by commodity and individually to the extent possible. Emanation mechanisms, methane volumes emitted, and safety and health implications of such methane occurrences are noted. Chemical composition of gas occurrence is discussed in detail where the composition of gas is unusual or noteworthy from an explosibility standpoint. General recommendations are made for use of the data presented. Some conclusions are drawn relative to safe mining practices in gassy metal and nonmetal mines.

1. INTRODUCTION

The data presented by the authors were collected and assembled during day-to-day activities of the Mine Safety and Health Administration and during special assignments undertaken by the authors for purposes of updating and streamlining MSHA regulations for gassy metal and nonmetal mines. Standards development activities have been a priority assignment in MSHA in recent years and during the course of these activities it became apparent that there was widespread interest in conditions related to methane occurrence in gassy metal and nonmetal mines. This paper was completed to condense into one publication some of the knowledge about gassy mines in the United States that had heretofore been available only in scattered MSHA files. Another goal in completing this paper was to provide data which could be used in mine design where hazards could be anticipated and to suggest some appropriate safety procedures.

Other related reports have been written about gassy mines in recent years. A report by Thimons (1) (numbers in parentheses refer to items in the list of references at the end of this report) described a proposed system for forecasting methane hazards in metal and nonmetal mines. Lumsden (2) presented data collected to describe various methods used by foreign countries to classify metal and nonmetal mines as gassy.

2. BACKGROUND

The mines described in this report were placed into the gassy classification by the Mine Safety and Health Administration or by the state in which the mine was located. The following regulations (30 CFR Parts 57.21-1 and 57.21-2) are the mine classification standards used to designate the mines as gassy.

57.21-1 Mandatory. A mine shall be deemed gassy, and thereafter operated as a gassy mine, if:
(a) the state in which the mine is located classifies the mine as gassy; or
(b) flammable gas emanating from the orebody or the strata surrounding the orebody has been ignited in the mine; or
(c) a concentration of 0.25 percent or more, by air analysis, of flammable gas emanating only from the orebody or the strata surrounding the orebody has been detected not less than 12 inches from the back, face, or ribs in any open workings; or
(d) the mine is connected to a gassy mine.
57.21-2 Mandatory. Flammable gases detected only while unwatering mines or flooded sections of mines or during other mine

reclamation operations shall not be used
to permanently classify a mine gassy. During such periods that any flammable gas is
present in the mine, the affected areas of
the mine shall be operated in accordance
with appropriate standards in this Section
57.21.

At the time of the writing of this report, MSHA was actively engaged in developing revised standards for classification
of and for operation of gassy metal and nonmetal mines and for that reason, near-future
standards could differ significantly from
those cited above.

3. DISCUSSION

Table 1 shows active and recently active
metal and nonmetal mines in the United
States which have been classified as gassy.
So far as is known, table 1 includes all
gassy mines which were active at the time
of the writing of this paper. As noted on
the table, other mines in the United States
have been classified as gassy but have been
inactive for many years and are not included. Currently inactive mines in table 1
have been included because they illustrate
certain points made in this paper. Table 1
shows the particular criterion (items a, b,
c, or d) applied in classifying each mine
as gassy. Frequently, although an ignition
of flammable gas has occurred (item "b" in
table 1), the mine is formally classified
as gassy based on the results of a gas
sample analysis which shows that 0.25 percent or more of methane was present not
less than 12 inches (0.30m) from the back,
face, or ribs of a mine opening (item "c"
in table 1). Although table 1 shows only
one reason for the gassy classification,
sometimes more than one of the conditions
in standard 57.21-1 may have been met.
Three of the table 1 mines were classified
gassy as a result of being connected to another gassy mine. Only one mine is included which was classified gassy under safety
standard 57.21-2.

3.1 Description of Incidents

It was felt that further elaboration under
a commodity grouping of mines would provide
valuable information on the conditions, the
accidents, or the disasters that led to the
gassy classification. Where a date is spe-
ified for classifying a mine gassy, the date
corresponds to the date on the classifying
document and not necessarily to the date
the sample was collected or the date on
which the incident occurred.

3.1.1 Oil Shale Mines

The U.S. Bureau of Mines Shaft No. 1 (also
known as the Horse Draw Shaft experienced
an ignition which occurred on November 1,
1978. Methane gas was ignited during an
oxyacetylene cutting operation being conducted at the 5 level shaft station; the
flame reportedly persisted for only about
5 minutes and there were no injuries. The
mine was classified gassy on November 1,
1978, as a result of the ignition. On
December 6, 1978, in another incident,
methane gas was ignited and an oil shale
fire resulted during blasting in a 2080-ft-
level drift. Flooding of the shaft was
utilized to extinguish the fire.

The Rio Blanco Mine experienced an ignition
of methane which occurred on November 26,
1979. The ignition occurred when hot slag
from an oxyacetylene cutting operation ignited methane in a shaft, the resulting
flame was of short duration and no persons
were injured. The mine was classified
gassy on November 30, 1979, as a result of
the ignition.

The Cathedral Bluffs Mine was classified
gassy on January 2, 1980, as a result of the
detection of 0.25 percent or more of methane
at the 960-ft-station of the Ventilation-
Escape Shaft. On at least eight occasions
during 1980, ignitions were reported to
MSHA, having occurred in the Production and
in the Ventilation-Escape Shafts. The
sources of ignition generally were falling
slag from cutting torch operations.

The White River Shale Oil Project experienced an ignition of methane and a resulting oil shale fire which occurred on December 5, 1983. Methane was apparently ignited in the ventilation shaft bottom at the
time of blasting and the burning methane set
fire to oil shale. The resulting oil shale
fire was extinguished by flooding the shaft
bottom. The project was classified gassy
on December 7, 1983, as a result of the detection of 0.25 percent or more methane in
the return air from the ventilation shaft.
The White River Shale Oil Project subsequently experienced another ignition and
fire which occurred in an ore-pass raise
at the time of blasting on June 9, 1984.
The fire reportedly was extinguished by
drenching with an estimated 100,000 (378 m^3)
gallons of water.

Methane release in oil shale mines generally
can be expected to be somewhat unpredictable. Oil shale formations are expected
to contain less methane near outcrops, but
increased methane content is generally to
be expected as distance from mine workings
to the outcrop increases and as depth of
cover increases. In those mines where

methane can be anticipated, releases can be expected to occur whenever oil shale is fragmented such as by blasting, continuous mining, or drilling; methane volumes released will generally be in proportion to volume of shale broken, and duration of release will correspond to the time required to fragment the shale. Significant methane releases may also occur through methane feeders which are usually associated with faults, fractures, fissures or porosity. Slow releases of methane may occur continuously from seemingly impermeable wall rock along mine entries.

Although Bureau of Mines studies at the Horse Draw Mine showed methane contents up to about 150 cu ft per ton (4.68 m^3/tonne) of shale broken, most Horse Draw data showed methane content in the range from about 5 to about 100 cu ft per ton (0.15 to

Table 1. Gassy metal/nonmetal mines in the U.S.[1]

Mine Name, location	Commodity	Reason(s) for Classification[2]				Mine Name, location	Commodity	Reason(s) for Classification[2]			
		a	b	c	d			a	b	c	d
Alchem Mine, Green River, WY.	Trona	✳				Kerr-McGee Mine, Eddy Co., NM.	Potash			✳	
B-38 Mine, Bonanza, UT.	Gilsonite				✳	Lisbon Mine, San Juan Co., UT.	Uranium			✳	
B-42 Mine, Bonanza, UT.	Gilsonite				✳	Mississippi Chemical Corp Mine, Eddy Co., NM.	Potash			✳	
Belle Isle Mine, St. Mary Parish, LA.	Salt		✳			North Tisdale Gravity Drainage Project, Johnson Co., WY.	Petroleum	✳			
Belle Mine, Bellefonte, PA.	Limestone			✳		Rio Blanco Mine, Rio Blanco Co., CO.	Oil Shale		✳		
Calloway Mine, Copperhill, TN.	Copper		✳			Stauffer Big Island Mine, Green River WY.	Trona	✳			
Cathedral Bluffs Mine, Rio Blanco Co., CO.	Oil Shale			✳		Tenneco Mine, Green River, WY.	Trona	✳			
Cote Blanche Mine, St. Mary Parish, LA.	Salt			✳		Texasgulf Mine, Granger, WY.	Trona	✳			
FMC Mine, Green River, WY.	Trona	✳				Weeks Island No.2 Mine, Iberia Parish, LA.	Salt			✳	
I-16 Mine, Bonanza, UT.	Gilsonite			✳		White River Shale Oil Project, Uintah Co., UT.	Oil Shale			✳	
I-24 Mine, Bonanza, UT.	Gilsonite				✳	U.S. Bureau of Mines Shaft No. I, Rio Blanco Co., CO.	Oil Shale Nahcolite Dawsonite	✳			
International Salt[3] Co. B-Shaft, Retsof, NY.	Salt										
Jefferson Island Mine, Iberia Parish, LA.	Salt			✳							

[1] Table consists of many, but not all gassy metal/nonmetal mines. Some mines included are not active but are enumerated as examples of specific conditions

[2] (a) Classified gassy by the state (b) an ignition of flammable gas occured (c) 0.25 % or more flammable gas was detected (d) mine connected to a gassy mine

[3] Classified gassy under 57.21-2.

3.12 m^3/tonne). Moreover the Bureau data may not truly reflect what has been or will be experienced at other sites. Some engineering firms have provided site specific design criteria for projecting ventilation requirements in oil shale mines. For example, methane emission rates for one proposed site were projected as follows:

Shaft sinking:	40 cu ft/ton
	(1.25 m^3/tonne)
Raise boring:	60 cu ft/ton
	(1.87 m^3/tonne)
Raise slashing:	40 cu ft/ton
	(1.25 m^3/tonne)
Lateral drifting:	40 cu ft/ton
	(1.25 m^{3}tonne)
Wallrock emission:	0.001 cfm/sq ft
	(5 x 10^{-6} m^3/s/m^2).

Little if any data on daily emissions are available from operating gassy oil shale operations due to depressed conditions in the industry which have severely curtailed the scale of oil shale operations.

3.1.2   Gilsonite Mines

Gilsonite is a solid hydrocarbon which is naturally occurring in near-vertical veins which extend to the surface in the state of Utah. Table 1 shows how these mines reportedly were classified as gassy. One gilsonite mine, I-16, was reportedly classified gassy as a result of the detection of 0.25 percent or more methane. Three gilsonite mines shown on table 1 were reportedly classified gassy because they were connected to another gassy gilsonite mine.

Over the years there have been several fires and explosions in gilsonite mines and mills. Gilsonite dust presents a serious explosion hazard, the minimum explosive concentration being about 0.02 ounces per cu ft (0.02 kg/m^3); in comparison, coal dust has a minimum explosive concentration of about 0.05 to 0.07 ounces per cu ft (0.05 to 0.07 kg/m^3). Gilsonite mines generally use extraction methods that rely upon compressed-air power so that electricity or explosives are not normally utilized underground for gilsonite fragmentation. Explosives, however, are used for rock fragmentation and a recent gilsonite dust explosion was ignited while blasting in the wall rock of the vein to provide space in the shaft for an air-lift pipeline. The cause of the explosion was presumed to be failure to properly stem the charged holes prior to blasting.

3.1.3 Domal Salt Mines

The Belle Isle Mine experienced a methane explosion which occurred on June 8, 1979. The explosion resulted in the deaths of five persons and resulted in injuries to a number of others who were present underground at the time. The disaster occurred when an outburst of methane and salt was triggered by a blast round in a development heading. About 10 minutes following release of the estimated 600,000 cu ft (17,000 m^3) of methane, the body of methane and air was ignited, most likely by an electrical source such as an electrical arc or spark, or by burning electric cable insulation. The mine was declared gassy on June 12, 1979. The Belle Isle mine voluntarily ceased operations on January 31, 1984. The cessation of operations was a result of ground stability problems which made it impossible to conduct mining operations and to assure the safety of employees.

Following the Belle Isle disaster, on June 18, 1979, an intensive examination was begun by MSHA to establish whether methane was present in other Louisiana domal salt mines. The Cote Blanche salt mine was classified gassy on July 14, 1979, as a result of the detection of 0.25 percent or more methane.

The Jefferson Island salt mine was classified gassy on June 27, 1979, as a result of the detection of 0.25 percent or more methane. The Jefferson Island mine was flooded on November 20, 1980, and is no longer in operation.

The Weeks Island salt dome is the location of two salt mines; the older mine has been converted to a National Strategic Petroleum Reserve and is no longer an active salt mine. It was classified gassy on July 7, 1979, as a result of the detection of 0.25 percent or more methane. The newer mine is termed the Weeks Island Mine No. 2 and is in no way connected to the older mine. The Weeks Island Mine No. 2 was classified gassy on December 15, 1980, following notification of MSHA by the shaft excavation contractor that 0.25 percent or more methane had been detected or October 17, 1980, emanating from a vertical drillhole adjacent to a slot raise. On October 6, 1982, an outburst was triggered when a development blast was made in the mine; the resulting body of methane in air was safely ventilated out of the mine and no persons were injured.

The occurrence of methane in the above-named domal salt mines is well known. When methane is released into the mine atmosphere it almost always accompanies a fragmentation process such as drilling, blasting, mechanical mining, or undercutting. Although most areas of the mines are relatively free of methane, other areas are susceptible to large releases of methane from unstable zones of methane-enriched salt. Observed methane releases occur from vertical and horizontal boreholes, from kerf cuts (undercuts), from feeders where bubbling may be visually observed accompanying methane release from fissures or from porous zones, and from outbursts.

Thimons (1) reported methane concentrations up to 165 ppm in the return air from domal salt mines, and so far as is known, little other data are available on levels of methane in the return air of domal salt mines. The largest volumes of methane are released during outbursts which may be defined as the sudden, violent release of gases and solids from a working face. Outbursts in domal salt mines most often occur during blasting of development faces although some lesser outbursts have occurred during bench blasting. The approximate range in size of the conical cavities that generally result following an outburst is up to 100 ft (30 m) in diameter and up to 278 ft (85 m) in height based on direct measurements. Outbursts are known to have occurred in all four domal salt mines described herein.

The final MSHA report on the Belle Isle disaster (3) estimated that a volume of 600,000 cu ft (17,000 m^3) of methane was released into the mine atmosphere by the June 8, 1979, outburst. This volume translates into an instantaneous methane release of about 40 cu ft per ton (1.25 m^3/tonne). Methane content of domal salt samples was reported by Schatzel and Hyman (4) and the maximum methane content reported therein translates to about 2.4 cu ft per ton (0.075 m^3/tonne). The authors report however that actual releases observed from outburst zones are apt to produce higher volumes of methane per ton due to the porosity and permeability which seem to characterize outburst-prone zones in domal salt mines and perhaps due to scale effects which have not yet been quantified. Iannacchione (5) reported a range of methane emissions of from 10 to 70 cu ft/ton (0.31 to 2.18 m^3/tonne) while continuously mining through an anomalous zone in a domal salt mine. The term "anomalous" has been defined (4) and generally means zones of visibly impure salt which could be defined by field observation.

3.1.4 Trona Mines

Five trona mines are located in the vicinity of Green River, Wyoming. Trona is a naturally occurring sodium sesquicarbonate which occurs in bedded deposits and which is mined by traditional coal mining methods such as room-and-pillar, shortwall, or retreat longwall mining. All five trona mines were classified gassy by the State of Wyoming at about the time when each commenced development and all have operated to date as gassy mines. Typically, the trona seam being mined is sandwiched between two oil shale beds although in places the bed overlying the trona is very lean oil shale, mudstone, or marlstone and shale interlaminated with thin stringers of trona. The beds above and below the trona are known to be sources of methane and appear to liberate the bulk of the methane encountered. However, MSHA studies have shown that methane also emanates to a lesser extent from the face and ribs as well. Methane from the faces and ribs may be associated with relatively thin (up to about 3-in.-thick) bands of oil shale which occur within the trona beds being mined. The MSHA studies estimated that 15 to 20 percent of face area methane liberation originates in thin bands of oil shale that may be present within the trona bed. Measurements made by MSHA during the last 3 months of 1984 showed that the total methane liberation for the five trona mines ranged from 55,000 to 2,071,000 cu ft (1,557 to 58,650 m^3) per 24-hr period.

Thimons (1) reported methane concentrations of up to 0.2 percent in the return air from trona mines. Lumsden (2) reported that the occurrence of larger volumes of gas from a longwall panel at the Alchem Trona Mine substantiated the belief that most flammable gas emanated from the roof and floor. Lumsden further reported that violent outbursts and gas blowers have not been encountered at that mine and that most gas enters the mine by steady-state emission from the roof.

Those trona mines operating in trona Bed 17 routinely drill pressure-relief holes into the mine roof at intersections and at additional regular intervals along entries. The Alchem, FMC, and Tenneco mines drill pressure-relief holes for purposes of roof control. The other two trona mines do not require such pressure-relief holes as gas is reportedly primarily in the mine floor.

A number of ignitions of methane have been experienced in the trona mines and some ignitions have resulted in injuries to personnel. In one accident at the Alchem Mine on June 3, 1977, and accumulation of methane was ignited by a faulty electrical splice during retreat mining of a room-and-pillar panel. Four persons suffered minor burns but reportedly did not require hospitalization.

3.1.5 Potash Mines

The Mississippi Chemical Corp. potash mine was classified gassy on November 4, 1980, as a result of the detection of 0.25 percent or more methane. The Kerr-McGee Chemical Corp. potash mine was classified gassy on February 8, 1981, also as a result of the detection of 0.25 percent or more methane. Both mines are shaft mines located in the Carlsbad, New Mexico, potash district. At the time of the writing of this paper, only the Kerr-McGee potash mine was operating. Future operational status of the Kerr-McGee and Mississippi Chemical Corporation mines could not be projected due to uncertain conditions on world fertilizer markets.

The Carlsbad district is located in a marine evaporite basin, the Southern Permian Basin. The total thickness of evaporites ranges from 1,000 ft (304 m) in the northern mines to 3,000 ft (914 m) in the southern mines. Commercial production of potash has been exclusively from the McNutt Member of the Salado Formation. The Lower Salado Formation extends to at least 400 ft (122 m) below the McNutt and this 400 ft minimum (122 m) of salt reportedly separates the potash mining levels from underlying sandstone and porous carbonate beds that contain oil and gas. The halite in the Upper Salado averages 500 ft (152 m) in thickness. Gases are present, primarily in mudstone and polyhalite seams within the McNutt Member, and the gases are composed primarily of methane and nitrogen (6). When such seams occur immediately above the mine roof it has become standard practice in the Carlsbad district to drill 10- to 30-ft-long, 1-5/8-in. diameter (3- to 9-m-long, 41-mm-diameter) pressure relief holes at all intersections to minimize roof falls.

With the exception of one known sample, all the gas samples collected from roof-relief holes in the Carlsbad potash district have exhibited a gas composition which was

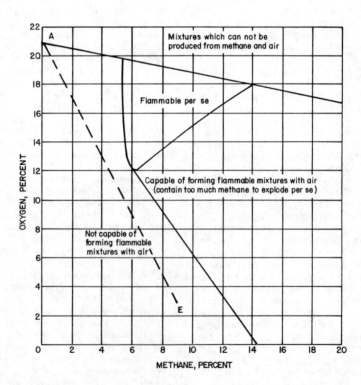

Fig. 1. Relation between quantitative composition and flammability of mixtures of methane, air, and nitrogen (Coward and Jones, 1952).

incapable of forming explosive mixtures in the mine atmosphere. Figure 1 shows an example of the dilution of a typical gas sample from a roof-relief hole. The sample which initially has the composition designated by point "E" dilutes to point "A" without passing through the flammable zone (7).

No ignitions of flammable gas are known to have been documented in writing in regard to the Carlsbad district mines. Nevertheless numerous gas incidents (other than ignitions) including injuries and fatalities have been documented in regard to the Carlsbad district mines. The composition of gas which is commonly released is typically very low in oxygen and high in nitrogen, rendering the gas a simple asphyxiant; at least one fatality is recorded which occurred when such gas was released in a volume sufficient to displace the mine atmosphere in a small area of a potash mine. Over the years other gas-related incidents, variously termed air-blast, blowouts, or outbursts, have caused injuries and fatalities in potash mines. On December 13, 1983, a continuous miner operator was fatally injured and another employee was injured due to an outburst of gas and fragmented rock that occurred when the continuous mining machine cut into a gas-filled fissure zone. Analyses of the gas released showed that although methane was present, the gas was not capable of forming flammable mixtures in the mine air. Although no instances of methane ignitions are known to have been documented in writing, potash operators should remain vigilant and continue routine gas analyses to provide prompt detection if or when gas conditions change.

3.1.6 Petroleum Mines

A few petroleum mines have been or are currently active in the United States. Only one petroleum mine has been classified gassy to date; it is a vertical shaft which is a part of the North Tisdale Gravity Drainage Project. The shaft was classified gassy by the State of Wyoming on May 13, 1981, reportedly about the time when shaft sinking commenced. The shaft reportedly intersected the petroleum bearing formation and petroleum was freely entering the atmosphere in the shaft. This shaft is separate from and not to be confused with the adit described in the next paragraph.

The North Tisdale Project experienced an incident which resulted in the death of one worker on January 25, 1978. On that date, an explosion occurred when a sump round was blasted in an adit on the project. An in-

vestigation of the explosion and fatality indicated the probable misuse of explosives in the presence of crude oil. It was concluded that placing and detonating unconfined, or partially confined explosives, in intimate contact with crude oil while blasting a sump provided conditions conducive to causing an oil-mist explosion. Because none of the criteria of 30 CFR 57.21-1 were met (methane in a concentration of 0.25 percent or more was not detected and there is no provision for an oil-mist explosion) the adit was not classified gassy either by the State of Wyoming or by MSHA. At the present time, the end of the adit which intersected the petroleum bearing formation reportedly has been sealed off and development workings have been extended beneath the petroleum reservoir rock to eliminate seepage of petroleum directly into the mine workings. The present operator reportedly proposes to capture the product in pipe lines which will protect the mine environment from contamination by petroleum. Thimons (1) reported that the methane level in the return airway of a petroleum mine was 115 ppm. Further information on the subject of petroleum mines is contained in reports by Energy Development Consultants Inc. (8) and by Golder Associates (9).

3.1.7 Miscellaneous Mines

Four of the gassy mines shown in table 1 do not lend themselves to a commodity grouping and therefore will be discussed individually under the "miscellaneous mines" heading. The four mines are the Belle Mine, the Calloway Mine, the Lisbon Mine, and the International Salt Co. B-Shaft.

The Belle Mine experienced an ignition of methane which occurred on October 26, 1976. The ignition occurred in a development raise on the 960 west level of the mine and two raise miners received second- and third-degree burns as a result of the ignition. The ignition occurred when an accumulation of methane was ignited when one of the workers struck a cigarette lighter (another source says a spark from a scaling pick ignited the methane). As a result of the ignition, the mine was classified gassy by MSHA on November 9, 1976. The Belle Mine had operated since about 1937 and had experienced three ignitions prior to the ignition which resulted in the gassy classification. Little information is available on the source of methane, the emanation mechanism or the emanation volumes. All the ignitions have reportedly occurred in blind raises which were not

339

thoroughly ventilated prior to commencing work. Thimons (1) reports methane concentrations ranging from 1 to 783 ppm in the returns of gassy and nongassy limestone mines.

The Calloway Mine experienced an ignition of methane on March 17, 1958. The ignition occurred in and about the base of a 50-degree inclined raise which extended 63 ft (19 m) above a sublevel drift. The sequence of events leading up to the accident began when gas emanating from blast holes being drilled at the face of the raise became noxious and the two raise miners exited the raise and remained in the subdrift for about an hour while the raise was ventilated with compressed air. The air was turned off and one miner reentered the raise. At that time an ignition occurred in the drift and raise. One miner died after travelling a short distance and the other miner was injured as a result of the ignition. No Federal gassy metal/nonmetal mine classification standards existed at the time but one source states that the mine was classed gassy by the State of Tennessee at the time of the ignition. Once applicable Federal standards were promulgated, the Calloway Mine was classified gassy by MSHA on about February 8, 1971, based on the ignition of flammable gas which occurred in 1958.

Methane liberated into the Calloway Mine is sometimes accompanied by a recognizable foul odor which normally alerts personnel to leave the immediate area and to take corrective action to remove the gas. Methane gas seems to be associated with waste rock in both the footwall and hanging wall which consist of metasediments including metagreywackes and schists. The vein material consists of iron and copper sulfides which do not contribute any methane. The waste rock minerals are almost exclusively silicates, although some carbonates and a very small amount of graphite are found. The graphite is reported to be the apparent source of methane and methane is most often released during the drilling of boreholes; release from "feeders" or "blowers" has not been reported. There are no coal measures within 50 miles (80 km) of the mine. Methane was first detected in 1949 during drilling an exploratory diamond drill hole horizontally into the footwall on 16 level. MSHA gas analyses of Calloway Mine return air in 1973 showed a methane concentration of 0.002 percent. The Calloway Mine ceased operating in about May or June of 1983 due to instability of the shaft pillar of the main production and service shaft (B-Shaft).

The Lisbon Mine experienced an ignition

of methane on April 23, 1973. The ignition occurred in a 10-ft by 12-ft by 23-ft-high (3.0-m by 3.7-m by 7.0-m-high) chamber which had been excavated adjacent to a 9-ft-high (2.7-m-high) haulageway. The chamber had been provided as a target for a proposed large-diameter borehole being drilled down from a higher level; the 23-ft (7.0-m) height was needed to provide space to attach a reaming bit to the drill string once the pilot hole intersected the chamber. Methane apparently accumulated in the chamber and was ignited by a worker using an oxyacetylene torch. The worker was burned about the hands and face and required hospitalization. The mine was classified gassy on April 24, 1973, as a result of the detection of 0.25 percent or more methane on the day following the ignition.

Other incidents involving the release of methane have occurred at the Lisbon Mine since 1973. The most notable incidents were methane and rock outbursts which occurred in March, April, and May of 1979. The largest outburst occurred on March 29, 1979; the second and third outbursts occurred on April 6, 1979, and May 17, 1979, respectively. The outbursts occurred in development headings at the time the headings were blasted. The March 29 outburst reportedly resulted in ejection of an estimated 250 tons (227 tonnes) of pulverized sandstone. Mining crews detected 3.5 percent methane upon entering the area following the outburst. No further data were recorded regarding the April 6 outburst. The May 17 outburst reportedly expelled 40 tons (36 tonnes) of material. MSHA investigators surmised that natural gas from oil and gas bearing formations beneath the Lisbon mine orebody, migrated along nearby faults to the contact between the Chinle and Cutler Formations. The trapped gases remained there under extreme pressure until released by development blasting. No persons were reported injured in any of the outburst incidents described above.

The International Salt Co. B-Shaft (formerly called the Sterling B-Shaft) experienced an ignition of flammable gas on April 18, 1975. The B-Shaft had been and still is separated from the present day Retsof Salt Mine and is not utilized in ventilation of the Retsof Mine. The ignition in the 1,100-ft-deep (355-m-deep) B-Shaft occurred while activities were underway relative to correcting a water quality problem. The International Salt Co. had initiated a program to properly dispose of brine wastes in response to an order from the New York State Dept. of Environmental Conservation. Activities underway at the time of the accident were aimed at removing debris from the B-Shaft.

Flood lamps which had been lowered into the shaft for illumination most likely ignited flammable gases when the lamps were broken by falling debris. As a result of the ignition, three persons were killed and another died on May 5, 1975. In addition, a number of persons were injured. Those persons killed or injured were gathered about the collar of the shaft and were hit by flying or falling objects.

The B-Shaft was not classified gassy at the time because no further activity was planned in B-Shaft. Recently, however, activities have been underway in B-Shaft and in the Sterling Mine related to mine dewatering and sump development and the affected areas of B-Shaft and the Sterling Mine were classified gassy by MSHA under the provisions of 30 CFR 57.21-2. MSHA memoranda issued to International Salt Co. on September 5 and 19, 1984, detailed the extent of the shaft area affected. The 57.21-2 classification was based on the 1975 ignition and on the more recent detection of methane in excess of 0.25 percent. Standard 57.21-2 provides for flammable gases detected during mine reclamation or mine dewatering and is not intended to provide a permanent gassy classification.

3.2 Methane Liberation

Figure 2 was prepared from data on gassy mines by David M. Hyman of the U.S. Bureau of Mines who granted permission to present the figure herein. As may be noted in the legend, methane emission data are presented by emissions from sections of mines, by total annual emissions, and by methane release during outbursts. The figure makes it possible to compare methane emanation data for metal and nonmetal mines with similar data from coal mines. Perhaps the most significant information to be gleaned from figure 2 is the extreme variability of methane emission from any given commodity group.

Generally, the greatest volume release of methane per ton shown in figure 2 is for outbursts and for annual basis emissions. Annual basis emissions are computed by dividing the volume of methane released in one year of mining by the tonnage mined in a year. Such an annual basis emission will seem high because it includes all the steady-state methane emission from roof, ribs, faces, and floor of the entire mine in addition to the methane emission from development and extraction areas. Such figures are useful, however, because they

provide values which may be used for mine ventilation system designs.

Face-area methane liberation figures are extremely useful in the design of the face ventilation systems necessary to handle the methane liberated during face blasting or during continuous mining. However, such face-area data are frequently difficult to obtain and generally are developed only during special ventilation studies made with specialized equipment in face areas. Some such data are shown in figure 2 under mine-section emissions.

The diagonal lines in figure 2 are lines of equal methane emission per ton of commodity. Mines which lie along the same diagonal line are characterized by the same emissions of methane per ton of commodity. Figure 2 is by no means a final summary of the methane liberation volumes which may occur in mines in general or in specific mines; persons applying such data should exercise due diligence in view of the extreme unpredictability of methane liberation in mines.

4. CONCLUSIONS

During the preparation of this paper, it became apparent that certain types of accidents have occurred with sufficient frequency to draw attention to preventing reoccurrence. It follows that present or future safety regulations must address certain mining-related activities if worker safety is to be assured.

Welding and cutting are examples of activies which continue to cause ignitions and it is apparent that such activities need to be conducted in a manner which will prevent future ignitions. Cigarettes, lighters, and other smoking materials have often been implicated as a source of methane ignition and smoking practices and materials require controls. A number of ignitions have occurred in raises where methane was released and accumulated in a high spot; specific operating practices are needed to assure safe operations in such locations. Electrical equipment failure, nonpermissible equipment, or improperly maintained equipment have been responsible for ignitions of methane and need to be guarded against. Improper blasting practices or inappropriate blasting materials have ignited methane and volatile dusts and proper practices need to be developed to prevent such occurrences. When ignitions occur, frequently there has been inadequate mine ventilation in addition to the improper practices already described, and adequate ventilation of working areas needs to be assured. In addition to those precautions already recommended, there is a need to specify the instruments and procedures which will assure early detection of methane so that hazardous accumulations can be dealt with promptly. The mining activities discussed above are by no means the only activities which need to be controlled and are presented as typical examples of activities which in the past have been closely associated with the ignition of methane.

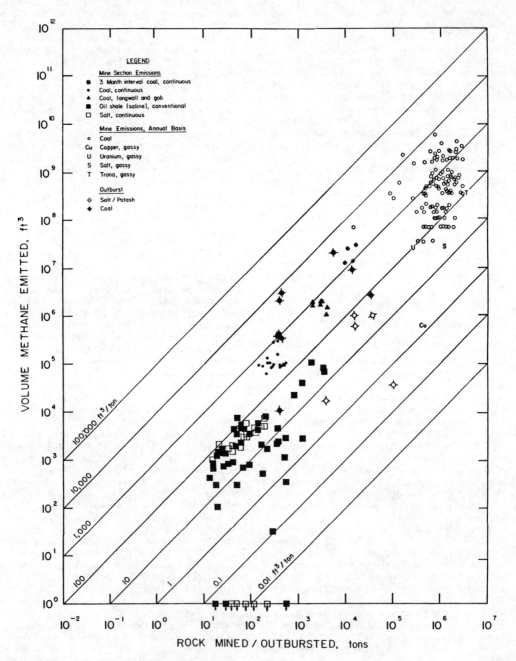

Fig. 2. Methane volume emitted per ton mined. (Courtesy of David M. Hyman, U.S. Bureau of Mines)

5. REFERENCES

1. Thimons, E. D., Vinson, R. P., and Kissell, F. N. Forecasting Methane Hazards in Metal and Nonmetal Mines. BuMines RI 8392, 1979, 9 pp.

2. Lumsden, A. M., and Talbot, R. Develop Data for Review of Metal and Nonmetal Mine Methane Hazard Classification. BuMines Contract J0100060 by Golder Associates Inc., 1983, 518 pp.

3. Plimpton, H. G., Foster, R. K., Risbeck, J. S., Rutherford, R. P., King, R. F., Buffington, G. L., and Traweek, W. C. Final Report of Mine Explosion Disaster. MSHA Accident Report, June 8, 1979, 135 pp.

4. Schatzel, S. J., and Hyman, D. M. Methane Content of Gulf Coast Domal Rock Salt. BuMines RI 8889, 1984, 18 pp.

5. Iannacchione, A. T., Grau, R. H., Sainato, A., Kohler, T. M., and Schatzel, S. J. Assessment of Methane Hazards in an Anomalous Zone of a Gulf Coast Salt Dome. BuMines RI 8861, 1984, 26 pp.

6. Griswold, G. B. Geology of the Carlsbad Potash Mining District. Exhibit J of Comments and Recommendations of the Carlsbad Potash Industry to the Mine Safety and Health Administration on the Gassy Mine Classification Procedure and Safety Standards. Presented in Carlsbad, New Mexico on March 31, 1982.

7. Coward, H. F., and Jones, G. W. Limits of Flammability of Gases and Vapors. BuMines Bull. 503, 1952, 144 pp.

8. Energy Development Consultants Inc. Mining For Petroleum: Feasibility Study. BuMines Contract No. J0275002 (Technical and Economic Feasibility Study of Oil Mining), July 1978, 361 pp.

9. Golder Associates. Oil Mining: A Technical and Economic Feasibility Study of Oil Production by Mining Methods. BuMines Contract No. J0275018, 1978, 317 pp.

An overview of coal and gas outbursts

ALAN A.CAMPOLI, MICHAEL A.TREVITS & GREGORY M.MOLINDA
US Department of the Interior, Bureau of Mines, Pittsburgh, PA, USA

ABSTRACT: An outburst in a coal mine environment is the rapid ejection of coal and gas from the face. Outbursts are caused by a combination of high coalbed gas pressure and structural stress, the load placed upon mine workings by the overburden. A review of the outburst problem in general is followed by a presentation of prediction and prevention techniques employed worldwide to mitigate outbursts.

1 INTRODUCTION

An outburst in a coal mine environ- ment is the rapid ejection of coal and gas from the face. Outbursts result in the formation of a cavity ahead of or to one side of the working place, usually at or near the top of the coalbed. During an outburst event, coal is pulverized and large quantities of gas (predominantly methane and/or carbon dioxide) are emitted, usually followed by a rapid reduction in the gas emission rate with time (Shepherd, et al 1979).

Outbursts can vary in magnitude, displacing from a few to thousands of tons of coal, with the volume of gas emitted generally equal to the gas content of the coal disturbed. For example, on October 16, 1981, at the Yubari-Shin Colliery in Japan, an outburst ejected 4000 m^3 of coal from the face area and 600,000 m^3 of methane gas, resulting in 93 fatalities (Sato 1982). This is not to say that large quantities of coalbed gas cannot be released into the mine workings without coal displace- ment. Such an event could be caused by the penetration of an impermeable strata, a clay vein for instance.

2 OUTBURST CONDITIONS

The thickness of overburden and the magnitude of the resultant vertical stress exerted on the coalbed are major factors relative to outbursts. Over- burden thicknesses in excess of 400 m with methane as the primary coalbed gas constituent and in excess of 150 m with carbon dioxide as the dominant gas are considered sufficient to produce coal and gas outbursts (Kolmsov & Bolsminskii 1981; Kidybinski 1980; Lama 1968; Leighton 1978).

Coalbed gas content is also a major factor in outbursts. A coalbed with a methane gas content greater than 9 m^3/mt is considered coal and gas outburst prone (Kowing 1981; Paul 1981; Janas 1979).

The presence of geological anomalies, such as fault zones, clay veins, or igneous intrusions, greatly increases the probability of an outburst event. A structurally weak section in a coalbed, combined with high gas content and pressure, could present an outburst hazard. An anomaly need not be major; very often the anomaly is not noted until detailed postoccurrence investigations are conducted (Thomas 1962). It has been observed that even insignificant tectonic disturbances can cause outbursts in gassy coalbeds (Kowing 1981).

Outburst frequency and magnitude have been compared in three different mining situations in Hungarian coal mines. Outburst frequency and magnitude were reported lowest at longwall faces, of moderate intensity in development head- ings, and greatest in cross-measure drift workings as a coalbed is approached (Shepherd, et al 1981).

Australian research indicates that coal rank plays an important role in out- bursts. Low- to medium-volatile bitum- inous coals have been identified as especially prone to outbursts (Shepherd, et al 1981). These outburst conditions are summarized in table 1. Additionally, the compressive strength, porosity, pore size distribution, gas desorption characteristics, and fracture density of the coalbed are outburst related characteristics (Kowing 1981).

upon the limited data compiled to date, it appears that more than the few coalbeds presently known to have a demonstrated propensity for outbursts will be mined in the future. Table 1 lists considerations in addition to gas content and overburden

TABLE 1. - Worldwide Summary of Outburst Conditions

Parameter	Characteristic Range
Overburden	>400 m with CH_4 dominant. >150 m with CO_2 dominant.
Gas content	>9 m^3/mt.
Geologic anomalies	Fault zones, clay veins, igneous intrusions, extreme fracture density.
Mining method	Cross-measure drift workings, longwall gateroad development.
Coal rank	Low to medium volatile bituminous.
Coal strength	Structurally weak coalbeds or weak areas within an outburst-prone coalbed.

All of the parameters listed in table 1 do not have to be present to cause an outburst. The characteristic ranges presented represent the opinions of many researchers around the world. They are not absolute numbers, but do provide a valid indication of the general conditions causing outbursts.

3 OUTBURSTS AND U.S. COAL MINES

About of 17,000 outbursts have been documented around the world, causing hundreds of fatalities (Jackson 1984). Outbursts are rare occurrences in U.S. coal mines or are poorly documented. However, with mining operations being conducted under increasing overburden thicknesses, the problem is likely to become more prevalent. Table 2 lists U.S. coalbeds meeting the overburden (>400 m) and gas content (>9 m^3/mt) criteria (LaScola 1983). Based

thickness that may determine whether an outburst hazard may be present.

One U.S. mine with a known outburst hazard is the Dutch Creek No. 1 Mine operating in the Coal Basin "B" Coalbed near Redstone, CO. On April 15, 1981, a methane and coal dust explosion occurred. Fifteen miners died as a result of the explosion (Elam, et al 1981). Investigators concluded that a coal and gas outburst had occurred, releasing large amounts of methane and fine coal, creating an explosive methane air and coal dust mixture. This mixture was ignited by a faulty lighting switch on a continuous mining machine (Elam, et al 1981). Outbursts occurred frequently during mining, especially on development sections. According to statements of the miners and mine officials, outbursts occurred as often as two or three times per shift. They described these events as coal and methane gas "flowing" rapidly from the face into the workings. The

TABLE 2 - Potential Outburst Coalbeds in the U. S.

State	Coalbed group or formation	Coalbed	County where sampled	Number of samples	Average gas content, m³/mt	Average depth of overburden, m
Alabama	Black Creek	Black Creek	Tuscaloosa	3	12.9	805
		Jefferson	Tuscaloosa	4	11.7	785
		Lick Creek	Tuscaloosa	2	12.7	905
	Mary Lee	Blue Creek	Tuscaloosa	2	13.2	795
		Mary Lee	Tuscaloosa	11	13.1	630
		New Castle	Tuscaloosa	1	17.5	650
	Pratt	American	Tuscaloosa	2	9.7	565
		Gillespy	Pickens	1	9.4	510
		Pratt	Tuscaloosa	4	11.7	535
	Cobb	Cobb	Tuscaloosa	2	11.0	505
Colorado	Mesaverde	Cameo	Mesa	4	10.5	840
		Undifferen- tiated	Mesa	1	11.6	1,460
	Williamsfork	Anderson	Garfield	4	10.6	1,015
		Coal Basin B	Pitkin (4)	--	--	825(4)
		Undifferen- tiated	Moffat	6	10.0	1,425
	Fruitland	Undifferen- tiated	La Plata	5	12.7	860
	Vermejo	Undifferen- tiated	Las Animas	3	12.5	480
Oklahoma	Hartshorne	Hartshorne	Le Flore	2	16.5	440
Utah	Sunnyside	Kenilworth	Emery	1	9.8	750
			Carbon	1	11.0	970
	Spring Castle	Castlegate A	Carbon	1	9.4	810
		Castlegate C	Carbon	1	10.6	1,000
Virginia	Pocahontas	Pocahontas No. 3	Buchanan	24	15.3	550
West Virginia	New River	Beckley	Raleigh	1	10.8	370

TABLE 3. - Worldwide Summary of Prediction Methods

Test boring	Volume of cuttings measurements for stress identification. Desorption analysis of cuttings. Adsorbtion analysis of cuttings. Gas pressure monitoring. Gas emission rate analysis.
Microseismics	Passive systems. Active systems.
Geologic investigations	In situ point load coal strength tests. Coal jointing evaluation for fault prediction. Reconnaissance drilling for tectonic irregularities. Induced fracture recognition.

majority of these outbursts discharged 30 to 40 mt of coal. But there were outbursts that expelled enough coal from the face to cover the continuous mining machine and the shuttle car behind it. The volume of methane gas accompanying these outbursts varied considerably in quantity. At times hazardous methane gas concentrations were found in intake aircourses, 30 m or more outby the faces, indicating that sufficient quantities of methane gas were released to overpower the intake air current (Elam, et al 1981).

4 PREDICTION METHODS

The methods used to predict outbursts can be classified as test boring, micro-seismic, and geologic investigations. Table 3 provides a summary of these general methods; application may vary according to geologic setting. It is suggested that no one method be relied upon for full prediction accuracy and that many or all of the methods be considered.

4.1 Test Boring

Drilling test boreholes in advance of the mine face increases the probability of identifying hazard areas prior to mining. Techniques here include observa-tion of the volume of cuttings produced, gas desorption and adsorption from cuttings, gas pressure, and borehole gas emission rate. Measurement of cuttings volume from test boreholes has been used to predict highly stressed areas in advance of the face (Davies 1980). Drill cuttings far in excess of the volume of the borehole being drilled indicate potential hazard areas.

Borehole gas emission rate analysis has shown promise in outburst prediction (Matuszewski & Kozlowski 1979). A reduction in incremental gas flow per unit depth indicates the intersection of a highly stressed compressed zone, while a large increase in incremental gas flow indicates a crushed zone. Both observations point to a possible outburst zone.

Gas pressure is one of the major factors in outbursts. Two methodologies have been described in the literature. One method measures gas pressure directly, with in-hole instrumentation (Matuszewski & Kozlowski 1979). The other establishes gas pressure indirectly

by measuring the desorption of drill cuttings (Paul 1981). Desorption testing involves indirect measurement of the in situ gas content, where a sample is enclosed in a sealed container and the gas released is measured. Analysis of these measurements can provide estimates of the original coalbed gas content and pressure.

Adsorption indices are a measure of the gas storage capacity of the coal. Laboratory samples are exposed to gas under pressure, and the amount of gas adsorbed is measured. Changes in gas capacity may be a function of changes in maceral composition, internal surface area, mining-induced strain, or geologic structure in the sample area (Williams 1982).

4.2 Microseismics

When a rock mass is subjected to changing structural stress conditions and/or gas pressure, movement within the strata releases seismic energy. This seismic energy is "rock noise" when it is within humanly audible frequencies. Rock noise has long been a warning to miners of impending danger. Microseismic monitoring of subaudible seismic energy was first developed by the Bureau of Mines to predict rock bursts in hard rock mines in the 1940's (Blake 1974). Systems can be passive (listening only to rock-gener-ated noise) or active (listening to artificially generated sonic pulses). Single-channel, one-geophone systems are used to measure the relative amount of noise, while multichannel systems are used to locate each rock noise source and to compile rock noise rates for stability estimates.

Research has shown that in-seam seismic techniques are capable of detecting a 0.9-m fault in a 2.7-m seam at a range of 260 m with a resolution of plus or minus 40 m (Greenhalgh & King 1980). Acoustic emissions activity has been recorded prior to and associated with several outbursts (Leighton 1978). Results indicate that outbursts are comprised of several failures in rapid succession and that anomalous microseismic emissions are generated 15 to 45 min prior to failure. Researchers claim that microseismic technology has demonstrated its potential as an outburst prediction technique, but state that further work is necessary to achieve acceptable reliabil-ity.

4.3 Geologic Investigations

In situ point-load strength testing has been employed as an outburst prediction tool. A portable cone penetrometer was developed which measures the distance a pointed cone can be forced into the coal face under a uniform force. Variations in the depth penetrated provide a relative indicator of coal strength. Weak coal zones have been correlated with outburst occurrences (Kidybinski 1980).

A forewarning of strike-slip faults in Australia has been obtained by measuring the increased frequency of fractures (cleat) in the coalbed. In the mine studied, outbursts only occurred close to such faults. This method did not predict the precise location of an outburst, but did give warning of increased hazard as far as 40 m in advance of mining (Shepherd & Creasy 1979). A certain amount of induced jointing in the coalbed is expected, given a set of overburden, mining method, and in situ coal strength conditions. An increase in jointing above that normally induced by mining has also been associated with outburst episodes (Hanes & Shepherd 1981).

5 PREVENTION METHODS

Mining operations at the Yubari-Shin Colliery were suspended indefinitely following the outburst disaster (Hirota, et al 1983). The closing of outburst mines is a radical last step that can be avoided, through the development of outburst prevention methods. Table 4 summarizes the outburst prevention methods under development worldwide. The basis of these prevention methodologies is to reduce the structural stress and/or the gas pressure values to below the outburst threshold. Their application may vary in effectiveness from coalbed to coalbed. All of these procedures can be used concurrently with mining operations and should be employed with extreme caution.

5.1 Mine Planning

Longwall mining has been demonstrated to be less hazardous than narrow entry development (Davies 1980). But sharp corners, less than right angles, on longwall headings and pillars in room and pillar mining cause concentration of stress due to a reduction in load-bearing area and should therefore be avoided.

Protective seam mining is a technique employed in multiple-seam mines. The mining of a coalbed in close proximity to an outburst-prone coalbed can greatly reduce the danger of outbursts. The protective seam, the one least liable to outburst, is mined first. It has been found in France that the relieving effect can extend to 55 m below and 90 m above the protective seam (Thomas 1962). The relieving effect is a combination of gas drainage and stress relief in the area of the outburst-prone coalbed.

5.2 Gas Drainage

Premining gas drainage is an effective

TABLE 4. - Worldwide Summary of Prevention Methods

Method	Technique
Mine planning	Mining method selection. Protective seam mining.
Gas drainage	Horizontal borehole systems. Cross-measure borehole systems. Vertical borehole systems.
Stress relief	Large-diameter borings. Slot cutting. Water infusion. Inducer shot firing.

method for reducing or eliminating outbursts. Whether accomplished through horizontal boreholes, cross-measure boreholes, or vertical boreholes, gas drainage in advance of mining can reduce coalbed gas pressure. Boreholes drilled from within the mine, whether horizontally into the coalbed or inclined into the adjacent strata, can provide a mechanism for gas removal ahead of the working face.

5.3 Stress Relief

Stress relief techniques serve to reduce outburst potential, by reducing the ability of the immediate face to store the energy necessary for outbursts. The fracturing of the immediate face insures that high gas pressures or structural stress conditions do not exist in the next cut. Structural stress is shifted to the solid coal inby the face. Gas pressure is reduced by coalbed gas flow through the induced fracture system. Stress relief boreholes, 95 to 140 mm in diameter, are drilled into the face to mechanically destress the coalbed by removing large volumes of drill cuttings (Paul 1981). Stress relief slot cutting on longwall faces is reported to be highly successful for reducing outbursts (Kolmsov & Bolsminskii 1981). Relaxation slots are systematically cut from 0.2 to 1 m deep into the face, destressing the immediate face area.

Water infusion, the injection of water at high pressure (>703,000 kg/m^2), has been explored as a stress relief mechanism to prevent outbursts (Kuschel 1973). The water pressure fractures the coal in advance of the face and provides for slow relaxation of stress through lateral movement enhanced by the presence of the water. However, the effectiveness of water infusion decreases as the depth of cover increases, owing to the reduction in coalbed permeability with depth.

Inducer shotfiring can be very effective in reducing outburst hazards. Blast holes are drilled into the face and loaded with charges (Davies 1980). The charge delivers a short intense shock wave to the strata, which may trigger an outburst. Stress relief is accomplished whether an outburst occurs or not by the fracturing of the blast area, transferring high stress concentrations to an area farther in advance of the face. Danger from outbursts occurring under such circumstances is minimal since mine personnel and equipment are evacuated from the area.

6 SUMMARY

Coal and gas outbursts affect only those few mines in the United States that presently operate under extreme overburden depth and coalbed gas pressure. The problem is likely to become more widespread as mining is conducted at greater depths. Various methods have been developed to predict the occurrence and minimize the effects of outbursts.

7 REFERENCES

Blake, W. "Microseismic Techniques for Monitoring the Behavior of Rock Structures," U.S. Bureau of Mines, Bulletin 665, 1974, 65 pp.

Brauner, G. "Recognition and Removal of High Rock Pressures in the Seam," Gluckauf, v. 109, No. 23, Nov. 8, 1973, pp. 1133-1136.

Davies, A. W. "Available Defenses Against Outbursts in the United Kingdom in 1980," The Aus. I.M.M. Southern Queensland Branch, The Occurrence, Prediction and Control of Outbursts in Coal Mines Symposium, Sept. 1980, pp. 203-215.

Elam, R., C. Lester, and A. O'Rourke. "Report of Investigation, Underground Coal Mine Explosion, Dutch Creek No. 1 Mine - I. D. No. 05-00301," Mine Safety and Health Admin., 1981, 55 pp.

Greenhalgh, S., and D. King. "In-Seam Seismic Methods for the Prediction of Outbursts in Coal Seams," The Aus. I.M.M. Southern Queensland Branch, The Occurrence, Prediction and Control of Outbursts in Coal Mines Symposium, Sept. 1980, pp. 139-150.

Hanes, J., and J. Shepard. "Mining Induced Cleavage, Cleats and Instantaneous Outbursts in the Gemini Seam at Leickhart Colliery Blackwater, Queensland," Proc. Australian, Inst. Min. Metall., No. 277, March 1981, pp. 17-26.

Hirota, T., Y. Watanabe, and M. Seismo. "Disaster of Coal and Gas Outburst at Yubari Shin Colliery," 20th Int. Conf. of Safety in Mines Research Institutes, Sheffield, United Kingdom, Oct. 3-7, 1983, 14 pp.

Jackson, L. "Outbursts in Coal Mines," International Energy Agency, Feb. 1984, 105 pp.

Janas, H. "Improved Method for Assessing the Risk of Gas/Coal Outbursts," Second International Ventilation Congress, Reno, NV, Nov. 4-8, 1979, pp. 22-27.

Kidybinski, A. "Significance of In Situ Strength Measurements for Prediction of Outburst Hazard in Coal Mines of Lower Silesia," The Aus. I.M.M. Southern Queensland Branch, The Occurrence, Prediction and Control of Outbursts in Coal Mines Symposium, Sept. 1980, Australia, pp. 193-201.

Kolmsov, O. A., and M. I. Bolsminskii. "The Development of Methods of Dealing with Sudden Outbursts of Coal and Gas at Mines in the Donesk Basin," XIX Int. Conf. on Mine Safety Research, Oct. 6-14, 1981, Katowice, Poland, 8 pp.

Kowing, K. "Geological Preconditions for Gas Outbursts in Hard Coal Mining," Gluckauf, July 9, 1981, pp. 340-346.

Kuschel, K. H. "Measures for the Prevention of Rock Bursts," Gluckauf, v. 109, No. 23, Nov. 8, 1973, pp 1136-1139.

Lama, R. D. "Outbursts of Gas and Coal," Colliery Engineering, March 1968, pp. 103-109.

LaScola, J. C. "Computer-Implemented Coalbed Methane Data Base," Ch. 29 in Proc. 1st SME-AIME Conf. on Use of Computers in the Coal Industry, Aug. 1-3, 1983, pp. 251-257.

Leighton, F. "Microseismic Activity Associated with Outbursts in Coal Mines," U.S. BuMines, 1978.

Matuszewski, J., and B. Kozlowski. "Control of Gas and Rock Burst Hazard Demonstrated by the Example of the Bituminous Coal Mine 'Nova Ruda,'" XVIII Int. Conf. of Mining Safety Research, Cavtat, Yugoslavia, Oct. 8-12, 1979, 5 pp.

Paul, K. "Further Development of Methods for Predicting and Preventing Gas Outbursts," Gluckauf, July 9, 1981, pp. 334-337.

Sato, K. "Abstract of the Report on Gas Outburst Disaster at Yubari New Colliery," Coal Mining Research Centre, Japan, 1982, 7 pp.

Shepherd, J. R., and J. H. Creasy, "Forewarning of Faults and Outbursts of Coal and Gas at Westcliff Colliery, Australia," Colliery Guardian Coal International, Oct. 1979, pp. 13-22.

Shepherd, J., L. K. Rixon, and L. Griffiths. "Outbursts and Geological Structures in Coal Mines: A Review," Int. J. Rock Mech. Min. Sci. and Geomech., v. 18, 1981, pp. 267-283.

Thomas, D. A. "Instantaneous Outbursts of Coal and Gas," Proceedings of the National Association of Colliery Managers, Sept. 25, 1962, pp. 25-32.

Williams, R. J. "Prediction Techniques Currently in Use and Being Researched in Australian Coal Mines," The Coordinating Committee on Outbursts Related Research, May 1982, 73 pp.

Characterization of the occurrence of methane

in Gulf Coast domal salt mines

ANTHONY T.IANNACCHIONE & STEVEN J.SCHATZEL
US Department of the Interior, Bureau of Mines, Pittsburgh, PA, USA

ABSTRACT: Within the last 10 years, faster extraction rates, deeper mining levels, and "complete" mining of several salt dome levels contributed to the increased methane emission hazards, especially those associated with outbursts. The U.S. Bureau of Mines is actively involved in identifying the parameters governing these hazardous conditions, which may lead to a methane explosion. These parameters are better understood when the ranges of methane contents, emission rates, and the gas pressures are known.

Data obtained in Louisiana domal salt mines have shown that variations in methane contents, emission rates, gas flows, and pressures are dependent upon the type of salt encountered. Normal production-grade pure salt was found to have a methane emission rate of less than 0.1 m^3/t and methane contents generally less than 0.1 $cm^3/100$ g. Emission rates from an adjacent anomalous zone ranged from 0.2 to 2.7 m^3/t and methane contents as high as 2.6 $cm^3/100$ g were determined. Extremely low gas flows and lack of pressure buildup were typical during drilling of exploration boreholes in normal salt. Gas flows of 0.33 l/sec and pressures of more than 6.2 MPa were observed from two boreholes when an anomalous zone was encountered. Localized permeability within this zone was observed by measuring tracer gas migration between two drill holes 7.3 m apart. Methane hazard characterization of a localized area within a Gulf Coast salt mine should consist of the following: (1) methane content determination, (2) mapping of geological features, (3) gas emission surveys, and (4) in situ tests of physical properties related to gas.

1 INTRODUCTION

Although methane has long been recognized as a hazard in coal mining, the danger it poses in domal salt mines has become apparent only recently due to the explosion at the Belle Isle salt mine (Plimpton, et al 1980) and the subsequent Mine Safety and Health Administration proposed gassy mine regulations for domal salt mines. Throughout the history of mining salt in the Gulf Coast region, hazardous methane occurrences have been recognized only infrequently. The inconsistency of the methane occurrences has detracted from the seriousness of the hazard and, consequently, hindered the development of control strategies and procedures. Analogies can be drawn with the early coal mining experience in the United States. Prior to passage of the Federal Coal Mine Health and Safety Act of 1969, coal mines could be classified as either gassy or nongassy. However,

history has shown that some of the worst methane-induced coal mine explosions have occurred in mines originally classified as "nongassy" (U.S. Senate 1969).

Recently, deeper mining levels and higher extraction rates have contributed to increases in the frequency of hazardous methane occurrences. These occurrences have sometimes been manifested as violent outbursts of methane and salt. Outbursts are generally associated with blasting, but they occur only in a small fraction of the total number of faces. However, these hazardous methane occurrences have often happened in groups and are sometimes associated with specific zones within several of the domal salt mines.

2 CHARACTERIZATION OF THE METHANE HAZARD

2.1 Methane Content Testing

One element of the Bureau of Mines

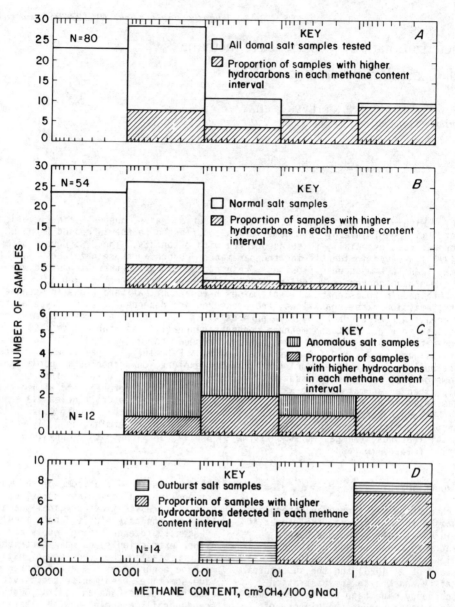

Figure 1. Methane content distribution with the number of samples containing higher hydrocarbons in each interval for each salt type.

methane control program for salt mines was the development of a test to quantify the methane content of solid salt samples. Qualitative data show that the distribution of methane in a salt dome is nonuniform. A classification system of salt "types" was sought to delineate relatively high-methane enriched salt from salt with lower methane contents. Three types of salt were described to

facilitate this classification:

Normal salt - White to gray, transparent to translucent, very low in impurities, commonly with grain sizes of 5-15 mm.
Anomalous salt - A relatively continuous matrix of striated crystalline salt with argillaceous and/or arenaceous material incorporated into an

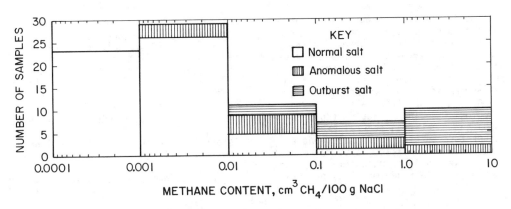

Figure 2. Salt type composition for each methane content interval.

opaque mass of rock. Often brine and gas seeps are associated with edge and central anomalous zones (Kupfer, 1980). This salt type also includes lithologies such as sandstone breccias and sylvite bands, which can be relatively continuous or may be discontinuous and pinch out along strike distances of 30 to 70 m. Some other impurities included are potassium salts, anhydrite, gypsum, and quartz.

Outburst salt - The essential criterion for all outburst salts is that they are physically associated with an outburst. Outburst salts were sometimes found to have enlarged and/or randomly oriented and striated crystals with various degrees of intercrystalline gas bubbles. However, the same characteristics were sometimes observed in areas not experiencing outbursts.

Methane content distributions are different for various salt types. Normal salt samples tested ranged from 0.0003 cm^3 CH4/100 g NaCl [the estimated lower limit of experimental resolution (Schatzel & Hyman, 1984)] to 0.31 cm^3 CH4/100 g NaCl. Anomalous salt samples range from 0.003 to 2.6 cm^3 CH4/100 g NaCl; outburst salt ranges from 0.014 to 7.4 cm^3 CH4/100 g NaCl (fig. 1)

Most of the anomalous salt samples presented in this paper were from a thick continuous zone of argillaceous salt from a domal salt mine. This zone had been associated with releases of methane and brine and had functioned as a migratory pathway for the transport of fluids. An increase in the percentage of higher order hydrocarbons other than methane was associated with this zone of anomalous salt. The occurrence of measurable amounts of higher hydrocarbons in gas

samples collected from methane content testing may therefore act as an indicator of hazardous gas and brine emissions (fig. 1).

Outbursts in salts have been associated with rapid emissions of large quantities of methane (Plimpton, et al 1980). Although many unique physical features can be found in outburst salt, many test samples were found to be essentially indistinguishable in appearance from normal salt. The methane content distribution for outburst and normal salts is shown in figure 2. Over 90 pct of the normal salts contained less than 0.01 cm^3 CH4/100 g NaCl, and all but one sample contained less than 0.1 cm^3 CH4/100 g NaCl. These 54 samples were collected from various locations in the Jefferson Island, Weeks Island, Cote Blanche, Avery Island, and Bell island salt mines. All outburst salts contained more than 0.01 cm^3 CH4/100 g NaCl, and all but 2 of the 14 outburst salt samples included in this study contained more than 0.1 cm^3 CH4/100 NaCl. This distribution shows that methane content testing can provide information for distinguishing the more hazardous potential outburst salts from the generally nongassy normal salts.

It should be noted that extreme ranges of gas contents for each salt type do overlap (fig. 2). For this reason, multiple samples collected in a statistically valid fashion can best verify the local gas content characteristics of the salt.

2.2 Geologic Mapping

Before a methane control plan can be formulated, the geology of problem areas must be identified. A gassy zone can be identified by geologic mapping and an exploration drilling program. Identifying salt rock types is important because previous Bureau research has shown the association of high methane content and methane emissions data (Iannacchione, et al 1984) with specific salt rock types.

Geologic mapping by Kupfer (Kupfer, 1980) has identified several examples of outbursts occurring in or near continuous zones of anomalous salt. Bureau of Mines research found one anomalous zone that emitted 80 times more gas per unit volume than the adjacent normal salt. This zone of anomalous salt was distinguished from the much more common normal salt by the occurrence of fluid flows, predominantly in the form of methane gas and brine, and the increased amounts of clastic material causing the overall salt mass to appear dark in color. The anomalous zone was mapped on the 366-m, 427-m, and decline levels at the Belle Isle salt mine and projected horizontally and vertically into the surrounding areas (fig. 3).

Geologic in-mine mapping should include

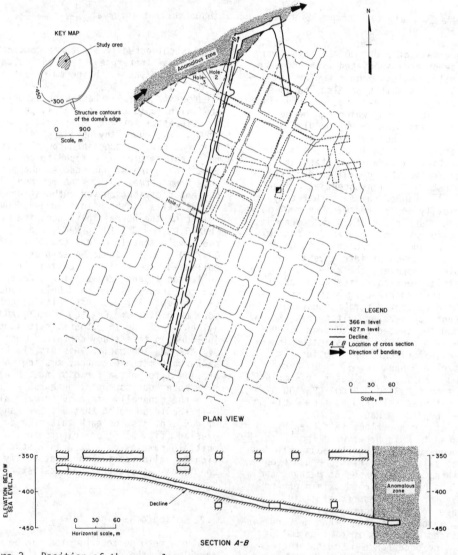

Figure 3. Position of the anomalous zone and location of 3 exploration boreholes at the Belle Isle salt mine, Louisiana, USA.

an inspection and description of salt types, with emphasis on the presence of brine or gas seeps, fluid or gaseous inclusions, grain size, striations, and orientation of banding. Once the mine has been sufficiently mapped, trends of anomalous and outburst prone zones of salt can be established for the entire mine. This will allow projections of these zones into unmined portions of the salt dome close to the current working sections. Also, because the domal salt is often nearly vertically bedded, these trends can be projected into superjacent and subjacent mining levels.

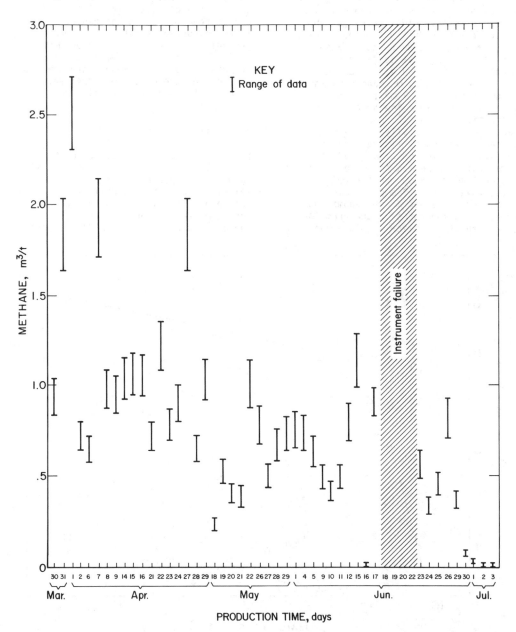

Figure 4. Calculated methane emission rates per ton of salt produced from the decline face.

2.3 Gas Emission Surveys

Once the zones of gassy salt have been recognized, accurate measurements of gas emission problems encountered during mining of the face through these areas can be collected routinely. In an effort to identify a range of methane emissions in domal salt mines during mining, the Bureau of Mines has measured emission rates in an advancing decline at the Belle Isle Mine during a 6-month period. Measurements were taken at an advancing decline face where a continuous-mining machine was in operation. During production periods, methane emissions ranged from 0 to 2.7 m^3/t of salt mined (fig. 4). Within a few hours after production ceased, methane emissions declined to essentially zero and remained at that level until mining resumed. The emission rate in normal salt adjacent to an anomalous zone was found to range from 0 to 0.1 m^3/t. Much greater emission rates (averaging 0.7 m^3/t of mined salt) were observed in the anomalous zone. Emission rates in the anomalous zone were almost twice as high when

mining advanced at an angle to salt banding as compared to when face advancement occurred parallel to banding (Iannacchione, et al 1984). This suggests some degree of directional permeability.

Similar emission surveys of other gassy zones will enable the operator to quantify the emission rate and composition of these gases. This information can be transferred to the geologic map. If anomalous zones determined from geologic mapping are associated with regions of significant emission rates, projections of these zones into unmined portions of the dome could delineate potentially hazardous emission zones.

2.4 Drill Hole Testing

In order to further investigate the physical characteristics associated with known and suspected areas of high gas emissions, exploration drill holes should be cored into each face advancing into unmined portions of the dome. These drill holes will yield data pertain-

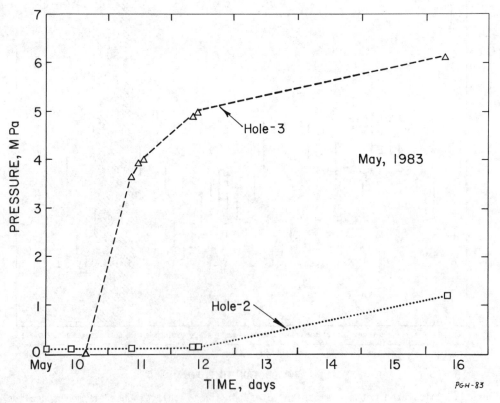

Figure 5. Gas-pressure buildup through time for holes 2 and 3.

ing to the composition, flow, pressure, and in situ borehole communication rate of gas from the salt rock penetrated. Cores collected from these holes can be inspected and described, delineating the physical and compositional characteristics of the salt and tested for methane content prior to mine face advancement. Gassy zones can be further delineated by changes in pressures and flow rates of gas as the hole is advanced. If two or more holes are drilled in the face, cross borehole communication can better define the extent and rate at which gases can migrate to the face area.

The Bureau of Mines drilled three exploration holes into the anomalous and adjacent normal salt described in the preceding sections using a brine solution. The first hole was drilled entirely in normal salt; the second and third holes were drilled through normal salt into the anomalous zone (fig. 3). The normal salt that was penetrated in each of the three holes had extremely small gas contents, low pressures, and low flow rates. However, in holes 2 and 3, gas flows in excess of 0.33 l/sec were measured when anomalous salt was encountered.

In situ gas pressures were measured in hole 2 and 3 by sealing the bottom 6.1 to 9.1 m of hole. Gas pressure in hole 3 rose to over 6.2 MPa during a 6-day period (fig. 5). However, in hole 2, which had been partially drilled 9 months prior to hole 3, gas pressure did not exceed 69 kPa until pressure in hole exceeded 4.8 MPa. At this point, hole 2 began to buildup pressure over a 4-day period, reaching approximately 1.2 MPa (fig. 5). These two holes were 7.3 m apart at their end points.

A gas injection test was conducted to determine if the salt formation between holes 2 and 3 had any measurable permeability. Cross-borehole permeability was determined by injection neon tracer gas into hole 3 at a pressure of 7.9 MPa while observing the tracer gas movement into hole 2. Approximately 1.98 m^3 of neon was forced into hole 3 and less than 0.001 m^3 migrated into hole 2. The gas required from 5 to 23 hours to migrate 7.3 m through the anomalous zone. This indicates the gas was traveling at more than 8.3×10^{-5} m/sec, but less than 4.2×10^{-4} m/sec within the formation. This relatively slow rate indicates some permeability within the anomalous zone and confirms the potential for gas migration to the mine opening from considerable distances within the formation. It should also be pointed out

that the sealed-off portion of the drill hole in the anomalous zone produced noticeable quantities of clay and brine, which could have resulted in decreased permeability by crusting the inside of the borehole.

The most important point to be made here is that localized permeability may also be associated with the same geologic conditions responsible for high gas content of salt. Therefore, drilling exploratory holes will not only allow inspection prior to mining and gas content testing, but will also provide data on gas pressure and flow characteristics and relative local permeabilities. This can provide useful information to a mine operator owing to the apparent nonuniformity of methane distribution within Gulf Coast salt domes.

3 CONTROLLING THE METHANE HAZARD

After the methane hazard has been characterized, several techniques and/or procedures can be employed to control this potential problem. These techniques and procedures would include the following: remote gas monitoring; use of permissible equipment; development of a special mine plan for gassy zones with special emphasis on ventilation and production criteria; employing gas drainage techniques to reduce in situ pressure; and if all else fails, possible avoidance of potentially very hazardous areas.

Remote gas monitoring and the use of permissible equipment are intended to decrease the chance of an explosion if a large gas emission has occurred. Currently, all miners must be withdrawn from the mines while blasting. Remote-area gas sensors placed near the production faces and within return airways would alert the mine operator of hazardous gas emissions. The use and maintenance of permissible equipment in known gassy areas would eliminate a source of ignition for this hazardous condition.

Mine layout should consider the location of these gassy zones to minimize their effect on high production areas and maximize the potential for large quantities of ventilating air currents. The establishment of many working faces, the ability to regulate large quantities of air to a gassy zone, and the blasting of a potentially hazardous face at desirable times, such as the last shift of a work week, would minimize health, safety, and production-related problems. If large emissions did occur, the mine ventilation system could be altered to more effectively dilute the methane in a select area of the mine over an entire weekend.

Bureau of Mines data have shown that high gas pressures, flows from drill holes, and flows between drill holes can exist within one gassy zone in a domal salt mine. Experience in coal mine gas drainage has identified these conditions as essential for successful drainage of face areas. Even though the magnitudes of the parameters governing these conditions are much smaller than typical coalbed conditions, the presence of these conditions support the potential use of gas drainage as a local control procedure in anomalous zones. Drainage holes drilled into high gas pressure areas with some degree of permeability will eventually reduce the gas pressure locally, thereby lowering the flow rates of gas from the salt into the face area. Finally, if conditions are serious enough, total avoidance of the gassy zone may be appropriate.

4 CONCLUSIONS

Data included in this report supports the concept of an erratic distribution of high gas content salt in Gulf Coast domal salt mines. Within many of these salt domes, there exist anomalous zones which may disect or totally surround the salt stalk. One such anomalous zone was investigated at the Belle Island salt mine and was found to contain the following abnormal characteristics: (1) measurable gas flows between boreholes, (2) directional permeability controlled by bands within the salt, and (3) gas pressures higher than the calculated hydrostatic head. All of these characteristics suggest the potential for hazardous gas emissions.

Characterization of the occurrence of methane from Gulf Coast salt mines is based on data from a combination of methane content testing, geologic mapping, gas emission surveys, and gas flow and pressure characteristics of in situ salt from gassy zones. A control strategy would include identification, delineation, and orientation of gassy salt zones; determination of the severity of the potential hazard; use of special mine-through procedures and techniques; and, if necessary, total avoidance of the zone.

5 REFERENCES

Iannacchione, A. T., R. H. Grau III, A. Sainato, T. M. Kohler, and S. J. Schatzel. 1984. Assessment of Methane Hazards in an Anomalous Zone of a Gulf Coast Salt Dome. U.S. Bureau of Mines RI 8861, 26 pp.

Kupfer, D. H. 1980. Problems Associated With Anomalous Zones in Louisiana Salt Stocks, USA. Paper in 5th Symp. on Salt, North. Ohio Geo. Soc., Cleveland, Ohio, v. 1, p. 119-134.

Plimpton, H. G., R. K. Foster, J. S. Riebeck, R. P. Rutherford, R. F. King, G. L. Buffington, and W. C. Traweek. 1980. Final Report of Mine Explosion Disaster, Bell Island Mine, Cargill, Inc., Franklin, St. Mary Parish, Louisiana. MSHA Accident Investigation Report, 156 pp.

Schatzel, S. J. and D. M. Hyman. 1984. Methane Content of Gulf Coast Domal Salt. U.S. Bureau of Mines RI 8889, 18 pp.

U.S. Senate. 1969. Hearings Before the Subcommittee on Labor of the Commission on Labor and Public Welfare. 91st Congress, 1st Session, Part 5, Appendix.

Determining face methane liberation patterns during longwall mining

ANDREW B.CECALA, ROBERT A.JANKOWSKI & FRED N.KISSELL
Pittsburgh Research Center, Bureau of Mines, PA, USA

ABSTRACT: As deeper seams are continually mined, methane liberation will continue to increase and must be monitored effectively. To effectively monitor and develop appropriate control technology, methane liberation patterns must be known. The Bureau of Mines recently completed a study to identify specific face methane liberation patterns during longwall mining. Two longwall faces were surveyed. Both had high methane liberation rates. At one longwall face, most of the methane was liberated during cutting of coal by the shearer mining machine. At the second face, a significant portion of methane was from total face and floor liberation. An effective methane monitoring system would be different for each longwall panel because of the differences in liberation.

1 INTRODUCTION

The Bureau of Mines conducted a study to measure and record methane levels in an attempt to identify the specific liberation patterns on longwall faces. Ignitions on longwall panels have been on the increase over the past few years. To effectively monitor methane, it is necessary to know the specific liberation areas. Once liberation patterns are known for a longwall panel, effective monitoring systems can be installed, and effective control procedures can be implemented.

Methane ignitions are still one of the most serious hazards facing the coal mine operator. Over the past few years, there have been a number of coal mine fatalities due to methane ignitions. Records for the past 10 years indicate an average of approximately 50 reported methane ignitions a year in coal mines in the United States. Over the latter part of this period the number of ignitions has increased due in part to ignitions occurring on longwall panels where the high rate of extraction liberates methane at a higher rate. As mining continues to expand to greater depths, it is estimated that methane levels will also increase (1).

Methane is liberated at the face in two ways during longwall mining. The first is gas liberated by the cutting action of the shearer mining machine during the cutting sequence. The second is gas emitted from the exposed coal along the total face, floor, and roof; this is also known as face bleeding.

Ventilation is the primary means used to control face methane liberation. In room and pillar mining, increased gas levels are usually handled by increasing the airflow. In an effort to utilize this control method, face airflows on some longwall panels have exceeded 2500 m^3/min with air velocities of over 300 m^3/min. In a few cases, these airflows have not been sufficient to dilute and disperse the methane liberated during longwall mining.

2 TEST SET-UP FOR METHANE MONITORING

Both remote-sensing and handheld methane monitors were used. Two remote sensing methane monitors were used, one located on the shearer and the other at the tailgate end of the face. Handheld monitors were used to monitor methane levels downwind of the shearer.

The remote sensing monitors were CSE 180R Monitors which have a remote sensor head, with cables available in lengths of between 3 and 33 m. (Reference to specific manufacturers is for information only and does not imply endorsement by the Bureau of Mines).

The sensor head uses a catalytic diffusion-type sensor to monitor methane. From the temperature differential across a wheatstone bridge in the sensor head, the instrument calculates methane concentrations in air from 0 to 5 pct (%). The concentration is recorded continuously on an internal strip chart recorder and is displayed on the the monitor face.

The monitor located on the shearer was used to determine methane liberated during face cutting. This unit monitored a point on the face side of the shearer body near the tailside drum (fig. 1). The monitoring location at panel 1 was at the tailside shearer splitter arm. Since the shearer at panel 2 did not have a splitter arm, the monitoring location was on the body of the shearer. Because of the amount of water and coal thrown by the tail drum at these locations, the sensor head was not located at the sampling point, but was housed in a sensor chamber on the walkway side of the shearer machine. This chamber was 15 cm in on all sides. Hard tubing extended from the chamber to the sampling point from which air was drawn into the chamber at about 4 L/min by two sampling pumps. This monitor had a response time of approximately 35 s.

The second monitor, located at the tail end of the face, was used to determine total face liberation. This monitor was strapped to a hydraulic support, approximately five supports from the tailgate. The sensor head was extended up and out to the front of the support. The sampling point was approximately

15 cm in from the roof, and 60 cm from the face.

Handheld monitors were used to determine methane decay levels downwind of the shearer machine. Readings were taken near the roof a few feet from the face, for a distance of 9 m downstream from the shearer, at 1.5 m intervals.

In addition, Federal regulations require that all working sections in coal mines have a methane monitor at the face to measure methane levels. In some cases, these values were correlated with values obtained from the monitor at the tail end of the face.

3 TESTING

Tests were performed at two different longwall faces to determine liberation and flow patterns during longwall mining. These faces were known to have high methane liberation, and each panel had completely different geological conditions (Table 1). Both of these longwall panels were ventilated from head to tail. These faces are considered to be at the extremes for methane liberation for longwall mining today, but as deeper seams are mined these could someday become the norms. Testing was performed at each panel by two Bureau personnel for one shift a day for one week.

4 RESULTS

The significant liberation and flow patterns of each longwall face was obtained. The results from testing at each face will be listed separately because of the major differences in the methane liberation and flow for each panel.

Longwall Panel 1:
    There were four significant findings:
    * Methane levels were lowest during the tail to head pass, (even though cutting was bidirectional).
    * Substantial methane dilution was occurring downstream from the shearer.
    * Methane levels increased during bumps.
    * Methane levels were highest during the headgate cutout.
Each of the findings will be described:
    * Methane Levels Were Lowest During the Tail to Head Pass.

Methane liberation is the same,

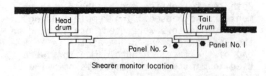

Figure 1. Sampling location on shearer mining machine for both longwall faces tested.

TABLE 1. Geological conditions of each longwall panel

Approximate values	Panel 1	Panel 2
Seam height.........m...	2.5	7 (mining top 3 m)
Overburden..........m...	500	600 m
Roof....................	Medium: sandstone, siltstone.	Medium: shale.
Cutting direction.......	Bidirectional.	Unidirectional.
Longwall type..........	Retreating.	Advancing.

independent of the cut direction, because a unit volume of coal contains a certain unit volume of methane gas. As the coal is cut, the methane gas is released. Methane levels around the shearer are determined from the rate of dilution by the primary airflow. Figure 2 shows typical methane levels at the shearer monitor for the head to tail (A), and the tail to head (B) pass, respectively. The average methane concentration for the head to tail pass was 0.72 pct; the average concentration for the tail to head pass was 0.53 pct. These reading were also supported by handheld measurements taken at the tail end of the shearer. The differences can be accounted for by the airflow patterns around the shearer for the two cut directions (fig. 3). On the tail to head pass, the airflow coming down face flowed directly to the drums, and was then forced out around the cowl, into the midsection of the work area. More air reached the drums, there was more turbulence, and methane levels were lower.

On the head to tail pass, the cowl partially blocked airflow to the drums, and methane liberated during cutting was not diluted with as much air at the shearer. Thus methane levels were higher at, and immediately downstream, from the shearer.

4.1 Substantial Methane Dilution Was Occurring Downstream from the Shearer.

This was supported by the decay rates downstream from the shearer taken by the handheld monitors, and from the remote methane monitor measurements. Table 2 shows the average decay rate downstream from the shearer for the head to tail pass. At 9 m downstream from the shearer, the methane concentration was 29 pct less than at the shearer.

The methane level measured at the tail end of the face is a combination of the face liberation rate (bleed off along the entire face), and the liberation during coal extraction. Table 3 compares the methane levels measured at the shearer and at the tailgate section of the face, during one half shift.

The methane levels at the tailgate at this longwall panel were throughly diluted and mixed with the face airflow and normally remained low. Also, methane levels at the tailgate varied relative to those measured at the shearer with a certain lag time which depended on the location of the shearer on the face. Methane levels at the shearer at this panel were in all instances higher than those measured at the tail, and under certain conditions, these levels were four to five times higher.

4.2 Methane Levels Increased During Bumps

Many times when mining deep seams, the overburden pressure builds and is spontaneously released through bumps. This occurs even more often on longwall panels because of the pressure created by the gob. When bumps occur, the excess pressures are transferred to the face. This causes additional fracturing of the coal seam, which liberates additional methane. When a substantial bump occurs, the methane levels can increase signifi-

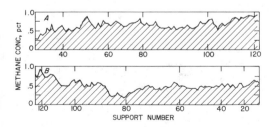

Figure 2. Shearer methane levels for both cut directions.

TABLE 2. Average methane decay downstream from shearer

Distance downstream of the shearer	Methane concentration, pct
1.5 m........	0.78
3.0 m........	.68
4.5 m........	.67
6.0 m........	.63
7.5 m........	.59
9.0 m........	.55

TABLE 3. Methane level simultaneous at shearer and tailgate

Time	Support number	Methane concentration, pct	
		Shearer	Tailgate
8:40..	115	0.2	0.2
9:00..	120	.7	.3
9:20..	87	.3	.3
9:40..	55	.3	.3
10:00..	20	.5	.4
10:20..	0	1.1	.4
10:40..	25	.2	.3
11:00..	35	.2	.2
11:20..	35	.3	.2
11:40..	35	1.2	.3

cantly. This can be seen in figure 4, which represents the remote monitor on the shearer. The tail to head pass was proceeding as normal up to support 70, at which time three major bumps occurred within a few minutes and the methane concentration jumped from an average 0.52 pct to an average value of 1.03 pct. Fracturing by the bumps had released additional methane trapped within the seam, which would ordinarily not have been released until the coal was cut.

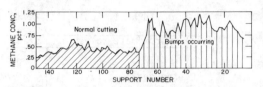

Figure 4. Increased methane liberation during bumps.

4.3 Methane Levels Were Highest During the Headgate Cutout

As previously mentioned, the methane liberation rate is fairly constant under normal conditions for the entire face area. However, measured methane levels vary due to the extent of dilution with ventilation air. During the headgate cutout, the liberated methane was not adequately mixed and diluted by the primary face airflow (fig. 5). Due to blockage by the shearer and the 90° turn, a good portion of air, often leaked into the gob. Because of this, the first 5 to 10 shields were often poorly ventilated, and the methane levels increase (fig. 6). Typical methane levels measured during the tail to head pass averaged 0.53 pct. From support 10 to the headgate, methane concentration at the shearer increased significantly, with a peak value of 2.3 pct near support 5.

Longwall Panel 2:
There were three significant findings during the testing performed at this longwall panel:

* Methane built up gradually along the face from headgate to tailgate.

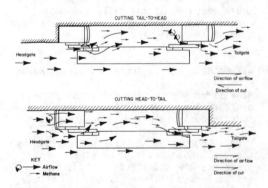

Figure 3. Airflow patterns around longwall shearer.

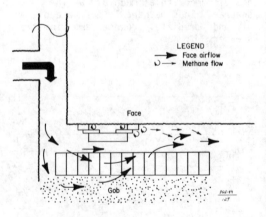

Figure 5. Typical airflow patterns in headgate area.

364

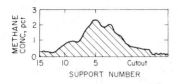

Figure 6. Measured methane concentrations during headgate cutout.

* Methane built up gradually as the day
  progressed.
* Methane was not liberated significantly by the cutting action of the
  shearer, but was emitted along the
  total face and floor.

4.4 Methane Built Up Gradually Along the
Face From Headgate to Tailgate.

Methane levels recorded at the shearer
indicated a gradual buildup of methane
along the face from the headgate to the
tailgate. Figure 7 shows this gradual
buildup of methane at the shearer mining
machine for one pass from cutting from
tail to head. At the beginning of the
pass, the shearer's methane concentration
was 0.8 pct and decreased continually to
the headgate, where the methane
concentration was 0.35 pct. This gradual
reduction in the methane concentration
from tailgate to headgate was seen for all
tests at this longwall panel.

4.5 Methane Built Up Gradually as the Day
Progressed.

This gradual buildup of methane during the
work shift was supported by all the
monitors used for testing. In figure 8,
the stationary methane monitor at support
159 shows the buildup for one pass. The
methane concentration initially was
0.65 pct at the beginning of the pass.

This value increased over the tail to head
pass to a peak concentration of 0.95 pct.
This is also supported by figure 9, which

is the measured methane concentrations at
the shearer mining machine. The arrows on
the graph indicate the direction of
cutting. On both the tail to head pass
and the head to tail pass, the methane
concentration was higher at the tailgate
than at the headgate. Also, the methane
concentration at the tailgate increased
0.4 pct from 9:00 am to 1:00 pm,
supporting the point that there was a
gradually build up of methane as the day
progressed.

4.6 Methane was not Liberated
Significantly By the Cutting Action of the
Shearer, But Was Emitted Along the Total
Face and Floor.

It is common to observe substantial
methane dilution as the distance increases
downstream from the shearer because in
most cases, a portion of methane is
liberated from the coal cut by the
cutting drums. However at this longwall
panel, this was not the case because
there was no significant methane
liberation by the cutting of the shearer.
The methane concentration was the same at
the shearer as it was 7.5 m upstream or
downstream from it.

5 DISCUSSION

The results of the two longwall surveys
were totally opposite. These mines are at
the extremes for methane liberation today,
but in the future as mining continues to
go deeper, similar panels may someday be
the norm for longwall mining. At the
first mine, most of the methane was
liberated during cutting of the coal by
the shearer. At the second mine, most of
the methane was from total face and floor
liberation. Because of the differences in
the liberation at each panel, an effective
methane monitoring system would need to be
different for each longwall panel.

At longwall panel 1, the significant part
of the liberation was from the coal being
cut by the shearer. Although the amount

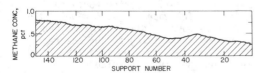

Figure 7. Gradual decrease in methane
from tailgate to headgate.

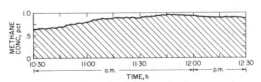

Figure 8. Gradual increase in methane at
tailgate as day progressed.

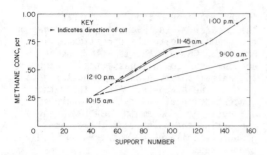

Figure 9. Gradual increase in
methane from headgate to tailgate,
and increase as day progressed.

of methane liberated is independent of the
cut direction or location on the face,
recorded methane levels were 26 pct higher
on the head to tail pass than on the tail
to head pass, due to the additional
turbulence and mixing of the ventilating
air during cutting from tail to head.
Recorded methane levels were highest
during the headgate cutout, during which
the liberated methane was not adequately
mixed and diluted by the primary face
airflow, mainly due to the blockage by the
shearer and to the 90° turn at the
headgate. During the head gate cutout, a
major portion of the air is forced back
into the gob for the first 10 to
15 supports. It was also observed that
when overburden pressure builds
and spontaneously releases through bumps,
methane concentrations increase
significantly.

Methane levels at the tailgate varied
relative to those measured at the shearer
with a lag time which depended on the
shearer's face location, and although the
tailgate monitor did vary relative to the
shearer's monitor, the changes were not
proportional to those at the shearer. The
actual concentration changes at the
tailgate were very minor except when the
shearer was cutting out at the tailgate.
This was due to the dilution of liberated
methane with the primary airflow, as the
distance increased downstream from the
shearer. This dilution downstream from
the shearer was observed with the handheld
monitors.

To effectively monitor a longwall panel of
this nature, the monitor should be located
on or near the shearer mining machine.
Hazardous methane levels could be
encountered at the shearer and not
substantially increase the methane level
at the tailgate end of the face.

At longwall panel 2, the significant part
of the methane was from total face and
floor liberation. There was a gradual
buildup of methane along the entire face
from headgate to tailgate. Also there was
a gradual buildup of methane as the day
progressed. This was due to the
additional face and floor exposure as the
day progressed. Coal was mined on this
face for only one shift. During the other
two shifts the face had time to bleed off
a portion of the methane. As cutting
progressed through the day, new coal was
exposed and methane liberation increased.
No significant methane gas was liberated
by the cutting of coal by the shearer
mining machine, as evidenced by the fact
that concentration at the shearer was the
same as that 7.5 m upstream or downstream
from the shearer.

For this longwall panel the most effective
monitor location is at the tail end of the
face. Since the methane is not being
liberated by cutting of coal by the
shearer, a monitor located on the shearer
would not be of benefit, unless conditions
change.

6 CONCLUSION

As deeper seams are continually being
mined, methane liberation will continue to
increase. To effectively monitor methane,
liberation patterns must be known. At the
two longwall panels surveyed for this
study, these patterns were totally
different. At the first panel, most of
the methane was liberated during cutting
of coal by the shearer. There was a
substantial variation in methane levels
during the mining cycle dependent on the
effectiveness of dilution by the primary
airflow. Methane levels were higher
during the head to tail pass than tail to
head, and were highest during the headgate
cutout. At the second panel, most of the
methane was from total face and floor
liberation. Because of this, methane
levels at the tailgate end of the face
were substantially higher than at the
headgate area, independent of the
shearer's face location. Because of the
liberation at each panel, an effective
methane monitoring system would need to be
different for each longwall panel. More
effective ways to dilute this methane,
both from the cutting by the shearer
mining machine and from face liberation,
must continually be developed.

7 REFERENCE

Irani, M. C., E. D. Thimons,
  T. G. Bobick, M. Duel, and
  M. C. Zabetakis. Methane Emissions
  from U.S. Coal Mines, A Survey.
  BuMines IC 8558, 1972, p. 57.

Effect of gas pressure on permeability of coal

SATYA HARPALANI
University of Arizona, Tucson, USA

MALCOLM J.McPHERSON
University of California, Berkeley, USA

ABSTRACT: An investigation was carried out to study the influence of gas pressure on flow characteristics of coal. Results indicated that the capacity of coal to hold methane is strongly dependent on gas pressure. For pressures upto $4·14$ MPa (600 psi) the gas content of coal increased continuously. Gas pressure – permeability experiments were carried out on cylindrical coal specimens under constant effective hydrostatic stress conditions. Results indicated that coal, unlike other rocks, does not exhibit the Klinkenberg effect – decrease in permeability with increase in gas pressure. A reverse Klinkenberg effect was observed giving increasing permeability with rising mean pressure suggesting enlargement of capillaries and pores within the flow paths. Also, the gas pressure history of coal has a marked influence on subsequent permeability.

1 INTRODUCTION

Over the past 30 years, emissions of methane into coal mines have increased significantly. This has occurred because of greater comminution of the coal by mechanized procedures, higher productivity, faster moving faces and a trend towards deeper workings (McPherson and Hood, 1981). In order to plan the ventilation requirements of a mine, it is important to have an estimate of the rate at which methane is emitted from source beds and migrates through the strata towards the workings.

Numerous computer models have been devised to simulate the release and migration of methane in strata surrounding mine workings, and to assist in the design of methane drainage systems. Most of these models rely upon a knowledge of the permeabilities of the coal and the associated strata. In order to achieve satisfactory simulations of gas migration, it is necessary to have a knowledge of the permeability of coal, its capacity to hold methane and the variation of these factors with time.

Associated with any flow of gas from coal is a resulting drop in the pressure of gas in the pores. The gas pressure, therefore, varies continuously as mining

activity progresses. This paper examines the influence of gas pressure on flow characteristics of coal. It also describes the variation of the capacity of coal to hold methane with respect to pressure and how this affects the gas emission into mine workings. The results obtained have been used to interpret the relationship between gas pressure and the capacity of coal to hold methane, as well as coal permeability.

2 PERMEABILITY SPECIMEN PREPARATION

One of the most important factors that influences the accuracy of permeability measurements is specimen preparation. Many rocks can be cored easily, cut and machined to the required size, but not coals which are friable. Various methods have been tried in the past to give cylinders of coal (Harpalani, 1985) but chipping has been a problem with each one of them. Patching the core with some kind of glue or cement has, therefore, usually been necessary to give smooth surfaces resulting in an error due to variations in the actual cross sectional area of the specimen. A new technique was, therefore, investigated for this study.

A brick cutting bandsaw was used to slab the coal lump along the desired planes to give roughly rectangular prisms which were then shaped to approximate cylinders on a horizontal belt grinder. Hard plastic tubing, about half an inch in length, was glued to either end of the specimen and these ends were clamped in a lathe machine. The specimen was then machined to the required diameter using a compressed air grinder, with a silicon carbide wheel, giving an excellent cylindrical surface. Lastly, the core faces were ground on a surface grinder to generate parallel and perpendicular faces.

It has been a routine procedure in the past to commence permeability tests by evacuating the samples to "clean" the internal voidage of any water vapour and adsorbed gases that might interfere with the permeability measurement (Somerton, 1974). However, during this investigation, it was observed that evacuation of coal at normal temperatures caused partial clogging of the flow paths and a consequent reduction in permeability. Evidence suggested that this may be due to liquefaction or sublimation of volatile materials at very low pressures. A technique to evacuate coal at low temperatures (-150 deg C) was developed, which avoided, or minimized this fall in permeability (Harpalani, 1984).

3 PERMEABILITY EXPERIMENTAL APPARATUS

For this investigation, it was necessary to design an experimental rig that permitted simultaneous measurement of applied stress, applied gas pressure and gas flowrate through the specimen. The experimental equipment is shown diagramatically in Figure 1.

Independent control of the axial stress was maintained by mounting the cell in a stiff testing machine with fine load control, and of the radial stress by a hydraulic system exerting oil pressure on a 4 mm thick silicone rubber sheath around the coal. Gas was supplied via a regulator, and pressure gauge to the top of the triaxial cell. Gas was led away from the lower end of the specimen to the soap bubble flow meter.

Three flowmeters, with different ranges were used making it possible to measure any flowrate between 0.1ml/min and 100ml/min.

4 PROCEDURE FOR TRIAXIAL PERMEABILITY TESTS

The specimen to be tested was inserted into the cell. Steel perforated discs were fitted to the two flat ends of the specimen in order to distribute the gas

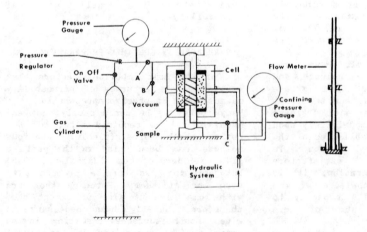

Figure 1. Schematic diagram of the permeability apparatus.

uniformly. Platens were then inserted into the ends of the cell. The upper one was connected to the gas cylinder and the lower one to the bubble meter. The entire cell assembly was placed in the center of the testing machine.

Using the fine control of the machine a small axial load was applied to the specimen. The specimen was then subjected to a confining pressure sufficient to seal the membrane against the solid surfaces. The regulator of the gas cylinder was adjusted to give a gas pressure of about 275 kPa (40 psi). Both confining and axial stresses were then increased simultaneously until their desired levels for that particular experiment were reached.

The downstream pressure was always atmospheric and the upstream pressure was varied, thus changing the mean gas pressure. The confining and axial stress levels were to be kept constant for one complete cycle of increasing and decreasing pressure. However, since gas pressure in the pores has a significant effect on the actual stress level, it was decided that effective stress, rather than the applied stress, should be kept constant.

Effective stress is defined as the difference between the external stress and pore pressure. Mathematically, effective stress, σ', is

$$\sigma' = \sigma - P$$

where σ = applied stress
         P = pore pressure

Pore pressures are equal in all directions – axial and radial. Therefore, for hydrostatic stress of $\sigma_a = \sigma_r$, the applied axial stress was modified every time the inlet gas pressure, P_1, was changed. For an increase in gas pressure from P_1 to $P_1 + \Delta P$, the axial stress increased by an amount ΔP. The load was adjusted to bring the axial stress back to σ_a. However, for radial stress, the variation was more complex due to the fact that gas pressure varied along the length of the specimen. The gas pressure, at the upper end of the specimen was P_1, and that at the lower end was atmospheric. As a result, the effective confining radial stress at the upper end was $(\sigma_r - P_1)$ and at the lower end, σ_r. Since it would have been very difficult to vary the applied radial stress along the length of the sample during a test, an approximation to constant effective

radial stress was obtained by using the log mean gas pressure.

$$\text{log mean } P = \frac{P_1 - P_2}{\ln (P_1 / P_2)}$$

P_2 was always atmospheric and P_1, the absolute pressure at the upstream gas regulator. The radial stress was increased, (or decreased) every time P_1 was increased (or decreased) by the value equal to the lograthmic mean of the two pressures.

In other words, instead of keeping the applied stress constant at a hydrostatic value for an experiment, it was the effective stress that was kept constant, axially as well as radially.

Commencing from an initial upstream pressure of .275 MPa (40 psi), the pressure was increased in increments and flowrate measured at every stage. After the maximum pressure of 2.75 MPa (400 psi) had been reached, it was then decreased in steps, taking flow measurement at regular intervals until the initial level was re-attained. Effective stress was maintained constant for every cycle.

5 VARIATION OF PERMEABILITY WITH MEAN GAS PRESSURE

A series of tests were carried out to investigate the behavior of coal permeabilities with respect to mean gas pressure within the samples. The effective stress was held constant whilst the inlet methane pressure was varied from 0.345 to 2.75 MPa (50 to 400 psi). This corresponded to mean gas pressures (absolute) of 40 to 215 psi (0.275 to 1.48 MPa).

For most rocks, the variation of permeability follows the Klinkenberg equation:

$$k = k_L + \frac{b}{P_m}$$

where k_L = "absolute" permeability
         P_m = mean gas pressure
         b = constant dependent upon
             the gas used

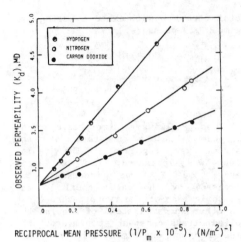

RECIPROCAL MEAN PRESSURE $(1/P_m \times 10^{-5})$, $(N/m^2)^{-1}$

Figure 2. Variation of permeability with mean gas pressure.

Figure 2 shows this relationship for most rocks (Klinkenberg, 1941). Permeability is normally plotted against $1/P_m$ in order to produce a linear relationship. The intercept of the line on the permeability axis indicates the permeability of the material to a single phase incompressible fluid. The Klinkenberg phenomenon is ascribed to slip flow at the solid-gas boundaries within the pores and capillaries and presumes that the pressure of the gas has no effect on the physical structure of the material. The layer of gas at the interface becomes immobile at high pressures thus decreasing the permeability. Several researchers have reported unusual effects produced by coal. An extrapolation of the line to the $1/P_m$ = 0 axis has sometimes indicated a negative liquid permeability - a physical impossibility. Others have reported a negative slope of the k v $1/P_m$ line (Gawuga, 1974). This is most difficult to explain in terms of boundary layer flow.

Figure 3 shows the result of the gas pressure tests on a coal sample. The three sets of curves are for effective hydrostatic stresses of 2.75, 4.14 and 5.52 MPa (400, 600 and 800 psi). Each test consisted of incremental increases in inlet gas pressure (and, hence, mean gas pressure), then similar decreases in gas pressure. Despite considerable scatter, the points for the 2.75 and 4.14 MPa (400 and 600 psi) hydrostatic loading exhibit clear hysteresis effects, the permeability remaining

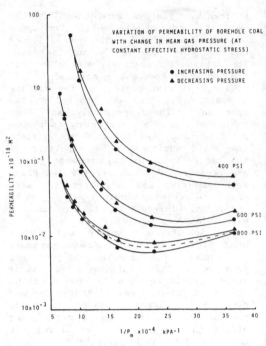

Figure 3. Variation of coal permeability with change in mean gas pressure at different stress levels.

generally a little higher during the second part of the cycle, i.e., decreasing gas pressures. The 5.52 MPa (800 psi) curve is better defined with no observable hysteresis effect. Similar results were obtained for three other experiments.

An examination of the curves given on Figure 3 indicates that for the 4.14 and 5.52 MPa (600 and 800 psi) tests, a Klinkenberg-like behavior is exhibited at low gas pressures (below 500 kPa or 72.5 psi), i.e., $1/P_m$= 20 x 10^{-4} kPa^{-1}. At higher pressures, a reverse Klinkenberg effect is observed giving sharply increasing permeabilities with rising mean gas pressures. For the 2.75 MPa (400 psi) tests, the curves rise continuously - slowly at low gas pressures then accelerating upwards at the higher gas pressures.

These findings seem to suggest an alteration in the structure of the flow paths caused by the pressure of the gas within those flow paths. The effect is accentuated at low confining stress and high gas pressure. There are several possible explanations. Enlargement of the capillaries and pores within the flow paths may have taken place at the expense of non-connected or isolated pores within the coal matrix. There may even have been some contraction of the solid material. However, this would be negligible for the range of stress values applied (2.07 to 6.21 MPa). On the other hand, the experimental technique may have exaggerated the effect. The employment of a log mean in order to maintain a constant effective hydrostatic stress (applied stress – pore pressure) is certainly an improvement on earlier similar tests. However, at any given measuring point, the radial stress was uniform over the outside surface of the cylindrical specimen whereas the gas pressure varied axially along the sample. Hence, only at the position of logarithmic mean gas pressure in the sample would the applied stress be truly hydrostatic. At the upper (high pressure) end the radial stress would be less than the axial stress leading to the possibility of "barrelling" of the sample and tensile microfracturing. This would give the increased permeability observed in the tests. Further work is suggested with a modified design of triaxial cell to investigate gas pressure as a function of permeability relationship for coal.

6. RETENTION AND RELEASE OF METHANE IN COAL MINES

Before proceeding with the investigation to study the variation of coal permeability with mean gas pressure, it was felt appropriate to determine its capacity to hold methane as a function of gas pressure. This section briefly describes the calculation of the components of gas content of coal, the mechanism of its release and its subsequent flow into the mine workings.

6.1 Gas content of coal

The gas content of a coalbed is made up of two components – free gas compressed in pore spaces and gas adsorbed on the internal surface of the pores. Figure 4 is a representation of methane molecules inside a coal pore.

Free gas is the mobile gas stored in the coal pores. To determine this quantity, the porosity of the coal must be measured. (Porosity is the ratio of void volume to the bulk volume of a porous material). The quantity of free gas can then be calculated at different

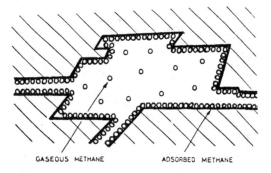

GASEOUS METHANE   ADSORBED METHANE

Figure 4. Pictorial representation of methane molecules inside a coal pore.

pressures and expressed as the volume of gas, at S.T.P., per unit weight of coal. However, the greater part of the gas content retained within the coal is held on the surfaces of coal pores and microfractures in adsorbed form as a monomolecular layer. Since the internal surface area of coal may be as large as $90m^2/gm$, the quantity of adsorbed gas can be extremely high. This quantity is a function of gas pressure in the pores, the temperature, moisture content of coal and its composition (carbon and ash content). However, for a given coal, adsorption experiments are carried out on moisture free coal at a constant temperature. Corrections for moisture and temperature are subsequently applied.

6.2 Experimental Results

The Gas Expansion method was used for the experimental work to determine the gas content of coal. Using helium, the porosity of coal was determined and the amount of free gas calculated for different pressures. Following this, the quantity of adsorbed methane at different pressures was determined for up to 4.14 MPa (600 psi). The capacity of coal to hold methane (sum of free gas and adsorbed gas) was calculated for values up to this pressure since the reported gas pressure in underground coal mines seldom exceeds this value.

Figure 5 shows the quantities of free and adsorbed gas for one coal sample, as well as the total gas content at 17^0C. Four other experiments were carried out on coal samples taken from different areas of the same coal seam.

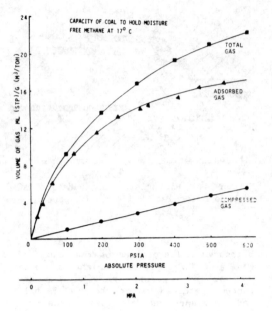

Figure 5. Adsorption isotherm for one coal sample.

The results for all of these samples were similar with some variation in their capacities to hold methane. Figure 6 shows the average gas content of the seam at 18°C (average of the 5 samples taken from the same seam, all corrected to 18°C). Curve fitting procedures indicated that this curve followed the equation.

$$G = 9.63 \ P^{0.87} \ m^3/ton$$

where G = Total gas content
P = Gas pressure in MPa

On examination of the results, a few points become evident:
1. Free gas is a small fraction of the total gas content.
2. The coal has a fairly high adsorptivity for methane.
3. The volume adsorbed continues to rise at a pressure of 4.14 MPa (600 psi) but at a reducing rate.
4. The average porosity of this coal seam is 20% - a value much higher than usually reported in the literature.

6.3 Mechanism of desorption in mines

Methane which remains adsorbed in the coal over geological time will have reached a stable equilibrium with the compressed gas. The approach of an underground mine working changes the stress pattern and results in a stress envelope around the face. This causes microfracturing and weakening of the strata. These induced microfractures, coupled with cleavage planes result in a very large increase in permeability in the destressed zones above and below mined out areas. This provides the paths along which the free gas can migrate. The flow of gas results in a fall of gas pressure within the pores. From the shape of the adsorption isotherm, it is clear that if the gas pressure is reduced, then the coal becomes less capable of holding methane in an adsorbed state. Methane molecules detach themselves from the internal surfaces of the pores. Thus the process of desorption (reverse of adsorption) is initiated. Increasing quantities of methane become gaseous and free to expand out through the flow paths towards the mine workings. The process of desorption in-situ is a very slow one and methane emissions from coal mines have been reported even after some years of working.

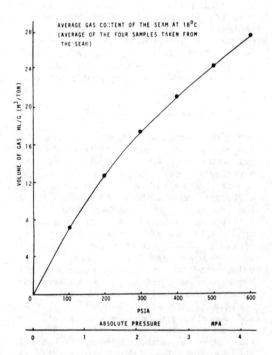

Figure 6. Average capacity of the coal seam to hold methane at differing gas pressures.

7. CONCLUSION

From the experimental work described in this paper, the following conclusions can be made:

1. The variation of coal permeability with respect to reciprocal mean gas pressure within the coal samples was found not to follow the linear relationship exhibited by other rocks. Considerable deviations from the Klinkenberg phenomenon were observed. The permeability decreased initially as the gas pressure increased, then rose dramatically as the gas pressure further increased (reverse Klinkenberg). The phenomenon may be explained by a combination of slip flow, gas adsorption and dilation of the internal flow paths provided by bedding planes in coal. The effects of variation in the first two are dominant at low gas pressures and the latter at higher gas pressures.

2. The effect of applied stress on coal permeability was found to be quite significant. The dramatic decrease in permeability with increase in hydrostatic stress can be explained by the fact that high stresses close the pores and microfractures in the coal matrix reducing the flowpaths for passage of gas. More interesting was the observation that at higher stress values (5.52 MPa) coal begins to behave more like other rocks. An extrapolation of this observation is that in very deep mines, coal might exhibit the normal Klinkenberg effect – increase in permeability with decrease in pressure. This would mean that once the mining activity begins, coal permeability would gradually increase causing a regular, if not higher, flow of methane into workings.

3. The coal permeability is strongly dependent on pressure history, the permeability remaining higher with the pressure decreasing. Once again, this indicates that if a coal seam at any time in the past had high gas pressure, then the permeability would be greater than expected.

4. The amount of free gas, which is dependent on the coal porosity, constitutes a small portion of the total gas content even for highly porous coals. The value of porosity determined is much higher than the values reported in the past. This could be due to the usage of helium and the gas expansion technique as compared to mercury injection method used more frequently.

5. For the range of gas pressures considered during adsorption experiments, the majority of methane is contained within the coal structure as a monomolecular adsorbed layer. All samples showed that at 4.14 MPa (600 psi) the amount of adsorbed gas was still rising and monomolecular saturation was not reached for the tests conducted.

ACKNOWLEDGEMENTS

This work was performed by the authors at the Mine Ventilation Laboratory of the University of California, Berkeley, as a part of a more general project to study the gas flow characteristics of coal. The project was conducted under contract to the U.S. Steel Corporation and funded by the Gas Research Institute. The authors wish to thank these organizations for their support.

REFERENCES

McPherson, M.J. and M. Hood 1981. Ventilation planning for underground coal mines. Annual report, U.S. Department of Energy. Contract W-7405-ENG-48.

Harpalani, S. 1985. Gas flow through stressed coal. Ph.D. thesis. University of California, Berkeley: 26-27.

Somerton, W. H., I. M. Soylemezoglu and R. Dudley 1974. Effect of stress on permeability of coal. Final report, U.S. Bureau of Mines Contract H0122027.

Harpalani, S. and M. J. McPherson 1984. The effect of gas evacuation on coal permeability. Intl. Jour. Rock Mech. Min. Sci. Vol. XXI, No. 3: 161-164.

Gawuga, J. 1979. Flow of gas through stressed carboniferous strata. Ph.D. thesis. University of Nottingham, U.K.

Klinkenberg, L. J. 1941. The permeability of porous media to liquids and gases. Drilling and Production Practice, A.P.I.

McPherson, M. J. 1975. The occurrence of methane in mine workings. Journal of the Mine Ventilation Society of South Africa. Vol 28, No. 8: 118-125.

10. Case studies

The most disastrous explosion in Turkish coalfields

N.BILGIN
Istanbul Technical University, Turkey

ABSTRACT : This Paper summarizes the causes of, and circumstances attending, the explosion which occured at Armutçuk Colliery, on the 7th March 1983. 102 miners were killed and 86 were injured. A description of the colliery, the state of the district before the accident and the results of the laboratory experiments already carried out on the explosiveness of coal dust collected from the same area are given briefly. The principal underground evidences such as direction of forces from displaced roof supports, the caking of dust, the nature of the deaths and injuries of the victims are detailed. The causes of the accident were discussed and some recommendations in order to prevent a repetition are given.

1 DESCRIPTION OF THE COLLIERY

Armutçuk Colliery is located on the coast of the Black Sea approximately 45 km west of the town of Zonguldak, it covers an area of 38.5 sq.km.and has coal reserves of some 94.10^6 t. The average annual output is 490.000 t with 2797 men employed underground and 1195 on the surface. There are 5 workable coal seams in the area changing from 3m to 30m in thichness. The coal which is very liable to spontaneous combustion has a mean calorific value of 6400 kcal/kg. The average moisture content is 8 %, ash content 11 %; sulfur 1 %, and volatil matter 33 %. The coal can be classified as sub-bituminuous with a free swelling index (F.S.I) of 1.5. The general exploitation method used in the area is short wall with top-slicing, 40 % of the coal is usually left in the place and this makes spontaneous combustion of the coal very common in the colliery. In average 10 working panels per year are closed for this reason (Arıoğlu 1982).

Steel archs and wooden supports are used and the faces are not mechanised. Electric cap lamps are in use with flame safety lamps issued as firedamp detectors. Certain of-ficals are also issued with methanometers.

The explosion occured at approximately 5.30 p.m on 7th March 1983. 102 miners were killed and 86 were injured. Rescue operations were almost completed whithin few days.

2 THE STATE OF THE DISTRICT BEFORE THE ACCIDENT

Figure 1 shows the location of the working panel 12003. The panel is 450 m. from the main shaft no 14. The Büyükdamar seam has a tichness of 10-14 m. which includes a few cm of dirt band. The panel lies some 330-350 m below the sea lavel. Production commenced in January 1983, coal being won on three shifts per day and at the time of the accident the average weekly output was 3.500 tons. The main gate (-346) was equipped with a panzer conveyor and two auxiliary fans. The 12000 A band roadway was being used as a return airway of the working panel 12003. The eastern part of the panel 12003 was closed because of the spontaneous combustion of the coal. There is 20 m of pillar between return airway of the closed panel and working panel.

The following methane contents for

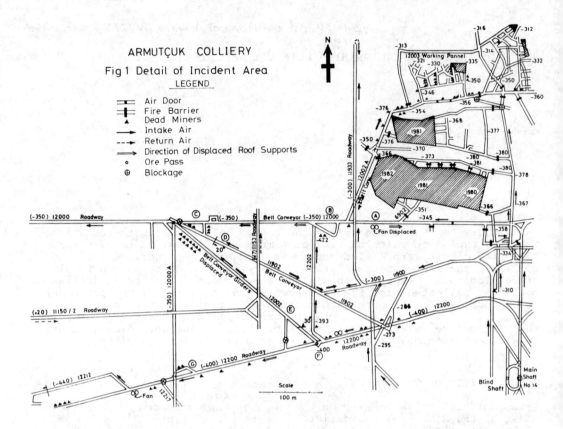

ARMUTÇUK COLLIERY
Fig 1 Detail of Incident Area
LEGEND

- Air Door
- Fire Barrier
- Dead Miners
- Intake Air
- Return Air
- Direction of Displaced Roof Supports
- Ore Pass
- Blockage

panel 12003 were recorded in the official notebook.

Table 1. Methane measurements in Pannel 12003.

Date	Shift	Methane %
8.2.1983	16-24	1.2
9.2.1983	8-16	1.5
12.2.1983	8-16	2.0
14.2.1983	8-16	0.5
15.2.1983	8-16	1.8
16.2.1983	8-16	0.7
17.2.1983	8-16	0.8
25.2.1983	8-16	0.6
2.3.1983	8-16	0.5
3.3.1983	8-16	3
4.3.1983	8-16	0.1~1
5.3.1983	8-16	0.1~1
6.3.1983	8-16	0.1~1

3 LABORATORY EXPERIMENTS CARRIED OUT ON THE EXPLOSIBILITY OF COAL DUST COLLECTED FROM ZONGULDAK COALFIELD

A cloud of fine dust, heated at any points to its ignition temperature, rapidly ignites throughout its volume. An explosive burning takes place with increasing pressure and speed of propogation of the flame. The most disastrous explosion on record, that at Courrières in France in 1906 in which 1100 men killed, was caused by a blown-out shot in heading, raising and igniting a cloud of dust. In order to investigate the explosibility of coal dust collected from Zonguldak Coalfield, an extensive laboratory experiments were carried out at Berggewerkschoftlicke Versuchsstrecke research establishment (Saltoğlu 1970). Coal dust specimen were collected from Armutçuk, Kozlu and Üzülmez Collieries. The results already published by training branch of E.K.I now T.T.K, indicated that, the coal dust specimen collected from Armutçuk Colliery was 2 and 3 times more reliable to dust explosion than the others. Some recommadations in order to prevent a coal dust explosion were included in this report.

4 OBSERVATIONS

Table 2 indicates the areas where the bodies were found and the nature of the deaths. It is clearly seen from this table that two explosions occured during the accident, one in level (-318/-350) and other in level (-400).

Table 2. The nature of the deaths.

Working Area	Total Numb. of Deaths	Violence %	Burning and Violence %
In all	102	16	18
1	36	16	26
2	3	0	0
3	17	11	11
4	12	0	0
5	39	26	24

Working Area	Poisining %	Burning %	Poisining, burning %
In all	20	6	40
1	2	0	56
2	0	0	100
3	25	0	53
4	100	-	-
5	6	-	44

Working areas could be described as :
1. Panel 12003 and 12002A Band Roadway, -318/-350.
2. 12000 Band Roadway, -350.
3. Band Roadways, -350/-300,-350/-400.
4. 12000 Roadway, -350, 1400 m west from panel 12003.
5. 12200 Main Roadway, -400.

It is generally believed that after a coal dust explosion, the ash content, the volatil matter content and free swelling index of coal dust would be much different than the original. The term "free-swelling" is applied to the behaviour of some bituminous coals when heated under specified conditions. The softened coal can be expand, for example due to gases been released on pyrolisis, in the direction away from the heating surface. The degree of swelling can be expressed in numbers of 1 to 9 based on cross-sectional areas profiles of the swollen sample. The free-swelling index is used to estimate the caking properties of coals during combustion on grade, to dif-

ferentiate between agglomerating and non agglomerating coals, and to detect oxided coals (Abernethy 1971).

The analysis of the dust specimen collected from incident area gave the results tabulated in Table 3, in this table the letters A,B,C,D,E, F and G,H,I are the areas where the samples were collected and they are indicated in figures 1. The big difference between the values obtained from dusts collected from incident and non-incident areas strongly emphasizes a coal dust explosion.

Table 3. The results of coal dust analysis.

The incident area	Ash Content %	Volatil Matter content %	F.S.I.
A	64	22	0
B	60	21	0
C	66	17	0
D	60	17	0
E	61	17	0
F	37	24	0
G	53	31	0
H	5	39	4
I	28	26	1

The areas where dust specimen were collected could be described as :
A. Band Roadway 2000(-350).
B. Motor Grage 2000(-350)
C. Power Station 2000(-350)
D. Band Roadway 1902 (-350/-300)
E. Band Roadway 12002 (-350/-400)
F. Ore Fass (-400)
G. Motor Garage -400
H. Unbroken coal specimen from non incident area
I. Dust speciment collected from non incident area.

In the following day of the explosion, detailed examinations were made of the incident area and indications of the directions of forces from displaced roof supports were carrefully examined. They are shown in figure 1.

5 CONCLUSIONS

102 miners were killed and 86 were injured during the explosion which occured at Armutçuk Colliery, on the March 1983. Varying views were expressed during the inquiry about

what caused or contributed to the
incident. Hovewer its beyond doubt
that the firedamp accumulated in
12002 A(-375/-350) band roadway
caused the first explosion. The advancing wave of the explosion.
Stired up the dust on the 1200(-350)
11900 (-350/-300); 12002 (-350/-400)
band roadways; the initial ignition
of the dust cloud followed with the
continued propogation of flame away
from the point of ignition and a
second explosion took place in
12200 (-400) main roadway. The
analysis of coal dust specimen collected from different areas, nature
of deaths, direction of forces from
displaced roof supports, observations made in the incident area
strogly emphasize that two explosions occured in the Colliery.
Although a detailed report was published by training branch of Ereğli
Coal Establishment (now T.T.K)
about coal dust explosions, precautions such us stone dust barriers
were not built in Armutçuk Colliery.
The maintenance of a good standart
of ventilation to minimize the risk
of accumation of methan and the immediate detection of this if it
should occur, the removal or rendering harmless of accumulation of
coal dust are strongly recommended
in Zonguldak Coalfield.

REFERENCES

Arıoğlu, E., and Yüksel, A. 1983
    Statistical investigation of
    spontaneous combustion of coal in
    Armutçuk Colliery, the possible
    application of hydraulic stowing.
    Mining Congress. Ankara.
Saltoğlu, S. 1970, The explosiven-
    ess of coal dust collected from
    Zonguldak Coalfield and the use
    of stone dust barriers in order
    to prevent coal dust explosions.
    E.K.I training branch, no 31,
    Zonguldak.
Abernethy, R.F., and Synder, J.P.
    1970, A small scale test for
    free swelling index of coal.
    U.S.B.M, R.I 7 534, p.p.17.

A laboratory evaluation of spray-applied rigid urethane foams

ROBERT J.TIMKO & EDWARD D.THIMONS
Pittsburgh Research Center, Bureau of Mines, PA, USA

MERVIN D.MARSHALL
Mine Safety Appliance Research Corporation, Evans City, PA, USA

ABSTRACT: The objectives of this research were to examine and document the laboratory performance of several different brands of rigid urethane foam. Eighteen brands, produced by 13 manufacturers, were examined. All testing was performed in a laboratory under controlled conditions.

Preliminary lab tests included flame spread evaluations, flame penetration, ignition temperatures, air permeability, and adhesion. Having established arbitrary cut off values, the original 18 candidates were narrowed to eight. More specialized tests followed. The effects of water immersion and dry aging on flame spread performance and other physical properties were then examined.

This publication was written to provide information for those concerned with using rigid urethane foams as a sealant. Its intent is to describe the various laboratory evaluations performed and to delineate the performance of each candidate. This report neither promotes nor refutes the use of rigid urethanes. It presents results that will enable those responsible for using, as well as enforcing the use of rigid urethane foams in underground mines to make informed decisions.

1 INTRODUCTION

Sealants used in the underground mining industry are vitally important to maintain a safe and productive working environment. Coatings on stopping and overcast faces enable ventilation air to follow its prescribed course through a mine. Applying sealants to the ribs and roof of intake airways greatly reduces the deleterious effects of temperature and humidity on exposed surfaces.

In most coal mines cementitious sealants are the prevalent types of coating used. These are usually hand applied with a brush or trowel. Cementitious sealants are inorganic and, thus, lack the ability to flex or compress as the substrates undergo external compressive forces. This inability to move with the substrate means that continuous maintenance must be performed to replace damaged sealant.

Sealants do exist that can deform with their substrates. These include urethanes, sodium silicates, and to some extent, cementitious sealants containing solid, flexible additives. Unfortunately, most contain some fraction of organic components. Because organics can burn, these have not been readily accepted underground.

Flexible sealants are divided into two categories of application: by hand and spray. The hand-applied usually contains some type of latex additive to ensure flexibility when cured. Spray-applied flexible sealants are, for the most part, urethane foam.

Urethane foams are commonly referred to as rigid foams because they become rigid when cured. Urethane foam was introduced several years ago, initially as insulation in the construction industry and as flotation in the ship building industry.

Because of its excellent adherence and flexibility, it was immediately accepted by the mining industry as a sealant. This evaluation of rigid urethane foams was part of a larger contracted effort

TABLE 1. Rigid urethane foam candidates[1]

Product identity	Supplier	Flame spread index[2]	Density kg/m^3	Pct closed cell	.03 m thick application cost ($/m^2)
CSI 9120...........	Chemetics Systems	20	32	96	2.37
CSI 9152...........	Chemetics Systems	20	32	95	2.26
Corofoam G325......	Cook Paint & Varnish	30	32	>90	2.04
FS-24..............	Foam Systems Co.	25	32	90	2.37
FS-25..............	Foam Systems Co.	25-30	32	90	2.37
FS-234.............	Foam Systems Co.	25	35	90	3.23
Chempol 30-2124....	Freeman Chemical	25	32	94	1.94
Rigimix E/F........	Mine Safety Appliance Co.	25	32	>90	2.69
X-156..............	Mine Safety Appliance Co.	20	32	>90	3.55
Polysystem 7622-02.	Olin Chemical Corporation	25	35	ND	2.37
FMS-20.............	Polymir	20	34	95	2.04
Texthane 220-20....	Texas Urethanes	25	32	95	3.44
UFS-250............	United Foam	25	32	ND	2.37
Isonate CPR 468....	Upjohn Co.	25	32	92	2.04
USC-230............	Urethane Systems	25	32	95	1.72
FMS-20[3]............	Utah Foam Prod. Co.	25	32	94	2.26
SS-0640............	Witco Chemical Co.	25	32	>90	2.26
SS-0768............	Witco Chemical Co.	25	32	>90	2.15

[1]Data provided by MSAR from manufacturer specifications.
[2]Flame spread values determined by ASTM E-84 test method.
[3]No relation to Polymir FMS-20. In the report this foam will be referenced as FMS-20U.

with the Mine Safety Appliance Research Corporation (MSAR), which included urethane foams as well as other types of rigid foams. All laboratory work was done in MSAR facilities, with Bureau researchers overseeing the effort.

Spray-applied urethane foams are the organic sealants most often used underground. They are best suited for underground use because of their physical properties, ease of application, and relatively low cost. Eighteen different spray-on rigid urethane foams produced by 13 different companies were selected. A complete list, including specific characteristics, is shown in Table 1.

2 BACKGROUND

2.1 CHEMICAL COMPOSITION OF RIGID URETHANE FOAMS

Rigid urethane foams are two-part chemical systems which must be metered in specific proportions for successful application. The A or activator component typically contains a polymeric isocyanate, diphenylmethane diisocyanate (MDI), which has a recommended threshold limit value

- time weighted average (TLV-TWA) of 0.02 parts per million parts of air (ppm). This means that 0.02 ppm is the average concentration to which workers may be exposed for a normal 8 h workday and 40 h work week without adverse effects. The principal hazard is respiratory irritation following inhalation of the vapor.

The average concentration of MDI immediately above an open, full container has been found to be less than 0.01 ppm at 316 K. The activator should, therefore, present no problems as a vapor.

The B component of the rigid urethane foam system contains polyols, a blowing agent, fire retardant, surfactant, catalysts and a flushing agent. The polyols or polyalcohols have very low toxicities. The blowing agent, a fluorocarbon, has a TLV-TWA ceiling of 1,000 ppm. The flushing agent, generally methylene chloride, which has a TLV-TWA of 500 ppm, is used to clean the sprays following application. These chemicals are used in sufficiently small quantities to preclude most exposure problems.

When applying rigid urethane foams,

FIGURE 1. Applying rigid urethane foam to an underground stopping.

hazardous concentrations of certain vapors and mists could be generated, especially in poorly ventilated areas. For this reason, most regulatory agencies recommend the use of positive pressure air masks, including full eye protection, for anyone in the vicinity of the application.

The two components of rigid urethane foams have an unmixed density between 1,041 and 1,282 kg/m^3. They are delivered via pump to a spray gun, blended, and sprayed on surfaces at densities approaching 32 kg/m^3 (fig. 1). The expansion ratio is between 32 and 40 to 1. To obtain a .03-m thick coating, a 0.8-mm thick unexpanded coating should be applied.

3 LABORATORY EVALUATION OF RIGID URETHANE FOAMS

3.1 FOAM SELECTION

All of the 18 candidate foams selected for evaluation had the following characteristics:

o Densities between 32 and 35 kg/m^3,
o Flame spread ratings of 20 to 30, based on ASTM E-84 tests,
o Spray-on application,
o Greater than 90 pct closed cell,
o Less than $3.77/m^2 application cost (.03 m nominal expanded thickness).

The foams chosen all had similar densities because of constraints associated with vertical surface application. The closed cell density must be above 26 kg/m^3 to prevent collapse. Because of heat dissipation problems, foams with densities above 48 kg/m^3, when applied in expanded thicknesses of .03 m or more, may crack or scorch.

The ASTM E-84 test determines burning characteristics of materials as a function of flame spread over their surfaces. The test equipment includes a 7.62 m long rectangular tunnel, .45-m wide by .30-m high. Materials are suspended horizontally at the roof and ignited. Flame spread rate, heat evolved, and smoke emitted are measured. The Mine Safety and Health Administration (MSHA) accepts only those materials with a flame spread index of 25 or less.

Only sprayable candidates were selected, since this is the most common application method underground. Two application systems are available--a trailer mounted rig and a portable, manually operated system.

None of the candidates had closed cell contents of less than 90 pct. This is important because closed cell content relates directly with air permeability. With smaller closed cell values, air can flow more easily through the foam coating, rendering it worthless as an air barrier sealant.

All candidates were evaluated with expanded spray-on thicknesses of .03 m and had application costs of less than $3.77/m^2. This made some foams approximately three times more expensive than hand applied cementitious sealants, which average $1.29/m^2 with a nominal thickness of 3.2 mm. The added expenses are mainly due to: 1) higher material costs, 2) more complex application equipment, and 3) additional maintenance required on equipment. However, the added application costs are recoverable because: 1) rigid urethanes seal more effectively, reducing initial leakage through stoppings, and 2) they retain their sealing abilities much longer than cementitious sealants.

3.2 FLAME SPREAD INDEX EVALUATION

The ASTM E-162 Radiant Panel Test, was used to determine the flame spread index of the foams (figs. 2A and 2B). In this test, a sample is placed in front of a heat radiating panel of a standard heat flux and ignited by a sample ignition source. Flame propagation with respect to time is measured, along with the heat evolved. These two values are multiplied to give the flame spread index. Average flame spread indices for the 18 urethanes are shown in figure 3.

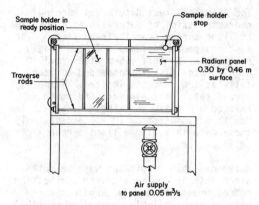

FIGURE 2A. ASTM E-162 radiant panel test apparatus (front view).

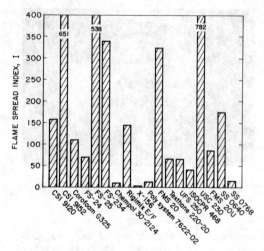

FIGURE 3. Flame spread indices for 18 urethanes.

to the bottom of the specimen so that the flame front now moves up the panel. The exhaust stack is recentered over the new panel location.

Results of the modified E-162 radiant panel tests are shown in figure 5. With the ignition pilot at the bottom of the

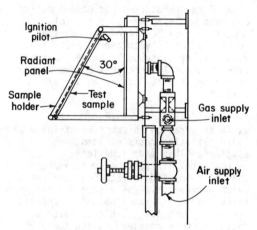

FIGURE 2B. ASTM E-162 radiant panel test apparatus (side view).

After this initial series of tests, an arbitrary cutoff value of 150 was chosen. This value has no real meaning. Looking at the results in figure 4, two different performance characteristics are apparent. One hundred and fifty was chosen because it appeared to reasonably separate the two foam classes.

3.3 MODIFIED FLAME SPREAD INDEX EVALUATION

A modified E-162 radiant panel test was devised by MSAR as a more severe criterion of foam flammability. In this test, the angle that the foam specimen makes with the radiant panel is reversed (fig. 4), bringing the sample closer to the radiant panel at the bottom rather than the top. The ignition pilot is moved from the top

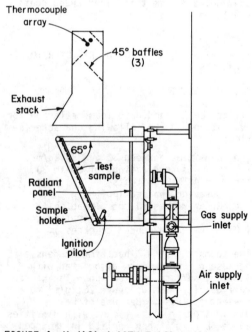

FIGURE 4. Modified ASTM E-162 radiant panel test apparatus (side view).

specimen, the flame front spread more rapidly up the foam sample. Flame spread versus time increased for all samples an average of 450 pct, with a corresponding increase in the flame spread index.

Although the same testing apparatus is used, the E-162 test and modified E-162 test, not accepted by ASTM, give very different results. The intention was not to see how well the samples correlated between the tests, but merely to expose samples to a more severe examination.

In these tests the arbitrary cutoff point chosen was 2,000, leaving eight candidates for further testing. Again, this value had no real meaning but was chosen solely as a means of classifying the foams in a group whose properties are more closely aligned.

3.4 FLAME PENETRATION

Rigid urethane foams are used primarily as stopping and overcast sealants to restrict or eliminate air leakage in underground mines. Urethane foams must not only have acceptable flame spread characteristics,

but also they must be able to limit or prevent flame penetration.

Resistance to direct flame exposure is measured by a flame penetration test. This examination was conducted on the eight remaining candidates using a method developed by the Bureau of Mines. The test apparatus is shown in figure 6. A .15 m x .15 m x .03 m sample is inserted into the holder. A pencil-point flame is adjusted to .05 m above the foam and set so that a .04-m blue cone flame is emitted. The actual foam surface temperature is measured by a thermocouple. The mirror beneath the test apparatus is used to determine if burn-through takes place within 7 min. None of the samples permitted burn through within the allotted time.

3.5 IGNITION TEMPERATURES

Two temperatures were determined: the flash-ignition temperature (FIT) and the self-ignition temperature (SIT).

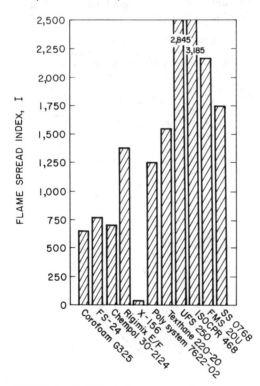

FIGURE 5. Modified flame spread indices.

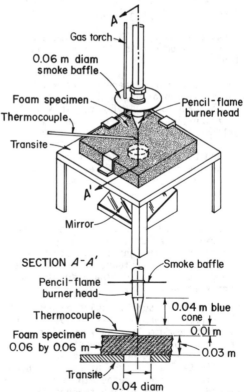

FIGURE 6. Flame penetration test apparatus.

387

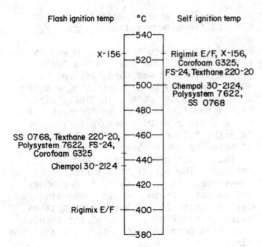

Flash ignition temp | °C | Self ignition temp

X-156 — 520 — Rigimix E/F, X-156, Corofoam G325, FS-24, Texthane 220-20

500 — Chempol 30-2124, Polysystem 7622, SS 0768

SS 0768, Texthane 220-20, Polysystem 7622, FS-24, Corofoam G325 — 460

Chempol 30-2124 — 440

Rigimix E/F — 400

FIGURE 7. Flash and self ignition temperatures of rigid urethane foams.

Before ignition, foams must undergo sufficient external or internal heating for flammable gases and decomposition products to be released. If the proper fuel-air ratio exists, an ignition may take place. When a flame exists external to the foam, the products given off may flash and ignite. The temperature at which this occurs is the flash ignition temperature.

There is also a temperature at which the decomposition products themselves ignite, without any external flame. This is the self-ignition temperature. In most cases, self ignition temperatures are higher than flash ignition temperatures. The ASTM Method D 1929 determines the self- and flash-ignition temperatures. Figure 7 shows the temperatures required to ignite rigid urethane foams.

These results show at what temperatures problems could occur with rigid urethanes. For example, electric arcs can greatly exceed the flash ignition temperature required to ignite foam; care should be exercised in using urethane foams near high-voltage wiring.

Self ignition usually occurs by applying a thick (more than .05 m expanded) coating of urethane foam to a surface. The foam undergoes an exothermic reaction as it cures. Depending upon component reactivity and the type of catalyst used, some foam internal temperatures will exceed the self-ignition temperature. The foam then begins to decompose rapidly, ignite, then burn from the inside.

3.6 AIR PERMEABILITY

Since the main use for rigid urethane foams underground is as sealants on stoppings and overcasts, they must be impermeable to air. The permeability of the eight candidate foams was determined at .03, .13, and .25 m wg. Figure 8 shows a schematic of the test device.

All samples tested had leakages of less than 2.5×10^{-4} m^3 s^{-1}/100 m^2 at each pressure drop. The closed cell content of each foam was at least 90 pct. As expected, the high closed cell content allowed a high pressure differential to be exerted, with no breakdown of the individual closed cells.

3.7 ADHESION

Foams that have successfully passed all previous tests may still be unacceptable for underground use if they do not adhere adequately to various substrates. To evaluate this, candidate foams were sprayed on different substrates, permitting them to adequately cure, and measuring the force required to pull a representative sample from the substrate.

Five substrates were used: coal, wood, slate, concrete block, and plastic brattice. Each substrate was used in four different conditions: dry, dry and rockdusted, wet, and wet and rockdusted. For wet tests, water was applied to each surface until the surface was completely saturated. Rockdust was applied until the substrate was completely covered.

An adhesion test, modified from earlier Bureau research, was developed. A pull-tab, consisting of a .05 m x .05 m, flat, perforated metal sheet, with an eyebolt centered in the tab, was placed on

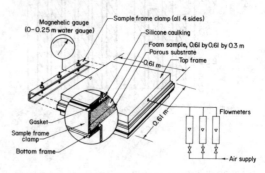

FIGURE 8. Test device for measuring air permeability.

388

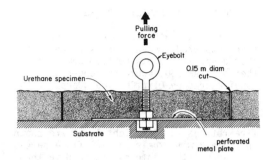

FIGURE 9. Test device for determining adhesion.

each substrate (fig. 9). Approximately .05 m of foam was sprayed onto the substrate and pull-tab. After curing, a .15 m diameter hole-saw, centered over the eyebolt, was used to cut a circle ensuring that the sample was free of the adjacent foam.

Each sample was placed on a tensile evaluation device, the substrate secured, and the foamed tab pulled until it separated from the substrate. Data for overall average foam strength on each substrate are shown in figure 10, A through E. These were averaged over all substrate conditions and presented as a single result. The averaged values represent realistic information of each brand's performance. Two foams, Rigimix E/F and FS-24, consistently outperformed the others. X-156 and Chempol 30-2124 were the least adhesive.

4 WATER-IMMERSION TESTS

Rigid urethane foams used as sealants underground should be capable of withstanding both high relative humidities and, at times, actual water contact. Placing urethane foams in water could cause: 1) structural weakening due to the absorption of water, 2) increased flammability due to leaching out of flame retarding chemicals, and 3) increased air permeability due to internal cell rupture, reducing the closed cell content.

Four .15 m x .46 m x .03 m rigid urethane foam samples of each brand were immersed in distilled water for 96 h. The samples were then removed from the water, weighed to determine water retention, and then reweighed after 48 h to determine a post immersion dry weight.

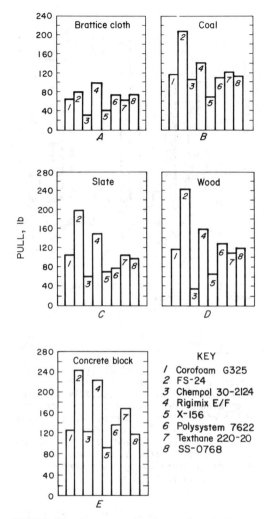

FIGURE 10. Average adhesion values of rigid urethane foams.

The eight final candidates were then again subjected to the E-162 radiant panel test (fig. 11). Most foams showed a significant increase in flame spread indices after water immersion, due to leaching out of the fire retardant additives.

Tests comparing pre-immersion values of closed cell content, compressive strength (determining what pressures are required to initiate specimen compression), and density showed that the differences were small enough to be considered insignificant.

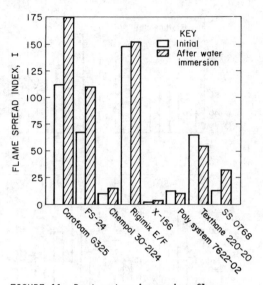

FIGURE 11. Post water immersion flame spread test results.

5 DRY-AGING TESTS

In some mines heat, not moisture, is an important reason for poor sealant performance. This is especially true in deep mines. Heat also has a tendency to drive off certain components of some rigid urethane foams. This can alter the chemical composition and, ultimately, the fire-retardant properties of the foam. Accordingly, dry-aging tests were conducted in parallel with the water-immersion tests.

Nine .15 m x .46 m x .03 m samples of each brand were weighed, then stored in a 373 K oven for 28 days. After removal, the samples were equilibrated to room temperature and humidity, then reweighed. Weight loss due to dry aging for all candidates was less than 7 pct.

Next, each foam was subjected to an E-162 radiant panel test to determine if fire retardancy has been compromised by dry aging. Results of the eight urethane foam samples are seen in figure 12. The results were similar to those found after water immersion. Two separate groups are noted: those consistently having a flame spread less than 25, and those with an index above 25. Since 25 is the maximum value MSHA will permit for use on stoppings or overcasts underground, it was apparent there were foams that could not meet this strict criterion.

6 COMPONENT AVAILABILITY PROBLEMS

After the laboratory examinations had been completed, researchers were advised that a specific urethane foam component, Thermolin RF 230, would no longer be manufactured by Olin Chemical Corp., because of a lack of demand. This polyol was used in five of the eight brands of foam evaluated and was a high performance fire retardant. The eight foams are shown in Table 2 with their initial flame spread and rank. It is interesting to note that the foams containing the discontinued polyol, especially Chempol 30-2124 and X-156, performed less well in the adhesion testing than did other foams with a different polyol.

At this time, Thermolin RF 230 is not being manufactured by anyone, nor are any plans known to restart production. Deleting the foams having this component, none of the remaining rigid urethane foams tested had a flame spread index of 25 or less, according to the ASTM E-162 radiant panel test.

7 SUMMARY

Sealants are used on stoppings and overcasts in underground mines to decrease leakage and increase the quantity of air available in working sections. Most coal

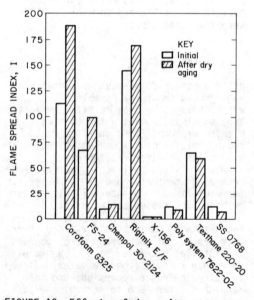

FIGURE 12. Effects of dry aging on flame spread.

TABLE 2. Rigid Urethane Foam ASTM E-162
flame spread index and rank

Foam	Flame spread index	Rank
Chempol-30-2124......	10	2*
Corofoam G-325.......	112	7
FS-24................	68	6
Polysystem 7622-02...	12	3*
Rigimix E/F..........	144	8
SS-0768..............	13	4*
Texthane 220-20......	65	5*
X-156................	2	1*

*Contained Thermolin RF 230, a fire
retardant component.

mines use cementitious sealants. These are inorganic and lack the ability to flex when convergence occurs at the coated surface. The mortar face cracks or spalls, creating a low resistance leakage path for air.

A flexible organic sealant, used predominantly in metal/nonmetal mines, is rigid urethane foam. This is a two-part sealant that is sprayed onto a stopping or overcast in thin coats that expand to between 32 and 40 times their original volume.

This research was part of a larger contracted effort in which several different types of foams were evaluated. The evaluations were done by the contractor with Bureau personnel overseeing the work. Because only spray applied rigid urethanes appear feasible for underground sealant work, only sprayed-on rigid urethanes were evaluated.

Eighteen different foams were subjected to two initial flame spread evaluations. Arbitary minimum performance specifications were devised for each flame spread test. After a second, modified flame spread test, only eight candidates remained.

The effects of water immersion and the corresponding changes in flame spread were examined, as were the effects of dry aging. Again, flame spread differences were noted. Post-immersion samples were examined for closed cell content, compressive strength, foam area increase and density and showed little difference from virgin samples. Dry aged samples were evaluated for weight changes and flame spread. Several candidates had a weight loss. In addition, four foams exhibited an increase in flammability.

Virgin foam samples were also tested for flame penetration, self- and flash-ignition temperature, air permeability and adhesion. Performance in each test was satisfactory.

One urethane foam component common to five of the eight candidates was a high performance fire retardant. The component was discontinued shortly after the laboratory testing was completed. This effectively eliminated the only foams evaluated that had a flame spread index of 25 or less, according to the ASTM E-162 radiant panel test.

Those responsible for applying urethane foam must be thoroughly trained in proper application techniques, as well as equipment maintenance. Proper metering of the two components is essential for successful application. This requires thorough cleaning of all equipment after spraying. Probably the most important rule in applying rigid urethane foams is to never exceed .05 m expanded thickness for any one coating. The curing process involves an exothermic reaction which can raise internal temperatures of some urethane foams above the self-ignition temperature and cause a fire within the foam body.

8 REFERENCES

American Conference on Governmental Industrial Hygienists. Threshold Limit Values for Chemical Substances and Physical Agents in the Work Environment with Intended Changes for 1983-1984. Cincinnati, OH, 1983, 93 pp.

American Society for Testing and
Materials. Standard Test Method for
Ignition Properties of Plastic.
D-1929-77 in 1977 Annual Book of ASTM
Standards: Volume 08.02, Plastics.
Philadelphia, PA, 1977, p. 627.

American Society for Testing and
Materials. Standard Test Method for
Surface Burning Characteristics of
Building Materials. E 84-81A in 1981
Annual Book of ASTM Standards:
Volume 04.07 Building Seals and
Sealants; Fire Standards; Building
Construction. Philadelphia, PA, 1981,
p. 824.

American Society for Testing and
Materials. Standard Test Method for
Surface Flammability of Materials Using
a Radiant Heat Energy Source. E 162-81A
in 1981 Annual Book of ASTM Standards:
Volume 04.07 Building Seals and
Sealants; Fire Standards; Building
Construction. Philadelphia, PA. 1981,
p. 930.

Grantham, R. J. Use of Polyurethane Foam
in Mines. Mine Safety Appliance
Co., Evans City, PA, Jan. 1980, 49 pp.

Mine Safety Appliance Research Corp.
Evaluating Rigid Foams for Constructing
and Repairing Mine Stoppings
(contract J0308006). BuMines OFR 40-85,
1985, 151 pp.; NTIS PB 85-187656.

Mitchell, D. W., J. Nagy, and E. M.
Murphy. Rigid Foam for Mines. BuMines
RI 6366, 1964, 37 pp.

Mitchell, D. W., E. M. Murphy, and J.
Nagy. Fire Hazard of Urethane Foam in
Mine. BuMines RI 6837, 1968, 29 pp.

Timko, R. J., and E. D. Thimons. New
Techniques for Reducing Stopping
Leakage. BuMines IC 8949, 1983, 15 pp.

The Upjohn Company, Industrial Health
Services. Engineering, Medical Control
and Toxicological Considerations.
Tech. Bull. 105, Feb. 15, 1970, 52 pp.

Wilde, D. G. Combustion of Polyurethane
Foam in an Experimental Mine Roadway.
Great Britian Safety in Mines Research
Establishment. Res. Rep. 282, 1972,
41 pp.

Designing mine ventilation systems – Case studies

FLOYD C.BOSSARD
F.C.Bossard & Associates Inc., Butte, MT, USA

ABSTRACT: An overview of mine ventilation design principles, along with their application in general case histories are presented. The design of mine ventilation systems for four underground mines are reviewed. The analysis covers (1) occupational safety and health aspects, (2) design criteria established, (3) mine ventilation balance requirements, (4) primary ventilation fan specifications and electric horsepower requirements, and (5) other pertinent aspects developed by the application of computerized mine ventilation modeling techniques.

1. PLANNING A MINE VENTILATION SYSTEM

The most critical period in the design of any mine ventilation system is that time when the engineer is establishing the basic criteria. Specifications, basic ventilation engineering factors, and even required assumptions go to make up the criteria that serve as design parameters for the proposed mine ventilation system.

First of all, any engineer in the United States responsible for designing a mine ventilation system has to become familiar with the federal and state occupational safety and health regulations for the proposed operation. The federal regulations pertaining to mine ventilation system design, installation, and operation are found in the Code of Federal Regulations, Title 30 - Mineral Resources. Title 30 is updated annually and is available for purchase through the U. S. Government Printing Office, Washington, D. C. 20402. Throughout the year, impending changes in the regulations are announced in the Federal Register.

Most tunnel construction regulations (enforced by the Occupational Safety and Health Administration) are outlined in CFR 29.

The primary objectives of designing the mine ventilation system for an underground mining operation are to provide a suitable volume-quality supply of fresh air to the working areas while simultaneously diluting airborne toxic gases and dust contaminants

generated; diluting combustible gases (i.e., methane); diluting radioactive contaminants (i.e., radon-daughters); and exhausting the contaminated air from the mine. The air delivered underground may require heating during cold weather. Underground air volumes may become excessively warm and require cooling prior to delivery to the workplace. Mine primary ventilation systems should afford reasonable resistance pressure losses (efficient electrical energy consumption).

Mine ventilation circuitry has to be established and controlled to contribute to the protection of underground personnel in the event of a mine fire. In the analysis of a mine fire control program and associated worker health protection, the mine ventilation plays a dominant role in providing safeguards and protection for the workmen and the physical facilities of the mine. Whether a new mine ventilation system is being designed or an engineering evaluation is being conducted of an operating mine ventilation system, the utilization of computer model simulation techniques should be investigated. Mine ventilation design and analysis are ideal applications of computer model simulation. What other area in mine design has so many variables influencing the final results? Mine ventilation design is very much a problem of optimization because of the alternatives involved. The potential for effective economic optimization of a mine ventilation engineer utilizes mine ventilation computer software.

Following is a discussion of criteria that have to be addressed by the engineer responsible for the design and operation of a mine ventilation system. These major criteria are not listed in any ranked order of importance. It should be pointed out also that the list of topics discussed does not exhaust the subject. Typically, the mine ventilation system for various phases of the mine operation's life (i.e., 1, 2, 5, 10 and 20 years) is incorporated in the mine ventilation analysis.

1.1 Mine Planning Criteria

1. Short-term, near-term, and long-term mine production capacity.
2. Maximum physical size.
3. Installed electro-mechanical equipment. The ventilation of stationary electromechanical equipment (substations, crusher/conveyor systems, pump station, etc.) is normally planned for and accomplished by incorporating industrial ventilation systems. Methane liberated during crushing and storage of gassy products and dust emissions caused by material handling and transport are evaluated.

The ventilation requirements for rubber-tired diesel equipment are complex. Sufficient ventilation air has to be provided to every workplace where diesel equipment is operating to meet the diesel equipment nameplate ventilation volume requirements or to meet the threshold limit values for diesel toxic exhaust emissions and the airborne concentration of dusts generated. The diesel units are utilized because of their mobility and the flexibility they afford the mine. The mobility is the very reason why many mechanized mines are under-ventilated. The diesel equipment is utilized in more than one location, but the ventilation air cannot normally be shifted to follow and service the diesel equipment.

4. Primary airway cross-section and support.
5. Mining methods.

1.2 Mine ventilation criteria

1. Ventilation velocity criteria. The maximum recommended air velocity in primary airways will vary, dependent upon the type of mining operation and the length of airway. The following are generally accepted ranges of air velocity in primary airways.

2. Thermodynamic Criteria. All sources of heat entering and leaving the mine air

Table 1. Design range of air velocities in primary airways.

Air Velocity, fpm (m/s)	Type of Airway
1000-1500 (5.1-7.6)	Service Shaft
1500-2000 (7.6-10.2)	Production Shaft
500-1500 (2.5-7.6)	Main Entry
500-1000 (2.5-5.1)	Conveyor Tunnels, Declines.
±2000 (±10.2)	Sills and Raises (no production) activities)
500-1000 (2.5-5.1)	Sills & Raises (with production activities)
2000-3000 (10.2-15.2)	Exhaust Mains
3000-4000 (15.2-20.3)	Exhaust Shafts (concreted)
2000-3000 (10.2-15.2)	Exhaust Shafts (rock section)

should be evaluated. Major sources of heat in an underground mine include adiabatic compression, electromechanical equipment, groundwater, and wallrock. In a potentially hot mine, air cooling requirements will have to be evaluated. The mine inlet air may have to be heated during the winter season.

3. Type of primary mine ventilation system. An early decision has to be made as to the type of primary mine ventilation system that will be incorporated in the mine. Major alternative primary systems include:
   a. Primarily exhausting system.
   b. Primarily blowing system.
   c. Push-pull ventilation system.

4. Radon gas emission criteria. The rate of radon gas emissions in a uranium mine will vary with the porosity, permeability, and U_3O_8 grade of the ore.

5. Coefficient of friction criteria. Most published airway coefficient of friction factors were established from research conducted in airways of relatively modest size (i.e., 50 to 150 square foot area, 4.6 to 13.9 m^2) and incorporating timbered, rough rock or smooth-lined sections. Generally, the K factor reduces in airways with comparable surface lining conditions as the cross-sectional area increases.

6. The effect of the proposed mine ventilation system on mine fire control has to be analyzed. Or, vice versa, the ventilation design engineer has to appreciate what effect a mine fire may have upon the

proposed mine ventilation system.

7. Transient air-loss estimates have to be incorporated in the calculation of the mine ventilation balance. This uncontrolled leakage of fresh air into the mine exhaust air system prior to delivery to the production area workplace can be very significant (over 50 percent).

8. Combustion gas emission criteria.

9. A minimum average air velocity of 50 to 100 fpm (0.25-0.51 m/s) is required to minimize stratification in mine workings and a resultant buildup of toxic diesel exhaust emissions and combustible methane gas. The minimum average air velocity required increases as the cross-section of the mine opening increases.

In summary, the establishment of realistic and effective criteria for designing a mine ventilation system is the key to providing a safe and healthy workplace underground.

2 OIL GRAVITY DRAINAGE PROJECT

A mine ventilation analysis of a potential underground oil gravity drainage project is discussed. Basic mine ventilation design was based upon both a non-gassy mine designation as well as a gassy mine classification by the U.S. Mine Safety and Health Administration. This paper summarizes the findings of the gassy mine study and the associated mine ventilation analysis.

The project mine ventilation system was designed to (1) dilute combustible stratagases encountered to low, safe concentrations; (3) provide dilution air volumes in association with dust suppression techniques to reduce the respirable dust content in ambient air to below the threshold limit value; (4) to provide protection for the underground personnel in the event that a mine fire occurs; and (5) to maintain an ambient environment in which productive work can proceed safely, efficiently, and in reasonable comfort.

2.1 Strata gas

A potential problem that may arise during the mining of oil is hydrogen sulfide gas. Hydrogen sulfide is a colorless, flammable gas. The primary hazard associated with the inhalation of hydrogen sulfide gas is pulmonary irritation and respiratory paralysis. The Mine Safety and Health Administration (MSHA) enforces a time-weighted threshold limit value (TLV) of ten parts per million for hydrogen sulfide.

The methane gas content of the shale is 2 to 16 cf/ton. This methane content is less than that found in the Piceance/Uinta oil shales (up to 90 cf/ton). The methane content in gassy coal mines producing from different coal seams will vary from 100 to 500 scf/ton of coal. The methane gas content of the shales determined by the USBM Direct Method are relatively low, indicating the possibility that the project may not become classified as a gassy mine. The actual emissions of methane encountered when underground excavation activities commence will determine whether the project is non-gassy or gassy.

An in-mine continuous methane detection system, hardwired to a methane recording and warning system located on surface, will be installed. Hand-held, portable methane sampling equipment should also be available and utilized by underground personnel.

A remote, continuous monitoring system is not required in a gassy or non-gassy mine by any state regulatory group or by the Federal Mine Safety and Health Administration. Federal regulations do require that in gassy mines, development equipment be equipped with methane monitors that alarm at 1% methane and de-energize the equipment automatically when the concentration reaches 1.5% (30 CFR 57.21-80).

Installations for the sampling of the following gases should also be considered.

- Hydrogen Sulfide. Detectors should be installed in the exhaust air from long, dead-end development headings that are areas of groundwater influx.

- Carbon Monoxide. A continuous detector for monitoring carbon monoxide should be installed in the air exhausting from the operating panel. The purpose is to determine any potential build-up of carbon monoxide, a product of combustion of oil as well as diesel fuel. This equipment would provide early detection of a mine fire.

The most serious underground emergencies result from underground fires. The most serious fire that could occur in the project would be located in the primary inlet airway, fumigating all areas downstream serviced by this air unless fire doors are installed in critical locations.

An intricate system of control doors will be required between parallel, adjacent inlet and exhaust airways systems to effectively control ventilation flows. All doors should be constructed of inflammable materials (i.e., metal and concrete).

When the mine ventilation balance is 100,000 cfm, a methane influx of less than 250 cfm is required if an average methane concentration of 0.25 percent is not to be exceeded in the exhaust air. Five hundred cfm of methane influx will result in a 0.5

percent methane concentration in the exhaust air. If more than 1000 cfm of methane enter the mine, sections of the mine will contain more than one percent methane. Men are prohibited from performing production duties in areas when the methane concentration exceeds one percent. Additional primary inlet and exhaust airways would have to be driven to alleviate this condition

2.2 Mine ventilation design criteria

The following list of criteria include specifications and assumptions that served as basic parameters for mine ventilation design work.

1. The project has a mine ventilation balance of 30,000 cfm during the development phase and 100,000 cfm during the production phase.

2. The effects of methane emissions on the mine ventilation system were analyzed for methane emission rates of
- 0.005 cfm/ft^2, wallrock surface
- 0.001 cfm/ft^2, wallrock surface
- 0.002 cfm/ft^2. wallrock surface
- 0.1 cfm/production drill hole
- 4 ft^3/barrell of groundwater
- 10 cfm at the oil/gas/water collection station
- 15 ft^3/ton of excavated shale rock

3. Excavation parameters
- 6' φ service shaft (downcast)
- 3' φ borehold (exhaust)
- 6' φ air shaft (exhaust)
- 10' wide x 11.5' high tunnels

4. The ventilation air distribution patterns are designed to rpovide a minimum air velocity of 100 fpm in major tunnels and 50 fpm air velocity in the connecting tunnels.

5. A mine groundwater inflow of 25 gpm is anticipated during mining operations.

6. Two 8-inch diameter pipes will provide 16,000 cfm of compressed air at the bottom of the Service Sahft to supply ventilation air for the initial excavation breakthrough to the 3-foot diameter borehole. The pipes will be embedded in the shaft concrete lining or serve as guides for the service cage.

7. The oil drained from the reservoir will be totally contained in a collection piping system with the contained pressure equal to the reservoir gas pressure.

2.3 Mine ventilation system

The initial primary circuit would be

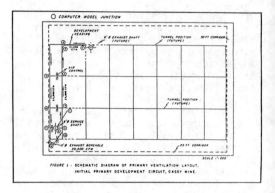

FIGURE 1 - SCHEMATIC DIAGRAM OF PRIMARY VENTILATION LAYOUT, INITIAL PRIMARY DEVELOPMENT CIRCUIT, GASSY MINE.

established by a "break out" crosscut driven from the bottom of the 6' φ Service Shaft to intersect the 3' φ Borehole. Sixteen thousand scfm of ventilation air will be provided via two 8' φ pipes in the Service Shaft to ventilate the driving of this crosscut.

Following completing of the holing, an exhaust fan capable of exhausting 30,000 cfm of air will be installed at the top of the 3' φ Borehole.

A double entry development system will be driven from the Service Shaft to connect with the Exhaust Shaft (See Figure 1). Holings between the two entries will be made on 250-300 foot intervals. A request for a variance for approval of this extended holing distance will be presented to MSHA. Current MSHA regulations state that holings of two entries driven in a gassy mine have to be made on 100-foot intervals.

Air control doors will be installed in all crosscut holings, except the last holing, between the two entries. Under this configuration 25,000 cfm of fresh inlet air courses down one entry and crosses over through the last crosscut holing; exhaust air passes down the second entry and is exhausted from the mine via the 3' φ Exhaust Borehole. Air controls noted on Figure will have to be installed to accomplish this routing of air flow.

The development headings beyond the last break-through will be ventilated with an auxiliary fan and rigid ductwork. One auxiliary fan located in fresh air ahead of the last break-through would service both advancing dead-end headings with about 2,5000 cfm of fresh air apiece. The auxiliary ventilation system will consist of a 40 HP axivane fan and a 42-inch diameter duct, providing fresh air to dead-end headings.

Following completion of the holing between the Service Shaft and the Exhaust Shaft, an exhaust fan system capable of

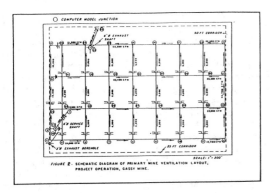

FIGURE 2. SCHEMATIC DIAGRAM OF PRIMARY MINE VENTILATION LAYOUT, PROJECT OPERATION, GASSY MINE.

increasing the mine ventilation balance to 100,000 cfm will be installed at the top of the Exhaust Shaft. The 100,000 cfm of air will provide 4.5 to 5 air changes per hour, with an air transit time of 10 to 30 minutes, depending upon the circuit traversed.

The bulk of this fresh air (±95,000,000 cfm) will course through the tunnels connecting the Service Shaft and the Exhaust Shaft. Auxiliary fans will be installed in this fresh air base. These fans in series with ventilation duct will provide the air necessary for driving the tunnel and crosscut system proposed in Figure 2. The fans should be capable of providing 20-25,000 cfm of air. The grid system of tunnels will be developed by completing each block outlined in a systematic pattern. One-third (one row) of the rectangles will require that three sides have to be developed before a holing is made. This development will be pursued from both ends over a total distance of about 650 feet. This means that a dead-end heading will be driven 325-350 feet before a holing is made. This holing distance will also require a variance from current MSHA gassy mine regulations.

After these holings are completed, air controls (ventilation doors) will be installed to control the ventilation air flows and advance the fresh air base through the last holing. Periodically, the auxiliary fans will be advanced and installed near these last crosscut holings.

Figure 2 is a schematic plan view of the final primary mine ventilation system during the operation of the underground oil gravity drainage project. The drainage holes have been drilled. An oil/gas collection storage and pumping system has been installed.

The total mine ventilation balance is 100,000 cfm. All of this fresh air enters through the 6'φ Service Shaft. Two thousand cfm is exhausted up the 3' φ Ex-

haust Borehole. The bulk of the vitiated air will be discharged to surface via the 6'φ Exhaust Shaft. The primary ventilation system is designed so that the fresh air courses away from the Service Shaft in two tunnels and is returned to the Exhaust Shaft in the second pair of tunnels.

A minimum volume of air (6,000 cfm) is allowed to circulate through each crosscut between two tunnels. It is necessary to provide the air volume for diluting and removing methane emissions in each of the crosscuts. Computer modelling incorporating a wallrock methane emission rate of 0.001 cfm/ft^2 indicate that the methane concentration in air will range up to 0.3 percent during the early development phase and up to 0.8 percent during the project operation. When a wallrock methane emission rate of 0.002 cfm/ft^2 is incorporated in the model, the methane concentration in air will range up to 0.6 percent during the initial development stage and in excess of one percent during the project operation. Any methane concentrations in excess of one percent will necessitate workmen removal from that area.

The air flow is controlled by an array of ventilation doors. Additional fire control doors will be installed in the main tunnels to compartmentalize the mine in anticipation of any combustion episode that might occur.

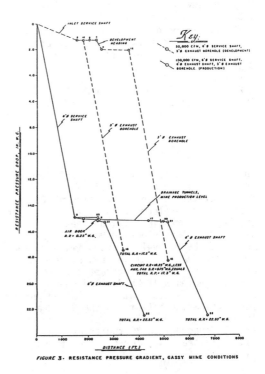

FIGURE 3. RESISTANCE PRESSURE GRADIENT, GASSY MINE CONDITIONS

397

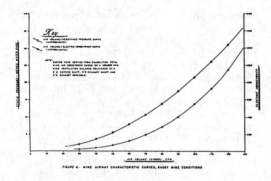

FIGURE 4. MINE AIRWAY CHARACTERISTIC CURVES, GASSY MINE CONDITIONS

Figure 3 is a schematic diagram illustrating the mine resistance losses for the primary mine ventilation circuits. The airway resistance pressure drop through the mine is plotted versus length of airway.

A plot of mine static pressure versus the quantity of air passing through the primary mine ventilation circuits was constructed. These mine airway characteristic curves (Figure 4) are useful in predicting a realistic limit of air delivery for the various airways versus mine static pressure requirements and electric horsepower consumption.

The curves were derived from computer calculated total mine air horsepower data, based on a 100,000 cfm mine ventilation balance, delivered through the mine via a 6' φ Service Shaft, 6' φ Exhaust Shaft, and a 3' φ Exhaust Borehole.

The primary fan specifications include:
1. Initial development - 31,000 cfm, 17.5 in. w.g. available static pressure, 0.069 lbs/cf air density, 150 electric horsepower.
2. Project Operation - 105,000 cfm, 22.5 in w.g. available static pressure, 0.066 lbs/cf air density, 550 electric horsepower.

3 PRECIOUS METAL MINE

The mine, capable of producing 2000 tpd of ore, is accessed by a spiral ramp and a 13-foot diameter concreted hoisting shaft. Underground development consists of horizontal levels approximately 100 feet apart, with a draw level located 900 feet below surface. Each level is approximately 1000 feet long. An end-slice open stope mining method is utilized to extract the ore. Vertical holes will be drilled down between levels. Ore will be blasted into the stope which bottoms on the 900 draw level. On the 900 draw level, broken ore will be loaded into 26 ton trucks and hauled to the shaft for hoisting to surface.

3.1 Mine ventilation system

The 13-foot diameter hoisting shaft serves as the primary inlet airway from surface to the 900 draw level. Old mine workings are upcast from the open stope on the 800 level to surface. An exhaust fan on the 300 level provides the energy for directing 60,000 cfm of exhaust air through this circuit to surface. Ninety thousand cfm of air on the 900 level will travel west and exhaust to surface via the incline. A second primary exhaust fan will be located on the surface, exhausting from a fan drift holed into the decline. An air lock would be installed in the incline, between the portal and the fan drift.

Major mine ventilation design criteria include
1. The mine ventilation system should minimize the impediment of air lock installation to operating diesel haulage equipment.
2. Two thousand tpd of ore will be hauled on the 900 draw level. Simultaneously, production activities will be taking place on up to six other levels.
3. The 900 draw level dead-end mucking crosscuts will be ventilated by auxiliary fans or fan ejector systems.
4. Maximum airway velocity
    3000 fpm - air raises
    1500 fpm - hoisting shaft
    1000 fpm - incline and haulageways
5. A minimal air leakage factor has been included in the mine ventilation balance calculations.
6. The operating diesel equipment totalled 650 bhp.
7. The mine ventilation balance requirements totalled 150,000 cfm of air.
8. The cross section of the 900 draw level and the incline in 15.5 ft. x 13 ft.

A plot of the mine characteristic curves is shown in Figure 5.

The primary ventilation fan performance and electric horsepower requirements are:
1. Incline portal: 90,000 cfm, 3.3 in.

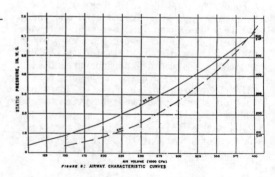

FIGURE 5: AIRWAY CHARACTERISTIC CURVES

w.g. available static pressure, 0.075 lbs/
cf air density, about 100 electric horsepower.
  2. 300 Level: 60,000 cfm, 2.5 in. w.g.
available static pressure, 0.075 lbs/cf air
density, about 50 electric horsepower.

4 BASE METAL MINE

A ventilation analysis of a proposed deep-
level mining operation has been conducted.
Development and mining operations are to
be conducted on eight levels, between 2000
and 3500 feet below surface. Approximately
10,000 tpd of ore are to be mined by a
modified room-and-pillar mining method.
At any one time, up to 10,000 bhp of diesel
equipment may be operating in the mine. The
mine ventilation design was based on the
delivery of 2,000,000 cfm of air underground to the production areas. The
following figure 6 is a plot of the
resistance pressure loss curves through
planned major circuits in the mine.
  Figure 7 includes airway characteristic
curves of proposed primary airways.
Resistance pressure is plotted against
volume and electric horsepower.
  The effective area of the primary inlet
and exhaust airways from surface range
from 250 to 400 square feet.
  The proposed primary ventilation fan
specifications are:
  1. No. 1 inlet - 1,000,000 cfm at 7.9
in. w.g. static pressure
  2. No. 2 inlet - 950,000 cfm @ 7.8 in.
w.g. static pressure
  3. No. 3 exhaust - 1,000,000 cfm @ 10.8
in. w.g. static pressure

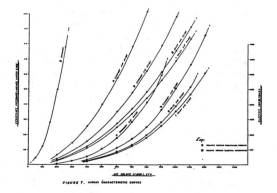

FIGURE 7. AIRWAY CHARACTERISTIC CURVES

  4. No. 4 exhaust - 760,000 cfm @ 10.3 in.
w.g. static pressure
  5. No. 5 exhaust - 250,000 cfm @ 7.9 in.
w.g. static pressure
Actual air density is 0.071 lbs/cf.

5 OIL SHALE MINE

The ventilation system design for a typical
oil shale mine producing along the outcrop
of the Mohagony ledge in Colorado is discussed. The proposed mine is to operate
in the range of 80,000 tpd, utilizing a
modified room and pillar mining method.
An upper bench is intially driven, followed
by removal of the lower bench.
  The principal environmental problems that
will be encountered underground and require
controlling in the work environment are:
(1) diesel combustion products of carbon
monoxide, carbon dioxide, nitrogen dioxide,
oxides of nitrogen, and unburned hydrocarbons; (2) airborne concentrations of
dust (including free silica) in the respirable dust fractions; (3) methane gas
released from the interstices of the oil
shale and the surrounding strata; and (4)
potentially high ambient air temperatures.
  The large production rooms in room-and-
pillar mining operations will present
special mine ventilation problems from
potential air stratification and build-up
of toxic contaminants. If a healthy
working environment is to be maintained in
these large rooms, a number of actions must
be considered.
  1. A minimum air velocity of 100 fpm
should be maintained.
  2. A system of large, flexible brattices
between pillars for controlling and routing
of ventilation air flows should be utilized.
  3. Install large propellor ejector fans
within the room area to direct ventilation
airflows to active work faces.
  4. Utilize auxiliary ventilation systems

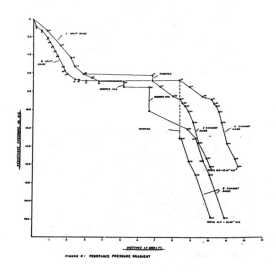

FIGURE 6: RESISTANCE PRESSURE GRADIENT

for ventilation of deadend headings.

The methane gas content of oil shale is less than that found in gassy coal beds. Available information indicates that the methane content of oil shale may range from zero scf/ton at the outcrop, to 5-10 scf/ton one to two miles from the outcrop, and 60-80 scf/ton within the deeper portions.

5.1 Mine ventilation analysis

The design mine ventilation balance is 7,000,000 cfm. The transient air losses have been estimated at 15 percent. This mine ventilation balance was based on providing an air velocity of 100 fpm in the main and panel entry development areas, and the upper bench operations in the panel. Maintaining this average air velocity will reduce the potential for stratification.

The installed diesel equipment totalled about 42,000 horsepower. Haulage trucks constitute almost 50 percent of the total diesel equipment and account for more than 50 percent of the diesel fuel consumed. It is estimated that an average of 850-975 gph of diesel fuel will be consumed.

The maximum concentration of gases will be encountered in production areas where operating diesel equipment is concentrated. Maximum predicted levels of carbon monoxide will range up to approximately 7 ppmv. The threshold limit value (TLV) is 50 ppmv. The carbon dioxide levels will range from 300 to 1400 ppmv (TLV = 5000 ppmv). Maximum predicted levels of nitric oxide will range up to 15 ppmv (TLV = 25 ppmv). Nitrogen dioxide levels will range up to 2.5 ppmv (TLV - 5 ppmv, ceiling level).

Ventilation analysis were conducted periodically over the operating life of the mine. The maximum resistance pressure losses in the primary circuits in the various ventilation plans studied will range from 0.5 to 5 inches water gage.

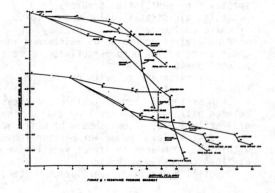

FIGURE 8 : RESISTANCE PRESSURE GRADIENT

Figure 8 is a plot of typical resistance pressure loss curves through planned major circuits in the mine.

A half dozen large primary fans (± 1,000,000 cfm) will operate at all times. Three or four smaller primary fans (250,000 cfm to 600,000 cfm) will also be required.

The electric energy consumed by the primary ventilation system will range from 1,000 to 7,000 horsepower throughout the life of the mine. Additional electric energy will be required for auxiliary fan and jet fan ventilation systems, and primary fans exhausting from mined-out areas.

A computer analysis was conducted to predict methane concentrations throughout the mine based upon methane emission rates of 5 and 10 cubic feet of methane released in the mine per ton of shale rock broken.

Low levels of methane gas will be released during mining of the oil shale. About 50 to 75 percent of this methane will be released during blasting operations. The bulk of the remaining methane will be released prior to and subsequent to the blasting. Maximum methane concentrations in the panel production areas following blasting can range from

- 0.8 to 1.2 percent, for rock releasing 5 cf of CH_4/ton
- 1.6 to 2.4 percent, for rock releasing 10 cf CH_4/ton.

The background methane levels in the exhaust airways should average less than 0.01 percent for that timeperiod that doesn't include the blasting operation. The potential for methane buildup, to dangerous levels in the mined-out panel or section areas, may present a potentially hazardous situation. The control of methane in abandoned areas of the mine will require substantial efforts.

The ventilation air-transit times will be approximately 30 to 70 minutes for panel preparation and panel production areas, and almost three hours for main entries under development.

6 OBSERVATIONS

Future world energy requirements will dictate the need to recover oil products from underground sources that cannot be claimed by present day recovery techniques. The principal enviornmental and physical factors facing the recovery of oil by gravity drainage or mining of oil shales are the occupational safety and health aspects and potential rock mechanic problems. The mine ventilation design requirements of such underground opera-

tions will have to be more sophisticated
than present-day ventilation practices
in hard rock and coal mines. The implications of (1) mining in areas adjacent to
oil and gas reservoirs and (2) removing
50 to 75-foot thick seams of oil shale
over tens of square miles are obvious.

The persons responsible for mine ventilation design of oil gravity drainage or oil
shale mines are in a unique position to
make technical history or create tragedy.

Selection of proper ventilation system in the Jui-San coal mine

SHENG-SHYONG HUANG
EMRO, ITRI, Taipei, Taiwan

ABSTRACT: Mine ventilation system consists of two basic steps: the first step includes all the intake airways and return airways; the second step covers the local ventilation system. Obviously the selection of a mine ventilation system is very important to the operation of a mine with complicated and long airways. The ventilation system not only directly affects the ventilation condition but is also related to the operating cost. The following illustrates the successful selection of a proper ventilation system carried out in the Jui-San coal mine. This renovated ventilation system consists mainly of using the additional inclined shaft as a return-air shaft and installing a suitable 300 horsepower main fan to replace the two units of 120 horsepower main fans installed at slope 3 and slope 4. The mine then has gained substantial advantages from the ventilation improvement.

1 THE SELECTION OF VENTILATION SYSTEMS IN THE JUI-SAN COAL MINE

The Jui-San coal mine is the largest coal mine in Taiwan. Its annual coal production is about 210,000 metric tons. The coal production comes mainly from four operating blind inclined shafts. In order to decrease the high air temperature of the working faces in slope 3 and slope 4 due to poor ventilation, a new inclined shaft (A: $5.78M^2$ L: 1382m) was considered to be utilized for increase the total through-put air to the various working faces. However this change of ventilation system has raised considerable problem because the inclined shaft was connected to the old system. The ventilation engineer of the Energy and Mining Research and Service Organization (EMRO), ITRI was assigned to help the Jui-San Coal Mine to solve this ventilation problem. After a comprehensive ventilation survey of the mine they presented a report of improvement of ventilation system to the Jui-San Coal Mine. The length, size and resistance of airways is shown in table 1. The ventilation system of the Jui-San coal mine is shown as figure 1.

In this report the following analyses and suggestions of ventilation problems were adequately accepted by the mine and put into action. The mine then had much substantial benefits from the ventilation improvement.

1.1 The disadvantages and difficulties to use the new inclined shaft for the intake-air shaft:

1. If the new inclined shaft is used for intake-air shaft it is helpful to decrease the air temperature of working faces during winter time. However, in summer time the air temperature on the surface is high and the intake air temperature is high also. As the high temperature of intake air can reach the working face in two minutes, an adverse effect may occur harmful to the environmental condition of the working faces.
2. The horizon haulage way now

Table 1. The length, size and resistance of airways.

Name of Airway	Length (m)	Area (m^2)	Resistance (Weisbach)	Resistance per 100 meters (Weisbach)	Resistance per 100 meters (Murgue)
horizon-haulage way (1)	756.4	5.38	0.04189	0.00554	5.54
horizon-haulage way (2)	2528.0	4.46	0.20749	0.00821	8.21
main slope	429.3	4.74	0.03340	0.00778	7.78
slope 1	773.2	3.96	0.13349	0.01726	17.26
slope 1-2 (I)	905.9	3.98	0.15144	0.01672	16.72
slope 1-2 (R)	704.7	2.89	0.42449	0.06024	60.24
slope 2	780.0	4.45	0.12400	0.01590	15.90
slope 2-2 (I)	655.0	3.92	0.16631	0.02539	25.39
slope 2-2 (R)	444.0	3.26	0.12539	0.02822	28.22
slope 3	795.0	4.14	0.10227	0.01286	12.86
slope 3-2 (I)	477.0	4.11	0.09267	0.01943	19.43
slope 3-1 (R)	646.0	3.70	0.13394	0.02073	20.73
slope 3-2 (R)	467.0	3.26	0.18746	0.04014	40.14
slope 4	699.0	4.34	0.07507	0.01074	10.74
slope 4-1 (R)	593.0	4.21	0.06454	0.01088	10.88
I-Tung slope	647.5	4.69	0.04475	0.00691	6.91
middle slope	832.0	3.75	0.20932	0.02516	25.16
San-Tung slope	1013.0	4.27	0.04729	0.00467	4.67
Chung-Tung slope	326.0	3.90	0.04764	0.01461	14.61
Wu-Tung slope	570.0	4.64	0.05064	0.00888	8.88
level of slope 1	730.0	3.10	0.22646	0.03102	31.02
coal face of slope 1	387.0	-	0.55771	0.14411	144.11
level of slope 2	1456.0	2.37	1.08220	0.07433	74.33
coal face of slope 2	467.0	-	1.91865	0.41085	410.85
level of slope 3	987.0	2.65	0.67087	0.06797	67.97
coal face of slope 3	197.0	-	0.67307	0.34166	341.66
level of slope 4	2534.0	2.80	1.06207	0.04191	41.91
coal face of slope 4	590.6	-	10.96899	1.85725	1857.25

used for air intake is about 3,300 meters long. The rock temperature is rather low due to closeness to ground surface and a long period of air cooling has already taken place. In summer time the rock of low temperature will absorb some heat from the air current to cool down the air temperature. Therefore, the horizon haulage way should be kept as an intake-air way. The heat coming from virgin rock will release heat much faster in a new opening where the rock has not been cooled down by the air current. Therefore the new slope is not suitable to be used for air intake.

3. The ventilation system as the new slope is used as an return-air slope shown as figure 2.

4. The quantity of air should be 1,200 m^3/min more when the new slope is used for air intake as compared to its use for air return. The additional air quantity is needed to compensate for the permittable (planning) air leakage that may occur in slope 3 and slope 4. If the total demand of air quantity of slope 3 and slope 4 is 3,000 m^3/min, then the intake air quantity of the new slope should be 4,200 m3/min. The air velocity of the new slope is 727 m/min. It is far above the permittable maximum air velocity of 450 m/min. in the intake-air way as required by the mine law.

5. It is very difficult to arrange the layout of ventilation system and air quantity control in slope 3 and slope 4.

If the new slope is used as an intake-air slope, and the third and the fourth return inclined shafts are used as return-air slope, then an air regulator should be installed in slope 3 and slope 4 to control the air splitting and make sure all the working faces can get sufficient quantity of air. However, it is a serious problem to install air regulator in the haulage slope because it may hinder the haulage and may intro-

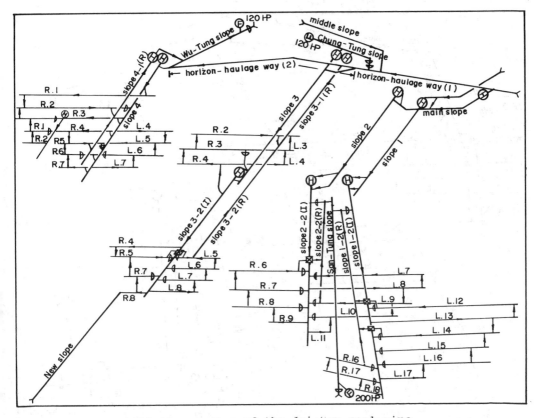

Fig. 1. The ventilation system of the Jui-San coal mine.

duce safety problems. If booster fan is installed to control the air distribution of slope 3 and slope 4 the ventilation electric power consumption would be increased. On the other hand it is very difficult to maintain a suitable air quantity control.

1.2 According to the result of ventilation networks analysis, the comparison of the fans used at present with the fans to be used when the new slope is used as air return (plan) is shown in table 2. The data of Table 2 show that: (a) the total fan horsepower of the plan is 38 horsepower less than that used at present. About U$13,658 can be saved annually if the unit cost of electric power is U$0.055 per 1 Kw-hr.. (b) It can save some supervision and maintenance cost of fans and enhance the safety because the total

number of fans used in the plan is 6 sets only, and the number used at present is 20 sets. (c) The comparison of effective air quantity obtained at present and in the plan is shown in table 3. It is obvious from the data of Table 3 that the air quantity of each slope in the plan is much more than that can be attained at present, especially the third slope.

1.3 In the early stage to use the new slope as return-air slope, the new slope and slope 4 have not yet been connected together. It is used only as the return-air slope of slope 3. Then the 300 HP main fan of the new slope needs not be operated with full load. When we reduce the rotating speed of the 300 HP main fan by 14.5%, its loading capacity will be lowered to 200 HP, and the effective air

405

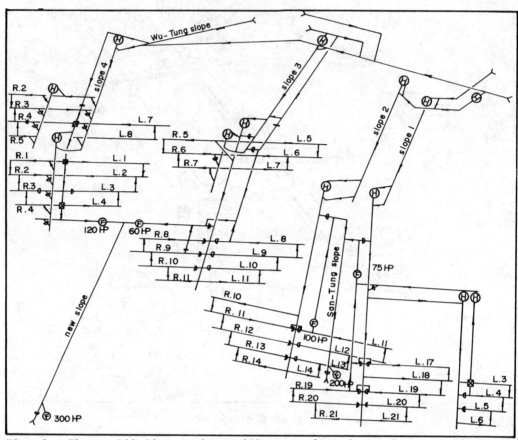

Fig. 2. The ventilation system as the new slope is used as an return-air slope.

Table 2. Comparison of fans used at present and in the plan.

	Present				Plan		
Name of slope	Fans Used		Remarks	Name of slope	Fans Used		Remarks
	Set	HP			Set	HP	
San-Tung	1	200	main fan	San-Tung	1	200	main fan
Chung-Tung	1	120	"	New Slope	1	300	"
Wu-Tung	1	120	"	Slope 1	1	75	booster fan
Slope 1	4	115	booster fan	Slope 2	1	100	"
Slope 2	7	145	"	Slope 3	1	60	"
Slope 3	4	143	"	Slope 4	1	120	"
Slope 4	2	50	"				
Total	20	893		Total	6	855	

Table 3. Comparison of effective air quantity at present and in the plan.

Unit: m^3/min.

Name of slope	(1) Present	(2) Plan	Comparison of (2) and (1)	
			Air Quantity	%
Slope 1	1359	1500	+141	+10.4
Slope 2	1270	1300	+30	+2.4
Slope 3	1147	1659	+512	+44.6
Slope 4	1227	1553	+326	+26.6

406

Table 4. Comparison of effective air quantity attained at present and in the plan.

Unit: $m^3/min.$

Name of slope	(1) Present	(2) Plan	Comparison of (2) and (1)	
			Air Quantity	%
Slope 1	1359	1367	+8	+0.6
Slope 2	1270	1271	+1	+0.1
Slope 3	1147	1630	+483	+42.1
Slope 4	1227	1373	+146	+11.9

quantity of slope 3 and slope 4 will be increased by 483 $m^3/min.$ (42%) and 146 $m^3/min.$ (11.9%) respectively. The comparison of effective air quantity at present and in the plan is shown in table 4.

2 PRACTICAL BENEFITS OF VENTILATION IMPROVEMENTS OF JUI-SAN COAL MINE

The Jui-San coal mine adopted a modified ventilation system recommended by the EMRO and improve its ventilation performance considerably. This renovated system consists mainly of using the new inclined shaft as a return-air shaft and installing a suitable 300 horsepower main fan on the surface to replace the two units of 120 horsepower main fans installed at slope 3 and slope 4. The 300 horsepower of main fan is now operated only to the capacity of 200 horsepower because the slope 4 and the new slope have not connected together yet and the 300 horsepower fan is used only as the main fan of slope 3. The practical benefits of ventilation improvements are shown as follows:
1. The intake air quantity of slope 3 and slope 4 has increased 55.6% (729 $m^3/min.$) and 15.7% (181 $m^3/min.$) respectively.
2. The average dry-bulb air temperature of the coal faces at the intake-air side and the return-air side is decreased from 33.3°C and 34.4°C to 31.1°C and 31.8°C respectively. The decrease of temperature is 2.2°C and 2.6°C.
3. The average wet-bulb air temperature of the coal faces at the intake-air side and the return-air side is decreased from 30.5°C and 31.4°C to 28.5°C and 29.6°C respectively. The decrease of temperature is 2°C and 1.8°C.

4. The air-recirculation at slope 3 has been eliminated completely.

407

Ventilation experiences with longwall mining
at Macquarie Colliery

S.BATTINO & P.B.MITCHELL
BHP Collieries Division, Wollongong, Australia

ABSTRACT: It is generally accepted that the introduction of longwall extraction is often associated with complex ventilation planning problems. These problems are made much more severe when the colliery introducing this mining system is already experiencing high gas emissions even during the development phase and the coal seam worked and/or adjoining seams also exhibit a proneness to spontaneous combustion. This paper presents the ventilation experiences at Macquarie Colliery, which has encountered and overcome difficulties with both these hazards but where ventilation systems have to be carefully and continually reassessed.

1 INTRODUCTION

Macquarie Colliery, in the Newcastle region of the Sydney Coal Basin (refer to Fig.1), extracts the Dudley Seam at a depth of 280 m. The seam has a gas composition of approximately 99% methane and 1% CO_2 (air-free basis), a gas content of 3 m^3/tonne and a relatively low gas macropermeability of the order of 40 mdarcy.

Access to the seam workings is via a 7.5 m diameter intake shaft used for men and materials and a 5.5 m diameter return shaft which has a 10 man capacity winder installed for second egress purposes. Both shafts are fully concrete lined.

The initial layout for the mine intended to incorporate the use of shortwall extraction techniques which had been very successful at other BHP collieries. The envisaged production levels and face unit displacement were such that ventilation could be achieved by using twin centrifugal fans similar to existing units at other BHP mines. The fans were initially operated at 520 rpm producing 125 m^3/s at 2.5 kPa.

Specific ventilation problems were experienced from an early date in colliery development, mainly associated with the presence of faults and dykes. The geology of the strata is complex with five overlying and two underlying seams believed to contribute to gas emissions into the working seam especially as strata relaxation takes place after longwall mining.

Prior to the introduction of longwall extraction, seam gas, ventilation and geological investigations were carried out. These included working seam gas content and composition tests, gas balance and ventilation surveys during the development phase and the assessment of gas drainage application. From these data, predictions were made of gas emissions for a range of daily production levels and the required air quantities to maintain these emissions within statutory limits. However during the extraction of the No. 1 longwall block significant difficulties were encountered with high methane emissions from the unsealed goaf as well as disturbing levels of carbon monoxide in return airways. A reappraisal was necessary of the ventilation system and regular gas monitoring was conducted in the returns and at the goaf edge.

On the basis of the experiences with the No. 1 longwall mining, the ventilation system was reassessed for the No. 2 longwall. The main alterations incorporated a single entry maingate isolated from the No. 1 longwall roadways, goaf sealing from the returns and the introduction of gas post-drainage of the longwall goaf. Notwithstanding the recent extraction of the No. 2 longwall block, the gas data substantiated the necessity to continue close monitoring of longwall ventilation.

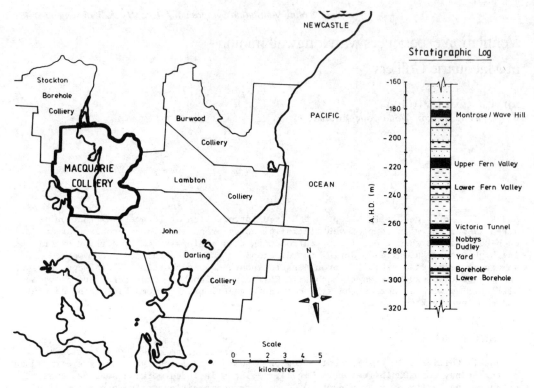

Fig. 1. B.H.P. Collieries in Newcastle Region of Sydney Coal Basin.

2 EARLY GAS AND VENTILATION EXPERIMENTS AND SURVEYS

From the early stages of pit development, gas and ventilation monitoring programmes were undertaken to determine the source(s) of high methane readings recorded at Macquarie Colliery (Battino, 1980). The investigations included mine air sampling on a daily basis from the upcast shaft during a full production week to establish the variation in the methane-make with coal output, gas balance surveys in a working panel where fault zones had been intersected and the assessment of the relative gas contributions of adjoining seams into the coal seam extracted. Gas emission prediction studies were also carried out in preparation for longwall mining and the ventilation requirements to maintain methane levels below statutory limits were established.

Typical variations of gas-make with coal production and time in the development phase of Macquarie Colliery are shown in Fig. 2. The graph indicates that the methane-make resulting from the colliery output during the week of investigation is approximately 0.27 m^3/sec. In general,

there is a positive correlation between the calculated methane-make and the mine production during the previous 24 hours. It should be noted that the ventilation and percentage methane readings were taken each time during dayshift. The low methane make obtained for the Monday, can be explained in terms of the weekend non-production period, indicating that reasonable continuity of coal extraction and face exposure is necessary to sustain the above level of gas-make.

The results of the survey conducted to establish the influence of the presence of fault zones on resulting gas emissions and make in a development district are presented in Fig. 3. Generally, there is close correspondence between the peaks of methane make figures and the locations of the faults intersected in the heading investigated.

Gas balance examinations undertaken during various production periods indicated gas-make figures in the order of 0.1 m^3/sec from the working face in solid developments.

The gas emission prediction studies carried out prior to the extraction of the first longwall block were based on several

recognised world-wide methods, namely those
developed by Winter, Stuffken, Schulz,
Barbara Experimental Mine in Poland and
the European Coal Committee. The reasons
for the choice of these particular theo-
ries were the existing local geological
and mining conditions, coal seam characteristics, gas content, permeability and
diffusion parameters as well as specific
gas emission data obtained from Macquarie
and other Australian collieries.

Table 1 (Lunarzewski and Battino, 1983)
presents the variation in methane emissions
for production levels in the range of 1000
to 10000 tonnes. This allowed the prediction of the ventilation levels required to
keep methane concentrations within statutory limits separately for longwall mining
and for the complete mine workings. The
dilution levels of 1.25 and 2.25% CH_4 were
computed on the basis of permissible
methane concentrations in the ventilation
current, while the 0.5% CH_4 values were
included for upcast shaft ventilation
requirements. Examination of the gas
emission rates obtained by the five different methods revealed that variations of
5 to 40% could be expected. However,
based upon the data available at that time,
the calculations were considered reasonably reliable for Macquarie Colliery
conditions. Nevertheless, it was strongly
recommended that regular gas emission and
gas balance measurements be programmed
soon after the start of the No. 1 longwall
extraction in order to verify the accuracy
of the methods used and to correct any
major variation which may have occurred.

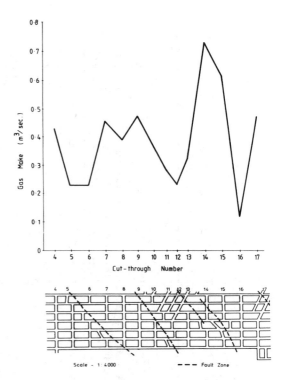

Fig. 3. Influence of Fault Zones on Gas Make in a
Development Panel, Macquarie Colliery.

3 LONGWALL NO. 1 EXPERIENCES

3.1 Ventilation system

As a result of the mine surveys the fan
speed was increased from a belt driven
520 rpm to 740 rpm. This gave a considerable increase to the total mine air
quantity to allow longwall mining to
commence.

The first longwall face which commenced
in March 1983 was 141 m wide with a panel
length of 1500 m. Initially ventilation
proposals were to split the return air at
the tailgate, with part going to the 2
North East section and the remainder
returning to the 1 North East return to
assist in the dilution of the methane
bleed from the goaf (Fig. 4). Because all
the air in the return gate panel traversed
the face, quantities in the order of only
15 m³/s were being used to dilute the
methane in the goaf return. It soon became
apparent that this method of ventilation
was inadequate to maintain methane levels
within statutory limits.

A revised ventilation layout having all
air return via the bleed roadway supplemented by additional air entering the tail-

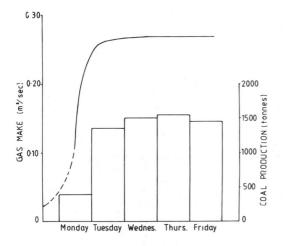

Fig. 2. Plot showing variation of gas make,
coal production and time, Macquarie Colliery.

411

TABLE 1

Prediction of Gas Emission Levels and Ventilation Requirements
for Different Outputs at Macquarie Colliery
for methods of Barbara Experimental Mine & European Coal Committee

| | LONGWALL MINING | | | ALL UNDERGROUND WORKINGS | | | |
| Coal output (tonnes/day) | Gas emission calculated (m³/min) | Gas emission including irregularity co-efficient* (m³/min) | Air quantity necessary for dilution to 1.25%CH₄ (m³/sec) | Total gas emission calculated (m³/min)*** | Total gas emission including irregularity co-efficient** (m³/min) | Air quantity required for dilution to 2.25%CH₄ (m³/sec) | Air quantity required for dilution to 0.5%CH₄ (m³/sec) |
1	2	3 = 2xcoefft.	4 = 3xcoefft.	5 = 2+coefft.	6 = 5xcoefft.	7 = 6xcoefft.	8 = 6xcoefft.
1000	7.07	10.61	14.15	51.07	84.27	62.42	280.9
2000	9.11	13.67	18.23	53.11	87.63	64.91	292.1
3000	10.68	16.02	21.36	54.68	90.22	66.83	300.7
4000	12.00	18.00	24.00	56.00	92.40	68.44	308.0
5000	13.17	19.76	26.35	57.17	94.33	69.87	314.4
6000	14.22	21.33	28.44	58.22	96.06	71.55	320.2
7000	15.19	22.79	30.39	59.19	97.66	72.34	325.5
8000	16.09	24.14	32.19	60.09	99.15	73.44	330.5
9000	16.94	25.41	33.88	60.94	100.55	74.48	335.2
10000	17.74	26.61	35.48	61.74	101.87	75.46	339.6

* Irregularity coefficient = 1.5 due to changes in barometric pressure and mining conditions for longwall.
** Irregularity coefficient = 1.65 due to changes in barometric pressure and mining conditions for whole of pit.
*** These values include longwall gas emission and all developing panels gas emission (ribside and extraction). For development, 1500 t/d av. production was assumed and av. methane emission 44 m³/min.

Note: Statutory regulations at time of the study stipulated a 2.5% methane limit in return headings. Present regulations are now 2.0% maximum.

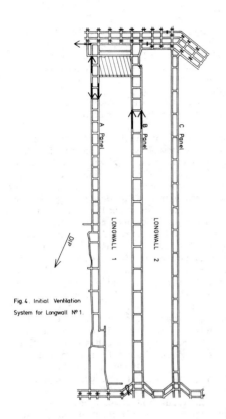

Fig. 4. Initial Ventilation System for Longwall Nº 1.

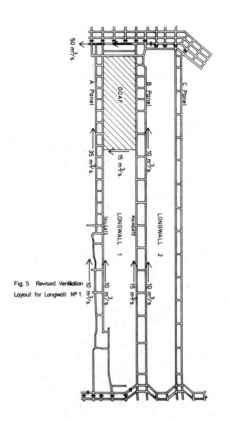

Fig. 5. Revised Ventilation Layout for Longwall Nº 1.

gate section was adopted. Quantities up to 35 m³/s flowed along this single return roadway to 1 North East where combined with development panel return air, gave a total of 50 m³/s in the main return. The revised ventilation layout is shown in Fig. 5. this proved successful, until the face had retreated approximately 500 m when unacceptably high methane concentrations were again recorded. Gas emissions from the adjacent seams were higher than predicted. Even though methane readings along the face were of the order of 0.3%, return concentrations exceeded statutory limits. The dip of the seam was working to the advantage of face operations, but the high concentrations flowing out from the goaf cavity through the open cut-throughs was of serious concern.

3.2 Gas emissions recorded

During the extraction of the No. 1 longwall block, gas balance surveys were conducted around the working face and goaf areas, intake and return gateroad headings, individual development panels and main colliery returns. At each specific test location,

air quantities, CH₄ concentrations and roadway area measurements were taken. The methane levels were ascertained by D6 methanometer readings as well as by gas sample analysis in the laboratory. It was stressed that in all gas balances undertaken, the number of test points investigated was commensurate with the prevailing conditions and that a larger number of these sites had to be studied to achieve greater accuracy.

The results of the gas balance measurements carried out on various occasions after the start of the No. 1 longwall mining are presented in Table 2, which also shows the daily longwall production at the time of sampling. In general, all measured gas-makes were found to be much greater than those predicted. Examination of the data revealed that the gas emission levels recorded for all underground workings were reasonably close to those predicted, whereas the emissions registered uniquely for the longwall area varied more significantly (refer to Tables 1 and 2). On the basis of these new findings, a modified prediction of gas emission levels and ventilation requirements for different outputs was made.

The sources of methane and contribution

413

TABLE 2

Gas Balance in
No. 1 Longwall Area
- Macquarie Colliery -

Date	Pro- uction (tonne/) (day)	Intake emission*	Return emission	CH₄ make
		m³/min		
28.4.83	3770	18.2 + 2	43.0	22.8
19.5.83	4000	19.4 + 3	46.4	24.0
20.5.83	4000*	20.5 + 4	50.0	25.5
9.8.83	6850	16.0 + 7	59.5	36.5
25.8.83	5900*	14.2 + 6	47.0	26.8
26.8.83	5900	15.9 + 8	54.0	30.1

* On these dates, there was no production
but the figures shown are based on the
previous day's output.
** The values added in this column refer
to the calculated gas emissions along the
gate roads in the goaf area.

of each seam to the absolute gassiness of
the pit are summarised in Table 3. It is
clear that the immediate roof caving areas
provided the most significant contribution
to the absolute gassiness of the longwall
area, although the contribution of the
overlying adjacent seams in the roof was
not to be discarded. One cross-measure
and two in-seam boreholes were drilled at
this stage to assess the viability of gas
pre and post-drainage as a measure of
maintaining the methane emission levels
within statutory limits. The use of a
Pemberthy type steam ejector was adopted
to apply suction levels to 50 mm Hg.

Although limited experimentation was
conducted, it was encouraging to find that
even with relatively low vacuum levels,
some improvement in gas flowrates was
achieved.

3.3 CO detection and subsequent
     ventilation considerations

Approximately six months after the commencement of Longwall No. 1, abnormally high
carbon monoxide readings were recorded at
the goaf edge and in the longwall return
airway. It had been known from laboratory
samples that seams adjacent to the working
seam had a slight propensity for spontaneous combustion although no conclusive
indications had occurred in workings at
nearby collieries. Because of the unusually high carbon monoxide levels, a number
of air sampling stations were established
as shown in Fig. 6. Bladder samples were
regularly taken, initially on a weekly
basis but more frequently with increased
levels of oxidation in the goaf as the face
retreated. Tables 4a and 4b show the
results from stations established at the
goaf edge and the longwall return. The
carbon monoxide to oxygen deficiency ratio
remained quite high at the goaf edge but
did not indicate conclusively that a
definite heating had developed. Due to
the subsequent levelling out of the readings
recorded, it was felt that the oxidation
rate of the coal in the goaf was deceler-
ating as the goaf compacted, effectively
reducing the migration of the air across
the goaf to the return airway. This was
substantiated by the higher carbon monoxide
readings taken at the goaf edge approximately 50 m behind the face and at the

TABLE 3

Sources of Methane and Contribution of
each seam to the absolute Gassiness of the Longwall area

Source	Coal Seams	Seam Thickness (m)	Distance From Working Seam (m)	Contribution to Gassiness %	m³/min
Working seam and immediate roof caving	Dudley	1.9	-	63	24
	Nobby	1.0	2.4		
	Victoria Tunnel	3.4	6.1		
Roof	Lower Fern Valley	0.8	36.3	21	8
	Upper Fern Valley	5.6	46.1		
	Montrose/Wavehill	4.5	87.6		
Floor	Yard	1.1	14.0	16	6
	Borehole	6.8	32.0		

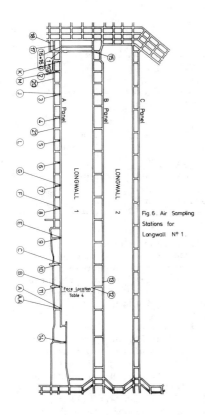

Fig.6. Air Sampling Stations for Longwall N°1.

face start line (sample points B and 2) where the caving strata was not fully compacted. By maintaining a steady rate of extraction it was predicted that no rapid increase of oxidation would occur.

The immediate sealing of the longwall goaf was seriously contemplated but the following important considerations were in favour of completing the longwall under the existing method of ventilation:

(a) The carbon monoxide readings and the Graham Ratio determined at the indicative sampling points gave no conclusive evidence of increased heating;

(b) The main body of the goaf exhibited no signs of heating activity;

(c) Certain goaf migration paths had

been established for the air and appeared to be in a reasonable state of equilibrium. Any alteration of pressures across the goaf would change the atmospheric condition within the goaf and cause renewed heating;

(d) Because of the high methane emission rate from the working and surrounding seams, any 'U" system would require a back return at the tailgate to keep the tailend of the face free of gas. This would still result in air passing through the fringe of the zone of maximum heating activity; and

(e) A period of only two months was required to complete the longwall block.

In conjunction with the decision to maintain the present system of ventilation, a programme was initiated to increase the air sample monitoring stations around the goaf. Site preparations for seals around the No. 1 longwall area commenced.

The extraction of Longwall 1 was successfully completed on 13th March 1984 with an overall average weekly production for the 54 weeks of 10415 tonnes.

4 LONGWALL 2 EXPERIENCES

4.1 Revision of ventilation system

The results recorded during the extraction of the No. 1 longwall block made it clear that a different longwall layout and method of ventilation was required which satisfied the following:

(a) To reduce the incidence of heatings which developed within the goaf, minimal migration of air through this area was necessary; this ruled out any concept of a bleed system;

(b) Because of the high methane emission rates experienced and evidence of progressive increases in emission levels, a simple "U" ventilation system would not be satisfactory;

(c) To reduce the exudation of the goaf gas into the returns and maintain an extinctive atmosphere on completion of the face, the minimum number of entries and hence seals had to be established; and

TABLE 4a

Gas Analysis of Samples - Longwall 1, 18th November 1983

Sample point	Location	CO_2	O_2	CO	CH_4	H_2	N_2	CO/O_2
18	Main Return	0.08	20.2	0.0012	2.30	<0.001	77.4	0.40
2	Longwall Return	0.17	19.4	0.0053	3.5	<0.001	76.9	0.55
5	Longwall Return	0.13	19.3	0.0036	4.6	<0.001	76.0	0.44
8	Longwall Return	0.08	19.7	0.0027	3.8	<0.001	76.4	0.50
1G	Goaf Cavity	0.39	1.90	<0.001	84	<0.001	13.7	<0.01

TABLE 4b

Gas Analysis of Samples – Longwall 1, 17th December 1983

Sample points	Location	CO_2	O_2	CO	CH_4	H_2	N_2	CO/O_2
18	Main Return (in 55 m³/s of air)	0.08	20.30	0.0009	2.05	.001	77.60	0.37
2	Longwall Return (in 15 m³/s of air)	0.18	19.30	0.0054	3.70	.001	76.90	0.52
8	Longwall Return	0.10	18.55	0.0029	3.70	.001	76.6	0.39
10	Longwall Return	0.05	20.35	0.0009	1.65	.001	77.9	0.31
B	Behind Face	0.12	17.70	.0067	8.8	.001	73.4	0.39
1G	Goaf Cavity	0.33	1.40	0.001	91.0	.001	7.3	0.01

(d) A methane drainage system needed to be incorporated since ventilation alone proved insufficient in diluting the high gas emissions.

A single entry maingate roadway was driven between 1 North East and 2 North East leaving a solid coal barrier between longwalls. The face ventilation direction was reversed for this arrangement. This reduced the maingate seals to two for this longwall. Because the tailgate panel had already been driven, the number of seals required for this section was already established.

To comply with the above features, three proposals were considered for ventilating the second face:

Proposal 1 – Simple "U" system
Although this system for ventilating the face was ideal as far as minimising the potential to goaf heatings, the high methane emissions ruled against this method. Methane drainage would be limited since roadways behind the face would not be maintained, and hence gas dilution at the face would be insufficient.

Proposal 2 – Back return system
The method utilises the concept of a limited bleed system at the tail end of the face whereby the air splits, allowing part to go beside the goaf to the first open cut-through. This migration beside the goaf maintains a relatively gas-free condition at the return end of the longwall. The concept of this system has been well documented (Highton, 1979). This method of ventilating the face (Fig. 7) would improve the control of gas at the tail end and reduce the risk of spontaneous combustion but there was still some doubt about the levels of methane in the air returning via the goaf edge, particularly as this section was only to be supported by timber props. Any reduction in flow via the "back return" could lead to explosive mixtures along this

pillar length. Also, the migration paths through the goaf to the cut-through in this back return method were very near to the active zone. Careful control of the regulator in the tailgate intake airway and good seal construction in the return were essential.

Proposal 3 – Goaf seals with Y Ventilation
This is a variation of the ventilation system used for No. 1 longwall. All access to the goaf is sealed except for the cut-through immediately behind the face, as shown in Fig. 8. Return air splits, as for proposal 2, to draw the goaf gases away from the tail end of the face. Although a similar concern existed with

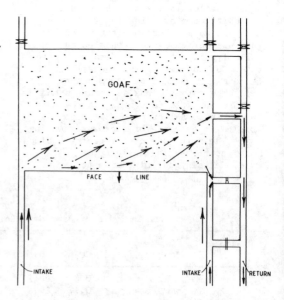

Fig. 7. Back Return Ventilation System for Longwall N° 1.

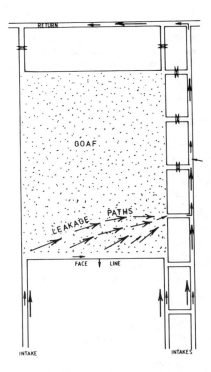

Fig. 8. The Y Ventilation System

the limited goaf migration to this "back return" cut-through, with a steady rate of extraction and continuous sampling, some control could be maintained over any possible heating. This method was amenable to a methane drainage installation, and enabled access for thorough sampling of the goaf. It was accepted that high differential pressures would exist across the goaf and excellent seal construction was required.

As the gate roads had already been developed and limitations imposed by the main ventilation arrangement for the mine made proposal 2 difficult to apply, proposal 3 was adopted for Longwall No. 2.

4.2 Gas drainage application

The second longwall block extraction commenced in July 1984 with very satisfactory results. However, after three months of production and a reasonably sized and well established goaf, unacceptably high methane levels (up to 1000 l/sec) were recorded in the return headings. An underground methane drainage plant was immediately commissioned to intercept and extract the excessive and hazardous gas from the goaf and adjoining coal seams before its entry into the ventilation system. This was

undertaken by drilling cross-measure holes firstly from the back of the goaf (main returns) and subsequently in the tailgate panel to intersect the various overlying and underlying seams. These boreholes were then sealed with standpipes connected to a 100 mm methane range and the gas drained under vacuum using two Nash Hytor pumps installed in the 1 North East headings. Total effective gas flowrates from these boreholes were in the range of 6 to 8 m^3/min of CH_4.

Subsequent investigations revealed that the longwall No. 2 return seals were ineffective in both keeping the oxygen level down through the goaf area and controlling the methane for drainage purposes. A campaign was therefore undertaken to reseal the walls, ribs, roof and floor with tekcrete. On the basis of all available data, a relationship was established between longwall production, methane volumes drained and gas levels in return headings. The necessity to adhere to this relationship in order to satisfy statutory requirements meant, on certain days, that longwall extraction could not be carried out for the full four shift period.

Prior to high methane emissions being recorded, the best weekly longwall production achieved was approximately 38000 tonnes. Subsequent emission problems required further cross-measure drainage holes to be drilled from the tailgate and more effective sealing to be undertaken. On completion of the No. 2 longwall block in March 1985, the average weekly production for the 33 weeks of its extraction was 15140 tonnes.

4.3 Effectiveness of seals

Before the commencement of the second longwall face, stoppings were erected across the entrance to the face start line adjacent to the 1 North East return roadway. The cut-throughs in the tailgate section (C Panel) had been stopped off during the development phase, and, to allow the longwall ventilation circuit to be completed, the stoppings at locations B and C were breached (see Fig. 9). As the wall retreated, the stoppings along the tailgate section were successively removed and then rebuilt when the face had extracted to the next cut-through. Because an increasingly higher pressure differential was developing between the face and the 1 North East return as the face retreated, leakage of methane through the tailgate side seals was a constant problem. After high production periods, face shutdowns were necessary due to high methane levels

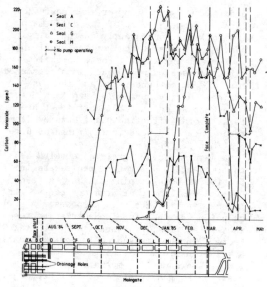

Fig. 9   Carbon Monoxide Readings from Behind Stoppings
Adjacent to Longwall 2 Goaf

in the return. This was ameliorated to a
certain extent when the methane drainage
plant was commissioned in late October,
1984 but this in turn exacerbated the
problem of increased air migration through
the goaf and particularly alongside the
goaf edge behind the seals on the tailgate
side where roadway compaction was not
complete. The re-established air circulation resulted in higher carbon monoxide
readings being recorded behind these seals.
The influence the methane drainage plant
had on the rate of oxidation can be seen
in Fig. 9, where during holiday periods,
the carbon monoxide levels decreased at
seals adjacent to the drainage points.

It was later found that seal C, which
was a big bag stopping, was placed on
loose coal and subsequent resealing and
grouting around the perimeter decreased
the measured CO level.

Although carbon monoxide readings as
high as 225 ppm were recorded, the oxygen
deficiency ratio rarely exceeded 0.5 and
definitely not on any sustained basis. No
conclusive evidence of a heating was established but it was apparent that the
stoppings along the tailgate section were
not as effective as anticipated in creating an extinctive atmosphere.

5 CONCLUSIONS

Information gained from the extraction of
the first two longwall blocks has revealed
methane emission levels which cannot be

controlled by ventilation alone. The
application of methane drainage to the
goaf and adjoining seams is essential to
maintain safety and productivity during
longwall mining at the colliery. To this
end a higher capacity surface drainage
methane plant is being installed to replace
the existing underground exhausters.

The revised ventilation system for Longwall 2 appears to have provided improved
control over the problem of spontaneous
heating in the goaf and, in turn, has
resulted in increased productivity.

To maintain higher longwall outputs
further experimentation is necessary on
sealing techniques to achieve more effective drainage and control gas outflow into
the return headings. It is expected that
during the extraction of future longwalls
frequent re-evaluation of ventilation
quantities will be required to keep the
faces operational on a continuous basis.

6 ACKNOWLEDGEMENTS

The results of the investigations are
published with the permission of The Broken
Hill Proprietary Company Limited, although
the views expressed are those of the
authors and not necessarily those of the
Company. Recognition is made of the
significant contribution by Mr. B. McKensey,
Manager, Macquarie Colliery for valuable
discussions and suggestions at the various
stages of investigations.

7 REFERENCES

Battino, S. 1980. Gas investigations at
    Macquarie Colliery, BHP Collieries
    Research internal report, Wollongong.
Highton, W. 1979. The case against bleed-
    er entries and the reasons for a safer
    and more efficient alternative, Second
    International Mine Ventilation Congress,
    Reno, Nevada.
Lunarzewski, L., and Battino, S. 1983.
    Predictions of gas emissions and venti-
    lation requirements for longwall mining
    at Macquarie Colliery, BHP Collieries
    Research internal report, Wollongong.

11. Underground heat flow and climate

The computation of heat loads in mine airways using the concept
of equivalent wetness

M.G.MACK & A.M.STARFIELD
University of Minnesota, Minneapolis, USA

ABSTRACT: The computation of heat and moisture transfer rates into mine airways
is relatively easy if (i) the airway can be approximated, in cross-section, by a
circular hole, and (ii) the conditions of moisture and temperature around that
hole are uniform. In practice, however, moisture is seldom distributed uniformly
around the perimeter of an airway and the symmetry of the solution is destroyed.
In a recent paper, Starfield and Bleloch (1983) developed the concept of
replacing the non-uniformly wet airway by a uniform airway of 'equivalent
wetness', and showed that the value of that equivalent wetness was not
intuitively obvious. This paper discusses how the equivalent wetness can be
calculated and demonstrates the usefulness of the concept. In particular, it is
shown how it can be used to develop relatively fast solutions on microcomputers
for a number of problems of practical interest. These include predicting the
effects of daily or seasonal fluctuations in the inlet air temperatures.

1 INTRODUCTION

A common problem in the ventilation of
deep mines is the need to calculate the
heat and moisture transfer occuring in
long airways or shafts, where the main
source of heat is the surrounding rock.
Heat flows through the rock and heat and
moisture transfer occur from the rock
surface to the air in the airway. Axial
heat flow in the rock can usually be
neglected, so that it is possible to
compute the heat flow through the rock at
one cross-section of the airway at a time.
Heat and moisture transfer rates at that
cross-section can then be calculated and
used, with the appropriate psychrometric
equations (Hemp 1982), to predict air
temperatures further along the airway.
These in turn are used in the boundary
conditions for the heat conduction problem
at the next cross-section, and so on. The
difficult part of the calculation is to
solve the time-dependent heat conduction
equation at each cross-section.

There are certain simple problems for
which this solution has been found and
tabulated. Goch and Patterson (1940)
presented the solution for a circular (in
cross-section) airway, the surface of
which is held at a constant temperature.

Jaeger and Chamalaun (1966) and Starfield
(1966a) tabulated the solution for a
circular airway in which there is
convection to air at a constant
temperature. These solutions can be used
to calculate heat and moisture transfer
rates, and hence air-temperatures in an
airway in an isotropic homogeneous medium,
provided the entire surface of the airway
is completely dry or else uniformly wet.

In real airways the restrictions under
which the tabulated solutions apply seldom
hold. Air temperatures are not constant,
but fluctuate from day to day and over the
course of a year. Geometrically, airways
are seldom circular in cross-section, and
the surface wetness is not uniform. The
surrounding rock may be neither
homogeneous nor isotropic. Geometry is the
least important of these considerations,
unless the cross-section of the airway is
very far from circular or square.

Moisture and heat transfer rates can be
calculated for the real problem using
methods such as finite elements. In
practice, such methods will not be used by
ventilation engineers because of the cost,
in both time and money. The challenge we
face is to provide methods that account
for the important effects on air
temperatures but that are sufficiently

quick to be implemented on microcomputers. This paper will describe a concept called the equivalent wetness, and examine how it can be used to obtain the solution to a number of practical problems in mine ventilation.

2 THE CONCEPT OF EQUIVALENT WETNESS

At any cross-section, sensible heat is transfered from the rock surface to the ventilating air, moisture is evaporated from wet portions of the rock surface, and there may be radiant heat transfer between rock surfaces (e.g. from a dry roof to a wet floor) and from the surfaces to the air. The relevant equations are

$$S = h.A.(T_{surf} - T_{air}) \qquad (1)$$

and

$$M = f.e.A.(p_{sat}(T_{surf}) - p_{vap}) \qquad (2)$$

where S = sensible heat transfer rate
 M = moisture transfer rate
 h = convective heat transfer coefficient
 A = area of surface
 T_{surf} = temperature of surface
 T_{air} = drybulb temperature of the air
 f = wetness factor
 e = moisture transfer coefficient
 p_{sat} = saturated vapor pressure of air (at the rock surface temperature)
and p_{vap} = actual vapor pressure of the air.

Although radiant heat transfer is governed by the equation

$$R = \sigma(T_1^4 - T_2^4)$$

it can be simplified to

$$R = k_{rad}(T_1 - T_2)$$

where σ = Boltzman's constant
 k_{rad} = radiative heat transfer coefficient (k_{rad} depends on the temperatures T_1 and T_2 but can usually be assumed to be constant in mines)
and
 T_1, T_2 = temperatures of surfaces (or surface and medium) between which radiation occurs.

The wetness factor f in equation 2 was introduced by Starfield (1966b). It can vary from 0 for a dry surface to 1 for a completely wet surface. It can be thought of as a correction factor for the fact that, at the microscopic level, part of a surface that appears wet may in fact be dry. The wetness factor can vary along an airway and also around its circumference.

Most airways, for instance, are wetter on the floor than on the roof.

Consider, for example, a circular cross-section which is wet (with wetness factor f) over part of the surface and completely dry over the rest of the surface, as shown in Figure 1. Obviously, the isotherms in the surrounding rock will be distorted near the floor (Starfield and Dickson 1967), and heat flow will not be radially symmetric. This complicates the solution of the heat conduction equation. It follows that a useful (if somewhat wishful) question to ask is whether it is possible to find an equivalent problem, with uniform wetness ϕ over the entire surface, which leads to exactly the same heat AND moisture transfer rates as in the problem of Figure 1. We would call ϕ the equivalent wetness factor. Intuitively, one might expect that the equivalent wetness could be found by simply multiplying the actual wetness factor by the fraction of the surface which is wet. Thus if the bottom quarter of the surface is wet with f = 0.4, one might expect the equivalent wetness factor ϕ to be 0.1.

The only way to explore the validity of this concept is via numerical experiments. In a recent paper Starfield and Bleloch (1983) developed an extremely fast solution for heat conduction in the rock surrounding an airway as represented in Figure 1. They called their method the 'quasi-steady' method and used it to show that the concept of equivalent wetness was viable, but that there was no simple way of estimating the equivalent wetness factor ϕ; the intuitive calculation, based on areas, was totally incorrect.

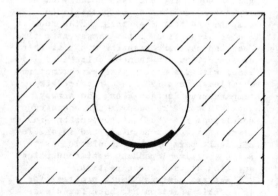

Figure 1. Cross-section through typical airway. (The heavily shaded portion of the surface is wet, the rest of the circumference is dry.)

422

The work described in this paper began with a thorough investigation of the concept of equivalent wetness, using the quasi-steady algorithm to perform a large set of numerical experiments. The method was used to calculate the heat and moisture transfer rates from many different partly wet airways exposed to air at constant temperature. The heat and moisture transfer rates for each case were then calculated for the same conditions except that the real partly wet surface was replaced by a uniformly wet surface. The equivalent wetness factor that gave the same, or similar, heat and moisture transfer rates as the real airway was found by trial and error for each case.

This investigation showed that the main factors to which the equivalent wetness is sensitive (other than the actual wetness of the surface and the proportion of the surface which is wet) are ha/k_{rock} (where a is the radius of the airway and k_{rock} the thermal conductivity of the surrounding rock), k_{rad}, the wetbulb depression (i.e. the difference between the wet and drybulb temperatures), and the drybulb temperature. Table 1 presents approximate values for equivalent wetness that depend only on the actual wetness and the proportion of the surface that is wet, as well as more complicated (and accurate) empirical formulae that account for the factors listed above.

3 A SIMPLE APPLICATION

Suppose that the inlet air temperatures for an airway do not change with time. Suppose too that at each cross-section the surface of the airway is partly wet and partly dry (as in Figure 1). The wet and drybulb temperatures can then be predicted quickly and easily on a microcomputer using the 'quasi-steady' algorithm. They can also be calculated using the concept of equivalent wetness. This can be accomplished as follows.

Combining equations (1) and (2) the total heat transfer from a wet surface can be written

$$F = S + M.L = h.A.(T_{surf} - T_{eff})$$

where L = latent heat of water and the effective air temperature is defined by

$$T_{eff} = T_{air} - \frac{f.L.e}{h}(p_{sat}(T_{surf}) - p_{vap}) \quad (3)$$

This definition of an effective air temperature can be used to treat combined heat and moisture transfer as convective heat

transfer to air at the effective temperature. The latent and sensible heat transfer contributions are required separately to calculate wet- and drybulb temperature changes and they can be obtained from

$$S = h.A.(T_{surf} - T_{air})$$
$$\text{and} \quad M = \frac{F - S}{L}$$

If we have a surface that is only partly wet, then the above equations can still be used if we replace the wetness factor f by the equivalent wetness factor ϕ. In this way the tables of Jaeger and Chamalaun or Starfield that were developed for convective heat transfer for radially symmetric heat flow into dry airways can in fact be used to predict heat and moisture transfer rates for airways that are wet on only part of their circumference.

There is no advantage to be gained by calculating heat and moisture pick-up along an airway in this way on a microcomputer: the quasi-steady algorithm is both fast and reliable for this purpose. However this application of equivalent wetness could be used on a calculator, together with the equivalent wetness formulae from Table 1. We can in fact test how accurate these calculations would be by comparing them with solutions obtained via the quasi-steady algorithm. Table 2 shows some of the results of these comparisons. Use of the equivalent wetness gives reasonable results for low wetnesses (say 0.2) but the accuracy deteriorates as the actual wetness increases. The last result in Table 2 represents one of the worst possible situations. Although use of the formula produces a 19% error in the heat transfer, the absolute error is only 12 W/m, which is acceptable.

4 FLUCTUATING INLET AIR TEMPERATURES

Fluctuations in the inlet air temperature (either daily or seaonal) can often be important. The quasi-steady algorithm cannot be used to predict the effect of these fluctuations except in the following very crude fashion: to predict the average temperatures at the end of an airway, we could assume constant inlet air temperatures that are the average values of the fluctuating inlet air temperatures, while to predict the maximum (or minimum) temperatures at the end of the airway, we could assume constant inlet air temperatures that are equal to the maximum (or minimum) values of the fluctuating inlet air temperatures.

Table 1. Formulae for equivalent wetness.

Fraction Wet	Actual Wetness Factor	Approximate Equivalent Wetness	Accurate Expression for Equivalent Wetness
0.25	0.2	0.035	$0.033 + \dfrac{\ln(10/H)}{500} + \dfrac{(db-wb)-5}{5000} + \dfrac{(k_{rad}/h)^{\frac{1}{2}}}{200} + \dfrac{35-db}{2500}$
0.25	0.5	0.070	$0.056 + \dfrac{[\ln(10/H)]^{1.6}}{250} + \dfrac{3[(db-wb)-5]}{5000} + \dfrac{3(k_{rad}/h)^{\frac{1}{2}}}{100} + \dfrac{3(35-db)}{2000}$
0.25	1.0	0.080	$0.074 + \dfrac{7[\ln(10/H)]^{1.6}}{1000} + \dfrac{3[(db-wb)-5]}{2000} + \dfrac{(k_{rad}/h)^{\frac{3}{5}}}{20} + \dfrac{3(35-db)}{1000}$
0.50	0.2	0.080	$0.074 + \dfrac{[\ln(10/H)]^{1.6}}{500} + \dfrac{(db-wb)-5}{2500} + \dfrac{(k_{rad}/h)^{\frac{1}{2}}}{110} + \dfrac{35-db}{1400}$
0.50	0.5	0.15	$0.14 + \dfrac{\ln(10/H)}{100} + \dfrac{(db-wb)-5}{500} + \dfrac{3(k_{rad}/h)^{\frac{2}{5}}}{100} + \dfrac{3(35-db)}{1000}$
0.50	1.0	0.22	$0.19 + \dfrac{3[\ln(10/H)]^{1.5}}{200} + \dfrac{(db-wb)-5}{250} + \dfrac{(k_{rad}/h)^{\frac{1}{2}}}{11} + \dfrac{3(35-db)}{500}$

These crude approximations ignore the effect of heat stored in the rock near the airway surface during fluctuations. In the rest of this section we will show how the concept of equivalent wetness, together with Duhamel's theorem, can be used to compute this storage effect, and we will then compare the results with the crude approximations (neglecting peripheral heat storage) described above.

4.1 Duhamel's theorem for variable air temperatures

Duhamel's theorem, as stated by Carslaw and Jaeger (1959) is: if $F(x,y,z,t)$ represents the temperature at (x,y,z) at a time t in a solid in which the initial temperature is zero, while its surface is kept at unit temperature, then the solution of the problem when the surface is kept at temperature $\Theta(t)$ is given by

$$v = \int_{0}^{t} \Theta(\tau) \cdot \frac{\delta}{\delta t} F(x,y,z,t-\tau)\,d\tau \qquad (4)$$

For a cross-section through an airway $F(x,y,z,t)$ corresponds to the problem addressed by Goch and Patterson (1940). However, although the air temperature at a cross-section is known, the rock surface temperature is unknown because convection occurs from it to the air. Approximating the surface temperature Θ in (4) by a continuous piecewise linear function, integrating by parts, and equating the

conduction through the rock to the convection from the rock, leads to the following algorithm:

(i) Set $\Theta_0 = 0$

(ii) Calculate Θ_1 from

$$h[\Theta_1 - v(\Delta t)] = \Theta_1 \frac{Q(\Delta t)}{\Delta t}$$

(iii) for all $j > 1$ calculate Θ_j from

$$h[\Theta_j - v(j\Delta t)] = \Theta_0 G(j\Delta t)$$
$$+ \frac{1}{\Delta t}(\Theta_1 - \Theta_0)Q(j\Delta t)$$
$$+ \frac{1}{\Delta t} \cdot \sum_{n=1}^{j-1}(\Theta_{n+1} - 2\Theta_n + \Theta_{n-1})Q[(j-n)\Delta t]$$

where $\Theta_j = T_{surf} - VRT$ at $t = j\Delta t$

$v(j\,t) = T_{air} - VRT$ at $t = j\Delta t$

and VRT = virgin rock temperature. It should be noted that these equations are explicit. The calculation of Θ_j requires the values of Θ at previous steps only. G and Q are, respectively, the flux at time $j\Delta t$ and the total heat flow up to time $j\Delta t$ for the problem adressed by Goch and Patterson. Both of these functions have been tabulated by Goch and Patterson and approximating formulae (for use on a computer or calculator) have been obtained by Starfield and Bleloch (1983) and Whilllier and Thorpe (1971).

Table 2. Examples of the use of equivalent wetness

Description	Heat Transfer W/m			Moisture Transfer mg/m/s		
	Quasi-Steady	Approximation	Formula	Quasi-Steady	Approximation	Formula
$h=15.6$ W/m^2 °C	52.83	65.63	56.88	46.18	40.92	44.92
$k_{rad}=5$ W/m^2 °C						
radius=1m		Error = 24.2%	Error		Error = -11.4%	Error
VRT=50°C			=			=
DB=35°C			7.7%			-2.7%
WB=30°C						
age=1 year						
f=0.2 over ¼ of surface						
$h=10.4$ W/m^2 °C	176.2	193.4	181.4	30.75	25.09	30.83
$k_{rad}=5$ W/m^2 °C						
radius=1m		Error = 9.8%	Error		Error = -18.4%	Error
VRT=50°C			=			=
DB=25°C			3%			0.3%
WB=20°C						
age=1 year						
f=0.2 over ¼ of surface						
$h=10.+$ W/m^2 °C	239.5	304.5	270.8	116.1	95.73	110.5
$k_{rad}=5$ W/m^2 °C						
radius=5m		Error = 27%	Error		Error = -17.5%	Error
VRT=50°C			=			=
DB=25°C			13.1%			4.8%
WB=20°C						
age=1 year						
f=0.2 over ¼ of surface						
$h=10.4$ W/m^2 °C	444.9	477.3	471.4	22.7	20.1	22.68
$k_{rad}=5$ W/m^2 °C						
radius = 5m		Error = 7.3%	Error		Error = -11.5%	Error
VRT=50°C			=			=
DB=25°C			6%			0.1%
WB=25°C						
age=1 year						
f=0.2 over ¼ of surface						
$h=10.4$ W/m^2 °C	59.73	138.5	71.22	88.72	51.54	83.94
$k_{rad}=5$ W/m^2 °C						
radius=1m		Error = 132%	Error		Error= -41.9%	Error
VRT=50°C			=			=
DB=25°C			19.2%			-5.4%
WB=20°C						
age=1 year						
f=1 over ¼ of surface						

4.2 Seasonal fluctuations in airways

The algorithm developed from Duhamel's theorem allows one to predict heat transfer for fluctuating inlet air temperatures in a dry airway. It can easily be extended to the problem of an airway that is uniformly wet by replacing T_{air} by the effective air temperature T_{eff} of equation (3). Finally, it can be extended to an airway that is only wet over a part of its circumference by using the equivalent wetness factor ϕ instead of the actual wetness factor f.

This approach leads to a quick solution on a microcomputer that takes into account storage effects at the periphery of the airway. The first question to consider is how accurate that solution is. Comparisons were made with finite difference solutions and these showed that the two solutions were almost identical.

The next question to consider is whether it really is necessary to take the heat storage effects into account: is it possible to make good predictions via the crude use of the quasi-steady algorithm as described at the beginning of this section? The answer can be found by comparing the results obtained from the two algorithms. This was done and it was found that the accuracy of the algorithm neglecting the storage capacity of the rock was independent of the wetness and

425

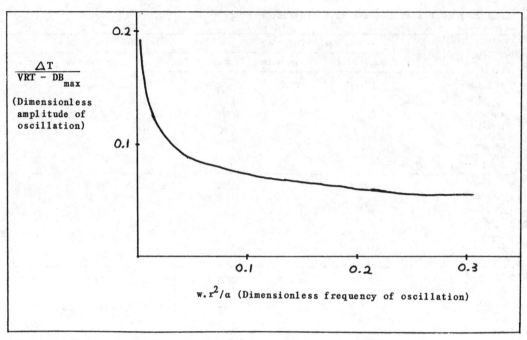

$\dfrac{\Delta T}{VRT - DB_{max}}$

(Dimensionless amplitude of oscillation)

$w \cdot r^2/\alpha$ (Dimensionless frequency of oscillation)

Figure 2. Showing the combination of dimensionless parameters that leads to a 10% error (maximum) if one neglects heat regeneration at the surface of an airway when computing the effect of fluctuating air temperatures.

was in fact only sensitive to two factors:

    i) the ratio of the amplitude of the air-temperature oscillation to the difference between the maximum drybulb temperature and the virgin rock temperature, (a dimensionless measure of the amplitude of the air temperature fluctuation), and

    ii) the ratio wr^2/α (effectively a dimensionless frequency) where w is the seasonal frequency.

Figure 2 shows for what values of these parameters a maximum error of 10% (in the total heat flow from the rock at a cross-section) occurs in simulating a four year period by neglecting the heat storage capacity of the rock.

4.3 The effect of daily fluctuations

Figure 2 suggests that while it may often be permissible to neglect Duhamel's theorem when investigating the effect of seasonal fluctuations in the inlet air temperatures, the heat regeneration mechanism in the rock at the periphery of the airway is likely to be important at higher frequencies, as in daily fluctuations. This suggests that problems involving daily temperature fluctuations should be solved (using Duhamel's theorem,

effective air temperatures, and the concept of equivalent wetness) as described in section 4.2. There is however a practical problem: in order to accurately approximate the sinusoidal temperature fluctuation at least 10 time-steps (Δt) are required per cycle. To represent 4 years of seasonal fluctuations thus requires 40 steps. To represent 4 years of daily sinusoidal fluctuations however, 14600 steps would be required!

The problem can be resolved by breaking the solution down into two parts: first the solution obtained under average, constant inlet air conditions and then a daily, sinusoidal fluctuation superimposed on that solution. The first solution can be obtained via the quasi-steady algorithm, and the second using Duhamel's theorem as in section 4.2. In practice it was found that if daily cycles were superimposed on the average solution at any time, then all transients associated with the introduction of the cycles died out within five or six days, leaving a recurring fluctuation. Duhamel's theorem could thus be implemented using only 50 or 60 time steps.

In the case of daily fluctuations, it was found that the rock heat storage capacity did indeed have a significant effect on the temperatures in the airway

426

and that it could not be neglected. The algorithm using Duhamel's theorem with the concept of equivalent wetness thus provides the only viable quick estimate of the effect of daily fluctuations.

5 CONCLUSIONS

We have shown how the concept of equivalent wetness can be used to simplify the solution of many practical problems associated with the prediction of temperature and humidity increases along mine airways. In particular, we have suggested that the concept can be used:
(i) on pocket calculators, for the case of constant inlet air temperatures, and
(ii) on micro-computers for fluctuating inlet air temperatures.

Moreover, we have shown that the heat storage capacity of the rock may be ignored for seasonal fluctuations, but cannot be ignored when predicting the effect of daily fluctuations in the inlet air temperatures.

ACKNOWLEDGEMENTS

This work was carried out under contract to the Chamber of Mines of South Africa Research Organisation by the Itasca Consulting Group, Inc. It forms part of the Research Organisation's program on the reduction of heat flow in mines.

REFERENCES

Carslaw,H.S. and J.C.Jaeger 1959. Conduction of heat in solids. London:Oxford University Press.

Goch,D.C. and H.S.Patterson 1940. The heat flow into tunnels. J.Chem.Metall.Min. Soc.S.Afr. 41:117-121.

Hemp,R.1982.Psychrometry. In J.Burrows (ed.),Environmental engineering in South African mines. Johannesburg:Mine Vent. Soc.S.Afr.

Jaeger,J.C. and T.Chamalaun 1966. Heat flow in an infinite region bounded internally by a circular cylinder with forced convection at the surface. Aust. J. Phys. 19:475-488.

Starfield,A.M. 1966a. Tables for the flow of heat into a rock tunnel with different surface transfer coefficients. J.S.Afr.Inst.Min.Metall. 66:692-694.

Starfield,A.M. 1966b. The computation of temperature increases in wet and dry airways. J.Mine Vent.Soc.S.Afr 19:157-165.

Starfield,A.M. and A.L.Bleloch 1983. A new method for the computation of heat and moisture transfer in a partly wet airway. J.S.Afr.Inst.Min.Metall. 84:263-269.

Starfield,A.M. and A.J.Dickson 1967. A study of heat transfer and moisture pick-up in mine airways. J.S.Afr.Inst. Min.Metall. 67: 211-234

Whillier,A. and S.A.Thorpe 1971. Air temperature increases in development ends. Internal research report of the Chamber of Mines of South Africa.

Simplified method to calculate the heat transfer between mine air and mine rock

XINTAN CHANG & RUDOLF E.GREUER
Michigan Technological University, Houghton, USA

ABSTRACT: Totally dry airways are rare. Even seemingly dry rock contains liquid water and water vapor, which by migration, evaporation, or condensation exert considerable influence on rock and air temperatures.

Whereas many mathematical solutions for the dry heat transfer exist and are widely used, only isolated attempts to describe the wet case have become known.

The Mining Engineering Department of Michigan Technological University approached this problem in connection with efforts to provide transient-state simulations of ventilation systems. A rigorous mathematical approach was used, and it was proved that general solutions can be obtained. Examples are given and the pertinent computer programs are presented.

For short time intervals, like those of interest in mine emergency planning, simplified solutions can be useful. They are derived and their applicability is discussed.

INTRODUCTION

The heat transfer between mine air and mine rock has always been of great concern to mine ventilation engineers. Ventilation planning for hot mines as well as predictions of natural draft effects in ventilation systems require temperature precalculations. Considerable efforts have therefore, for more than half of a century, been invested into developing methods for these precalculations.

There exist by now a large number of approaches to predict the combined dry heat transfer by conduction within the rock and by convection from the rock to the air. None of these approaches can, however, take the parallel mass transfer of water and its temperature effect satisfactorily into account.

This is regrettable because it is recognized that even seemingly dry airways are to a considerable extent influenced by water migration, evaporation, or condensation. Small water quantities can, due to the large latent heat of water, have great effects.

The fact that no convincing mathematical descriptions of the simultaneous heat/mass transfer exist indicates the complexity of the problem. The Mining Engineering

Department of Michigan Technological University made an attempt to close this gap, and it shall be described in the following paper.

PHYSICAL MODELING

Past attempts to describe the effects of water on mine air temperatures fall roughly into two categories. One of them uses statistical tools to interpret systematical field-measured data for empirical relationships between temperature, humidity and other ventilation parameters. It does not attempt to explain the nature of their dependence or to draw generally valid conclusions. Rather it offers useful equations of localized significance. The other one is a semi-empirical technique which tries to derive functional relationships between temperature and other ventilation parameters. A number of loosely defined factors, coefficients, and more or less justified relations are introduced for this purpose to provide for general applicability. There exist, however, so many assumptions, which are neither theoretically sound nor universally tenable, that the usefulness of this approach has to be limited in range or a lack in accuracy has to be expected.

Like several predecessors, the authors tried to follow a rigorous analytical approach. To simplify the presentation, airways surrounded by isotropic and homogeneous rock and with cylinder symmetry shall be discussed. No energy changes of the water other than the enthalpy change of evaporation is considered, because it is in magnitude many times larger than all other energy components.

For the problem under discussion it seems to be reasonable to assume a common temperature for both the dry and the wet portion of the wall surface. Then, by applying Fourier's equation of thermal conduction to the heat transfer process in a section perpendicular to the airway axis, the governing equation and initial boundary conditions read

$$\frac{\partial T_r}{\partial t} = a \left(\frac{\partial^2 T_r}{\partial R^2} + \frac{1}{R} \frac{\partial T_r}{\partial R} \right) \tag{1}$$

subject to

$$t=0 \quad \text{all } R \quad T_r = T_{rv} \tag{2}$$

$$t>0 \quad R \to \infty \quad T_r = T_{rv} \tag{3}$$

$$R=R_o \quad \lambda \frac{\partial T_r}{\partial R} = h(T_r - T_a) + \psi L D_t (X_{ww} - X_w) \tag{4}$$

where

T_r = rock temperature
T_a = air temperature
T_{rv} = virgin rock temperature
T_{rw} = temperature on rock surface
t = time
R = coordinate in radical direction
ψ = wetness (area ratio)
a = thermal diffusion coefficient
L = latent heat of evaporation at T_{rw}
D_t = mass transfer coefficient
X^t = absolute humidity of air
X_w = absolute humidity near wet wall
R_{ww} = hydraulic radius of airway
h^o = heat transfer coefficient

Its solution describes the temperature distribution within the rock under the restrictions of the prescribed boundary initial conditions. Consequently, the temperature and humidity variation of the passing air can be determined.

TECHNIQUE OF SOLVING THE GOVERNING EQUATION

As can be seen, the major difficulty in solving the governing equation lies in the non-linear boundary conditions, Eq. (4). Since the absolute humidity X can be a very complicated function of temperature and pressure, it seems that the equation can only be solved through a numerical method.

It is possible to overcome this problem in the following way.

For a section under consideration, X_w is a constant. As the effect of atmospheric pressure changes on the variation of saturation pressure of water has only secondary significance, X_{ww} can be considered as a function of T_r at $R=R_o$ only. Therefore, the boundary condition in Eq. (4) can be rewritten as

at $R=R_o$

$$\lambda \frac{\partial T_r}{\partial R} = h(T_r - T_a) + \psi L D_t \{f(T_r) - X_w\} \tag{5}$$

set

$$T_r(t,R) = T_o(t,R) + T_1(t,R) \tag{6}$$

The governing equation can then be resolved into two sets of equations, i.e.

$$\frac{\partial T_o}{\partial t} = a \left(\frac{\partial^2 T_o}{\partial R^2} + \frac{1}{R} \frac{\partial T_o}{\partial R} \right) \tag{7}$$

subject to

$$t=0 \quad \text{all } R \quad T_o = T_{rv} \tag{8}$$

$$t>0 \quad R \to \infty \quad T_o = T_{rv} \tag{9}$$

$$R=R_o \quad \lambda \frac{\partial T_o}{\partial R} = h(T_o - T_a) \tag{10}$$

and

$$\frac{\partial T_1}{\partial t} = a \left(\frac{\partial^2 T_1}{\partial R^2} + \frac{1}{R} \frac{\partial T_1}{\partial R} \right) \tag{11}$$

subject to

$$t=0 \quad \text{all } R \quad T_1 = 0 \tag{12}$$

$$t>0 \quad R \to R_1 \quad T_1 = 0 \tag{13}$$

$$R=R_o \quad \lambda \frac{\partial T_1}{\partial R} = hT_1 + \psi L D_t \{f(T_o + T_1) - X_w\} \tag{14}$$

430

As can be easily verified, Eqs. (7) to (14) are equivalent to the previous governing equation and the associated restrictions. Moreover, T_o exactly fits the same definition as the temperature distribution in the rock under "dry" conditions. One solution which has remained popular since the fifties is

$$\theta = \frac{T-T_a}{T_{rv}-T_a} = 1 - \sqrt{\frac{R_o}{R}\frac{B_i}{B_i'}}\left(1-\mathrm{erf}(y_1)\right.$$

$$\left.-e^{B_i'(\frac{R}{R_o}-1)}\ e^{(B_i')^2 Fo}\ \{1-\mathrm{erf}(y_2)\}\right) \quad (15)$$

where

B_i = Biot number
B_i' = B_i + 0.375
Fo = Fourier number
y_1 = $(R-R_o)/(2\sqrt{at})$
y_2 = $B_i\sqrt{Fo}+ (R-R_o)/(2\sqrt{at})$

T_1, the temperature correction due to the phase change of water, is governed by Eqs. (11) to (14), in which the semi-infinite domain has been replaced by a finite domain from R_o to R_1, an arbitrarily prefixed radius of the influential circular. This serves to facilitate the process of solution and does not cause any numerical difference since R can be any number, as large as wanted.

The mathematical term of $f(T_o+T_1)$ in Eq. (14) can be expressed in various forms in accordance with different accuracy requirements. It can, without difficulty, be obtained from steam tables for saturated vapor pressure under atmospheric conditions. Consequently, with the aid of Taylor's expansion, $f(T_o+T_1)$ can be approximated as

$$f(T_o+T_1) \simeq f(T_o) + f'(T_o)T_1 \quad (16)$$

provided T_1 is reasonably small, or in other words, when the relative humidity of air is not too low.

When considering an air current of very low relative humidity, Taylor's expansion takes the form

$$f\{(T_o+T_1)_i\} \simeq f\{(T_o+T_1)_{i-1}\} +$$

$$f'\{(T_o+T_1)_{i-1}\}\ \{(T_o+T_1)_i - (T_o+T_i)_{i-1}\}(17)$$

where the subscript indicates time intervals. This is a more general form, yet more difficult to handle. Eq. (17)

indicates that the boundary condition for T_o, Eq. (10), has to be rewritten to accommodate the T_o term in the Taylor's expansion. Consequently, no simple solution like Eq. (15) can be obtained for it. In fact, in this case, T_o has a similar solution as that for T_1. Though the corresponding solution has also been obtained and verified, to keep this paper as concise as possible we will stay on the condition inferred by Eq. (16), i.e. small T_1.

On the surface of the wall, $R=R_o$, we have

$$T_o(t,R) = T_o(t,R_o) = T_o(t) \quad (18)$$

which indicates that $f(T_o)$ and $f'(T_o)$ are functions of t only. By substituting the above relations back to boundary condition (14), the following equation results.

$$\lambda\frac{\partial T_1}{\partial R}\Big|_{R_o} = hT_1 + \psi LD_t\Big(f\{T_o(t,R_o)\}$$

$$+ f'\{T_o(t,R_o)\}\ T_1-X_w\Big) = \xi T_1+\eta \quad (19)$$

where

$$\xi = h + \psi LD_t f'\{T_o(t,R_o)\} \quad (20)$$

$$\eta = \psi LD_t[f\{T_o(t,R_o)\}-X_w] \quad (21)$$

The newly converted boundary condition is a function of time t. Recalling the implicit requirement imposed by the approximation, Eq. (16), one realizes that the whole time span must be divided into a series of small time intervals. Each of the time intervals should be reasonably small to allow the temperature correction T_1 to suitably fit the linear approximation.

One can set: time = $\Sigma(\Delta t)$, where every time interval is not necessarily equal to all others as long as the accommodated temperature correction T_1 and the variation of T_o are small. Furthermore, to assure standing on a theoretically sound basis, it is imperative that the temperature distribution patterns in the successive time intervals perfectly match each other.

If one takes one of the time intervals as an example, the governing equation of T_1 reads:

$$\frac{\partial T_1}{\partial t} = a \left(\frac{\partial^2 T_1}{\partial R^2} + \frac{1}{R} \frac{\partial T_1}{\partial R} \right) \tag{22}$$

subject to

$$t=0 \quad \text{all } R \quad T_1 = F(R) \tag{23}$$

$$t>0 \quad R=R_1 \quad T_1 = 0 \tag{24}$$

$$R=R_o \quad \lambda \frac{\partial T_1}{\partial R} = \xi T_1 + \eta \tag{25}$$

in which the time at the beginning of the present time interval is set as time zero, and ξ and η are constants corresponding to the relevant conditions at $t=0$. $F(R)$ is the distribution pattern of temperature corrections by the end of the last time interval. Obviously, for the first time interval, the equation $F(R)=0$ has to be used in accordance with the initial condition for T_1 in Eq. (12).

The solution process is quite complicated, involving rigorous mathematical derivations and verifications. Fortunately, no further assumptions or approximations are needed. For conciseness, the solutions are directly stated below.

The following definitions are used.

$$\nu_i(t,R) \equiv \frac{T_1 + \sigma_i}{T_{rv} + \sigma_i} \tag{26}$$

$$A_i \equiv \frac{R_o \tau_i \epsilon_i \ln(\frac{R}{R_o}) + \tau_i}{(1+\tau_i)\{R_o \epsilon_i \ln(\frac{R_1}{R_o})+1\}} \tag{27}$$

$$B_i \equiv \frac{\pi \tau_i \epsilon_i \{1 + R_o \epsilon_i \ln(\frac{R_1}{R_o})\}}{(1+\tau_i)\{R_o \epsilon_i \ln(\frac{R_1}{R_o})+1\}} \tag{28}$$

$$C_{i,j} \equiv \frac{J_o^2(k_j R_1) e^{-ak_j^2 t_i}}{(KJ)^2 - (k_j^2 + \epsilon_i^2) J_o^2(k_j R_1)} \tag{29}$$

$$\{V_o(k_j R)\}_i \equiv J_o(k_j R)(KY) - Y_o(k_j R)(KJ) \tag{30}$$

where

$$KJ = k_j J_1(k_j R_o) + \epsilon_i J_o(k_j R_o)$$

$$KY = k_j Y_1(k_j R_o) + \epsilon_i Y_o(k_j R_o)$$

$$D_{i,j} \equiv \frac{\tau_{i-1}(\epsilon_{i-1} - \epsilon_i)}{k_j^2(1+\tau_i)\{R_o \epsilon_{i-1} \ln(\frac{R_1}{R_o})+1\}} \tag{31}$$

$$E_{i,j} \equiv \frac{\epsilon_i(\tau_{i-1} - \tau_i)}{k_j^2(1+\tau_i)} \tag{32}$$

$$Z_n \equiv k_n^2 C_{2,n}\{D_{2,n} + E_{2,n} -$$

$$\frac{2\tau_1 \epsilon_1}{R_o(1+\tau_2)} \sum_{m=1}^{\infty} \frac{C_{1,m}(\epsilon_2 - \epsilon_1)}{k_n^2 - k_m^2} \tag{33}$$

$$\sigma_i \equiv \frac{\eta_i}{\xi_i} \tag{34}$$

$$\tau_i \equiv \frac{\sigma_i}{T_{rv}} \tag{35}$$

$$\epsilon_i \equiv \frac{\xi_i}{\lambda} \tag{36}$$

$J(k_j R)$ = Bessel function of the first kind
$Y(k_j R)$ = Bessel function of the second kind
k_j = solution of the characteristic equation (40)

Then the solution of T_1 is:

By the end of the first time interval, which is represented by subscript 1,

$$\nu_1(t,R) = A_1 - B_1 \sum_{m=1}^{\infty} C_{1,m}\{V_o(k_m R)\}_1 \tag{37}$$

By the end of the second time interval,

$$\nu_2(t,R) = A_2 + \pi \sum_{n=1}^{\infty} Z_n\{V_o(k_n R)\}_2 \tag{38}$$

By the end of the third time interval,

$$\nu_3(t,R) = A_3 + \pi \sum_{p=1}^{\infty} [k_p^2 C_{3,p}\{V_o(k_p R)\}_3$$

$$\{D_{3,p} + E_{3,p} + \frac{2(1+\tau_2)}{R_o(1+\tau_3)} \sum_{n=1}^{\infty} \frac{Z_n(\epsilon_3 - \epsilon_2)}{k_p^2 - k_n^2}\}] \tag{39}$$

In all following time intervals the solution has exactly the same appearance as that of Eq. (39) except for the

432

subscripts which should be properly adjusted to reflect the current time interval under consideration.

Once $\nu(t,R)$ has been found, T_1 can be easily obtained through the definition of ν, Eq. (26). From Eq. (6), the determination of the temperature distribution in the wet case can finally be completed.

PRACTICAL WAYS TO CALCULATE THE WATER CAUSED TEMPERATURE CHANGE T_1

Although an analytical solution has been obtained, it is still a formidable task to calculate numerical values for it. Scrutinizing the solution one sees that one deals with a large number of infinite series, which do not necessarily converge rapidly. In most cases, a good convergence can be expected when the time span of an interval is relatively large. Theoretically speaking, the solution of the modified governing equations (7)-(10) and (22)-(25) approach the true solution of the original governing equations (1) to (4) when the temperature correction in each time interval approaches zero. This presents no real problem to practical consideration, since temperature variations encountered in normal ventilation planning are not too drastic. However, one may in some cases want to consider the situation immediately after rock of virgin temperature is exposed. For this extreme case, a compromise between accuracy and computer time is necessary. This problem will be discussed in the following section.

Scrutinizing the solution further, one notices the need to solve the characteristic equation

$$\begin{vmatrix} J_o(k_jR_1) & Y_o(k_jR_1) \\ KJ & KY \end{vmatrix} = 0 \qquad (40)$$

A numerical method was employed to calculate the eigenvalue k.

The most important feature is the necessity to reduce the calculation load. For instance, the solution of the characteristic equation itself can be a time consuming procedure while one cannot afford to miss an eigenvalue before reaching a satisfactory result. Fortunately, their eigenvalues are constant in each time interval. When the temperature distribution is being calculated, one does not need to calculate them over and over again for each point. These and other possibilities in time savings must be analyzed carefully to achieve a successful program.

Comparing the solution of the wet case correction, Eqs. (37) to (39), with the solution for the corresponding dry case, Eq. (15), it is easily realized how much more complicated the simultaneous mass/energy transfer process is compared with the dry heat transfer. It is difficult to present the data in a concise way, as the number of controlling dimensionless numbers for the wet case are larger than for the dry case.

From the solution we can see that, besides Fourier number and Biot number, the dimensionless groups ε and τ are significant in characterizing the heat transfer process of a wet case. Loosely speaking, the former indicates the relative capacity of heat transfer on the surface compared to that of rock, the latter represents the relative heat transfer capacity due to evaporation.

A computer program was written to test the foregoing solutions. A flowchart of the program is shown in Fig. 1. One set of the typical results is shown in Table 1 and Fig. 2 for which the saturated vapor pressure $f(T)$ was converted from a steam table between $0^{\circ}C-40^{\circ}C$ and 1 atm to a polynomial. The mass transfer coefficient was obtained by setting Lewis' relation equal to 1.

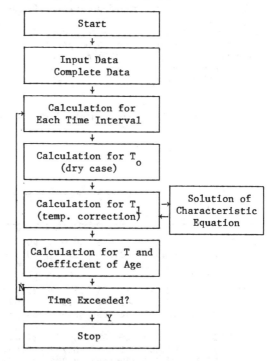

Fig. 1. Flowchart of program for the solution of temperature.

SIMPLIFIED SOLUTION

With the previous solution, we can simulate the temperature variation in the rock around an airway from time zero till the establishment of an equilibrium at time infinity. In some cases, the moisture may play no role until the time t_o, which is a finite non-zero value. In other cases it may only be desired to observe the temperature variation for a very short period of time. The problem of dealing with the first few hours after virgin rock is exposed to ventilation was briefly described before.

The solution, Eqs. (37) to (39), can in principle handle these cases. But, as noticed in the last section, the convergence rate is not satisfactory if time intervals are too short. In one test run of the developed program, a time interval was deliberately set to zero. The results were still not close enough to the reality, even though over 500 terms from the infinite series were summed up. This compares with 3 to 40 required terms if time periods are longer than about one hour. In fact, any further increase in the number of calculated terms of the infinite series in an attempt to gain a better accuracy will unavoidably encounter the difficulty with round-up errors of the computer.

To surmount the obstacle to accommodate the simulation of a short time period, the governing Eqs. (22) to (25) must be handled in a different way. By rewriting Eqs. (22) to (25) as

$$\frac{\partial T_1}{\partial t} = a \left(\frac{\partial^2 T_1}{\partial R^2} + \frac{1}{R} \frac{\partial T_1}{\partial R} \right) \tag{41}$$

subject to

$$t=0 \quad \text{all } R \quad T_1=0 \tag{42}$$

$$t>0 \quad R \to \infty \quad T_1=0 \tag{43}$$

$$R=R_o \quad \lambda \frac{\partial T_1}{\partial R} = \xi(T_1+\sigma) \tag{44}$$

one knows that they are basically the same as the governing equations in the dry case except for the fact that ξ and σ are functions of t rather than constants. Similar results are expected. Let's use the approximate equations

$$K_o(x) \approx \sqrt{\frac{\pi}{2x}} e^{-x} \tag{45}$$

$$K_1(x) \approx \sqrt{\frac{\pi}{2x}} e^{-x}(1 + \frac{3}{8X}) \tag{46}$$

where: K = Bessel functions

Omitting detailed derivations, we can get the result:

$$v(t,R) = v_o - \sqrt{\frac{R_o}{R}} \frac{R_o \varepsilon v_o}{R_o \varepsilon + 0.375} \{1-erf(y_1)$$

$$-e^{pq} e^{p^2 t} [1-erf(y_2)\}] \tag{47}$$

Table 1. Calculated temperature distribution corresponding to the conditions defined in Fig. 2.

thermal conductivity = 10.0 KJ/hr m.k.; relative humidity = 90%

Time (hr)	T ($^\circ$C)	R (m) 2 (wall)	3	4	5	6	7	8	9	10	30
10	T_o	29.239	39.879	40.0	40.0	40.0	40.0	40.0	40.0	40.0	40.0
	T_1	-5.539	0.0	0.0	0.0	0.0	0.0	0.0	0.0	0.0	0.0
	T_o+T_1	23.700	39.879	40.0	40.0	40.0	40.0	40.0	40.0	40.0	40.0
100	T_o	22.544	33.923	38.463	39.729	39.938	39.995	40.0	40.0	40.0	40.0
	T_1	-3.661	-1.634	-0.522	-0.052	0.0	0.0	0.0	0.0	0.0	0.0
	T_o+T_1	18.883	32.289	37.941	39.677	39.938	39.995	40.0	40.0	40.0	40.0
1100	T_o	17.143	24.488	29.174	32.448	34.804	36.501	37.705	38.539	39.100	40.0
	T_1	-1.430	-1.048	-0.782	-0.584	-0.431	-0.313	-0.222	-0.152	-0.100	0.0
	T_o+T_1	15.713	23.440	28.392	31.864	34.372	36.188	37.483	38.387	39.000	40.0

TEMPERATURE DISTRIBUTION

The curves represent temperature distribution of
wet cases in 10, 100 and 1100 hours, respectively.

reference atm. pressure   101.35 kpa
specific heat of air   1.0035 kj/kg.k
virgin rock temperature   40.0 deg. c.
thermal diffusivity   0.01 m²/hr.
thermal conductivity   x kj/hr.m.k

reference air density   1.20 kg/m³
enthalpy of evaporation   2430. kj/kg
air temperature   15.0 deg. c.
conv. heat trans. coeff. 20. kj/hr.m².k
rel. humidity of air   y per cent

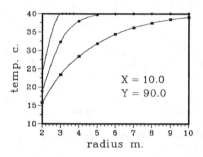

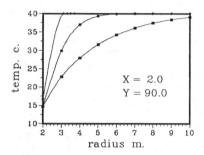

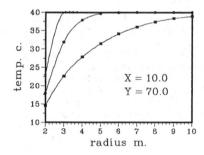

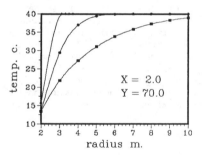

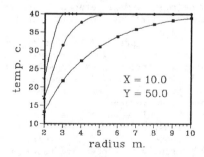

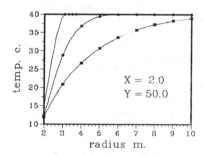

Fig. 2

435

where

$$y_1 = \frac{R-R_o}{2\sqrt{at}} \qquad (48)$$

$$y_2 = \frac{(\varepsilon R_o + 0.375)\sqrt{at}}{R_o} + \frac{R-R_o}{2\sqrt{at}} \qquad (49)$$

$$p = \frac{\sqrt{a}(\varepsilon R_o + 0.375)}{R_o} \qquad (50)$$

$$q = \frac{R-R_o}{\sqrt{a}} \qquad (51)$$

This solution has a satisfactory convergence rate for short time intervals.

Eq. (47) provides a very useful tool to simulate the wet case if heat transfer occurs over a short period of time and the temperature correction is small. Recalling that the previous solution, Eqs. (37) to (39), can be used to handle temperature calculations for any time period without limitation, provided that the time span is properly divided into intervals and temperature correction T_1 is small, we can then understand the limitation inherent in the present solution, Eq. (47). It can only tackle one period of time with a small temperature correction. It cannot be extended to multi-intervals of time because the initial conditions in each time interval other than in the first one are too involved to allow analytical solutions.

A computer program was also written to verify this solution and it gives consistent results as the previous method.

APPLICATION OF SOLUTION

It is easy to visualize, and miners know from experience, that water evaporation and condensation affects airway wall temperatures to a significant extent. The calculation results, shown in Fig. 2, verify such experience.

The importance of knowing temperatures of airway walls and their changes with time and ventilation conditions is very obvious for ventilation planning in hot mines. The solutions presented here may be helpful for this purpose. They were, however, derived with the intention to develop improved methods to simulate ventilation systems in emergency conditions, in particular fire conditions.

Efforts for such simulations have been pursued at Michigan Technological University for more than a decade (Greuer 1973, 1977). A correct assessment of temperature distributions is, under such conditions, vitally important to pre-calculate thermal forces and resulting airflow distributions. A large number of different case studies have proved that the simulation efforts were basically correct.

Unfortunately, the influence of moisture in these simulations could only be approximated. It is hoped that the above described approaches, which have by now been incorporated into the existing programs, will mean an improvement.

Present efforts are directed towards a full transient-state simulation of ventilation systems taking even minute time intervals such as those surrounding airflow reversals into account (Greuer 1983). Event and interval oriented simulation techniques are used. Both adopt the concept of quasi-equilibria in the simulation time intervals and rely on the correct calculation of the heat exchange between air and airway walls. A basic flowchart of the transient-state simulation program is shown in Fig. 3.

CONCLUDING REMARKS

Being aware of the great influence of water migration, evaporation and condensation on airway wall temperatures, the authors hope that the calculations presented here are a useful contribution for assessing this influence.

Eqs. (37) through (39) present "exact" solutions of this influence provided that the time intervals were properly set and the correction for the water influence was reasonably small, say, within 2 or 3 degrees Centigrade. As mentioned previously, such a requirement implies that the foregoing solution is suitable for the case of high relative humidities or low degrees of wall wetness. A computer program was written to realize this solution.

When a very strong influence of water evaporation on the wall temperature variations is introduced by large differences in the initial rock/air temperatures and humidities, Eq. (47) can be employed for a better convergence, if the transient state for only a short time interval is of interest. This approach has already been adopted in one of the programs which are presently developed and tested at Michigan Technological University.

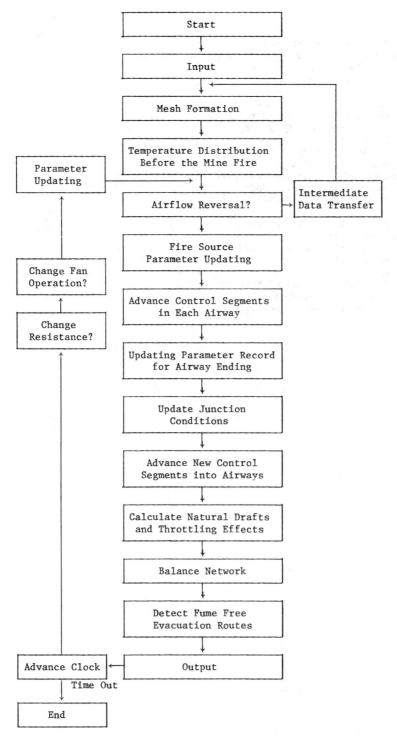

Fig. 3. Flowchart of the transient-state ventilation
simulation program.

437

REFERENCES

Amano, K., Y. Mizuta & Y. Hiramastsu.
1982. An improved method of predicting
underground climate. Int. J. Rock
Mech. Min. Sci. & Geomech. Abstr.
19:31-38.
Greuer, R. E. 1973. Influence of mine
fires on the ventilation of underground
mines. USBM Contract No. S0122095,
NTIS PB 228 834, 179 pp.
Greuer, R. E. 1977. Study of mine fires
and mine ventilation - Computer
simulation of ventilation systems under
the influence of mine fires. USBM
Contract No. S0241032, NTIS PB 299 231,
162 pp.
Greuer, R. E. 1983. A study of the
precalculation of effects of fires on
ventilation systems of mines. USBM
Contract No. J0285002, NTIS PB 84-159
979, 293 pp.
Luikov, A. V. 1968. Analytical heat
diffusion theory. Translated from
Russian by J. P. Hartnett. Academic
Press.

The computer simulation of mine climate on a longwall coal face

I.LONGSON & M.A.TUCK
University of Nottingham, UK

ABSTRACT: Exhaustion of the shallower coal seams in Western Europe is resulting in a shift of coal production to seams at greater depths with a consequential rise in the virgin rock temperature. This factor coupled with the nature of longwall operations and the steps taken over recent years to increase face output is making the maintenance of a satisfactory working climate very demanding.

The paper describes the development of a computer model to simulate the heat and moisture transfers occurring on a longwall coal face and to predict the resultant psychrometric conditions of the ventilating airstream. The approach and techniques used to achieve this are discussed.

Comparison between predicted and actual measured conditions is given. The overall aim is the production of a user friendly predictive program which can be used for planning purposes.

1 INTRODUCTION

In the U.K., as existing face capacity is replaced, many of the new faces will have higher installed powers and be located in zones of higher strata temperature. This will result in more faces experiencing environmental difficulties, particularly with respect to climatic conditions. In many cases these problems are unlikely to be solved by increasing face air quantities as face velocities are already approaching the recognised upper limit and so more refrigeration will be required. This will require major capital and operational expenditure.

It is easy to spend large capital sums and incur high operational expenditure without achieving the desired amount of climatic control. There is, therefore, a great need to have reliable prediction programs in order to aid in the making of sound, informed decisions which will have a significant influence in the long term on the level of production costs.

A number of techniques have been devised to predict the heat flow and resultant effect on the psychrometric condition of the air in mine workings. These can be classified, due to their important physical and operational differences, under two headings, namely tunnels and working zones.

In the former it is usual to assume a cylindrical tunnel allowing the theory of radial heat transfer to be used, like in the work of Goch and Patterson (1940), Starfield (1967), Gibson (1976) and more recently Starfield and Bleloch (1983). In the latter case for stopes or faces it is possible to apply the theory of heat flow from a semi-infinite solid, such as in the work of Starfield (1966a,1966b) and Middleton (1979). Simulation programs have also been written for complete longwall districts, (Voss 1971) (Middleton 1979).

The work described in this paper on the simulation of mine climate on a longwall coal face forms part of an on going research program within the Department of Mining Engineering at the University of Nottingham concerned with the computer prediction of climatic conditions underground. The paper describes the approach used to model the heat flows into the air on a longwall coal face, the techniques of computing the heat and moisture transfers from the various heat sources and the methods of calculating the resultant psychrometric conditions of the air.

The overall aim is to produce a program which can be used to plan ventilation/refrigeration requirements and which can be used to compare various cooling strategies for hot workings.

2 SOURCES OF HEAT AND MOISTURE ON A LONG-WALL COAL FACE

The major heat and moisture inputs to the ventilation airstream on a longwall coal face are from the following sources;

    (a) exposed strata surfaces
    (b) conveyed coal
    (c) waste or goaf region
    (d) machinery
    (e) sprays

there are other sources which are of little significance, for a fuller description see Hemp (1982).

The strata temperature varies with locality but, for the conditions being modelled, this invariably exceeds the air temperature. This results in heat and moisture entering the airstream through evaporation from the newly exposed wet strata surfaces produced as the face advances.

Once the coal has been cut it will continue to transfer heat/moisture as it is conveyed outbye.

Air leakage through the caved goaf also picks up heat/moisture and leakage of this air back into the face airstream will also constitute a major heat input.

Longwall coal faces are highly mechanised and all the power will ultimately appear as heat in the airstream unless work is done against gravity.

The use of water sprays for dust suppression can be a source of heat or coolth as well as moisture depending on the temperature of the water supplied.

All the sources described above interact in a complex manner. The heat flow at a particular time on a longwall face is not purely a function of the present conditions, but requires that previous activity on and around the face is considered. The chosen approach in this paper is essentially steady state but with reliance on some empirically based mean values.

3 THE COAL FACE SIMULATION MODEL

3.1 The elemental approach

In the model the face is represented as a rectangular opening within the strata and an elemental approach (consistent with the finite element approach used by Gibson 1976)) is used to model the complex heat flows that occur. The coal face is considered to be formed, along it's length, by a number of elements of relatively small length. All parameters are assumed to remain constant within an element and any changes are applied at the boundaries between successive elements. This produces

a stepped effect as illustrated in figure 1. However, if the elements are suitably short in length, the predicted route should closely follow the observed variation.

An element length of ten metres has been found to be suitable as this results in a reasonable computer mill time and allows the cutting machine and associated support advance operations to be defined within the same element.

3.2 Strata heat: roof, face and floor

In order to model the heat flow within the program the contributions from the roof, face and floor are considered separately. The thermal parameters of the strata and the inlet air condition are known therefore the problem is reduced to one of calculating the heat flow contribution from the strata and using it to predict the outlet air condition for a particular element.

To do this the face is considered as a single slab and the roof and floor split into slabs of width equal to the advance increment, (Starfield 1966a,1966b). A division of the elements in this way, coupled with appropriate assumptions, allows the equations of one-dimensional heat flow from a semi-infinite solid to be employed. The assumptions necessary are;

    (a) The roof and floor slabs are perfectly insulated from each other
    (b) Geothermic gradient effects are ignored
    (c) The strata units are homogeneous and isotropic
    (d) The roof and floor strata does not contribute to the face heat flow prior to their exposure to the ventilating airstream.

The face has a uniform thickness slice removed from it cyclically at time increments Δt. Details of a face element and the generation of the slabs are shown in figures 2a and 2b. In order to calculate the exposure time of the slabs a subprogram is used which models the assumed constant speed of the cutting machine along the face. Knowing the machine position, physical dimensions of the face and the time taken to cut a shear enables the exposure times to be calculated.

3.2.1 Theory

The partial differential equation governing one-dimensional heat flow from a semi-infinite solid (Carslaw and Jaeger 1959), using $T(x,t)$ to denote the temperature at distance x from the exposed surface at time

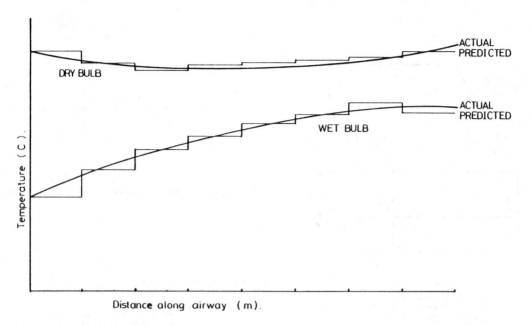

Figure 1. The elemental approach.

t is

$$\partial^2 T/\partial x^2 = 1/\alpha \cdot \partial T/\partial t \qquad (1)$$

α = thermal diffusivity.
   Solving equation 1 subject to the boundary and initial conditions;
   (a) $T(x,0) = Tvr$ = virgin rock temperature
   (b) as $x \rightarrow \infty$ $T(x,t) = Tvr$
   (c) flow of heat into air at dry bulb temperature Tdb is controlled by a surface heat transfer coefficient h, where

$$k \cdot \partial T/\partial x = h(Ts - Tdb) \qquad (2)$$

Ts = $T(0,t)$ = dry surface temperature
   k = thermal conductivity of rock W/mC;
and simplifying the solution to give the surface temperature gives

$$(Ts-Tdb)/(Tvr-Tdb) = \exp(h^2 \cdot \alpha t/k^2) \cdot \\ erfc(h\sqrt{\alpha t}/k) \qquad (3)$$

erfc = complementary error function.
The complementary error function can be calculated from a series (Hemp 1982).
Equation 3 enables the dry surface temperature of any particular slab within an element to be calculated. The heat flow can then be calculated using

$$Q = h(Ts - Tdb) \qquad (4)$$

   In this analysis the surface heat transfer coefficient is assumed to consist purely of a convective term (Whillier 1982).

3.2.2 Influence of wetness

The previous analysis assumed the existence of dry conditions but as longwall coal faces are rarely dry due to inherent moisture present in the strata and the use of dust suppression sprays, the resultant effects of moisture have to be included in the model. In order to evaluate the effect of wetness, a wetness factor, η, is used, (Starfield 1967)

$$\eta = \text{area of wet surface/total area} \qquad (5)$$

   All wet areas are considered together in one zone and it is assumed that there is no conduction between dry and wet areas. At a wet surface sensible heat transfer, Qc, and latent heat transfer, Qℓ, heat transfer will occur. At equilibrium the algebraic sum of these must equal the heat supplied by the strata to the surface, Qs, so

$$Qs = Qc + Q\ell \qquad (6)$$

441

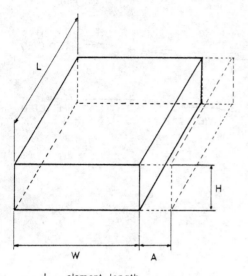

L = element length
W = ventilated width
A = advance per shear
H = extracted height

Figure 2a. Face element details.

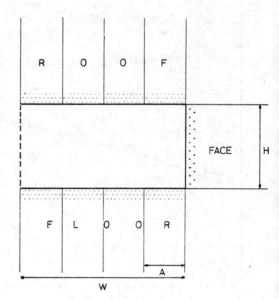

Figure 2b. Generation of roof, face and floor slabs for elemental heat flow calculation.

The sensible heat transfer can be expressed as

$$Qc = A.h.\eta.(Tws - Tdb) \qquad (7)$$

where   A = area of surface
        Tws = wet surface temperature

and the latent heat transfer as

$$Q\ell = A.H.\eta.0.7.L.(es - e)/p \qquad (8)$$

where   L = latent heat of vapourisation of water
        es = saturated vapour pressure at Tws
        e = vapour pressure of air
        p = absolute pressure

finally the strata heat as

$$Qs = A.k.\eta.dT/dx \qquad (9)$$

where dT/dx = temperature gradient at wet surface

As the variables contained in the above equations are mathematically interdependent the heat balance expressed in equation (6) cannot be applied directly. However an equivalent iterative technique has been used to find the equilibrium wet surface temperature. The total flow from a particular slab is then calculated from a sum of the sensible heat flow

$$Qc = A.h.((Ts-Tdb).(1-\eta)+\eta.(Tws-Tdb) \qquad (10)$$

and the latent heat flow

$$Q\ell = A.h.\eta.0.7.L.(es - e)p \qquad (11)$$

3.3 Strata heat flow from the goaf

The goaf or waste area left behind as a coal face advances can be a major heat source. Air is able to leak into the goaf from the intake airway and to the return due to the pressure differential across the face. The amount of leakage depends on a number of factors but notably the quality of the gateside packs and the state and composition of the goaf. Browning (1980) reported that 22% of the district airflow at a particular colliery passed through the goaf. Measurements made by the authors at a colliery in Nottinghamshire support this observation with values around 20%.

A major problem in modelling the goaf area is presented by its continually changing condition and more work is required in order to gain a fuller understanding of the variations in airflow patterns and associated heat transfer phenomena that occur.

A large number of variables need to be determined and presently an analogy with a packed bed is used (Eckert and Gross 1963). Currently the assumptions used within the face model are;

442

(a) 20% of the district airflow passes through the goaf area. It is assumed to enter at the intake end of the face. Leakage emanates back onto the face at the cutting machine and the return lip.

(b) The goaf consists of randomly packed uniform spheres, through which plug flow occurs.

(c) The flow regime is laminar.

(d) The sensible heat flow is purely convective.

(e) The temperature of the rock particles is the local virgin rock temperature.

(f) The relationships used to calculate the sensible and latent heat flow from strata have been applied to the goaf.

Measurements conducted on two faces in the Nottinghamshire coalfield showed that under normal production conditions with a VRT of 40° C, the air emerged at the return end of the face at a dry and wet bulb temperature of 38 and 34° C respectively. These measurements were used in order to quantify all the necessary parameters within the model. Other necessary assumptions are;

(a) A porosity value of 39% is attributed to the goaf.

(b) An elemental approach is used.

(c) The area through which the air in the goaf is confined to is equal to a distance ten metres back from the waste edge of the supports and twice the extracted height. Pickering (1969) observed that the majority of the airflow in the goaf flows close to the back of the face supports.

(d) A wetness factor of 0.5 is attributed to the goaf.

The above assumptions allowed a model to be developed which correlated well with measured goaf outlet conditions. The model is simple and strongly empirical in nature, but it is hoped that a more theoretically based solution can be produced at a later date.

3.4 Conveyed coal

In Western Europe the predominant method of underground coal transport is by conveyor. The heat released as coal is conveyed outbye can be considerable (Gracie and Matthews 1976, Browning 1980, Watson 1981). It can be regarded to comprise of two components, that of the original strata and heat from the breaking process. This component is considered separately because of it's magnitude and because it does not directly reduce the ability of the exposed strata to transfer heat. The ability of coal to transfer heat is greatly enhanced on a conveyor because it is crushed, wetted and there is a high relative velocity

between the moving conveyor and the air.

The model used to simulate heat transfer from coal on a conveyor is based on work conducted by Watson (1981) on dry coal. For the purposes of the analysis the coal is treated as a multi-layered, continuous plane parallel slab and the equations derived are for steady state conditions. Only heat flow from the upper surface of the conveyed coal bed is considered and the coal is assumed to be at VRT when cut and loaded at the face machine. This assumption has been verified by measurement.

The total heat exchange at the upper surface of the coal is made up of convective, qc, evaporative, qe, and radiative, qr, heat exchanges. At equilibrium the balance

$$q = qc + qe + qr \qquad (12)$$

must hold, where q is the heat transferred to the ventilating airstream. This treatment results in the following equations for the sensible and latent heat transfers, which are

$$Q_{sen} = A.(hc + 2.0412).(Tu - Tdb) \qquad (13)$$

$$Q_{lat} = A.hc.0.7.L.(eu - e)/p \qquad (14)$$

Q_{sen} = sensible heat transfer
Q_{lat} = latent heat transfer
A = surface area of coal bed
hc = surface heat transfer coefficient
Tu = temperature of upper surface of coal
eu = saturated vapour pressure at Tu

From laboratory studies Watson (1981) shows that the value of hc can be calculated from;

$$hc = 0.025.U^{0.7} \ kW/m \ C \qquad (15)$$

where U is the relative velocity between the conveyor and the airstream. The upper surface temperature of the conveyed coal is calculated in the same way as the dry surface temperature of strata, using thermal constants obtained experimentally by Watson for run of mine coal with 5% moisture. The exposure time for a particular element of the conveyed coal can be calculated knowing the machine position and the speed of the conveyor.

This method allows only the surface of the bed to cool as the coal is conveyed outbye but not the bulk and is therefore not consistent with reality. Other methods have been tried but to date no satisfactory method of representing the simultaneous cooling of the surface and bulk has been found.

3.5 Machinery

UK longwall coal mining operations are highly mechanised and this constitutes a major heat source. All the electrical power supplied to a longwall district must either ultimately appear as heat at various locations or do work against gravity. The heat flow paths around underground machinery are complex and so an empirical approach is necessary in order to predict the percentage of their rated power which appears as heat. An added complication is that the machinery works intermittently and at varying rates of output.

In order to estimate the heat flow from machinery one of two approaches can be used. Firstly a fixed percentage of the rated power can be used or secondly predictive equations based on tonnage produced and rated machine power can be used, as by Voss (1971). The first method is used within this program, the user specifying the corrected machine power. This is based on work reported by Verma (1984), and for a face machine an average value of 60% of the rated machine power is generally used.

Within the program, machinery is classed as either a spot or linear type source with refrigeration designated as a negative heat input.

3.6 Sprays

Sprays using either normal or chilled service water are used extensively underground, primarily for dust suppression but also for air conditioning. Sprays can significantly affect the face environment.

The effect of a chilled or normal service water spray is based on work by Bluhm (1976), Bluhm, Van der Walt and Whillier (1976,1977) and Whillier (1977). The spray is considered as a single stage crossflow spray chamber through which a proportion of the face airflow passes. Within the model the spray is considered to act in an extra control volume added onto the element in which the spray is acting (Middleton 1979). Within this control volume no allowance is made for heat and work entering the envelope and the entry conditions are the exit conditions for the element in which the spray is working.

3.7 Miscellaneous routines in the program

There are a number of smaller but still essential routines used within the program. These include an autocompression routine, a routine to calculate the variation in VRT with a change in level. There is also a mixing routine which is used to mix goaf air with the face airflow or the airflow affected by a spray with the unaffected face airflow. It is based on the law of conservation of mass and the law of conservation of energy with regards to moisture content and enthalpy.

3.8 Calculation of outlet conditions

Using the sensible and latent heat flows together with other energy exchanges that take place within an element, such as autocompression, enables the total changes in enthalpy and moisture content to be calculated.

$$\Delta h = Q_{sen}/\dot{M} + \text{other enthalpy terms} \qquad (16)$$

$$\Delta x = Q_{lat}/(\dot{M}.L) + \text{other moisture content terms} \qquad (17)$$

Δh = change in specific enthalpy
Δx = change in moisture content
$\dot{M}$ = mass flow of air

The outlet dry bulb can be calculated from;

$$Tdb_2 = Tdb_1 + \Delta h/(Cpa + x_2.Cpv)$$

and the outlet moisture content from;

$$x_2 = x_1 + \Delta x$$

Subscripts 1 and 2 refer to inlet and outlet.

Cpa = thermal capacity of dry air at constant pressure.
Cpv = thermal capacity of water vapour at constant pressure.

Knowing the outlet moisture content, which is directly related to vapour pressure the outlet wet bulb temperature can be calculated using an iterative technique which balances vapour pressures until a set tolerance is achieved (Gibson 1976).

Using standard psychrometric equations, (McPherson 1979), it is now possible to calculate all the remaining psychrometric parameters of the air. The outlet pressure of the air can also be calculated by applying Atkinson's law and the equation of polytropic flow to the steady flow energy equation. The final parameter to be calculated is the air cooling power (Mitchell and Whillier 1972).

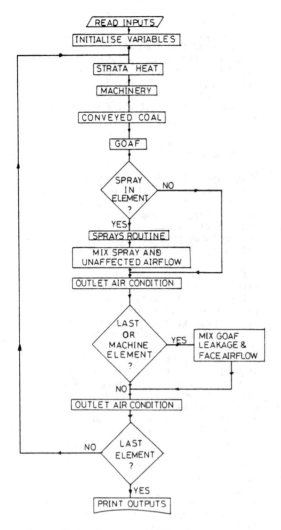

Figure 3. Basic program flowchart.

4 THE COAL FACE TEMPERATURE PREDICTION PROGRAM

The present program is stored and run on the ICL 2977 mainframe computer at the University of Nottingham. Initial development of the main body and some of the subroutines was undertaken on APPLE II microcomputers within the Department of Mining Engineering.

The program is written in FORTRAN77 and is in a modular form to allow easy insertion of new or amended routines. A basic flow chart is shown in figure 3.

A single data file is currently needed in order to run the program. The input requirements are;

1) State of air at inlet: dry/wet bulb temperature, air quantity, absolute pressure.

2) Thermal parameters of the strata: conductivity diffusivity, virgin rock temperature and geothermal gradient.

3) Face details: length, extracted height, ventilated width, depth at intake and return ends, web cut per shear, time to cut a shear, present cutting machine position, speed of face conveyor, type of ventilation, direction of machine travel on present cut, friction factor and wetness factor.

4) Machinery details: spot or linear sources, rated power, location, length.

5) Spray details: water flow rate and temperature, location.

The output is presented in tabular form and gives the psychrometric state of the air at ten metre intervals along the face. Graphical output is presently under development. The program can be run interactively or as part of a batch queue system.

5 CORRELATION WORK

In order to test the quality of the predictions produced by the program it has been and is continuing to be tested against measured values. Measurements have been taken on two underground districts at a local colliery of the National Coal Board. Dry bulb temperatures and humidity have been monitored continuously at several positions on both districts using thermohygrographs. Weekly surveys undertaken on differing days are used to measure the variation in airflow and pressure with time and also dry and wet bulb temperatures are measured at selected sites for correlation purposes.

Both districts are located in the Deep Soft seam and are of the advancing U-type with homotropal ventilation. The faces are 250 m in length and the average extracted height is 2 m. Both work a two shift system and saleable production is currently running at 8-9000 tonnes per week.

5.1 3's district

This is a high production, highly mechanised district currently advancing in excess of 15 metres per week. The face equipment consists of a heavy duty AFC, the drives however are considered to lie outside the face zone, and a double ended ranging drum shearer (150 kW). The face supports are 4 legged shield supports. To suppress dust whilst advancing shield supports, sprays have been fitted but observation showed

that these are rarely used. There are dust
suppression sprays on the ranging drum
shearer and the measured water discharge
temperature from these was 30° C.

The roof strata is shale and the floor
consists of seatearth with shale underlying it. The virgin rock temperature has
been measured at 40° C.

The correlation exercise described is for
the 2 May 1984. The face survey was undertaken during the dayshift, details are
given in Table 1.

5.1.1 Computer simulation

Three computer runs were undertaken corresponding to machine positions at 100, 54 and
3 metres from the intake end, which should
correlate with the measurements made at the
return lip, 40 chock and 80/124 chock
respectively. The shear time used in each
exercise was two hours and the wetness
factor 0.5.

The comparison between the measured and
predicted results is illustrated graphically in figure 4.

From figure 4 a good correlation is
obtained. With the machine at 100 metres
the correlation obtained at the return end
is acceptable, the same can be said for
the correlations at 187.5 and 125 metres
with the machine at 54 and 3 metres respectively. This is not true at 56.25 metres
with the machine at 3 metres, however this
coincided with the machine cutting or
sumping into the intake end when measuring
at 56.25 m. This is an intermittent operation and the coal flow rate on the face
conveyor is markedly reduced and these
conditions are not taken into account by
the program.

The correlations obtained are acceptable
especially when one takes into account the
limitations of the surveying technique
used. Spot measurements are taken along
the face over a distinct time period during
which conditions can vary appreciably. For
example air intake conditions can change
over a short period of time depending on
activities occurring outbye of the intake
lip and especially with respect to the
input from the ripping machine. Fortunately
during the survey activity in the intake
was limited and no ripping or campacking
operations were undertaken at the intake
end of the face.

An interesting feature shown in figure 4
is the great effect which machine position
has on the thermal conditions encountered
on a coal face and hence the influence
which conveyed coal has on the face
climate.

From the results of other correlation
runs it can be broadly stated that under
normal production conditions the predictions obtained by the program are to within
±1° C of the measured values. Overall the
correlation with the dry bulbs is better
than that for the wet bulbs. This is understandable but is a feature which requires
further investigation.

6 DISCUSSION

This paper records the current state of
development of a working coal face heat
program. At this stage it is directed
particularly towards simulating UK conditions of an advancing, highly mechanised,
conventionally ventilated, longwall coal
face with virgin rock temperatures not
exceeding 45° C.

The chosen approach has been to closely
study and monitor climatic conditions at
a local colliery during the development of
the computer model. This has been beneficial because it has enabled a better appreciation of particular process mechanisms,
interactions and temporal variations.

A steady state analysis has been adopted,
although it is recognised there are actually very significant deviations from this.
However by using reasoned assumptions a
detailed approach to the modelling has been
attempted and satisfactory predictions
obtained for a range of power loader positions and various face operating conditions.
This is encouraging and is the justification
for moving on to test predictions against a
wider correlation base.

This detailed approach has identified the
positions in the extraction cycle which
produce the worst climatic conditions and
variations that occur with time have been
established. Currently the sensitivity of
the solution to input variations is being
investigated and this is likely to have an
important influence on the future data
input and also on data management within
the program.

Further work is required to model more
effectively several of the contributing
heat and moisture exchanges between the
sources and the ventilating airstream in
order to improve some of the techniques
described earlier in the paper. However,
the program has been assembled to enable
future advances in the modelling of particular transfer mechanisms to be incorporated easily, thus ensuring that program
development can proceed in an evolutionary
manner.

Once satisfactory correlation has been
accomplished for a wider selection of face

Table 1 Observed survey data 2-5-1984

Measurement position	Tdb/Twb C	Comments
Return lip	31.1/27.1	Ripping, campacking, coaling, machine at 90 – 100 chock
40 chock	30.2/26.1	Coaling, machine at 125 chock to intake
80 chock	29.5/25.2	Coaling, chocking upstream, machine almost at intake end
124 chock	27.7/21.3	Coaling, sumping, product flow on AFC reduced
Intake lip	27.2/18	Machine sumping into intake end, pack on dosco off

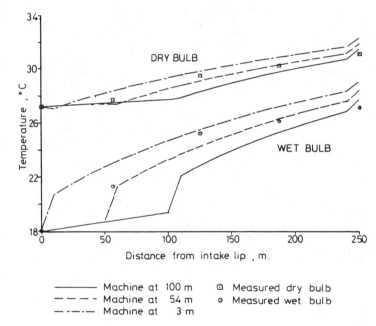

Figure 4. Correlation trials 3's, 2 May 1984.

installations it is hoped that an inter-active, user friendly, version of the program can be produced which will be suit-able for industrial use.

7 CONCLUSIONS

The paper has reported on the development of a longwall coal face climate prediction program as part of on-going mine environ-mental research at the University of Nottingham. The program has been developed specifically to predict the variation in and also ultimately to guide in the effective control of climate on a UK advancing longwall coal face within a relatively high VRT zone. Initial correlat-ion trials have shown an acceptable agree-ment between observed conditions and program predictions.

The simulation work described is aimed at

providing reliable climatic prediction to enable ventilation/refrigeration planning alternatives to be realistically compared as an aid to selecting the optimum scheme.

ACKNOWLEDGEMENTS

The authors wish to acknowledge the assistance of the following:
 Professor T Atkinson for the provision of the facilities within the Department of Mining Engineering.
 The Science and Engineering Research Council for financial assistance.
 Mr J E Wood, Area Director, NCB North Nottinghamshire for permission to make underground measurements.
 Mr R Aldred, Ventilation Engineer, NCB North Nottinghamshire Area for his interest and advice on underground work.
 The Manager and ventilation staff of Harworth Colliery for assistance with the underground correlation work.
 Members of the Environmental Group within the Department of Mining Engineering, University of Nottingham for assistance with underground measurements and their contribution to discussions.

REFERENCES

Bluhm,S.J.1976.Predicting the performance of spray chambers for cooling air. Heating, air conditioning and refrigeration.13:27-39.

Bluhm,Van der Walt and Whillier.1976. Performance tests on horizontal spray chambers. Chamber of mines report 42/76.

Bluhm,Van der Walt and Whillier.1977. The design of spray chambers for bulk cooling air in mines. Chamber of mines report 62/77.

Browning,E.J.1980.The emission of heat in longwall coal mining. First session working group No 1, International Bureau of Mine Thermophysics.Katowice.

Carslaw and Jaeger.1959.Conduction of heat in solids. Oxford press.London.

Eckert and Gross.1963.Introduction to heat and mass transfer. McGraw-Hill.

Gibson,K.L.1976.The computer simulation of climatic conditions in underground mines. Ph.D. thesis, University of Nottingham.

Goch,D.S. and Patterson,H.S.1940.The heat flow into tunnels. J.Chem.Metall.Min.Soc. S.Afr.41:117-121.

Gracie,A. and Matthews,R.1975/76.Strata, machinery and coal in transit; their relative roles as heat sources.Min.Engr. pp181-86.

Hemp,R.1982.Sources of heat in mines.

Chapter 22. Environmental engineering in South African mines.Mine Ventilation Society of South Africa. Johannesburg.

McPherson,M.J.1979.Psychrometry: The measurement and study of moisture in air. Mining Department Magazine,University of Nottingham.31:41-51.

Middleton,J.N.1979.Computer simulation of the climate in underground production areas. Ph.D. thesis, University of Nottingham.

Mitchell,D. and Whillier,A.1972.Cooling power of underground environments.J.Min. Vent.Soc.S.Afr.25:140-51.

Pickering,A.J.1969.Airflows and methane emissions on fully caved, mechanised coal faces in South Nottinghamshire.Min.Engr. 110:93-107.

Starfield,A.M.1966a.Heat flow into the advancing stope.J.Min.Vent.Soc.S.Afr.19: 13-29.

Starfield,A.M.1966b.The computation of air temperature increases in advancing stopes. J.Min.Vent. Soc.S.Afr.19:189-99.

Starfield,A.M. and Dickson,A.J.1967.A study of heat transfer and moisture pickup in mine airways.J.S.Afr.Inst.Min.Metall.67: 211-34.

Starfield,A.M. and Bleloch,A.L.1983.A new method for the computation of heat and moisture transfer in a partly wet airway. J.S.Afr.Inst.Min.Metall.83:263-69.

Verma,Y.K.1984.Control of mine climate.Min. Engr.pp315-23.

Voss,J.1971.Prediction of climate in production workings.Gluckauf.Yr 107:412-418.

Watson,A.G.1981.The contribution of conveyed coal to mine heat problems.Ph.D. thesis, University of Nottingham.

Whillier,A.1977.Predicting the performance of forced draught cooling towers.J.Min. Vent.Soc.S.Afr.

Whillier,A.1982.Heat transfer.Chapter 19. Environmental Engineering in South African mines.Mine Ventilation Society of South Africa.Johannesburg.

A prediction of mine climate around a longwall district

J.FIALA, J.KOUT & A.TAUFER
Mining Institute of the Czechoslovak Academy of Sciences, Ostrava, Czechoslovakia

ABSTRACT: In this paper there is introduced the brief description of the calculation method of airflow temperature changes in longwall face district. As the sources of heat there are considered surrounding rock massif, energetic systems and mined material in transport on belt conveyers. The main attention has been intented on the influence of surrounding rock massif and humidity. The basic knowledge is that airflow gets to connection with underground workings walls surface temperature which is calculated on the basis of knowledge of the virgin rock temperature in a given district and of the age of workings. The airflow humidity is included into the calculations by means of the moisture coefficient that has been determined statistically for different types of underground workings. In the example of the method practical using the measured and calculated temperature changes of airflow in the intake haulage road and longwall face are compared.

1 INTRODUCTION

Czechoslovak underground black coal industry is characterized nowadays by reaching down and in many cases by exceeding of depth levels that are critical for the most of mining branches. This fact concerns with mine ventilation, too where the more difficult mine-geological conditions take out solving the very consequential problems.

One of the main problem are unacceptable climatic conditions at many underground workplaces by that not only economy is unfavourably influenced but also safety of work. That is a well known fact that climatic conditions influencing by means of additional mine ventilation adaptation is not only expensive but often operationally and technically uneasy. Therefore there is a requirement to perform the mine ventilation adaptation in the project phase namely on the basis of exact prediction bases e.g. mine climate prediction calculations on projected workplaces.

Thermodynamics department has solved the problems of the airflow temperature changes prediction with a view to the most exposed parts of mines - longwall face districts. In these districts the most intenzively function the main heat sources i.e. surrounding rock massif, installed energetic systems and mined material on belt conveyers.

The paper introduces the brief description of the worked up calculation method of the airflow temperature changes in intake haulage road and longwall face and the level of the achieved results is evident from the practical example.

2 EVALUATION OF CLIMATIC CONDITIONS

The accurate determination of the individual climatic factors has the influence on workers and their labour productivity and it is a basic presumption for the judgement of the suitability of working environment and for the determination of the admissible limits in which there is possible to work under non-limited labour time. Climatic conditions are characterized by dry-bulb temperature of airflow, its humidity and velocity. The reciprocal combination of these parameters defines four categories in accordance with the maximum admissible labour time (see Table 1). With respect to the productivity and safety of work there is advisable to influence the mentioned parameters so that workplaces could be classed with the category of non-limited labour time.

There is necessary to remind that the Czechoslovak regulations take the heat production of workers into account.

Table 1. The climatic conditions evaluation at the airflow rate above 2 m/s (the heat production of workers is above 96 W/m^2).

Category	Relative humidity (%)	Dry-bulb temperature (°C)
I. Non-limited labour time	70	27.8
	80	26.8
	90	25.9
	100	25.1
II. 6 hours labour time	70	30.3
	80	29.3
	90	28.5
	100	27.7
III. 5 hours labour time	70	31.4
	80	30.4
	90	29.6
	100	28.9
IV. 4 hours labour time	70	32.5
	80	31.6
	90	30.8
	100	30.0

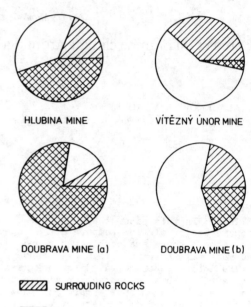

HLUBINA MINE   VÍTĚZNÝ ÚNOR MINE

DOUBRAVA MINE (a)   DOUBRAVA MINE (b)

▨ SURROUDING ROCKS

▭ ELECTRICAL EQUIPMENTS

▨ MINED MATERIAL

Fig. 1. The share of the main heat sources on the airflow enthalpy change in intake haulage roads in different mines.

3 HEAT AND MOISTURE SOURCES

Rock massif, energetic systems and mined material on belt conveyers have been considered as the main sources of heat in the Ostrava - Karvina mines. Their influence on the airflow enthalpy increasing is not constant but it depends on the specific conditions of single mines and often of individual workplaces (see Figure 1).

The secondary heat sources e.g. low-temperature oxidation, lighting, body metabolism etc. are considered to be negligible in the comparison with the above heat sources.

Water in mines from various water sources has the great influence on the state of the climatic conditions, both service and natural. The change of the liquid state of water causes the reduction of dry-bulb temperature of airflow but on the other hand the specific humidity and the enthalpy of airflow are increased.

3.1 The heat released by surrounding rock massif

The heat quantity that is released from rock massif into mine airflow depends on the virgin rock temperature, the age of workings, thermophysical parameters of rocks, roughness of the surface of workings and the quantity and temperature of airflow. This heat can be expressed by means of the Newton's relation

$$dQ_1 = \alpha \, 0 \, dL \, (\, T_h - T_v \,) \qquad (1)$$

where dQ_1 = the heat released from rock massif, W
α = the heat transfer coefficient, W/m^{2}K
0 = the perimeter of a working, m
dL = the length of a working, m
T_h = virgin rock temperature, °C
T_v = airflow temperature, °C

and the heat that is received by airflow by the relation

$$dQ_2 = C_p \, Q_m \, dT \qquad (2)$$

where dQ_2 = heat received by airflow, W
C_p = specific heat, J/kg K
Q_m = airflow rate, kg/s
dT = temperature difference, K

If there is accepted a presumption that every released heat passes to airflow, then $Q_1 = Q_2$ and consequently

$$Q_m C_p dT = \alpha \, 0 \, dL \, (T_h - T_v) \qquad (3)$$

At boundary conditions $L = 0$ and $T_v = T_0$ the resultant relation is

$$T = T_h - (T_h - T_0) \exp - \frac{\alpha \, 0 \, L}{Q_m \, C_p} \qquad (4)$$

The heat transfer coefficient, , is the complex function of many parameters and its expression is very difficult. The following relationship has been accepted for calculations

$$\alpha = 3.65 \, \varepsilon \, v^{0.8} \, D^{-0.2} \qquad (5)$$

where ε = roughness of walls
v = airflow velocity, m/s
D = equivalent diameter, m

The equation (4) describes the heat transfer from rock massif under the steady-state conditions which do not correspond with the real state. Unsteady-state process of the interaction between rock massif and airflow can be expressed from the equation in which the non-dimensional surface temperature of working has been used.

$$F_t = \frac{T_h - T_p}{T_h - T_v} \qquad (6)$$

where F_t = non-dimensional surface temperature
T_p = temperature of wall surface, $^\circ C$

$$T_p = T_h + (T_v - T_h) \, F_t \qquad (7)$$

The non-dimensional surface temperature, F_t, can be approximated (Scerban, Kremněv 1959) as follows

1. $Fo < 10$

then $F_t = \dfrac{Bi}{Bi + 0.375} (1 - \exp(z^2) \mathrm{erfc}(z))$

where $z = (Bi + 0.375) \sqrt{Fo}$
2. $Fo \geq 10$

then $F_t = 1 - 0.8 Fo^{-0.21} Bi^{-1}$, $Bi > 25$

or $F_t = 1 - 0.5 Fo^{-0.2} Bi^{-0.85}$, $Bi \leq 25$

where Fo = Fourier's number
Bi = Biot's number

Since airflow is contiguous with working surface temperature of which is not identical with the virgin rock temperature, the equation (4) can be re-written into the form

$$T = T_p - (T_p - T_0) \exp - \frac{\alpha \, 0 \, L}{Q_m \, C_p} \qquad (8)$$

3.2 The heat from energetic systems

With regard to facilities of the Ostrava - Karvina mines only electrical equipments have been taken into account because pneumatic and Diesel systems are used only exceptionally.
If there is fulfilled a presumption that all electrical energy is changed into the heat in the final effect and it is transferred into airflow, then is accepted the relationship $Q_m = P$. Since the electrical equipments are not in the operation continuously and the presumption of the immediate heat transfer is not fulfilled there is necessary to use the correction coefficients of time loading, b, and the usage of power input, k. The relationship is as follows

$$Q_e = P \, k \, b \qquad (9)$$

where Q_e = heat released by energetic systems, W

The values of the correction coefficients k and b are introduced in Table 2.

Table 2. The values of the coefficients k and b in the Ostrava - Karvina mines conditions.

Kind of electrical equipment	b	k
Belt conveyer	0.38	0.35
Flight conveyer	0.38	0.47
Coal-getting and driving machines	0.29	0.85
Transformers	$Q_e = P(1 - \eta) \, 0.8$	

3.3 The heat from mined material on belt conveyers

With regard to the increasing of output, its concentration and mining depth levels (increasing of the virgin rock temperatures) mined material has became a significant factor that influences the airflow temperature changes, especially when a heterotropal ventilation system is used. The intensity of the heat transfer in the system mined material - airflow is increasing as the mined material proceeds to the area with a relatively lower temperatures.

The quantity of the released heat from the mined material can be expressed by a relation

$$Q_u = M C_u \Delta T_u \qquad (10)$$

where Q_u = heat from mined material, W
M_u = mined material rate, kg/s
C_u = specific heat of mined material, J/kg K
ΔT_u = mined material temperature drop, K

The value of the mined material temperature drop, T_u, can be determined approximately from the relation

$$T_u = 0.0024 L_d^{0.8} (\overline{T}_u - T_v) \qquad (11)$$

where L_d = lenght of conveyers, m
$\overline{T}_u$ = temperature of mined material below a face, $^\circ$C

3.4 The influence of moisture

Untill the influence of moisture has been considered, although moisture is very significant factor in the thermodynamics processes in mine conditions.

The using of the moisture coefficient, A, is one of the possibilities how to get hold of the influence of moisture in calculations. The moisture coefficient value can be calculated from relationship

$$A = \cfrac{1}{1 + \cfrac{l_v}{C_p} \cdot \cfrac{dx}{dT}} \qquad (12)$$

where A = moisture coefficient
l_v = evaporation heat, J/kg
dx = specific humidity change, g/kg

Since there is impossible to determine the value of the quotient dx/dT in (12)

in projected mine workings in advance, there was necessary to determine the value of the coefficient A experimentally on the basis of presumption that the value is constant for a given working. The practical measurements were realised in the Ostrava - Karvina mines in various mine - geological conditions. There was found that the value of A is mostly in the range from 0 to 1; in intake haulage roads from 0.12 to 0.28 and in faces from 0.15 to 0.47.

4 THE METHOD OF TEMPERATURE CHANGES CALCULATION

Considering that walls´surface temperature, T_p, is not only the function of time but it is changed by the influence of airflow that is why the working is divided into short intervals in each of them the surface temperature is considered to be constant and it is calculated separately together with the temperature change. The calculated airflow temperature in the end of the each interval becomes the intake temperature of the next one. The distribution of the working into the intervals makes possible the better expression of the influence of point heat sources e.g. transformers and the actual state of the thermodynamics processes in a given interval.

The equation (3) that contains the heat sources introduced in the part 3 can be re-written into the form

$$Q_m C_p dT + Q_m l_v dx = (\propto O(T_p - T_v) + \Gamma) dL \qquad (13)$$

where $\Gamma = (Q_e + Q_u)/L$, W/m

After the arrangement there is possible to write

$$Q_m C_p dT = A(\propto O(T_p - T_v) + \Gamma) dL \qquad (14)$$

The airflow temperature in the end of each interval is calculated from the expression (14) by means of following relation

$$T_{i+1} = T_i + \frac{A l (\propto O(T_p - T_i) + \Gamma)}{Q_m C_p} \qquad (15)$$

where i = number of interval
l = length of interval, m

Note: If there is a point heat source in an interval (e.g. transformer) then it is necessary to include the term in the right side of the relation (15)

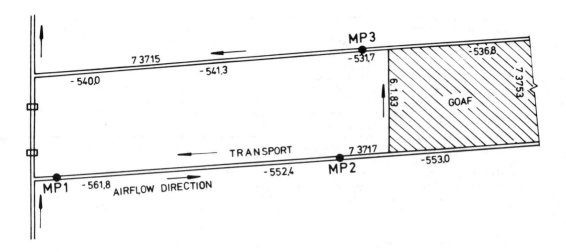

Fig. 2. The situation in the longwall face district of the Doubrava Mine.

$$\frac{P \cdot A}{Q_m C_p}$$

where P = name-plate rating of equipment, W.

5 THE EXAMPLE OF THE TEMPERATURE CHANGES CALCULATION

In the frame of the investigations and for the verification of the calculation method in practice there were made great numbers of measurements of the temperature and moisture changes of airflow in the Ostrava - Karvina mines conditions. The attention was intented on longwall face districts where the retreating mining method had been used and that is the prevailing mining method in our conditions. As the example of the calculation method practical using there is introduced the longwall face district of the Doubrava Mine.

5.1 The description of the situation in the face district

The longwall face (see Fig. 2) is demarcated by the intake haulage road, by the face and by the outlet road. The coal seam, the thickness of which is 3.9 m, is mined by using of the retreating mining method with the controled caving of the roof. The district is heterotropally ventilated. The shearer KWB-3 RDU was used as the coal-getting machine. The transport of mined material from the district is ensured by the flight conveyer Rybnik 73 in the face and by the flight conveyer TH 600 C and by the belt conveyers TP 650 in the intake haulage road. The average depth level of the district is 550 meters under the sea level. The daily average output from the district is 1100 tons. The coal seam was wetted along the whole length of the face.

5.2 In-situ measurements

At the measuring points MP1, MP2 and MP3 (see Fig. 2) there were measured thermomoisture parameters of airflow, its velocity and atmospheric pressure at the same time in 15 minutes intervals in the 3 hours period and at points MP1 and MP2 in addition there were measured the temperatures of mined material on belt conveyers. The consuption of electrical energy of energetic systems and their time loading were recorded continuously. The virgin rock temperature was determined by using of the calculation method (Fiala, Taufer, Benes, Kohut 1984).

The using of instrument measuring technique was limited by the requirement of their flameproof construction. Temperature and moisture measurements were made by using of the common Assmann´s psychrometers with thermometers with 0.2°C division.

The airflow velocity was measured with the semi-automatic anemometers Rosenmüller and atmospheric pressure with the Ascania instruments.

The mined material temperatures were measured on samples that had been collected into the insulated containers in which the temperature of the samples was measured immediately by using of the flameproof thermometers G1-A (Knejzlik et al. 1982). The sampling at the measuring point MP1 was made from the same point as at the measuring point MP2.

The data of the measurements then were used as the input data for the following checking calculations of the temperature changes in the intake haulage road and in the face and they are introduced in Table 3.

Table 3. The data concerning with the intake haulage road 7 3717 and the face 7 3753 of the Doubrava Mine.

Parameter	Intake road	Face
Length, L (m)	444	141
Section, S (m^2)	10.7	7.25
Age, (day)	995	1
Thermal conductivity of rocks, k (W/m K)	2.963	0.22
Thermal diffusivity, a (m^2/s) 10^{-6}	1.406	0.138
Moisture coefficient, A	0.3	0.2
Specific humidity mean change, dx (g/kg)	0.963	5.722
Airflow rate, Q_m (kg/s)	26.8	
Atmospheric pressure B (Pa)	106653	
Depth level (m)	-552	
Virgin rock temp., T_h (^{o}C)	32.5	
Power input:		
Conveyers (kW)	270	360
Shearer (kW)	-	250
Output level (kg/s)	26.6	

The thermal conductivity of carboniferous rocks and of coal was measured by using of the divided-bar instrument and by Shotherm QTM-D2 apparatus on samples that had been taken from the watched intake haulage road and from the face.

5.3 The comparison of the calculated and in-situ measured results

The calculations were made on the basis of the parameters from Table 3 and from the measured course of dry-bulb temperatures of airflow at the measuring point MP1. There is made the comparison of the measured and calculated temperatures of airflow at the measuring points MP2 and MP3 in the following Table and diagram.

Table 4. The measured and calculated dry-bulb temperatures of airflow at the measuring points MP2 and MP3 in different time.

Time	Dry-bulb airflow temperature				
	MP1	MP2	$\overline{MP2}$	MP3	$\overline{MP3}$
7.30	20.2	21.9	22.0	25.2	25.2
7.45	20.4	22.3	22.2	25.3	25.5
8.00	20.4	22.3	22.2	25.6	25.5
8.15	20.6	22.5	22.4	25.8	25.7
8.30	20.4	22.1	21.4	25.7	22.2
8.45	20.2	21.9	21.2	25.4	22.0
9.00	20.2	21.7	21.2	25.4	21.8
9.15	20.2	21.7	21.2	25.5	21.8
9.30	20.2	22.1	22.0	25.8	25.3
9.45	20.3	22.3	22.1	25.7	25.5
10.00	20.3	22.2	22.1	25.6	25.4
10.15	20.2	22.1	22.0	25.6	25.3

Note: The calculated temperatures are introduced in columns $\overline{MP2}$ and $\overline{MP3}$, the measured temperatures in columns MP1, MP2 and MP3.

6 DISCUSSION OF RESULTS

In this part of the paper the authors should like to mention some results that follow from the measurements and the calculations.

1. The increasing of temperatures in the intake haulage road 7 3717 and in the face 7 3753 is essentially constant (1.2 and 3.25^{o}C, respectively) regardless of the stopping of the mining and drawing process in a short period of time (from 8.00 to 9.15). This is probably caused by the emission of the heat that was accumulated in steel elements in the district e.g. supports, conveyer´s frame, motor´s body etc.).

2. From the course of the curves in the Figure 3 and from the records of the electrical equipments operation (shearer, flight and belt conveyers) and of the output there is evident that the variance of the intake airflow temperatures into the road and the face is influenced by the heat emission from electrical equipments and from mined material in transport on belt conveyers.

From the mentioned Figure there is also seen that the calculated temperatures at

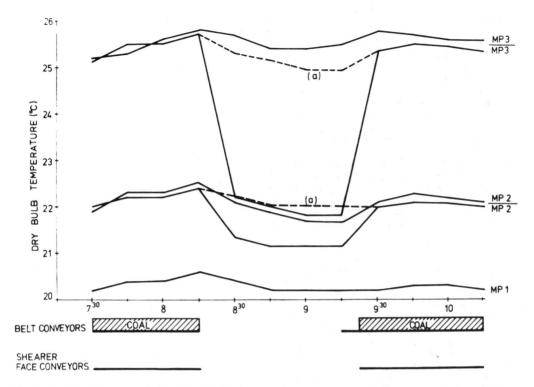

Fig. 3. The diagram of the dry-bulb temperatures course at the measuring points
MP1, MP2 and MP3.

the points MP2 and MP3 well correspond
with the measured temperatures. The only
exception is the period of time from 8.15
to 9.15 because the releasing of the a-
ccumulated heat in steel elements, while
the production is stopped, is not included in calculations. The curves (a) simulate the state of continuous activity
and with regard to the mentioned heat
accumulation effect the real state is
better expressed.

3. The mean deviations of the measured
and calculated temperatures (variant a)
are in the road and in the face 0.11 and
0.21, respectively. This deviations are
functioned by the neglection of some additional heat sources, such as low-temperature oxidation, lighting, body metabolism
etc. and also there is necessary to say
that the exact expression of the considered heat sources in real conditions is
very difficult. Therefore it was necessary the using of some simplifications that
have to do with the determination of the
dimension of workings, physico-thermal
properties of rocks, the expression of
the heat released by energetical systems,

humidity etc.

4. The determination of the heat that
is released by mined material in transport on belt conveyers and/or flight conveyers is the difficult problem of the
prediction calculations.

Practical measurements in the Ostrava-
Karvina coalfield conditions have shown
that the mined material temperatures below the faces have been from 4 to 8 °C lower then the virgin rock temperatures.
For the calculations of the heat released
by mined material in accordance with (10)
and (11) there was used the mean value of
the decreasing of mined material and the
average wet-bulb temperature of airflow
at the measuring point MP1. The calculated decreasing of the temperature of mined material corresponds to measurements.

5. By using of this calculation method
there is also possible the determination
of the airflow relative humidity at the
end of workings by means of the moisture
coefficient, A. Since this parameter has
been determined statistically for various
types of workings the resulted relative
humidity value is also only of orientation.

7 CONCLUSIONS

The introduced example of the calculation of the temperature changes was selected from the extensive set of the measurements that had been made in the frame of the investigation program in the Ostrava-Karvina mines. The comparisons of the measured and the calculated temperature changes confirm that the worked up method of the calculation is acceptable for the predictions of the airflow temperature state in longwall face districts and the results can become the qualified basic material for proposal of the ventilation adaptation.

As was introduced above the method is acceptable as well as in this state for practical using. In the next investigation there will be necessary to extend its using at next types of workings e.g. shafts, crosscuts etc. and also to take into account workings that are ventilated by using of auxiliary systems. It will be important to intent on more qualify expression of the heat sources especially on mined material, goaf etc.

The working up of the calculation method that affords not only the data about prediction state of mine airflow on proposed workplaces but that will contains also the data concerning with the necessary adaptations of the mine ventilation is the main aim.

8 ACKNOWLEDGEMENT

The authors wish to thank Doc.ing.A. Otahal, PhD from the Mine Ventilation Department of the Mining and Metallurgical University in Ostrava for the valuable reminders and Mrs.J.Horakova and Mrs.M.Velebilova for the preparation of the paper.

9 REFERENCES

Fiala,J., Taufer,A., Benes,M., Kohut,R. 1984, Determination of virgin rock temperatures in prospective areas of underground mining. Proc. the 3rd Int. Mine Vent. Congress, pp. 369-373, Harrogate.

Knejzlik,J. et al., Investigation and development of instruments for laboratory and in-situ measurements. Research report II-6-1/02.4, Min. Inst. of the Czech. Ac. Sci., Ostrava, Czechoslovakia, 1982.

Scerban,A.N., Kremnev,O.A. 1959, Scince foundation of the calculation and control of the heat régime of deep mines. 1st edition, Kijev, USSR.

The reduction in heat flow due to the insulation of rock surfaces in mine airways

P.BOTTOMLEY
Chamber of Mines of South Africa Research Organization, Johannesburg, South Africa

Abstract: A theoretical and experimental study of the effects of insulating a mine airway is presented. It was theoretically found that for a totally insulated airway the possible reduction in heat flow was in the region of 50 per cent, whereas for an airway with the footwall remaining uninsulated the reduction in heat flow was only in the region of 25 per cent. Initial results from an insulated experimental test site support these findings.

1 INTRODUCTION

As the South African gold mining industry is working to greater depths and, hence, to greater virgin rock temperatures, it is of crucial importance to be able to maintain an acceptable working environment. Present workings have extended to approximately 3 600 m below surface, with virgin rock temperatures up to 60 °C being experienced, and serious consideration has been given to extending operations down to depths of 5 000 m, into rock temperatures of approximately 70 °C. The current and expected increase in heat load on the mine ventilation systems has resulted in a rapid, almost exponential growth in installed refrigeration capacity since 1976[1]. A similar trend has occurred in the cost of primary power and it has become imperative to develop more cost effective methods for combating the underground heat load problem and in fact to reduce the flow of heat into the environment.

The many kilometers of intake airways contribute a large proportion of the overall mine heat load. This proportion grows with the life of a mine due to the airways extending to reach the workings further from the shaft. The practice of surface or station bulk air cooling also results in an increase in the heat load contribution by the airways. This is due to an increased temperature driving force between rock and air. One possible method of reducing the heat load is by the insulation of the rock surfaces. This paper presents a study of some of the factors to be considered when contemplating insulating airways in order to reduce heat flow. Nomograms for predicting the reduction when insulating typical airways are presented, and some details and initial results from an experimental field trial are shown.

The study has been divided into three phases:

1. The theoretical prediction of the overall heat flow benefits. This involves the mathematical solution of the two-dimensional heat flow equation to determine the heat flow at a single cross-section, and then the step-wise calculation of the changing air conditions along the length of the airway. While the latter part of this computational process is relatively simple, the former presents some difficulties. These are overcome by using finite element techniques or where possible, simplified analytical approaches, such as a quasi-steady method[2]. The accuracy of these calculations is limited due to the difficulty in realistically modelling the practical situation.

2. The actual measurement of heat flow in an underground insulated airway. Data are presently being collected at an insulated airway in a deep gold mine.

3. The identification of suitable insulation materials. The benefits of coating the rock surfaces lie not only in

the insulation effect but also in the support properties of the material. In fact, some mines are considering the complete shotcreting of all rock surfaces in the haulages and cross-cuts as they are developed. These measures are being considered from a rock stabilization point of view without thought of the insulating benefits. Materials suitable for both support and insulation purposes would have obvious benefits.

Not all of this programme of work has been completed, but this paper serves to present the major influences on heat flow reduction when considering the effects of insulation in mine airways. Following a brief look at the history of mine insula-tion some of the findings of the theoreti-cal work and predictions of heat flow reduction are presented. The under-ground experimental test site at a deep gold mine and the techniques being used are described and some early experimental results are given. Finally some of the desired properties of the insulating material are described.

2 HISTORY OF MINE INSULATION

During the period 1962 to 1964 an insu-lation field trial was conducted at Loraine Gold Mines Limited. This project utilized two types of insulation, namely polyurethane foam and slag wool. The heat flow was measured by drilling holes to a depth of about 3 m into the rock mass and measuring temperatures at 300 mm inter-vals. Although a number of measurements were made over a period of more than two years and it was found that the heat flux into the airway was reduced, the results were never published. Nor was the wide scale insulation of airways ever imple-mented. Presumably either the results were not conclusive or the benefits not sufficiently attractive at that time.

In 1967, the Environmental Engineering Laboratory attempted, by means of a resistance paper analogue to quantify the effect of an insulated airway with the footwall uninsulated. It was argued that most of the benefits of insulating an airway will be negated by the increased heat flow through the footwall which in practice would probably not be insulated. The decreased benefits of insulation with a finite length of tunnel were shown. The benefit decreases because insulation, by reducing heat pick-up, promotes a smaller air temperature gradient along the airway, thereby maintaining a higher driving force. It was considered that the overall

benefits of insulating were small and it was concluded that the insulation of mine airways was not justifiable at that time. It is considered that the methods used are open to criticism and the results presented are in error.

In 1968 Gould[3] used a numerical method, based on the Goch-Patterson approach[4], to compare the temperature rise of air through 600 m of tunnel when either completely insulated or uninsulated (with varying degrees of wetness for the uninsulated case). The results show a significant decrease in the outlet air temperature due to insulating when compared to the uninsulated case, typi-cally by 1,5 °C wet-bulb. His overall conclusion was that there are many practi-cal problems which require attention before insulation becomes feasible, and that the higher the virgin rock tempera-tures, the more justifiable becomes the expense of insulating. However, at the virgin rock temperatures being encountered at the time (1968) there was no merit in its application.

In more recent times Hughes[5] point-ed out some advantages of insulating air-ways. He attempted to quantify the econo-mic benefits due to the reduction in heat flow and reduction in resistance to air flow. Break-even cost figures for the insulation were evaluated. He presented a positive case for the use of insulation. Unfortunately his economic justification is erroneous due to, amongst other things, the assumption of steady state heat transfer rather than the transient problem existing, as well as his consideration of only one cross-section of tunnel rather than a finite length.

The past work on mine insulation has generally resulted in negative conclusions with regard to its wide-spread implementa-tion. However most of this work was done about 20 years ago. Since then the environmental control practices in South African gold mines have changed radi-cally. Significantly higher virgin rock temperatures are being encountered and the widespread use of chilled service water and the natural partner - the bulk cooling of air - are well established. The avail-able insulation materials have changed radically over these two decades. Furthermore the increased depths and higher rock stresses have increased the possible benefits with regard to rock support. All these factors have indicated the need for a careful re-consideration of the insulation of mine airways.

3 THEORETICAL PREDICTION OF THE EFFECT ON HEAT FLOW OF INSULATING MINE AIR WAYS

In order to make a full theoretical study of the effects of insulation, all the various parameters which influence heat flow in tunnels must be considered. Some of these parameters are:
Condition of ventilation air.
Virgin rock temperature.
Tunnel age.
Rock specific heat.
Rock density.
Rock conductivity.
Tunnel radius.
Tunnel length.
Surface heat transfer coefficient.
Ventilation air quantity.
Heat transfer due to radiation.
Dampness of the rock surface.
Insulation thickness.
Insulation conductivity.
Equivalent conductivity of footwall ballast.
Percentage covering of insulation.
Time of rock exposure prior to application of insulation.
In this paper many assumptions and typical values have been used in order to reduce the number of variables and simplify the results. Further work to be published will include a complete analysis of all the relevant parameters. In this paper only dry conditions are considered, the heat transfer due to radiation is assumed to be negligible and the conductivity of the footwall ballast is assumed to be the same as the rock conductivity.
The theoretical assessment of the reduction in heat flow due to insulation is reported here for five cases:
1. A single cross-section of airway with the entire perimeter uniformly insulated from the time of excavation.
2. A single cross-section of airway with with the perimeter partially insulated from the time of excavation.
3. A single cross-section of airway with the entire perimeter uniformly insulated 6 months after excavation.
4. A finite length of tunnel with the entire perimeter uniformly insulated from the time of excavation.
5. A finite length of tunnel with the footwall uninsulated. The insulation on the sidewall and hangingwall were assumed to be present from the time of excavation.
It must be stressed that the first three cases are hypothetical and merely serve to give a comparative initial estimation. The more realistic models are those used in cases 4 and 5.

The following specific set of conditions has been used in these analyses:
Air dry-bulb temperature :25 °C
Virgin rock temperature :45 °C
Circular tunnel radius :2 m
Rock thermal conductivity:6,0 W/mK
Rock thermal diffusivity :$2,4 \times 10^{-6} m^2/s$
Insulation conductivity :0,03 W/mK
Surface heat transfer coefficient :10 W/m^2K
(this corresponds to an air velocity of about 4 m/s)
Airway assumed to be dry.

3.1 Single cross-section of airway completely insulated from time of initial excavation

An estimation of heat flow into a cylindrical tunnel, with uniform boundary conditions, situated in an infinite solid, may be calculated from tabulated values given by Jaeger and Chamalaun[6].
The effect of the insulation layer on the surface is accounted for by neglecting its thermal capacity (which is small compared to that of the rock), and calculating an equivalent overall surface heat transfer coefficient, h_e, through:

$$1/h_e = 1/h + \frac{\text{Insulation Thickness}}{\text{Insulation Conductivity}}$$

where h is the actual surface heat transfer coefficient.
The effect of varying the insulation thickness is therefore reflected in a resistance to heat flow term at the rock surface.
Figure 1 shows the variation of heat flow into the tunnel with time, for varying thickness of insulation. Figure 2 shows the percentage reduction in heat flow, compared to the uninsulated case, as

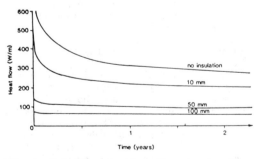

Fig.1. Reduction in heat flow due to varying thicknesses of insulation.

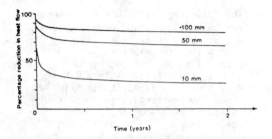

Fig.2. Percentage reduction in heat flow
due to varying thicknesses of insulation.

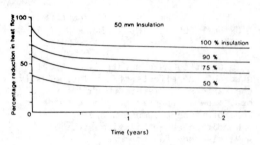

Fig. 4. Percentage reduction in heat flow
due to varying the percentage covering of
insulation.

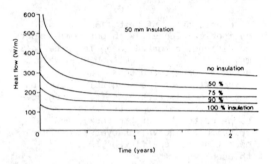

Fig.3. Reduction in heat flow due to varying the percentage covering of insulation.

derived from Figure 1. Three points
should be noted at this stage. Firstly,
the reduction in heat flow diminishes
rapidly during the first few months and
after that tends to level out. Secondly,
there is a diminishing return effect with
regard to the thickness of insulation.
Thirdly, the magnitude of the reduction in
heat flow for a 50 mm insulation thickness
is about 60 to 70 per cent, even after the
initial rapid decay has damped out.

3.2 Single cross-section of airway
    partially insulated from time of
    initial excavation

In practice it will probably be difficult
to insulate the footwall effectively.
Although in reality the ballast in the
footwall would have some insulating effect
this has been ignored and the effect of
partially insulating the rock surfaces has
been examined. The estimation of heat
flow into a cylindrical tunnel with nonuniform boundary conditions may be calculated using a quasi-steady method developed by Starfield and Bleloch[2].

Figures 3 and 4 show the effects of
varying the percentage covering of the
insulation. The thickness of insulation
used in these calculations was 50 mm. The
case where 75 per cent of the perimeter is
insulated corresponds to the case of an
uninsulated footwall. The point to note
from these figures is that when the rock
surfaces are completely covered there is
approximately a 65 per cent reduction in
heat flow, however when the footwall is
left uninsulated this reduction was
reduced to about 40 per cent.

3.3 Single cross-section of airway which
    is completely insulated some time
    after excavation

The previous analysis does not show the
effect of insulating an airway some time
after it has been excavated, where the
rock has been cooled to a certain extent
prior to the insulation being applied.
Consider the previously described tunnel
conditions with an age of 6 months before
the application of a complete covering of
50 mm of insulation. Prior to insulating,
the rock surface had been extensively
cooled to a temperature of about 28 °C
(recall that 6 months before it had been
at the virgin rock temperature of 45 °C).
Thus, on applying the insulation layer
there is initially a large reduction in
heat flow due to the small driving force
between the insulation surface and the air
stream. After the insulation has been
applied the temperature of the rock surface increases and after an infinite time
period the heat flux from the airway would
approach that from a tunnel that had been
insulated in a similar manner immediately
after being excavated. The variation in
heat flux before and after the airway is
insulated is shown in Figure 5.

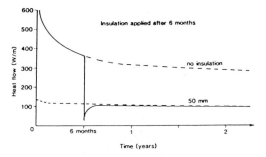

Fig. 5. The reduction in heat flow due to applying insulation 6 months after excavation.

3.4 Finite length of tunnel with the entire perimeter uniformly insulated from time of excavation

The examination of a single cross-section of tunnel overestimates the overall heat reduction effects of insulation when observed along a finite length of tunnel. By reducing the heat flow, the insulation helps to maintain a higher temperature driving force than would be experienced in an uninsulated tunnel.

In order to evaluate the heat flow into a length of tunnel it is necessary to use a stepping procedure which calculates the air temperatures leaving a segment of tunnel, for subsequent input temperatures into the next segment. This procedure is easily implemented on a micro-computer. The results of using such a procedure on varying lengths of tunnels, with differing air flow rates, are shown as a nomogram in Figure 6. The insulation thickness was 50 mm and similar conditions to those previously outlined were assumed, except that the surface heat transfer coefficient was allowed to vary (the heat transfer coefficient used previously corresponded to an air velocity of 4 m/s).

The results of insulating the tunnel are expressed as a percentage reduction in heat flow compared with an uninsulated tunnel. The nomogram in Figure 6 may be used to estimate the heat reduction for varying lengths of tunnel and varying ages. As an example comparable to the earlier results, consider an air velocity of 4 m/s, for a tunnel length of 2 000 m; the heat flow reduction would vary from 58 to 51 per cent as the age varies from 1 to 10 years. Note that these savings are not as high as those indicated by investigating a single cross-section; see Figure 2 where the corresponding value for the age of 1 year is 70 per cent.

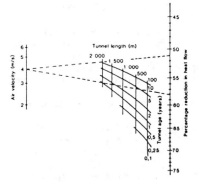

Fig.6. Percentage reduction in heat flow for tunnels with a complete covering of 50 mm thick insulation.

3.5 Finite length of tunnel insulated from time of exposure, with the footwall uninsulated.

To evaluate the heat flow reduction in a partially insulated finite length of tunnel, the stepping procedure outlined in Section 3.4 was applied.

Again for varying tunnel lengths, and differing air flows the results are set out as a nomogram, in Figure 7. The assumed parameters were the same as those of Section 3.4.

For the same example as before, that is an air velocity of 4 m/s and a tunnel length of 2 000 m the heat flow reduction would vary from 28 to 23 per cent as the tunnel ages from 1 to 10 years. These values are approximately half those computed for a totally insulated airway, and significantly different from those values evaluated at a single cross-section.

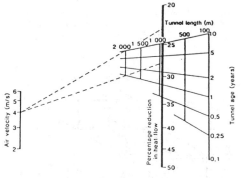

Fig.7. Percentage reduction in heat flow for tunnels with a partial covering of 50 mm thick insulation.

461

4 THE AIRWAY INSULATION EXPERIMENT

A section of airway at Western Deep Levels gold mine is being used to investigate experimentally the effects of insulation. The test site is divided into three 30 metre sections, which were treated as follows:

1. Section A - An uninsulated control section.

2. Section B - The sidewalls, hanging-wall and foot wall were insulated with approximately 50 mm thick polyurethane foam.

3. Section C - The side wall and hangingwall were insulated with approximately 50 mm thick polyurethane foam. The foot wall was left untreated.

This is shown schematically in Figure 8.

Before treatment with insulation a comprehensive set of resistance temperature measuring devices was installed in the centre of each section in the pattern shown in Figure 9.

Each probe hole was drilled 6 meters deep and more of the resistance thermometers were distributed near the surface than deep in the rock. Some details of the test site are given below:

Air dry bulb temperature :30 °C
Virgin rock temperature :43 °C
Depth below surface :2 700 m
Square cross-section :3,5x3,5 m^2
Air velocity :2,0 m/s
Rock diffusivity :2,44x10^{-6}m^2/s
Rock thermal conductivity :5,48 W/mK
Date of excavation :August 1983.
No fissure water or drain water.

The variation in heat flux around the perimeter of each section was calculated from the collected temperature data by numerically evaluating the temperature gradient at the rock surface. A typical example is shown in Figure 10. The application of insulation was completed on the 22nd of March, 1985 and temperatures recorded at weekly intervals thereafter. For each of the four weeks since insulation was completed, the total heat flux

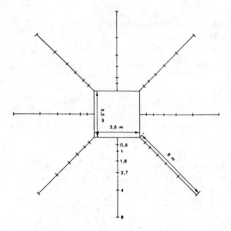

Fig.9. Distribution and pattern of temperature measuring points.

from each section was computed, and is shown in Figure 11.

At this early stage in the experiment the results show a marked decrease in heat flow in the insulated sections, as predicted in Figure 5. The heat flow in the total insulated section is less than half that in the uninsulated section, and that in the partially insulated section is approximately two thirds of the uninsulated section. At this time the sharp decrease in the heat flow during the first two weeks in the uninsulated section cannot be explained. It is expected that temperature logging will continue for a further 6 months in order to obtain a complete picture of the heat flow pattern.

5 INVESTIGATION OF SUITABLE MATERIALS

It was the original intention to coat the test site with a phenolic formaldehyde

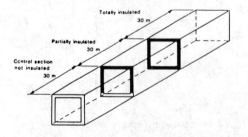

Fig.8. Schematic of insulated test site.

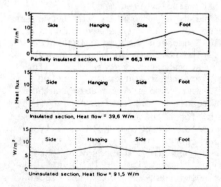

Fig.10. Variations in heat flux around perimeters of test sections.

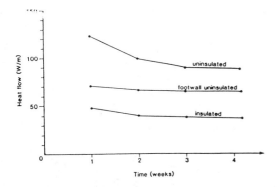

Fig.11. Variations in heat flow with time from the experimental test sections.

foam. However, problems were encountered with the application of this material to the rock surface. The test site was eventually insulated with a polyurethane foam which, for a number of reasons, is not suitable for full scale practical use. There are however a number of materials to which consideration should be given; these may be divided into cement based materials and plastic foams (and in some instances a combination). A material suitable for both support and insulation purposes would have obvious benefits.

There is thus a need to examine the suitability and costs of various materials that are presently available and if necessary to encourage the development of more suitable materials. Some of the properties that should be considered in assessing the suitability of materials for the large scale insulation of mine intake airways are as follows:

1. Thermal Conductivity
The lower the conductivity, the less the amount of material needed to have the same insulating effect. As a reference value it should be noted that polyurethane has a conductivity of 0,03 W/mK and it has been tentatively estimated that thicknesses of the order of 50 mm would be suitable for insulation purposes.

2. Toxicity
During application and curing, a number of insulation foams emit harmful gases, in particular formaldehyde gas which is carcinogenic and for which concentrations of greater than 2 ppm should be avoided.

3. Fire Resistance
It is extremely important that the insulation product will not support combustion. Both the fire hazard and the emission of toxic fumes are extremely serious considerations.

4. Composition
The insulation material should not be susceptible to attack from bacteria and should be unaffected by running water or high humidity.

5. Structural Strength and Durability
A high structural strength would be beneficial with regard to the shoring of rock and the prevention of weathering, as well as for good durability of the insulation surface. The product should have a good adherence to the rock. An expected life of the order of 10 years with a limited patching up should probably be the aim.

6. Ease of Application
The insulation material should be simple to apply in large quantities. The hangingwall can cause difficulties due to the insulation falling off before setting. Product density, conductivity and application thickness should be easily controlled in an underground environment.

7. Price
The price of supply and application of the insulation, combined with its conductivity will eventually determine the feasibility of insulating intake airways.

CONCLUSIONS

While some work has yet to be done on finalising the prediction of the change in heat flow due to insulation, it is evident that substantial savings in heat flow can be realised, but only if the footwall can be adequately insulated. A reduction in the region of 50 per cent can typically be expected for a fully insulated tunnel. This drops to only 25 per cent for a tunnel with an uninsulated footwall.

The experiment at Western Deep Levels gold mine, although in its very early stages, is providing an expected pattern of heat flow. Substantial reductions in heat flow are evident between the insulated and non-insulated sections.

REFERENCES

1 Sheer,T.J. Burton,R.C. & Bluhm,S.J. Recent developments in the cooling of deep mines. The South African Mechanical Engineer Vol 34 Jan 1984.
2 Starfield,A.M. & Bleloch,A.L. A new method for the computation of heat and moisture transfer in partly wet airways. Journal of S.A. Inst. of

463

Mining and Metal Vol 83. Nov/Dec.
1983.

3 Gould,M.J. The determination of parameters concerned with heat flow into
underground excavations. Dissertation
for MSc (Eng) University of the
Witwatersrand, March, 1968.

4 Goch,D.C. & Patterson,H.S. The heat
flow into tunnels, Journal of the
Chem. Metal and Mining Soc. of S.A.,
Sept. 1940.

5 Hughes,R.O. A few thoughts on heat
loads in intake haulages. Journal of
Mine Ventilation Society of S.A. Vol
31 March 1978.

6 Jaeger,J.C. & Chamalaun,T. Heat flow in
an infinite region bounded externally
by a circular cylinder with forced
convection at the surface. Aust. J.
Phys., Vol 19 1966.

Mathematical modelling of diffusive and heatmass transfer processes in ventilating mining workings of an arbitrary form

A.A.BAKLANOV & G.V.KALABIN
Kola Branch of the Academy of Sciences of the USSR, Apatity, USSR

ABSTRACT: The problems dealing with large chambers ventilation are discussed taking into account the basic physical processes governing dilution and removal of harmful impurities from confined space of an arbitrary form.

1 INTRODUCTION

Mathematical modelling has been widely used recently in solving problems of mine ventilation and is considered a most promising method of investigation. As to the chambers of an arbitrary form an additional and rather complicated problem arises to study the distribution of air flow in workedout area. Besides, it is very important to take into account heat and sometimes humidity processes in mine aerology models as they can cause significant air movement. These problems have not been completely studied.

This work deals with the way of solving the problem of large chamber ventilation taking into account the basic physical processes and suggests a mathematical model of dynamics of mine atmosphere with allowance for heat processes and interaction with rock and a model of transfer and diffusion of multicomponent impurities in chambers of an arbitrary geometry. In constructing mathematical models there appears the necessity of experimental check of the models and determination of some parameters. For these reasons some comparisons are given with the results of physical modelling of chamber ventilation with rectangular workings taken as an example (Kalabin, 1981, 1983).

2 MODEL OF MINE ATMOSPHERE DYNAMICS

To describe thermohydrodynamics of mine atmosphere the following system of equations is suggested:

$$\frac{dU}{dt} = -\frac{1}{\rho}\frac{\partial P}{\partial z} + \frac{\partial}{\partial x}\tau_{11} + \frac{\partial}{\partial y}\tau_{12} + \frac{\partial}{\partial z}\tau_{13} + J_x \quad , \quad (1)$$

$$\frac{dV}{dt} = -\frac{1}{\rho}\frac{\partial P}{\partial y} + \frac{\partial}{\partial x}\tau_{21} + \frac{\partial}{\partial y}\tau_{22} + \frac{\partial}{\partial z}\tau_{23} + J_y \quad , \quad (2)$$

$$\frac{dW}{dt} = -\frac{1}{\rho}\frac{\partial P}{\partial z} - g + \frac{\partial}{\partial x}\tau_{31} + \frac{\partial}{\partial y}\tau_{32} + \frac{\partial}{\partial z}\tau_{33} + J_z(3)$$

$$\frac{dT}{dt} = \frac{L}{C_p}\Phi + \frac{\partial}{\partial x}H_1 + \frac{\partial}{\partial y}H_2 + \frac{\partial}{\partial z}H_3 + J_T \cdot (4)$$

$$\frac{1}{\rho}\frac{\partial \rho}{\partial t} + \frac{\partial U}{\partial x} + \frac{\partial V}{\partial y} + \frac{\partial W}{\partial z} = 0 \quad , \quad (5)$$

$$P = \rho R T \quad . \quad (6)$$

where $\dfrac{d\varphi}{dt} = \dfrac{\partial \varphi}{\partial t} + U\dfrac{\partial \varphi}{\partial x} + V\dfrac{\partial \varphi}{\partial y} + W\dfrac{\partial \varphi}{\partial z},$

t – time;
U, V, W – the components of velocity vector $\vec{U}$ in the directions of axes of Cartesian coordinates x, y, z;
ρ, P, T – density, pressure and temperature of the air;
q – free fall acceleration;
L – latent evaporation heat of condensation;
Φ – velocity of liquid phase formation;
C_p – heat capacity of the air at P=const;
R – gas constant of the air;

$I_\alpha(\alpha=x,y,z,T)$ — members characterizing the capacity of ventilation installations and heat sources;

$\tau_{ij}(i,j=\overline{1,3})$ — symmetric Reynolds stress tensor:

$H_i(i=\overline{1,3})$ — turbulent heat flow.

As boundary conditions at the chamber inlet (S_1) and outlet (S_2) we assume Neimans conditions depending on the concrete problem

$$\frac{\partial U}{\partial n}=0; \; \frac{\partial V}{\partial n}=0; \; \frac{\partial W}{\partial n}=0; \; \frac{\partial T}{\partial n}=0 \quad (7)$$

when $\overline{x}\in S_i$

or Dirichlet

$$U=U_\phi; \; V=V_\phi; \; w=W_\phi; \; T=T_\phi \quad (8)$$

when $\overline{x}\in S_i$,

where n — the normal to the boundary of the region of integration.
The boundary conditions for pressure are assumed (Bram van Leer, 1977).
At the boundary with the massif (G) we assume the adhesion conditions for the velocity components

$$U=V=W=0 \quad (9)$$

when $\overline{x}\in G$,

and the surface temperature will be determined from the heat balance equation at the air — massif interface.

$$c_s\rho_s\left(K_s\frac{\partial\tilde{T}}{\partial n}\right)_G - \rho\, c_p\left(D_T\frac{\partial T}{\partial n}\right)_G = Q_T , \quad (10)$$

where T — massif temperature;
c_s, ρ_s — specific heat and soil density;
K_s — coefficient of the soil temperature conductivity;
D_T — coefficient of molecular heat transfer;
Q_T — function characterising artificial heat sources on the massif surface.

To calculate the massif temperature $\tilde{T}$ using the corresponding system of equations of heat conductivity of massif rocks we construct a numerical model analogous to the one described in (Baklanov, 1983);

To complete the system of equations (1)-(6) we use the form described by Smagorinsky (1963) for tensor by Reynolds

$$\tau_{ij}=K_M\, D_{ij} , \quad (11)$$

where D_{ij} the tensor of deformation

$$D_{ij}=\frac{\partial U_i}{\partial x_j} + \frac{\partial U_j}{\partial x} + \frac{2}{3}\delta_{ij}\frac{\partial U_h}{\partial x_h} \quad (12)$$

and K_M — the effective viscosity as shown in (Clark, 1977)

$$K_M=(K\Delta)^2/Def/ \quad (13)$$

$$K_M=(K\Delta)^2/Def/ (1-R_i)^{1/2} \quad (14)$$

when $R_i \leqslant 1$,

$$(Def)^2=\frac{1}{2}\left(D_{11}^2+D_{22}^2+D_{33}^2\right)+D_{12}^2+D_{13}^2+D_{23}^2 \quad (15)$$

where R_i — Richardson number;
$\Delta = \Delta x\cdot\Delta y\cdot\Delta z$ — the volume of an elementary-net cell;
K — the nondimensional parameter determined experimentally.

For the turbulent heat flow

$$H_i=K_H\frac{\partial T}{\partial x_i} \quad (16)$$

where $K_H = K_M$.

To take into account the real three-dimensional geometry of the workings let us use in numerical realization the method of fictitious regions (Marchuk, 1977) the essense of which consists in addition of fictitious regions to the actual region of integration till obtaining a more suitable form and in redefining of the model equation system by special conditions of its expansion on fictitious region. In the given paper we shall use modifications (Aloyan et. al., 1982) with variational—difference method (Penenko, 1981). With such an approach the conditions on the boundaries between the main and fictitious regions are natural for variational functional and are taken into account automatically when recording the conditions of stationarity of the functional summary analoque at the points of net region of the corresponding surface.

Construction of finite—difference approximations is made for a joint system "air—massif" by discretization of the basic integral identity resulted from the conditions of stationarity of summary functional. Such approach to solving the hydrodynamic type problems has been developed by Penenko (1981). We construct numerical algo—

rithm solving discrete equations on the basis of method of splitting (Marchuk, 1977, Penenko et.al., 1976).

3 MODEL OF DIFFUSION AND TRANSFER OF IMPURITIES IN WORKINGS

Depending on the effluents composition of impurities in the workings may be rather diverse. It may consist of some dozens of components and undergo various chemical transformations. That is why we shall construct a model of transfer and transformation of multicomponent impurity. Let us consider the impurity consisting of n components and denote with $\vec{C} = \{C_i\}$ $(i = \overline{1, n})$ — volumetric concentrations of impurities averaged by a time interval t. Then the following system of equations is taken as the model basis:

$$\frac{\partial \vec{C}}{\partial t} + U \frac{\partial \vec{C}}{\partial x} + V \frac{\partial \vec{C}}{\partial y} + \left(W - \omega\right) \frac{\partial \vec{C}}{\partial z} + B\vec{C} = (17)$$
$$= \Delta_c \vec{C} + \vec{I}_c$$

where
$$\Delta_c \vec{C} = \frac{\partial}{\partial x} D_x \frac{\partial \vec{C}}{\partial x} + \frac{\partial}{\partial y} D_y \frac{\partial \vec{C}}{\partial y} + \frac{\partial}{\partial z} D_z \frac{\partial \vec{C}}{\partial z}$$

$D_x = \text{diag} \{D_{xi}\}$, $D_y = \text{diag} \{D_{yi}\}$,

$D_z = \text{diag} \{D_{zi}\}$, $(i = \overline{1, n})$ — the coefficients of turbulent diffusion in the directions of the corresponding coordinate axes;
$\omega = \text{diag} \{W_{gi}\}$ $(i = \overline{1, n})$ — diagonal matrix, the elements of which are equal to the velocities of sedimentation of the corresponding substances;
$B(\vec{x}, t) = \{B_{ij}(\vec{x}, t)\}$ $(ij = \overline{1, n}; \vec{x} = (x, y, z))$ — matrix operator describing the interaction of various impurities (in the case of passive impurities $B = 0$);
$\vec{I}_c = \{I_{ci}(\vec{x}, t)\}$ $(i = \overline{1, n})$ — the sources of impurity release.
As the initial conditions we assume

$$\vec{C} = \vec{C}_o(\vec{x}) \quad \text{when } t = 0. \tag{18}$$

Let us define the boundary conditions. At the region outlet let us take arbitrary conditions

$$\frac{\partial \vec{C}}{\partial n} = 0 \quad \text{when } \vec{x} \in S_2 \tag{19}$$

at the inlet we assume

$$\vec{C} = \vec{C}_s(\vec{x}, t) \quad \text{when } \vec{x} \in S_1 \tag{20}$$

if impurities enter through S_1.

Balance equation for each impurity component on the boundary with the massif surface is

$$D \frac{\partial C_i}{\partial n} - W_{gi} C_i \sin \alpha - \beta_i C_i = Q_{ci} , \tag{21}$$

where D — the molecular diffusion coefficient;

α — angle of inclination to the horizontal plane;
β_i $(i = \overline{1, n})$ — coefficients of absorption of impurities by the surface;
Q_{ci} — power of surface impurity sources.

The methods of obtaining discrete analogs and numerical realization are similar to those mentioned above. However, the impurity concentration is a strictly non-negative value, hence, in solving equations (17) there appears an additional requirement of monotonicity of numerical schemes. At the impurity transfer stage we shall use the modification of Fromm scheme (Bram van Leer, 1977).

4 CALCULATION RESULTS

Let us briefly consider some results of numerical experiments of modelling of aerodynamic, diffusive and heat processes in chamber-like workings.

Let us consider the alteration of aerodynamic characteristics of ventilation flow when the through workings sharply widen, i.e. when a chamber is formed on the mine shaft axis. In this case we assume that heat factors are not decisive, therefore the surface heat balance is considered to be zero, and the vertical air stratification in the working — to be indifferent. Fig.1 shows the calculation results as velocity vector fields in the horizontal section at various degrees of horizontal flow confinement $m = d_o/\sqrt{hB_k}$ (d_o — the size of the inlet shaft, h — the chamber height, B_k — the chamber width) and the jet rates U_B at the inlet. As it is seen from fig.1a a recircular scheme of chamber ventilation with the formation of two stable large vortexes along the whole chamber length is formed with m = 0.26. Numerical experiments showed that at m-increase there forms a parallel-flow recircular scheme of ventilation with the formation of four local vortexes located in the chamber corners (fig.1b). A further increase of m-parameters and U_B-decrease result

467

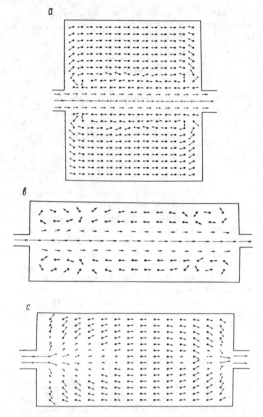

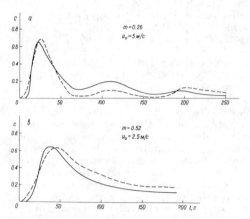

Fig.2. Alteration of concentration vs time at the outlet of a flat chamber: a) m = 0.26, U$_B$ = 5 m/s; b) m = 0.52, U$_B$ = 2.5 m/s.

Fig.1. Vector fields of current velocities in workings with different degree of horizontal jet confinement: a) m = 0.26, U$_B$ = 5 m/s; b) m = 0.52, U$_B$ = 2.5 m/s.

in a gradual vortex degeneration and in formation of a parallel–flow ventilation scheme. Figure 1c depicts the vector fields of horizontal rate at m = 0.52 and U$_B$ = 2.5 m/s.

The solution of the equation of the impurity transfer and diffusion for such flows shows that at contamination at the chamber inlet the impurity removal at the outlet occurs in different ways depending on the chamber ventilation scheme. With the parallel–flow scheme there observed a gradual decrease of concentration at the outlet vs time. With a recircular ventilation scheme there formed additional maxima of concentration caused by impurity diffusion and transfer by vortex flows. Figure 2 gives graphs of variations of concentration at the chamber outlet vs time for a parallel–flow and a recircular ventilation schemes. To compare the numerical modelling results the figure al-

so gives concentration curves obtained on a physical model (Kalabin, 1981, Kalabin, 1983) for analogous conditions.

Diffusion coefficient calculation has been developed for flat chamber–like workings and is based on mathematic and physical modelling of diffusion processes. We suggest the following method of calculation. At first the fields of flows in chambers are calculated on the base of a mathematical model. Then physical modelling of chamber ventilation processes is carried out (Kalabin, 1981) for analogous conditions and the concentration dependence on time at the chamber outlet is measured. Then with the help of a mathematical model of impurity diffusion and varying the diffusion coefficients we succeed in getting the time dependence on the concentration analogous to that obtained on a hydromodel (Kalabin, 1983). Having obtained such a regime we get the values of real diffusion coefficients D$_x$, D$_y$ for a flat chamber with a certain degree of confinement.

On the base of multiple numerical experiments and comparison with the results of hydromodelling there have been obtained the dependences of change of turbulent diffusion coefficients D$_x$ and D$_y$. Figure 3 shows the calculation results of D$_x$ for a chamber of the following parameters: the chamber length is 40 m, the chamer height and the dimensions of the inlet and outlet channels are h = d$_o$ = 2m with the change of the chamber width B$_K$ and a degree of confinement m from 0.25

468

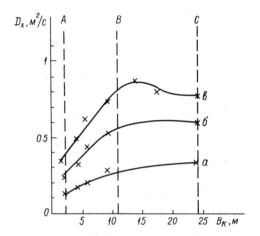

$D_x, \text{м}^2/\text{c}$

Fig.3. Dependences of the turbulent diffusion coefficient D_x on the degree of confinement m at different flow velocities.

to 1 respectively. The velocity at the chamber inlet was also set equal to n. Figure 3 shows that D_x -change at the AB -section is quite normal and is rather precisely described by the following dependence:

$$D_x = \alpha U_0 \left(d_0 + \beta B_\kappa \right) , \quad (22)$$

where $\alpha = 0.0175$; $\beta = 0.403$ (experimental data). In the BC -section an increase of D_x-coefficient is slowing down, then becomes stable and in some cases it even decreases. This fact is caused by the formation of recircular zones along the side zones of a chamber and by compression of the main flow. The BC -section coincides just with the transition from a parallel-flow recircular scheme to a recircular one at $m < 0.33$.

The determination of the main turbulent characteristics allows to find an analitical solution for practical calculations of change of impurity concentrations in a flat chamber. Let us imagine the chamber volume in a form of the sum $V = V_1 + V_2$, where V_1 is the flowing part, V_2 is the section uniting the stagnant zones with reverse flows. For such a system let us write down the following equations, describing the impurity transfer (Kafarov et al., 1976)

$$\frac{\partial C}{\partial t} + U \frac{\partial C}{\partial x} = D \frac{\partial^2 C}{\partial x^2} - \frac{K}{H_1} \left(C - C_1 \right) \text{ in } V_1 \quad (23)$$

$$H_2 \frac{\partial C}{\partial t} = K \left(C - C_1 \right) \text{ in } V_2, \quad (24)$$

where C, C_1 – impurity concentration in the flowing and stagnant zones;
H_1, H_2 – a part of the volume of the flowing and the stagnant zones;
K – coefficient of exchange between the flowing and the stagnant zones;
D – coefficient of longitudinal mixing (dispersion).

By integrating the system of equations (23)-(24), making the procedure of convolution and some calculations we obtain the following solution:

$$c(t,x) = \frac{\exp\left[-\frac{(x-Ut)^2}{4Dt}\right]}{\sqrt{4\pi Dt}} + \int_0^t \frac{1}{\sqrt{4\pi D(t-\tau)}} * (25)$$

$$* \exp\left[-\frac{(x-U(t-\tau))^2}{4D(t-\tau)}\right] \exp\left[-\frac{K}{H_1}(t-\tau)\right] C_1(\tau,x) d\tau ,$$

$$C_1(t,x) = \int_0^t \frac{K}{H} \exp\left[-\frac{K}{H_2}\tau\right] \frac{1}{\sqrt{4\pi D(t-\tau)}} \exp\left[-\frac{(x-U(t-\tau))^2}{4D(t-\tau)}\right] d\tau \quad (26)$$

The integrals are computed without any difficulty with the help of standard programs. The determination of D- and K -coefficients is the most complicated problem but in case we have a numerical model of aerogasdynamics of chambers, the calculation of the coefficients becomes quits real. The coefficient of mixing may be determined by D_x from the numerical experiments and dependences shown in fig.3 or by the formula (22). The coefficient of exchange between the zones may be represented in the form (Kafarov et. al., 1976):

$$K = \frac{Q_{ich}}{V} + \frac{H_2 D_m}{\Delta \ell^2} , \quad (27)$$

where Q_{ich} – volumetric air rate from the flowing zone into the stagnant one and in the opposite direction;
V – full volume of the system;
D_m – coefficient of molecular diffusion;
Δl – characteristic dimention of the stagnant zone.

The value of the K -coefficient may be obtained on the base of comparison of the outlet concentration profiles. These profiles are obtained with the help of a numerical model with the analagous solutions by the formulae (25)-(26). When the assumed coicidence is obtained we shall get an un-known value of the coefficient of exchange K. Thus, using the analitical and nume-

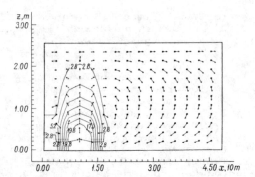

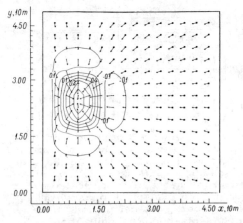

Fig.4. Air circulation in the chamber
under a heat source effect ($T' =$
$20^{\circ}C$): a) vector fields of velocity and
isolines of the vertical velocity (m/s)
in the horizontal section $(z = 0.7 \ m)$;
b) vector fields of velocity and isolines of deviation of temperature $(^{\circ}C)$
in the vertical section $(y = 24 \ m)$.

rical solutions of the problem it becomes possible to determine the impurity
concentration vs time and space at
the assumed air amount applied into
the chamber.

In conclusion let us consider air
circulation in an isolated chamber-like
working under the influence of a heat
sources of an artificial origin. Let us
assume a relative overheat of a local
surface section at the chamber bottom
($T' = 20^{\circ}C$). The results of numerical
modelling for the case are given in
fig.4.

These investigations have been conducted in cooperation with the Laboratory of Hydrodynamic Problems of the
Environment, Computation Center of the
Sibirian Departnmet of the USSR Ac.
of Sci.

5   REFERENCES

Aloyan, A.E., Baklanov, A.A., Penenko,
V.V. 1982. Application of methods of
fictitious regions in problems of numerical modelling of quarry ventilation. Meteorology and hydrology
no.7, p.42–49.

Baklanov, A.A. 1983. A numerical model of hydrometeorological regime of
quarries with allowance for the processes of moisture exchange and
heterogeneous cover of slopes. In
B. Methods and mathematical modelling in hydrodynamic problems of
the environment. Novosibirsk: CC
Sib. Depart. of the USSR Ac. of
Sci.

Bram van Leer. 1977. Towards the ultimate conservative difference scheme 1 V A approach to numerical
convection. In J. Comp. Phys.
23(7):267–299.

Clark, T.L. 1977. A small–scale dynamic using a terrein–following transformation. In J. of Comp. Physics:
24:184–215.

Kafarov, V.V., Dorokhov, I.N. 1976.
Systems analysis of the processes
chemical technology. The bases of
strategy. Moskva: Nauka.

Kalabin, G.V. 1981. Experimental investigation of opening (distribution) of
confined turbulent flows. Novosibirsk,
J. ETPRPI. 3:106–111.

Kalabin, G.V. 1983. The character and
regularities of processes of gaseous
products dilution and removal in
slot–like chambers. Novosibirsk, J.
FTPRPI. 1:77–82.

Marchuk, G.I. 1977. Methods of computing mathematics. Moskva: Nauka.

Penenko, V.V., Aloyan, A.E. 1976. Numerical method of calculation of fields
atmosphere boundary layer meteoro-
logical elements. J. Meteorology and
hydrology. 6:11–25.

Penenko, V.V., 1981. Methods of numerical modelling of atmosphere processes. Leningrad: Gidrometeoizdat.

Smagorinsky, J. 1963. General circulation experiments with the primitive
equation. J. Mon. Wea. Rev. 91(3).

Taylor, T.D., Ndefo, E. 1971. Calculation of flow of viscous fluid in the
channel with the help of splitting. In
B. Proceedings of the second international conference on numerical
methods in fluid dynamics. Berlin–
Heidelberk–New York. p.218–229.

12. Microcomputers

Mine ventilation computer modeling
More practical, economical, and available

CHUCK R.McLENDON
Allied Corporation, Green River, WY, USA

ABDUL J.KUDIYA
Hercules Incorporated, Salt Lake City, UT, USA

ABSTRACT: The accomplishments of this paper enable a new simulation of mine ventilation networks to be performed on mini or micro computers. The resulting computer program has been derived from the condensation and modification of existing programs designed for large mainframe computers. In addition, this program has expanded capabilities and uses.

One additional capability is the option to, in effect, smooth out air resistance data. For example, in the simulation of mine air branches, significant differences are often found between the measured and computer-calculated values of air quantities. With this program, an optional subroutine can be called to analyze the complete mine network on the basis of known air quantities. The subroutine makes necessary resistance adjustments which correlate with measured quantities. This technique eliminates the trial and error efforts previously necessary to adjust resistance data.

After a correct resistance/quantity relationship has been established for a mine, this program is capable of running with only fan quantities. That is, for changing main fan input, the program will output mine air flow quantities and patterns.

This program has been field tested, with excellent results, in a large room-and-pillar mine. However, this program should have near-universal applicability in any underground mine that has access to a mini or micro computer.

1 INTRODUCTION

At the Alchem Mine, near Green River, Wyoming, a need exists for the solving of novel, complex ventilation problems practically and efficiently. The mine covers an area of approximately 13 sq miles (34 sq km) and consists of a flat-lying trona (sodium sesquicarbonate) seam at an average depth of about 1 550 ft (1 610 m). Four shafts and four axivane surface fans are utilized to maintain a total mine air flow of 1.2 Mcfm (570 m^3/s).

The location of the mine shafts present a great design opportunity for maintaining a long and short range ventilation. Three shafts are within a couple hundred yards (meters) of one another, while the fourth shaft is about two miles (three kilometers) away.

The deferring of construction of a fifth ventilation shaft, for at least a decade, has been largely possible through crafty ventilation planning, computer modeling, and the excellent cooperation from operations people in carrying out unconventional ventilation schemes. It is the computer modeling which has proven to be a valuable tool in aiding ventilation planning. First only usable with large mainframe computers, ventilation modeling is now more practicable with microcomputers.

2 TRANSITION FROM MAINFRAME TO MICRO-COMPUTER HARDWARE

The usefulness and predicting capabilities of the pioneer computer models depend upon a large mainframe computer and a considerable amount of good data. These two factors have been addressed and improved in this study.

Many mines do not have access to a large, expensive mainframe computer. Computer modeling on a small desk-top microcomputer can be more economical and allow many program runs in a single time frame. Microcomputers, like the one at the Alchem Mine, can be purchased for less than $20K (U.S. 1985 dollars), which includes peripheral devices. The cost of microcomputers is currently dropping. The validity of many computer models depend upon many accurate air measurements. Such measurements can be time and resource consuming. Furthermore, the accuracy of measured values is sometimes difficult to obtain due to interference from ore and traffic conveyances and other numerous factors. This study has made substantial inroads into minimizing the input data needed to achieve valid model output.

3 ADAPTION AND MODIFICATION OF EXISTING PROGRAMS

A Pennsylvania State University program (Ramani, et al; 1975) had worked well at the Alchem Mine and was used as a starting basis for this study. Electric analogue computer programs and basic ventilation theory were also utilized. The University program was adapted and streamlined to run on a Micro PDP-11, manufactured by Digital Equipment Corporation (See Appendix A). This microcomputer has 10 megabyte of memory plus peripheral diskettes (floppy disks).

An optional subroutine was added to the program to smooth out estimated air resistance in air branches (See Appendix B). A branch is a roadway segment which connects any two air junctions. The subroutine also improves the prediction of air quantities in the networks. The improved prediction capability also allows the simulation of mine ventilation networks using only main fan quantities as a fixed input. The solution of problems involving main fan shutdowns, modifications, reversals, and blade changes are some of the practical applications of this modification.

4 USING AND OPERATING THE MODEL

A basic knowledge of the University program (Ramani, et al; 1975) or other similar program would be helpful in using this program. Many variable names have been utilized which are consistent with the base program.

An example run of the program with the optional subroutine might use routine air flow measurements taken in the mine. The program smoothing-run involves three basic steps:

First, the model is executed with fixed (measured) air quantities (Q) and estimated resistance values in the unfixed branches, which need to be adjusted. The pressure head values (P), are then converted to resistance values (R), for the fixed branches, using the equation $R=H/Q^2$.

Second, the resistance values obtained from the previous equation are used as input. Also the fixed fan quantities are inputted. In this step, all the fixed quantities from step one are treated as variables.

In the final step, the resistance values are refined from the air quantities obtained from steps one and two.

5 FINAL COMMENTS

At the Alchem Mine, the calculated values of branch air flow have been in close agreement with historically measured quantities. This modified program should have near-universal application in room-and-pillar mining.

6 REFERENCES

Hartman, H.L. 1982. Mine ventilation and air conditioning. New York: John Wiley and Sons.

Jorgensen, R. 1970. Fan engineering. New York: Buffalo Forge Co.

Ramani, R.V; Owili-Eger, A; and Manula, C.B. 1975. A master environmental control and mine system design simulator for underground coal mining. Washington, D.C.: Government Printing Office.

APPENDIX A

```
      PROGRAM VENT3
      VIRTUAL NA(3000),BRANCH(220),OUT(220),HE(250),SUMNVP(220),
     1IND(40),DD(80),C(40,6),EQ(40),HH(220),XS(10),YS(10),
     2DDTAB(9,9),ABS1(220),ABS2(220),AN(220),FF(220),
     3FX(20,40),HEIGHT(220),LENGTH(220),NVP(220),R(220),
     4QMAX(40),QMIN(40),QUANT(220),TEMP1(220),TEMP2(220),
     5WIDTH(220),X(20,40),IPRES(220),
     6J1(220),J2(220),KEY(39),NP(40),SUMH(500)
      REAL SPECNAG
      REAL NVP,LENGTH
      INTEGER BRANCH
      CHARACTER*50 TITLE
      OPEN(UNIT=81,FILE='MINE.DAT',STATUS='OLD')
      TITLE=' MINE VENTILATION SIMULATION'
    1 WRITE(6,605)
      WRITE(6,610) TITLE
      WRITE(6,615)
      DO 11 I=1,220
      QUANT(I)=0.
      ABS1(I)=0.
      ABS2(I)=0.
      AN(I)=0.
      FF(I)=0.
      HEIGHT(I)=0.
      LENGTH(I)=0.
      NVP(I)=0.
      R(I)=0.
      QUANT(I)=0.
      TEMP1(I)=0.
      TEMP2(I)=0.
      WIDTH(I)=0.
      IPRES(I)=0
   11 J1(I)=0
      J2(I)=0
      DO 22  I=1,40
      QMAX(I)=0.
      QMIN(I)=0.
      EQ(I)=0.0
      NP(I)=0
   22 X(J,I)=0.
      DO 33  I=1,39
      DO 337 I=1,40
      DO 337 J=1,16
  337 C(I,J)=0.0
      E=0.
      IBAL=0
      IFANCO=0
      INPRT=-1
      IRES=0
      ITPRES=0
      MAXIT=0
      MAXJ=0
      NB=0
      NE=0
      NFIXB=0
      NJ=0
      DO 44  I=1,80
   44 READ(81,1170)  (J1(I),I=1,220)
      REWIND 81
      READ(81,1170)  (J1(I),I=1,220)
      WRITE(6,1170)  (J1(I),I=1,220)
      WRITE(6,1170)  (J2(I),I=1,220)
      READ(81,1170)  (J2(I),I=1,220)
      WRITE(6,1171)  (R(I),I=1,220)
      READ(81,1171)  (R(I),I=1,220)
      READ(81,1171)  (QUANT(I),I=1,220)
```

```
      WRITE(6,1171)  (QUANT(I),I=1,220)
 1167 FORMAT(3F10.4)
 1168 FORMAT(3F10.4)
 1169 FORMAT(4I15)
 1170 FORMAT(10I5)
 1171 FORMAT(10F10.4)
      READ(81,*) E,
      READ(81,1169) IBAL,IFANCO,INPRT,IRES,ITPRES,MAXIT,MAXJ,NB,NF.
     1NFIXB,NJ,NP(1),NP(2),NP(3)
      WRITE(6,1169) IBAL,IFANCO,INPRT,IRES,ITPRES,MAXIT,MAXJ,NB,NF.
     1NFIXB,NJ,NP(1),NP(2),NP(3)
      DO 1172 J=1,3
      READ(81,1168)   (FX(I,J),I=1,5)
      WRITE(6,1168)   (FX(I,J),I=1,5)
 1172 CONTINUE
      READ(81,1167)  (QMAX(I),I=1,3)
      WRITE(6,1167)  (QMAX(I),I=1,3)
      READ(81,1167)  (QMIN(I),I=1,3)
      WRITE(6,1167)  (QMIN(I),I=1,3)
      DO 1173 J=1,3
      READ(81,1168)   (X(I,J),I=1,5)
      WRITE(6,1168)   (X(I,J),I=1,5)
 1173 CONTINUE
      CLOSE (UNIT=81)
      NM=NB-NJ+1
      NFBPE=NFIXB+NF
      WRITE(6,502) NB,NJ,MAXJ,NM,NF,MAXIT,NFIXB,E
      WRITE(6,503) IFANCO,IRES,ITPRES,INPRT,IBAL
      IF( INPRT .LE. 0 )   GO TO 17
      IF( IRES .EQ. 2 )   WRITE(6,532)
      IF( IRES .EQ. 1 )   WRITE(6,533)
      IF( IRES .EQ. 0 )   WRITE(6,534)
   17 DO 15 I=1,NB
      BRANCH(I)=I
      IF( IRES .EQ. 1 )   GO TO 16
      IF( IRES .EQ. 2 )   GO TO 13
      IF( INPRT .GE. 1 )   WRITE(6,5042)  I,J1(I),J2(I),R(I),
     1QUANT(I)
      GO TO 15
   16 IF( ABS(QUANT(I)) .LE. 1.E-10 )   GO TO 14
      DELIP=ABS(I) - ABS2(I)
      R(I)=ABS(DELIP) / (QUANT(I) * QUANT(I))
      IF( INPRT .GE. 1 )   WRITE(6,5041)  I,J1(I),J2(I),R(I),
     1ABS1(I),ABS2(I),QUANT(I),DELIP
      GO TO 15
   13 T=HEIGHT(I) * WIDTH(I) * AN(I)
      R(I)=FF(I) * (HEIGHT(I) + WIDTH(I)) * AN(I) * LENGTH(I)/
     1(2.6 * T * A * T)
      IF( INPRT .GE. 1 )   WRITE(6,504)  I,J1(I),J2(I),R(I),
     1FF(I),HEIGHT(I),WIDTH(I),LENGTH(I),AN(I)
   15 CONTINUE
      CALL DETBRA(ABS1,R,NFBPE,NB,BRANCH,MAXJ,OUT,J1,J3,NM,NJ,NF,NFIXB
      CALL DETMSH(NB,BRANCH,OUT,NA,J1,J3,NF,NVP,SUMNVP,INPRT,NM)
      IF(NF .GT. 0) CALL DETFAN(NVP,LENGTH,BRANCH,NF,KEY,NP,NFIXB,
     1J1,J3,IFANCO,ND,IDEG,DD,QUANT,C,QMAX,QMIN,X,FX)
      CALL ITERAT(IBAL,NFIXB,MAXIT,NE,NFBPE,ND,QUANT,EQ,C,R,NA.
     1NF,QMAX,QMIN,DD,IFANCO,NB,NM,E,SUMNVP)
      WRITE(6,521)
      DO 111 I=1,NB
      HH(I)=R(I) * ABS(QUANT(I)) * QUANT(I)
      IF( IFANCO .EQ. 2 )   QUANT(I)=QUANT(I) * .01
      WRITE(6,522)   I,J1(I),J2(I),R(I),QUANT(I),HH(I)
  111 CONTINUE
      IF( NF .LE. 0 )   GO TO 300
      DO 200 I=1,NF
  200 CALL POLYEV(QUANT(I + NFIXB),EQ(I),I,ND,C)
      IF( NFIXB .LE. 0 )   GO TO 400
  300 WRITE(6,600)
```

475

```
          JE=0
          DO 370 I=1,NFIXB
          JS=JE+1
          JE=ME(I)
          SUMH(I)=-SUMNVP(I)
          K=IABS(NA(J))
          PMI=NA(J)/ K
360       SUMH(I)=SUMH(I) + PMI * HH(K)
370       WRITE(6,522) I,J1(I),J2(I),R(I),QUANT(I),SUMH(I)
400       IF( ITPRES .EQ. 0 )  GO TO 450
          CALL TOTPRS(NJ,NB,J1,J2,NFIXB,NEBPF,HH,SUMH,EQ)
          IF( NF .LE. 0 )  GO TO 1
450       WRITE(6,523)
          WRITE(6,524)  (I,J1(I + NFIXB),J2(I + NFIXB),
         1QUANT(I+NFIXB),EQ(I),I=1,NF)
          STOP
500       FORMAT(A1,A7.9A8)
502       FORMAT(///,6X,2HNB,6X,2HNJ,4X,4HMAXJ,6X,2HNM,6X,2HNF,
         $3X,5HMAXII,3X,5HFIIX,9X,1HE,/,7I8,F10.4,/)
503       FORMAT(2X,6HIFANCO,4X,4HIRES,2X,6HITPRES,3X,5HINPRT,
         $4X,4HIBAL,/,5I8,//)
504       FORMAT(I7,214,F15.6,5F7.1)
5041      FORMAT(I7,214,F15.6,2F10.2,F10.5,E10.3)
5042      FORMAT(I7,214,F15.6,5X,F10.2)
521       FORMAT(///,7HBRANCH,2X,2HJ1,2X,2HJ2,14X,9X,1H0,9X,
         $1HH,/)
522       FORMAT(I7,214,F15.6,F10.4,F10.3)
523       FORMAT(///,4HFAN,2X,2HJ1,2X,2HJ2,9X,1HQ,8X,2HEQ,/)
524       FORMAT(3I4,F10.4,F10.3)
532       FORMAT(7HBRANCH,2X,2HJ1,2X,2HJ2,9X,1HR,5X,2HFE.
         $7HHEIGHT,2X,5HWIDTH,7HLENGTH,5X,2HAN,/)
533       FORMAT(7HBRANCH,2X,2HJ1,2X,2HJ2,9X,1HR,6X,4HABS1,6X,
         $4HSUMH,/)
605       FORMAT(1H1,//,1X,80(1HA),/)
610       FORMAT(10X,'TITLE',5X,1A50//)
615       FORMAT(/,1X,80(1HA),//,1H1)
          END
          SUBROUTINE DETBRA(ABS1,R,NEBPF,NB,BRANCH,MAXJ,OUT,J1,J2.
         1NM,NJ,NF,NFIXB)
          VIRTUAL ABS1(220),R(220),BRANCH(220),OUT(220),J1(220).
         1J2(220),JC(999)
          REAL NVP,LENGTH
          INTEGER BRANCH
          DO 10  I=1,NB
10        ABS1(I)=R(I)
          GO TO 304
303       KLINE=1
304       KLINE=0
          IS=NEBPF + 1
          JE=NB - 1
          L=0
          DO 18  I=IS,IE
          DO 17  J=IS,JE
          IF( R(J + 1) - R(J))  17,17,16
16        T=R(J)
          R(J)=R(J + 1)
          R(J + 1)=T
          T=BRANCH(J)
          BRANCH(J)=BRANCH(J + 1)
          BRANCH(J + 1)=T
          L=1
17        CONTINUE
```

```
18        IF(L)  19,19,18
          JC-JE - 1
19        DO 20  I=1,MAXJ
20        JC(I)=0
          I=NB + 1
          L=0
          N=0
          DO 31  IJ=IS,NB
          I=I - 1
          OUT(I)=0.
          K=BRANCH(I)
          JA=J1(K)
          JB=J2(K)
21        IF(JC(JB) - JC(JB))  26,28,21
22        IF(JC(JA) - JC(JB))  22,25,22
          JA=JC(JA)
          JB=JC(JB)
          DO 24  J=1,MAXJ
23        IF(JC(J) - JC(JA))  24,23,24
24        CONTINUE
          GO TO 31
25        JC(JB)=JC(JA)
          GO TO 31
26        IF(JC(JA))  22,27,22
27        JC(JA)=JC(JB)
          GO TO 31
28        IF(JC(JA))  29,30,29
29        OUT(I)=1.
          N=N+1
          GO TO 31
30        L=L+1
          JC(JA)=L
          JC(JB)=L
31        CONTINUE
          IF(N - NEBPF - NM)  32,33,32
32        WRITE(6,505) I                       32,33,32
          NJ=NB + 1 - NFIXB - NF - N
          NM=NB - NJ + 1
          WRITE(6,534) NJ,NM
33        IF(KLINE .EQ. 0 )  GO TO 303
          IF( NEBPF .LE. 0 )  GO TO 50
          DO 35  I=1,NEBPF
35        OUT(I)=1.
50        DO 55  I=1,NB
55        R(I)=ABS1(I)
          RETURN
505       FORMAT(//,27H04* NO. OF BASIC BRANCHES =,I4,5H + NF)
534       FORMAT(//,5X,19HNUMBER OF JUNCTIONS,I6,10X,10HNUMBER OF ,
         $6HMESHES,I5)
          END
          SUBROUTINE DETFAN(NVP,LENGTH,BRANCH,NF,KEY,NP,NFIXB,J1,J2,
         1IFANCO,ND,IDEG,DD,QUANT,C,QMAX,QMIN,EX)
          VIRTUAL KEY(39),NP(40),J1(220),J2(220),ND(40),QUANT(220),
         1C(40,6),DD(80),QMAX(40),QMIN(40),X(20,40),EX(20,40)
          REAL NVP,LENGTH
          INTEGER BRANCH
          IF( NF .LE. 0 )  RETURN
          WRITE(6,513)
          DO 87  L=1,NE
          KEYL=KEY(L)
          NPL=NP(L)
          JJ=L + NFIXB
          WRITE(6,514)     L,J1(JJ),J2(JJ),KEYL
          IF( KEYL .GT. 0 )  GO TO 846
          WRITE(6,516)   (I,X(I,L),EX(I,L),I=1,NPL)
          IF( IFANCO .NE. 2 )  GO TO 351
          DO 350  I=1,NPL
350       X(I,L)=X(I,L) * 100.
```

```
351   DD(JJ)=QUANT(JJ)
      IFANC1=IFANCO + 1
840   CALL OLDFAN(L,NPL,ND,IDEG,C,FX,X)
      IF( KEYL .EQ. 0 )        GO TO 850
      ND(L)=ND(KEYL)
      IDEG1=IDEG + 1
      DD(JJ)=QUANT(JJ)
845   DO 845  J=1,IDEG1
      C(L,J)=C(KEYL,J)
      QMAX(L)=QMAX(KEYL)
      QMIN(L)=QMIN(KEYL)
850   WRITE(6,518)    (C(L,I),I=1,IDEG1)
      CALL POLYEV(QMAX(L),FQMAXL,L,ND,C)
      CALL POLYEV(QMIN(L),FQMINL,L,ND,C)
      WRITE(6,536)  QMAX(L),FQMAXL,QMIN(L),FQMINL
87    CONTINUE
      RETURN
513   FORMAT(/)
514   FORMAT(//,4HFAN ,8X,2HJ1,4X,2HJ2,7X,3HKEY,5X,5HPOINT,
     $11X,1HX,8X,2HFX,//,I4,4X,2I6,I10)
516   FORMAT(40,2X,2F10.4)
518   FORMAT(//,2X,12HCOEFFICIENTS,//,2(3X,1P4G15.6,/))
536   FORMAT(10X,4HQMAX,F11.3,3X,5HFQMAX,F11.3,/,10X,4HQMIN,
     1F11.3,3X,5HFQMIN,F11.3)
      END
      SUBROUTINE DETMSH(NB,BRANCH,OUT,NA,J1,J2,ME,NVP,SUMNVP,INPRT,NM)
      VIRTUAL BRANCH(220),OUT(220),NA(3000),J1(220),J2(220),
     1ME(250),NVP(220),SUMNVP(220)
      REAL NVP,LENGTH
      INTEGER BRANCH
      JK=0
      L=0
      DO 54  I=1,NB
37    K=BRANCH(I)                54,54,37
      L=L + 1
      JK=JK + 1
      NA(JK)=K
      JA=J1(K)
      JB=J2(K)
      N=I + 1
38    IF( OUT(J) )               43,39,45
39    K=BRANCH(J)
40    IF( JB - J1(K) )           41,40,41
      JK=JK + 1
      JB=J2(K)
      NA(JK)=K
      GO TO 43
41    IF( JB - J2(K) )           45,45,45
42    JB=J1(K)
      JK=JK + 1
      NA(JK)=K
43    IF( JB - JA )              44,51,44
44    OUT(J)=-1.
      GO TO 50
45    CONTINUE
46    IF( NA(JK) )               47,46,46
      JB=J1(K)
      GO TO 48
47    JB=J2(K)
48    JK=JK - 1
49    K=BRANCH(I)                49,49,38

      WRITE(6,525)     K
      STOP 50
50    GO TO 38
51    DO 53  J=N,NB              52,53,53
52    OUT(J)=0.
53    CONTINUE
      ME(L)=JK
      JE=JK
54    CONTINUE
      DO 55   I=1,NB
      IF( ABS(NVP(I)) .GT. 1.E-10 )    GO TO 56
55    CONTINUE
      GO TO 57
56    WRITE(6,507)
      DO 62   I=1,NB
      IF( ABS(NVP(I)) .LE. 1.E-10 )    GO TO 62
      WRITE(6,508)     I,NVP(I)
62    CONTINUE
57    IF( INPRT .GE. 2 )      WRITE(6,509)
      JE=0
      DO 61
      JS=JE + 1
      JE=ME(I)
      SUMNVP(I)=0.
      DO 60  J=JS,JE
      K=IABS(NA(J))
      PM1=NA(J)
60    SUMNVP(I)=SUMNVP(I) + PM1 * NVP(K)
      L=JE
      IF( INPRT .GE. 2 )       WRITE(6,510)  I,L,SUMNVP(I)
61    IF( INPRT .GE. 2 )       WRITE(6,511)  (NA(J),J-JS,JE)
      IF( INPRT .GE. 2 )       WRITE(6,512)  JE
      RETURN
507   FORMAT(//,4X,6HBRANCH,11X,4HNVP/  )
508   FORMAT(I11,F14.7)
509   FORMAT(///,5HMESH,2X,6HNO. OF,6X,6HSUMNVP,6X,
     $3HBRANCHES CONSTITUTING THE MESH,/,7X,8HBRANCHES,/)
510   FORMAT(I4,I10,F10.5)
511   FORMAT(30X,8I5)
512   FORMAT(/18X,7HTOTAL =,I5)
525   FORMAT(//,27H ** NO MESH FOUND AT BRANCH,I4)
      END
      SUBROUTINE INITER(NFBPF,DD,NFIXB,QUANT,IFANCO,NB,ME,NA)
      VIRTUAL NA(3000),QUANT(220),ME(250),DD(80)
      INTEGER BRANCH
      IF( NFBPF .GT. 0 )         GO TO 60
      DD(1)=10.
      IE=1
      GO TO 80
60    IE=NFBPF
      IF( NFIXB .LE. 0 )        GO TO 80
      DO 65  I=1,NFIXB
      DD(I)=QUANT(I)
      IF(IFANCO .EQ. 2 )       DD(I)=DD(I) * 100.
65    CONTINUE
80    DO 85  I=1,NB
85    QUANT(I)=0.
      JE=0
      DO 93  I=1,IE
      JS=JE + 1
      JE=ME(I)
      DO 93  J=JS,JE
      K=IABS(NA(J))
      QUANT(K)=QUANT(K) + DD(I)
91    QUANT(K)=QUANT(K)          92,91,91
      GO TO 93
```

477

```
 94 QUANT(K)=QUANT(K) - DB(I)
 93 CONTINUE
    RETURN
    END
    SUBROUTINE ITERAT(IBAL,NFIXB,MAXIT,ME,NEBPE,ND,QUANT,
   1FO,C,R,NA,NE,QMAX,QMIN,DD,IFANCO,NB,NM,E,SUMNVP)
    VIRTUAL NA(3000),ME(250),ND(40),DD(80),C(40,6),FQ(40),
   1R(220),QMAX(40),QMIN(40),QUANT(220),SUMNVP(220),
    REAL NVP,LENGTH
    INTEGER BRANCH
    IF(IBAL .EQ. 1) CALL INITER(NEBPE,DD,NFIXB,QUANT,IFANCO,
   1NB,ME,NA)
    IB=NFIXB + 1
    DO 108 II=1,MAXIT
    IF( NFIXB ) 820,820,810
810 JE=NE(NFIXB)
    GO TO 830
820 JE=0
830 L=0
    SUMD=0.
    DO 107   I=IB.NM
    IM=I - NFIXB
    JS=JE + 1
    JE=ME(I)
    SUMNVP=-SUMNVP(I)
    SUNDH=0.
    DHE=0.
 94 N=ND(IM)
    IF( NEBPE - I )   98,94,94
    CALL POLYEV(QUANT(I),FQL,IM,ND,C)
    FQ(IM)=FQL
 95 IF( N - 1 )   97,97,95
    J=J-1
    DO 96   II=2,N
    DHE=(DHE + FLOAT(J) * C(IM,J + 1)) * QUANT(I)
 96 SUMH=SUMH - FQ(IM)
 97 SUMH=SUMH - FQ(IM)
    DHE=DHE + C(IM,2)
 98 DO 101   J=JS,JE
    K=IABS(NA(J))
    DH=R(K) * ABS(QUANT(K))
    H=DH * QUANT(K)
    SUNDH=SUNDH + DH
    IF( NA(J) )   100,99,99
 99 SUMH=SUMH + H
    GO TO 101
100 SUMH=SUMH - H
101 CONTINUE
    SUMDH=SUMDH + SUMDH - DHE
102 D=-SUMH / SUMDH
    IF( ABS(SUMDH) - 1.E-20 )   106,106,102
    IF( ABS(D) .LE. 1.E-20 ) GO TO 107
    IF( NF .EQ. 0 ) GO TO 120
    DO 115   IF=JS,JE
    K=NA(IF)
    KA=IABS(K)
    IF( KA .GT. NEBPE )   GO TO 115
    IF( KA .LE. NFIXB )   GO TO 115
    KE=KA - NFIXB
    QT = QUANT(KA) + FLOAT(K/KA)*D
    IF( QT .GT. QMAX(KE)   D=.5*(D/ABS(D))*(QMAX(KE)-QUANT(KA))
    IF( QT .LT. QMIN(KE)   D=.5*(D/ABS(D))*(QUANT(KA)-QMIN(KE))
115 CONTINUE
120 DO 103   J=JS,JE
    K=IABS(NA(J))
    IF( NA(J) )   104,103,103
    GO TO 105
103 QUANT(K)=QUANT(K) + D
104 QUANT(K)=QUANT(K) - D
105 CONTINUE
    SUMD=SUMD + ABS(D)
    IF( ABS(D) - E )   107,107,106
106 L=L 1
107 CONTINUE
    IF( L )   109,109,108
108 CONTINUE
    WRITE(6,519)   IT,SUMD
    WRITE(6,520)
    RETURN
109 WRITE(6,519)   IT,SUMD
    RETURN
519 FORMAT(///,20X,18H* * * OUTPUT * * *,///,I8,11H ITERATIONS,
   $9X,6HSUMD =,E14.7)
520 FORMAT(10X,25H* * D STILL GREATER THAN E)
    END
    SUBROUTINE OLDFAN(L,NPL,ND,IDEG,C,FX,X)
    VIRTUAL ND(40),C(40,6),X(20,40),FX(20,40)
    VIRTUAL ALPHA(6),BETA(6),SIGMA2(6),P(7,20),S(6),W(6),Z(6)
    REAL NVP,LENGTH
    INTEGER BRANCH
    IF( NPL - 2 )   910,910,970
910 ND(L)=1
    IDEG=1
    IF( NPL - 1 )   920,930,940
920 STOP   1020
930 C(L,1)=FX(1,L)
    C(L,2)=0.
    RETURN
940 IF( X(1,L) - X(2,L) )   960,950,960
950 STOP   1050
960 C(L,1)=(FX(1,L) * X(2,L) - FX(2,L) * X(1,L)) / (X(2,L) - X(1,L))
    C(L,2)=(FX(2,L) - FX(1,L)) / (X(2,L) - X(1,L))
    RETURN
970 DSQ=0.
 63 DO 64   J=1,NPL
    DSQ=DSQ + FX(J,L) * FX(J,L)
    P(1,J)=0.
 64 P(2,J)=1.
    W(1)=NPL
    BETA(1)=0.
    IF( NPL - 7 )   66,65,65
 65 IDEG=5
    GO TO 67
 66 IDEG=NPL - 2
 67 IDEG1=IDEG + 1
    DO 73   I=1,IDEG1
    K=I + 1
    Z(I)=0.
    DO 68   J=1,NPL
 68 Z(I)=Z(I) + FX(J,L) * P(K,J)
    S(I)=Z(I) / W(I)
    DSQ=DSQ - S(I) * S(I) * W(I)
    SIGMA2(I)=DSQ / FLOAT(NPL - I)
 69 IF( SIGMA2(I) - 1.E-6 )   78,78,69
    ALPHA(I)=0.
    DO 71   J=1,NPL
 71 ALPHA(I)=ALPHA(I) + X(J,L) * P(K,J) * P(K,J)
    ALPHA(I)=ALPHA(I)/ W(I)
    DO 72   J=1,NPL
    P(K+1,J)=(X(J,L) - ALPHA(I)) * P(K,J) - BETA(I) * P(I,J)
 72 W(K)=W(K) + P(K+1,J) * P(K+1,J)
 73 BETA(K)=W(K)/ W(I)
 74 SMALL=SIGMA2(I)
    I=1
```

```fortran
      GO TO 190
120   IF( NEIXB .LT. I .AND. I .LE. NEBPF  )   GO TO 136
      TPRS(J1(I))=TPRS(J2(I)) + HH(I)
      IF( I .GT. NEIXB )  GO TO 190
      TPRS(J1(I))=TPRS(J1(I)) - SUMH(I)
      GO TO 190
130   TPRS(J1(I))=TPRS(J2(I)) - EQ(I - NEIXB)
      GO TO 190
140   IF( ABS(TPRS(J2(I))) .GE. 1.E-10 )  GO TO 190
      IF( J2(I) .EQ. 1 )  GO TO 190
      IF( NEIXB .LT. I .AND. I .LE. NEBPF  )   GO TO 150
      TPRS(J2(I))=TPRS(J1(I)) - HH(I)
      IF( I .GT. NEIXB )  GO TO 190
      TPRS(J2(I))=TPRS(J2(I)) + SUMH(I)
      GO TO 190
150   TPRS(J2(I))=TPRS(J1(I)) + EQ(I - NEIXB)
190   CONTINUE
      IF( IDONE .EQ. 1 )  ITPII
      WRITE(6,210)  ITPII
210   FORMAT(///, ALL TOTAL PRESSURES HAVE BEEN CALCULATED.,/,
     $ (THE PROCESS REQUIRED,I4, ITERATIONS.),/.
     $ THE NON-ZERO VALUES ARE PRINTED BELOW.,/.
     $ (JUNCTION NO. ON 1ST LINE,TOT. PRES. BELOW IT ON 2ND),)
215   I=0
219   J=0
220   I=I + 1
      IF( I .EQ. 1000 )  GO TO 230
      IF( ABS(TPRS(I)) .LT. 1.E-10 )  GO TO 220
      J=J + 1
      JOUT(J)=I
      TOUT(J)=TPRS(I)
      IF( J .EQ. 6 )  GO TO 230
      GO TO 220
230   IF( J .EQ. 0 )  GO TO 300
      WRITE(6,235)
235   FORMAT(/)
      WRITE(6,240)   (JOUT(K),K=1,J)
240   FORMAT(6(5X,I3,.2X))
      WRITE(6,250)   (TOUT(K),K=1,J)
250   FORMAT(6F10.3)
      IF( J .LT. 6 )  GO TO 300
      GO TO 219
300   RETURN
      END
```

```fortran
      IDEG1=IDEG + 1
      DO 77  J=2,IDEG1
76    IF( SMALL - SIGMA2(J) )    77,77,76
      J=J
      SMALL=SIGMA2(J)
77    CONTINUE
78    WRITE(6,517)    SIGMA2(I)
      IDEG=I - 1
      IF( IDEG )    780,780,79
780   ND(L)=1
      C(L,1)=S(I)
      C(L,2)=0.
      RETURN
79    ND(L)=IDEG
      IDEG1=IDEG + 1
      DO 81  I=1,IDEG1
80    P(I,1)=0.
81    P(I,I+1)=1.
      DO 82  J=1,IDEG
82    P(I,J+2)=P(I,J+1) * ALPHA(J) - P(I,J) * BETA(J)
      DO 83  I=2,IDEG
      DO 83  J=1,IDEG
83    P(I,J+2)=P(I-1,J+1) * ALPHA(J) - P(I,J+1) * ALPHA(J) - P(I,J) * BETA(J)
      DO 84  I=1,IDEG1
84    C(L,I)=C(L,I) + P(I,J+1) * S(J)
      RETURN
517   FORMAT(/,18X,8HSIGMA2 =,1PE14.7)
      END
      SUBROUTINE POLYEV(QL,EQL,L,ND,C)
      VIRTUAL ND(40),C(40,6)
      REAL NVP.LENGTH
      INTEGER BRANCH
      NDL=ND(L)
      EQL=0.
      IF( NDL .EQ. 0 )  GO TO 10
      EQL=C(L,NDL+1) * QL
      IF( NDL .EQ. 1 )  GO TO 10
      JL=NDL
      DO 9  IL=2,NDL
      JL=JL - 1
9     EQL=(EQL + C(L,JL)) * QL
10    EQL=EQL + C(L,1)
      RETURN
      END
      SUBROUTINE IOTPRS(NJ,NB,J1,J2,NEIXB,NEBPF,HH,SUMH,EQ)
      VIRTUAL EQ(40),HH(220),J1(220),J2(220),SUMH(500)
      VIRTUAL TPRS(999),JOUT(6),TOUT(6)
      REAL NVP.LENGTH
      INTEGER BRANCH
      DO 10  I=1,999
10    TPRS(I)=0.
      ITPII=0
      IDONE=0
100   IDONE=0
      ITPII=ITPII + 1
      IF( ITPII .LE. NJ )   GO TO 105
      WRITE(6,102)
102   FORMAT(///,MAX. ITERATIONS REACHED FOR ABS. PRESSURE.,/,
     $ SOME JUNCTIONS POSSIBLY NOT CALCULATED.,//)
      GO TO 215
105   DO 190  I=1,NB
      IF( ABS(TPRS(J1(I))) .GE. 1.E-10 )   GO TO 140
      IF( J1(I) .EQ. 1 )   GO TO 140
      IF( ABS(TPRS(J2(I))) .GE. 1.E-10 )   GO TO 120
      IF( J2(I) .EQ. 1 )   GO TO 120
      IDONE=1
```

479

```
      SUBROUTINE RESIST(QUANT,NB,R,RN)
      VIRTUAL R(300),QUANT(300),QFIX(300),
      RN(300)

      READ(82,*) (QFIX(I), I=1,300)
      DO 25 I=1,NB
      RN(I) =(R(I)*QUANT(I)*QUANT(I))/(QFIX
      (I)*QFIX(I))
   25 CONTINUE
      RETURN
      END
```

In the Main Program:

```
C     ISC IS THE FLAG TO ACTIVATE SUBROU-
      TINE RESIST


      IF(ISC .LE. 0) GO TO 996
      CALL RESIST(QUANT,NB,R,RN)
      DO 997 I= 1,NB
      R(I) = RN(I)
  997 CONTINUE
      ISC=0
      GO TO (Re-run the program)
  996 CALL TOTPRS.
```

Ventilation surveys and networks using microcomputers

C.J.HALL
University of Idaho, Moscow, USA

ABSTRACT: Two major programs have been written for a microcomputer, specifically a TRS 80 Model 3 to allow simulation of a mine ventilation system. Input for the first program is the data measured at the mine. Some of the output from this program is used as input to the network program. Output from the first program include resistances, natural ventilation pressures and inlet and outlet specific volumes of the air. The specific volumes allow calculation of the compressibility of the air. The network program is pseudo-compressible, so that it can output airflows likely to be measured at the mine regardless of changes in the specific volume of the air.

1 DETERMINATION OF BRANCH RESISTANCE

The field measurements are taken to allow branch resistances to be calculated from:

$$\Delta P = R \dot{V}_m^2 \rho_m \dots\dots\dots\dots\dots\dots(1)$$

in which: ΔP = the pressure drop due to friction, shock and changes in kinetic energy, in Pa

R = the branch resistance, in m^{-4}

$\dot{V}_m$ = the mean volumetric flow rate, in m^3/s

ρm = the mean density, in kg/m^3

The computer program used to process the field data is written for barometric pressures at the beginning and end of the branch. Any time there is a change in elevation in an airway, there is a change in pressure caused by the change in potential energy, or by the weight of the column of air. This pressure change must be subtracted from the total pressure change, as measured by the barometers, in order to determine the ΔP in equation 1. As the air moves through the branch, its properties follow some sort of path. The most commonly used path is a straight line on a density/depth diagram - that is, the mean density is the mean of the entering and leaving densities. This is used to determine the no-flow pressure change, and

the difference between this and the measured pressure change is ΔP. Unfortunately, this path equation is accurate for changes in elevation of no more than 500 - 600 m. A path equation that is good for 2 - 3000 m is:

$$Pv_a^n = C\dots\dots\dots\dots\dots\dots\dots\dots(2)$$

in which: P = the barometric pressure, in kPa

v_a = the specific volume of dry air, in m^3/kg

n and C = constants

If we look at a Pv diagram, Figure 1,

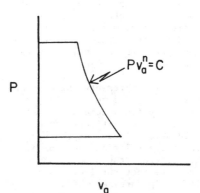

Fig. 1: Pressure/Specific Volume Diagram for Air

we can see that the area to the left of the curve, $-\int v_a\,dP$, is equal to the sum of the changes in potential and kinetic energy, the work and the lost work per kilogram of dry air plus its associated water stuff, or

$$\frac{n(P_e v_{ae} - P_i v_{ai})}{1 - n} =$$
$$\Delta pe^* + \Delta ke^* + {_1}w_2^* + {_1}lw_2^* \ldots(3)$$

in which: n = the constant in equation 2

Δpe^* = the change in potential energy per kilogram of dry air plus associated water stuff, in kJ/kg

Δke^* = the change in kinetic energy per kilogram of dry air plus associated water stuff, in kJ/kg

${_1}w_2^*$ = the work per kilogram of dry air plus associated water stuff, in kJ/kg

${_1}lw_2^*$ = the lost work per kilogram of dry air plus associated water stuff, in kJ/kg

i = the state of the incoming air

e = the state of the exiting air

The term water stuff is used to cover the water vapor and condensed water associated with the air. Thus, in an upcast shaft with condensation, provided the condensed water is moving upwards, it is included in the potential energy term, even if it is not travelling at the same velocity as the air.

The work term, ${_1}w_2^*$, would typically be a fan input into the system. It is apparent that if we ran a branch through a fan, there would be a sudden increase in pressure, and the path equation $Pv_a^n = C$ would not hold for that branch. Thus the branch would be split in two, one ending at the fan inlet and the next starting at the fan outlet. Thus the work term may be omitted from equation 3.

The lost work term represents the product of the pressure drop due to friction or shock and the specific volume of the air at that position. Thus it represents the work we could have gained from the system if, instead of the friction/shock pressure loss, we had obtained the same pressure drop through a perfect turbine, and got the work from this turbine. If we add the kinetic energy term to the lost work term, we can develope the pressure drop required in equation 1.

The potential energy term varies with evaporation and condensation. The simplest equations are:

For evaporation

$$\Delta pe^* = g(1 + [\omega_i + \omega_e]/2)(Z_e - Z_i)\ 10^{-3}\ldots(4a)$$

For condensation

$$\Delta pe^* = g(1 + \omega_i)\ (Z_e - Z_i)\ 10^{-3}\ldots\ldots(4b)$$

in which g = the gravitational constant, in m/s^2

ω = the humidity ratio, kg water vapor/kg dry air

Z = the elevation above some datum, in m

Equation 4a implies that the evaporation per foot of depth is constant. This, of course, is not correct, but since the humidity ratio is small, the potential energy term error will be small. If we consider a downcast shaft that is dry at the top, but has water entering the shaft half way down, it is obvious that the path cannot be represented by $Pv_a^n = C$, and the shaft should be split into two branches, one dry and the other wet.

Equation 4b implies no evaporation before condensation starts, and that all condensed water is carried out of the shaft. If there is evaporation before condensation starts, again the shaft should be split into two branches. Not all condensed water is carried out of the shaft - some hits the shaft furniture and drops down the shaft. Very rarely, a significant amount of water forms in large enough drops in the air that it falls down the shaft.

We can now rewrite equation 3 as

$$\frac{n(P_e v_{ae} - P_i v_{ai})}{1-n} =$$
$$g(Z_e - Z_i)\ (1 + \omega)\ 10^{-3} + {_1}lw_2^* \times 10^{-3} (5)$$

in which ${_1}lw_2^*$ includes friction, shock and kinetic energy terms, and ω is the appropriate water stuff potential energy term.

The remaining equations that we need to solve equation 1 for the branch resistance are:

$$n = \ln(P_e/P_i)/\ln(v_{ai}/v_{ae})\ldots\ldots\ldots(6)$$
$$v_a = R_a T/P_a \ldots\ldots\ldots\ldots\ldots\ldots(7)$$

in which:

R_a = 0.28700 kJ/kg K

T = the absolute dry bulb temp., in K

P_a = the partial pressure of the air, in kPa

$$P_a = P - P_v \ldots\ldots\ldots\ldots\ldots (8)$$

in which :

P_v = the partial pressure of the water vapor, in kPa

$$\omega = 0.62186 P_v/P_a \ldots\ldots\ldots\ldots (9)$$

$$\Delta P = {}_1 1 w_2 / v_{am} \ldots\ldots\ldots\ldots\ldots (10)$$

The value of P_v is obtained from measurements of barometric pressure and wet and dry bulb temperatures. It is usually acceptable to obtain the mean specific volumes and densities as the mean of the entering and leaving values.

$$\rho = (1 + \omega + \omega_w)/v_a \ldots\ldots\ldots\ldots (11)$$

in which

ω_w = the mass of water associated with one kilogram of dry air.

Thus in our survey we have to take the necessary measurements to allow us to solve equations 5 through 11.

2 DETERMINATION OF NATURAL VENTILATION PRESSURE

The natural ventilation pressure is determined by balancing the weight of air in the branch against the weight of a column of standard air of the same height. The direction of flow is taken into account as well as the weights, to give positive or negative values of NVP. For branches with less than 500 m difference in elevation, the branch density used is the mean of the entering and leaving densities. When the difference in elevation of the ends of the branch exceed 500 m, the mid point conditions are calculated thermodynamically, using half the heat exchange, half the evaporation and half the lost work. The mean density is then calculated using Simpson's rule,

$$\rho_m = (\rho_i + 4\rho_{mp} + \rho_e)/6 \ldots\ldots (12)$$

If it is decided to use natural ventilation in the network, and if the collars of the mine openings are at different elevations, then natural ventilation pressures have to be applied to all surface branches. The computer program gives two surface branch natural ventilation pressures, one based on density, as already described. The second natural ventilation pressure is based on the calculation of resistance, and in fact represents a difference in atmospheric pressures at the same elevations over the openings. Such a difference in pressure would cause a wind on surface, and in the absence of

natural effects underground, air to flow through the mine in the same direction.

3 SELECTION OF BRANCHES AND NODES

The data required for input to the field data reduction program are the values of the parameters at the beginning and end of the branch. The most direct method of obtaining this information is to measure these parameters simultaneously, in a method referred to as "leap frogging". This method takes longer, by 50 - 100%, than the "base and roving instrument" technique. Errors in measured pressure drops could be up to 15 Pa with presently available equipment, whereas errors of several hundred Pascals can be made using the latter technique.

If air quantities are not known with sufficient accuracy, a quantity survey should be run before the survey is planned. Ventilation plans and sections are then studied, and branches and nodes identified. Initially the network should be kept as simple as possible. Once the model is satisfactory (that is, the computer simulated flows are close to the actual ventilation quantities) it may be elaborated as necessary. There should, however, be good reasons for adding each branch.

The selection of a branch must satisfy two major requirements. First, the mass rate of flow must be substantially constant throughout the branch, so that equation 1 may be solved for R. Second, the actual path followed by the air must conform reasonably to the mathematical model. Thus a branch must be terminated and a new branch started where:
- two airflows join together
- one airflow splits into two
- at a main or booster fan
- when a horizontal airway becomes inclined, or vice versa
- in a shaft where
    - there are significant lengths where there is evaporation and no evaporation. The adiabatic lapse rate can be negative with evaporation.
    - there are significant lengths where there is condensation and no condensation. The adiabatic lapse rate changes from 10 to about 4 C/km.
    - there is a major addition of heat at one location, such as at a pump station
    - there is a significant amount of leakage of air into the shaft at a significantly different state from the air in the shaft. This is often

a major problem at older multilevel mines.

The above list is not exhaustive.

Only those branches with reasonable pressure drops need be surveyed for resistances. A pressure drop of 150 Pa (0.6" WG) should be accurate to ±10%. Thus at many mines, 50% or less of the airways need be surveyed. Some of the other airways might be surveyed for natural ventilation pressures.

In order to determine the resistances of the branches not surveyed, standard airways should be surveyed and the resistance per unit length determined. When possible, airflows should be adjusted to give the maximum possible through the standard when it is surveyed. Standards can be used in shafts in multilevel mines, where there is a large airflow through the top of the shaft, but progressively less further down. A pressure survey of the lower part of such a shaft is rarely justified.

4 MEASUREMENTS AND PROCEDURES

The measurements taken for a branch are the barometric pressure and wet and dry bulb temperatures at each end of the branch simultaneously, the volumetric flow rate at either or both ends of the branch, and the height of the barometers above the drift floor. The pressure and temperature readings can be taken every minute over a ten minute interval by one observer at each station. The volumetric flow rate can be taken several times during this period by a third observer, or by the two P/T observers immediately before and after the ten minute period.

The pressure may be measured with aneroid barometers, because they are as accurate as mercury and glass instruments and easily portable. Since the instruments are accurate to say, 5-15 Pa (equivalent to 0.5-1.5 m of air), the calculations are carried out to 1 J/kg, and the height of the barometer above the floor should be estimated to 0.25 m. The actual elevation is obtained from a survey office estimate of the floor elevation, good to say 0.1 m, and the recorded height over the floor.

Wet and dry bulb temperatures should be measured using battery operated psychrometers, to allow the observer to take readings at one minute intervals. The temperatures can be estimated to a tenth of a degree, although this accuracy is not required to obtain the lost work, and in

fact, the thermometers are probably not that accurate.

The air velocity is measured with an anemometer. The areas where the velocities are taken should have been measured before the survey, and in fact are probably part of the mine ventilation control system. Although the nodal pressures are required, the barometers are usually set up at a more convenient location. If it is suspected that the pressure at the node is significantly different from that at the barometer, the pressure difference can be measured with a tube and water gauge.

The temperature and humidity are required at the beginning and end of the branch, not at the node. Thus, if we consider a upcast shaft station at a multilevel mine, the node is in the shaft. Two branches terminate at the node, one horizontal and one vertical. One vertical branch starts at the node. Three separate temperatures and humidities are required, one for the level air just before it enters the shaft, a second for the air coming up the shaft before it mixes with the level air, and a third for the air mixture proceeding up the shaft.

5 THE FIELD DATA REDUCTION PROGRAM

Input to the program can be in SI or US units, with output in the same units as input. The parameters are barometric pressure, wet and dry bulb temperatures and elevation at inlet to and exit from the branch, the volumetric flow rate at inlet, exit or both, and the water humidity ratio (the mass of condensed water per unit mass of dry air) at entrance to the branch. The computer immediately outputs these input values. The standard density to be used to calculate natural ventilation pressure is input, as is the selection of the branch as an underground or surface branch.

The computer calculates in SI units. It obtains the saturation vapor pressure at the wet bulb temperature from the equations published by ASHRAE, which are for saturated steam over water rather than the psychrometric value. The vapor pressure is calculated using Ruben and Stroude's equations.

For underground branches, the computer calculates the heat exchange per unit mass of dry air. If there is evaporation, the computer uses:

$$_1q_2^* = 1.0045(t_e-t_i) + \omega_e h_{ge} -$$

$$\omega_i h_{gi} - (\omega_e-\omega_i-\omega_{wi})h_{fe} -\omega_{wi}h_{fi}$$

$$+9.80665(Z_e-Z_i)(1+[\omega_i+\omega_{wi}+\omega_e]/2)\ 10^{-3}\ ..(13)$$

For condensation,

$$_1q_2^* = 1.0045(t_e-t_i) +\omega_e h_{ge} +\omega_{we} h_{fe} -\omega_i h_{gi} +$$

$$9.80665(Z_e-Z_i)\ (1+\omega_i)\ 10^{-3} \ldots\ldots\ldots.(14)$$

in which:

$_1q_2^*$ = the heat exchange per unit mass, kJ/kg dry air

h_g = the enthalpy of saturated steam, kj/kg

h_f = the enthalpy of saturated water vapor, kJ/kg

Straight line equations are used for both h_g and h_f.

The computer outputs the mean volumetric flow rate, the heat exchange per unit mass, the evaporation or condensation per unit mass, the lost work per unit mass, the resistance, the product of the resistance and mean density, and the natural ventilation pressure.

The reasonableness of the resistance can be judged from a knowledge of the branch and the lost work. The heat exchange and evaporation or condensation allows an assessment of the validity of the measurements, and also some idea of the amount of leakage into a branch.

The computer also outputs the specific volume of the air at inlet, exit and midpoint, if that is calculated, and the mean specific volume; the density at inlet, exit and mean; the humidity ratios at inlet and exit, and the mass rates of flow of dry air at inlet and exit, depending on which volumetric flow rates were measured.

For surface branches, the only output other than the input values are the natural ventilation pressures due to density and atmospheric pressure.

6 THE NETWORK PROGRAM

The network program is a standard incompressible program using US units, with the addition of variable leakage to each branch. The input consists of identification of the network, fan characteristic curves, airway geometry, and individual branch data. Each branch is made up of three parts, inlet, branch and exit. Three resistances may be specified, for the three parts of the branch. Two types of leakage may be specified for each branch, a constant leakage either into or out from the branch, and a variable leakage. The latter is specified as a factor applied to the quantity flowing in the branch, so that leakage is proportional to the rate of flow. The constant leakage could represent compressors or compressed air. The variable leakage could represent leakage from intake to return in parallel airways connected by crosscuts, or the change in specific volume of the air as it moves along an airway. Both types of leakage may be applied to the air entering the branch, to obtain the amount of air in the branch, and to the quantity in the branch, to obtain the amount of air leaving the branch. The three air quantities act on their respective resistances to give up to three pressure drops in a branch. If a fan is specified for a branch, it can be specified to act on any of the three quantities in the branch.

The inlet and outlet resistances have to be specified as resistance values, and they could represent shock losses. The branch resistance may be specified as a resistance value, or as the geometric parameters of the branch. In addition, the resistance of a branch with a number of identical airways in parallel may be obtained by specifying the number of parallel airways, with a default value of one.

The computer selects the minimum number of meshes to ensure the unique solution of the network. It balances quantities at the nodes, calculated unbalanced heads around each mesh, and the quantity adjustment to be applied to each branch in each mesh. This process is iterative, to close on a specified maximum quantity adjustment at a node or for the mesh.

It is apparent that the quantity leaving a branch does not have to equal the quantity entering the branch, so at one node the quantity entering and leaving will not balance. This is obviously a surface node, and is node 1. What is not so obvious is that the balancing of quantities at a node has to be an iterative process if leakage factors, as opposed to constant leakage, are used. Solution of a network with leakage factors specified, therefore, takes considerably longer than when no factors are specified. It is possible to get the process to diverge. However, the most common error made when using this program is to place the unbalanced node (node 1) in a position other than at the surface.

7 A MINE SURVEY

The mine surveyed had two shafts and two levels, as shown in Figure 2, the ventilation network. The mine was being developed, but was not in production. Branches 1 and 2 were the main downcast shaft, and branch 11 the upcast shaft. Branch 4 was a raise with a fan at the top, and branch 5 a storage area with a fan at the end of it to ensure no recirculation from branch 4. The main development was in branches 3 and 9. A small amount of development was being carried out in branch 8, which was heavily regulated to ensure the main airflow through branches 9 and 10. Branch 12 contained the main mine fans. Branch 10 was isolated because it had a large airflow, was relatively long, and could therefore be used as a standard for other branches should their pressure drops be too small for the determination of reliable resistances. There were some booster fans in branch 9, so that at a later date this branch should be split into a network of tens of branches and several fans. Thus the survey presented here is a typical initial survey.

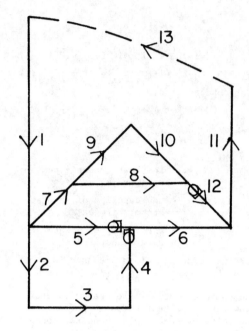

Fig. 2: Ventilation Network

BRANCH 1   (FIGURE 3 - UNDERGROUND BRANCH)

PI	PE	TI	TE	TWBI	TWBE
10.09	10.981	56	47.7	34.1	40.6

ZI	ZE	QI	QE	OMWI
10445	8089	0	377000	0

F
UNDERGROUND BRANCH
MEAN VOLUME FLOW RATE, CFM, IS 395485
HEAT EXCHANGE, BTU/LBM, IS-.0940683
EVAPORATION, LBM/LBM, IS 4.56351E-03
CONDENSATION, LBM/LBM IS 0
LOST WORK, FT LBF/LBM, IS 42.7992
RESISTANCE*10 POWER 10 IS .0392651
RESIST.CORR.TO ACT.DEN. IS .0290374
NVP(STD.DENSITY= .055 LBM/FT3), SWG, IS .21035

RHOI	RHOE	RHOMEAN	OMI	OME
.052792	.0582332	.0554645	9.08049E-04	5.47156E-03

VI	VE	(VE+VE)/2	VM	MAI	MAE
18.9595	17.2663	18.1129	18.095	0	21833.6

BRANCH 13   (FIGURE 4 - SURFACE BRANCH)

PI	PE	TI	TE	TWBI	TWBE
10.031	10.09	-30	-30.4	-29.5	-29.9

ZI	ZE	QI	QE	OMWI
10325	10445	0	0	0

F
SURFACE BRANCH
NVP IN SURFACE BRANCH, IN WG, IS-3.28184 ;-.189734 BY DENSITY AND-3.0921 BY ATMOSPHERIC PRESSURE

The output of the field reduction program for branches 1 and 13 is shown in Figures 3 and 4. The units are U.S. The underground branch, Figure 3, gives a lost work of 43 ft lbf/lbm, or 130 J/kg, so that the pressure drop due to friction should be correct to ±10%. The survey was run in winter, with the downcast air heated by direct fired heaters, so that the small heat exchange (-200 J/kg) and significant evaporation (0.0046 kg/kg) are both reasonable.

The first two lines of output in Figure 3 are the input data, the values being at branch inlet and outlet. Pressures are barometric, in lbf/in^2, temperatures in F, elevations in feet and airflow rates in ft^3/minute. The last two lines are densities, in lbm/ft^3, humidity ratios in lbm/lbm, specific volumes in ft^3/lbm dry air, and mass rates of flow in lbm dry air/minute. The change in specific volume from 18.96 to 17.27 ft^3/lbm dry air allows a variable leakage factor of -3.4% to be specified for each end of the branch.

A possible source of major errors in the survey was the measurement of volumetric flow rates. It was not possible to measure these in branches 1 and 2. The quantity in branch 2 was estimated from measurements at the exit from branch 3, using the appropriate specific volumes. The quantity measurement was made within an hour of the pressure/ temperature measurements. The quantity flowing in branch 1 was estimated

from the branch 2 quantity and quantity measurements in branches 5 and 7, duly adjusted for specific volumes. Unfortunately, the branch 5 and 7 quantities were not measured on the same day as the branch 1 pressure/temperature survey. This was obviously a planning error. Conditions remained much the same over the three day period that the survey was run, but the error in determining the resistance of branch 1 could well be more than the ±10% related to the pressure drop. The resistances used in the network program are the product of the resistance and density, since the network program does not correct for density.

The surface branch, Figure 4, was not measured at the mine, but it had to be analysed in order to get the density NVP. Because the shaft collars were relatively close together in a valley, it was considered that the weathr related NVP would be insignificant. Thus the density NVP only is used in the network simulation. The gross relative pressure error in the surface branch, which is responsible for the large weather related NVP, was caused by the pressure measurements being taken at the top of each shaft on different days. This incorrect pressure relationship will cause an error in the density NVP, but the magnitude is insignificant. The effect can be reduced by increasing PI by 0.108 lbf/in^2, which changes the density NVP by less than 0.01" WG whilst reducing the weather related NVP from -3" to -0.1" WG.

TABLE 1

Branch No	Lost work ft lbf/lbm	Resistance x Density x 10^{10}	NVP "WG	VI	VE
				ft^3/lbm	
1	42.8	0.029	0.21	18.960	17.266
2	3.2	0.015	0.40	17.284	17.138
3	25.3	0.100	0.	17.165	18.655
4	145.8	0.471	0.05	18.656	19.060
5	1.3	1.179	0.	17.434	17.880
6	63.9	0.200	-0.01	19.245	19.198
7	87.2	0.200	0.	17.758	17.795
8	-0.7	-0.064	0.	17.883	19.060
9	21.3	0.064	0.	17.937	18.882
10	34.0	0.093	0.	18.888	18.900
11	179.0	0.089	1.30	19.055	20.341
12	254.5	0.517	0.	18.768	19.616
13	–	0.	-0.19	–	–

TABLE 2

Branch	QI cfm x 10^{-3}		QE cfm x 10^{-3}	
1	400.66		373.88	(377)
2	153.55		152.43	(154)
3	152.43		173.49	(172)
4	173.49	(176)	176.97	
5	11.75		11.75	(11.3)
6	188.73	(179)	188.73	
7	208.46		208.46	(217.7)
8	27.51		29.36	(34)
9	180.95		198.08	(191)
10	198.08	(193)	198.08	
11	419.54	(429)	448.56	
12	227.44		230.81	(227)
13	448.56		448.56	

The temperatures used for this branch must be the actual branch temperatures, that is atmospheric temperatures. These were not measured, because the air at entry to branch 1 had been heated, and the air at exit from branch 11 is mine exhaust air, not atmospheric air. Because temperatures were about -30 F (-35 C), a value of -30 was used at the top of the upcast shaft, and -30.4 (corresponding to 3.3 F/1000 feet, or 7.0 C/1000 m) at the top of the downcast shaft.

Some of the output from the field data reduction program is listed in Table 1, and from this, input to the network program was selected.

The lost work for branch 2 is too low to be significant, so the resistance for the branch was obtained from branch 1 adjusted for length. This resistance, together with the actual airflow, would give a lost work of 1.8 ft lbf/lbm, for an error of 1.4 ft lbf/lbm, or a combined barometric pressure error of less than 1.5' or air (0.5 m of air).

Branch 10 was used as a standard when necessary, and it is obvious that this is required in branches 5 and 8. The lost work errors in these branches were less than 2 ft lbf/lbm, for combined barometric pressure errors of less than 2' of air (say 0,5 m air).

The branch 12 lost work was grossly in

error. If the field data reduction program had been run at the mine this would have been spotted and the branch remeasured. Since it was discovered too late for remeasurement, branch 10 was used as a standard. In addition, the variable leakage factor was adjusted.

Variable leakage factors were used in branches 1,2,3,4,8,9,11 and 12. In addition, constant leakages representing compressed air, were used in branches 3 and 9.

The fan curves were input from the measured values in the field and the manufacturer's curves adjusted for density. The measured positions were close to the manufacturer's curves.

The computer generated quantities are shown in Table 2, along with the quantities used to determine the branch resistances in parentheses. The largest differences are less than 10,000 ft^3/minute(5 m^3/s), or say 5%, so that the survey may be considered satisfactory.

One final point that is apparent from Table 2 is that about 50,000 ft^3/minute (25 m^3/s) more air leaves the mine than enters it. Of this, 15,000 ft^3/minute (7.5 m^3/s) is compressed air and 35,000 ft^3/minute (17.5 m^3/s) due to expansion of the ventilating air.

Modelling and prediction of climatic conditions in a deep level development tunnel

M.A.OUEDERNI & E.P.DÉLIAC
Centre Mines Infrastructures, Ecole Nationale Supérieure des Mines de Paris, Fontainebleau, France

P.CASSINI
Centre d'Etudes et Recherches des Charbonnages (CERCHAR), Verneuil-en-Halatte, France

ABSTRACT : Although prediction of climatic conditions in deep mines has made considerable progress over the past years, particularly in South Africa and in Germany, a practical, easy -to-use and accurate model of development ends environment does not appear to exist as yet due to the complexity of heat exchanges in a single end, with two airflows in the same tunnel and various pieces of machinery.

Consequently the "Charbonnages de France" group and the "Centre Mines Infrastructures" of the "Ecole des Mines de Paris" have developed their own model, together with a micro-computer program called CLIMEND, based on their experience of the German approach to the problem. One of the main problems to be solved was the calculation of the rock surface temperature; a new method, based on the Rees' theory, has been devised, which takes into account air temperature fluctuations and the evaporation of wet areas, and uses a new concept of a boundary layer around the tunnel. Temperature and moisture data have been collected in a few places of some roadways driven in hot strata at relatively deep levels. They allow the calibration of the various models and give an idea of the accuracy of heat predictions. CLIMEND models the various heat exchanges (convection and radiation) and mass exchanges of the air along segments of the tunnel using the calculation of one segment to initiate that of the next one. It requires the empirical knowledge of the heat generated at the development front, based on statistical information. It enables an accurate adjustment of the required parameters (rock wetness, thermodynamic properties) by comparison with preliminary results. It is also very easy to use by an engineer with little or no computer experience. The program is used on micro-computers such as IBM PC

INTRODUCTION

Research on the prediction of climatic conditions in deep underground workings has been carried on for over twenty years, particularly in South Africa, one of the pioneers in this field. Yet the problem of modelling the environment of a deep single end tunnel has not been solved satisfactorily up to now, although this could prove to be of critical importance, since mine personnel has to work in such anexcavation before a holing to the primary ventilation can be established. Among the reasons is the complexity of the various heat exchanges, as illustrated on figure 1; in fact they can be listed as follows :

- q_c convective heat flux density
- q_r radiative heat flux density
- q_{ev} evaporation heat flux density
- H_b heat exchange between intake and return air, through the air duct
- H_e heat flow from the broken rock to the air
- H_m heat exchange at the end of the tunnel

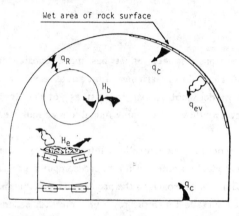

Legend :

➤ conv. or rad. heat transfer

⤳ evaporation heat transfer

Figure 1 : the various heat exchanges in a single development end

Two approaches to model the heat balance in such an excavation are well documented; they are the South African one (Chamber of Mines), mainly consisting of empirical correlations based on extensive statistical data (Whillier and Ramsden, 1975), and the German one (Bergbau Forschung), using a computer program to calculate the various unknowns from one section of the tunnel to the next one (Voss, 1980). The South African method has shownits usefulness in assessing the overall heat load in a mine; it is however too inaccurate to precisely predict temperatures in a particular tunnel and, furthermore, it is only adapted to twin development ends with one airflow for each

tunnel. It is being updated with a new numerical method (Starfield et al., 1983). The Bergbau Forschung approach is currently used in France, where the Virgin Rock Temperature (VRT) exceeds 45°C in some places of the Lorraine collieries. Yet, it relies on two parameters which have to be empirically adjusted, called wetness factor and, equivalent conductivity; the calculation is firstly done in averaged conditions of air intake and face end heat exchange and then, once the adjustments are made, the second step consists of calculating the climate in instantaneous conditions (machines stopped, for instance). It has been found that, unless the wetness factors and equivalent conductivities are adjusted again (which is not realistic), significant differences are observed between predicted and actual values. Furthermore, it requires a main frame computer in its present use.

Consequently, it has been decided to investigate the possibility to accurately model and predict the climatic conditions in a deep development end through a micro-computer program, starting from the German method of calculation. The work has been part of a common research project between the Charbonnages de France company and the mining department of the Ecole Nationale Supérieure des Mines de Paris (CMI). The micro-computer software has been called CLIMEND and is the main object of this paper.

In the first part, theoretical considerations are given, followed in the second one by a short description of CLIMEND; then experimentaldata is detailed in the third part and test-runs of CLIMEND are discussed in the fourth one, compared with calculations using the Bergbau Forschung program.

THEORETICAL PROBLEMS

The calculation of heat flow in a deep level development end has to account for three major points (Ouederni, 1984)

i) evaluation of rock surface temperature from totally dry to totally wet conditions

ii) determination of heat and mass exchange coefficients (mainly convection and evaporation)

iii) assessment of changes induced by the pile of broken rock at the face, by its transport along the tunnel, or by the face itself.

The calculation of the radiative heat transfer and of the heat exchanged between the intake and the return air is now well mastered (Rohsenow and Hartnett, 1973; Ouederni, 1984) and will not be detailed here. Suffice to say that it is accounted for in this model.

Heat Flux Density from the Surrounding Rock

General equations : The total flux density is $q = q_c + q_{ev} + q_r$. The convective heat contribution is given by the classical relation

$$q_c = h_c \, (RST - DBT)$$

with

h_c : convective heat transfer coefficient (in W/m2.°K)

RST : rock surface temperature (see below)

DBT : dry bulb air temperature

When the rock surface element is wet, q_{ev} must be calculated and is given by

$$q_{ev} = KL_v \, (P_s - P_v)$$

with

K : mass transfer coefficient (in kg/m2.s.Pa)

L_v : evaporation latent heat (in J/kg)

P_s : vapour pressure at saturation (in Pa)

P_v : partial water vapour pressure (in Pa)

Rock surface temperature : Despite simplification using the assumption of an axially symmetric tunnel, the heat conduction equation remains mathematically unsolved. In addition, classical numerical methods are difficult to use and take no account of air temperature fluctuations, i.e. the temperature is assumed constant. However, an empirical solution can be found easily and rapidly using Rees'theory. It is based on the following formula

$$RT\,(r,t) = VRT - (VRT - RST) \exp\,(-C(r-D/2)/\sqrt{at})$$

with

r : distance to the axis of the tunnel (in m)

t : age of the tunnel at the point of investigation (in hours)

RT : rock temperature

a : rock diffusivity (in m2/hr)

D : hydraulic diameter of the tunnel (diameter of a circular tunnel with the same cross-section area) (in m)

C : Rees' coefficient, depending on the rock type (dimensionless)

and on the classical Newton boundary condition

$$q_c = \lambda \left(\frac{\partial RT}{\partial r}\right)_{r = D/2}$$

with

λ : mean thermal conductivity of the rock (in W/m°K)

It is necessary to adjust the Rees'coefficient using experimental data, in order to use the above equation.

In addition to this, an accurate prediction of the tunnel climate over a short period requiresto account for fluctuations of the temperatures, moisture or wetness of the walls, and/or air flow. To be able to solve this problem, the concept of a rock boundary layer around the tunnel has been used (figure 2). It is then assumed that the above fluctuations are limited to this layer, of a given thickness e; inside the layer, the stationary conduction equation is solved to calculate the temperature gradient at any time t, with the following boundary conditions :

- $r \geq e + D/2$: RT is given by the Rees' formula under mean conditions
- $r = D/2$: the heat flux density is that given in the above Newton condition

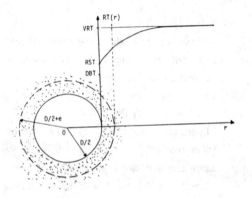

Figure 2 : the boundary layer model
(from Ouederni, 1984)

Heat and mass exchange coefficient : The most commonly used equation is that of Mc Adams, empirically estalished as follows

$$hc = 4,32 \ f \ V^{0,8} \ D^{-0,2}$$

with V : air velocity (in m/s)

f : roughness factor (dimensionless)

When the air velocity is less than 1m/s, the natural convection can no longer be neglected and V is replaced by a resultant velocity V' which depends on the difference RST-DBT (Froehlicher, 1980).

The mass transfer coefficient K, depends mainly on hc and the dry bulb temperature.

Broken Rock and Tunnel Face

The broken rock : it is assumed to be well divided, so that it releases its energy rapidly, mostly by evaporation. To ascertain this would require delicate measurements; in this model only an approximate evaluation is used, derived from the measurements described further in the paper.

The tunnel face : it is virtually impossible to accurately model the heat exchange at the end of the tunnel because it depends on the mining method, on the equipment, on its working rate, on the interaction between machines and the rock (heat storage, for instance) on the distance from the discharge of the duct to the face, etc... At this stage, it has therefore been decided to assume this heat exchange experimentally known and to simultaneously collect a large amount of field data, to enable statistical correlations at a later stage.

The heat exchanged by the broken rock and at the end of the tunnel is therefore cumulated as a localized heat load, later referred to as the front-end heat pick-up.

CLIMEND

Organization of the Program

The purpose of this program is to assemble the above described model with the help of a micro-computer, in order to provide a calculation of the climatic conditions in a tunnel (temperature, moisture content) as well as of the various heat or mass exchange parameters. Because of the large amount of calculations required, it has been written in FORTRAN and developed on an IBM-XT; its organization is illustrated on figure 3. As can be seen on the figure, CLIMEND is an iterative program. When starting the calculation, it divides the tunnel into sections; for each iteration it computes the conditions in the air duct (from the intake to the face), then brings in the face contribution to the heat exchange, and eventually works out the climate in the tunnel itself backwards, from the face to the entrance.

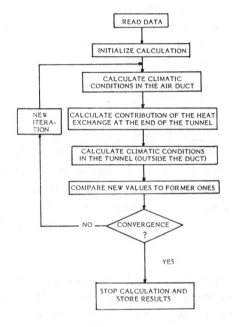

Figure 3 : simplified organization of a CLIMEND calculation

```
*****************************************************
*                                                   *
*            CLIMEND : MAIN MENU                     *
*                                                   *
*  1 : READ THE TUNNEL DATA FROM A DISKETTE         *
*  2 : INTRODUCE THE TUNNEL DATA                    *
*  3 : RETRIEVE OR MODIFY THE TUNNEL DATA           *
*  4 : SAVE THE TUNNEL DATA ONTO A DISKETTE.        *
*  5 : MODIFY CALCULATION PARAMETERS                *
*  6 : CALCULATIONS                                 *
*  7 : RESULTS                                      *
*  8 : QUIT                                         *
*                                                   *
*****************************************************

What is your option number ? :
```

Figure 4 : main menu of CLIMEND

Menu of CLIMEND : Figure 4 shows the main menu of the program; it can be split in three main functions ; data file management, calculations, output of results.

Data file management : in order to make the program as interactive as possible, the data has been organized in "sub-files" which can be accessed directly. It is possible to have any of them displayed for potential modifications, instead of creating a complete new file at each change. The various sub-files are basically :

- tunnel specifications
- duct specifications (including a regular leakage distribution)
- rock parameters
- fan specifications
- intake air
- face description (including the broken rock)
- machines specifications (including working rate)

It is of interest to note that

i) the Rees' coefficient is not taken as a rock parameter, its value being adjusted at the present stage; when more information is available it will be incorporated in the rock sub-file

ii) the three last sub-files are subject to fluctuations

iii) to account for variations in humidity, CLIMEND allows to divide the tunnel in up to four sections where the wetness factor can be set at different values

iv) it is possible to introduce a standing time in the face description, in order to correct the age of the tunnel if the development has been stopped for some time.

Calculation : it has been explained that fluctuations of certain parameters should be accounted for. The program therefore performs a "mean" calculation with the values given in the data file. Those values must therefore be input as average values (particularly for the power consumed by the machines) in this calculation, there is no use of the boundayer layer model. Then, after this initial calculation, CLIMEND asks whether an "instantaneous" one is necessary. If so, it asks for the instantaneous values of the parameters contained in the three last data sub-files, and performs a second calculation, similar to the first one. The accuracy of the calculation can easily be changed by modifying, through a special option (see figure 4), either the Rees' coefficient, the boundary layer thickness, the wetness factors, the residual error on temperatures, moisture content, or on the convection coefficient h_c.

Output of results : there is no graphical output of results at this stage; the program offers achoice of tables, the most important one being obviously that of the temperatures and the moisture content of the air.

EXPERIMENTAL DATA

Description of the Sites

Experimental data was collected in the Vouters section of the French collieries of Lorraine (Houillères du Bassin de Lorraine), belonging to the Charbonnages de France Group. In this mine the coal seams dip from 60° to 90°, so thatthe access cross-cuts meet different types of rock. Two sites were selected for the measurements, at the same depth of 1146 m below surface. They consist of a main haulage (main drift North 1146) and a cross-cut (third auxiliary North East 1146), both with a forced ventilation system. The general layout is shown on figure 5. The developing method is drill and blast, with a pneumatic multi-jumbo and a 57 kW diesel LHD.

Ventilation is forced through an air duct to as near as 15 m from the face. Altogether, six measurements were carried out between 1983 and 1984, with the length of the tunnels from 240 to 600 m. Each measurement lasted more than one week, thus enabling to calculate mean climatic conditions as well as fluctuations due to either an unusually high or low activity, or different characteristics of the intake air.

The Data Collected

Virgin rock temperature : it has been measured at different depths in the rock mass with a geothermometer, intrinsically safe device developed by Cerchar (Charbonnages de France's Research organization). It has an accuracy of ± 0.2 °C, with a digital display sensitive to 0.1 °C.

Temperature and moisture : five to six SINA thermohygrographs were placed along the tunnels, as shown on figure 5. They continuouslyrecord the dry bulb temperature and the relative humidity of the air. They have been selected because of their accuracy and their safety in an explosive atmosphere (Cassini, 1984). Calibrations with brines have yielded accuracies of ± 0.3 °C on DBT and $\pm 2\%$ on relative humidity. In addition, dry and wet bulb temperatures have been measured with classical hygrometers in order to complete the SINA recordings and detect possible heterogeneities.

Airflow : airflows were carefully measured at different points, as well as leakages through the air duct.

First Conclusions from the Measurements

Two important results have been obtained from this field data collection, notwithstanding the results of CLIMEND runs. Firstly it has been found that in the face area, the dry bulb temperature does not increase (except for one period, when the development was stopped); the heat given to the air results in an increase in the moisture content temperature (this is obviously due to the important water flow). Secondly the rock mass has an important regulating effect. During the same measurement, at different time intervals, the variations in temperature and moisture content decreased from the face to the entrance of the tunnel.

DISCUSSIONS

In order to test the validity of the theoretical model outlined hereabove and that of the CLIMEND program, test-runs were performed on one of the six measurements mentioned in the previous section, namely the May 1984 measurement of the 1146 level cross-cut. As explained in the description of CLIMEND, the program starts with a calculation of mean conditions, where adjustment parameters are the Rees' coefficient and the wetness factor in up to four sections of the tunnel (whereas, in the Bergbau Forschung program the Rees' coefficient is replaced by the equivalent conductivity of the rock); thereafter, it is possible to calculate instantaneous climatic conditions, the adjustment being done through the boundary layer thickness and, to some extent, through the wetness factor (mostly in the section at the front end of the tunnel). This

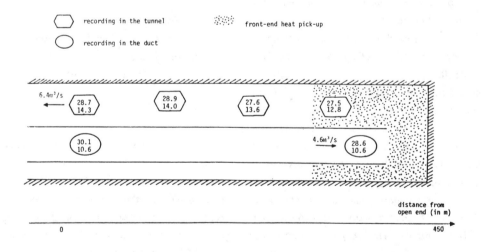

(Nota: top figure is DBT,in °C,and bottom one is moisture content,in g H_2O / kg air)

Figure 5 : layout of survey points during the May 1984
measurement in the North East 1146 Cross-cut.

495

section illustrates this, with comparisons of experimental measures, CLIMEND and the Bergbau Forschung program predictions.

Calculation of Mean Conditions

General tunnel description : table I gives the detail of the tunnel description

TOTAL LENGTH (m)	VR? (°C)	INTAKE AIR WBT DBT (°C)	AIR FLOW (m³/s)	MASS FLOW (kg/s)	LEAKAGE DENSITY (per 100m)	FRONT-END HEAT PICK-UP DRY EVAP. (kW)
448	33.53	21.3 30.1	6.4	7.9	8%	0.5 20.5

Table I : description of 1146 level
cross-cut in May 1984.

The tunnel was divided in three sections for the CLIMEND calculation (front end, intermediate, main section), and two sections for the German one, the wetness decreasing from the face to the open end.

Calculation and adjustment : table II shows the result of the calculations, compared with experimental evidence.

As can be seen on this table, the agreement is excellent, provided the adjustment is that given in table III.

It should be noted that the value of 0.25 found for C is in excellent agreement with that calculated by Simode (1976) in the same mine (C = 0.255), in fact if this value is changed by 0.05, the total heat load in the tunnel varies in the same direction by slightly less than 10%. Similarly the calculated equivalent conductivity (2.75 W/m.°K) checks with the experimental one (2.7 W/m.°K). The roughness factor has been chosen after the Bergbau Forschung experimental results. In fact, the few (first value is experimentally measured, second one is predicted by CLIMEND, bracketed one is given by the Bergbau Forschung program)

DISTANCE FROM OPEN END (m)	DUCT AIR DBT (°C)	TUNNEL AIR DBT (°C)	TUNNEL AIR MOISTURE CONTENT (g H₂O/kg)	RST DRY WET (°C)
10	30.1 30.1 (30.1)	28.7 28.7 (28.7)	14.3 14.3 (14.3)	29.6 24.3 (29.9) (24.2)
135	——— 29.8 (29.6)	28.2 28.2 (28.1)	14.0 14.0 (14.0)	29.0 23.9 (29.2) (23.8)
280	——— 29.3 (29.1)	27.6 27.6 (27.6)	13.6 13.6 (13.7)	28.4 23.5 (28.5) (23.4)
385	——— 28.9 (28.7)	27.5 27.5 (27.5)	12.8 12.7 (13.2)	28.2 23.0 (28.3) (22.9)
405	28.6 28.8 (28.6)	27.8 (27.6)	12.4 (12.9)	28.5 23.0 (28.4) (23.0)

Table II : calculations of mean conditions
in 1146 level cross-cut

	REES' COEFFICIENT	WETNESS FACTOR SECTION 1	SECTION 2	SECTION 3	EQUIVALENT CONDUCTIVITY (W/M°K)	ROUGHNESS FACTOR
CLIMEND	0.25	0.1 (200)	0.15 (197)	0.45 (51)	—	1.7
BERGBAU FORSCHUNG	—	0.19 (343)	0.3 (105)	—	2.75	1.7

Table III : adjustment parameters for the
mean conditions calculation
(bracketed values are lengths of sections
of tunnel, in m)

differences between the two numerical simulations are mainly found in the rock surface temperature (see Table II) and in the calculation of the convective heat transfer coefficient. The Bergbau Forschung program does not appear to calculate the effect of natural convection and uses a fixed value for h_C when the air velocity is low (h_C = 4.65 W/m2.°K), whereas CLIMEND calculates an actual value for h_C (here h_C = 3.26 W/m2.°K).

Calculation of instantaneous fluctuating conditions

The example chosen here is when the development is in full activity. Table IV gives the results of the calculations compared with experimental recorded values.

The intake air has not changed, but the front end heat pick-up has been increased as follows
- increase in DBT : 0.9 kW
- increase in humidity : 30.1 kW.

It should be reminded here that the mining method was drill and blast, hence the relatively low difference between peak and mean values for this heat exchange. Had the development been driven with a cutting machine, this difference would have been much higher. The new adjustment is detailed in table V. The boundary layer thickness can be taken as 0.65 m in this tunnel; the same value was found valid when the development is stopped and, in fact, it has been found that a variation in this parameter does not affect the heat exchanged much. As shown on tables IV and V, CLIMEND predicts reasonably accurate values of temperatures and moisture contents, with an additional adjustment limited to the boundary layer and to the wetness close to the face. The Bergbau Forschung program is also accurate, but the wetness factors and the equivalent conductivity have to be re-adjusted, like for a new tunnel calculation. In addition to this, empirical corrections on DBT and moisture content are input at the front end; for example, in the values given in table IV an additional 1.3 g H_2O/kg air has been used, which explains the slight difference with CLIMEND results when getting close to the face. It is felt that this lack of physical meaning could explain the difficulty to predict conditions in a new tunnel

DISTANCE FROM OPEN END (m)	DUCT AIR DBT (°C)	TUNNEL AIR DBT (°C)	TUNNEL AIR MOISTURE CONTENT (g H_2O/kg)	RST DRY WET (°C)	
10	30.0 30.1 (30.0)	28.7 28.8 (28.7)	14.8 14.7 (14.7)	29.6 (29.9)	24.5 (24.4)
135	--- 29.8 (29.6)	28.2 28.2 (28.2)	14.6 14.5 (14.6)	29.0 (29.3)	24.1 (24.2)
280	--- 29.3 (29.0)	27.6 27.6 (27.7)	14.5 14.0 (14.5)	28.4 (28.5)	23.7 (24.0)
385	--- 28.9 (28.7)	27.7 27.4 (27.6)	13.6 13.0 (13.8)	28.1 (28.4)	23.1 (23.5)
405	28.6 28.8 (28.6)	--- 27.7 (27.8)	--- 12.7 (13.7)	28.3 (28.8)	23.1 (23.9)

Table IV : calculations of "peak" conditions in 1146 level cross-cut)

(same comments as for table II)

	BOUNDARY LAYER THICKNESS (M)	WETNESS FACTOR			EQUIVALENT CONDUCTIVITY (W/M.°K)
		SECTION 1	SECTION 2	SECTION 3	
CLIMEND	0.65	0.1 (200)	0.15 (197)	0.45 (51)	---
BERGBAU FORSCHUNG	---	0.19 (343)	0.55 (105)	---	2.25

Table V : adjustement for instantaneous conditions calculations

(same comments as for table III)

still to be developed, when using the German program. However CLIMEND has not been used yet to simulate conditions without experimental evidence and its accuracy as a predictive tool still has to be demonstrated.

CONCLUSIONS

A fundamental investigation into the theoretical calculation of the heat balance in a single end development tunnel at depth has shown that it was possible, with a few assumptions, to predict the climatic conditions along the tunnel, thus confirming previous work done in Germany. After some modifications of the Bergbau Forschung's program, particularly by introducing a Rees' coefficient and by calculating the natural convection heat transfer at all air velocities, it has been possible to derive a seemingly powerful micro-computer program. The concept of a boundary layer around the tunnel seems to hold very promissive results as far as fluctuations of either intake air characteristics or face heat pick-up are concerned, although it is still early to draw definite conclusions. A strong emphasis is presently placed on further refinement of the output of CLIMEND results and on the creation of an important and reliable experimental data base, to obtain accurate values of the adjustment parameters (mainly Rees' coefficient and boundary layer thickness) and statistical correlations to predict arbitrarily chosen heat loads (broken rock and face end). Measurements will include mechanized development ends and deeper workings.

ACKNOWLEDGMENTS

The authors wish to thank management of CERCHAR for permission to publish this paper. They are very grateful to the Houillères du Bassin de Lorraine and particularly to the personnel of the Vouters section and of the Ventilation department for their invaluable help to obtain the experimental evidence presented here, as well as to the Direction of HBL for the permission to publish these experimental results. The views given here are those of the authors and not necessarily those of Charbonnages de France.

REFERENCES

Cassini P. (1984) "La mesure de l'humidité en mines grisouteuses et son enregistrement", Bulletin d'information du Bureau National de Métrologie, n°57, Paris, July 1984.

Froehlicher (1982) "Contribution à l'étude du climat minier", DEA thesis, INPL, ed. by ENSMIMN, Nancy.

Ouederni M.A. (1984) "Prévision des conditions climatiques dans un traçage profond", DEA Thesis, ENSMP, ed. by Centre Mines Infrastructures, Paris.

Rohsenow W. and Hartnett J. (1973) "Handbook of heat transfer", pub. by Mc Graw Hill, New York.

Simode E. (1976) "Contribution à l'étude du climat dans les voies en creusement". Internal report, H.B.L.-C.C.A. n°226/75, Merlebach.

Starfield A.M. and BLELOCH A.L.(1983) "A new method for computation of heat and moisture transfer in a partly wet airway", Journal of South African Institute of Mining and Metallurgy, Vol.83, pp 263-269.

Voss J. (1980) "Le climat dans les creusements mécanisés de voie et sa prévision", Journées d'information C.C.E., Luxembourg (Nov. 1980), ed. by Société de l'Industrie Minérale, Saint-Etienne.

Algorithm for fast simulation of mine ventilation using dual microcomputers

YUUSAKU TOMINAGA, HIROAKI MATSUKURA & KIYOSHI HIGUCHI
Hokkaido University, Sapporo, Japan

ABSTRACT: In order to plan a mine ventilation network by micro computer and to control airflow rate in a network in real time with a mini computer, a novel algorithm for mine ventilation simulation were developed. The principal results of the study are 1: the correction value of airflow rate for a basic mesh can be estimated as a root of an equation for a parabola derived from Kirchoff's second law, 2: convergent time of ventilation simulation is dependent on the order of applying basic meshes for correction airflow rate, 3: a dual micro computer system is available to speed up the simulation of a ventilation network.

1 INTRODUCTION

Recently with growth of microelectronics, micro computers have been used for mine ventilation simulation. As a micro computer takes more time to calculate than a large computer, development of software for micro computer is needed.

Three methods have been considered for mine ventilation simulation, 1: Hardy Cross method (Scott & Hinsley 1951), 2: Δ-Y transformation method (Kumazawa 1945) and 3: node potential method (Gunther 1967 and Isobe et al. 1979). In general, simulation time by Hardy Cross method is shorter than by other methods.

Hashimoto (1966) and Oka et al.(1967) pointed out the factors for converging time of the Hardy Cross method, such as a) choice of chord branch, b) total branch number in a basic mesh, c) chance of correcting airflow rate and d) modification of correction value of airflow rate.

In the present paper information is given on two of these factors and a dual micro computer system for rapid simulation of ventilation network.

2 CORRECTION VALUE OF AIRFLOW RATE

Initially, let the airflow rate through all the branches in a network be assigned to satisfy Kirchoff's first law. Kirchoff's second law is expressed by equation 1 in the case of a closed circuit shown in fig. 1.

$$\sum_{\ell=1}^{j} d_{\ell} R_{\ell} (Q_{\ell} + \Delta Q_{\ell})^2 - F = 0 \qquad \ldots (1)$$

Where R is aerodynamic resistance, Q_{ℓ} is airflow rate assumed initially and ΔQ_{ℓ} is correction airflow rate. F is natural and fan ventilation pressure in the circuit. d_{ℓ} is a coefficient with values of 1 and -1. If the airflow direction coincides with the predetermined direction, d_{ℓ} is 1 and if it is in the opposite direction on the circuit then d_{ℓ} is -1.

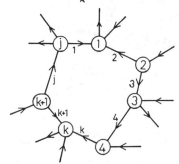

Fig. 1. A typical closed circuit.

If inlet and outlet airflow rates on the circuit illustrated in fig. 1 are assumed correct, ΔQ_{ℓ} is given in equation 2.

$$\Delta Q_{\ell} = d_1 d_{\ell} \Delta Q_1 \qquad \ldots (2)$$

Where ΔQ_1 is the correction airflow rate in the first branch as a chord branch in the circuit.

Substitution of equation 2 into equation

1 results in equation 3.

$$A \Delta Q_1^2 + B \Delta Q_1 + C = 0 \qquad \ldots (3)$$

$$\text{Where } A = \sum_{k=1}^{j} d_k R_k$$

$$B = 2 \sum_{k=1}^{j} R_k |Q_k|$$

$$C = \sum_{k=1}^{j} d_k R_k Q_k^2 - F$$

Equation 3 has two roots, but, when assumed airflow rates are near to the correct values, the correction airflow rate is given by equation 4.

$$Q = (-B + \sqrt{B^2 - 4AC})/2/A \qquad \ldots (4)$$

A ventilation network used in this study is illustrated in fig. 2.

Simulation time using equation 4 is about 70% of the time taken by Hardy Cross method.

3 EFFECT OF BASIC MESH ORDER ON CONVERGENT TIME

In this study chord branches were chosen

in consideration of a branch incident to a fan, high aerodynamic resistance and total branch number in a mesh as few as possible. The chord branches with parallel in fig. 2 are numbered arbitrarily and the numbers are shown in parentheses. As twenty basic meshes are included in the network, there are twenty conbinations of starting mesh for initial correction airflow rate. Following mesh for correction airflow rate is decided at first stage in such a way that common branches between former mesh and following one, are as many as possible occur. At following stages, correction airflow rates are estimated under the mesh order decided at first stage untill all the correction airflow rates converge less than 0.001 m^3/sec.

Simulation time on each mesh order is shown in first turn in fig. 3 where rectangles give these cases and circles are results of one hundred combinations chosen at random from the twenty factorial combinations of mesh order to display appearances of simulation time. The numbers with and without parenthses are stage numbers till airflow rate convergence and typical

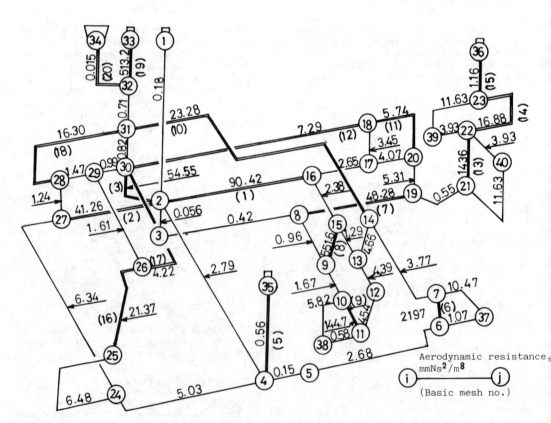

Fig. 2. A typical ventilation network

500

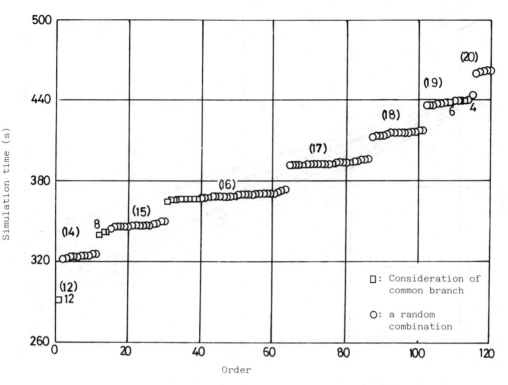

Fig.3. Simulation time on each mesh order, number without parentheses is of starting mesh number and number with parentheses is maximum stage number at convergence.

starting mesh number, respectively.

From the figure it is clear that mesh order is one of the factors to speed up ventilation simulation. The method choosing mesh order according the number of common branch is available, if the appropriate starting branch was given.

To search out the appropriate starting mesh, the property of correction airflow rate was investigated. In the case that mesh'12' is selected as starting mesh, the logarithm of summation of all the correction airflow rate at each stage is shown in fig. 4. A line is a regression line with a correlation coefficient -0.997.

Fig. 5 shows all the regression lines with typical starting mesh number concerning twenty starting meshes.

From fig.'s 3 and 5, it is clear that the steeper a regression line is, the more rapid airflow rate converges.

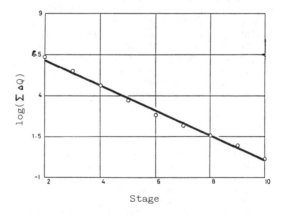

Fig. 4. A summation of all the correction airflow rates at each stage when starting mesh'12' is selected.

4 A DUAL MICRO COMPUTER SYSTEM FOR RAPID SIMULATION OF VENTILATION NETWORK

Two dual micro computer systems are discussed in this paper to improve simulation speed. One is that each computer shares in the estimation of correction airflow rate for half part of the basic meshes. After all the correction airflow rate are estimated, airflow rate in the branch is corrected to satisfy Kirchoff's first law.

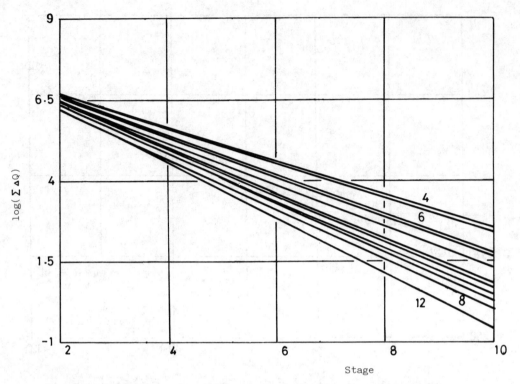

Fig. 5. Regression lines concerning each starting mesh.

The other is that computer'A' shares in the decision of the optimum mesh order for rapid simulation and computer'B' calculates correction airflow rate for a mesh. The first mesh order in the computer'B' is selected arbitrarily. After airflow rates are converged under an assumed fan working point, apparent mine charactaristic curve can be drawn by airflow rate through fan and a summation of pressure loss in the branches from intake to exhaust. An intersection of both characterristic curves of fan and mine gives a fan working point. When difference of fan working points between new and assumed exceeds error allowance, 0.1 mmAq, new fan working point is taken the place of assumed one. Afterword, computer'A' sends a signal to computer'B' and recieves the optimum mesh order at this time from the computer'B'. Following calculation to estimate fan working point, computer'A' calculates correction airflow rate with the optimum mesh order. The procedure mentioned above is repeated untill fan working point converges. Flow chart of the procedure is illustrated in fig. 6.
Simulation times on the network shown in fig. 2 are compared among three cases, such as 1: by a dual micro computer system, one shares in selecting the optimum mesh

order and the other shares in the estimation of correction airflow rate, 2: by uni-micro computer system and 3: by a dual micro computer system, each computer shares in the estimation of correction airflow rate by half. From the fig. 7, a dual micro computer system with combination of selection of the optimum mesh order and estimation of correction airflow rate is most available in three systems.

5 CONCLUSION

More effective methods to speed up simulation of a ventilation network have been introduced. Main results in the study are as follows,
1. The correction airflow rate for a basic mesh is given as a root of an equation for a parabora derived from Kirchoff's second law.
2. The optimum mesh order for rapid simulation exists as convergent time is dependent on the mesh order for correction airflow rate.
3. When first mesh is given, following mesh should be chosen so that common branches between first mesh and following mesh become as many as possible.

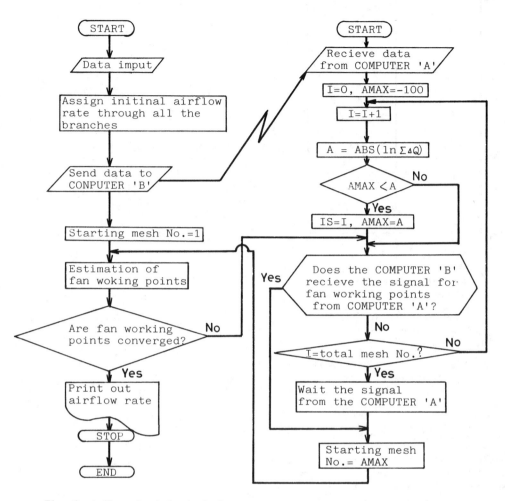

Fig. 6. A flow chart for a dual micro computer system.

4. A dual micro computer system in which one computer searches the optimum mesh order and other computer estimates correction airflow rates for meshes is available.

ACKNOWLEDGEMENT

The authers are indebted to Professor Pierre Mousset-Jones, University of Nevada-Reno, for his assistance in the English translation of the part of this paper, and wish to thank Professor R. V. Ramani, Pennsylvania State University, for his encouragement to the study.

This study was supported by the Grant-in-Aid for Scientific Research of the Ministry of Education, Science and Culture of Japan.

REFERENCES

Scott, D.R. and Hinsley, F.B. 1951. Ventilation network theory, Colliery Engineering, Vol.28.
Hashimoto, B., Asaeda, E., and takahashi, K. 1966. Analysis of mine ventilation networks by digital computer, JMMIJ, Vol.82, No.933.

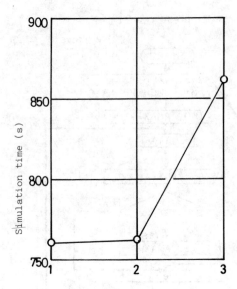

1: A dual micro computer system with
   combination of mesh order search
   and calculation of correction
   airflow rate.
2: A uni-micro computer system.
3: A dual micro computer system
   sharing by half.

Fig. 7. Simulation time on a network
illustrated in fig. 2.

Oka, K., Kiyama, H. and Hiramatsu, Y. 1967
   Analysing ventilation Network problems
   by a Digital Computer, JMMIJ, Vol.83,
   No.945.
Kumazawa, Y. 1945. Trigonal transformation
   in mine ventilation Network, Journal of
   the Mining Institute of Hokkaido, Vol.1,
   No.5.
Gunther, J. 1967. Une nouvelle methode de
   calcul des reseaux D'aerage, REVUE DE
   L'INDUSTRIE MINERALE, 11.
Isobe, T., Nohara, H. and Tominaga, Y.
   1979. An analytical investigation of
   mine ventilation network using a graph
   theory, JMMIJ, Vol.95, No.1096.

An interactive microcomputer program for mine ventilation network analysis

ZACHARIAS G.AGIOUTANTIS & ERTUGRUL TOPUZ
Virginia Polytechnic Institute and State University, Blacksburg, USA

ABSTRACT: VENTSIM is an interactive microcomputer program for analysis of mine ventilation networks. The program is comprised of two parts. The first part is a menu driven data program which allows the user to input data or edit data files with minimum effort. The second part solves a given ventilation network calculating the air quantities and the corresponding head losses of each airway by using the Hardy-Cross iterative procedure.

The program creates a fan directory where different fan types and curves can be stored, as well as a job directory where different case study files can be kept. The output includes a listing of the active airways and nodes of the network, the special constraints set on airways and the solution of the network.

1 INTRODUCTION

Most of the programs available for analysis of mine ventilation systems are designed for mainframe computer systems with limited access to production engineers. The continuous improvements in performance and prices of microcomputers, however, have created a demand for interactive and user-friendly microcomputer programs that may take longer to run, but provide online error trapping for input data and immediate answers.

VENTSIM is divided into two parts. The first part is composed of a fully interactive menu driven program for data preparation and storage. It permits the user to keep a directory of different network layouts as well as a directory of fan curves. The second part uses an already established solution procedure, the Hardy-Cross method, for balancing a ventilation network. It has some built-in error trapping routines that will print an error message and abort execution in case the user exceeds the limitations of the program.

The output of the program is displayed on the screen and at the same time an output file containing a detailed listing of the input data as well as the solution of the network is created and stored.

The program is written in BASIC and runs on the IBM-PC, IBM-XT and the IBM portable computer as well as on a series of compatible systems. Minimum system configuration includes 128K of main memory, one disk drive and a CRT.

2 THE MODEL

The first part of the program uses a tree-structured menu that includes five levels of imbedded sub-menus as shown in Figures 1 and 2. The user may choose valid option numbers of sub-menus to access the various editing modes, solve the network or access a help screen. Within the sub-menus the user may choose different field numbers (options) to modify one or more data fields in each screen. In most cases input data is checked for validity at the time of input. The whole data set is automatically re-saved after completion of each step.

The second part of the program balances the network. To conserve memory space and allow for larger networks, the solution routine is included as an independent segment. When this routine is loaded into memory, the input data have to be re-read from the disk. The job parameters as well as the fan curves are loaded at this point. The user is prompted for the name of the output file and at this stage he may choose to eliminate booster fans that

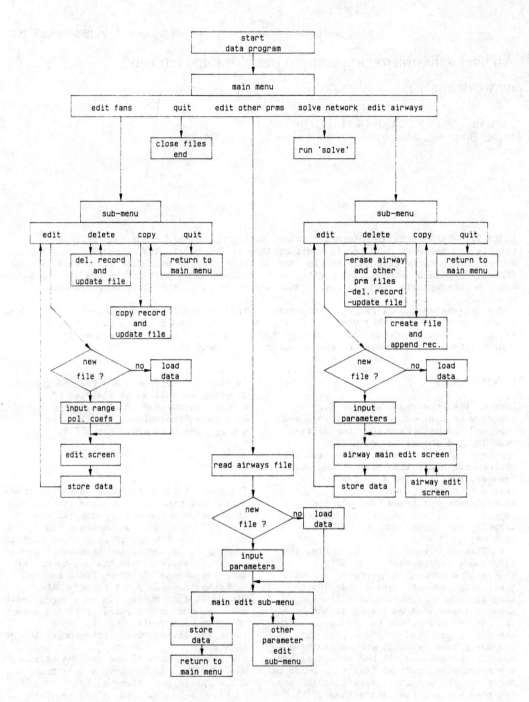

Figure 1. Data program

1 = Edit/Input fan parameters
2 = Edit/Input airway characteristics
3 = Edit/Input other parameters
4 = Solve network
5 = Help
6 = Exit to DOS

Enter option:

Figure 2. Main menu

may be assigned to fixed flow airways by
the solution routine. Then the total
number of airways, nodes and meshes
present in the current network will be
calculated and displayed.

Scott and Hinsley, (1951), suggested
that a network may be balanced in less
iterations if the meshes are chosen so
that high resistance airways are not
common to two or more meshes if that is
possible through the formulation of the
network. This leads to a unique method of
constructing the meshes which may be
considered as an advantage of the method.
The program, which is a modified code
based on Boumis (1980), sorts the airways
starting from fixed flow airways, airways
with fans, and then other airways in order
of their decreasing resistances. The last
condition implies that the leakage paths,
which are considered high resistance
airways will be higher in the list. The
meshes are then formed by starting at the
bottom of the above list and by placing
those airways back in the network in a way
that does not form a closed loop (mesh).
The remaining airways are omitted at this
stage, since they are the basic airways.
When the basic airways are added, the
loops are closed and the meshes are
formed. The airways that have directions
opposite to that of the basic airways are
assigned negative signs.

The program then calculates the natural
ventilation pressure for each mesh as well
as the initial flow in the airways. The
assignment of the initial flow is done as
follows: first the inflow of gases is
added to the airways; then the required
flow in the fixed flow airways is added,
followed by the maximum permissible flow
in the fan airways.

The program checks for adverse operating
situations, such as fans acting against
each other, and then the network is
balanced through the iterative procedure.
Iterating will stop either when a limiting

number of iterations is reached or when
the balancing error is below a specified
upper bound.

After balancing the network and
displaying the solution on the screen the
program will calculate the free-split
paths for each fan, if requested. If no
main fans are assigned to the network,
then, the program will provide the option
of assigning a fan to an airway and
calculating the corresponding free split.
In other words, the program will find a
directed path from intake to exhaust with
the maximum total head loss. The operating
point of a fan under these conditions will
then be calculated. A dynamic programming
routine was implemented at this stage. The
routine will search the network for the
sequence of paths resulting in the highest
head losses regardless of mesh formulation.
If, however, any one - directional flow
closed loop is encountered, calculation
will stop with a warning message. A macro
flow-chart of the program is given in
Figure 3.

2.1 Assumptions

Although the program can be applied to
many situations there are some assumptions
made that the user should be aware of:

1. Leakage paths must be entered by the
user as high resistance airways. These are
going to be treated as regular airways in
the iterative procedure. However, they
will not participate in the calculation of
the free split head losses.

2. The fan types assigned to the airways
are represented by one operating curve.
Different curves of the same fan may be
kept in the directory and compared in
individual runs of the program.

3. Node number (1) corresponds to an
atmospheric pressure node and there may be
more than one such node in a network. All
intake and exhaust nodes should be
numbered (1).

4. The resistance of an airway is
calculated according to the following
priorities:

a. As a function of a surveyed air
quantity and head;

b. Directly as given by the user in
which case a corresponding friction factor
is calculated;

c. As a function of airway geometry and
a given friction factor.

5. Geometric data for the airways such
as the length or the perimeter are used
together with a given friction factor to
calculate the resistance. When such data
are unknown or not important for the
analysis the program will assign the value

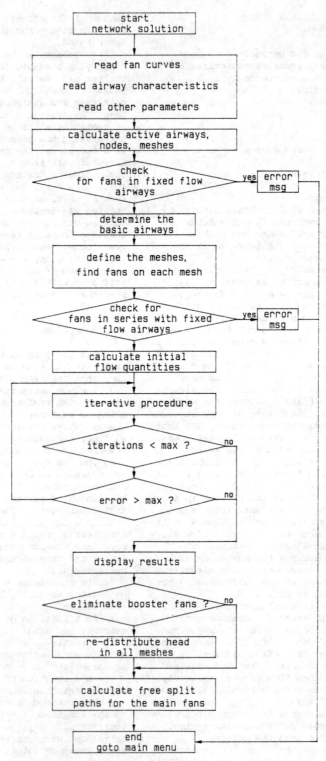

Figure 3. Solution routine

of '0' to the particular data field.

6. The flow and head units are assumed to be cubic feet per minute and inches of water gage, respectively.

7. Fan curves are approximated by polynomials. The user may specify his own interpolation function (up to a ninth degree polynomial) or the program will calculate either a parabolic (second degree) or a cubic (third degree) polynomial from the pairs of the flow and the head values for a particular fan curve.

8. Fans are also characterized by maximum and minimum flow values that denote the operating range of the fan. If the calculated flow for a fan lies outside this region the fan will automatically be excluded from the analysis.

2.2 Limitations

The number of independent meshes in a ventilation network is given by:

Number of meshes =
Number of airways - Number of nodes + 1

The program chooses the meshes so that only one fixed flow airway or fan airway is present in each mesh if that is possible. The program has several limitations that are dictated by the principles used to balance the network.

1. The number of the fixed flow airways should not be more than the number of independent meshes or an error will occur.

2. All active airways should be included meshes. If non-participating nodes or meshes are found in the preliminary analysis, the program will stop with an error message and the user will have to go back and change the activity status of the corresponding airways or add leakage paths.

3. If fans acting opposite to each other are assigned to airways in a series the program will stop with an error message.

4. In the following situations involving the setting and operation characteristics of fans, the program will not abort execution but the corresponding fan will be rendered inactive in the iterative procedure:

a. Installation of a fan in an airway designated as fixed flow airway;

b. The required flow through a fan airway lies outside of the operating range of the fan;

c. If a fan is operating around the end points of its operating range the message "FAN OUT: at border of operating range" will be printed and the fan will be excluded from the analysis;

d. Because of the memory limitations,

the program can analyze networks up to 400 airways and permits up to three digit node numbers. Up to 15 other parameters per category (such as inflow of gases, fixed flow airways, etc.) are allowed. These limitation can be relaxed by modifying the dimension statements. The fan directory can hold up to 20 different fan curves.

3 RUNNING THE PROGRAM

VENTSIM consists of one diskette that includes two programs, VMENU and VSOLVE with total disk space requirements of 150K for the compiled version. Data files are stored on the same disk and include:

- FANLIST, containing a set of fan curves and their characteristics.

- <name>.ADT, containing the geometric characteristics of the airways in the network while <name>.OTH contains the set of other properties assigned to certain parts of the network for case study <name>. A set of such files is created for every new job. A file <name>.OUT will be created by default after each run. For consecutive runs of the same network the user should be aware that the new output sequence will overwrite the old file unless a different name is specified.

- JOBLIST, containing a list of the existing job names.

Help screens are also stored on the disk and include all the files with the extension '.SCR'.

To run the program the user should first boot up the system. With the prompt (A>) in the display, type VMENU and press RETURN. The title of the program will be displayed on the screen. Whenever a "more..." appears in the lower right hand corner of the screen the user should press any key to move to the next screen. The main menu will be the third screen to appear after a brief explanation of the program capabilities and limitations. Data preparation as well as execution of the solution routine are controlled by this menu.

3.1 Preparation of Data Files

In order to set up a network for computer analysis the following steps are necessary:

1. Update the fan directory with all the fan types that will be used in the analysis. It is advisable to set different fan files for different settings (pitch or speed) of the same fan.

2. Create a file by defining the airways

of the network.

3. Create a file with all the control parameters of the network.

The first option the main menu will introduce is the sub-menu shown in Figure 4. The first option of this sub-menu "Edit old file or create new file," will prompt for a name assigned to a fan curve file. If the name appears in the fan directory which is displayed on the screen, the old file will be edited. If the name does not appear in the directory a new file will be created and the user will be prompted with the input sequence. Input requirements include maximum and minimum flow permitted through the fan and the number of coefficients (<=10) of the polynomial to be used in approximating the fan curve as shown in Figure 5. Entering '0' for the number of coefficients will result in a request for discrete pairs of data points (i.e., quantity of airflow and the corresponding head loss) to be inputted.

```
           Edit / Input fan parameters
                 Directory Entries

F1      F2      F3


1. Edit old file or create new file
2. Delete a file from directory
3. Copy a file under another name
4. Help
5. Return to main menu


              Enter Option:
```

Figure 4. Sub-menu for fan parameters

```
F1       OPERATING DATA -- ref number: 1

1. Maximum flow :          160
2. Minimum flow :           15
3. Number of coefs <=10:     4

                Coefficients

   4 .          coef # 1 : 11.856
   5 .          coef # 2 :-2.234146E-02
   6 .          coef # 3 :-1.108609E-04
   7 .          coef # 4 :-4.607172E-07
```

Figure 5. Fan data

The program will calculate the coefficients of a parabolic or third degree polynomial approximation for a fan curve given any number of data points by using the second or third degree least squares approximation, respectively. The user must input at least three pairs of points for the parabolic approximation and four for the cubic approximation. The number of data points does not affect execution time significantly but more points will provide a better approximation for the fan curve. The data points will not be saved in a file; the user, however, has an option to edit these points before the polynomial coefficients are calculated (Figure 6).

The second option of the fan sub-menu will delete a file from the directory. The third option allows the user to duplicate a file under a different name which will be appended to the directory. This option is particularly useful when setting up different operating curves for the same fan. The fourth option will access a help screen while the fifth option returns control to the main menu.

The second option of the main menu, "Edit/Input airway characteristics," will produce a sub-menu similar to the above (Figure 7). Again the user can edit an existing job file or create a new one. All parameters are summarized in a sample shown in Figure 8, and each airway characteristic can be edited individually

Field	Quantity	Head
1	50.000	10.500
2	80.000	9.120
3	100.000	8.000
4.	120.000	6.720
5.	150.000	4.500

Select field number for changes; if NONE hit 'RETURN'

Figure 6. Fan curve data

```
        Edit / Input airway characteristics
                  Directory Entries

WANG


1. Edit old file or create new file
2. Delete a file from directory
3. Copy a file under another name
4. Help
5. Return to main menu


                Enter Option:
```

Figure 7. Sub-menu for airway characteristics

AIRWAY CHARACTERISTICS FOR WANG

1. Total airways 10   Max Iterations 10   Maximum Error 10 cf/min

Fld	Act	S	E	A	P	L	R	K	Q	H
2¦	1	1	2	0.0	0.0	0	0.9E+00	0.0E+00	0.00E+00	.00E+00
3¦	1	2	3	0.0	25.0	360	0.6E+00	0.4E+02	0.00E+00	.00E+00
4¦	1	3	4	0.0	0.0	0	0.2E+01	0.0E+00	0.00E+00	.00E+00
5¦	1	2	5	0.0	0.0	0	0.4E+01	0.0E+00	0.00E+00	.00E+00
6¦	1	4	5	0.0	0.0	0	0.9E+01	0.0E+00	0.00E+00	.00E+00
7¦	1	4	6	0.0	36.0	1500	0.2E+01	0.8E+02	0.00E+00	.00E+00
8¦	1	5	7	0.0	0.0	0	0.8E+01	0.0E+00	0.00E+00	.00E+00
9¦	1	6	7	0.0	0.0	0	0.3E+01	0.0E+00	0.00E+00	.00E+00
10¦	1	7	1	0.0	0.0	0	0.4E+01	0.0E+00	0.00E+00	.00E+00
11¦	0	6	1	0.0	0.0	0	0.2E+01	0.0E+00	0.00E+00	.00E+00

Select field number for changes; if none, hit 'RET', if MORE enter '0'

Figure 8. Table of airway data

Airway : 2 - 3

1. Activity   :   1
2. Starting node   :   2
3. Ending node   :   3
4. Section (sq.ft) :   0
5. Perimeter (ft) :   25
6. Length (ft)   :   360
7. Resistance (10^{-10}): .5538461
8. K-factor (10^{-10}):   40
9. Surveyed quantity (10^3): 0
10. Surveyed head (inWG) :   0

Select field number for changes; if NONE
hit 'RETURN'

Figure 9. Airway data

OTHER PARAMETERS FOR WANG

1. Number of gas inflow junctions   0
2. Number of fixed flow airways   2
3. Number of fan airways   1
4. Number of natural pressure airways   1

Select field number for changes; if NONE
hit 'RETURN'

Figure 10. Sub-menu for airway parameters

by selecting the appropriate data field
number (Figure 9). Field number 1 modifies
the total number of airways, maximum
number of iterations and maximum error
permitted. Field numbers 2 to 10 modify
the individual airway characteristics.

The airway activity parameter has the
following function. When the value of '0'
is assigned to the parameter, the airway,
although present in the network layout,
will not participate in the formation of
the meshes or the analysis. This is
particularly useful when designing a
network in which the user can quickly
introduce or dispense with an airway. A
value of '1' describes an active airway
while a value of '2' signifies a leakage
path. Assigning a value of '1' to a
leakage path will not affect the solution
but the airway will not have that
designation in the output file.

An airway is defined by its two end
nodes. Optionally the user can input flow
quantities and head losses obtained from a

ventilation survey for any given airway.
If both of these quantities are non-zero
the resistance of the airway will be
calculated from these values. Otherwise,
the directly inputted resistance value
will be used. If this value is entered as
'0', then the cross-sectional area, the
perimeter, the length of the airway and
the corresponding friction factor will be
used to calculate the resistance of that
airway. In this case the Atkinson equation
will be used:

$$R = K\, L\, P/(5.2\, A^3)$$

where:
R = Resistance lb min^2/ft^8
K = Friction Factor lb min^2/ft^4
L = Length (ft)
P = Perimeter (ft)
A = Cross-sectional area (sq.ft)

The third option of the main menu will
prompt for the job name and display the
sub-menu shown in Figure 10.

This step can not be accessed for a
given network unless a file for airways
has already been created through the
second option of the main menu. Some other
ventilation characteristics may be
assigned to certain network components
through this step. These characteristics
may include the inflow of compressed gases

511

in some specific nodes (Option 1), air quantities in fixed flow airways (Option 2), fan types on fan airways (Option 3), and natural pressure head (Option 4).

Inflow of compressed gases such as methane emissions, compressed air, etc., may be added to the air quantities in some cases. These additional quantities are assumed to be acting at the beginning or ending node of that airway. It is possible to fix air quantities in certain airways (Figure 11). In this case, the program will solve the network and keep the quantities in these airways fixed. The number of the fixed-flow airways cannot be more than the number of the basic meshes. The sub-menu for inputting the fan airways includes a listing of the fans. If a fan reference number is not included in the fan directory (or it is zero) that fan will be deleted. The natural ventilation pressure is inputted in inches water gage and the sign should denote whether it is acting on the assigned direction of flow or not.

1. NUMBER OF PARAMETERS

FIELD	AIRWAY	FIXED AIR FLOW (X1000)
2.	4- 5	12.00
3.	4- 6	50.00

Select field number for changes; if NONE hit 'RETURN'

Figure 11. Fixed flow airways

In all of the above sub-menus assigning a zero quantity (i.e., zero inflow of compressed gas or zero for a fan type) to an airway or node will result in the deletion of the corresponding input data.

When inputting the other properties of the network the program checks for invalid or conflicting assignments and non-existent or non-participating airways and nodes.

3.2 Output

The fourth option of the main menu will load the solution routine. The user will be prompted for the name of the output file as well as whether he wants to eliminate potential booster fans. Before solving the network and while determining the basic airways and forming the meshes, the program runs checks for fans in fixed flow airways, for fans in series with fixed flow airways, and for fans acting opposite to each other within a mesh. The meshes are formed and initial quantities are assigned to the airways. The active number of nodes and airways as well as the full compositions of the meshes formed will be displayed.

The iterative procedure will stop either when the preset number of iterations is exceeded or when the error between two consecutive iterations is below the limit set. Then the program will regulate the fixed flow airways. The corresponding meshes will be balanced either by inserting a pressure head increase (booster fan) or a pressure drop (regulator). At this stage the booster fans may be eliminated by redistributing the corresponding pressure increase to all meshes except those that contain main fan units. If no regulator can be assigned to the mesh on which the booster fan is acting, the booster will not be eliminated and a warning message will appear. Then the nodes with compressed air intake and the airways with natural ventilation pressure, if any, will be displayed. A list of all the active airways with the corresponding flow quantities and pressure heads will then be displayed. Special airways such as leakage paths and regulated airways are distinguished. The program will then calculate and display the free split head losses for all the main fans or for optional fan locations if the initial network does not provide for main fans. All information that appears on the screen except for some of the warning messages is automatically saved in a file that can be printed at a later stage (Figure 12).

3.3 Example

The example problem analyzed by Wang (Figure 13) in Hartman et al, (1982), was solved using VENTSIM. The input data as well as the output sequence presented in Figures 8 to 12 correspond to the above example problem. Results are similar to those presented in the reference, though presented in a different format.

4 CONCLUDING REMARKS

The program is a powerful tool for analyzing existing networks as well as designing network expansions as mining advances. The best way to solve such a problem would be to establish a simulated network by inputting the measured air quantities and head losses and then adjusting some airway characteristics until the operating conditions meet those

```
*****************  Mesh table  *****************

              Mesh 1 : the 4 airways are
  [ 4, 5]  -[ 2, 5] [ 2, 3] [ 3, 4]
              Mesh 2 : the 6 airways are
  [ 4, 6] [ 6, 7] -[ 5, 7] -[ 2, 5] [ 2, 3] [ 3, 4]
              Mesh 3 : the 4 airways are
  [ 7, 1] [ 1, 2] [ 2, 5] [ 5, 7]
```

Network solution

```
          The network is balanced after  5  iterations
          The maximum error is less than    10.00  cf/min
```

Fixed Flow Airways
Regulator or booster fan conditions

```
Airway    Air-flow (X1000)        Adjustment
 4-  5         12.00     Booster fan for   0.386  inWG needed
 4-  6         50.00     Resistance (regulator) for   0.529 inWF needed
```

Commends on fan operation

```
Airway    Fan-type    Airflow (X1000)          Remark
 7-  1      F1            99.96         at 8.054934  inWG pressure
```

Active airways with natural pressure
Airway   Natural pressure (inWG)
7- 1   -0.35

```
                 ****  Airway table  ****
     Airway    Flow (cf/min X1000)    Head inches WG    Type
      1-  2          99.959               0.899
      2-  3          62.000               0.213
      3-  4          62.000               0.650
      2-  5          37.959               0.999     reg. at   0.386 inWG
      4-  5          12.000               0.136
      4-  6          50.000               0.792     reg. at   1.529 inWG
      5-  7          49.959               1.897
      6-  7          50.000               0.713
      7-  1          99.959               4.296
```

```
               ****  Airway characteristics  ****
   Airway    Act.   Section   Length    Perim.     Resistance
    1-  2     1       0.0       0.0       0.0      0.9000 E-10
    2-  3     1       0.0      360.0      0.0      0.5538 E-10
    3-  4     1       0.0       0.0       0.0      1.6900 E-10
    2-  5     3       0.0       0.0       0.0      4.2500 E-10
    4-  5     1       0.0       0.0       0.0      9.4500 E-10
    4-  6     3       0.0      1500.0     0.0      1.6226 E-10
    5-  7     1       0.0       0.0       0.0      7.6000 E-10
    6-  7     1       0.0       0.0       0.0      2.8500 E-10
    7-  1     1       0.0       0.0       0.0      4.3000 E-10
    6-  1     0       0.0       0.0       0.0      1.5000 E-10
```

```
              Free-split paths for fans
      Airway          Head      Airflow (X1000)
      7 - 1          8.441          99.959
```

Figure 12. Example of output file

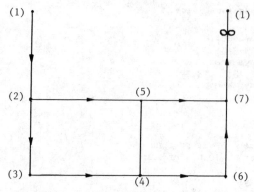

Figure 13. Example network

observed in the real network. Then one may
experiment by adding, deleting or
modifying airways and/or other network
parameters. The elimination of booster
fans is of particular interest in
underground coal mines where only main
fans are permitted. Elimination of the
booster fans will increase the overall
operating head of the main fans and a new
operating point will be calculated. By
using the appropriate fan curve that
includes that operating point and by
balancing the network again, slightly
different operating conditions may be
obtained. The procedure may be repeated
until satisfactory results are obtained.

5 REFERENCES

Boumis T.P., (1980), "Computer Solution of
Mine Ventilation Networks," Mineral
Resources, Vol. 1, Greece.
Hartman H.L., et al, (1982), "Mine
Ventilation and Air Conditioning," John
Wiley, New York.
Scott D.R. and Hinsley F.B., (1951),
"Ventilation Network theory," Colliery
Engineering.

13. Face ventilation I

Characterization of the performance of large capacity face ventilation systems for oil shale mining

C.E.BRECHTEL, M.E.ADAM & J.F.T.AGAPITO
J.F.T.Agapito & Associates Inc., Grand Junction, CO, USA

E.D.THIMONS
US Department of the Interior, Bureau of Mines, Pittsburgh, PA, USA

ABSTRACT: The performance of two large capacity face ventilation systems was measured using sulfur hexaflouride tracer gas tests in a large dead-end heading at the Colony shale oil mine. The test room was 17 m wide by 9 m high and 98 m long. Testing compared the effectiveness of a free-standing, jet fan and a reversible fan with rigid duct. The jet fan performance was similar to the ducted system while using less power. Tracer gas was used to simulate the production of mine air pollutants including blasting fumes, hot diesel emissions, free methane from surrounding strata, and methane desorbing from rubblized shale. The tests showed that both systems could provide effective ventilation during oil shale mining.

1 INTRODUCTION

The large openings required for room and pillar oil shale mines, coupled with the very large diesel powered equipment required for production are expected to create substantial ventilation problems at the working face. Projected rates of air pollutant production will require face ventilation air quantities that are as large or larger than those required for an entire panel in a typical coal mine. Conventional ventilation equipment is capable of supplying the face air flow rates, but the effects of room dimensions and large-scale turbulence on ventilation effectiveness are unknown.

These concerns motivated several oil shale mining companies, through the Colorado Mining Association, to sponsor a design and testing project in conjunction with the Bureau of Mines and the Department of Energy. The primary tasks of the project were to:

o Review current industrial practice to identify data pertinent to the design of face ventilation systems.

o Design and fabricate two test systems with the required air flow capacity.

o Measure the ventilation effectiveness in full size openings in an oil shale mine.

Review of ventilation literature indicated that there was ample design data on the operation of air moving systems, but little data to describe the motion of air or the pollutant dilution effectiveness of air in large rooms. Analytical techniques to describe air motion and turbulence, and their effect on the dilution of air pollutants on the room scale are currently not available in the mining industry. Ventilation network models determine average properties of the air stream, such as velocity, temperature, density, or pollutant concentrations, but cannot provide information about the distribution across the excavation. Advances in fluid flow modeling have produced computer codes capable of describing turbulent flow on the room scale, but application has been mainly to problems associated with the aerospace industries (Patankar and Spalding, 1972; Rodi, 1980).

Sulfur hexaflouride (SF_6) tracer gas provides a means of quantifying the average performance of ventilation systems at different locations in a test room. SF_6 has been used in underground ventilation studies by Thimons et al. (1974), Thimons and Kissell (1974) and Matta (1978). The gas is inert, does not occur naturally and is detectable in concentrations as low as one part per trillion (1×10^{-12}). The dilution of the tracer gas provides a direct measurement of the action of the face ventilation system. The low concentrations can be linearly ex-

trapolated to operational levels of pollutant production so that actual fresh air
requirements can be calculated. Different
types of pollutant production can be
simulated by altering the mode of tracer
gas release.

2 FACE VENTILATION SYSTEMS

Seven conceptual designs of face ventilation systems, based on current ventilation
practices, were generated. Comparison and
evaluation of the designs identified the
following two systems as being the best
approaches:

o <u>A Free-Standing Jet Fan</u>; consisting
  of a 1.4 m diameter fan (Spendrup AMF
  1400-70-12) with 2-speed, 75 kw
  motor, mounted on a scissors lift to
  allow elevation to 5 m above the
  floor.
o <u>A Reversible Fan with Rigid Duct</u>;
  consisting of a 1.4 m diameter, 2-
  stage fan (Spendrup AMF 2400-70-12)
  with two 93 kw motors connected to
  1.4 m diameter round steel duct.

Both systems were designed to deliver 47.2
m³/s which was based upon the projected
diesel horsepower during oil shale loading
operations. The jet fan system, shown
in Figure 1, was ranked very highly
because of its operational compatibility
and low capital and operating cost. It

Fig. 1. Photograph of the jet fan.

was considered unacceptable in gassy
mining conditions, because its operation
would cause some recirculation in the face
due to entrainment along the expanding
jet. The ducted fan system in Figure 2,
was selected for gassy oil shale mining.
Rigid duct was chosen because it would
allow comparison of the system's effectiveness in both blowing and exhausting
modes.

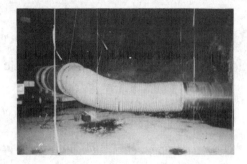

Fig. 2. Photograph of the ducted fan
system.

Operation of the two systems is compared
in Figure 3. For the jet fan, the turbulent jet of air issuing from the fan outlet forms the channel or duct to conduct
the air to the face. Near the fan, the
air jet is approximately the diameter of
the fan, but expands rapidly until it is
approximately one-half the opening. The
air reaches the face with a small portion
of its initial velocity, then sweeps the
face and returns down the opposite side of
the heading. The low friction loss allows
the jet fan to operate with lower power
consumption.

The ducted system is shown for both
blowing and exhaust modes. Positioning of
the fan inlet is a critical parameter in
the elimination of recirculation. In the
blowing mode, the fan inlet must be around
the corner and upstream in the last open
crosscut. Air is carried to the face by
the duct and sweeps it at high velocity.
This results in very good mixing of air
pollutants in the face area. The exhausting air flows back to the last open
crosscut along the heading. In the exhaust mode, fresh air travels up the heading to the face area and exhaust air flows
back to the last open crosscut through the
duct. Face air sweep velocities are low,
resulting in reduced mixing energy. The
area of capture is very small, and studies
by Luxner (1969) and Haney et al. (1982),
have shown that increasing the distance
between the face and duct inlet reduces
the ventilation effectiveness dramatically. Positioning of the outlet must be
such that the exhaust air cannot be recirculated. Reversible operation of this
particular system requires that the fan
and duct be located on the wall at the
upstream side of the last open crosscut.
In the exhaust mode, the exhaust air jet
would project the air downstream in the
last open crosscut thereby minimizing re-

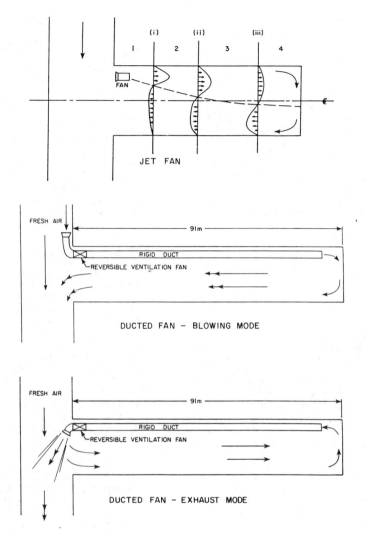

Fig. 3. Schematics of fan system operating in a dead heading.

circulation. If this is not possible, the exhaust would have to be ducted up to the roof to allow the outlet to be located downstream.

3 FIELD TEST SITE

Field tests were conducted at the Exxon Colony pilot mine near Parachute, Colorado. Figure 4 is a map of the mine. Crosscut 7 was used as the test room because it was the only part of the mine where ventilation air could be drawn past the room to simulate a last open crosscut.

The total mine ventilation air was 58.5 m³/s and was channeled past Cross-cut 7 by constructing the brattice wall shown in the figure. The channel width was made 4.6 m to generate flow past velocities of 1.4 m³/s, similar to what is projected for an operating oil shale mine. The room was closed at a distance of 98 m by constructing another brattice wall.

A grid system was constructed in the test room so that comparative measurements could be made for both fan systems. Figure 5 shows the layout of the grid system in both plan and side views. At each point, small pulleys were attached at the roof and floor with a 9 mm rope traveling through them. Orange plastic streamers were attached to the ropes at 1.5 m spacings to show the direction of air

519

motion. Air velocity, air direction and tracer gas sampling were restricted to the grid points.

Locations of the fan systems during testing are shown in Figure 5. Air velocity and air flow direction studies were used to select the optimum operational position for the tests. Tests of the jet fan at various heights above the floor showed that the fan's ability to project air to the face was reduced by increasing the height. The optimum position was as close to the bottom corner as possible. The ducted fan was placed around the corner for blowing mode tests. In the exhaust mode test, the ducted fan outlet passed through the brattice wall so that outlet air was dumped directly into the mine exhaust channel.

4 TRACER GAS SIMULATIONS

The tracer gas tests were designed to simulate different mechanisms of pollutant production. Oil shale loading operations at the face were expected to be the pollutant source that governed face air quantity requirements. The combined effect of carbon monoxide and nitrogen oxides was generally thought to be the governing factor; however, reports by Breslin et al. (1976) and Daniel (1984) suggest that diesel particulates could be the key design factor. Respirable oil shale dust has been found to contain between 10 and 13 percent quartz and is a potential problem during loading operations. Face blasting in oil shale mining requires the detonation of up to 900 kg of ANFO, which produces high levels of gaseous pollutants and dust in the face area. Methane is known to occur as both a free gas in solution with the ground water and in solid solution with the kerogen. Although not expected to be a problem in oil shale mines on the rim of Piceance Creek basin, mines in the central Piceance Creek and Uinta basins have been classified gassy during development mining operations.

Tracer tests, designed to simulate each of these potential mine air pollutants included:

o Simulation of hot diesel exhaust: This test was designed to simulate the systems' ability to dilute diesel emissions (gaseous and particulates). A 53,000 k-joule/hr kerosene space heater was placed in the face area with the exhaust routed through a vertical stack to dump 5 m above the floor. Tracer gas flowing at a constant rate was mixed in the hot gas

stream before the outlet. The space heater generated a stream of hot gases with a buoyancy similar to engine emissions. The mine ventilation and face ventilation systems were started and the steady-state concentration of SF_6 measured.

o Simulation of blast clearing: This test was designed to simulate the fans' effectiveness at clearing a heading after blasting. The test room was sealed, and SF_6 gas released to give a uniform concentration of approximately 1000 ppt (1000 x 10^{-12}). The fan was run for a short period of time to mix the gas uniformly. The mine ventilation system was then started, and the fans were used to clear the tracer gas from the room.

o Simulation of methane layering: SF_6 was mixed with 52.4 mole percent helium in air to simulate the density of methane gas. It was released from very small holes along a 15 m long pipe that was suspended at the roof. The pipe would simulate the intersection of a crack that was conducting methane gas into the mine at roof level. The tracer gas was released at a uniform rate for 45 to 60 min., and gas samples were taken to show if the tracer would form a roof layer similar to methane. The fans were then started to test their effectiveness at breaking up the layer.

o Simulation of methane emissions from a muckpile: In this test, the mixture of air, helium and SF_6 was released from a group of pipes laid out in the face area to simulate methane desorbing from a freshly blasted muckpile. The tracer gas was released for 45 to 60 min., and then the fans were started. The steady-state concentration was measured to establish the effectiveness of the two systems.

In addition to these tests, the inlet recirculation of each fan was measured using the SF_6.

The locations of tracer gas release and the tracer gas sampling points are illustrated in Figure 6 for each type of test. Automatic, programmable syringe samplers were used to collect tracer gas samples. Each sampler was loaded with nine - 25 cc syringes. Both the time to begin sampling and the time interval over which the sample was drawn into the syringe could be programmed. Samplers were suspended at different locations with respect to the roof to provide a measurement of the variations in SF_6 concentration in the verti-

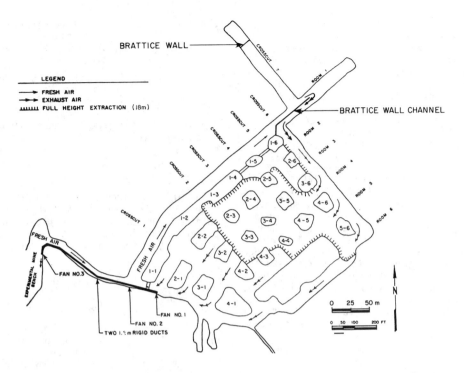

Fig. 4. Map of the Colony pilot oil shale mine showing the location of the test room.

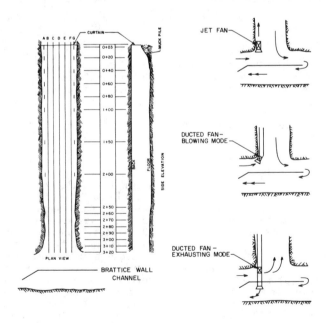

Fig. 5. Test room grid system showing the location of the fans during the tracer gas tests.

cal as well as horizontal plane. Tracer gas release was controlled by electronic mass flow meters that were calibrated for the mine air density (1.009 kg/m^3).

This study employed two types of SF$_6$ release:

o Steady-state, continuous release of SF$_6$.
o Release of a fixed quantity of SF$_6$ and mixing that quantity uniformly throughout the test room.

Analysis of both types of test was designed to yield a uniform measure of the ventilation effectiveness, which is designated as the dilution efficiency (E_D). Dilution efficiency is defined as the ratio of the actual quantity of air that is diluting the tracer gas divided by the fan outlet flow rate. A value of 1.00 indicates that all of the air flowing through the fan is being perfectly mixed with the tracer gas. This type of efficiency index was used previously by Haney et al. (1982) and called the face ventilation index. In a steady-state type of release, equation 1 would be used to determine the dilution efficiency.

$$E_D = C_{ideal}/\overline{C}_{ss} \qquad (1)$$

where: E_D = Dilution efficiency

$\overline{C}_{ss}$ = Time weighted average concentration of SF$_6$

C_{ideal} = Ideal concentration of SF$_6$ assuming perfect mixing with the fan outlet flow rate

The ideal concentration is calculated using equation (2):

$$C_{ideal} = q_{SF_6}/Q_{fan} \qquad (2)$$

where: q_{SF_6} = Flow rate of SF$_6$
Q_{fan} = Fan outlet flow rate

For dilution of uniformly mixed tracer gas in a room of volume (V), Matta et al. (1978) used equation (3) to determine the effective flow rate of air.

$$Q_E = V \frac{(\ln C_o - \ln C)}{t_2 - t_1} \qquad (3)$$

where: Q_E = Effective air flow rate
V_E = Room volume
C_o = Initial SF$_6$ concentration at time t_1

C = Concentration of SF$_6$ at time $t_2 < t_1$

t_2, t_1 = time

The dilution efficiency is then calculated by equation (4):

$$E_D = Q_E/Q_{Fan} \qquad (4)$$

Fan inlet recirculation was measured by releasing the tracer gas at a constant rate at the fan inlet and measuring the concentration in the outlet air. Equation 5 is used to determine the recirculated flow.

$$Q_R = \frac{Q_{Fan} \, C_{OT} - q_{SF_6}}{C_R} \qquad (5)$$

where: Q_R = Flow rate of air being recirculated
Q_{Fan} = Fan outlet flow rate
C_{OT} = Concentration of SF$_6$ in the outlet
q_{SF_6} = Flow rate of SF$_6$ injected into the fan inlet
C_R = Average concentration of SF$_6$ in air around the fan inlet

The recirculated flow rate divided by the fan outlet flow rate is the proportion of air being recirculated at the inlet.

5 TRACER GAS RESULTS

5.1 Blast clearing test

In the blast clearing tests, the test room was closed by sealing the brattice wall channel. Gas containing 101 ppm (101 x 10^{-6}) SF$_6$ was released at a rate 5 l/m for 30 minutes. In the blowing mode tests, the fan was then used to mix the air throughout the room, resulting in an average concentration of 947 parts per trillion (ppt). The samplers were allowed to take a background sample, then the brattice wall channel was opened and the mine fans were activated. Figure 7 shows the time-concentration data for the nine samplers located Section 40 (12 m from the face) during the jet fan testing at a flow rate of 41.7 m^3/s. The curves

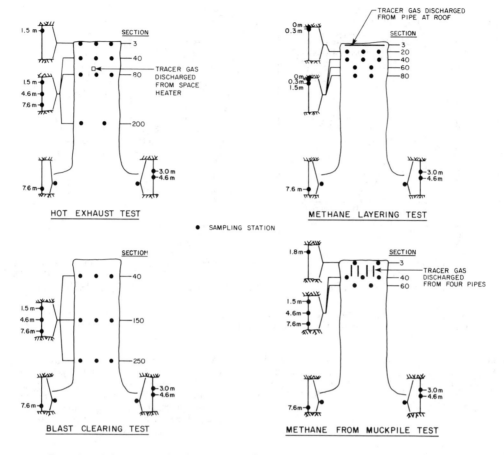

HOT EXHAUST TEST

METHANE LAYERING TEST

● SAMPLING STATION

BLAST CLEARING TEST

METHANE FROM MUCKPILE TEST

Fig. 6. Schematic of tracer gas release points and sampler locations.

are typical of all the blast clearing tests. Dilution efficiencies for the different tests are compared in Table 1.

Table 1. Comparison of dilution efficiencies from the blast clearing tests

Test	Fan Flow rate (m^3/s)	Dilution Efficiency at section			Average
		40	50	250	
Jet fan	41.7	0.77	0.65	0.83	0.75
Jet fan	28.3	1.00	1.00	1.00	1.00
Ducted fan-blowing	42.8	1.00	0.94	0.97	0.98
Ducted fan-exhausting	34.5	0.71	0.78	0.88	0.79

The dilution efficiencies in all of the tests were high and showed that the sys-

tems delivered between 71 and 100 percent of the fan outlet volume to the immediate face area. The ducted system achieved the highest efficiency in the blowing mode with a 19 percent reduction of dilution efficiency in the exhaust mode. The jet fan system was effective at delivering 77 percent of its outlet flow at a distance of 98 m from the fan outlet. The efficiency increased when the outlet flow rate was reduced to 28.3 m^3/s. In fact, the tracer gas measurements indicate that it delivered as much air to the face at the lower flow rate as it did at the higher flow rate. This suggests some interaction between the jet stream and room dimensions that is not well understood. Similar results were reported by Volkwein (1978), in measurements of jet fan effectiveness at Union's Parachute Creek shale oil project.

The distance that air will circulate into the test room as a result of the last

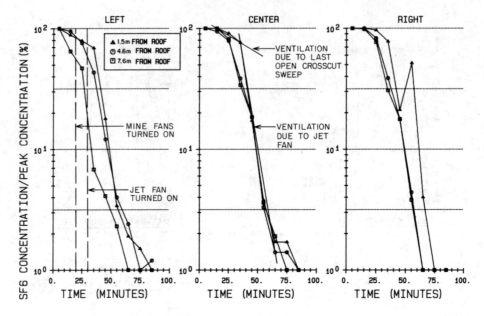

Fig. 7. Concentration of SF$_6$ versus time in the blast clearing test at section 40 (12 m from the face) - jet fan.

open crosscut flow was measured during these tests. Air flow was detected as far as Section 150 with flow rates of 8.3 m^3/s and 5.3 m^3/s measured at Sections 250 and 150, respectively.

5.2 Diesel exhaust simulation

Diesel exhaust emissions were simulated by releasing a steady stream of release gas containing 101 ppm SF$_6$ into a kerosene space heater connected to a 4.6 m vertical stack. The release gas was injected at a rate of 5.44 l/min. The exhaust stream temperature was 148.9°C, well below the temperature at which SF$_6$ begins to break up.

The mine fans and face ventilation systems were in steady operation for 20 minutes before the tracer gas release began. Figure 8 shows the time concentration curves for samples at Section 40 during the ducted fan test in the blowing mode. The SF$_6$ concentration built up to a steady-state value within 40 minutes. Time-weighted average concentrations were calculated for the last three sample points and used to determine the dilution efficiencies in Table 2.

The space heater was used to increase the buoyancy of the exhaust stream similar to diesel emissions. This produced a small stratification of the tracer gas,

with the SF$_6$ concentration averaging 7.5 percent higher at 1.5 m from the roof than at 7.6 m from the roof. This was significantly less than stratifications measured during actual loading operations, where concentrations were between 36 and 72 percent higher near the roof.

In this test, the jet fan was marginally more effective than the ducted fan in the blowing mode with an average of 7 percent higher dilution efficiency at the face.

5.3 Methane layering simulation

Methane layering occurs because methane bleeding from fissures in the roof is less dense than air. The density contrast causes most of the methane to float near the roof and makes dilution of the methane more difficult. In order to correctly simulate this phenomenon and make a realistic measurement of the fan systems' capability to break up a methane layer, a release gas mixture of 52.4 mole percent helium in air with 1.09 ppm SF$_6$ was employed.

The gas was released from a 15.0 m long pipe suspended at the roof at section 3. Small holes were drilled in the pipe at 30 cm spacings, so that the gas would be released uniformly along the roof line. Gas was released at a rate of 23.6 l/min. for 120 minutes to develop the layer. The gas

524

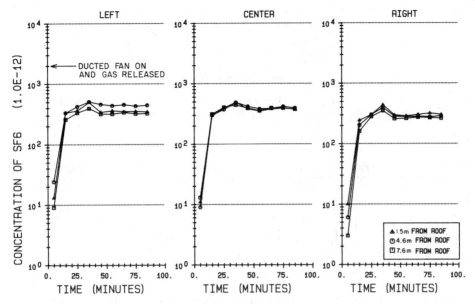

Fig. 8. Concentration of SF_6 versus time in the hot exhaust test at cross section 40 (12 m from the face) - ducted fan blowing.

Table 2. Comparison of dilution efficiencies from the hot exhaust simulations

Section	Distance from roof m	Dilution Efficiency	
		Jet fan @ 41.7 m^3/s	Ducted fan blowing 42.8 m^3/s
0+03	1.8	0.76	0.97
0+40	1.5	0.69	0.63
	4.6	0.70	0.59
	7.6	0.71	0.67
0+80	1.5	0.76	0.74
	4.6	0.83	0.75
	7.6	0.83	0.77
Overall Average		0.78	0.74

was then left flowing and the fan system started.

Time-weighted average concentrations were calculated during the 50 minute build-up period to evaluate the extent of the layer. Figure 9, shows the layering developed in the test of the ducted fan. After 40 minutes of tracer gas release, the fans were started and the time to

break up the layer and steady-state concentration determined. Time-concentration curves for section 20 during the jet fan test are presented in Figure 10. Table 3 lists the dilution efficiencies. The ducted fan system in the blowing mode showed very high efficiency in this test because it delivered its full flow of air directly at the tracer gas release point.

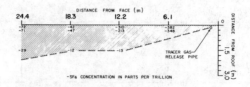

Fig. 9. Schematic illustrating degree of SF$_6$ layer produced during the methane layering test - ducted fan blowing.

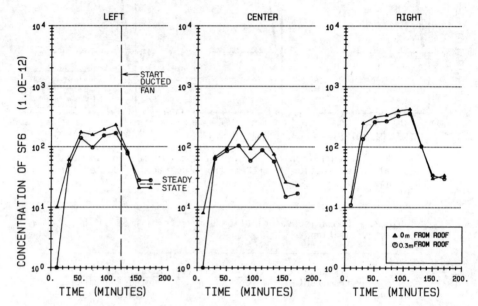

Fig. 10. Concentration of SF$_6$ versus time in the methane layering test at cross section 20 (6 m from the face) - jet fan.

Table 3. Comparison dilution efficiencies for the methane layering tests

Section	Distance from roof m	Dilution Efficiency	
		Jet fan @ 41.7 m^3/s	Ducted fan blowing 42.8 m^3/s
0+20	0	0.40	1.00
	0.3	0.41	1.00
0+40	0	0.57	0.77
	0.3	0.54	0.77
	1.5	0.64	0.83
Overall Average		0.59	0.83

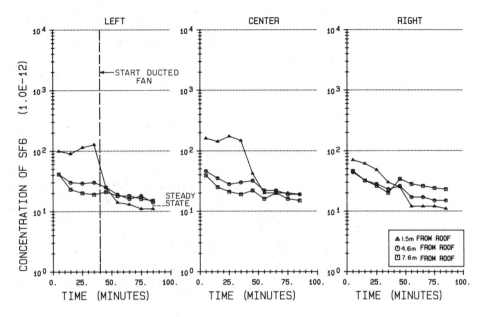

Fig. 11. Concentration of SF_6 versus time in the methane from muckpile test at section 40 (12 m from the face) ducted fan blowing.

Table 4. Comparison of dilution efficiencies for the methane from muckpile tests

Section	Distance from roof m	Dilution Efficiency	
		Jet fan @ 41.7 m³/s	Ducted fan blowing 42.8 m³/s
0+03	1.5	0.45	0.67
0+40	1.5	0.74	0.71
	4.6	0.79	0.59
	7.6	0.69	0.52
Overall average		0.79	0.60

If the gas release point was moved, the effectiveness would probably drop to near the level of the jet fan. Both systems broke up the layering effectively in a period of 40 minutes.

5.4 Methane from muckpile simulation

The 52.4 mole percent helium in air was used to simulate methane desorbing from a freshly blasted muckpile. The gas containing 1.09 ppm SF_6 was released at a rate of 23.59 l/min. for a period of 60 minutes to build a tracer gas cloud in the face. The gas was released from four pipes laid out in the configuration of a muckpile. Small holes were drilled in the pipes at 30 cm spacings to allow uniform release over the face area. The fans were started after 60 minutes and the steady-state concentration measured.

Time concentration data for section 40 in the ducted fan test are shown in Figure 11. The curves show the buildup of SF_6 and the fact that the tracer gas is being transported up through the still air

527

because of the lower density of the gas mixture. The SF_6 concentration 7.6 m from the roof averaged 20 percent of the concentration at 1.5 m from the roof.

The fans were then started and the SF_6 concentration quickly reduced to steady-state values. Dilution efficiencies calculated from the steady-state values are compared in Table 4. The jet fan had an average of 19 percent higher dilution efficiency in this test.

5.5 Fan inlet recirculation measurements

Tracer gas measurements of fan inlet recirculation were made for both systems in the blowing mode. Recirculation of the fans was 24 and 28 percent of the outlet flow for the jet fan and ducted fan, respectively. This was surprisingly high for the ducted fan, and was due to the location of the fan and the restriction of flow area by the brattice wall channel.

Elimination of the recirculation would improve the dilution efficiencies of the ducted fan to a range of 0.81 to 1.0.

6 DISCUSSION

The tracer tests indicated that both of the systems are capable of providing effective face ventilation during oil shale mining operations. The jet fan gave similar levels of ventilation effectiveness at lower power consumption than the ducted system. This is because the jet fan does not use ducting, and thus, does not incur appreciable friction losses. Properly used, the jet fan has many advantages over the ducted system for oil shale mining, including:

o low capital and operating cost
o ease of movement and positioning
o higher air velocities throughout the heading

Present MSHA regulations prohibit the use of the jet fan in gassy situations because of recirculation. Inlet recirculation could be eliminated by attaching ducting to the inlet and drawing air into the fan from upstream in the last open crosscut. Studies of controlled recirculation reported by Leach and Slack (1969) indicate that recirculation by entrainment along the jet will not increase the methane concentration as long as the inlet fresh air quantity remains constant.

The results of the fan tests are highly dependent upon fan inlet location, recirculation, room dimensions, fan outlet location and air flow in the last open

crosscut. These tests sought to characterize the fans in what was considered to be good simulations of normal operation. The tests were successful in that they showed the fans were able to provide effective ventilation under field conditions.

7 CONCLUSIONS

Major conclusions resulting from this work are:
o SF_6 tracer gas testing is an effective method of characterizing face ventilation performance.
o The results of the tests performed in this study are dependent upon recirculation, fan inlet and outlet location, pollution source, room dimensions, and last open crosscut flow.
o Recirculation reduced the dilution efficiencies measured in this work by a range of 17 to 27 percent.
o Both systems showed high dilution efficiencies and were effective in ventilating the face at a distance of 98 meters.
o The jet fan delivered similar performance at with less power consumption than the ducted system. It was obviously advantageous for application to oil shale mining because of its maneuverability and simpler operation.

REFERENCES

Breslin, J.A., A. J. Strazisur, and R. L. Stein, 1976. Size distribution and mass output of particulates from diesel engine exhausts. BuMines RI 8141(r).
Daniel, J.H., 1984. Diesels in underground mining - a review and an evaluation of an air quality monitoring methodology. BuMines RI 8884.
Haney, R.A., S.J. Gigliotti and J.L. Banfield, 1982. Face ventilation systems performance in low-height coal seams. Proceedings of the 1st Mine Ventilation Symposium, Reno.
Leach, S.J. & A. Slack, 1969. Recirculation of mine ventilation systems. Colloquium on firedamp measurement and control. Institute of Mining Engineers. Isleworth.
Luxner, J.V., 1969. Face ventilation in underground bituminous coal mines. BuMines RI 7223.

Matta, J.E., E.D. Thimons and F.N. Kissell, 1978. Jet fan effectiveness as measured with SF_6 tracer gas. BuMines RI 8310.

Patankar, S. V. and D. B. Spalding, 1971. A calculational procedure for heat, mass and momentum transfer in three-dimensional parabolic flow, Int. J. Text Mass Trans., 15.

Rodi, W., 1980. Turbulence models and their application in hydraulics, State-of-the-Art Paper, IAHR.

Thimons, E.D. and F.N. Kissell, 1974. Tracer gas as an aid in mine ventilation analysis. BuMines RI 7917.

Thimons, E.D., R.J. Bielicki and F.N. Kissell, 1974. Using sulfur hexaflouride as a gaseous tracer to study ventilation systems in mines. BuMines RI 7916.

Volkwein, J.C., 1978. Letter to Mr. L. Pyeatt, Union Oil Company of California, Parachute Creek Shale Oil Project, Parachute, Colorado. SF_6 measurements of jet fan effectiveness.

The resistance to airflow on a longwall face

MALCOLM J.McPHERSON
University of California, Berkeley, USA

ABSTRACT: This paper describes one of a series of tests that were carried out to characterize the behavior of ventilation in a longwall district of a coal mine in the Western United States. A detailed survey was carried out to measure the variations of airflow and frictional pressure drop along the longwall and at the face ends. The corresponding values of resistance were determined and found to be concentrated at the face ends. The measured data were utilized to evaluate a range of friction factors for mechanized longwalls. These enable the resistance of a faceline to be determined. A separate procedure is given for estimating the additional resistance at face ends and across power loaders. The method of determining the total resistance of a longwall face is illustrated by a worked example.

1 INTRODUCTION

During the summer months of 1982 the Thompson No.1 Mine, owned and operated by the Snowmass Coal Company, Carbondale, Colorado, was placed on standbye except for development and essential maintenance. The mine included a longwall face, fully equipped with an Anderson–Boyes 500hp radio-controlled chainless shearer. A Dowty-Meco 150hp armored flexible chain conveyor and 350 ton Hemscheidt Troika shield supports. The coal seam was 7ft (2.1m) thick and dipped at 30°. The face operated on full dip to give level airways. Ventilation was ascentional and antitropal.

It is difficult to conduct detailed ventilation surveys on an active longwall. The steady-state conditions, freedom of access, and utilization of sensitive and delicate instrumentation that are necessary for accurate ventilation measurements are neither available nor welcome on an operating face. During the standby period, the company offered to make the mine available for field studies funded by the U.S. Department of Energy. This provided a rare opportunity to conduct detailed ventilation measurements on a longwall face.

Friction factors and values of resistance have been well tabulated in ventilation literature for shafts and airways. There is little comparable data for fully mechanized longwall faces. Four major tests on ventilation were made at Thompson No. 1. This paper describes one of them, and is intended to provide the data that enables the total resistance of a longwall face to be assessed.

2 EXPERIMENTAL PROCEDURE

The district operated on a twin entry system. As the face retreated, the inner airways were allowed to collapse while the outer airways were maintained as bleeder entries. This system provided a positive ventilation pressure across the gob and kept the faceline free from emissions of methane from the waste.

Over the lower half of the face length at Snowmass, the armored flexible conveyor was pushed close to the coal face and the conveyor jacks fully retracted. The distance from the coal face to the front legs of the chocks varied between 7.5 and 10 ft. (2.3 and 3m) Face conditions were good. About half way along the face, however, resin grout injection had been used to stabilize the ground around a small fault in the seam. Above that point, the face had become wider due to sloughing from the coal front. The uneven surface left on the face combined with the broken coal to give an aerodynamically rougher flowpath. The shearer was parked at the upper end of the face (tailgate) where the width of the face had increased to 14 ft.(4.3m).

The test commenced by erecting a brattice in the lower bleeder airway and opening the regulating brattice in the maingate leading to the face. These steps were taken to produce a relatively high airflow along the face line and, hence, to improve the accuracy of airflow and pressure drop measurements. More than 50 per cent of the air supplied to the panel still entered the caved area from the lower bleeder airways. Nevertheless, as a safety precaution, one of the team patrolled the cross-cuts at the top edge of the caved waste throughout the day, monitoring methane concentrations. Most of these cross-cuts showed 0.1 or 0.2 per cent methane. Higher concentrations were found towards the starting-off line of the face. There was no significant increase in gas concentration in any one cross-cut during the day.

Stations A1 through A10 were set up at

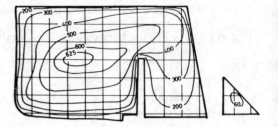

Figure 2. Velocity contours at one of the airflow measuring stations. (feet per minute)

locations indicated on Figure 1. The face was divided into 58 ft.(17.7m) lengths with shorter increments where the face widened and across the shearer. Additional lengths of 70 ft.(21.3m) in the conveyor road and 58 ft. in the upper haulage road were included in order to determine the resistance of these airways close to the face ends.

Airflows were measured at midpoints between stations by the "point traverse" method. This involved taking anemometer readings at a number of points on a grid covering the cross-section of the measuring station. The readings enabled velocity contours to be established in a manner similar to those illustrated in Figure 2. These, in turn, were used to determine the volume flow of air.

The frictional pressure drops between stations were determined by the gauge and tube technique. A length of 1/8 inch (3mm) internal diameter plastic tubing was laid out between each pair of consecutive stations in turn. The ends of the tubing were each attached to the total head connections of a 2 ft pitot tube. A 0 to 0.25 inch w.g. (0 to 60Pa) magnehelic diaphragm gauge was connected

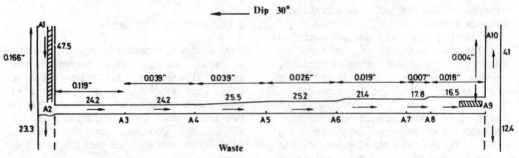

Figure 1. Pressure drops (inches w.g.) and airflows (kcfm) measured on the face.

into the line in order to measure the pressure difference between the two pitot tubes.

The pressure tubing was suspended from roof supports and tested for pressure integrity at each setting. The pitot tubes were held facing into the airflow. Care was taken to prevent the bodies of the observers from affecting the readings, a precaution that is particularly important in the cramped confines of a longwall face. The observers holding the pitot tubes lay flat on their backs on the face conveyor during each reading and traversed the pitot tubes slowly over the central part of the cross-section, but maintaining the orientation of the pitot head into the airflow. This technique avoided spurious readings arising from any anomolous local turbulence.

For each reading, the magnehelic gauge observer sited himself between a pair of chocks, well out of the main airstream – again in order not to affect the airflow or associated pressure drop. The pressure gauge was observed until a stable reading was established. All anemometer and magnehelic readings were corrected according to the relevant calibration.

3 TEST RESULTS

The measured pressure drops and corresponding airflows are shown in Table 1. Pressure drops are given in thousandths of an inch of water gauge (milli inch w.g.) and airflows in thousands of cubic feet per minute (kcfm). The corresponding resistances, calculated from the square law have units of

$$R = \frac{p}{Q^2} \quad \frac{milli\ in.\ w.g.}{(kcfm)^2} \quad (1)$$

$$or \quad \frac{(in.w.g.) \times 10^{-3}}{(ft^3/min)^2 \times 10^{+6}}$$

$$= \frac{(in.w.g.)}{ft^6} \quad min^2 \times 10^{-9}$$

This is sometimes called the Practical Unit (P.U.) of mine resistance as it derives directly from the square law without the need for any multiplying constant. For conversion to S.I.units, 1 P.U. = 1.1183 Ns^2/m^8.

In order to compare resistances for the various sections of the face, Table 1 also gives the resistances in P.U. per foot length. These values are further illustrated on Figure 3. The peaks of resistance at the face ends show very clearly. These arise from the concentration of equipment causing obstruction to the airflow, and from the shock loss as the airflow changed direction at the junctions of airway and face.

The larger cross-sectional area through the upper half of the face is revealed as a lower resistance per unit length. The slight shock loss as the face widens in section A7–A8 shows as a small increase in resistance. There is a further rise due to the presence of the shearer at the upper end of the face, but this is not so great as would normally be expected from such a major

Table 1. Resistances on the longwall face

Station Length From To		ft	Pr. Dp. m.in.w.g.	Airflow kcfm	Resistance P.U.	Resistance P.U./ft	Comments
A1	A2	70	166	47.523	0.0735	0.00105	Stageloader & Electrics
A2	A3	58	119	24.181	0.2035	0.00351	Maingate end of face
A3	A4	58	39	24.230	0.0664	0.00114	
A4	A5	58	39	25.536	0.0598	0.00103	
A5	A6	58	26	25.170	0.0410	0.00071	
A6	A7	58	19	21.415	0.0414	0.00071	
A7	A8	20	7	17.797	0.0221	0.00111	Face widens
A8	A9	45	18	16.452	0.0665	0.00148	Shearer
A9	A10	58	4	4.120	0.2356	0.00406	Tailgate

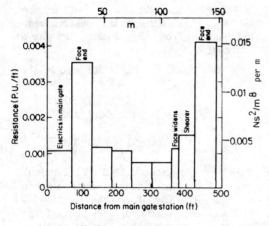

Figure 3. Variation of resistance (per unit length) throughout the longwall face.

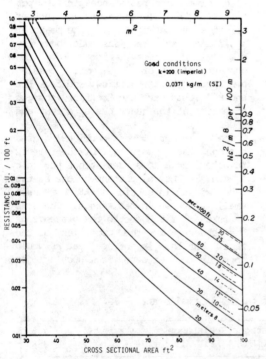

Figure 4. Nomogram for faceline resistance (good conditions on the face)

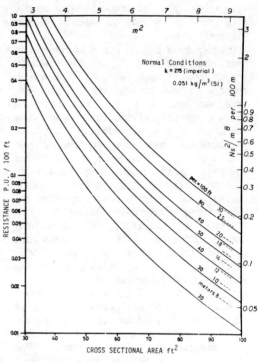

Figure 5. Nomogram for faceline resistance (average conditions on the face)

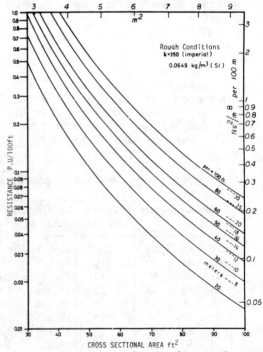

Figure 6. Nomogram for faceline resistance (rough conditions on the face)

obstruction. The reason for this was that the face was much wider and the cross-sectional area considerably larger at the shearer than along the rest of the face.

The cumulative increase in resistance from a point 70 ft. (21.3m) outbye the maingate end of the face to a point 58 ft. (17.7m) outbye the tailgate end was 0.81 P.U. (0.91 Ns^2/m^8). A summary of the breakdown is given in Table 2.

Table 2. Summary of face resistances

	Resistance		Percentage of total
	P.U.	Ns^2/m^8	
Stage loader/ electrics	0.0735	0.0822	9.1
Face ends and tailgate	0.4391	0.4910	54.2
Face line	0.2972	0.3324	36.7
Total	0.8091	0.9056	100.0

4 ESTIMATION OF THE RESISTANCE OF A LONGWALL FACE

The results given in the previous section show that more than half the total resistance occurred at the face ends. It is clear that previous practices of estimating the resistance of a longwall face simply as a function of length and height could give misleading results. Shock losses must be assessed separately and the appropriate resistance combined with the face line resistance.

In this section of the paper a more detailed analysis of the Snowmass face resistance is undertaken and a method is proposed for the estimation of the resistance of other longwall faces. This involves asessing the resistance of (i) the unobstructed face line (ii) face ends and (iii) shearer or other power-loader.

4.1 Face line friction factors

The four 58ft (17.7m) sections between stations A3 and A7 (Figure 1)

represented an unobstructed face line bounded by the shield supports, the armored conveyor and the coal face. The upper half of this 232 ft (70.7m) length had a greater cross-sectional area but rougher conditions due to sloughing of the coal face.

The resistance of any airway may be expressed in Imperial Units as

$$R = \frac{k}{5.2} \frac{L}{A^3} \frac{O}{} \quad \frac{lbf \, min^2}{ft^4} \frac{ft \, ft \, ft^2 (in.w.g.)}{ft^6 \quad lbf}$$
(2)

where k = friction factor ($lbf.min^2/ft^4$)
L = length (ft)
O = Perimeter (ft)
A = cross-sectional area (ft^2)

In this system of units, numerical values of the friction factor are very small and lists of k values given in the literature are normally multiplied by the factor 10^{10}. Following this conversion, the units of R become

$$\frac{(in.w.g.) \, min^2}{ft^6} \times 10^{-10}$$

It will be noted that this differs by a factor of 10 from the resistances (P.U.) calculated from the observed airflows and pressure drops (equ.1).

In order to calculate k factors from the measured resistances (P.U.), the relationship becomes

$$k = \frac{R \times 5.2 \times A^3}{L \times O} \times 10 \quad \frac{lbf.min^2}{ft^4} \times 10^{10}$$
(3)

Table 3 gives the k values calculated from equation 3 and also in SI units (kg/m^3), where

$$k = \frac{(lbf.min^2)}{ft^4} \times \frac{4.4482 \times 60^2}{(0.3048)^4} = \frac{Ns^2}{m^4} = \frac{kg}{m^3}$$

Hence,
k (Imperial Units $\times 1.855 \times 10^6$) = k (SI)

Two points emerge from the k factors calculated for the longwall face. First, the values are considerably higher than those listed in the literature for shafts, airways or slopes. This is a direct and inevitable consequence of the aerodynamic drag

535

Table 3 Calculation of friction factors for four consecutive lengths of face.

Section	Length ft	Q kcfm	p milli. in.w.g.	A ft^2	O ft	R P.U.	k factors lbf.min^{2}x10^{10} ft^4	kg m^3
A3–A4	58	24.230	39	46.24	31.5	0.0664	187	0.0347
A4–A5	58	25.536	39	57.41	37.0	0.0598	274	0.0509
A5–A6	58	25.170	26	61.67	36.0	0.0410	212	0.0394
A6–A7	58	21.415	19	72.64	41.0	0.0414	347	0.0644

caused by the powered supports, added to the high d/D values of the roof beams, conveyor structure and chocks (where d is the distance protruded into the airway and D is the hydraulic mean diameter of the face). The rubbing surfaces along a mechanized longwall face are aerodynamically much "rougher" than other airways underground.

Second, the measured k factors show a significantly higher value in the upper section of the face. This is entirely in accord with the deteriorating appearance of the face in this section.

4.2 Prediction of face-line resistance:

The longwall face at Snowmass varied in cross-sectional area, perimeter and surface roughness. This variability has proved to be advantageous to this project as it has given ranges of friction factors and resistances that may be used to estimate face-line resistances of other longwalls.

Three friction factors have been chosen, to represent (i) good, (ii) normal and (iii) rough conditions for a longwall face equipped with power supports. The values are given in Table 4.

Table 4. k values for a mechanized longwall face.

lbf.min^{2}x10^{10} ft^4	kg m^3	Condition
200	0.0371	Good
275	0.0510	Normal
350	0.0649	Rough

The resistance of a face-line may be estimated by choosing a k value from Table 4 and using one of the following two equations:-

Imperial units:

$$R = \frac{k \ L \ O}{52 \ A^3} \qquad P.U. = \frac{milli.in.w.g.}{(kcfm)^2} \qquad (4)$$

where k is the value read directly from Table 4 in lbf.min^{2}x10^{10}/ft^4

L = length (ft)
O = perimeter (ft)
A = cross-sectional area (ft^2)

S.I. Units: In the more rational SI system, no constants are necessary and the equation is simply

$$R = \frac{k \ L \ O}{A^3} \qquad Ns^2/m^8 \qquad (4a)$$

where k is read from Table 4 (kg/m^3)
L = length (m)
O = perimeter (m)
and A = cross-sectional area (m2)

In these calculations, A should be taken as the full cross sectional area available for airflow including open flow paths between chock legs. In the majority of modern installations of power supports, there will be relatively little flow between the chock legs. The perimeter, O, may be taken as a line that traverses the coal face, the conveyor structure (including spill plates), the front of the chocks and the underside of the roof beams.

Example

A mechanized longwall is 350 ft. long, the cross-sectional area is 60 ft^2 and the perimeter is 40 ft. Using an average condition k factor of 275 (Imperial Units), the face line resistance is:

$$R_f = \frac{275 \times 350 \times 40}{52 \; (60)^3} = 0.343 \; \text{P.U.}$$

At a mean airflow of 20 kcfm (20,000 cfm) this will give a pressure-drop along the face-line of

$$p = R_f Q^2 = 0.343 \times 20^2 = 137 \; \text{milli in.wg}$$

$$\text{or} \quad 0.137 \; \text{inches wg}$$

4.3 Nomograms for rapid estimation of face-line resistance:

In order to provide a rapid means of estimating face-line resistances three nomograms have been constructed to represent good, normal and rough conditions. These are reproduced on Figures 4, 5 and 6 and give the face line resistance per 100 ft (or 100m) for any given cross-sectional area and perimeter.

4.4 Face Ends:

The ends of longwall faces differ considerably in the resistance they offer to airflow. This is due to variations in equipment, geometry and roof supports. The shock losses that occur when an airstream is required to change direction are often described in terms of 'velocity heads':

$$P_{sh} = \frac{X \rho u^2}{2g} \qquad \frac{\text{lbf}}{\text{ft}^3} \frac{\text{ft}^2}{\text{s}^2} \frac{\text{s}^2}{\text{ft}} = \frac{\text{lbf}}{\text{ft}^2}$$

$$\text{or} \quad \frac{X \rho u^2 \times 1000}{5.2 \times 2g} \; \text{milli inches wg}$$

$$\qquad\qquad\qquad\qquad\qquad (5)$$

where
X = Shock loss factor
ρ = air density (lbf/ft^3)
u = air velocity (ft/s)

and
g = gravitational acceleration (32.2ft/s^2)

Rewriting the equation in terms of a volume flow, Q kilo ft^3/min, where

$$Q = u \times 60 \times A/1000 \qquad (6)$$

$$P_{sh} = \frac{X \rho}{5.2 \times 2g} \quad \frac{Q^2 \times 10^9}{60^2 A^2}$$

Taking a standard value of 0.075 lbf/ft^3 for air density gives

$$P_{sh} = \frac{62.21 \times Q^2}{A^2} \qquad (7)$$

This may be written in the form of a square law as

$$P_{sh} = R_{sh} \; Q^2 \qquad (8)$$

where the shock resistance is

$$R_{sh} = \frac{62.21 \times X}{A^2} \quad \text{P.U.} \quad (9)$$

In the SI system, the derivation of the R_{sh}, X relationship is much simpler.

$$P_{sh} = X u^2 \rho/2 \quad \text{N/m}^2 \; (\text{or Pa})$$

where
ρ = air density (kg/m^3)
and u = air velocity (m/s)

Taking standard density to be 1.2 kg/m^3,

$$R_{sh} = \frac{0.6 \; X}{A^2} \qquad \text{Ns}^2/\text{m}^8$$

$$\qquad\qquad\qquad\qquad\qquad (10)$$

In all these analyses, A has been taken to be the cross-sectional area of the unobstructed face.

It then remains to determine the shock loss factor for the often constricted conditions at a face end. The literature pertaining to both mine ventilation and the heating and ventilating industry contains lists of loss factors for many types of duct configurations. At the junction of an airway with a longwall face, there are at least three identifiable causes of shock loss; namely, (i) the sudden change in flow direction as the airflow turns into the face-line, (ii) the cross-sectional area of the face will often be different to that of the airway, particularly at the face end -

hence, there will be a shock loss due to the sudden contraction, and (iii) there will normally be a concentration of equipment at the face end resulting in a further obstruction loss. Each of these matters is further examined below with suggested means of estimating the corresponding shock loss factors.

Bend:

For a sharp 90° bend, the shock loss factor, X, is approximately 1.4. This drops rapidly as the radius of curvature increases. However, in the case of a junction between an airway and a longwall face, the turn will invariably be sharp and, in the great majority of cases very close to 90°. Hence, the value of X_b = 1.4 for the bend will be used for all such junctions.

Contraction:

At the sudden contraction as the air enters the faceline, the recommended shock loss factor is

$$X_c = \left(\frac{1}{C_c} - 1\right)^2 \qquad (11)$$

where the coefficient of contraction, C_c, depends upon the geometry of the face end. However, the effect of the contraction is fairly small compared with that of the bend and obstructions from equipment. A mean value of 0.7 for C_c is suggested, giving

$$X_c = \left(\frac{1}{0.7} - 1\right)^2 = 0.184 \qquad (12)$$

Obstruction:

In many cases, the entry to the faceline will be heavily obstructed by the face conveyor gearhead and transfer point. The shock loss is dependent upon the cross-sectional area of the face and how much of this is filled with equipment. Hence, in this case, there is no single value of shock loss factor that is generally appliable. The "obstruction formula" that may be employed is:

$$X_{ob} = \left(\frac{A}{C_c (A-a)} - 1\right)^2 \qquad (13)$$

where A = the full cross-sectional area of the face (ft² or m²)
    a = the cross-sectional area that is obstructed (ft² or m²)
    C_c = coefficient of contraction (take as 0.7)

Total shock loss factor at face-end:

The three shock loss factors X_b, X_c and X_{ob} are not independent. Shock losses may interfere one with another and it is possible that the integrated effect may be less than the sum of the individual components. However, a simple addition of the three shock loss factors gives a result which is in sensible agreement with the measurements made at Snowmass. It is suggested, therefore that the overall face-end shock loss is computed as

$$X_{fe} = X_b + X_c + X_{ob} \qquad (14)$$

4.5 Correlation with Snowmass intake face end:

At the intake end of the Snowmass face, the cross-sectional area of the face was 44 ft² of which 20 ft² were obstructed by the gear head of the conveyor. This combined with the transfer point on to the gate conveyor, causing considerable contraction of the airflow at the face-end. The high air velocity and excessive turbulence were all too obvious. The shock loss factors estimated from the relationships given above were

bend X_b = 1.4
contraction X_c = 0.184
obstruction X_{ob} = $\left(\frac{44}{0.7(44-20)} - 1\right)^2$
              = 2.621

Total shock loss factor at intake end:

X_{int} = 1.4 + 0.184 + 2.621 = 4.205

Equation (9) gives the equivalent resistance to be

$$\frac{62.21}{44^2} \times 4.205 = 0.135 \text{ P.U.}$$

Table 1 gives the total resistance of the first 58ft of the face (section A2 to A3, including the face end) to be 0.2035 P.U. Through this section, the cross-sectional area was 44 ft^2, the perimeter 31ft and the friction factor 187 (x10^{-10} lbf.min ^{2}/ft^4), similar to that of section A3-A4 (see Table 3).

Hence the resistance of this section of face line without the shock losses would be

$$R = \frac{k\,L\,Q}{52\,A^3} = \frac{187 \times 58 \times 31}{52\,(44)^3} = 0.076 \text{ P.U.} \qquad (4)$$

The equivalent resistance due to the shock losses is then given as the difference between the measured total resistance and that due to the 58 ft of face-line, i.e.

$$0.2035 - 0.076 = 0.128 \text{ P.U.}$$

This differs from the computed resistance of 0.135 P.U. by only 0.007 P.U., or 5 per cent. This is considered to be a satisfactory correlation.

4.6 Shearer:

The power loading machinery offers additional resistance to airflow on a longwall face. The pressure losses occur because of two distinct effects. First, wake losses occur at the downstream end of the shearer as the higher velocity airstream through the restricted area around the body of the machine is projected into the lower velocity downstream. Second, the increased velocity in the restricted zone along the shearer causes enhanced frictional losses in this area.

It is possible to calculate the theoretical shock loss caused by a symmetrical obstruction to airflow in a straight mine airway. However, such calculations give results which are less than half the losses measured across a shearer on a longwall face. This was also the case at Snowmass. The reasons for such discrepancies arise from the over-simplifications necessary for a theoretical treatment. In practice, the resistance offered by a power-loader

depends not only upon the dimensions of the machine and those of the face, but also upon:
* the position of the shearer within the cross-section.
* the location of the machine along the face, particularly if in proximity to the face ends and also with respect to positions of open flow paths in the waste behind the chocks.
* whether the machine is moving and, if so, in which direction relative to the airflow.
* the position and air-deflecting capabilities of water sprays.
* the position of the operator.
* the orientation of the cutting drums and any deflecting plates that may be fitted.
* the local friction (k) factor of the longwall.

One further indirect factor associated with a producing unit is the depth of broken coal on the face conveyor and its direction of travel relative to the airflow.

With such a diversity of variables, it is not surprising that a theoretical treatment based on a simplified geometry yields unsatisfactory results. An empirical method is, therefore, suggested and is based on the Snowmass measurements. However, it is also suggested that further investigations on a range of longwall faces are required to provide design data for shock losses across power-loaders. Table 1 shows that over the 45 ft (14m) long section, A8-A9, containing the shearer, the measured resistance was 0.0665 P.U. (0.074 Ns^2/m^8) The mean cross-sectional area of the face between stations A8 and A9 was 80 ft^2 (7.43m^2). The perimeter of the face cross-section, not including the shearer was 46 ft (14m).

The resistance of the 45 ft (13.7m) section, in the absence of the shearer is given by equation (4).

$$R = \frac{k\,L\,Q}{52\,A^3} \qquad \text{(Imperial Units)}$$

or

$$\frac{k\,L\,Q}{A^3} \qquad \text{(SI Units)}$$

Using the value of $k = 347$ lbf.min^2/ft^4 (0.0644 kg/m^3) for the upper part of the face (Table 3) gives

$$R = \frac{347 \times 45 \times 46}{52 \times (80)^3} = 0.0270 \text{ P.U.}$$

or

$$\frac{0.0644 \times 13.7 \times 14}{(7.43)^2} = 0.0301 \text{ Ns}^2/\text{m}^8$$

for the unobstructed face. Hence, the effective resistance of the shearer is

$$R_{shear} = 0.0665 - 0.0270 = 0.0395 \text{ P.U.}$$
$$(0.0442 \text{ Ns}^2/\text{m}^8)$$

The shock loss factor for the shearer is given by equations 9 or 10.

$$X_{shear} = \frac{R_{shear} A^2}{62.21} = \frac{0.0395 \times 80^2}{62.21} = 4.06$$

On the basis of this result, it is suggested that an estimate may be made of the resistance of a longwall shearer by assuming a shock loss factor of 4.

4.7 Caved Area:

The overall resistance of a longwall face will decrease considerably if there are open areas available for airflow in the waste zone behind the chocks. For planning purposes, however, the resistance of the face should be estimated on the basis of the conveyor being at its closest position to the coal front, the chocks fully advanced and the waste area caved up to the rear of the chocks. This situation will give the smallest cross-sectional area and highest resistance offered by the longwall under normal operating conditions.

5 SUMMARY OF PROCEDURE FOR ESTIMATING FACE RESISTANCE

1. Determine the length, L, mean cross-sectional area, A, and perimeter, O, of the face and estimate a friction factor, k, from Table 4. Then determine the face-line resistance, R_f, from Figures 4, 5 or 6, or from equations 4 or 4a.

2. For each face-end, determine the following shock loss factors where applicable.

Sharp right angled bend, $X_b = 1.4$
Sudden contraction at
intake end, $X_c = 0.184$
Face-end obstruction, $X_{ob} = \left\{ \frac{A}{0.7(A-a)} - 1 \right\}^2$

where A = cross-sectional area of face
and   a = cross-section of obstruction facing the airflow.

[For the majority of longwall faces in the United States, the height of the face is the same as that of the airways. For this reason, no allowance need be made for an expansion loss at the return end unless there is a concentration of equipment at this location.]

3. Use a shock loss factor of $X_{shear} = 4$ for each shearer on the face.

4. Sum all shock loss factors to give X_{tot}.

5. Determine the equivalent resistance of the shock losses.

$$R_{sh} = \frac{62.21}{A^2} X_{tot} \quad \text{(gives R in P.U. if A is in ft}^2\text{)}$$

or

$$R_{sh} = 0.6 \, X_{tot}/A^2 \quad \text{(gives R in Ns}^2/\text{m}^8 \text{ if A is in m}^2\text{)}$$

6. Determine the full face resistance

$$R = R_f + R_{sh}$$

7. The resistances of the main gate and tail gate should be determined separately from

$$R = \frac{k \, L \, O}{52 \, A^3} \quad \text{(P.U. Imperial Units)}$$

or

$$\frac{k \, L \, O}{A^3} \quad \text{(Ns}^2/\text{m}^8 \text{ SI units)}$$

using values of friction factor, k, appropriate to the conditions expected, and taking into account the location of equipment close to the face.

Having determined the full face resistance including those adjoining lengths of main and tail gates that contain equipment, then that resistance may be used in network simulations for ventilation planning purposes. Additionally, the frictional pressure drop, p, may be determined for any given airflow, Q, from the square law.

$$p = R Q^2$$

using the appropriate set of units, i.e.

Imperial:

        p in milli-inches w.g.

        R in P.U.

and   Q in kilo cfm.

S.I.:

        p in Pascals (N/m^2)

        R in Ns^2/m^8

and   Q in m^3/s

A worked example of the procedure is given in the appendix.

6 CONCLUSIONS

This paper provides data and a procedure for estimating the resistance to airflow offered by a modern mechanized longwall face. The technique is based upon a combination of theoretical analyses and empirical data gained from detailed measurements made on a fully equipped longwall. Face resistances obtained through this procedure may be incorporated into a network that also represents leakage through the caved area behind the face, in order to simulate the complete district. Another paper (Brunner) describes a quantified representation of leakage paths.

ACKNOWLEDGMENTS

The author and his colleagues are grateful to the United States Department of Energy for funding the project that included this study. We also thank the Snowmass Coal Company for lending us one of their mines in which to conduct the tests. In particular, we are most grateful to Mr. Jay Reynolds, mine manager, and Mr. Stephen Self, chief engineer, for their ready cooperation and very practical assistance in organizing and conducting the field work.

REFERENCES

Atkinson, J.J. 1854. On the theory of airflow in mines. Trans North of England Inst. of Mining Engineers, 3, 73-222.

A review of spontaneous combustion problems and controls with application to U.S. coal mines. D.O.E. report, contract No. AC03-768 F000 98, September 1982.

Bear, J. 1972. Dynamics of fluids in porous media. American Elsevier, N.Y.

Brunner, D.J. 1985. Ventilation models for longwall gob leakage simulation. U.S. National Ventilation Symposium 2, Reno Nevada, September.

McPherson, M.J. and Brunner, D.J. 1983 An investigation into the ventilation characteristics of a longwall district in a coal mine. Report to the U.S. DOE Contract No. DC AC03 768 F00098.

McPherson, M.J. 1971. The metrication and rationalization of mine ventilation calculations. The Mining Engineer, No.131, August.

Wallace K.G. Jnr. and McPherson M.J. 1982. Mine ventilation economics, DOE report. Contract No. DC AC03 768 F00098.

Whittaker, B.N. 1974. An appraisal of strata control practice. Trans. Inst. of Mining Engineers, October.

APPENDIX

Worked example of estimating face resistance

A mechanized longwall face has the following planned specifications.

Length:	120m
Cross-sectional area with chocks forward	6m
Corresponding perimeter	10m

The intake face end will have the cross-sectional area reduced to 2.5m^2 due to a conveyor transfer point. There is no additional equipment at the return face end. The face is at right angles to both main and tail gates. There is one shearer on the face.

Compute the face resistance and the frictional pressure drop at an airflow of 15 m^3/s.

Procedure:

1. Assuming normal conditions for powered supports, k is estimated to be 0.051 kg/m^3 (Table 4) The face-line resistance may be determined from either

$$R_f = \frac{k\,L\,O}{A^3} = \frac{0.051 \times 120 \times 10}{(6)^3}$$
$$= 0.283 \ \ Ns^2/m^8$$

or

read from Figure 5: at A = 6m^2 and perimeter = 10m, the nomogram gives a resistance of 0.23 Ns^2/m^8 per 100m.

Hence, $R_f = 0.23 \times \dfrac{120}{100} = 0.28 \ Ns^2/m^8$

2. At intake face-end, shock loss factors are

bend $\qquad X_b$ = 1.4
contraction $\ Xc$ = 0.184
obstruction $X_{ob} = \left\{ \dfrac{6}{0.7\ (6-2.5)} -1 \right\}^2$ = 2.10

Total shock loss factor at intake

X_{int} = 1.4 + 0.184 + 2.10 = 3.684

3. At return end, only the sharp bend contributes a significant shock loss, hence,

X_{ret} = 1.4

4. Shearer shock loss factor

X_{shear} = 4.0

5. Total of shock loss factors

X_{tot} = 3.684 + 1.4 + 4.0 = 9.084

6. Equivalent resistance for shock losses

$R_{sh} = \dfrac{0.6 \times 9.084}{6^2}$ = 0.151 Ns^2/m^8

7. Full face resistance

$R = R_f + R_{sh}$ = 0.283 + 0.151
$\qquad\qquad$ = 0.434 Ns^2/m^8

8. The frictional pressure drop across the face at an airflow of 15 m^3/s is given as

$p = R\,Q^2 = 0.434 \times 15^2 = 98$ Pa

Extended advance face ventilation systems

S.K.RUGGIERI
Foster-Miller Inc., Waltham, MA, USA

J.C.VOLKWEIN & F.N.KISSELL
Pittsburgh Research Center, US Bureau of Mines, PA, USA

C.McGLOTHLIN
Beaver Creek Coal Company, Price, UT, USA

ABSTRACT: Remotely controlled continuous miners and automated temporary roof support systems have opened the door to extended cuts in room and pillar mining. Cuts of up to 40 ft (12.20m) or more are possible between place changes which could significantly increase coal mine productivity. Application of these new developments, however, has been limited by the lack of adequate face ventilation technology.

The Bureau of Mines, Foster-Miller, Inc. (FMI), and Beaver Creek Coal Company (BCC) have recently evaluated a new auxiliary face ventilation system for extended advances of up to 40 ft (12.20m). The new system uses the recently developed Improved Sprayfan System. Additional forward and reverse pointing sprays were added to supply a clean split of air to the face and to direct contaminated air to the mouth of the exhaust curtain. An underground evaluation at BCC showed that the new system consistently provided effective face ventilation at curtain setbacks up to 40 ft (12.20m) from the face.

The success of this initial system evaluation, resulted in additional research to further improve extended advance auxiliary ventilation systems. Machine-mounted (continuous miner) systems are currently under development and testing in the full-scale continuous miner test facility <u>under a wide range of mining and ventilation conditions including varying seam heights and turning crosscuts</u> using curtain setbacks up to 60 ft (18.30m).

INTRODUCTION

Water sprays have been used for many years on continuous miners for dust control, bit lubrication, and cooling. In the mid-1970's, a new water spray system, called the Sprayfan system (Wallhagen 1977 & USBM Technology News 1983), was developed by FMI under contract to the Bureau. The development of the Sprayfan system was centered around the fact that water sprays can move air like small fans by transfering the momentum of the water droplets to the airstream.

The Sprayfan system consists of several spray manifolds strategically placed on the continuous miner. These sprays redirect the main ventilation flow to the face and sweep contaminated air (dust and methane) across the face toward the return. The Sprayfan system effectively eliminates short-circuiting of the ventilating air directly to the return. The layout of an improved Sprayfan system (Ruggieri 1984),

recently developed by FMI under a Bureau contract, is shown in figure 1. Although the figure depicts operation of

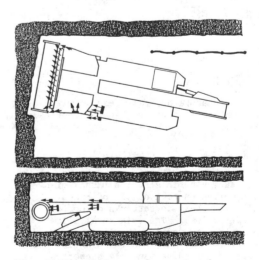

Fig. 1. Improved sprayfan system layout.

the system for a right-hand brattice return only, a dual system (mirror-image) is required for brattice returns on either side.

Since its original introduction in 1977, water spray systems, using Spray-fan principles have been installed on well over 300 continuous miners. The figure continues to grow as many oper-ators are now installing the system as part of their miner rebuild and replace-ment programs.

The Sprayfan system has proven capa-bility to control methane concentrations at the face (with the brattice 20 ft (6.10m) from the face) as well as or better than conventional water spray systems with the brattice 10 ft (3.05m) from the point of deepest penetration. This capability is illustrated in figure 2, which shows tracer gas concen-tration maps for conventional sprays with a 10 ft (3.05m) brattice setback and the Sprayfan system with a 20 ft (6.10m) brattice setback. This data was measured in a full-scale continuous miner test facility at FMI.

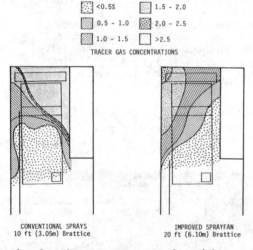

▒ <0.5%	▒ 1.5 - 2.0
▒ 0.5 - 1.0	▒ 2.0 - 2.5
▒ 1.0 - 1.5	☐ >2.5

TRACER GAS CONCENTRATIONS

CONVENTIONAL SPRAYS
10 ft (3.05m) Brattice

IMPROVED SPRAYFAN
20 ft (6.10m) Brattice

Fig. 2. Laboratory test results with conventional sprays and improved sprayfan system.

The proven capability of the Sprayfan system has resulted in the Mine Health and Safety Administration (MSHA) grant-ing variances to a number of mines allowing brattice setbacks of up to 20 ft (6.10m). This, in turn, allows operators to completely mine a standard 20 ft (6.10m) cut without advancing the curtain beyond permanently supported roof.

Technology in radio remote control and automated temporary roof support, has advanced to the point where 40 (12.20) to 60 ft (18.30m) cuts are now possible with machine operators remaining under permanently supported roof. In addition to improving safety, this technology could significantly increase coal mine productivity by reducing the number of place changes. Applications of this technology, however, have been limited because the technology to ventilate these deeper cuts has not kept pace.

In early 1984, a unique opportunity arose at a Utah coal mine (BCC) to develop and test a new ventilation system on a remotely controlled con-tinuous miner. The mine uses a remotely controlled continuous miner to make 40 ft (12.20m) cuts. The roof in this mine is competent enough to allow this depth of cut between roof bolting cycles. Temporary supports, however, were required to advance the ventilation line curtain. This requirement limited productivity and, more importantly, required personnel to set the temporary supports inby permanently supported roof. Mine management, therefore, was extremely interested in the capability of the sprayfan system to adequately ventilate the face with a brattice setback of up to 40 ft (12.20m).

Laboratory tests in the full-scale continuous miner facility were conducted which resulted in the development of a modified version of the Sprayfan system for ventilating cuts of up to 40 ft (12.20m).

Initial laboratory development

Initial testing in the continuous miner facility showed that the ventilating capability of the Sprayfan system was limited to brattice setbacks of approxi-mately 22 ft (6.71m). At brattice setbacks greater than 22 ft (6.71m), a recirculation cell developed over the miner body. This recirculation increased the tracer gas concentrations at the face, and was likely to cause visibility problems for the remote operator due to recirculation of dust and spray mist.

The Sprayfan, therefore, needed additional power along both sides of the miner to induce an adequate airflow split to reach the face area and to effectively channel contaminated air into the return. An additional spray manifold (known as "A-block") was installed on the off-curtain side of the miner located as far outby as practical (as shown in fig. 3). This manifold, containing two sprays, begins the airsplit process and pushes clean air up to the hingepoint region of the miner where the Sprayfan system continues to move the air up to and across the face.

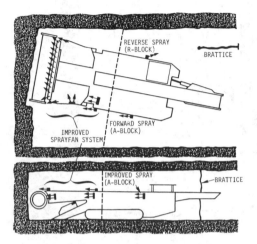

Fig. 3. Extended advance ventilation system.

To prevent the airsplit from "dying out" after sweeping the face area, a reverse spray manifold (known as "R-block") was installed on the curtain side of the miner aimed towards the curtain mouth. Again, the manifold was located as far outby as practical. This manifold provides a final "boost" of the contaminated air into the return. The R-block design was the result of earlier research conducted by the Bureau in which reverse sprays were used for continuous miner dust control.

Laboratory testing of the modified system showed that it could effectively ventilate the face with minimal recirculation at brattice setbacks of up to 40 ft (12.20m). Test results with the miner at the end of a 20 ft (6.10m) box cut, primary airflow of 8,000 cfm (227 m^3/m), a 40 ft (12.20m) brattice setback, and the system operating at 250 psi (1724 kPa) are shown in figure 4. Tracer gas concentrations were comparable to those obtained with the Sprayfan system at a 20 ft (6.10m) brattice setback and much lower than those obtained with conventional sprays and a 10 ft (3.05m) brattice setback.

Testing of the modified system at shorter brattice setbacks, however, showed excessive recirculation. At

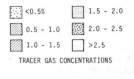

▨ <0.5%		▤ 1.5 - 2.0	
▥ 0.5 - 1.0		▧ 2.0 - 2.5	
▨ 1.0 - 1.5		☐ >2.5	

TRACER GAS CONCENTRATIONS

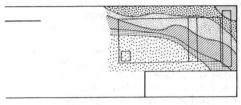

Fig. 4. Laboratory test results with extended advance ventilation system (40 ft (12.20m) brattice setback).

brattice setbacks of 20 ft (6.10m) or less, the increased power provided by the R-block, pushed contamination past the mouth of the brattice and allowed it to mix with the intake air. Turning off R-block eliminated the recirculation problem.

The extended advance face ventilation system developed in the laboratory, therefore, consisted of two subsystems. At brattice setbacks of 20 ft (6.10m) or less, the basic Sprayfan system presented in figure 3 is operated. When the curtain setback exceeds 20 ft (6.10m), A- and R-blocks are turned on.

This system was evaluated on a continuous miner in BCC's Gordon Creek No. 2 Mine.

Underground evaluation

In March of 1984, the extended advance face ventilation system, as developed in the laboratory, was installed by BCC. After a short period of operation underground, however, the mine modified the original R-block design.

The original manifold had consisted of two sprays oriented at an outward angle of 15° to the miner. The mine noticed that during mining of the on-curtain (sump) side of the entry, the R-block sprays prematurely impinged on the rib and caused some recirculation in front of the curtain mouth. Conversely, when mining the off-curtain (slab) side of the entry, the outward spray angle of 15° was not adequate to push contaminated air across the entry to the curtain mouth. Contaminated air was pushed backward toward the fresh air side of the curtain mouth resulting in some recirculation. The mine designed and installed a new R-block with two separate sets of two sprays each:

1. One set with an outward angle of 10° to be operated when cutting the on-curtain side of the entry

2. One set with an outward angle of 45° to be operated when cutting the off-curtain side of the entry.

Each set of sprays was valved separately to allow independent use.

Methane production in the Gordon Creek No. 2 Mine is extremely low and MSHA granted a variance to allow testing of the system with brattice setbacks to 40 ft (12.20m) which was the limit allowed by the roof control plan.

The basic strategy used for evaluating the performance of the Sprayfan underground was to determine the speed and effectiveness with which it was capable of sweeping contamination out of the face area and into the return under a variety of different face conditions. Generally such testing is conducted by monitoring actual face methane (CH_4) concentrations at several points on the miner and also in the return. However, since methane gas was not present at this mine in quantities sufficient for practical test procedures, the tracer gas, sulfur hexafluoride (SF_6), was used to evaluate the effectiveness of the new system. The basic method used is described in U.S. Bureau of Mines IC 8899 (Divers 1982).

The SF_6 gas was released on the off-curtain corner of the mining machine boom (fig. 5). Gas sample bottles were collected in the immediate return and the time required to remove two-thirds of the gas from the face was used as the ventilation effectiveness criteria.

Prior to the start of testing in a given cut, air velocity and volume measurements were made at the mouth of the line curtain and at the return sampling position. In addition, the layout and configuration of the cut as well as the distance from the curtain mouth to the return position were documented.

The evaluation of the extended cut system was completed over eight operating shifts and two idle shifts. A total of 39 static tests, representing a variety of different face conditions, were conducted during the idle shifts. Dynamic testing during active mining consisted of 40 separate SF_6 tests performed over 16 cuts. A representative cross section of the static and dynamic test results were chosen for inclusion in this paper.

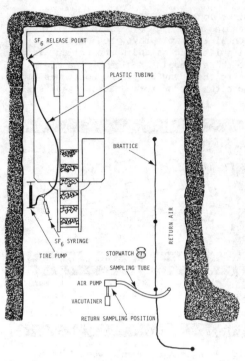

Fig. 5. Tracer gas setup.

The first test series conducted during the evaluation involved determining extended cut system face ventilation effectiveness at four brattice setback distances of 10 (3.05), 20 (6.10), 30, (9.15), and 40 ft (12.20m). To provide a basis for comparison, the following static tests were completed:

1. Primary ventilation tests (all sprays off) at a setback distance of 10 ft (3.05m)

2. Improved Sprayfan (R-block off) operating at setback distances of 10 (3.05) and 20 ft (6.10m)

3. Extended Advance System (A- and R-blocks on) operating at setback distances of 30 (9.15) and 40 ft (12.20m).

Figure 6 contains plots of percentage (%) of SF_6 captured versus time for the extended cut system operating at the four setback distances. Also shown is the plot of primary ventilation at the "best case" condition of 10 ft (3.05m) brattice setback. All four graphs of system performance are very similar: the plots "track" each other very closely regardless of brattice setback distance. The four graphs also represent a face ventilation effectiveness substantially better than that of the

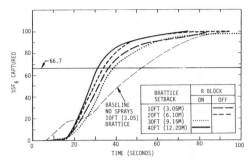

Fig. 6. Percent of SF$_6$ captured versus time - sprayfan at various setback distances.

10 ft (3.05m) primary ventilation test: average system time duration for 66.7% capture was 35 seconds versus 51 seconds for the primary ventilation test (a reduction of 31%).

Following is a summary of the conclusions drawn from the static test results:

1. Through use of the Extended Advance System, effective face ventilation was maintained at brattice setback distances of up to 40 ft (12.20m). The system was capable of ventilating the face at 30 (9.15) and 40 ft (12.20m) setback distances as well as the Sprayfan was at 20 ft (6.10m) setback distances

2. The Extended Advance System was capable of ventilating the face better at all brattice setback distances than primary ventilation airflow was at a 10 ft (3.05m) setback distance.

These conclusions were confirmed through dynamic testing. Figure 7 contains plots of percentage (%) of SF$_6$ captured versus time for three brattice setback distances with the Extended Advance System operating during mining of slab cuts. As observed in the static test results, the graphs are very similar, indicating comparable face ventilation effectiveness regardless of

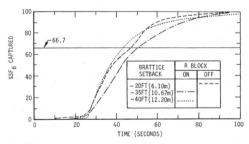

Fig. 7. Percent of sulfur hexafluoride captured versus time during actual mining (slab cuts).

setback distance. There were, however, several differences between the static and dynamic test results.

These include:

1. The dynamic tests contained an additional time delay of approximately 10 seconds before the start of significant SF$_6$ capture

2. The time delay contributed to a higher average time duration for 66.7% capture (35 seconds during static testing versus 47 seconds during dynamic testing).

Altered airflow patterns during actual mining and the back-mixing volume of the open sump were thought to be the cause of these increases. Even with these increases, system performance was still better than that measured during static tests with only primary ventilation airflow and a 10 ft (3.05m) brattice setback.

During mining, the system was also evaluated from a dust standpoint using instantaneous dust monitors. Results showed that very little dust generated by the miner affected either the shuttle car operator or the remote miner operator's position. Average dust concentrations at the shuttle car operator were only 4% higher than the concentrations measured in the primary ventilation upstream of the continuous miner face. Concentrations at the remote miner operator were only 10% higher than at the shuttle car operator. In addition, observations from the remote operator's position during cutting showed the cutterhead to be clearly visible even at brattice setbacks of 40 ft (12.20).

The performance of the system showed the potential for improved safety and increased productivity. Extended advances of up to 40 ft (12.20m) with good face ventilation were possible without sending miners under temporary supports to extend and maintain the brattice.

The demonstrated performance of the system has enabled BCC to obtain variances from MSHA and the system is now in use on two of their remotely controlled continuous miners. The success of this initial system evaluation, resulted in additional research to further improve the capability and performance of extended advance auxiliary ventilation systems.

Recent laboratory development

The success of the initial evaluation at

547

BCC has resulted in additional research to further develop and improve extended advance auxiliary ventilation systems. The objectives of this additional research are to:

1. Extend and evaluate system capability over a wider range of mining conditions
2. Simplify system operations
3. Extend system capability to brattice setbacks up to 60 ft (18.30m).

The original development and evaluation of the Extended Advance System was limited to conditions at BCC's mine. Before the system can be recommended for general industry use, additional development and evaluation is needed over a wider range of operating conditions. Current plans call for testing the system under the following conditions:

1. Seam heights up to 12 ft (3.66m)
2. Airflows ranging from 3,000 (85) to 16,000 cfm (453 m^3/m)
3. Exhaust brattice and exhaust tubing
4. Straight cuts (sump and slab positions) and turning crosscuts.

Plans also call for investigating alternative airmovers, particularly in the A- and R-block locations. The system evaluated at BCC successfully used water sprays as airmovers in these locations. In some mines, however, application of water on the rear of the miner may cause excessive bottom wetting and operational problems. Alternatives may use electrically or hydraulically driven fans or compressed air powered venturis.

The system, as evaluated at Beaver Creek, is moderately complex requiring three different systems operating during a 40 ft (12.20m) advance. For the first 20 ft (6.10m), the basic Sprayfan and A-block is operated; from 20 (6.10) to 40 ft (12.20m), the R-block is added. When the miner moves from the sump to the slab cut side the angle of the R-block discharge is changed. At BCC, this switching of spray systems was incorporated into the remote control system with solenoid valves on the miner. Optimum system performance, however, requires that the miner operator switch sprays at the appropriate times during the mining cycle. While this is not considered a critical problem, additional research is planned to determine if system operation can be simplified.

Testing is also being conducted to extend system capability beyond 40 ft (12.20m) brattice setbacks. Tests were started in the full-scale continuous miner gallery using the system evaluated at BCC with a brattice setback of 60 ft (18.30m). The initial tests were conducted with no shuttle car behind the miner. Results, illustrated in figure 8, showed that as much as 25% of the tracer gas in the return airstream was recirculating and mixing with the intake airstream. Even with this degree of recirculation and mixing, however, the system was able to pull a significant amount of clean air to the face. Both the peak and average tracer gas concentrations at the face were still below that measured with conventional water sprays and a 10 ft (3.05m) brattice setback.

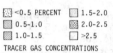

TRACER GAS CONCENTRATIONS

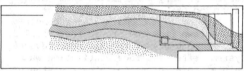

Fig. 8. Laboratory test results with extended advance ventilation system (60-ft (18.30m) brattice setback; no shuttle car).

When a model shuttle car was installed behind the miner, the recirculation was reduced by more than 50% as illustrated in figure 9. However, contamination was blown outby past the mouth of the curtain where it mixed with the intake air. While this degree of recirculation did not significantly affect tracer gas concentrations at the face, there was concern that it may affect the remote operator and shuttle car operator dust exposures and cause visibility problems from the remote operator's location.

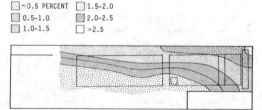

Fig. 9. Laboratory test results with the extended advance ventilation system (60 ft (18.30m) brattice setback; shuttle car in position).

Several concepts were evaluated to try to further reduce recirculation. These included:

1. Passive barriers – Passive barriers were mounted on the curtain side of the shuttle car and continuous miners. These barriers, which extended from the top of the models to within 6 inches (152.4 mm) of the roof, were intended to reduce mixing between the intake and return air streams. Test results, however, showed little improvement, indicating that most of the recirculation was occurring between the end of the shuttle car and the brattice line

2. Additional air moving power – The air moving power of the system was increased in several ways including:

    a. Increasing the spray pressure at A- and R-block

    b. Increasing the number of sprays at A- and R-block

    c. Adding air movers to the sides of the shuttle car

This increased power was intended to pull more fresh air in along the off-curtain side of the shuttle car and to force the return air into the mouth of the curtain. Test results, however, showed that increasing system power actually increased recirculation rather than reducing it while creating no further reduction in face concentrations

3. Bell mouth at the end of the brattice – Testing with a straight line curtain hung 2.5 ft (0.76m) from the rib showed significant amounts of contaminant blowing outby past the curtain mouth. To reduce this blowby, the end of the curtain was angled outward into the entry approximately 45° increasing the width of the brattice mouth to approximately 4.5 ft (1.37m). This bell mouth at the end of the curtain significantly reduced recirculation

4. Balancing system power and primary airflow – Previous research on the sprayfan system had shown that optimum system performance requires balancing the system power and the primary airflow. For example, operating with too high pressure at low primary airflows causes recirculation.

Preliminary testing of the extended advance system is showing similar results. At a primary airflow of 8,000 cfm (227 m^3/m), the system operating at 250 psi (1,724 kPa) recirculates approximately 20% of the contaminate in the return airstreams. When the system pressure was reduced to 150 psi (1,034 kPa), recirculation was reduced by more than 50%.

As of the publication date (May 1985) further testing, and development were in progress under the mining and ventilation conditions previously mentioned. Future plans call for a detailed underground evaluation of the newly developed extended advance ventilation system.

REFERENCES

Wallhagen, R.E. November 1977. Development of optimal diffuser and Sprayfan systems for coal mine face ventilation. USBM Contract No. H0230023, NTIS PB277 987.

Sprayfan aids in effectively controlling, methane. January 1983. USBM Technology News No. 162.

Ruggieri, S.K., D.M. Doyle, & J.C. Volkwein. April 1984. Improved sprayfans provide ventilation solutions. Coal Mining. 21(4):94-98.

Divers, E., N. Jayaraman, & J. Custer. 1982. Evaluation of a combined face ventilation system used with remotely operating mining machine. BuMines IC 8899. p. 7.

An operator's experience using antitropal and homotropal longwall face ventilation systems

J.W. STEVENSON
Jim Walter Resources Inc., Brookwood, AL, USA

1 INTRODUCTION

Jim Walter Resources, Inc., operates five underground coal mines with a sixth temporarily closed due to economic conditions. The mines are located in north central Alabama near Birmingham and in 1984 produced in excess of 8.5 million tons of low to medium volatile metallurgical coal that is used primarily by the Japanese Steel Industry or as a compliance coal for electric power generation. Full production of 9.5 million tons is scheduled to be on line in 1987. Continuous miners are used for development with longwall systems used for secondary or retreat mining.

2 MINE AND LONGWALL DEVELOPMENT

In most instances, longwall panels are developed with four entries and two splits of ventilating air. The outside entries serve as return air courses with the center two serving as a belt entry and the intake escapeway. Because we had previously petitioned MSHA and had installed an early warning fire detection system as an additional safeguard, we are permitted to use belt air in the faces. The track entry also serves as the intake escapeway since all track transportation equipment is battery powered except for a few diesel powered units - mainly personnel carriers.

At the depths we are mining, the Blue Creek Coal bed of the Mary Lee series is considered to be ultra-gassy, requiring large volumes of air to ventilate the face where coal is being cut, mined, and loaded. The minimum air quantity required at the end of the exhausting line curtain varies from 14,000 to 20,000 cfm depending upon the mine. It is not uncommon for a 6,000 foot-deep developing panel to require 160,000 to 200,000 cfm air to ventilate the faces and dilute

methane rib liberation. Total mine air quantity varies from 2 million to 3.6 million cubic feet a minute. Total methane exhausted by the ventilating air current varies from 15 million to 22 million cubic feet a day. The connected electric motor load to the fans ranges from 6,250 to 14,000 horsepower and at $325.00/horsepower year, the yearly power costs to ventilate the mines approximates 2.0 to 4.6 million dollars.

3 LONGWALL MINING

The first retreating longwall face under heavy cover started operating in No. 3 Mine during the first quarter of 1979. Since that time, we have systematically added longwall mining units; presently we are operating seven longwall faces with the eighth face scheduled to startup the first of June at No. 7 Mine (the second system in No. 7 Mine) and the ninth projected to start in September at No. 5 Mine (the first system at No. 5 Mine). When full longwall production is achieved, twelve longwall systems are scheduled with three in each of the Jim Walter Resources deep mines - No. 3, No. 4, No. 5, and No. 7 Mines. We are now mining our 23rd panel with face widths varying from 400 feet to 650 feet and panel depths exceeding 5,000 feet. Faces are projected to be 900 feet wide with panel depths of 6,000 feet. Continuous miner development production approximates 35% of the mine production; three longwalls at each mine produces 65% of the total production.

Of the 23 panels mined by longwall, 19 have been mined with conventional wrap-around or antitropal ventilation, and four faces partially or wholly extracted using homotropal ventilation. This presentation will discuss some of Jim Walter Resources experiences together with what we consider advantages and disadvantages between the two systems.

Although all of our shearers are bi-directional, they are mostly operated in a unidirectional mode cutting from the tailgate towards the headgate. In this mode the shearer mining rate is not restricted by material flow between the underframe of the shearer and the face conveyor chain. The shearer cleans up the face as it flits towards the tailgate to begin the second cut. During the clean-up or flit portion of the cycle, material flows under the shearer and if rock must be cut and loaded, an operations problem may occur with the armoured face conveyor (AFC) chain if it becomes fouled or broken as the rock passes under the shearer. However, this is a problem that will be solved as our efforts continue to be directed towards operating the shearer in a bi-directional configuration as faces extend from the present 650 feet width to a projected 900 feet.

One of the problems anticipated as the faces become wider is the quantity of methane released by cutting and liberated by the newly exposed coal face. Presently, the ventilating air current across the face to dilute methane to a 1% concentration may approach 1,000 to 1,200 feet a minute at the shearer. With wider faces, we anticipate additional methane, but realize that we have well exceeded the limit on air velocity as a dust control measure. In seam horizontal degasification will be required. We are now operating two long-hole horizontal drilling machines for our inseam capture of methane and piping the gas to the surface to sell as an unconventional energy source. A third drill will be delivered in June and a fourth is to be ordered this summer. It is anticipated that a horizontal drill program will become a part of the cycle for each longwall development panel. In addition to degasification of the coal bed, we plan to continue our program of infusing the horizontal drill holes with water to allay dust after the gas production decreases but before the longwall face intercepts the hole. The inseam degasification of the longwall panels will reduce the quantity and velocity of air necessary for methane control across the longwall face. Water infusion will lower respirable dust levels. Both approaches will be beneficial when mining wider faces with either face ventilation system.

In addition to horizontal long-hole inseam degasification, we maintain a pattern of 90 plus vertical holes in advance of mining to degasify the coal bed. Also,

two to four gob drainage holes are drilled above each longwall panel to remove methane from the voids that are created by the collapse of the roof strata in the cavity above the coal bed behind the longwall face. Obviously methane control of the Blue Creek Coal bed and the super and subjacent coal measures requires a considerable amount of effort at all levels of management and operations.

4 ANTITROPAL FACE VENTILATION

Figure 1 is the map of No. 3 Mine, our first longwall operation. The first panel was extracted in the southwest part of the property and all panels have been mined using antitropal longwall face ventilation. Experience has proven in every instance that the first two longwall panels in a block require the most ventilating air and may approach 250,000 cfm. With successive panels, lower quantities of air are required on the active panel and less air is needed to maintain methane concentrations at acceptable levels at the bleeder connectors. In the first longwall panel, we are mining our eighth consecutive face. From a ventilation viewpoint, the eighth face appears to require the same amount of air on the face but the least amount of total air to maintain compliance at the bleeder connectors and gob.

Figure 2 depicts an antitropal face ventilation plan, where the air flow direction is opposite the direction of coal flow on the longwall face conveyor. All four entries on the headgate side, which includes the belt entry, are utilized as intake air courses. The headgate bleeder connectors are tightly regulated, and depending upon caving and the tightness of the gob, additional plastic brattice checks may be required to direct the ventilating air current across the longwall face.

Normally a tightly regulated secondary intake ventilating air current is provided on the tailgate side of the longwall. The longwall face air current and the secondary intake air current meet in the vicinity of the tailgate. After merging, both splits of air flow into the gob and bleeder system. Care must be taken to insure that the secondary air current merges with the face ventilating air current in such a manner that the tailgate end of the longwall face remains ventilated. If the gob is open, plastic checks may be required between development pillars in the tailgate to insure that the ventilating air current does not prematurely enter the gob and leave a portion of the longwall face near the tailgate inadequately

ventilated. If the gob has caved so
tightly that a sufficient air quantity
cannot be provided to maintain acceptable
methane levels on the face, the secondary
intake air course on the tailgate side of
the longwall may be converted into a regulated return to maintain airflow on the
face. When large air volumes resulting
in high ventilating pressure losses are
required for methane control, regulator
settings are critical. Using antitropal
ventilation permits regulators to be
positioned in areas remote from normal
travel and therefore not subject to inadvertant change. In addition, the longwall block itself provides a barrier to
eliminate leakage into the gob until the
face is ventilated. It therefore con-
serves air.

At all transfer points, a system of
water sprays is used to allay and control
dust. If warranted, a steel framework
hood covered by plastic brattice cloth
is erected to shield the coal from the
air stream.

Methane recorders located in the belt
entry near the stage loader do not reg-

ister a significant elevation in concentration when the belt conveyor is
fully loaded as compared to empty.

5 HOMOTROPAL FACE VENTILATION

Figure 3 is the map of No. 4 Mine, the
second mine to install longwalls. We are
now mining our fifth panel. Panel No. 1
and No. 2 on the east side of the mine
and Panel No. 1 on the west side were
mined entirely or in part utilizing
homotropal face ventilation. Figure 4
is a ventilation plan where homotropal
face ventilation is used and the ventilating air flow is in the same direction
as the coal flow on the longwall face conveyor. After the first panel has been
extracted, the stopping line between the
primary intake air course and the gob is
critical. More critical, however, is the
fact that when using homotropal ventilation, one entry alone must supply the
longwall face with the same air quantity.
When the number of entries decreases from
four to one, the ventilating pressure

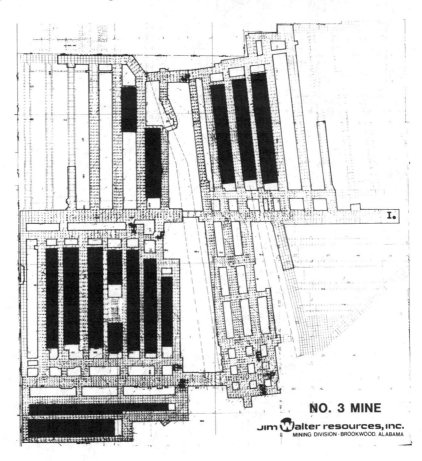

NO. 3 MINE

JIM Walter resources, inc.
MINING DIVISION - BROOKWOOD, ALABAMA

Figure 1
No. 3 Mine Map

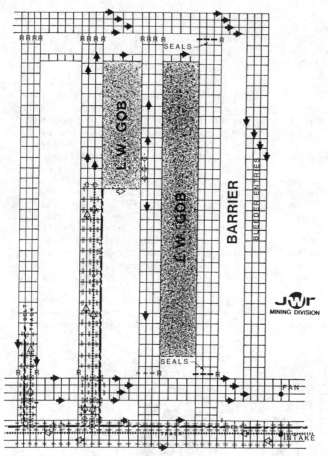

Figure 2
Antitropal Face Ventilation Plan

loss is sixteen times larger. This loss does not include any leakage loss through stoppings that are adjacent to caved gob areas and therefore usually subjected to strata convergence and floor heave. Both increase resistance to flow. In the four headgate entries, heavy regulation must be applied to insure that the secondary air current and the face ventilating current merge in the headgate near the corner of the longwall face. Although two of the entries can be easily regulated, the track entry requires an airlock door arrangement to provide necessary regulation. Futhermore, the belt entry must be regulated to provide the necessary air quantity to insure that the secondary air current in the headgate entries merge with the longwall face current in the vicinity of the headgate. After merging, both air currents flow into the headgate bleeders and the bleeder system. Care must be taken to insure that the secondary air current in the headgate merges with the

face ventilating air current in such a manner that the headgate end of the longwall face remains ventilated. Because regulators are located in travelways e.g. the track entry airlock, inadvertant changes can be made. In the belt entry, rock and coal moving on the conveyor belt will wear and change the size of the opening in the belt regulator resulting in a decrease in the amount of air going across the face or the two air currents merge at a location that shifts and may permit a portion of the longwall face near the headgate to be unventilated with the face ventilating air current entering the gob prematurely, or the stage loader area being ventilated with dust laden air from the headgate face. The tightness of the cave behind the No. 1 shield and the inby crosscut determines how much dust laden air passes over the stage loader. The belt regulator is another dust source generator.

Figure 5 is the mine map of No. 7 Mine

554

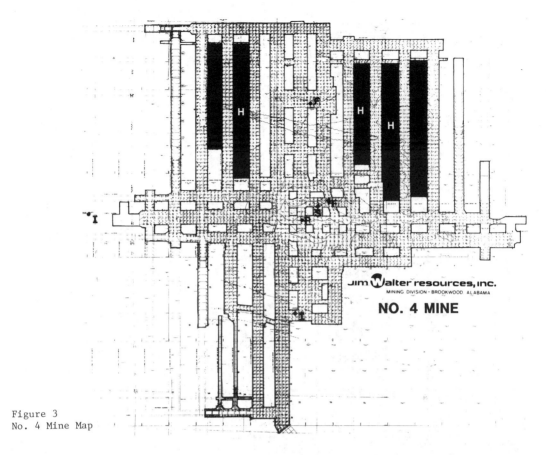

Figure 3
No. 4 Mine Map

where the first panel was partially ventilated by the homotropal system but was converted to antitropal because of methane and dust control considerations at the stage loader in the headgate belt entry as stated in the previous paragraph.

6 DUST CONTROL CONSIDERATIONS

To maintain compliance with respirable dust standards and to reduce float coal dust in the vicinity of the shearer is both challenging and frustrating. Unlike methane in air mixtures where airflow changes can be immediately evaluated by measuring methane concentrations in the air stream, respirable dust is elusive and it is therefore difficult to determine the success or failure of a change. Although instantaneous dust sampling equipment may be beneficial in evaluating the change, the key to the success of the dust control program lies in the experience of the person trained in dust control practices and techniques and translating his observations and findings into corrective practices.

Because of high velocity, the advantage of homotropal ventilation may be obfuscated. The cross sectional area towards the tailgate side of the shearer is wider by the depth of the drum cut resulting in a lower air velocity. At the headgate drum on the shearer, the air velocity must converge and increase approximately 20% in an area where dust is not only generated by where it may become airborne due to drum rotation. The increased ventilating air current velocity transports float dust to the headgate where the air velocity decreases at the stage loader and deposits float dust on the floor and coal ribs where it is not only readily apparent, but also in close proximity to heat and arcs caused by electric motors, controls and power cables to create a potential hazard. The Blue Creek Coal bed is soft with a Hardgrove grindability index approximating 100. When cutting the coal face, the zone in front of the lead cutter drum takes weight due to roof pressure and the face sloughs to create a dust cloud. When cutting homotropally, the dust cloud is in the narrower part of

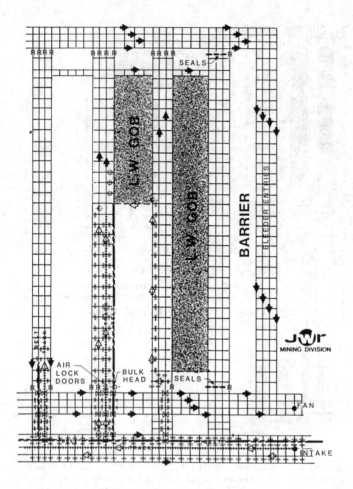

Figure 4
Homotropal Ventilation Plan

the face and in the elevated air velocity face zone so that additional dust is transported to areas where the velocity decreases to again drop dust in the headgate/stage loader vicinity.

Conversely, observations indicate that high velocity at the lead cutter drum may assist in controlling dust at the miners breathing zone due to the streamlining tendency of the air stream as it diverges downwind of the shearer when antitropal ventilation is used. In high air velocities, the shearer clearer concept, splitter and belt arm arrangement with properly directed water sprays can assist in diverting dust laden air towards the longwall face and again out of the miners breathing zone. The splitter and belt arm arrangement is somewhat effective in directing the dust cloud towards the face and out of the breathing zone of the shearer operator and helper depending upon the distance that the sloughage occurs upwind of the shearer.

7 CONCLUSIONS

At this time, the simplicity and the reproducibility of the longwall face ventilation system on succeeding panels makes antitropal the preferred system. However, because of the 2000 feet of cover over the mines, pillar configurations in continuous miner development panels are being re-evaluated to reduce the impact of front abutment pressure in both headgate and tailgate entries. It will require two to four more years to complete the evaluation of the yield-stable-yield pillar development mining plan. If two of the tailgate entries can be maintained in good condition when stable pillars are used, and an excessive number of wood or fibercrete cribs to control front abutment pressure are not required, the inherent dust control advantages of homotropal face ventilation will necessitate that we reassess our ventilation and dust control plan. In addition, the ability of our surface and

556

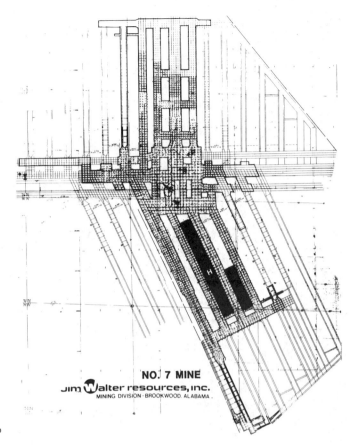

NO. 7 MINE
Jim Walter resources, inc.
MINING DIVISION - BROOKWOOD. ALABAMA

· Figure 5   No. 7 Mine Map

underground degasification efforts to
reduce the velocity of the ventilating
air current across the longwall face must
impact on existing face ventilation
practices and plans.

Jim Walter Resources is now operating
seven longwall systems and on every system,
antitropal ventilation is used. This is
attributed to the requirements of ventilating air currents in excess of 1000 fpm
for methane control and the inability to
maintain open and isolate from adjoining
gobs a sufficient number of air courses in
the tailgate resulting in high ventilating
pressure losses. An Engineer is scheduled
to work full time to evaluate ventilation
and dust control practices and techniques
on all longwalls and to standardize
successful practices on those walls not
employing such practices or techniques. He
works with suppliers and manufacturers to
evaluate new, innovative design changes
for implementation on all faces.

14. Air cooling and refrigeration

Geothermal air heating in a southwestern Wyoming trona mine

D.T.MOORE
Stauffer Chemical Company of Wyoming, Green River, USA

ABSTRACT: Since 1981, Stauffer Chemical Company of Wyoming has been utilizing an abandoned mining panel in their Big Island Mine to heat mine intake air. Ambient air enters the mine at temperatures as low as -45°F (-43°C) and is heated geothermally to temperatures in excess of 40°F (4°C) before being introduced to active portions of the mine. This method of heating intake air has proven to be economical since no fuel is required for heat generation. This paper presents Stauffer's experience with geothermal air heating and future expectations of this program.

1 INTRODUCTION

The Stauffer Chemical Company of Wyoming Big Island Mine and Refinery are located roughly 25 miles (40 km) northwest of Green River, Wyoming at an elevation of 6200 feet (1890 m) above sea level. Underground mining operations supply trona, a sodium-sesquicarbonate mineral, to a surface refinery where high purity commercial soda ash is produced. Two trona seams are being mined by conventional and continuous mining methods. The two seams are 800 feet (244 m) and 850 feet (259 m) below the surface and are accessed by three concrete lined, circular shafts.

From mine startup in 1962 until 1980, only two shafts existed. All intake air entered the mine through the service shaft (#1 Shaft), used also for transporting men and materials, and for routing mine utility systems such as water and power. The production shaft (#2 Shaft), divided down the center by a concrete curtain wall, also serves as the mine exhaust shaft. Two Jeffrey Aerodyne fans, each with a 4160 vac-300 HP motor, are operated in parallel at the collar of #2 Shaft and create a negative pressure throughout the mine.

In 1980, a second production shaft (#3 Shaft) was added to increase production capabilities and to improve the ventilation network in the mine. This shaft is similar to #2 Shaft in that it is also divided by a concrete curtain wall separating the ore hoisting and the ventilation compartments of the shaft. Roughly 60% of the mine intake air now enters through #3 Shaft thereby reducing the airflow in the service shaft and the congested shaft bottom area.

Steam is utilized as a heat source to raise the temperature of air traveling down the service shaft during the colder months of the year to provide a comfortable working environment in the shaft and shaft bottom areas, and to prevent the freezing of the water lines installed within the shaft. This mechanical heating is accomplished by fans pulling ambient air across steam heater coils and then forcing the heated air to the shaft collar where it is mixed with other ambient air and introduced into the mine.

2 LOCAL CLIMATE

Southwestern Wyoming experiences some of the most severe winter temperatures in the United States. The National Oceanic and Atmospheric Administration's (NOAA) Environmental Data Service provides climatography data at two locations near the Stauffer plant site. Fontenelle Dam, located 30 miles (48 km) north of the plant, experiences an average annual temperature of 38.8°F (4°C) according to the NOAA. The NOAA reports that Green River, Wyoming, located 25 miles (40 km) southeast of the plant, experiences an average annual temperature of 43.3°F (6°C). The Stauffer plant, being located roughly half way between these NOAA weather stations, experiences temperatures somewhere within this range.

Stauffer's plant Environmental Services Department (ESD) maintains a weather station that records temperatures as well as other climate related data. Since 1982, the ESD has measured surface temperature extremes ranging from 96°F (36°C) to -45°F (-43°C). The 1983 calendar year average surface temperature was 40.6°F (5°C) at the plant site.

For the purposes of this study, it has been assumed that the average annual surface temperature at the Stauffer plant site is 41°F (5°C).

3 THEORETICAL STUDY

During the late 1970's, the soda ash market expanded rapidly. It became apparent that a third shaft was necessary for Stauffer to keep up with the increased demand. In addition to increased production capacity, this new shaft would provide additional intake air for the rapidly expanding mine. During the planning stages of this new shaft, some provision for heating this intake air had to be provided. It was at this time that the concept of heating air geothermally was introduced.

After determining that the earth's natural warmth was indeed a potential source of low cost air heating, an evaluation was needed as to whether such a concept was economically feasible. Mr. A.J. Ziegenhagen of Stauffer's Process Development Department applied his heat transfer expertise to this concept by setting up a computer model to determine: (1) if intake air could be adequately heated by routing it through an abandoned area of the mine, and (2) the length of time that the system would remain effective.

Trying to model actual airflow conditions in a room and pillar panel would require mathematics far too complicated to be easily solved. Rather, the model developed by Mr. Ziegenhagen assumed an infinitely wide slit of length L and height H with air entering at one end at a constant inlet temperature (T in) traveling at a uniform velocity V (see Figure 1). The rock above and below the slit is assumed to extend infinitely in both the +x and -x directions and has an initial temperature (T rock) throughout. In other words, the slit is assumed to be well below the earth's surface so that changes in surface temperatures will not influence the rock temperature near the slit. When air travels through the slit, sensible heat is transferred from the rock to the air. Since this model represents an unsteady system, both the outlet air temperature (T out) and the rock temperature (T rock) will decrease with time. The basic problem is to determine the outlet air temperature (T out) as a function of time.

Many aspects of an actual situation could not be considered in this model. Sources of heat not considered were the heat provided by the intake shaft walls, adiabatic compression of the air, latent heat from moist shaft walls and natural recharging of heat along exposed surfaces from surrounding rock. Thermal conductivity of salt was used in the calculations since this is probably the closest figure available to trona.

The initial goal was to maintain an outlet air temperature (T out) of 60°F (16°C), a temperature comparable to the air heated by steam heaters in the service shaft. After plugging in values for the many variables used in the model, it was found that the outlet temperature (T out) could be maintained at 60°F (16°C) for a period of less than one month. It was immediately obvious that the time, effort and expenses required to set up a system useful for one month was not feasible.

The model was solved again, this time for maintaining an outlet temperature of 44°F (7°C). At this temperature, the model predicted the system would be useful for many years, a much more feasible time frame.

The next step was to determine the most efficient means of routing the intake air through one or more of the available panels. There were many flow patterns considered.

1) The entire airflow could be split and passed in parallel through any number of panels.
2) The entire airflow could be passed in series through these panels.
3) Partitions could be built within any one panel forcing the air to make several passes through the panel before exiting.

A tradeoff of more efficient heat transfer between the rock and the air versus increased ventilation pressure drops had to be considered.

In general, the following conclusions were made:
1) The rock to air heat transfer is more efficient utilizing three passes through a panel than allowing the air to pass in one unobstructed pass through a similar panel.
2) When utilizing numerous panels at once, series flow provides more efficient heat transfer than parallel flow.
3) For panel widths utilized in this mine, there appears to be no advantage to making the number of passes greater than three.

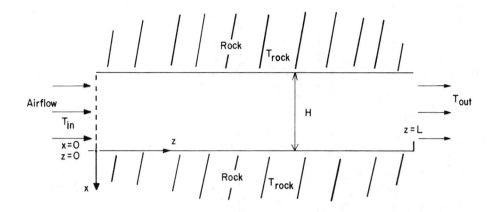

Figure I. Geothermal Air Heating Model

4 ACTUAL PROGRAM

From the results of Mr. Ziegenhagen's research, the program was begun by building partitions to route air through a panel three times. Since no pillars had ever been extracted in any of the previously mined panels, many abandoned panels were available for the program. The NW #1E Panel was selected for this program due to its close proximity to the new #3 Shaft (see Figure 2).

One of the disadvantages of setting up a panel with partitions for routing air is the cost of the partitions. However, during the upreaming and slashing operations used to install the new shaft, a considerable amount of waste rock had to be gobbed somewhere in the mine. This waste rock was placed in the NW #1E Panel to serve as partitions. By utilizing this waste rock as partitions, both gobbing the waste rock away from the #3 Shaft construction area and setting up the air heating panel were accomplished. In crosscuts where the gob could not be piled to the roof, wire mesh was hung from roof bolts to fill the gaps. The mesh was sprayed with a stopping sealant to provide an airtight partition. Four overcasts and 22 cinder block stoppings were built to reroute a mine return air route around and over this new mine intake.

Instrumentation was purchased and installed to monitor air and rock temperatures at different locations throughout the panel. Figure 2 shows instrument locations. Thermocouple probes, labeled as TC1 and TC2, were installed hanging from the roof near the bottom of #3 Shaft and near the east entrance of the NW #1E Panel. These two thermocouples are wired into an Omega

Model 400 Two Pen Chart Recorder so that our temperatures at these two points can be continuously monitored. A Honeywell seven-day circular chart recorder continuously monitors the temperature of the air leaving the air heating system and entering the active mine areas.

Two areas within the NW #1E Panel, labeled PP1 and PP2 on Figure 2, were selected for studying the temperature gradient developed within the rock adjacent to the panel. Figure 5 illustrates the basic setup of these areas. Two holes were drilled; one vertically into the floor and the other horizontally into a pillar. A wooden tamping stick with thermocouple probes placed through the stick at 0, 2, 4 and 6 foot depths were placed in each hole. The temperatures of the rock at each of these depths is measured monthly using an Omega Portable Thermometer Model 5800.

5 RESULTS OF PROGRAM

In effect, the entire mine, from the intake shafts to the exhaust shaft, is one large geothermal air heating system. Ambient air enters the mine at varying temperatures, circulates through the mine workings and exits the mine at a constant 62°F (17°C). We are concerned only with the part of the system necessary to warm the coldest intake air to a predetermined minimum comfortable working temperature. The NW #1E Panel, to date, has allowed the exhaust temperature to drop as low as 46°F (8°C).

The in situ temperature of the rock at mine level is estimated to be 65°F (18°C). When ambient air enters the air heating panel at temperatures below the rock temperature, which occurs roughly 7 months of

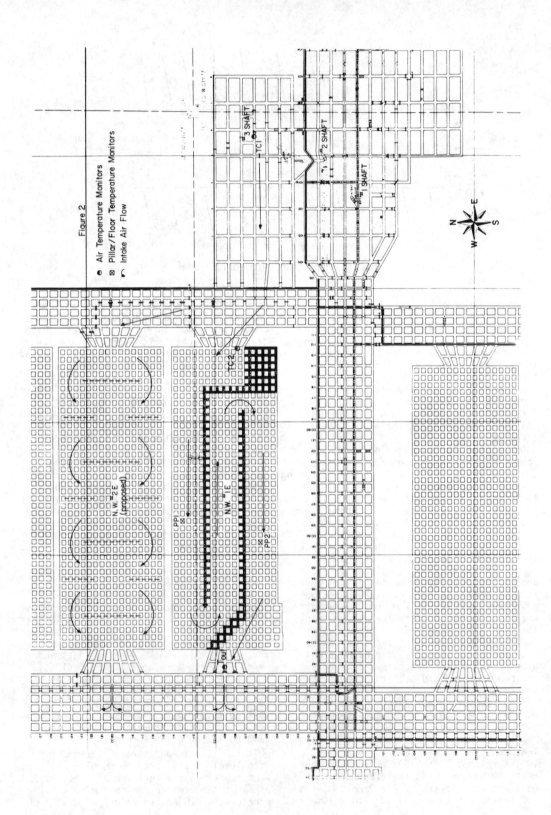

Figure 2

● Air Temperature Monitors
⊠ Pillar / Floor Temperature Monitors
↰ Intake Air Flow

the year in this mine, heat is transferred by conduction from the rock surfaces to the passing air. The air gets warmer while the rock surfaces get cooler. When very cold air passes by the rock surfaces, not enough heat is available at the surface to adequately heat the air. Heat then is drawn out from deeper within the rock to heat the air causing a temperature gradient to form well into the rock.

During the warmer months of the year when intake air temperatures are warmer than the pillar temperatures, heat is transferred from the air into the rock. This process is similar to recharging a storage battery. If enough warm air can be passed through the panel during the summer months to return to the rock all heat that was consumed during the winter months, the panel would realize no net heat loss and could be used indefinitely with the same efficiency.

However, far more heat is consumed during the winter than is returned in the summer. Since complete recharging fails to occur, the rock temperature immediately in and around the NW #1E Panel is gradually decreasing each year until a temperature is reached where complete recharging can occur. As a result of rock temperatures decreasing each year, the system reduces its capacity to heat air and the panel exhaust temperature slowly drops. We expect this decrease to continue until the average panel exhaust air temperature drops to approximately 41°F (5°C), which is nearly as low as the average ambient air temperature that enters the mine. This is the point of system equilibrium where complete recharging occurs and no further net heat losses occur from year to year.

A theory stating that a heat depleted panel can recharge itself when no longer in use has not yet been proven, but is being considered. If an air heating panel is closed off after a normal winter and no air is passed through it, it is believed that the rock within and adjacent to the panel, already somewhat depleted in heat, will gradually regain heat solely from the earth's natural warmth. The earth itself will act as an infinite heat source and will be little affected by our system. When heat consumption within the system ceases, heat migrating from within the rock mass will eventually return the system to its original temperature.

Figure 3 shows the fluctuations in temperature at two points within the system. The curve with the greatest range shows the average monthly temperatures at the bottom of #3 Shaft. Notice that the peaks and valleys on this curve are decreasing each year. Assuming each year has normal tem-

peratures, this gradual downslope represents the reduction in heating capacity of the shaft. The curve with the lesser range shows the average monthly temperatures of the air leaving the air heating panel and entering the active mine workings. It is when this curve drops below the minimum acceptable working temperature that a new panel must be utilized.

Figure 4 shows the heat transfer each month over the last four years. Positive heat transfer (air cooling) occurs when ambient air enters the mine at temperatures higher than the panel exhaust temperature. As this relatively warm air passes through the system, heat is drawn from the air and back into the adjacent rock mass. Negative heat transfer (air heating) occurs during those months when the ambient air temperature is lower than the panel exhaust temperature signifying that heat is removed from the adjacent rock.

The temperature gradient formed in the adjacent pillar and floor rock is shown in Figures 6 and 7, respectively. Similar to the air temperature curves (Figure 3), these curves have a gradual downward slope showing that net heat loss occurs within the adjacent rock each year. These curves clearly show that temperatures closer to the surface of the rock fluctuate greater from year to year since they are closer to the actual air flow. In theory, if temperature probes were installed at great enough depths within the rock, no change in temperature would be measured.

The temperature ranges of the probes placed in the pillar hole are greater than those in the floor hole even though they are located at the same location within the air heating panel. The pillar hole penetrates a 32 feet x 32 feet (10 m x 10 m) trona pillar, while the floor hole penetrates floor shales. Although a definitive explanation is not available, it is suspected that different rates of recharging account for the difference. The hole in the floor has an infinite body of heat behind it from which to draw heat. The pillar hole has only a 32 feet x 32 feet (10 m x 10 m) mass from which to draw heat. Since air flows entirely around the pillar, heat is consumed or returned to the pillar at a faster rate than from the floor. Thermal conductivity and specific heat capacity data obtained for trona and shale do not show enough variance to be significant.

Figure 8 illustrates the typical behavior of the air from when it enters the mine until it leaves the mine. Using data obtained from calendar year 1984, the average temperatures at different locations through

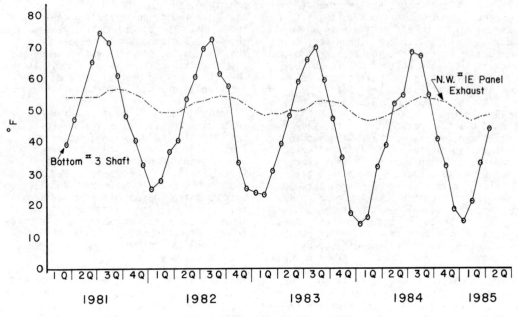

Figure 3 Average Monthly Air Temperatures

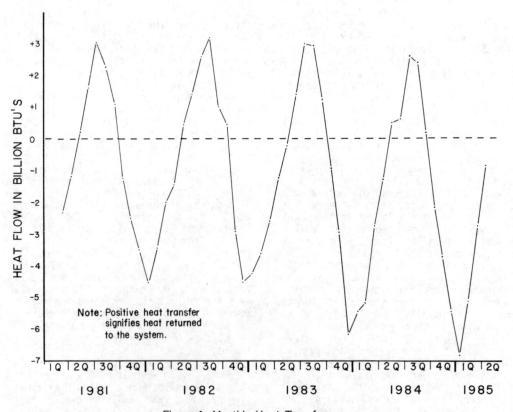

Figure 4 Monthly Heat Transfer

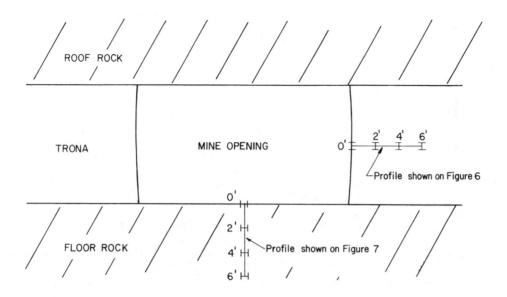

Figure 5 Location of Pillar/Floor Temperature Monitors

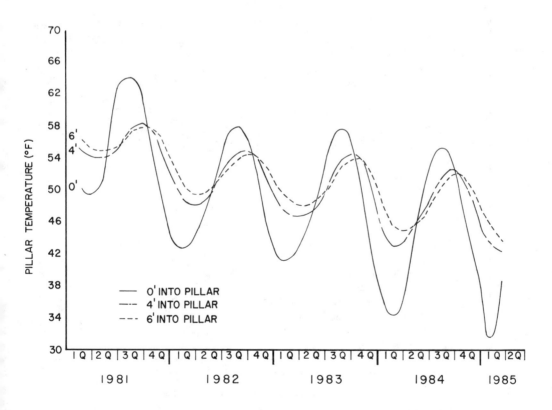

Figure 6: Temperature Profile Into A Trona Pillar

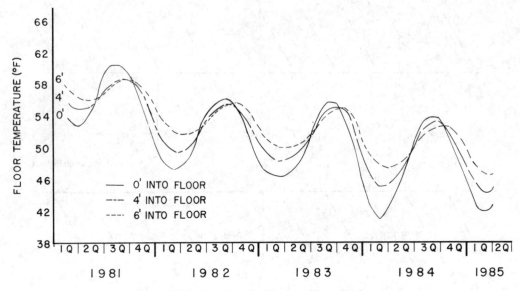

Figure 7 Temperature Profile Into Floor Rock

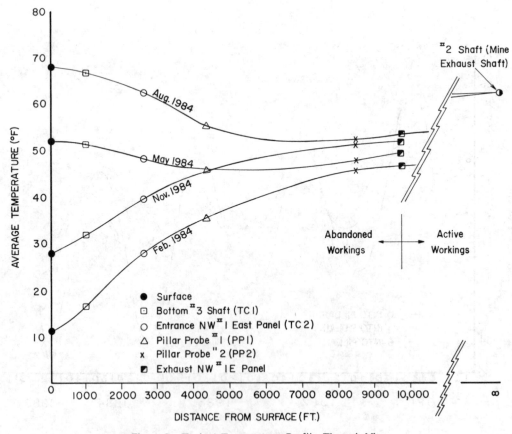

Figure 8 Typical Temperature Profile Through Mine

568

the mine for four typical months have been plotted. The mine is assumed to have an infinite length with the air heating system being the first 10,000 feet (3048 m). August is historically the warmest month each year with February the coldest. Air entering the mine with an average temperature differential of nearly 60°F (33°C) is reduced to a differential of roughly 7°F (4°C) by the time it enters the active workings. From there the differential is reduced even further until the air leaves the mine at a constant temperature.

6 RISK

All intake air entering the service shaft is heated on surface when ambient air temperatures are near or below freezing. Return air leaves the mine via the exhaust shaft at a constant temperature of 62°F (17°C). Therefore, the concrete liners of these two shafts have never been subjected to cycles of freezing and thawing. However, #3 Shaft has had to endure the full spectrum of ambient air temperatures. Temperature variances throughout each year have subjected the concrete lining in #3 Shaft to undergo up to 150 freezing and thawing cycles annually. Each of these freezing and thawing cycles cause a minute amount of damage to the concrete shaft liner. Collectively, these freezing and thawing cycles cause a gradual deterioration of the concrete liner. Compounding the problem, #3 Shaft has an inflow of up to 5 gallons per minute (19 /min) of water. The water does not enter the shaft at one point, but appears to seep out of the saturated liner throughout much of the shaft. Since water and concrete react differently when frozen or thawed, their relative forces acting against one another accelerate the rate of deterioration. If the air were heated on surface, damage to the concrete liner of #3 Shaft would probably have been minimal.

In addition to the repeated freezing and thawing of the concrete liner over the years, ice buildup within the shaft presents some problems. This ice buildup has a tendency to restrict the airflow traveling down #3 Shaft which increases the overall static pressure of the mine. In the spring when warmer temperatures begin to melt the ice, a falling ice hazard exists near the shaft bottom as basketball-size chunks of ice separate from the shaft liner.

In this mine, the trona ore is more competent than the shale layers above and below the ore body. Stauffer's mining plan requires that at least 6 inches (0.15 m) of trona be left in the roof to minimize

ground problems. In practice, leaving trona in the roof is not always possible, leaving patches of shale exposed to the mine atmosphere. Our experience indicates that the shale layers are much more reactive to temperature and humidity changes than trona. Due to the extreme air temperature and humidity changes in the geothermal air heating panel throughout the years, the exposed trona/shale interfaces become increasingly weaker, eventually leading to the separation of the bedding planes. The sloughing that results does not present a hazard to mine personnel since this panel is not used as a travelway.

Setting up a geothermal air heating panel requires a substantial amount of uncaved workings. Many mine operators rob pillars as they exit a mining area and allow the ground to cave behind. For many of these operations, committing a developed pillar area to an air heating program may not be economically feasible.

To effectively heat mine intake air, the overall mine static pressure head will be increased to account for the increased distance that the air must travel before actually being used. Although small, this increased pressure does represent an additional power cost for operating the main mine fans.

7 PROGRAM FUTURE

To date, only one panel has been utilized for geothermal air heating. The panel exhaust temperature has dropped to a low temperature of 46°F (8°C) and should continue to drop further as discussed earlier in this paper. Now that the panel exhaust temperature is approaching what is considered to be the minimum acceptable comfortable working temperature of 44°F (7°C), several options are being considered.

The most obvious option is to set up another complete panel for air heating. During the summer of 1985, the next mined-out panel to the north, NW #2E, is scheduled to be set up and used beginning the winter of 1985-1986. This panel will also be a one panel air heating system, but with solid brattice curtains used as baffles to force air throughout the width of the panel (see Figure 2). This panel layout will be relatively inexpensive to set up and instrument since the cost of solid partitions will be avoided. Pressure losses should be reduced over the previous panel since the new layout simulates parallel airflow circuits rather than a three pass series circuit previously used.

While utilizing this new panel, the previous air heating panel will be completely shut off from air flow and will be monitored to study the natural panel recharging rate. Knowing how quickly a depleted panel recharges itself when not in use will determine the number of panels necessary to maintain adequate continuous air heating.

After two or more panels have been set up for single panel use, parallel flow air heating can be used. The intake airflow can be split between any number of available panels. By routing less air though each panel, more efficient heat utilization will provide warmer temperatures for longer periods of time. The main drawback of parallel panel systems is that additional backup panels must be available when the parallel flow system reaches the end of its useful life.

Series flow through numerous panels provides the best heat utilization, but requires the most expensive setup cost and adds the greatest amount of pressure to our ventilation circuit. As with parallel flow systems, series flow systems also require backup panels when recharging is necessary.

The most practical system appears to be two to four panels set up in parallel so that the air can be switched from one panel to another simply by opening and closing ventilation control doors. One panel can be active at all times while the others are recharging. The frequency of panel changes will depend on recharging rates, the number of panels in the system, and volume of airflow. Once a sufficient number of panels are incorporated into the system, no further capital cost should be required for the remaining life of the property.

8 SUMMARY

Geothermal air heating is a successful means of low cost mine temperature control in locations where winter temperatures are extreme. The program utilizes an infinite and renewable resource that is available to most underground mines. This concept should be of interest for operations where mine intake air must be heated during the cooler months of the year and where there are abandoned workings that are both standing and accessible. In the four year life of this system, over $350,000 in natural gas costs have been averted due to this program. Once the complete system is developed, similar savings with very little cost should be realized for the duration of the mine's life.

Energy recovery turbines for use with underground air coolers

R.RAMSDEN & S.J.BLUHM
Chamber of Mines of South Africa Research Organization, Johannesburg, South Africa

ABSTRACT: In the past one of the major objections to the use of direct-contact air coolers in the South African gold mines has been the loss of energy associated with dropping the water pressure to the local barometric pressure. The recent introduction of turbines to recover part of this energy, normally less than 50 kW, in a number of air cooler installations has partly overcome this objection.

This paper describes some specific applications of these turbines and air coolers.

INTRODUCTION

The use of cooling is essential in the South African gold mines where mining takes place at depths up to 3500 metres below surface and virgin rock temperatures in excess of 60 °C are encountered. During the last eight years the installed refrigeration capacity on the South African mines has increased annually by an average 18 per cent and is at present approximately 1000 MW(R), or 284 000 refrigeration tons (Figure 1). Large refrigeration machines are installed both on surface and underground and large quantities of chilled water are used to distribute the cooling to the working places. The distribution of the cooling is achieved by chilling the normal mine service water and by cooling the ventilation air using water-air heat exchangers[1]. The underground cooling of the ventilation air takes place in large bulk air coolers situated in the main intake airways and, if necessary, also in smaller secondary air coolers which provide localized cooling near the workings.

The bulk cooling installations are usually of the direct-contact type[1]. These include horizontal multi-stage spray chambers in which the chilled water is sprayed vertically upwards into a horizontal air stream and units which make use of conventional cooling tower packing to create the wet surface area for heat transfer. The normal practice with the smaller secondary air coolers is to use finned tube heat exchangers known as cooling coils which are arranged in various combinations[1,2] in closed circuit cooling networks. Recently small mobile direct-contact units have been introduced for this application and will probably find wider use in the future.

Various designs and sizes of water turbines are being installed increasingly in mine chilled water and cooling systems. Firstly, there are relatively large Pelton wheel installations which typically generate about 2 MW of power while breaking the pressure of the full water flow rate

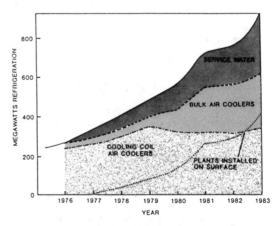

Fig. 1. Installed refrigeration capacity on South African gold mines.

down the shaft[5]. Secondly, there are
relatively small turbines which are used
in conjunction with the direct-contact air
coolers. The cold water pipe networks
supplying these air coolers are normally
integrated with the chilled service water
system and the water supply pressure at
the air cooling sites is seldom less than
1000 kPa. This is in excess of that
required at the spray nozzles and instead
of dissipating this pressure energy
through a valve, a small water turbine can
be used to generate useful power. This in
turn may be used to enhance the overall
performance of the heat exchanger by driving a pump, fan or other device.

Presently, in the South African gold
mining industry, there is a strong
interest in a concept termed hydro-power[3]
in which a considerable portion of stoping
machinery will be powered by water
supplied from surface under the full
gravity head. This water will be used to
power directly rockdrills and water jets,
as well as turbines to drive winches, fans
and other devices. Thus, there is an
interest in a third application for turbines, namely that of power generation
from an extremely high pressure supply of
water (typically 16 MPa).

When water pressure is reduced by a
throttling valve or through pipe friction
there is a water temperature rise[4] due to
the Joule-Thomson effect of $2,4 \times 10^{-4}$
K/kPa (or 2,3 K per 1000 m head). However
if useful energy is taken out of the water
stream with a turbine then there is a
corresponding reduction in the water
temperature rise. Thus the benefits of
using turbines are twofold; firstly there
is the production of power and secondly
there is a reduction in the apparent heat
load of the system. It has been shown[5]
that the savings in refrigeration costs
associated with the lower temperature
water being available can be more significant than those due to the power generated
through the turbine.

This paper is primarily concerned with
the use of relatively small turbines in
association with direct-contact air
coolers. However some discussion is also
presented on the other two applications
mentioned above.

TYPES OF TURBINES

Water turbines are divided into impulse
and reaction machines, depending on the
extent to which the water pressure head is
converted into kinetic energy within the
turbine. For the purposes of this paper

and the present underground applications,
these two divisions may be read as Pelton
wheel turbines and Francis turbines
respectively.

In a Pelton wheel turbine (see Figure
2), the total water pressure head is first
converted into kinetic energy through one
or more nozzles. The water jets from the
nozzles strike vanes or buckets attached
to the periphery of a rotating wheel. The
wheel rotates in air and the water leaves
the unit at the ambient pressure. The
leaving water stream thus has no residual
gauge pressure. Control of a Pelton wheel
may be achieved by needle valves which
change the flow characteristics of the
nozzles or by deflector plates which
deflect the water jet away from the
buckets.

Fig. 2. Pelton wheel turbine.

In Francis turbines (see Figure 3) the
water first passes through a ring of
stationary guide vanes in which only part
of the total head is converted into
kinetic energy. The guide vanes (also
termed nozzles) accelerate the water and
guide the water flow at the correct angle
into the runner. The pressure energy is
further converted into kinetic energy in
the runner and therefore a pressure
difference exists across the runner. The
actual pressure drop across a Francis
turbine is controlled by the configuration
of impeller, guide vanes and rotational
speed. The water flow rate may be controlled by the setting of valves upstream
or downstream of the unit. The general
design of a Francis turbine impeller is
similar to that of a centrifugal pump. In
fact pumps operating in reverse may be
used as energy recovery devices.

Fig. 3. Francis turbine.

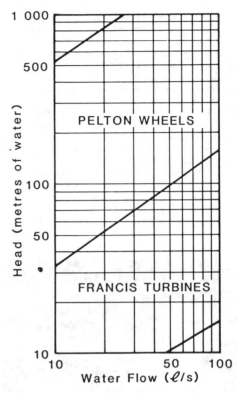

Fig. 4. Optimum operating conditions for single stage turbines.

All turbines are designed to meet a specific duty which is normally associated with the maximum efficiency of the machine. These operating conditions are classified by a term called the specific speed[6], which is given by:

$$N_s = NQ^{\frac{1}{2}}/(gH)^{\frac{3}{4}} \quad \text{(Radians)} \quad (1)$$

where N_s - specific speed (rad),
$\quad$ N - turbine speed (rad/s),
$\quad$ Q - turbine flow (m³/s),
$\quad$ g - gravitational constant (m/s²),
$\quad$ H - head (m of H_2O).

The specific speed is also used to categorize geometrically similar turbines and thus may be used to match a set of operating conditions with the most suitable type of turbine. Pelton turbines are reasonably efficient when operated with a specific speed (N_s) between 0,05 to 0,40 and Francis turbines with a specific speed (N_s) between 0,40 to 2,20 across each stage. For normal operating speeds, a map of water flow rate against pressure head, which shows the most suitable type of machine for various operating conditions, can be drawn (see Figure 4). Another general guide is that Pelton wheels perform optimally when the ratio of bucket velocity to the jet velocity is about 0,4 to 0,5.

There are numerous operating constraints that usually predetermine the choice of

turbine for a given cooler, as will be demonstrated in the next sections.

TURBINES USED WITH BULK AIR COOLERS

The performance of bulk air coolers, particularly spray chambers, can be greatly improved by collecting the water in a sump and re-spraying it into the air stream. These multistage units are arranged so that the water moves from stage to stage in a manner countercurrent to the air flow. For these coolers to operate satisfactorily, it is necessary for the spray nozzles to break the water stream into small droplets (typically 2 mm in diameter) and for the nozzle water pressure (normally about 200 to 300 kPa) to be sufficient for the water droplets to just reach the hangingwall and produce an even droplet distribution across the chamber. The water supply pressure required in the tower-type installations is normally about 50 to 100 kPa.

Frequently the water supply pressure is much higher than that required and, although the water pressure may be dropped across a throttling valve, there is often sufficient energy available to drive re-staging pumps through a water turbine. In a number of trial installations, Francis turbines have been used in this manner in association with multistage spray chambers. Three of these applications are described below.

Figure 5 shows a Francis turbine (nominal output 15 kW) which is used to drive a re-staging pump on a two-stage spray chamber having an air cooling duty of 2000 kW. The design specifications for this turbine installation are given in Table 1. The toothed belt drive (1,57 to 1 reduction ratio) between the turbine and

Fig. 5. Pump turbine unit installed on a two stage spray chamber.

the pump allows both the turbine and pump to operate at speeds close to their peak efficiencies. In many similar cooling installations it is found necessary to operate the turbine at a higher speed than the pump, and belt drives (V belts and toothed belts) have proved to be reliable in these mine applications. Even with the elevated turbine speed of 4700 rev/min, the specific speed (N_s) is 0,63 which is close to the lower limit for the efficient operation of Francis turbines.

A turbine can be used to drive more than one pump, as demonstrated in the next application where a Francis turbine is used to drive two re-staging pumps in a three-stage 750 kW spray air cooler (Figure 6). The design and performance of this three-stage spray chamber is fully described in Reference 7. The turbine and an electric motor are connected through centrifugal clutches to the drive shaft, thus providing a standby power supply in

Table 1. Design specifications for turbine and pump in a two-stage 2000 kW spray air cooler

Supply pressure to turbine	1000 kPa
Delivery pressure from turbine	300 kPa
Water flow through turbine	30 ℓ/s
Power taken out of water stream	21 kW
Turbine shaft power	14,1 kW
Turbine efficiency	67 per cent
Turbine speed	4700 rev/min
Pressure across 2nd stage pump	300 kPa
Water flow through 2nd stage pump	30 ℓ/s
Power supplied to 2nd stage waterflow	9 kW
Pump speed	3000 rev/min
Overall efficiency of system	43 per cent

the event of a turbine failure. In this application the speed of the turbine was determined by the speed of the electric motor (2900 rev/min) so that the specific speed was only 0,25. This speed is more suitable for a Pelton wheel; however a residual pressure was required for the first stage nozzles of the spray chamber. The turbine operated for two years without any major problems and a standby electric motor would not be fitted in future installations, thus removing the restraint on the turbine speed. The pay-back period for this turbine-pump installation is approximately one year and the turbine and

Table 2. Design specifications for turbine and re-staging pumps in a 3 stage 750 kW spray air cooler

Supply pressure to turbine	1000 kPa
Delivery pressure from turbine	175 kPa
Water flow through turbine	15 ℓ/s
Power taken out of water stream	12,4 kW
Turbine shaft power	7 kW
Turbine efficiency	57 per cent
Turbine speed	2900 rev/min
Pressure across re-staging pumps	175 kPa
Water flow through re-staging pumps	15 ℓ/s
Power supplied to water in each stage	2,6 kW
Pump speed	2600 rev/min
Overall efficiency of system	42 per cent

Fig. 6. Pump turbine unit with standby electric motor for a 3-stage spray chamber

pump specifications for the cooler are given in Table 2.

In an interesting air cooling installation, which is presently under construction, a total of 430 m^3/s of air will be drawn out of a main shaft and passed through two 10 MW spray air coolers in parallel before being returned into the shaft. These bulk air coolers are two-stage spray chambers each using 130 ℓ/s of chilled water. There is sufficient energy available in the primary water stream to drive a pump/turbine unit which provides water for the second stage. The pump-turbine units are directly coupled which gives obvious advantages in simplicity. The units have been selected to run at

Table 3. Design specifications for turbine and re-staging pump in a 2-stage 10 MW spray air cooler

Supply pressure to turbine	1000 kPa
Delivery pressure from turbine	300 kPa
Water flow through turbine	130 ℓ/s
Power taken out of water stream	91 kW
Turbine shaft power	70 kW
Turbine efficiency	77 per cent
Turbine speed	1450 rpm
Pressure across re-staging pump	325 kPa
Water flow through re-staging pump	140 ℓ/s
Power supplied to water stream	45,5 kW
Pump speed	1450 rpm
Overall efficiency	50 per cent

four-pole motor speed which will allow for the simple replacement of the turbine with an electric motor in the event of a failure. This results in a relatively low specific speed (N_s=0,41) which is however, acceptable as indicated by the relatively high efficiency (above 70 per cent). The specifications for the pump turbine unit used in this installation are given in Table 3.

As an alternative to turbines for taking energy out of the water stream, the high pressure water may be sprayed into the air and the drag on the air will induce an air flow. In one mine installation[1,8] cold water at 600 kPa is sprayed into the throat of ejectors, thus inducing air to be drawn into the venturi where the air and water become fully mixed. The efficiency of air entrainment (defined as: power absorbed by the air stream/power supplied in the water stream) is about 20 per cent for typical operating conditions in this type of cooler.

TURBINES USED WITH SECONDARY AIR COOLERS

It is generally accepted that the cost of distributing cooling through small water to air heat exchangers located at strategic positions[1] is high. Unfortunately, in many deep South African gold mines it is necessary to recool the air close to the workings and in recent years efforts have been made to reduce the costs of distributing cooling in this manner. Small direct contact air coolers with cooling duties of 300 kW or less have been developed. The use of turbines with these coolers has partly overcome the major objection of breaking the pressure in water networks. There are numerous different designs of these air coolers, making use of simple spray systems, cooling tower packings and knitted meshes[1]. Small turbines may be used in conjunction with these heat exchangers to drive fans or return pumps or both simultaneously. In a recent investigation it was demonstrated that a mine fan could be driven with a single stage Francis turbine over a wide range of water supply pressures and air resistances. A 610 mm (24 inch) diameter axial flow fan was coupled through a belt drive to a single stage turbine (all the parts used were off-the-shelf items). The fan-turbine unit (Figure 7) was assembled and its characteristics were measured in the Heat Exchanger Test Centre[9] of the Environmental Engineering Laboratory where the water and air flow rates through the unit could easily be varied. The range of operating conditions are given in Table 4.

Table 4. Range of operating conditions for fan turbine tests

Upstream water pressure to turbine	500-1200 kPa (1100 kPa)
Downstream water pressure from turbine	0-700 kPa (200 kPa)
Pressure drop across turbine	0-1100 kPa (900 kPa)
Water flow through turbine	3,1-4,6 ℓ/s (4,2 /s)
Air flow through fan	0,5-5,0 m^3/s (3,0 m^3/s)
Air pressure across fan	150-680 Pa (450 Pa)

(Design conditions in parenthesis.)

It was found that the unit operated satisfactorily over a wide range of water pressure. As the water pressure differen-

Fig. 7. Fan turbine unit.

tial across the turbine increased, so did the turbine-fan speeds , and the fan may be considered as being coupled to a variable speed motor. The conclusions which were drawn from this investigation are:

(i) the water flow through a given turbine with fixed nozzles is only a function of the pressure drop across the turbine and does not depend upon the power absorbed by the fan.

(ii) for a given water flow rate and water pressure differential the speed of the turbine/fan varies (about seven per cent for this particular fan and turbine) depending upon the fan power load.

(iii) the overall maximum efficiency for this type of fan - turbine unit which is constructed from off-the-shelf parts and

is operated under these ranges of conditions is likely to be about 40 per cent.

This unit demonstrated the versatility of the Francis turbine, and unlike the other examples which are described, this unit may be used in a wide range of applications.

At a particular mine installation the supply chilled water pressure to a mesh-type secondary air cooler was 1000 kPa whereas the required nozzle pressure was only 170 kPa. Instead of dropping this excess water pressure across a throttling valve a Francis turbine was used which drove the fan (Figure 8). The design specifications are given in Table 5.

It is necessary to operate the turbine at a much higher speed than the fan and a belt drive is used. In spite of the 2,76 ratio of speeds the specific speed of turbine is only 0,19 which is small, however practical limitations prevent the turbine from operating at higher speeds.

Table 5. Design specifications for a mesh cooler with a turbine driven fan

Water flow	3,77 ℓ/s
Water pressure upstream of turbine	1000 kPa
Water pressure downstream of turbine	170 kPa
Water pressure drop across turbine	830 kPa
Power taken out of water stream	3,1 kW
Tubine speed	4500 rev/min
Air flow	3,89 m^3/s
Fan pressure	276 Pa
Power given to airstream	1,1 kW
Efficiency of overall system	34 per cent
Fan speed	1643 rev/min

In a planned installation it is intended to use numerous small direct-contact air coolers in a cooling network which stretches over a number of levels. These coolers are to be fitted with Francis turbines which will drive pumps that return the water to the shaft area for repumping to the refrigeration machines. Although the water pressure at the air coolers will always be greater than that required by the spray nozzles, it will vary depending upon the cooler location and it is necessary to install a constant flow control valve upstream of each turbine. The water pump is directly coupled to the turbine.

Fig. 8. Turbine driven fan for mesh cooler.

Table 6. Design specification for air cooler with a turbine driven fan and return water pump

Water flow	9,4 ℓ/s
Water pressure upstream of turbine	720 kPa
Water pressure downstream of turbine	130 kPa
Water pressure drop across turbine	590 Kpa
Power taken out of water stream	5,54 kW
Turbine speed	4000 rev/min
Air flow	8,2 m^3/s
Fan pressure	240 Pa
Power given to airstream	1,97 kW
Return water pump pressure	98 kPa
Pump speed	2550 rev/min
Power given to return water stream	0,92 kW
Efficiency of overall system	52 per cent
Specific speed	1,09

In another recent application an air cooler was required as the only cooling load for an existing refrigeration plant[10]. Since the refrigeration plant was dedicated to the air cooler it was necessary for the characteristics of the refrigeration plant and those of the air cooler to be optimally matched. An analysis of the system showed that the optimum operating conditions could be achieved by making use of a simple cross-flow spray air cooler. The air cooler, which was constructed in the mine work-shops (Figure 9), used low pressure drop mist eliminators similar to those de-scribed by Jurani[11]. The air pressure drop across the air cooler at the design air flow rate was only 240 Pa. The circu-lating chilled water pump was supplied with the refrigeration machine and the supply water pressure at the cooler installation was much higher than that required at the spray nozzles. In fact there is sufficient energy available in the water stream for the turbine to drive both the fan and return water pump. This cooler was designed so that the turbine, return water pump and fan operated at different speeds (Table 6).

An obvious advantage of this arrangement is that no electric power is required at the air cooler site (additional power is pump at the refrigeration plant in order to provide sufficient pressure at the turbine). This unit was tested exten-sively on surface and found to work well, however because of water pressure limita-tions the unit has still to operate successfully in its underground applica-tion.

FUTURE DEVELOPMENTS

In the deep South African gold mines the heat produced at the working face may be as high as a third of the total mine heat load. Since the amount of cooling which is distributed through the water used for normal mining operations is limited, the balance of cooling must be provided through air coolers. At present a stope air cooler is being developed in which chilled service water at normal mine stope pressure (400 kPa) is to supply the cool-ing, as well as the power to drive air through the heat exchanger. For an air cooler which can be used at the working face to be practical it is believed that it must meet the design constraints presented in Table 7.

In a first prototype stope air cooler (Figure 10) the chilled water was directed in a jet onto buckets of a Pelton wheel which was directly coupled to a fan. The fan drew the ventilation air through a knitted mesh while the water from the Pelton wheel buckets fell through the mesh in a counterflow arrangement. This unit

Table 7. Design constraints for stope air cooler

Maximum mass	50 kg
Powered by normal stope water	400 kPa
Supply pressure of water flow	0,5-1 ℓ/s
Inlet water temperature	18 °C
Inlet air wet bulb temperature	30 °C
Cooling duty.	20 kW

Fig. 10. The first prototype stope air cooler.

performed satisfactorily as an air mover, however the heat transfer characteristics were poor. Improving the heat transfer characteristics would involve the use of a greater depth of mesh packing; this would in turn result in a larger power requirement at the fan. At the design conditions the specific speed of the turbine was only 0,07 and if the supply water pressure was increased the specific speed decreased further and the turbine operated at a lower efficiency so that the energy given to the airstream did not increase significantly. On the other hand if the water flow was increased, it was necessary to increase the overall size of the cooler which was also not practical. It was therefore decided to abandon the development of this prototype stope air cooler.

A second prototype is presently being constructed in which the air will be entrained through the cooler by spray nozzles. The air entrainment characteristics of various full cone and hollow cone nozzles which are mounted in a 610 mm diameter, 1,5 metre long section of ducting are being measured. Initial tests have shown that the air entrainment characteristics of the system at various water pressures agree with the predictions of McQuid[12]. When approximately 0,5 ℓ/s of water at 400 kPa is sprayed into the ducting with no restriction fitted at the end, the air entrained is approximately 1,3 m^3/s. In a stope air cooler it is envisaged that a mist eliminator would be fitted over the downstream end. The mist eliminator would remove the water droplets from the airstream and also provide an additional heat transfer surface. In preliminary tests, with low pressure drop mist eliminator plates fitted over the end of the unit, the airflow was reduced by about 30 per cent to approximately 0,8 m^3/s for 0,5 ℓ/s of water at 400 kPa, depending upon the type of nozzle used. At this stage, the initial development work is encouraging and a heat exchanger is presently being designed.

Fig. 9. Air cooler with turbine driven fan and return water pump.

ENERGY RECOVERY FROM WATER FLOW DOWN MAIN SHAFTS

In South African gold mines there is a widespread use of Pelton turbines to recover energy in relatively large installations, from the main chilled water flow down shafts. These installations are situated near the shaft at the middle levels and water is then fed further down into the mine, making energy recovery closer to the workings on the smaller scale discussed previously still possible.

Typically the power ratings for these installations would be 1 to 3 MW with supply pressures of 10 to 15 MPa. For normal motor speeds this implies a specific speed requirement of 1,5 to 3,0 Radians, which as was shown earlier is ideal for Pelton wheel type turbines; this is evident by the high efficiencies of over 80 per cent that are achieved. It is estimated that there are now about 40 such installations, with the first few being installed in the mid-1970s. These Pelton turbines may be loaded by being directly coupled to either mine de-watering pumps or synchronous generators. The former application has the disadvantage that the volumes of water available for pumping have to be controlled carefully in order that the turbine can be run when required.

USE OF TURBINES WITH MINE 'HYDRO-POWER' SCHEMES

As mentioned earlier there is presently a keen interest in the South African mining industry in the possibility of making use of the total hydrostatic pressure of the chilled water being fed from surface to power mining machinery. This concept has become known as hydro-power[3]. Presently this is being implemented on a small scale in a test stope where water from surface is being used to power water jets and positive displacement hydro-transformers, which then power hydraulic rockdrills using an oil-water emulsion. Work is also progressing well towards the development of a rockdrill which will be powered directly by pure water.

A range of turbines which may be used to drive fans, pumps and winch motors is presently being designed and evaluated. A typical output power being considered is 30 kW for a supply pressure of about 20 MPa. Thus there is an application for a machine with an extremely high supply pressure but having a relatively small power rating. The specific speed require-ment for this machine at two-pole motor speeds is 2×10^{-4} Radians which is well outside of the normal Pelton wheel range. However, an impulse type turbine for this purpose is presently under development in South Africa and is showing promise, having achieved efficiencies of above 40 per cent in surface tests.

CONCLUSION

The increase in refrigeration capacity installed on the South African gold mines has led to more attention being paid to methods of distributing cooling to the workings. The use of small turbines has partly overcome one of the major objections to direct contact air cooler, where there is a loss of energy associated with reducing the water pressure to atmospheric pressure at the cooler.

In the experience of the Environmental Engineering Laboratory the majority of the problems encountered in turbine/air cooler installations are due to the supply water pressure being less than the design value, resulting in a reduction in turbine output power. It is recommended that wherever practical, the available supply pressure should be measured before the design of the turbine installation is completed.

In installations where the correct supply water pressure is available, the turbines have proved to be reliable and may be considered as developed engineering appliances that can be acquired and used in a similar manner to other prime movers such as electric motors.

ACKNOWLEDGEMENT

This paper arises from work carried out as part of the research programme of the Research Organization of the Chamber of Mines of South Africa.

REFERENCES

1. BLUHM, S.J., RAMSDEN, R. and FERGUSON, D.W.B. 1984. Heat exchangers for cooling air in mines. The S.A. Mech. Eng. Vol. 34(10), pp.358-366.
2. CSATARY, C.J. and SMIT, J. 1975. Method of ventilating multi-reef workings and some aspects of the refrigeration plant design for Elsburg gold mine. In Proceedings International Mine Ventilation Congress Johannesburg 1975. Hemp R. and Lancaster, F.H. eds

(Johannesburg: Mine Ventilation
Society of South Africa, 1976),
pp.321-326.

3. MIDDLETON, N.T., VILJOEN, A. and
    WYMER, D.G. 1985.   Hydropower and
    its implications for the distribu-
    tion and use of energy in mines.
    Paper presented at Mine Vent. Soc.
    of S.Afr. Symposium on mine water
    systems, Sun City, Bophuthatswana,
    6-8 March.

4. RAMSDEN, R. 1983.   The temperature
    rise of chilled water flowing
    through pipes, J. Mine Vent. Soc. of
    S.Afr., Vol. 36(9), pp.85-93.

5. WHILLIER, A. 1977.   Recovery of
    energy from the water going down
    mine shaft, J. S.Afr. Inst. of Min.
    and Met., Vol. 77, pp.183-186.

6. DOUGLAS, J.F., GASIOREK, J.M. and
    SWAFFIELD, J.A. 1979.   Fluid
    Mechanics, Pitman, p.602.

7. FERGUSON, D.W.B. and BLUHM, S.J. 1974.
    Performance testing of an energy
    recovery turbine at a three stage
    spray chamber, J. Mine Vent. Soc. of
    S.A., Vol. 37(11), pp.121-125.

8. STROH, R.M. 1980.   Refrigeration
    practice on Anglo American gold
    mines. Paper presented at Mine
    Vent. Soc. of S.Afr. Symposium on
    Refrigeration and Airconditioning,
    Vanderbijlpark.

9. Chamber of Mines Research Organization
    Heat Exchanger Test Centre. 1983.
    Industry application brochure,
    Johannesburg.

10. THORP, N., and BLUHM, S.J. 1985.   The
    use of mini-cooling plants on Durban
    Roodepoort Deep. Paper presented at
    S. Afr. Inst. of Min. and Metal.
    colloquium 'Mining and Environment',
    Randburg.

11. JURANI, R.S. 1983.   Development,
    performance and application of high
    capacity portable bulk spray coolers
    at Mt. Taylor Mine. J. Mine. Vent.
    Soc. of S. Afr., Vol. 36(12),
    pp.123-125.

12. McQUID, J. 1975.   Air entrainment
    into bounded axisymmetric sprays.
    Inst. of Mech. Eng., Thermodynamics
    and Fluid Mechanics Group
    Proceedings 1975, Vol. 189, 28/75,
    pp.197-202.

Novel mine cooling methods for future deep level mines

C.J.CSATARY
Consultant, Johannesburg, South Africa

ABSTRACT: Mining experts are in doubt about the economics of ventilating and cooling very deep mines with surplus air and water from the surface. It was found, that a large portion of the heat production (above 50%) could be caused by autocompression energy increase of the downcast air and water. Loss of refrigeration capacity between the surface and the workings (leakage and heat transfer) will add to the inefficiency of the currently used mine cooling systems. It is proposed, that the basic design criteria for the planning of mine ventilation and cooling systems should be based on minimum permissable fresh air and water consumption from the surface. These minimums should carry maximum cooling capacities into the mine.

1 INTRODUCTION

We found that a large portion of the heat produced underground is caused by the downcast air and excess service water used for cooling the very deep mines. In addition leakages and heat transfer to chilled water pipes, uninsulated compressed air pipes and pumping out columns are adding further heat. Furthermore, electric or hydraulic power in the mine remains a direct heat source, while compressed air has considerable cooling power. My ideas to ventilate and cool mines are as follows:

1.1 Improved forms of air control.
1.2 Recirculation of mine air.
1.3 Low (subzero) temperature cooling of the downcast air on surface or in a sub-vertical shaft.
1.4 The 'Aircycle Refrigeration System' as applied to mines.
1.5 High pressure closed circuits.
1.6 Combinations of from 1.1 to 1.5.
   Others proposed:
1.7 Production of ice on the surface (11)
1.8 The 'Hydropower'
1.9 Desalination of water by freezing (15)

2 CURRENT MINE COOLING METHODS

2.1 Chilling water with underground plants

Advantages: High positional efficiency, low water pressure, good for scattered mining. Disadvantages: Limited heat rejection capacities, size limitation of machinery, low COP-o, expensive excavations, problems with supervision, repair and maintenance. COP is 3-4, condensing temperatures are up to 60°C, the Overall System Efficiency (OSE) is reduced to 2-2,5 depending upon the type of circuit. Capacity is approximately 21000 KWR with 600 kg/s upcast air.

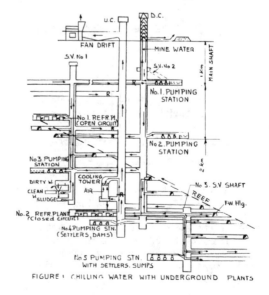

FIGURE 1 CHILLING WATER WITH UNDERGROUND PLANTS

2.2 Chilling water with surface plants

This method is a genuine financial penalty in spite of Pelton Wheels driving pumps or generators. The energy recovery is 60% with pumps. The OSE is only 1,5-0,5 at 3-4 km.
Example. Assumptions: Plant 40 MWR, depth 4 km, water flow 353,83 l/s, $\Delta t°C = 30 - 3 = 27°C$.
Basic Plant. Capital Expenditure at R1016/KWR is R40,64 million. COP = 6,5.
Power = 6153,8 KWE. At PV = R2500/KWE/10yrs PV = R15,384 millions.
Pumps. 7389,86 KWE, PV/10yrs, R18,474 millions.
Penalties. Autocompression of water: 353,83 x 9,79 x 4 x 0,4 = 5542,39 KWR. Loss of coldness: (50% true loss up to stopes):
(353,83 + 49,03) x 4,187 x 18-3-(4 x 0,93) x 0,5 = 8355,58 KWR. Autocompression of downcast air of 600 kg/s: 600 x 9,79 x 4 = 23496 KWR.
Total Penalties: 37393,97 KWR (93%).
Each type of loss of refrigeration above causes extra Capital Expenditure and extra electric power requirements for extra plant and pumps. This proportion is improving with increased installed capacity because the penalty for downcast air may remain constant.

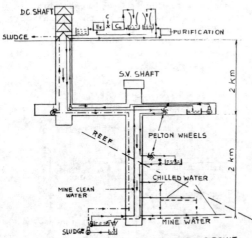

FIGURE 2. SURFACE PLANT OPEN CIRCUIT

OSE = KWR_{Net}/KWE_{Net}. If OSE = 0,5 at 4 km depth, the power is for 120 MWR system 120000/0,5 = 240 MWR. The power used by the 40 MWR plant is only 18465,1 KWE. Considering the effects of the penalties, the Capital Expenditure may climb to R200 million and the power on Present Value/10yrs basis may rise to 100 million.

(Total = R300 million)

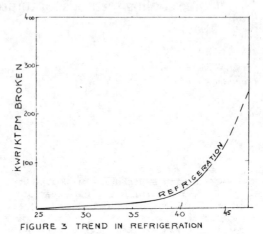

FIGURE 3 TREND IN REFRIGERATION

2.3 Cooling of the downcast air on surface

The current method is to cool to 5-8°C with 0-3°C chilled water. At 0°C water we may balance autocompression energy increase of 600 kg/s air. (23 MWR)

Table 1. Cooling capacity of d.c. air on the surface (Above 0°C).

Saturated, t			
	°C	10	0
Sigma heat,hs	KJ/kg	32,62	11,25
Cooling Capacity	KJ/kg	20,14	41,51
Total (600kg/s	MWR	12,084	24,906

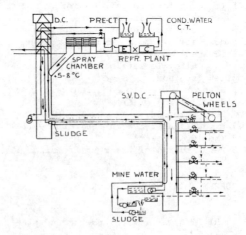

FIGURE 4 ABOVE ZERO DEGREE CELSIUS COOLING OF THE D.C. AIR ON THE SURFACE

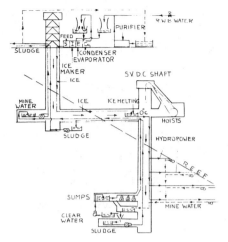

FIGURE 5 ICE FROM SURFACE.HYDROPOWER.

3 NOVEL MINE COOLING METHODS

3.1 Production of ice on surface (11)

Water (ice) may carry maximum coldness.
The system is penalised by the autocompression of air and water, pumping and fan-power. A trial plant is operating at ERPM and Rand Mines Ltd purchased another 1000 tons per day plant for Harmony Gold Mine. Ice can carry four times as much coldness as water.

Table 2. Cooling capacity of ice, KJ/kg

Ice	Water, t°C			
t°C	0	10	20	30
0	334,94	376,81	418,68	460,55
-5	345,44	387,31	429,18	471,05
-10	355,94	397,81	439,68	481,55
-15	366,44	408,31	450,18	492,05
-20	376,94	418,81	460,68	502,55

$C_p = 2,1$ KJ/kg, 144 Btu/lb = 334,94 KJ/kg
Btu/lb = 2,326 KJ/kg

Current disadvantages of ice production:
a) No energy recovery before melting is possible.
b) Penalised by autocompression of water, air.
c) Pumping is against full static head.
d) COP = 2 (at -5°C ice)
e) COP 3 with expensive plants.
f) Transport underground is difficult for long distances and for sub shafts.
g) Melting large quantities fast may be another problem.

Table 2 was calculated from:
$$Q = M_w \times \left[(\Delta t_w \times 4,187) + Q_L^{ice} + (t_{ice} \times c_{ice}) \right]$$
where M_w = mass flow kg/s, t_w = temperature difference of water above 0°C water.
Q_L^{ice} = latent heat of ice = 334,72 KJ/kg (144 Btu/lb)
t_{ice} = ice temperature in °C
c_{ice} = specific heat of ice KJ/kg (appr. 2,1 KJ/kg)
It is noted, that the density of ice is changing with temperature.
Carnat COP of ice making = $\dfrac{T_e - T_c}{T_c}$ where

T_e is equal to evaporating temperature °K

T_c is equal to condensing temperature °K

The economics of the present ice production is not so advantageous. Unless there is a breakthrough with a continuous ice production technique, the COP of the ice producing plant may remain below 2 and the OSE might remain similar to the chilled water scheme. Calculations (11) indicated, that the system may be 30% more economical, than chilled water in a 4 km deep mine from the surface.

3.2 Desalination with ice making and/or with freezing

This method was thought to help in the introduction of hydraulic power to drive the drills. Essentially it is a chilled water (ice) supply scheme from the surface, when Pelton Wheels would be replaced by the hydraulic machines. Ice in the system must be melted before the water gets to the drills. The quality requirement of water for hydraulic drills could be 500 - 1000 p.p.m. dissolved solids. This is why desalination by the freezing method is being considered using surface plants. Underground desalination may be possible with the Mine Aircycle Refrigeration System. It is expected, that the OSE of desalination by the freezing method will be less than of ice or chilled water supply schemes. Expenditures would be further increased. (15) If we use closed circuit chilled water supply schemes, a small capacity purification plant is sufficient to fill the system occasionally. It is too early to comment on the technical and economical feasibility of the various desalination schemes.

3.3 Subzero temperature cooling of the downcast air on surface

Figure 6 shows the solutions to downcast subzero temperature air from the surface to the mine (3,7,8). It is possible to use

Table 3. Cooling capacity of subzero temperature air on surface.

Saturated, t °C		-10	-20	-40	-60
Sigma heat, hs	KJ/kg	2,75	-4,76	-29,21	-50,13
Cooling	KJ/kg	50,01	57,52	81,96	102,89
Cooling of 600 kg/s air	MWR	30,0	34,51	49,176	61,734

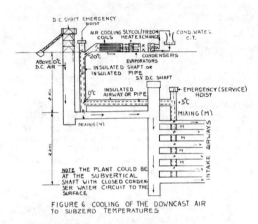

FIGURE 6 COOLING OF THE DOWNCAST AIR TO SUBZERO TEMPERATURES

an underground plant to do the same for a sub shaft using high pressure condenser water circuit from the surface.

The methods are using either a dry insulated shaft or an insulated pipe in a suitable downcast shaft. It is important to note that autocompression energy increase will increase dry-bulb temperatures up to 9,81°C/km (9,79 in SA). Thus -20°C air will be +0°C after 2 km and -40°C air will be +0°C after 4 km in the dry shaft or in the insulated pipe. There is extra heat transfer from the rock in spite of insulation. The depth of the "cooling shaft" depends upon the initial temperature.

3.4 The mine aircycle refrigeration.

I started to work on this concept as early as 1975. It was put to Dr. A. Whillier, Director of Environmental Engineering. He turned down the idea saying, that the COP of the compressor-expander aggregate is too low (0,5-0,7). He probably knew about the icing problem as well. Being very busy as Group Ventilation Engineer of J.C.I., I had to shelve such research project, though I could see its advantages. In December 1979 I joined Dr. Whillier, which gave me the opportunity to review this cooling cycle again as part of an overall invest-

igation on the economics of ventilating and cooling very deep mines.

The aircycle refrigeration concept is 150 years old. This was the cycle which produced ice for the first time in history, using mechanical refrigeration. The failure of not taking seriously this concept as against the vapour compression cycles is understandable. Chilling water as against cooling air had its explanation too. The aircycle is a ventilation and cooling cycle and no other medium than Refrigerant 729 (Air) is involved. After exploring the possibilities and the potentials of this system I submitted my reports to Dr. A. Whillier and his successor (3.9) proposing that a research project should be set up to develop the system.

The reversed Brayton cycle was a closed circuit (Figure 7)

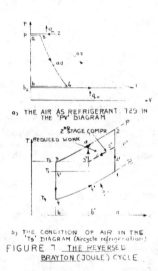

a) THE AIR AS REFRIGERANT 729 IN THE 'PV' DIAGRAM

b) THE CONDITION OF AIR IN THE 'Ts' DIAGRAM (Aircycle refrigeration)
FIGURE 7 THE REVERSED BRAYTON (JOULE) CYCLE

Figure 8 shows the first simplified 'Mine Aircycle' that I designed in 1981. A similar cycle was distributed to a number of manufacturers for comments at that time. Using this system, I expected to halve the Capital Expenditure and reduce the electric power requirements of a chilled water cooling system with plant on the surface, to a fifth or tenth.

In Figure 9 we show a mining layout using the aircycle mine cooling system in its simple form.

The air is compressed on the surface, intercooled, aftercooled, dried and drained of water and supplied to the shaft at 35°C approximately on the bank. It is then downcasted in the downcast or in the upcast shafts, cooled to approximately 35°C temp-

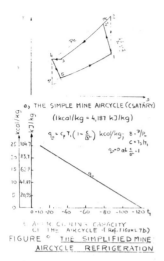

a) THE SIMPLE MINE AIRCYCLE (CSATARY)

(1kcal/kg = 4,187 kJ/kg)

$$q_o = c_p T_i \left(1 - \frac{c}{\theta}\right) \text{ kcal/kg}; \quad \theta = \frac{P_2}{P_1}$$
$$c = \frac{k-1}{k}$$
$$q_o = 0 \text{ at } \frac{c}{\theta} = 1$$

b. AIR COOLING CAPACITY
OF THE AIRCYCLE (Ref. 11, 6 and 7b.)

FIGURE 8. THE SIMPLIFIED MINE
AIRCYCLE REFRIGERATION

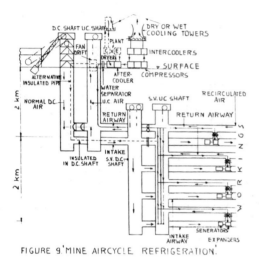

FIGURE 9. MINE AIRCYCLE REFRIGERATION.

erature at inlet to the expanders.

The expanders will drive induction generators working on the mine grid, and the 35°C temperature will be reduced through the expander by some designed 60-80°C. The air will leave the expander at pressure ratio of approximately 5 at subzero temperatures. The discharged low (subzero) temperature air can be directly mixed with mine air in the right proportions. It can cool air indirectly with the help of coils recirculating glycol. We can cool water, produce ice, and perhaps we can desalinate recirculated water underground. The temperature of the discharged air can be as low or lower than -50°C. In such conditions the air is practically dry and most of the moisture is frozen.

There are a number of large firms who are interested in developing this new mining technique. First to mention is Atlas Copco.

The main advantage of this mine cooling system would be that it nearly satisfies the basic design criteria of minimizing the supply of normal downcast air and water from the surface. None of the other systems or the ice system can do the same. In addition, this system improves the quality of the mine air with respect to wet-bulb temperatures, concentration of contaminants (dust, gases, radon).

In spite of the low value of the COP = 0,5-0,7 of the aggregate, the OSE could be 4-5 at 4 km depth against approximately 0.5-0,65 of chilled water or ice supplied from the surface. The technical feasibility of the mine aircycle depends on the expansion process, which must not be jeopardized by ice formation inside or outside the expander. The mines would not have objections if the subzero temperature air would be discharging ice or snow as well. Vapour in the return mine air which will freeze when mixed with subzero air may not contain salts from the mine water. It may be possible, that the subzero air (dry) can be used for desalination instead of other refrigerants or brine. There is no technology available to do desalination by freezing or ice making with subzero air. There is no "loss of coldness", if we adopt such technologies underground. I am working on a new underground desalination technique which might use subzero air directly.

3.5 High pressure closed circuit system using the 'Rock and Tubes' type of heat exchanger concept.

I proposed various versions of this promising mine cooling method including a joint project with the Hydropower scheme (3,4,10). The concept of closed circuit is well known in the deep gold mines and in the Ruhr for pressures up to 600m water. Attempts to introduce a high pressure system at Western Deep Level and Vaal Reef gold mines failed for several reasons, such as quality of water, small capacity (200 KWR). The tubes and shells were designed for the high pressure. (16)

The system could work with either a surface plant or with an underground plant or with combined surface and underground plants as it is shown in Figure 11.

The surface plant may be cooling a suitable glycol (proposed by A.W.T. Barenbrug) which would double the cooling capacity of a given pipe layout. The system is technically feasible only if my proposed "Rock and Tubes" type of heat exchanger can be developed. A vertical and a horizontal ver-

585

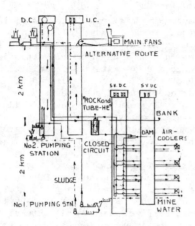

FIGURE 10 HIGH PRESSURE CLOSED
CIRCUIT MINE COOLING SYSTEM
(With Plant on the Surface)

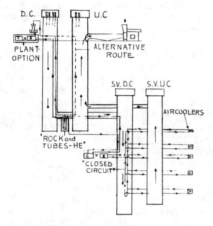

FIGURE 11. HIGH PRESSURE CLOSED
CIRCUIT WITH PLANT AT 3 Km DEPTH

sion of this proposed heat exchanger is
shown in Figure 12.

Depending upon the size of excavation,
one may be able to build units up to 50 MWR
capacity and 25 MPa pressure or more as
against the largest units designed for 8,5
MPa head with 200 KWR capacity in Western
Deep Level (16). I am working on a new design of high pressure heat exchanger using
the rock as a shell.

4 CONCLUSIONS

4.1 Currently used chilled water mine cooling methods are extremely expensive. The
exception is the cooling of the downcast
air to above zero temperatures on the surface.

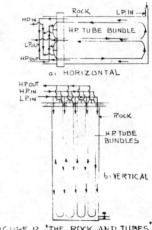

FIGURE 12 "THE ROCK AND TUBES"
UNDERGROUND HEAT EXCHANGER

4.2 Producing ice or the method of desalination by ice making or by freezing could
be an important improved technology to save
water but it may not be a great economical
achievement (11).

4.3 Cooling of the downcast air to below
zero temperatures has technical and economical potential, but the aircycle has higher positional efficiency and OSE.

4.4 The mine aircycle refrigeration system
is a very promising concept. This would be
a technological and economical breakthrough
in mine cooling practices.

4.5 High pressure closed circuits can upgrade Overall System Efficiencies of currently used chilled water systems, using the
"Rock and Tubes" type of heat exchangers.

5 REFERENCES

5.1 - 5.12 Reports are obtainable from Dr.
M.D.G. Salamon, Chamber of Mines S.A.

5.1 Csatary, C.J. Preliminary notes on the
subject of recirculating of air in mines.
Internal report no. 47/80, Jan 18/1980.

5.2 Csatary, C.J. An introduction to the
concept of controlled recirculation of
mine air. Internal report no. 124/1981.

5.3 Csatary, C.J. Notes on ideas of cooling deep level mines. Internal report no.
178/82, 15 September 1982.

5.4 Csatary, C.J. Notes on the possible design of high pressure heat exchangers to
discharge heat from the mines to the surface. Internal report no. 179/82.

5.5 Csatary, C.J. Heat produced by the
South African Gold Mines. Internal report
no. 180/82, 16 September 1982.

5.6 Whillier, A. Ice for cooling in deep
mines, An example. Internal report no.
190/81, 20 January 1981.

5.7 Csatary, C.J. A preliminary cost comparison between cooling of the downcast
air against cooling of extra service
water on the surface of deep mines.
Internal report no. 225/1982.

5.8 Csatary, C.J. An introduction to the
fundamentals of the possible use of the
aircycle refrigeration system to cool
deep level mines (Part 1 Technical aspects) Interim report no. 253/83.

5.9 Csatary, C.J. Closed circuit high pressure mine refrigeration system for energy
saving. Internal report no. 272 November
1984.

5.10 Sheer, J. Research into the use of ice
for cooling deep mines. Mine Ventilation
Congress, Harrogate. Paper 16 1984.

5.11 Burton, R.C., Plenderleigh, W.
Stewart, J.M. and Holding, W.
Recirculation of air in the ventilation
and cooling of deep gold mines.
The third Int. Mine Vent. Congress,
Harrogate, England. 13th - 19th June '83.

5.12 Csatary, C.J. Effects of leakages on
air distribution and fanpower. Internal
report no. 181/82.

5.13 Csatary, C.J. Farren, P. Blenkiron, J.
Practical considerations of changing over
existing air cooling to service water
cooling at Western Areas Gold Mine Co.
Ltd. Symposium of latent concepts in
cooling deep mines. S.A. I.M.E. 18th-19th
October 1977.

5.14 Shone, R.D.C. Harries R.C.
Desalination of mine water. Symposium.
Mine Ventilation Society of South Africa.
March 1984, Sun City.

5.15 Stroh, R.M. The refrigeration system
of Western Deep Levels Ltd. The Journal
of Mine Vantilation Society of S.A.
Vol. 27, No. 1, January 1974 p.15.

5.16 Sheer, T.J. Cilliers, P.F. Chaplain
E.J. and Correia, R.M. Some recent
developments in the use of ice for cooling mines. Symposium on mine water systems, Bophuthatswana. The Mine Ventilation Society of S.A. March 1985.

Power generation from mine refrigeration system

ROBERT TORBIN
Foster-Miller Inc., Waltham, MA, USA

EDWARD THIMONS
US Bureau of Mines, Pittsburgh, PA, USA

STEVEN LAUTENSCHLAEGER
Hecla Mining Company, Wallace, ID, USA

1 INTRODUCTION

The Bureau of Mines, as part of its
Health and Safety Research Program, has
investigated a technique for improving
the performance of mine refrigeration
systems. By converting the potential
energy of the incoming water to electrical energy, rather than to heat energy,
the water stays colder. This results in
cooler mine temperatures and a healthier, more comfortable working environment for miners. A primary spinoff
advantage resulting from this research
is an energy savings from employing such
a refrigeration technique. Recognizing
the value of recoverable potential
energy, the Bureau of Mines has undertaken a project to study the application
of energy recovery techniques to the
United States mining industry. To this
end, the Bureau has contracted with
Foster-Miller, Inc. of Waltham, MA to
investigate, evaluate and demonstrate
commercially available means of efficiently recovering the energy from any
incoming cooling water source. This
project has led to procurement, installation, and operation of a complete
thermal energy recovery system. The
test site is the Lucky Friday Mine,
owned and operated by the Hecla Mining
Company of Wallace, ID. The Lucky
Friday Mine is located in Mullan, ID in
the Coeur d'Alenes region of upper Idaho.

2 BACKGROUND

It is a well-established fact that as
mines go deeper, the mining environment
becomes hotter. The higher temperatures
and humidities encountered at large
depths make a comfortable and healthy
work environment difficult and expensive
to maintain. Therefore, large refrigeration systems are employed to condition
the working environment to help control
temperature and humidity. The refrigeration systems can be located on the
surface or on the underground working
levels of the mine. Regardless of
chiller location, large quantities of
water are required, either as direct
chilled water or as chiller service
water, to operate these refrigeration
systems. At the Lucky Friday Mine, the
chillers are located underground. The
large head offers an excellent opportunity for energy recovery from the available potential energy of the water.
This potential energy has been exploited
quite successfully in recent years by
the deep mining industry of the Republic
of South Africa. Energy recovery techniques developed by the Research Organization of the Chamber of Mines of South
Africa are commonly employed in several
mines. The primary means of energy
recovery is the use of hydraulic turbines to replace electric motors as
drivers for generators, pumps, or fans.

Current United States mining practices
have not placed a significant value on
this available energy. Several mines
allow the available potential energy to
be dissipated by friction and momentum
transfer within the pipeline and energy
dissipators. Using high pressure water
to operate a turbine can result in three
benefits to mines:

1. The turbine produces a significant
amount of power. Modern impulse turbines can convert as much as 85 percent
of the water's potential energy (at its
inlet) into useful mechanical energy.
The turbine can be coupled, with or
without a gear reducer, to a pump to
lift reject water out of the mine, or it
can be directly connected to an electric
generator.

2. The turbine also produces refrigeration savings equal to the recovered
mechanical energy. Without energy

recovery, the potential energy of the water is converted to heat. This conversion can be minimized by converting the available potential energy into the form of shaft work. The resulting lower water temperatures can be expressed directly as refrigeration savings.

3. A third benefit is that a turbine performs the necessary water pressure reduction and eliminates the need for kinetic energy dissipators in the mine. The pressure reduction allows the use of less expensive, low pressure downstream piping for the distribution of water throughout the mine. This facilitates installation of new piping extensions. This piping network also represents the vast majority of piping in the mine.

3 THERMAL ENERGY RECOVERY

Recovery of potential energy through hydropower principles is well known. Somewhat less well known is the fact that recovered power also produces thermal energy savings equal to the recovered mechanical energy. Typically, the potential energy of the service or cooling water for mines, obtained from the surface, is converted almost entirely to heat energy, resulting in an increase in the water temperature. Some of the conversion takes place through turbulence in the pipeline, while the major portion is usually converted through kinetic energy dissipation at the exit. The temperature rise is independent of any heat transfer in the mine shaft. The expected temperature rise is equal to 2.33°C for every 1000m drop (Whillier 1976). The temperature rise of the water across the turbine can be calculated from the following equation (see fig. 1 for the definition of the terms):

$$\Delta T = [f + (1 - \eta_T)(1 - f)](0.00233 \ \Delta H)$$

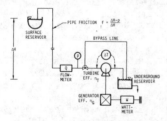

ΔH : CHANGE IN ELEVATION FROM THE SURFACE TO TURBINE (METERS)
Q : WATER FLOW (LITERS/SEC OR KG/SEC)
p : GAUGE PRESSURE UPSTREAM OF TURBINE (METERS)
f : PIPE FRICTION FACTOR
η_T : TURBINE EFFICIENCY
η_G : GENERATOR EFFICIENCY
ΔT : TEMPERATURE RISE ACROSS TURBINE (°C)

Fig. 1. Turbine-generator system diagram.

As a typical example, if the pipeline frictional losses represent 10 percent of the total head (f = 0.1), the turbine efficiency is 70 percent (η_T = 0.7) and the head is a 1000m, the temperature rise is 0.862°C. This represents only 37 percent of the original rise with no turbine (2.33°C). The lower water temperature resulting from turbine usage can be expressed directly in terms of refrigeration savings, either as kilowatts or tons.

To maximize the energy recovery by the turbine-generator, the pipeline should be sized to keep the frictional losses reasonably small.

The reduction in water temperature increase is represented by thermal savings. The magnitude of the thermal savings is exactly equal to the power extracted by the turbine. This savings is given by the following equation:

$$P_C = \eta_T(1 - f)Qg\Delta H$$

where: P_C = thermal energy savings (watts)
g = acceleration of gravity (meters per square second)

When retrofitting an energy recovery system, the installation of a turbine can help extend the capacity of an existing refrigeration plant by making it more efficient and reduce the water required to produce the same cooling capability. This means less reject water being pumped out of the mine, and therefore, a power savings can be accounted for in a reduction in required pumping horsepower. For a new installation, the thermal savings would be reflected in the sizing of the refrigeration plant capacity, and appropriate credit should be taken.

The power generated by a turbine-generator set is given by the following equation:

$$P_R = \eta_S gQ\Delta H$$

where: P_R = power savings (watts)
η_S = overall system efficiency (including friction, turbine, and generator losses)

The energy produced by the turbine generator set can result in significant savings due to a reduction in energy purchases.

While the initial investment in a thermal energy recovery system can be significant, the three benefits stated earlier are realized at little additional operating cost. Figure 2 graphically illustrates the potential operating savings (before taxes) to a mine when an impulse turbine generator set is installed under the following assumed conservative operating conditions.

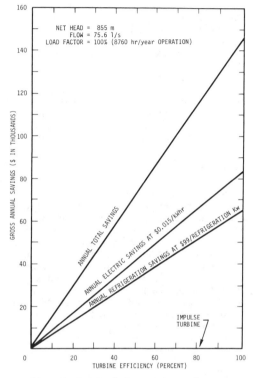

Fig. 2. Potential energy savings.

Electrical costs of $0.015/kWhr, refrigeration costs of $99/kW, and average water flow into the mine of 75.6 l/sec at a net head of 855m. It is also assumed that the thermal recovery system works continuously throughout the year. The graph shows that for a Pelton wheel efficiency of about 80 percent, the total annual savings are over $100,000.

4 AVAILABLE TURBINE HARDWARE

Generally, deep mines will start at 600m underground and can exceed 3000m in depth. The range of flow will vary from a few liters per second up to several hundred liters per second. This range of head and flow is generally outside the application envelope for Francis turbines, but is well suited for impulse turbines or high pressure multistage centrifugal pumps operated in reverse.

4.1 Impulse turbines

The impulse turbine offers excellent efficiency, in excess of 80 percent over a wide range of flow (40 to 120 percent of full load capacity). Figure 3 is a typical performance curve. Control is

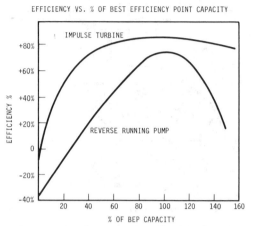

Fig. 3. Performance curves for impulse turbine and reverse pumps.

simple. A needle is positioned inside the nozzle either manually or by an automatic actuator operated from a feedback signal from the system. All of the system flow can be used to produce power. The runner assembly and the nozzle needle valve are the only two moving parts. The split casing design allows inspection and/or replacement with minimum downtime. There is no requirement for back pressure.

4.2 Reversed centrifugal pumps

Centrifugal pumps of various generic designs (from radial flow to axial flow) can be operated in reverse and used as hydraulic turbines. A comparison of the characteristics of normal pump operation with the characteristics of the same pump operated as a turbine at the same speed is shown in fig. 4 (Shafer 1982). The curves are normalized by the values of head, flow, efficiency, and power at the power best efficiency point (BEP). From this figure, one can draw some conclusions related to operation. First; the pump, operating as a turbine, can maintain higher efficiencies over a wider range of flows than when operating as a pump. Second; a relatively high flow rate (between 40 to 80 percent of the flow at the turbine BEP, depending on specific speed) is required when operating at zero power output. Reduction in flow below this value causes the turbine to begin absorbing power.

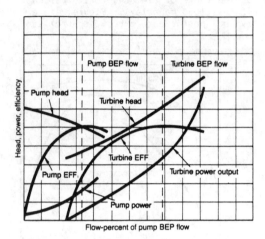

Pump and turbine RPM = constant

Fig. 4. Normalized performance curves for pump and pump in the turbine mode.

A standard centrifugal pump can usually be operated in reverse without any physical changes. However, the adequacy of the bearing design, the shaft stresses, the ability of threaded components coming loose, and the effects of increased pressure forces should be checked. To overcome some of these shortcomings, several companies have made internal modifications to standard multistage pumps. These changes have created true hydraulic turbines, not just pumps running in reverse.

Let us look at the advantages and disadvantages of the two types of generic machines. Reversed centrifugal pumps (standard units) offer low initial cost, off-the-shelf availability within a large range of heads and flows, and ability to operate with a high back pressure. Generally, pumps are familiar items to maintenance personnel. Pumps usually have relatively low absolute speeds, as well as relatively low runaway speeds. In comparison to a Pelton turbine, the reversible pumps have a 5 to 10 point lower operating efficiency. Pumps generally require somewhat higher maintenance due to erosion of impellers, internal bearings, and pressurized shaft seals. Pumps have a narrower operating range as efficiency drops off rapidly outside the range of 80 to 110 percent of design flow (see

fig. 3). Pelton turbines require no shaft seals (the casing operates at atmospheric pressure), do not require special overspeed protection, and can operate near maximum efficiency over a range of 30 to 120 percent of design flow. This is ideal for mining operations where large daily fluctuations in flow are typical. Although impulse turbines are highly reliable, they are relatively unknown to mine production and maintenance personnel. Their capital cost is greater than that of comparable pumps. Because of the outlet discharge characteristics, special foundation/mounting arrangements may be required in existing mines. Pelton turbines are available in the same head and flow ranges as pumps, however, the number of suppliers is not nearly as great.

5 MINE APPLICATIONS

Although turbine technology is well established, knowledge of mining operations is essential to instituting an effective energy recovery program. The critical needs of the mining operation must be met. The energy recovery technology must be applied so as to optimize the benefits. Although the ultimate "best" choice cannot be made until the detailed evaluations are completed, some typical installations can be used to illustrate the application of the concepts discussed. Four possible solutions are illustrated in fig. 5.

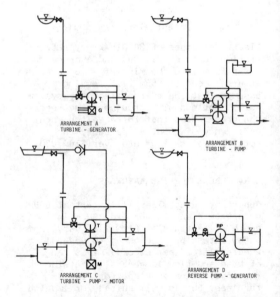

Fig. 5. Alternative system arrangements.

Arrangement A of fig. 5 is typical of a turbine-generator set. A Pelton turbine drives the generator, producing either ac or dc power. The turbine discharge is directed via a tailrace to the reservoir. As a safety feature, a turbine bypass is provided. The use of water from the reservoir varies and depends only on the mine operation. The turbine generator arrangement has some advantages over other arrangements due to its flexibility in siting and operation. The turbine generator set need only to be located near the cold water reservoir without concern for other water-related mine operations. The loss of the turbine or generator will have minimal impact on mine operations because of the availability of sufficient off-site power, and a turbine bypass piping arrangement.

Arrangement B depicts a turbine-pump set. The turbine drives a pump which takes suction from a hot dam (reject water reservoir). The source of this suction water is from the discharge of the refrigeration cooling water system and other drainage sources within the mine. The turbine discharge is directed to the cold reservoir where the water is ready for mine use. The pumping efficiency is reduced by the turbine efficiency. Therefore, one can either pump the same flow, but at lower head (as shown), or pump less flow at full head. This arrangement requires reasonable access to both hot and cold water reservoirs which is not always feasible. Furthermore, pump submergence also becomes a consideration if the roof height of the mine is restricted and sufficient NPSH is not available. Since the pump is used to return reject water and the turbine is used for incoming water, synchronization of these flows becomes important. Coordination of incoming and outgoing flows can be an operational problem depending on the mining operation, and its fluctuation during a 24-hr cycle.

Arrangement C is the same system as described under Arrangement B with the exception of an additional electric motor. The motor is used to make up the cumulative inefficiencies of the pump and the turbine, in order to return the full flow to the surface reservoir.

Arrangement D illustrates a reverse running pump driving a generator. In this arrangement, care must be taken to keep the discharge (formally the inlet) submerged to a minimum back pressure, to meet the NPSH requirements. The reverse running pump has the same advantages as the turbine-generator set described in Arrangement A in that it need only be located near the cold water reservoir, and it can be taken off line with minimal impact on mine operations. However, the reverse running pump has significant reductions in efficiency at flow rates other than its design flow rate, and its performance can vary drastically with the fluctuating flow. In some circumstances, the inlet flow must be throttled to control the performance within acceptable limits. This may require a design flow somewhat lower than peak flow, and the utilization of the bypass all the time to accommodate the excess flow above the design limit.

In another arrangement (not illustrated), the turbine could operate a compressor to supply local instrument and/or service air requirements. However, the air needs of the mine are not constant over a 24-hr day. Therefore, the system would be "off-line" over a large portion of the day if air receivers are used (excess storage capacity).

Economic considerations

Of the four applications just discussed, three were considered operationally and technically feasible at the Lucky Friday Mine site. These alternatives were the Pelton turbine-generator, the reverse running pump-generator, and the turbine-pump. An economic comparison of these three alternatives considered the following costs and revenues:
1. Capital costs
2. Installation costs
3. Operation costs
4. Maintenance costs
5. Major overhaul costs
6. Ownership costs
7. Income generated by the sale of power or savings from avoided costs
8. Savings in refrigeration costs.

A before tax rate of return analysis was performed because it was the most appropriate for Hecla's current tax situation. A rate of return (ROR) of 20 percent and a payback period of 3 years were considered as minimum requirements for each investment. The analysis produced the following results:

Turbine-generator   ROR = 34 percent
                      Payback: 3 to 4 years
Pump-generator   ROR = 18 percent
                      Payback: 5 to 6 years
Turbine-pump   ROR = 19.5 percent
                      Payback: 4 to 5 years

Based on this analysis and other technical considerations, the turbine generator system was chosen for the demonstration project.

6 RESEARCH PROGRAM AND DEMONSTRATION

The research program sponsored by the Bureau is composed of three phases. Phase I was a feasibility study. During this phase, existing water turbomachinery technology was reviewed. Alternative systems for using the high pressure water, such as shell and tube heat exchangers, high pressure closed circuit water lines, and hydrotransformer systems were evaluated and compared to the turbine system concept. In Phase II, various generic turbine pump and turbine generator system designs were developed. Complete systems were designed taking into account the sizing of required water reservoirs and backup pressure reducing systems needed during outages of the turbine-generator. Phase III involved the procurement, installation, operation, and testing of a complete thermal energy recovery system in a deep United States mine. The test site is the Lucky Friday Mine, owned and operated by the Hecla Mining Company of Wallace, ID. The primary minerals mined here are lead, zinc, and silver.

The turbine generator is rated to produce 210 kW of power at an average flow of 31.5 l/sec with an inlet head of 840m. The turbine efficiency at this design point is 87 percent. The acceptable flow range varies from 16 to 47 l/sec. The single nozzle Pelton turbine, supplied by Hayward Tyler of Burlington, Vermont, drives a Westinghouse supplied induction generator rated at 300 kW and operated at 3620 rpm. The system is operated continuously, recovering energy from incoming service water for the underground mechanical refrigeration units and other mining needs. The turbine generator system is situated in the mine on the first working level accessible from the Silver Shaft, 850m below grade as shown on the mine plan in fig. 6. The shaft continues to several lower levels, ultimately going over 1980m deep.

The incoming service water is stored in an underground reservoir from where it is redistributed to other areas of the mine through either pressurized or gravity fed piping systems. Due to the atmospheric discharge of the impulse turbine, the turbine generator set is

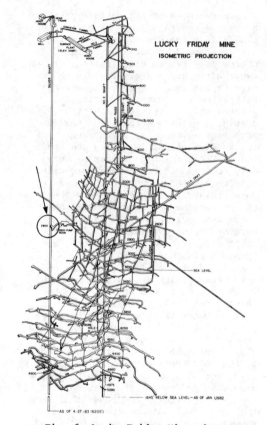

Fig. 6. Lucky Friday Mine plan.

mounted on a steel platform (3.3m high), which raises the turbine discharge above the dam crest. The system is provided with a piping bypass arrangement which allows continuous water service, regardless of the operating status of the turbine generator equipment.

Selection of the turbine generator system was dictated by the mine arrangement at the Lucky Friday Mine. The long distance between the existing reject water pumping station and the shaft was not conducive to the use of a turbine-pump system. It is desirable to locate the cold water reservoir close to the shaft to minimize the length of high pressure piping needed. Using the turbine to drive the local pumps (used for water redistribution) was also inadvisable, due to the wide fluctuation in daily flows. The flow fluctuations would be contradictory to the desire for constant pump output. In addition, the location of the water storage reservoir above the floor required elevation of the turbine, but certainly not the accompanying pump due to minimum NPSH requirements. The unfamiliarity of the Hecla personnel with turbine systems prompted a system selection that could easily be bypassed without extra expense of standby equipment and without detriment to the water distribution system. The turbine pump system violated these needs, and therefore, the turbine generator system was selected.

Figures 7 and 8 show the process diagram and the general arrangement drawing

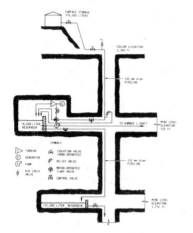

Fig. 7. Process diagram.

for the Lucky Friday site, respectively. Referring to fig. 7, with the system in the run mode the water:

1. Is supplied from the surface tanks
2. Passes through the downhole 6-in. butterfly valve
3. Passes through the three-way valve
4. Passes through the turbine needle control valve
5. Flows through the turbine, and is discharged to the cold water reservoir.

The cold water reservoir level is controlled by a water level transmitter and the modulating turbine needle valve. When the system is in the bypass mode, the water is diverted away from the turbine to a bypass pipeline by the three-way valve. The water flow through the bypass line is controlled by two restricting orifices and is discharged to the cold water reservoir. In this mode, the reservoir level is controlled by the modulating three-way valve with the turbine needle valve fully closed. A third mode of operation is the maintenance mode, during which the three-way valve is locked in the bypass position and the reservoir level is controlled by opening and closing the downhole 6-in. butterfly valve. The system has overpressure, overflow, overspeed, and high bearing temperature protection. An air

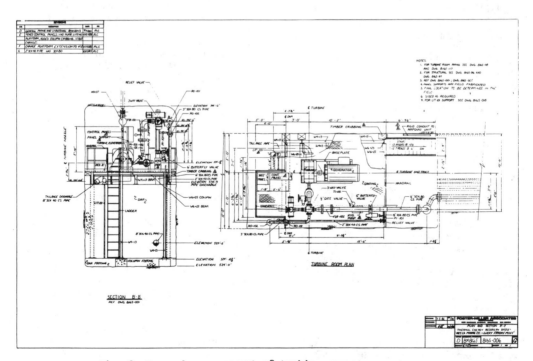

Fig. 8. General arrangement of turbine generator system.

Fig. 9. Turbine-generator and control panel.

Fig. 10. Turbine-generator and inlet piping.

release/vacuum breaker valve at the top of the shaft pipe column aids in filling and draining the pipeline. A relief valve is supplied to eliminate any dele- terious effects of water hammer on the pipeline in the shaft.

Figures 9 and 10 show the actual sys- tem hardware as installed in the mine.

7 PROJECT RESULTS

During the summer of 1984, the system installation was completed, and initial start-up tests were conducted to verify performance. During these tests, several problems became apparent which required remedial action to correct. The problems ranged from minor equipment malfunctions to major operating prob- lems. The following list itemizes some of the more significant problems:

1. High levels of turbine vibration
2. Poor instrumentation accuracy
3. Malfunctions of electrical equipment.

The most perplexing of the problems was the unexpected high vibration level and heavy splashing being emitted from

the turbine casing from around the shaft bearings. The vibration and splashing were occurring at the higher flow rates (>35 l/sec). Turbine vibration levels were exceeding acceptable levels, and were affecting the steel platform. The problem was initially evaluated as being caused by significantly lower than expected available head at the turbine nozzle. The loss of head was attributed to the shaft pipeline which was com- prised of both old 100-mm and new 150-mm diameter pipe. The portion of shaft pipeline with 100-mm pipe was replaced with 150-mm pipe. The head available at the turbine nozzle was, thereafter, very close to the predicted value. However, the vibration problem persisted.

At this point, the cause of the vibra- tion was attributed to the turbine buckets not dewatering properly at high flow rates. This problem was attributed to poor tailrace design, allowing par- tial vacuum conditions to exist inside the tailrace. It was surmised that tur- bine windage was possibly drawing froth from the tailrace water onto the turbine buckets. The solution to the problem was to simply install large vent holes (50 mm) through the casing on both sides of the turbine wheel. Once vented, the splashing and high vibration conditions were eliminated.

Another significant concern was the inability to get accurate performance data measurements. The compact design of the system, which was dictated by the size of the excavation, did not permit sufficient straight lengths of pipe for the turbine flow meters to measure flow accurately. The meters were only intended for gross indication of flow and were not meant for performance test- ing. Furthermore, the temperature sensors (RTDs) were also fine for gross temperature measurements, but could not provide the degree of accuracy needed for system thermal efficiency measure- ments. Since the system had a rela- tively small reservoir, the time avail- able to run a performance test was limited. Steady-state conditions were mostly impossible to attain during any of the performance tests. However, the data collected during the tests did com- pare quite well with the predicted values, which were based upon engineer- ing calculations of expected system performance.

There were many instances of improper electrical operation of key equipment. These problems were most notable in the main switch gear and instrumentation for the generator, and the motor operator

Table 1. Performance data.

	Head at nozzle (m)	Flow (l/sec)	kW output (net)	System efficiency (%)
Actual	865	25.3	166	0.74
Predicted	865	25.3	170	0.75
Actual	849	31.5	202	0.72
Predicted	849	31.5	212	0.75
Actual	826	37.9	240	0.71
Predicted	826	37.9	249	0.74

for the main turbine flow control valve. Several visits by manufacturers representatives were required before all equipment operations were corrected. Although finally corrected in the field, more attention to pre-installation inspection and testing was needed. Sophisticated diagnostic equipment and personnel are usually not available at the mine, incurring time-consuming delays for hardware and service representatives.

Although not a problem, there was concern about the methodology for the proper sizing of the underground reservoir. The cost of excavation from rock and construction of a concrete dam must be weighed against the need for adequate underground storage. the storage is needed to buffer the variations in demand and supply, and to provide some minimum quantity for potential loss of incoming flow. The minimum depth of the reservoir is a consideration for the minimum suction head requirements for any pumps using the reservoir for supply water. There are no currently known methods for use as guidance in sizing of these reservoirs.

Table 1 is a summary of the test data. In all three performance tests, the head and flow conditions for the actual and predicted data were identical. Therefore, the difference in performance between actual and predicted data was believed to be caused by the lower than expected efficiency of either the turbine, generator, or electrical system. It is impossible with the information on hand to pinpoint the loss of system efficiency with any certainty. However, the actual system efficiencies, over the range of flow, were deemed acceptable. As of the end of March 1985, the system was operated continuously for 6 months without any major problems, and has produced over 600,000 kWhr of power. This has resulted in a gross savings of over $15,000 to the Hecla Mining Company.

SITED REFERENCES

Whillier, A. July 1976. Energy recovery from water going down mine shafts. Report No. 46/76. Chamber of Mines of South Africa. Johannesburg, South Africa.
Shafer, L.L. November 1982. Pumps as power turbines. Mechanical Engineering.

ADDITIONAL REFERENCES

Van der Walt, J. & E.M. deKock July 1983. Mine ventilation and cooling projects. Engineering and Mining Journal.
Buse, F. 26 January 1981. Using centrifugal pumps as hydraulic turbines. Chemical Engineering.
Marks, J.R. November 1979. Refrigeration economics at the Star Mine. Society of Mining Engineers. 2nd Mine Ventilation Congress, Reno, NV.

Spray cooling developments

EDWARD D.THIMONS
Pittsburgh Research Center, Bureau of Mines, PA, USA

JOHN F.McCOY
Foster-Miller Inc., Waltham, MA, USA

JOHN R.MARKS
Homestake Mining Co., Lead, SD, USA

ABSTRACT: As part of a Bureau of Mines program, four limited bulk water spray coolers were designed and built. The first of these coolers, a single stage unit was tested in the Mt. Taylor Mine near Grants, NM, before it closed. The remaining three coolers, a two-stage cooler and two three-stage coolers were installed in the Homestake Gold Mine in Lead, SD. To date all the coolers have performed well under in mine conditions and have required only minimal maintenance. This paper details the design and performance of these coolers.

1 INTRODUCTION

The need for more efficient mine refrigeration equipment is greater than ever, as increasing numbers of U.S. mines are being operated at greater depths in order to reach minable ore deposits. The combination of high temperatures and high humidities encountered at these depths make a comfortable working environment difficult and expensive to maintain. Many ore bodies mined today could not be mined without cooling of the underground environment. Miners' safety, comfort, and productivity all suffer if air refrigeration needs are not satisfied. The water spray coolers described in this report are efficient and economical devices for dealing with the heat problem encountered in deep mines.

In deep, hot mines where virgin rock temperatures can exceed 130°F, mine cooling has conventionally been accomplished through the use of finned-coil coolers. These conventional coolers are similar in construction to automotive radiators. In these coolers, chilled mine water passes through the finned coils and cools the warm mine air as the air is blown past the coils. In the water spray coolers, however, heat exchange is accomplished by direct contact between the warm mine air and cold water sprays.

2 NOTES ON COOLER PERFORMANCE

Coolers are commonly regarded as having a specified cooling capacity of so many kilowatts; this can be misleading, since the cooling capacity of a unit depends on the inlet fluid temperatures and flow-rates. This may be illustrated by considering the extreme case of a cooler supplied with water at the same temperature as the input airstream. Obviously, the cooling capacity under these conditions is nil, and a better way is needed to characterize cooling performance.

2.1 FACTOR-OF-MERIT NOMENCLATURE

C = Thermal capacity of liquid water
    - 4.18 kJ/kg°C

C' = Average equivalent thermal capacity
    of humid air between t_{ai} and
    t_{wi} - kJ/kg°C

E_w = Effectiveness of the cooling water
    or water efficiency

F = Factor-of-merit or F-factor

G = Mass flow of air to the cooler
    - kg/s

L = Mass flow of cooling water to the
    cooler - kg/s

$$N = \frac{F}{1-F} /R^{0.4}$$

P = Cooling power or cooling duty of the cooler - kW (Refrigeration)

R = Capacity factor - LC/GC'

t_{ai} = Air wet-bulb temperature into the cooler - °C

t_{ao} = Air wet-bulb temperature out of the cooler - °C

t_{wi} = Temperature of cooling water into the cooler - °C

t_{wo} = Temperature of cooling water out of the cooler - °C

2.2 FACTOR-OF-MERIT

In this program, the performance of the cooling unit has been characterized by the factor-of-merit, or F-factor, originally developed by Whillier and improved by Bluhm. The F-factor of a cooler is independent of operating temperature and barometric pressure. It is equivalent to knowing UA for a conventional heat exchanger. The F-factor may also be thought of as the cooler efficiency when the capacity factor, R, is one.

If the water efficiency, E_w, is measured at some known arbitrary capacity factors, the following semi-empirical relationship permits calculation of the F-factor:

$$E_w = \frac{1-e^{-N(1-R)}}{1-Re^{-N(1-R)}}, \quad R \ne 1$$

$$E_w = F, R = 1 \text{ only}$$

where $N = \frac{F}{1-F} /R^{0.4}$

Once the F-factor is determined in this manner, then water efficiencies under other conditions may be predicted. The F-factor also allows you to determine when cooler maintenance is required.
Thus, if one knows the F-factor of a cooler and the input flow and temperature conditions, the cooling power may be calculated by:

$$P = E_w L C (t_{ai}-t_{wi})$$

where E_w is obtained from the first equation.

3 DESCRIPTION OF LARGE SCALE SPRAY COOLERS

3.1 SINGLE-STAGE COOLER

The original spray cooler was a single-stage unit (fig. 1). In this skid-mounted unit, an inlet transition and diffuser plate aids in spreading the inlet air jet to achieve more uniform flow. The air then passes through plates constructed of corrugated fiberglass roofing material and stacked as shown to produce channels (fig. 2). Air leaving these channels passes through the cooling water spray pattern shown in figure 2. Mist is removed from the cooling air in the wave plate eliminator before it exits through the transition section.
Cooling water is sprayed at a rate of 6.31 kg/s (100 gpm) from a manifold at 208 kPa (30 psi) through 40, 80° angle, solid-cone spray nozzles. The spray pattern produces a uniform water flux across the downwind end of the corrugated plates. Water entering the channels runs along the bottom of the channels, draining out of the upwind end of the cooler. This is achieved by installing the cooler so that it is pitched slightly downward at the upwind end. The water spray direction and channel flow direction are opposite to the airflow, producing a counter current cooling effect. Water is drained from the low end onto the floor.

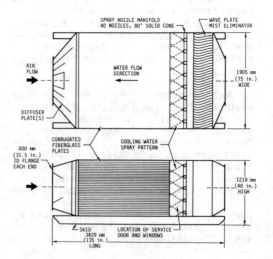

FIGURE 1. Schematic of single-stage cooler.

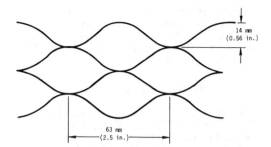

FIGURE 2. Corrugated fiberglass plates as stacked in single-stage cooler.

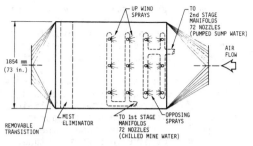

(a) Top View

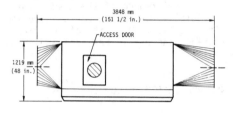

(b) Side View

FIGURE 3. Schematic of two-stage cooler.

3.2 TWO-STAGE COOLER

The Pneumafil Corporation of Charlotte, NC, adapted one of their bulk cooler concepts into a two-stage design with an overall length of 3,848 mm (151.5 in) (fig. 3). The downwind transition section can be removed during in-mine transportation. Increased cooler length permitted the use of a somewhat longer inlet transition which required no diffuser plates to achieve uniform flow.

Hot air entering the cooler encounters the second water stage first; producing a desirable counterflow effect. The second water stage consists of a pair of spray nozzle manifolds, holding 36 nozzles each. These manifolds sprayed in direct opposition to each other as shown in figure 3. Design flowrate for the 72 nozzles was 6.31 kg/s (100 gpm) at 138 kPa (20 psi). Each spray nozzle produces a 72°-angle, hollow-cone spray. Water is supplied to the second stage by a remotely located 2.2 kW (3 hp) pump, drawing water from the sump of the first water stage.

Air leaving the second water stage passes through the spray pattern of the first water stage. The first water stage is similar to the second water stage except that all 72 sprays are directed against the airflow. Cold service water is supplied at 6.31 kg/s (100 gpm) to the first stage at 138 kPa (20 psi) pressure.

Air is demisted by a wave plate eliminator before exiting the cooler. The water removed from the exiting airstream drains to a sump that collects water from the first stage (which is supplied to the second water stage as described previously). Water from the second water stage falls to a second water stage sump where it drains out onto the floor.

3.3 THREE-STAGE COOLER

Both three-stage cooler designs can be understood by reference to figure 4. These units incorporated dual 5.6 kW (7-1/2 hp) pumps on skids that are separable from the skids of the cooler for in-mine transport. The air inlet transition is identical in design to that used on the single-stage cooler. All

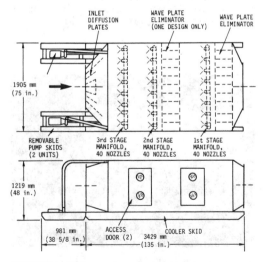

FIGURE 4. Schematic of three-stage cooler.

three water spray manifolds are identical. A uniform array of 40, 80° angle, full-cone spray nozzles are installed on each manifold. The 40 nozzles delivered 6.31 kg/s (100 gpm) at 207 kPa (30 psi).

In operation, hot air passes successively through the third and second water stages. In one design, a wave plate mist eliminator demists this air before it passes through the first water stage. The other design does not have this demist stage. The intent here was to evaluate any effect on thermal performance of the mist eliminator. Initial results indicate there is a positive effect. Downwind of the first water stage, the air is demisted and exits the cooler.

4 MINE APPLICATION OF COOLERS

4.1 MT. TAYLOR MINE

The single stage cooler was installed in the Mt. Taylor Mine in San Mateo, NM, in June 1982. The mine was a developing uranium mine, which was experiencing severe heat and humidity problems. It had a virgin rock temperature of 53°C (127°F) on the 3,200 level and was producing about 189.3 kg/s (5000 gpm) of water at approximately 49°C (120°F). The cooler was located on the 3,200 level in a drift where the dry bulb air temperature was 40.5°C (105°F). It was hung from the roof by chains to keep it out of the way of moving equipment and was pitched with the air inlet (water outlet) end 152.4 mm (6 in) lower than the air outlet end for proper water drainage through the channels in the corrugated plates (fig. 5).

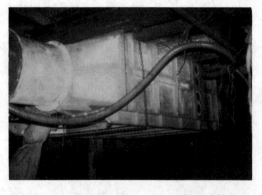

FIGURE 5. Single-stage cooler installed in Mt. Taylor Mine.

Mine engineers were pleased with the cooler, which remained in constant operation without any maintenance for four months until the mine was closed due to falling uranium prices.

4.2 HOMESTAKE GOLD MINE

The Homestake Gold Mine, located in the Black Hills of South Dakota, has been operating since 1876. Operations currently extend to the 8,000 level (2,440 m) where the virgin rock temperature is 56°C (133°F).

Air-conditioning was first used in 1963. Spot-coolers with reciprocating compressors delivered cooled air to development headings. In 1979, a 2.1 MW [580 tons(R)] plant was installed to chill service water for the deepest work areas. About 55 pct of the 60 kg/s flowrate (950 gpm) was used in cooling coils at stope entrances, but air and water side fouling significantly reduced airflow and heat transfer. Homestake then entered into a cooperative agreement with the Bureau of Mines and Foster-Miller Associates, Inc., to evaluate small direct-contact spray coolers. When connected to an 11 kW (15 hp) fan and ventilation bag, a cooler will deliver 2.36 m³/s (5,000 cfm) and 50 kW(R) [14 tons(R)] of cooling to a work area. The evaluation was very successful. Homestake currently uses 46 small spray coolers.

The steady move from cut-and-fill techniques to bulk-mining methods with increasing dependence on diesel LHDs prompted the search for a more compatible ventilation method. Late in 1982, the Bureau of Mines offered Homestake the chance to test three large spray coolers.

The coolers were originally designed for duct and bag, but have been installed as limited bulk-coolers at Homestake. Both limited and primary bulk cooling use airways to transport the refrigeration to work areas. This obviates the need for larger numbers of small fans, cooling devices, and ducting. Unlike primary bulk cooling, however, limited bulk cooling achieves higher positional efficiency and percent utilization. Work areas are assured a constant through-flow of cooled air.

The first three stage cooler with interstage mist eliminator, was installed on the 7,100 level in March 1983 (fig. 6).

FIGURE 6. Three-stage cooler installed in Homestake Gold Mine.

It services a 3-panel vertical crater retreat (VCR) drawpoint area, a charging chamber on the level above, and a development heading. Supplemental cooling is required on the level above due to the magnitude of heat load. The cooler gained immediate miner and supervisor acceptance, and has been operating continuously since installation.

The Pneumafil two-stage cooler was installed on the 7,400 level in February 1984 (fig. 7). It ventilates a large mechanized cut-and-fill stope and ramp system. Again, the large heat load necessitated supplemental cooling. The higher air flowrate effectively disperses diesel emissions and provides a much more uniform temperature profile within the large stope.

FIGURE 7. Two-stage cooler installed in Homestake Gold Mine.

The three-stage cooler without interstage mist eliminator is presently being installed 350 m downstream of the original chamber on the 7,100 level. It will service a VCR drawpoint area and a ramp system.

Maintenance on the first two coolers has been minimal. A few problems have resulted primarily from heavy blasting nearby and from water quality. The mine service water has a total hardness of 275 mg/L, and the sprays naturally wash out some of the dust, emissions, and fumes from the air, to the benefit of crews downstream. These effects combine to clog second and third water stage nozzles, many of which can no longer be cleaned and are scheduled to be replaced. The inter-cooler piping also needs cleaning. The circulation pumps have required minor repairs but have otherwise held up well. Scale deposited on the flow-straightening perforated plate in the Pneumafil cooler cut the airflow in half before the scale was removed with a chipping hammer. The mist eliminators have resisted scale build-up and have withstood heavy blasting most admirably.

Heavy blasting has taken its toll on the 7,100 level cooler, however. Blasts consuming 10 tons of powder have rocked the installation. Two windows were blown out and the inlet transition guide vanes were knocked loose and were removed in May 1984. Recent performance checks indicate that the F-factor has fallen significantly from its value at installation. The cooler will be fitted with a new inlet transition. This, along with a good cleaning and new nozzles, should return the F-factor to at least 0.60.

5 PERFORMANCE OF COOLERS

5.1 SINGLE-STAGE COOLER

This cooler, tested in the Mt. Taylor Mine, achieved an average F-factor of 0.525 with a standard deviation of 0.015. This is considered good for a single-stage spray cooler with only horizontal airflow, and was achieved in spite of evidence of temperature stratification due to nonuniform air-water contact throughout the cooler. Nonuniform air-water contact could be expected due to the mine limit on cooler length, which prevented the installation of an inlet transition to produce a flat velocity profile within the cooler. At the outlet of the mist eliminator, wet bulb temperatures ranged from 4.2 to 6.9°C (7.6 to 12.4°F) higher near the corners than near the center. Elimination of this stratification would improve cooler performance.

5.2 TWO-STAGE COOLER

Installation and operation of the two-stage cooler was somewhat different than called for by the original design. The outlet transistion was removed as it was not necessary to duct outlet air. This reduced static pressure loss across the cooler. Significantly, the mine did not need 6.31 kg/s (100 gpm) of chilled water for each cooler. For this cooler, approximately 3.5 kg/s (55 gpm) at 69 kPa (10 psi) of chilled water was supplied to the first water stage. Water to the second water stage was pumped out of the sump at 4.5 kg/s (72 gpm). This sump water was a mixture of water exhausted from the first water stage and recycled water which had been sprayed previously from the second stage. Excess water in the sump was drained into a tank at the 3.5 kg/s (55 gpm) inlet rate. This water is subsequently pumped through a shell and tube condenser. The purpose of using recycled water in the second water stage was to promote a higher water-to-air loading for better heat absorption from the hot entering air.

Table 1 presents the test results. Average F-factor is about 0.5. Temperature stratification of about 3.6°C (6-1/2°F) was observed in the outlet air and may explain the scatter of results. Since this cooler was not operated at the conditions or in the manner intended by the Pneumafil Corporation; performance somewhat below the predicted F=0.6 was to be expected.

5.3 THREE-STAGE COOLERS

Of the two three-stage coolers that were built, only the unit with two mist eliminating sections has been tested to date. The other three-stage cooler is only now being installed.

The cooler that was tested was modified for mine use. The outlet transition was removed and in this configuration, air pressure loss at about 12.3 m^3/s (26,000 cfm) was measured at 450 Pa (1.8 in water gauge) with the water sprays on and 250 Pa (1.0 in water gauge) with the sprays off. The water supply to the stages was:

a. First stage: chilled mine water; 3.1 kg/s (50 gpm) at 206 kPa (30 psi)

b. Second stage: mixed first and second stage sump water; 5.2 kg/s (82 gpm) at 241 kPa (35 psi)

c. Third stage: mixed second and third stage sump water; 5.5 kg/s (87 gpm) at 276 kg/s (40 psi).

Table 2 shows the results of performance measurements. The average F=0.606 is very good. As can be seen, each of the three stages is important to achieving this performance.

Table 1. Two-stage cooler tests

Test No.	Stages tested	Water temp In °C/(°F)	Water temp Out °C/(°F)	Air temp in Wb,db °C/(°F)	Air temp out sat. °C/(°F)	Water flow kg/s/(gpm)	Air flow m^3/s/(cfm)	Factor of Merit-F	Cooling kW/R-ton
1..	2	10.0 (50.0)	19.5 (67.2)	28.2 24.9 (82.8),(76.8)	20.7 (69.3)	3.5 (55)	7.73 (16,400)	0.500	137 (39)
2..	2	10.0 (50.0)	19.4 (66.9)	28.0 24.8 (82.4),(76.6)	20.8 (69.5)	3.5 (55)	8.02 (17,000)	0.489	134 (38)
3..	1	10.1 (50.1)	18.7 (65.6)	27.7 24.94 (81.8),(76.9)	21.6 (70.9)	3.5 (55)	8.49 (18,000)	0.433	123 (35)
4..	2	10.5 (51.0)	19.9 (67.9)	28.7 25.4 (83.6),(77.7)	20.8 (69.5)	3.5 (55)	6.89 (14,600)	0.510	134 (38)

Average F = 0.500, excluding test No. 3.
Air flows adjusted to obtain heat balance.

Table 2. Three-stage cooler tests

Test No.	Stages tested	Water temp In °C/(°F)	Water temp Out °C/(°F)	Air temp in Wb,db °C/(°F)	Air temp out sat. °C/(°F)	Water flow kg/s/(gpm)	Air flow m^3/s/(cfm)	Factor of Merit-F	Cooling kW/R-ton
1..	All	9.78 (49.6)	21.6 (70.9)	29.6, 24.8 (85.3),(76.7)	21.0 (69.8)	3.2 (50)	9.53 (20,200)	0.571	155 (44)
2..	All	10.2 (50.4)	22.3 (72.2)	30.2, 24.5 (86.4),(76.1)	20.9 (69.6)	3.2 (50)	10.4 (22,100)	0.612	158 (45)
3..	All	9.28 (48.7)	25.2 (77.4)	30.9, 26.7 (87.7),(80.1)	23.1 (73.5)	3.2 (50)	12.4 (26,200)	0.645	207 (59)
4..	1 & 2	9.22 (48.6)	24.2 (75.6)	31.5, 26.9 (88.8),(80.4)	23.6 (74.5)	3.2 (50)	12.9 (27,300)	0.572	197 (56)
5..	1	9.22 (48.6)	18.7 (65.6)	29.2, 24.2 (84.6),(75.6)	21.4 (70.5)	3.2 (50)	10.3 (21,800)	0.436	123 (35)
6..	All	9.28 (48.7)	21.9 (71.5)	28.9, 24.2 (84.0),(75.5)	20.9 (69.6)	3.2 (50)	12.1 (25,600)	0.596	165 (47)
7..	All	9.11 (48.4)	22.1 (71.8)	28.8, 24.3 (84.0),(75.8)	20.9 (69.6)	3.2 (50)	11.8 (25,000)	0.606	172 (49)

Average F for three-stages = 0.606.
All air flows adjusted to obtain heat balance.

6 DISCUSSION

The purpose of any ventilation system is to provide an intrinsically safe and physiologically tolerable underground work environment. Ventilation alternatives meeting these requirements are compared economically. Any characteristic of a given alternative can be eventually reduced to an economic equivalent. Some factors influencing ventilation alternatives analysis are the mining method, the contribution of various heat load sources to the total heat load, the surface ambient conditions, the labor required to maintain production quotas, the concentration of mining, and the weighted center of mining with due respect to the standard deviation. The age of the mine, including services available and condition of major airways, is also important.

Large spray coolers can be used in two ways. If used with duct as originally intended, their economics can be compared with spot-coolers. If used in a limited bulk cooling capacity, some specific conditions must be met. Water at 12°C (53°F) or lower is absolutely essential. A water quantity of 0.02 kg/s per kW(R) [1.1 gpm per ton(R)] must be available to meet projected heat load requirements. A number of headings can be serviced by a limited bulk cooler, but the areas must be in close proximity. If not, a larger number of smaller units may do a better job. The work area clusters themselves should be scattered. If not, primary bulk cooling may be more economical.

A high factor of merit is not always essential for primary bulk-cooling. The cooling duty may be met more effectively with a higher water flowrate and lower Δt. With limited bulk-cooling, however, a high factor of merit is essential. The water should remove as much heat from the air as practically possible when high pumping heads are involved, whether primary or limited bulk cooling is employed even if energy recovery on the service water side is employed.

7 CONCLUSIONS

The three large scale coolers tested in this program all performed well under in-mine conditions. The F-factor of the single-stage cooler of 0.525 and that of the three-stage cooler of 0.606 can be considered very good. The two-stage cooler F-factor of 0.500 may appear on the low side, but this results from the unit being operated under conditions not considered in its design. All of the coolers tested have performed for significant periods of time with only minimal maintenance requirements.

Experience with these coolers indicates the difficulty of achieving high performance (F=0.6) with only horizontal airflow. It took three stages of sprays with two electric pumps to reach this level with the three stage cooler. In any spray cooler, good performance requires uniform distribution and thorough mixing of the air and water. This usually demands both a uniform air velocity profile and uniform water loading throughout the contact region. None of the large coolers achieved good, flat air velocity profiles. This indicates temperature stratification existed, which resulted in loss of performance. Restriction on cooler size is the major problem in trying to achieve flat air velocity profiles with minimal loss of air pressure. However, size restriction is a fact of life for mining equipment, and solutions must be found within that constraint.

The use of the F-factor for characterizing cooler performance was invaluable in this program. This single number sums up the performance of a cooler, without regard for particular temperatures and flowrates. Laboratory testing is dramatically reduced and can be quickly correlated to mine test results and vice versa. F-factors also facilitate comparisons between different coolers in different mines. As a unitless number, there is no confusion in international work where different conventions of units are employed; one number suits all.

Finally, it is important to point out that any cooling technique employed must be tailored to the individual application for which it is intended. The large spray coolers discussed in this paper may not be appropriate for all mines with heat problems and, if they are employed, the manner in which they are used must be closely looked at in terms of the prevailing mining conditions.

8 REFERENCES

Bluhm, S. J. Performances of Direct-Contact Heat Exchangers. Journal of Mine Ventilation Society of South Africa, v. 34, Nos. 8 and 9, August and September 1981, pp. 155-174.

McCoy, J. F., A. Whillier, K. S. Heller, and E. D. Thimons. Development of an In-Line Water Spray Cooler. Journal of Mine Ventilation Society of South Africa, v. 35, No. 1, January 1982, pp. 1-6.

Whillier, A. Predicting the Performance of Forced-Draught Cooling Towers. Journal of Mine Ventilation Society of South Africa, v. 30, No. 1, January 1977, pp. 2-25.

15. Diesel

The effects of maintenance and time-in-service
on diesel engine exhaust emissions

ROBERT W.WAYTULONIS
US Department of the Interior, Bureau of Mines, Minneapolis, USA

ABSTRACT: Research was conducted under a U.S. Deparment of the Interior, Bureau of Mines contract to determine the effects of maintenance and time-in-service on diesel engine emissions and performance. Data from six Caterpillar 3306 PCNA and seven Deutz F6L 912 W engines from five underground mines are presented; their number of hours in service ranged from 485 to 9000. Also, a laboratory study induced faults in a new diesel engine to determine their effects on emissions. Analysis of the in-service engine data and the induced faults study revealed that harmful pollutants can be controlled to near new engine levels by sustained and proper maintenance. Maintenance activities causing the engine's fuel-air ratio to deviate from the factory setting have an immediate detrimental effect on most pollutants of concern. With proper maintenance, engine component wear affects emissions and performance only after several thousand hours in service.

1 INTRODUCTION

The increasing use of diesel-powered equipment in underground mines and concern for the health of miners led the Bureau of Mines, U.S. Department of the Interior, to embark on a project to obtain definitive data on the relationships between engine condition, maintenance practices, emissions, and time-in-service. Although it has been recognized for some time that properly adjusted engines and plentiful ventilation were necessary for safe operation of diesel-powered equipment (Grant 1973, Holtz 1960), little information existed to support this precept.

This report is based on data obtained from a study performed by Southwest Research Institute (SwRI) under U.S.B.M. contract H0292009 (Branstetter 1983). The objective of the SwRI study was to quantify typical mine diesel emission levels and relate this information to maintenance and time-in-service. This report discusses underground diesel maintenance, further analyzes the SwRI data on 13 in-service diesel engines, and presents the effects of induced faults on emissions.

2 DESCRIPTION OF METHODS

The SwRI project was divided into two major segments; the in-mine engine evaluations and a laboratory study of the effects of induced faults on engine emissions. Near identical emission and diagnostic instrumentation was used in both segments.

2.1 Instrumentation

Performance and diagnostic instrumentation were used to evaluate engine condition and measure exhaust emissions. The factors that affect emissions include engine speed, torque, fuel consumption, intake airflow, injection timing, nozzle crack pressure, compression, barometric pressure, and relative humidity. These parameters were measured in the mines and the laboratory using conventional instrumentation and techniques specified by the Environmental Protection Agency (EPA 1982, Fed. Reg. 1981) and described by Branstetter et al. Carbon monoxide (CO) and carbon dioxide (CO_2) were measured using non-dispersive infrared (NDIR) analyzers. Nitric oxide (NO) and nitrogen oxides (NO_x) were measured by a heated chemiluminescent analyzer, and hydrocarbons (HC) were measured using a flame ionization detector (FID). During the in-mine studies, a Bosch spot smoke meter was used to measure smoke density. The readout of the Bosch smoke meter or Bosch smoke number (BSN) has a scale of 0 to 10. Although correlations were not made in this study, the exhaust carbon and total particulate mass concentrations correlate well with the BSN (Alkidas 1984). In the laboratory, exhaust particulate measure-ments were accomplished with an exhaust dilution tunnel.

2.2 In-mine engine evaluations

Five mines supplied engines or complete vehicles with varying amounts of accumulated time-in-service to the field research team. During the testing of in-service engines it was difficult to obtain permission from mines to perform the in-depth and complete testing necessary to form quantitative relationships between time-in-servce, local maintenance practices, and emissions. Only mines having generally good maintenance programs were willing to allow the testing reported here; therefore it is assumed that the majority of the engines had received proper maintenance throughout their lives. Whenever possible mine operators supplied engines that had differing numbers of hours-in-service. Such a selection made it possible to attempt a qualitative correlation between engine performance and time. Reference to specific products in the text does not imply endorsement by the Bureau of Mines.

The six Caterpillar 3306 PCNA engines (designated "C") were six-cylinder, water-cooled, indirect-injection, four-cycle, naturally aspirated diesels rated 146 hp (109 kW) at 2200 rpm. The factory setting for fuel injection timing for these engines is 13 crank angle degrees before-top-dead-center (BTDC). These engines were all tested without removing them from the vehicle and were coupled to the dynamometer through the vehicle's torque converter. Full load-peak torque speed could not be attained owing to the torque converter's slip ratio and drive line mechanical losses. The load-speed test points used in the

Figure 1. A typical in-mine test set-up for the evaluation of diesel engines.

analysis of these engines include only 100 and 50 percent load-rated speed.

The seven Deutz F6L 912 W engines (designated "D") were all six-cylinder, air-cooled, indirect-injection, four-cycle, naturally aspirated diesels rated 84 hp (63 kW) at 2300 rpm. The factory setting for fuel injection timing is 24 degrees BTDC. The Deutz engines were tested apart from the mine vehicles; thus additional test data were obtained at 100% and 50% loads-peak torque speed. The format used to identify the engines in the forthcoming tables and graphs is "C or D - number of hours-in-service". For example, C-5000 designates a Caterpillar engine with 5000 hours-in-service. An in-mine engine test set-up is shown in figure 1.

2.3 Laboratory study of induced engine faults

A new Deutz F6L 912 W (D-0) engine was used to determine the effects of malad-justments or faults on the production of HC, CO, NO_x, and particulate matter. After initial tests to acquire the engine baseline data, faults were artificially induced at different levels of severity and the resulting changes in emissions were measured. These faults were intake combustion air restriction, exhaust-gas restriction, fuel injection timing adjustment, and overfueling. The first level of severity represents the limits of manufacturer-recommended specifications, and the second level was chosen to be deliberately excessive but within the realm of possibility through improper maintenance; intake restriction = 25 and 50 in. water (6 and 12 kPa), exhaust restriction = 3 and 6 in. Hg (10 and 20 kPa), and overfueling = 10% and 20% overrated. Fuel injection timing advance had three levels of severity: -4, +4, and +8 crank angle degrees from 24 degrees BTDC. The engine was tested with the individual faults and at the two or three levels of severity. Also, tests were conducted with two faults simultaneously. Finally, all four faults were induced simultaneously. Three baseline tests were performed on the engine during which all adjustments were set as recommended by the manufacturer.

2.4 Data analysis technique

Data were analyzed according to the protocol developed by Branstetter, using a normalized weighting equation. The mass emission rate for each pollutant was used in the data analysis. The data were reduced by combining load-speed test modes of each engine, for a given pollutant, into one number by weighting the emission value at the load-speed test points.

The weighting factors are based on the percentage of time the engine is assumed to be at a load-speed condition during a hypothetical operating cycle. Mass emissions were reduced and the units used in figures 2 through 4 are grams per brake horsepower-hour (gm/bhp-hr). So even though the C and D engine groups have different horsepower ratings, valid comparisons of the two engine groups can be made on a per-horsepower basis.

3 UNDERGROUND MINE MAINTENANCE

3.1 General

The objective of diesel engine maintenance is to keep engines in operating condition so that productivity continues and useful life is maximized. Preventive maintenance, periodic repairs, and adjustments are all part of a basic maintenance program. While main-tenance cannot guarantee hazard-free operation, it can prolong or restore near-original efficiency of the engine and vehicle (Springer 1975). Lack of basic maintenance decreases equipment availability and useful engine life and can increase engine failures. Underground mine maintenance programs must cope with unique environmental problems not common in other settings. Underground condi-tions vary but are usually very harsh and hard on equipment. Working areas are no larger than necessary, and often small poorly lighted cutouts serve as shops. Tools may consist of only essential items such as wrenches, a cutting torch, a welder, and large hammers. Some mines have large well-lighted shop areas where more complicated repairs can be undertaken and may even purchase specialized engine diagnostic equipment for preventive maintenance. Nevertheless, the underground mining environment is not conducive to performing diesel engine or any other maintenance activities.

The Mine Safety and Health Administration (MSHA), through certification or approval, assures that new diesel-powered equipment meets specified safety criteria. This includes a minimum ventilation requirement in loca-tions where the diesel unit is to be operated. The ventilation rate is calculated to dilute certain gaseous pollutants to one-half their individual Threshold Limit Values (30 CFR 32, 36, 1984). Once the equipment is put into operation, it is the responsibility of the mine operator to keep it in good condition.

3.2 Diesel engine maintenance

A brief discussion of the six major systems pertaining to the maintenance of diesel engines follows:

1. Intake air system- Dust-laden mine air causes intake air filters to become filled with dust, creating a restriction that may exceed the manufacturer's recommended limit. In-take air filters should be replaced when the pressure drop across the filter exceeds the manufacturer's specification, usually about 20 in. (5 kPa) of water. Not all air intake system failures can be detected by pressure drop indicators, e.g., a broken intake air duct or punctured filter will not be detected. The best method presently available for detection of these failures is a visual inspection of the air intake system. A failure, if not quickly repaired, will cause rapid engine wear and result in increased emissions and decreased performance.

2. Engine cooling system- The loss of engine cooling leads to scuffed cylinder walls and pistons, cracked heads, and burned valves. These conditions directly affect emission production and output power. A liquid-cooled engine relies on transfer of heat from the coolant to the radiator, and from the radiator to ambient air. Internal coolant passages of the engine and radiator must be kept free of mineral and rust deposits for effective heat transfer. Mine water is generally high in minerals and salts, rendering it unfit for use in engine cooling systems. It is recommended that a premix of a 50% mixture of distilled water and antifreeze be used.

Air-cooled engines reject heat via cooling fins which are an integral part of the engine. During normal operation these fins become coated with oil and dust, which bakes on to form an unsulating layer. If this layer is allowed to build on the engine, overheating will result. Periodic steam cleaning will prevent this situation.

3. Diesel fuel handling and quality- DF 2 diesel fuel should be used whenever ambient temperatures are above the cloud point of the fuel. This is because when compared to DF 1, DF2 possesses better lubrication properties and tends to extend fuel injection system component life. Additionally, DF 2 has a higher energy content per gallon. Low-sulfur (less than 0.5% by weight) fuel should always be used; the sulfur present in all diesel fuels directly affects the emissions of particulate sulfates and accelerates engine wear. Fuel contamination will also cause accelerated engine wear. It is important to minimize the number of fuel transfers and to store the fuel in tightly sealed containers which are clearly labeled.

4. Fuel injection system- The adjustment of the fuel injection system has considerable effect on the production of exhaust pollutants. The most critical adjustments are fuel flow rate and fuel injection timing. The engine fuel flow rate is usually set at the factory or at

the mine site, and this setting is based on the MSHA horsepower and ventilation rating. The fuel rate remains constant for long periods; however, it can be adjusted to yield higher output horsepower.

Unless manually adjusted, diesel injection timing generally remains constant over long service intervals. Timing could be improperly adjusted at the factory, improperly set by a serviceman, or otherwise altered to yield higher output horsepower. Engine manu-facturers usually allow a 1-degree deviation from the recommended setting. Although changes in fuel injection timing and fuel injec-tion rate can increase power output, increases in exhaust pollutants may accompany the power increase. The life of the injection system is greatly affected by the quality of fuel and lubricant. Contamination of the fuel erodes injector nozzle tips and the injection pump plunger and barrel. Like the rotating components of the engine, it is important that the lubrication oil in the injection pump to be clean.

5. Lubrication system- Failure of the lubrication system usually results in catas-trophic engine failure. System failures are often caused by a component failure, such as seized bearings, lubricant breakdown or contamination, or engine over-heating. To control these failures it is important to keep the crankcase lubricant free of solid and liquid contamination, and maintain the engine's cooling system. If an engine becomes excessively hot, the oil viscosity is lowered, resulting in loss of lubricity and accelereated engine wear.

6. Exhaust system- Excessive exhaust-gas restriction can result from either a partially plugged water scrubber or catalytic converter, or a dented exhaust pipe. Diesel engine manu-facturers generally consider 2 to 3 in. Hg (7 to 10 kPa) to be the acceptable limit. Excessive backpressure results in increases of some pollutants and decreases output power. Periodic inspection and cleaning of exhaust system components will preclude excessive backpressure.

4 TEST RESULTS OF IN-SERVICE CATER-PILLAR AND DEUTZ ENGINES

Table 1 contains diagnostic information and performance data for the 13 in-service engines, the D-0 engine, and partial information for a C-0 engine. The maximum horsepower is that developed by each engine at 100% of its maximum load at rated speed; brake specific fuel consumption (BSFC) in pounds per hour is also given at this condition. Fuel injector nozzle crack pressure, cylinder compression, and fuel injection timing are also listed to describe the internal condition of the engines. To identify possible reasons why emissions and performance vary with maintenance-related activities or time-in-service, the information

Table 1. In-service engine diagnostic information.

Engine	Max.hp	BSFC (lb/Bhp-hr @ max load/ speed)	Nozzle Crack Pressure (psi) Cylinder Number						Compression (psi) Cylinder Number						Injection timing (deg.BTDC)
			1	2	3	4	5	6	1	2	3	4	5	6	
C-0	146	n/a	400–800	"	"	"	"	"	no published data ————————————————→						13
C-485	139.9	0.3873	700	650	600	700	700	600	370	365	370	360	370	360	17
C-634	125.5	0.4231	750	750	750	650	750	800	330	350	375	n/a	350	355	15
C-2740	129.4	0.4066	700	700	700	700	800	750	n/a	295	n/a	300	290	280	n/a
C-3099	151	0.3774	600	700	700	700	600	700	350	360	345	350	355	320	18
C-3765	140.7	0.4052	900	700	800	700	700	700	340	350	345	330	330	325	n/a
C-5071	154.7	0.4085	750	800	800	800	800	800	n/a	n/a	n/a	n/a	385	375	n/a
D-0	83.5	0.4266	1850	1850	1850	1850	1850	1850	362 TO 435————————————→						24
D-600	78.2	0.4230	1750	1650	1650	1650	1700	1650	440	415	430	430	440	430	18
D-1720	86.8	0.4101	1550	1600	1675	1600	1650	1600	460	455	435	430	430	435	32
D-3300	79.7	0.4152	1400	1600	1500	1600	1500	1400	360	420	395	390	390	355	24
D-4500	77.5	0.4118	1700	1700	1600	1700	1600	1600	410	410	405	410	385	410	29
D-6500	78.7	0.4285	400	400	400	400	360	400	390	400	400	420	365	390	23
D-7375	63.4	0.4543	1600	1625	1600	1650	1700	V. LOW	320	340	360	330	335	330	n/a
D-9000	87.4	0.4141	1750	2100	2100	2000	1800	1800	390	420	450	420	450	440	21

contained in table 1 must also be examined in conjunction with the induced-faults data in table 2, and the emissions data in figures 2 through 5. Figures 2 through 5 are envelopes which contain all values for the respective engine group. Both engine groups are shown on common axes for visual comparison. Each engine can be identified by the legend and corresponding hours-in-service value. With the information contained in these tables and graphs, additional engine analysis can be performed by the reader.

Figure 2 shows the weighted average hydrocarbon (HC) emissions in gm/bhp-hr plotted against hours-in-service for all in-service C and D engines. The HC value for the C-0 engine is unknown. With the exception of C-485, C-3765, and D-6500, all the in-service engine values fall between about 0.28 and 0.50 gm/bhp-hr. Table 1 shows that C-485 had a 4-degree timing advance but otherwise good diagnostic indicators. The advanced timing accounts for these high HC emissions. The compression in C-3765 was slightly low, and the injection timing was not obtained. Low compression can cause inefficient combustion and an increase in HC emissions. Low fuel injector nozzle crack pressure results in a poor fuel spray pattern or dripping, which causes an incomplete mix of fuel and combustion air. D-6500 had very low fuel injector crack pressure in all six nozzles, which resulted in very high HC emissions. These data indicate a gradual increase in HC emissions for the first 4000 to 5000 hours-in-service.

In figure 3 the weighted average carbon monoxide (CO) emissions are plotted against time-in-service. Unlike the HC, NO_x, and Bosch smoke data, the emission envelopes for the C and D engines do not overlap for CO. The C engine group has the higher CO values. Excessive CO emissions result from incomplete combustion. For the C envelopes, data are grouped between about 1.9 and 3 gm/bhp-hr except for C-5071. Diagnostic information for this engine is incomplete; however, low compression or retarded timing will cause CO to increase. CO data for all the D engines fall between about 1.1 and 2.5 gm/bhp-hr. D-9000 is the highest in this engine group, and its slightly retarded timing contributes to a high CO value. In general CO emissions appear to increase slowly with time-in-service for the D engine group; it is difficult to distinguish a trend with time for the C group.

In figure 4, weighted average oxides of nitrogen (NO_x) emissions are plotted against time-in-service. For the C engine group, C-485 lies outside a decreasing trend with time; in the D engine group D-4500 also is an outlier. Both these engines have advanced fuel injection timing, which is the cause of these high NO_x emissions. D-600 has very low NO_x emissions

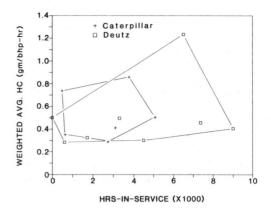

Figure 2. Weighted average hydrocarbon emissions vs. hours-in-service for the C and D engine groups.

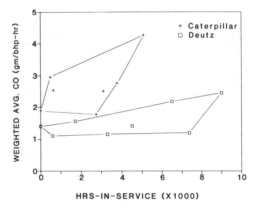

Figure 3. Weighted average carbon monoxide emissions vs. hours-in-service for the C and D engine groups.

resulting from this engine's retarded timing. Except for these three engines, the values for both engine groups fall between about 3.2 and 6 gm/bhp-hr, with the C group having the lowest values. This is possibly due to the lower combustion temperatures in the water-cooled C group, which tend to lower NO_x. These data indicate a gradual decrease in NO_x emissions with time in service, for both engine groups.

Figure 5 is a plot of the Bosch smoke number (BSN) at full load-rated speed vs. time-in-service for the C and D engine groups. BSN is an indication of combustion efficiency and particulate matter emissions. Within the C group, C-2740 had the lowest value at about 1. In table 1 it can be seen that this engine could only reach about 130 hp (97 kW) vs. its 146 hp (109 kW) rating, and this contributes to the low BSN; BSN is generally greater at higher

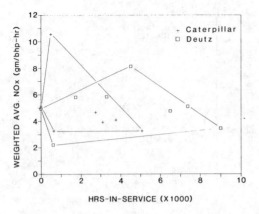

Figure 4. Weighted average oxides of nitrogen emissions vs. hours-in-service for the C and D engine groups.

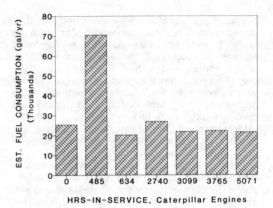

Figure 6. Estimated annual fuel consumption for the C engine group.

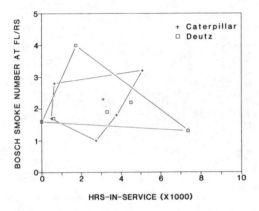

Figure 5. Bosch smoke number for C and D engine groups at full load-rated speed vs. hours-in-service.

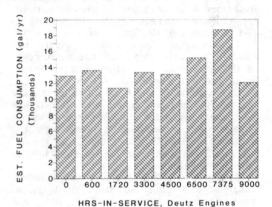

Figure 7. Estimated annual fuel consumption for the D engine group.

engine loads. C-634 and C-5071 both had high BSN's of about 3. C-634 could only attain about 126 hp (94 kW) and had slightly advanced fuel injection timing, which in combination increased BSN. C-5071 actually exceeded its rated power output by about 9 hp (6 kW), which could account for its high BSN. Within the D engine group, D-1720 had the highest BSN at about 4. A combination of greater-than-rated power output with advanced fuel injection timing was the cause of its high BSN. D-9000 had the lowest value (1.3 BSN) for the D group of engines. This was due to this engine's retarded timing. Additionally, the higher-than-baseline nozzle crack pressures effectively further retard fuel injection timing, and it was these conditions that counteracted the greater-than-rated power output and resulted in a very low BSN. Disregarding these extremes, BSN

for both engine groups fell between about 1.6 and 2.2 with a slight increasing trend over time.

Annual fuel consumption estimates are based on measured average brake specific fuel consumption for each in-service engine and 3456 hours per year operation. The engines are compared on the basis of equal work performed (horsepower-hours) in figures 6 and 7. Figure 6 indicates that a new C engine would consume about 25,000 gal (95 kl)/year of fuel; C-485 is seen to have extremely high fuel consumption at more than 70,000 gal (265 kl)/year. During the testing of this engine a leak in the high pressure fuel line occurred at the 50% load-rated speed test point. This very high estimate assumes that this condition was allowed to continue. C-634 would consume about 20,000 gal (76 kl)/year, slightly less than the C-0 value. Based on the information in

614

table 1, it cannot be determined why this engine was more fuel efficient. C-2740 had slightly higher than baseline fuel consumption at 27,000 gal (102 kl)/year, owing in part to its low cylinder compression. All other C group engines are estimated to consume near-baseline levels of fuel.

The annual fuel consumption estimates in figure 7 for the D engine group vary from about 11,000 to over 18,000 gal/year (42 to 68 kl/year). D-1720 consumes the least fuel of all D engines owing to its advanced injection timing, which enables it to get slightly more horsepower per unit of fuel consumed. Although this engine's output power was high, a penalty of high CO emissions as shown in figure 3 resulted. D-7375 is estimated to consume in excess of 18,000 gal (68 kl)/year owing to its one leaky injector and low cylinder compression. D-6500 is estimated to consume over 15,000 gal (57 kl)/year of fuel owing to very low injector nozzle crack pressures in all six cylinders. All other D group engines would consume about 12,000 to 13,500 gal/year (45 to 51 kl/year).

5 TEST RESULTS OF INDUCED FAULTS ON A NEW DEUTZ F6L 912 W ENGINE

A summary of the induced-faults tests is presented in table 2, which lists the percent deviation from baseline caused by the faults. The weighted baseline values for the gaseous emissions were HC = 0.31, CO = 1.15, NO_x = 3.15 gm/bhp-hr, and steady-state particulate matter at full load-peak torque speed = 569 milligrams per standard cubic meter (mg/scm), compared with 64 mg/scm at full load-rated speed.

The single fault that has the greatest influence on the production of hydrocarbons is retarded timing (+306%) and this is worsened when combined with restriction of intake air (+71% to +443%). Advancing injection timing also increases production of HC but levels off at 4 degrees advance to +107%. Exhaust-gas restriction tends to slightly increase HC by +17% at 3 in. Hg (10 kPa) and +2% at 6 in. Hg (20 kPa), while intake restriction and/or overfueling causes small decreases.

The test condition producing the highest CO emissions was the combination of intake air restriction and overfueling (+164% to +445%). Producing almost an equal amount of CO (+448%) was the combination of all faults at their most severe settings. Overfueling combined with exhaust restriction (+102% to +326%) followed by overfueling alone (+95% to +247%) produced less of an increase. Overfueling was the common element producing the highest CO. It appears that the advance in

Table 2. Percent deviation from baseline caused by induced faults in a new Deutz F6L 912 W engine.

Fault Description	Degree of Fault	HC	CO	NOx	Particulates * FL/RS	FL/PTS
Intake Air Restriction (in-H2O)	25	−28	+8	−15	+25	+44
"	50	−36	+28	−12	+75	+164
Exhaust Restriction (in-Hg)	3	+17	+1	+9	−15	−6
"	6	+2	+6	−3	−8	−11
Timing Advance (24 deg BTDC)	−4	+306	+53	−33	−4	−23
"	+4	+106	+4	+1	+2	−41
"	+8	+107	+12	+50	+29	+30
Overfueling (pct rated)	+10	−28	+95	0	+45	+125
"	+20	−20	+247	−24	+44	+173
Intake Air Restriction & Timing Adv.	25 −4	+71	+25	−27	+7	+150
"	50 −4	+443	+122	−38	+120	+94
Exhaust Restriction & Timing Adv.	3 +4	+59	+11	+2	−13	+5
"	6 +8	−29	+23	+53	+25	+59
Intake Air Restriction & Overfueling	25 +10	−3	+164	−14	+139	+194
"	50 +20	−21	+445	−35	+1038	+366
Overfueling & Timing Adv.	+10 +4	−24	+97	+11	+58	+118
"	+20 +8	−5	+112	+19	+102	+211
Intake & Exhaust Restriction	25 3	0	+8	+3	+8	+45
"	50 6	−38	+72	−8	+153	+91
Exhaust Restriction & Overfueling	3 +10	−33	+102	+6	+77	+368
"	6 +20	−46	+326	−13	+263	+324
Int.& Exh. Restrictions 0-fuel & Timing Adv.	25 3 +10 +4	−40	+72	+10	+132	+338
"	50 6 +20 +8	−44	+448	+6	+584	+352

* FL/RS – full load/rated speed
  FL/PTS– full load/peak torque speed

the injection timing in the absence of other faults somewhat compensates for overfueling and limits the production of CO.

The single fault most severely affecting

production of NO_x is fuel injection timing advancement (+1% to +50%). The greatest increase of NO_x occurred when timing advance and exhaust-gas restriction were combined (+2% to +53%). Retarded injection timing provided the greatest reduction in NO_x (-33%). Severely restricting intake air (-12% to -15%) and overfueling (-24%) both resulted in reduction of NO_x.

When the engine was operated at full load-rated speed, intake air restriction (+25% to +75%) or overfueling (+44% to +45%) had the most effect of single faults on particulate production. The combination of these faults at the most severe test point resulted in a 1038% increase above baseline. Combining all faults also greatly increased particulate production but at about half this condition (+584%). All combination faults except those with timing advance resulted in at least a 100% increase over baseline values.

Particulate values at full load-peak torque speed are somewhat similar to those at full load-rated speed. Of all the fault conditions that increased particulates, overfueling was a common element. The single faults that increased particulates were intake restriction (+44% to +164%) and overfueling (+125% to +173%).

In general, engine faults did not result in decreasing particulate matter reduction. Also, it was found that the level of emission caused by a pair of faults occurring individually is not as severe as the level when the same faults are induced simultaneously. For example, if an otherwise properly adjusted engine has an intake air restriction of 50 in. of water (13 kPa), particulate emissions increase 75%. If the engine is overfueled and otherwise adjusted properly, particulates increase by 44%. If the two fault conditions are combined, particulate emissions increase by more than 1000% over the baseline value. These maintenance-related faults effectively change the engine's fuel-air ratio, resulting in excessive exhaust pollutant production.

6 CONCLUSIONS

In the absence of severe faults or maladjustments, exhaust emission quality does not degrade excessively during the initial 4000 hours-in-service. After 4000 to 5000 hours-in-service the engines tested in this project typically developed the following trends: HC increased, CO increased, NO_x decreased, and particulates increased. Two explanations are offered for these qualitative trends. After time, engine component wear becomes significant and affects engine operation and composition of the exhaust. For example, a worn fuel injection system has the effect of retarding fuel injection timing and thus decreasing NO_x emissions. Also, it was observed that older engines are not as carefully maintained as newer engines, and minor faults are more prevalent. However, gradual component wear is believed to be the principal cause of the observed changes in exhaust emissions over time.

Induced faults tests revealed that intake air restriction, fuel injection timing, and overfueling had the greatest effect on emission rates, while certain combinations of faults had synergistic effects. It was observed that when the faults were removed, the engine emission characteristics returned to original levels. The production of HC is most affected by fuel injection timing maladjustments; retardation of the injection timing has the worst effect of any single fault. The combination of timing retardation and combustion air intake restriction promoted the greatest HC emissions. Overfueling the engine was the single fault that had the greatest effect on CO. Any other fault in conjunction with overfueling, with the exception of timing advance, increased the production of CO above the level caused by overfueling alone. The production of NO_x was most affected by injection timing. Retarding the timing decreased NO_x, while advancing the timing by more than 4 degrees substantially increased NO_x. The quantity of particulate matter produced while the engine was operating under full load was most affected by combining intake restriction with overfueling. These engine faults are caused by specific maintenance activities: (1) intake air filter change-out, (2) fuel injection timing adjustment, (3) fuel rate adjustment, (4) fuel injector nozzle cleaning and/or change-out, and (5) exhaust restriction monitoring.

The quantity of engine combustion products emitted into the mine atmosphere is in direct proportion to the amount of diesel fuel consumed. Proper engine maintenance results in lower fuel consumption and lower concentrations of exhaust pollutants, and this translates into cost savings and improved air quality. Diesel engine manufacturers have refined their product performance to be a balance between exhaust emissions, fuel efficiency, and durability. Any changes induced into a diesel engine that result in a fuel-air ratio deviation from the factory setting or accelerated engine wear will cause emissions, efficiency, and durability to degrade. Diesel engines in service in underground mines can be expected to perform several thousand hours with minimal degradation of exhaust-gas characteristics and performance if properly maintained.

REFERENCES

Alkidas, A.C. 1984. Relationships Between Smoke Measurements and Particulate Measurements. Society of Automotive Engineers Technical Paper Series No. 840412, 9 pp.

Branstetter, R., R. Burrahm, and H. Dietzmann. 1983. Relationship of Underground Diesel Engine Maintenance to Emissions. Volume I (contract H0292009, SwRI). BuMines OFR 110(1) - 84, 104 pp.; NTIS PB 84-195510.

_____. 1983. Relationship of Underground Diesel Engine Maintenance to Emissions. Volume II (contract H0292009, SwRI). BuMines OFR 110(2) - 84, 217 pp.; NTIS PB 84-195528.

Federal Register. 1981. (Fed. Reg.). Control of Air Pollution From New Motor Vehicles and New Motor Vehicle Engines; Particulate Regulation of Heavy-Duty Diesel Engines. Volume 46, No. 4.

Grant, B.F. and D.F. Friedman 1975 (comp.). Proceedings of the Symposium on the Use of Diesel-Powered Equipment in Underground Mining. Pittsburgh, PA, Jan. 30-31, 1973. BuMines IC 8666, 366 pp.

Holtz, H.C. 1960. Safety With Mobile Diesel-powered Equipment Underground. BuMines RI 5616, 87 pp.

Springer, K.J. 1975. Transportation, Trucks and Fuel Conservation. Presented at the IV Interamerican Conf. on Materials Technology, Caracas, Venezuela, 7 pp.

U.S. Code of Federal Regulations. 1982. Title 40. (40 CFR 86)—Protection of Environment; Part 86—Control of Air Pollution From New Motor Vehicles and New Motor Vehicle Engines: Certification and Test Procedures, Subpart N—Emission Regulations for New Gasoline Fueled and Diesel Heavy-Duty Engines; Gaseous Exhaust Test Procedures. U.S. Government Printing Office, Washington, DC, pp. 562-608.

U.S. Code of Federal Regulations. 1984. Title 30 (30 CFR 32)—Mineral Resources; Chapter I—Mine Safety and Health Administration, Department of Labor; Subchapter E-Mechanical Equipment for Mines; Tests for Permissibility and Suitability; Fees; Part 32--Mobile Diesel-Powered Equipment for Noncoal Mines. U.S. Government Printing Office, Washington, DC, pp. 209-221.

U.S. Code of Federal Regulations. 1984. Title 30 (30 CFR 36)—Mineral Resources; Chapter I-Mine Safety and Health Administration, Department of Labor; Subchapter E-Mechanical Equipment for Mines; Tests for Permissibility and Suitability; Fees; Part 36-Mobile Diesel-Powered Transporation Equipment for Gassy Noncoal Mines and Tunnels. U.S. Government Printing Office, Washington, DC, pp. 236-248.

Measuring diesel particulate emissions in underground coal mines

WILLIAM J.FRANCART & RICHARD T.STOLTZ
Mine Safety & Health Administration, Pittsburgh, PA, USA

The Dust Division, Pittsburgh Health Technology Center, has conducted three surveys to measure for the first time the quantity of diesel particulate matter emitted from diesel powered equipment during their regular duty cycles. These surveys were conducted in conjunction with the Ventilation Division as part of a program to determine air quantity ventilating requirements for dieselized coal mines.

Exhaust samples were collected directly from the exhaust manifolds of mobile diesel equipment to determine particulate emission rates. The measured rates were within ranges measured in the laboratory.

Respirable coal mine dust concentrations were determined throughout each section by collecting full shift respirable gravimetric samples. While it is not possible to distinguish between diesel particulate matter and coal dust by chemical analysis, an estimate of the contribution made by the diesel matter to the sample weight was made using time study and silica analysis data. Diesel particulate contents ranged from 33 to 99 percent of the respirable dust measurement.

This report presents the procedures employed to assess particulate concentrations and discusses the results obtained in the studies.

Heat emissions of diesel engines and their effects on the mine climate

P.BURGWINKEL
Aachen Technical University, FR Germany

1 INTRODUCTION

The development of German hard coal mining is characterized by the advancement of exploitation to greater depths and by a strong increase of face output. Both developments result in a significant deterioration of the climatic conditions in the mine which places the mine climate more and more into the foreground of operational planing.

Compared to electrically driven equipment with equal capacity, machinery materials operated by diesel engines release a heat current which is disproportionally more intense. It is often necessary to compensate this heat current by using additional cooling measures that have to be considered in an economic comparison between different haulage systems.

The study in hand has attempted to estimate nature and extent of the energy increase of the mine air caused by the use of diesel engines in a main haulage road. The investigation is based on measurements made with an 81 kW diesel-powered locomotive. First a series of measurements were executed on a locomotive test stand of the "Institut für Bergwerks- und Hüttenmaschinenkunde" of Aachen Technical University; subsequently, test runs were carried out with the same locomotive in an underground main road.

2 TEMPERATURE AND MOISTURE CONTENT OF THE AIR ABOVE AND BELOW GROUND

Air measurements took place above ground and at both ends of a main haulage road at a depth of about 780 m. The first measuring point in this haulage road was located at a distance of 700 m from the downcast shaft, the second point was 1700 m away from the shaft. The measuring section was a cross-cut with no branchways, 17,5 m^2 in cross-section, through which flowed a more or less constant ventilating current of 29 m^3/s while measurements were being taken.

The Result of an eight-days measuring period is shown in Figure 1, where T, the dry-bulb temperature, and x, the moisture content, are shown as functions of time. Fifteen values per hour were taken below ground and five values per day were taken above ground.

2.1 Dry-bulb temperature

Above ground, there are the normal, large daily fluctuations in temperature (T_{aG}). The largest difference in one day is 13,9 K; the difference between the highest and lowest temperature in the period of observation is 19,3 K.

The amplitude of the fluctuation in temperature declines considerably in the downcast shaft and the airways underground due to the heat exchange with the surrounding rock.

The daily variation at the 700 m measuring point is less than 0,6 K and the largest difference in temperature during the observation period is 1,5 K, whereas at the 1700 m measuring point this latter value has already decreased to less than 1,1 K. The more frequent daytime fluctuations are due to locomotive haulage.

The curves of the medium temperature above and below ground are similar. The fall in temperature above ground in the middle of the observation period is also clearly recognizable below ground.

It can be assumed that the rock surrounding the downcast shaft and main haulage road has cooled down; that is, it has reached an almost constant temperature with very little fluctuation throughout the year.

The rock absorbs heat when the air temperature is high and emits it when the tem-

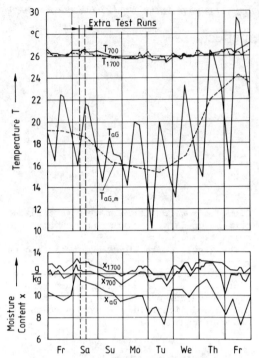

Fig. 1. Temperature and moisture concent above ground (T_{aG}, x_{aG}) and at the two measuring points below ground (T_{700}, T_{1700}, x_{700}, x_{1700}) as functions of time.

perature is low. Since the measurements were taken during summer, a slight fall in temperature was expected in the area around the shaft. When the summertime temperatures outside were high, like they were at the beginning and end of the measuring period, and the underground temperatures at a distance of 1700 m from the shaft were lower than those at 700 m. When the average temperature above ground sank to between 15° and 17°C during the test period, the temperature drop was reversed.

2.2 Moisture content

The humidity above ground varies between 7,3 and 12,0 g/kg during the test period. The air absorbs moisture continuously on its way to the working face. The large fluctuations above ground are decreased significantly less than the temperature fluctuations; that is, any changes in moisture content above ground can be observed throughout the whole intake ventilation roads. Generally speaking, 60 - 80% of rock heat in German coal mines is trans-

ferred by evaporation, hence by latent heat transfer.

Although the ventilation roads usually dry out after a longer period of ventilation, water seepage from the rock is mostly sufficient to sustain a continual latent heat transfer. The water evaporates on the rock surface; inside, transport takes place in liquid form (Voß, 1971).

The heat required for evaporation always comes from the rock, irrespective of surface and support of the gallery (Reuther, 1977). There is, to a large extent, latent heating of the air in the roadway during this investigation. The air enthalpy is higher at 1700 m than at 700 m during the entire test period; that is, the air continually absorbs energy from the rock. This remains true even for those days when the dry-bulb temperature of the air decreases in the test area.

3 A COMPARISON OF THE DIESEL ENGINE HEAT EMISSION WITH THE ENTHALPY INCREASE IN THE MINE AIR

The fuel energy driving the diesel engine is completely converted into heat, with the exception of the small amount of potential energy needed to climb a gradient, but which can normally be disregarded. This conversion partly takes place directly in the motor and transmission or indirectly through bearing and wheel friction in the wheel and axle sets of the entire train. Energy conversion and therefore heat generation in the diesel engine varies considerably during a haulage cycle. One complete cycle consists of a return trip from the shaft to the loading area, as well as stops and shunting at the loading area, shaft, and sometimes in between.

Only the energy emitted with the exhaust is given off directly and in full strength into the air in the gallery. Other energy flows first of all increase the temperature of the heat transfer media - the cooling water and transmission oil - and then heat up the whole engine and mine car bearings. The heat stored in this way is only partly emitted into the air in the gallery. A large percentage is given off during stops and shunting at loading area and shaft.

In order to keep the exhaust temperature below the maximum of 70°C stipulated by the mines inspectorate, water is sprayed into the exhaust pipe of the diesel engine. Consequently, the exhaust gases contain about 1,26 kg vapour per kg fuel consumption which develops during combustion, plus approximately 2,1 kg vapour as the result of the cooling process. A comparison

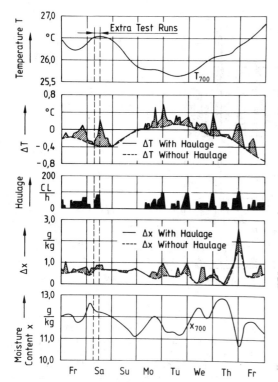

Fig.2 : Temperature T_{700} and moisture content x_{700} of the measuring point at 700 m; differences of temperature ΔT ($T_{1700} - T_{700}$) and moisture content Δx ($x_{1700} - x_{700}$), locomotive haulage in car-loads per hour (CL/h).

a clear correspondence between these hatched areas and the half-tone areas wich represent the number of car-loads transported per hour (CL/h). In the intervals between haulage cycles the increase in temperature and moisture content shows a significant minimum, whereas it reaches a distinct maximum in times of highest haulage intensity.

The dry and latent enthalpy increase of the air due to locomotive haulage should correspond in quality to the shaded areas of the T- and x-graphs.

A calculation of the enthalpy increase $\Delta \dot{H}$ of the air and a comparison between this increase and the heat emitted by the locomotive $\dot{Q}_L$ is illustrated in Figure 3.

The following conclusions can be drawn when comparing the curves in figure 2 and 3:

- There is a definite correlation between the heat given off by diesel locomotives and the increase in enthalpy in the mine air.
- On working days the locomotive releases an average heat flow of 25 kW in the test section. 84 % of this (21 kW) can be shown through the test methods to be dry and latent enthalpy increase. The remaining 16 % are counterbalanced by a lower heat transfer from the rock. There are peak values of over 50 kW for a number of hours nearly every working day.
- The heat given off by the locomotive (most of which is dry heat) causes only a slight increase in the dry-bulb temperature; basically, the additional heat causes the air to absorb considerably more moisture.
- Maximum values of temperature difference and moisture absorbtion coincide with the peak output times, whereas the minimum values of temperature and moisture difference are reached slowly. The time-delay of the moisture content is significantly longer than that of the temperature. A certain amount of the dry heat emitted by the locomotive is stored by the surrounding rock, causing increased evaporation.

4 ESTIMATING THE EXTRA COOLING COSTS MADE NECESSARY BY DIESEL-POWERED HAULAGE

In order to make this estimation it is assumed that e.g. because of high rock temperature and small seam thickness it is necessary to cool the air so as to achieve sufficient output in tolerable climatic conditions at the mine faces.

There is a very wide range of operating costs for air-cooling installations mentioned in technical literature, operating

between the heat needed for vaporisation and the total amount of heat released in the test section per haulage cycle shows that the latent heat emitted by the locomotive amounts to 23 %, with the remaining 77 % being emitted as dry radiated heat. Figure 3 illustrates the dry heat $\dot{Q}_{L,dry}$ and total heat emission $\dot{Q}_{L,tot}$ from the locomotive. The temperature difference can also be a negative value, because the dry-bulb temperature in the test section falls when temperatures above ground are high.

In order to be able to make a statement on the influence of locomotive haulage by means af the test data it is necessary to know how the graphs of temperature and moisture content would have looked without locomotive haulage taking place. It is not possible to make an exact calculation of these uninfluenced curves, yet a very near approximation is feasible with the help of a graphical method (Burgwinkel, 1983).

In figure 2 the hatched areas illustrate the increase in temperature and moisture caused by the locomotive haulage. There is

623

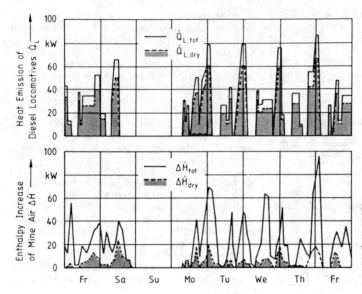

Fig. 3. Comparison of the dry and total heat emission of the locomotive ($\dot{Q}_{L,dry}$, $\dot{Q}_{L,tot}$) with the dry and total enthalpy increase of the air flow ($\Delta\dot{H}_{dry}$, $\Delta\dot{H}_{tot}$).

costs covering all the expenses arising from the running of a refrigeration plant.

Reuther (1977) estimates specific air-conditioning costs for cool air actually brought to the coal face as being 350 DM (110 US-$) per Gcal or 0,30 DM/kWh (0,10 US-$/kWh). This is an average value taken from plants in operation and includes all losses.

Dauber (1982) drew up a model calculation for high-powered cooling plants with specific costs of 0,10 to 0,14 DM/kWh (0,03 to 0,04 US-$/kWh). However, these costs do not include expenses for piping the cool air below ground, for which Dauber suggests an extra 50 % to 100 %. The possible range of costs can therefore vary from 0,15 to 0,28 DM/kWh (0,05 to 0,09 US-$/kWh) for cooling energy needed at the coal face.

Therefore specific costs for cooling of 0,20 to 0,30 DM/kWh (0,06 to 0,10 US-$/kWh) will be used in the following calculation.

The average fuel consumption of a diesel locomotive in the mine chosen for the investigation is a good 175 l per working day, which corresponds to 6,2 · 10⁶ kJ or 1720 kWh heat. About 84 % of this heat causes increased enthalpy of the air. As a rule, only 30 % to 40 % of the air current reaches the mining faces, roadway driving points or any other working points which possibly have to be cooled. This percentage will in all probability rise in the future. Therefore, the higher limit of 40 % will be used in the calculation.

Consequently, an extra energy load of a good 580 kWh is required for the cooling process each working day for each diesel locomotive in operation.

At a cost of 0,20 to 0,30 DM/kWh (0,06 to 0,10 US-$/kWh) each diesel locomotive produces extra cooling costs of 115 DM to 173 DM (36 US-$ to 54 US-$) per working day. This is in the order of the daily fuel costs for the locomotive. By improving the efficiency of the locomotive a reduction in fuel costs and - to the same extent - in cooling expenses can be achieved.

5 A CRITICAL LOOK AT THE COST ESTIMATION

The estimation carried out here can only be an attempt to ascertain reference values for the air heating by diesel locomotives or other diesel-powered engines, since too many other factors influence the correlation between diesel engines and mine air to such an extend that they cannot be ignored. The measurements underground showed that 84 % of the heat released in the test section by the locomotive made the mine air warmer. However, the actual extent to which the percentage falls the further the air continues through the tunnels and galleries to the working points, which are critical as far as air supply is concerned, could not be determined by this investigation. It can be presumed, however, that the percentage mentioned above will fall considerably according to the length of the airways which join the main haulage road.

The extra refrigeration energy as discussed here is small in comparison to the average total refrigeration energy required in mines with a high thermal load. Cooling plants with 2,5 MW to 10 MW rated power need from 50.000 to 240.000 kWh daily. Therefore, the 579 kWh required to deal with the heat caused by the locomotives is in a range that cannot be recorded by test methods or control engineering.

However, where a large number of diesel-powered engines are in operation, as would be the case if automotive technique (LHD-technique) was introduced, additional cooling measures should be considered. These possible extra costs should be taken into consideration when making economic comparisons since much less heat is emitted when electrically-powered engines are employed.

6 SUMMARY

The investigation described here is based on extensive test stand measurements made on a 16 t diesel locomotive with a rated power of 81 kW.

At full load, the locomotive emits 225 kW heat. In the 1000 m test section it could be proved by test measurements that 84 % of this heat flow added to the energy content of the mine air. The remaining 16 % are counterbalanced by a lower heat emission from the rock. The locomotive releases mostly dry heat which causes the air to absorb considerably more moisture. A calculation of additional ventilation expenses due to the use of diesel locomotives shows that a diesel engine may cost approximately 115 DM to 173 DM (36 US-$ to 54 US-$) per day for production and distribution of cool air. These extra costs are in the same range as the daily fuel costs.

When comparing the economy of various kinds of locomotives or haulage systems, these costs should be taken into consideration since electrically-powered locomotives, which are in competition with diesel locomotives, release only 5 % to 20 % of the heat of a diesel engine when doing the same work. This conclusion follows from a comparison with the performance characteristics of accumulator locomotives made by Fauser (1981).

REFERENCES

Voß, J. 1971, Klimavorausberechnung für Ab-baubetriebe. Glückauf 107, S. 412/18.
Reuther, E.-U., Dohmen, A., Billig, M. und Weuthen, P. 1977, Steinkohlenbergwerk der Zukunft. Teilprojekt Grubenklima, Bergbau-Forschung GmbH, Essen.
Burgwinkel, P. 1983, Experimentelle Untersuchung der Fahrdynamik, Energieumsetzung und Wärmeabgabe dieselhydraulischer Grubenlokomotiven. Diss. Aachen.

Dauber, C. 1982, Aufstellen von Kälteerzeugern großer Leistung auf Steinkohlenberg-werken. Diss. Aachen.
Fauser, H. 1981, Grubenlokomotiven und ihre Kennlinien. Glückauf-Forschungs.-H. 42, 1981, S. 98/111.

Mine planning with diesel-powered equipment:
Ventilation considerations

SUKUMAR BANDOPADHYAY
University of Alaska, Fairbanks, USA

RAJA V.RAMANI
The Pennsylvania State University, University Park, USA

ABSTRACT: Mathematical models have been formulated to evaluate diesel-powered equipment deployment schemes in a room and pillar mining system. Several planning exercises have been carried out to evaluate the impact of diesel exhaust on the section and on the ventilation network. The methodology described in this paper will permit determination of optimum deployment schemes, predict the effects of diesel exhaust in the mine air in either existing or new mine designs, and be useful to design and analyze mine ventilation plans.

INTRODUCTION

Although diesel-powered equipment in underground mining operation offers several advantages, the use of diesel powered units in underground coal mines in the United States has been very limited. This is, principally, because of the concern about potential health hazards from diesel exhaust.

In an effort to develop a better understanding of the health hazards associated with the exposure to diesel exhaust in underground mines, concentrated efforts are being made by the National Institute of Occupational Safety and Health (NIOSH), Environmental Protection Agency (EPA), the Mine Safety and Health Administration (MSHA) and the Bureau of Mines (USBM) to device analytical methods to characterize diesel emissions, their effects on health and control mechanisms to limit them at their sources. Diesel equipment manufacturers working through the American Mining Congress directed their efforts to conduct a comprehensive literature review, an evaluation and research gap analysis of the effects of diesel exhaust emission on health. As a result, present-day devices for treating exhaust gases are designed to lower the temperature of the exhaust gases, absorb some of the toxic gases, and remove most of the irritating constituents before discharge. All these improvements in the engine design, will lead to cleaner engines and reduce the emission quantity and toxicity. However, for the mine ventilation engineer, the engineering problem to ensure adequate quantities of air to reduce concentrations below the maximum allowable levels (MACS) for the various species in the exhaust will continue to exist, only with new and perhaps less restrictive constraints. In this paper, the development and use of mathematical models of mine production systems, the engine exhaust air quality and engine exhaust transport are discussed. The overall objective of this research effort is to evaluate the impacts of dieselization on production and ventilation systems and to aid in mine planning and design.

DIESEL EXHAUST FLOW MODELS

Diesel Exhaust Emission

Several aspects of a diesel engine affect the underground mine atmosphere. These include such engine dependent parameters as the type of the engine, the fuel used, the scrubbers and the operating mode. These determine the volume and quality of the engine exhaust. Important mine ventilation parameters include air velocities and cross-sectional areas of the mine air ways. Among the production related parameters are the number of diesel powered equipment deployed, their capacities, lengths and speeds of travel. These parameters are important for determining the concentration of exhaust gas species in the mine atmosphere.

When a diesel engine is travelling in an airway, engine load, engine speed and vehicle speed change with time as a function of the engine duty cycle. The engine exhaust volume and composition are a function of engine speed and BHP, making it necessary to estimate the changes of the pollutant volume with time and location as a function of engine speed and load. To estimate the source function, it is necessary to know the engine speed, engine load and vehicle speed at frequent intervals to account for variation in the exhaust flow rate. It would not be practical to instrument every teletram or similar diesel-powered vehicle to acquire this data. It is possible, however, to generate the engine duty cycle and exhaust volumes and composition on the basis of the simulation of a given production system (Ramani and Kenzy, 1978). One advantage of this approach is that, during planning, several diesel-powered equipment deployment schemes can be analyzed to generate values for the parameters of interest.

Diesel Vehicles in Mine Environment

Generally, diesel engine move faster than the air current in the face areas and, consequently, the air flowing through the haulage road is contaminated several times before it finally gets discharged into the return airways. This leads to a progressive rise in concentration in both spatial and time axes, due to the superposition of contaminants. The concentration profile will not reach a steady-state situation since the velocity vector is not uniform. The concentration front will not be fully developed in the face area because when a moving plane source is used, the distance required to obtain adequate mixing to reach a steady state situation is often greater than that found in the face area. The concentration gradient in the return airway, however, reaches steady state. Air velocity will affect the concentration growth in the return. The total contamination in the air at any point is the sum of the incremental contaminants from diesel engine(s) passed by the air. When the engine and the air are travelling in opposite directions, the predominant feature in the mass flow is the convective transfer, which causes deformation of both the front and rear profiles of the contaminant cloud. As the source moves further away from the face, the role of convective transfer gradually decreases along the length of the airway. In effect, the contamination concentration front is an increasing function of distance and time from the face.

Leakage Considerations

A parameter that will influence the concentration profile is the air leakage through the ventilation device. Because of air leakage, the velocity in the airway is not uniform over the entire length. Velocity at any distance x, however, can be represented by:

$$u(x) = u_o \exp(-a_o x)$$

where,

a_o is the leakage coefficent
u_o is the velocity at x = 0.
Usually, a_o is very small; therefore, $u(x)$ can be approximated by a truncated Taylor Series expansion of $u_o \exp(-a_o x)$ as $u(x) = u_o(1-a_o x)$.
For forcing-type ventilation, the x=0 plane is at the face and $u(x)$ can be approximated as $u(x) = u_o(1+b_o x)$ where b_o is the leakage coefficient for the forcing type ventilation system.

Single Engine Moving in an Airway with or Without Considerable Leakage

The mass transport equation obtained by taking a mass balance over an elemental volume in the space coordinate is represented by:

$$\frac{\partial c}{\partial t} + u(x) \frac{\partial c}{\partial x} - E_x \frac{\partial^2 c}{\partial x^2} =$$

$$g(x,t)\delta(1_f - x) \pm f(x,c,t) \qquad (I)$$

In Equation (I) the term $\frac{\partial c}{\partial t}$ is the rate of growth of concentration in the differential element, while $u(x) \frac{\partial c}{\partial x}$ is the net gain of material due to convective transfer. These two terms balance the total loss of material due to turbulent dispersion, which is represented by $E_x \frac{\partial^2 c}{\partial x^2}$. The source term is $g(x,t)$ and $f(x,c,t)$ is the loss or gain in pollutant volume due to leakage or other processes. The value for E_x is also dependent on the velocity of the source in the moving medium. In the case of diesel vehicles, the value of E_x is given by

$$E_x = E_o \frac{v \pm u(x)}{u(x)}$$

where E_o is the coefficient of turbulent diffusion due only to the mixing capability of the moving fluid, $u(x)$ is the velocity of the fluid, and v is the velocity of the vehicle.

For solving Equation (I), at least three conditions (one initial condition and two boundary conditions) are needed.

(1) Initial condition: Just before the diesel powered engine starts working, concentration of the pollutant in the mine roadway is zero. Mathematically, this condition is represented by:

$$c(x,t) = 0 \quad t \leq 0 \quad 0 \leq x \leq L$$

(2) Boundary conditions: The concentration at the air intake (x=0) and that at the outlet (x=L) or the main return is equal to zero. The physical situation corresponding to these conditions exists at the origin (x=0; the intake point) and at the end of the return drift, where intense mixing of the diesel-exhaust contaminated air takes place with the main return air. Mathematically, these boundary conditions can be expressed as:

$$c(0,t) = 0 \text{ and } \quad t \geq 0$$

$$c(L,t) = 0 \quad t \geq 0$$

In many cases, the parameters at the inlet cross section are known. For instance, the flow velocity distribution and the pollutant concentration can be measured. However, there is no certainty that hitherto unknown aerodynamic and diffusion processes occurring in the roadway do not influence the distribution of the parameters in the initial cross section. Such effects are usually assumed to be small, so that the boundary conditions in the inlet cross sections are given as $c(o,t) = 0$, $t \geq 0$. A more crucial problem is the imposition of boundary conditions in the outlet cross section. Here the distribution and the state of the pollutants are most often unknown. Often the objective of investigations and calculations is to find various parameters in the outlet section. A recognized method is to impose special conditions on the gradients, such as

$$\frac{\partial c}{\partial x}_{x=L} = 0 \quad t > 0$$

which implies that $c(L,t)$ is constant for $x>L$.

Source term: The engine moves with a finite velocity and, in effect, the spatial position of the contaminant source along the length of the airway changes with respect to time. The contaminant source function, $g(x,t)$, at any point x at a given time t can be calculated by the direct delta function as:

$$g(x,t) = u(t-T_1) A_1 + u(t-T_2) A_2$$
$$+ \dots u(t-T_n) A_n$$

where,
$u(t-T_i)$ is the direct delta function, and A_i is the volume of pollutant at engine mode i per unit volume per unit time.

It is necessary that the engine travel in the intake air and never in the return. The normalization function $\delta(1_f-x)$ ensures this condition and is defined by

$$\delta(1_f-x) = 1; \ 1_f \geq x$$
$$= 0; \ 1_f < x$$

where, 1_f is the length to the face from the origin.

Leakage terms: In Equation (I), $f(x,c,t)$ represents the loss or gain in pollutant volume due to air leakage, and can be calculated as follows:

$$f(x,c,t) = \{Q_{x+\Delta x} - Q_x\} \cdot c(x,t-1)$$

where $c(x,t-1)$ is the concentration of the pollutant at a location x at $(t-1)$ time period, and Q_x and $Q_{x+\Delta x}$ are the air quantities in locations x and x+Δx in the airway. The values for $f(x,c,t)$ are negative in the interval of $0 < x < 1_f$ and positive in the interval of $1_f < x < L$, where 1_f is the length to the face, and L is the length to the last point in the return.

Multiple Engines in a Single Roadway

A roadway with considerable leakage: taking a mass balance on a small element (Δx) of the roadway, the transport equation obtained is:

$$\frac{\partial c}{\partial t} + u_o (1-a_o x) \frac{\partial c}{\partial x} - E_x \frac{\partial^2 c}{\partial x^2} =$$

$$g(x,t) \delta(1_f-x) \pm f(x,c,t-1) \quad (II)$$

The initial and boundary conditions appropriate for this situation are:

Initial condition: $c(x,o) = 0$

$$0 \le x \le L$$

Boundary conditions:

$$c(x,t)\Big|_{x=0} = 0 \qquad t > 0$$

$$c(x,t)\Big|_{x=L} = 0 \qquad t > 0$$

In Equation (II) $g(x,t)$ is the source function due to multiple engines and is obtained by:

$$g(x,t) = \sum_{i=1}^{n} g_i(x,t)$$

where $g_i(x,t)$ is the source function for the ith engine.

A Roadway with Little Leakage: In this case, the leakage coefficient (a_o) is very small. Hence, the convective velocity can be reasonably

estimated by the average velocity over the entire length (L) of the roadway.

For exhaust type ventilation, the average velocity is given by:

$$u = u_o \left(1 - \frac{a_o 1_f}{2}\right)$$

where,

u_o is the velocity at x = 0.
a_o is the leakage coefficient.
1_f is the length to the face.

As before, taking a mass balance on a small element (Δx) of the roadway, the convection-diffusion equation obtained is:

$$\frac{\partial c}{\partial t} + u \frac{\partial c}{\partial x} - E_x \frac{\partial^2 c}{\partial x^2} =$$

$$g(x,t) \delta(1_f-x) \qquad (III)$$

The initial and boundary conditions appropriate for the situation are:

Initial condition: $c(x,0) = 0$

$$0 \le x < L$$

Boundary conditions:

$$c(x,t)\Big|_{x=0} = 0$$

$$c(x,t)\Big|_{x>L} = 0$$

Numerical solutions can be obtained (Bandopadhyay, 1982) for predicting the growth of the pollutant concentration for each combination of the roadway conditions and equipment deployment schemes.

MINE PLANNING DEMONSTRATION

Worldwide, diesel-powered equipment are used extensively in underground in both coal and non-coal mines. In the United States, the number of diesel units in underground hardrock mines varies as the number of operating mines changes with both metal prices and prevailing economic conditions.

Statistics are not available for this number, but it is estimated to be approximately 6,500 units in 1983, of which 650 units are in gassy, hardrock mines (Daniel, 1984). Statistics from MSHA show a rapidly increasing use of diesel equipment in underground coal mines over the last 5 years. Their numbers have increased from approximatly 200 units in 1977, to 578 units in 1980, to 943 units in 1982. As of March 31, 1983, MSHA records show 1,003 diesel units underground in operating coal mines. All diesel equipment used in underground coal mines, which is approved as permissible under CFR, title 30, is reported to the local MSHA districts. Hence, the number of diesel units in underground coal mines is well documented. There is a great potential for the application of diesel-powered equipment for face, intermediate, and main line haulage in coal mines. It is the intent in this section to create a complex realistic mine layout for studying a wide variety of diesel deployment schemes and demonstrate the application of mathematical models to mine planning.

The design of a ventilation system of the mine where diesel units are deployed is essentially an engineering problem. Ventilation requirements and system structure can only be defined on the basis of a production plan. However, the iterative and feed back nature of this process must be noted since production planning must consider the services required to support the operation. The methodology described here will permit determination of optimum deployment schemes, as well as, prediction of the effects of diesel exhaust in mine air in either existing or new mine designs.

DESCRIPTION OF A HYPOTHETICAL MINE

The mine layout chosen for the demonstration of the use of models in planning is in a 6-foot (1.8 m) thick coal seam under moderate cover. In the mine projection, nine main entries are planned with five center entries serving as intakes and the two on either side of the five forming the returns. There are a total of eight units projected in the mine, with only four in production at any time. Six-entry panels are employed in the sec-

tions, four intakes and two returns. The sections are equipped with a milling type continuous miner. The mine ventilation plan is represented by the network schematic shown in Figure 1.

The following diesel-powered equipment deployment schemes were evaluated.

1. Multiple Shuttle Car for the Face Haulage
2. Multiple Teletrams in the Section Haulage
3. Multiple Diesel Engines in a Network of Roadways

Multiple Shuttle Car for the Face Haulage

In the haulage configuration, two diesel teletrams were deployed in each production sections to transfer coal from the face to the section belt. The average haul distance (from dump to the face) for the teletrams is approximately 900 feet (274.3 m). These teletrams are designed to meet the requirements of schedule 31 of MSHA. The teletram has a 150 BHP $(112.5 \times 10^3 W)$ caterpillar (4 cycle) model 3306 NA engine.

The production system with the haulage scheme was simulated using a production system simulator (Manula and Albert, 1980). The simulator output provided information on elemental times in each mode of engine operation. In addition, engine rpm, vehicle speed, and tractive efforts developed at each multiple of the simulation time increments were also output by the model. All this data enables calculation of the dynamic parameters of the engine duty cycle. For each teletram, this information was averaged to produce values for the dynamic parameters for the entire mode of engine operation. These duty cycles in conjunction with the diesel engine emission characteristics for a range of loads and speeds (Figure 2) provide adequate information to generate both the volume and the analysis of engine emissions. To predict the effect of the diesel engine exhaust in the ventilating air in the roadway, the diesel exhaust flow model (Equation II) was used. Figure 3 is an example of the predicted carbon monoxide concentration at the face. At the beginning of the operation the concentration is

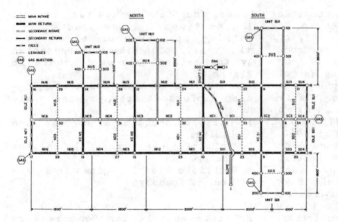

Figure 1.   Mine Network Schematic

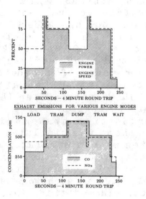

Figure 2.   Power and Speed Cycle for a Typical Shuttle Car Operation (Alcock, 1977).

CARBON MONOXIDE CONCENTRATION AT FACE (FACE HAULAGE)

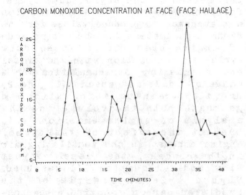

Figure 3.   Carbon Monoxide Concentration Profile at the Face.

less than 10 ppm. However, the maximum carbon monoxide concentration increases to about 26 ppm at the end of 32.40 minutes of engine operation. The large spike is due to the teletrams being very close to the face (engine 1 at the face and engine 2 at the change point).

The time averaged concentration plot is presented in Figure 4. The time over which the average was computed is the time required to complete one cut in the production system. Typically, this period is less than an hour. In this case, this time was 46 minutes. The time versus average concentration plot shows that the maximum time-averaged carbon monoxide concentration in the roadway is approximately 11.5 ppm, occurring at the face.

The concentration profiles of carbon monoxide along the roadway at various elapsed times of operation are shown in Figure 5. A comparison of the fixed point plots shows that at the start of the simulation the concentration is low. However, it rises to a maximum of 15 ppm at the end of 30.4 minutes of operation. It is important to note that the peaks occur at different points in the roadway at different elapsed times. This is so because of the convective transfer of the pollutants. It can also be observed that the concentration gradient $(\partial c/\partial x)$ in the return airway is approximately constant beyond some distance from the face.

MULTIPLE TELETRAMS IN THE SECTION HAULAGE

In an effort to predict concentrations for this haulage configuration, three diesel-powered teletrams were assigned to each of the production sections of the projected mine. The average one-way haul distances are 2600 feet (792.5 m) in section SU1, 3000 feet (914.4 m) in section NU2, 2000 feet (609.6 m) in section NU1, and 1800 feet (548.6 m) in section SD1. Here again, selection of the optimum number of teletram was obtained through simulation of the production system. The simulation of

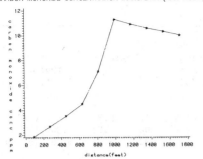

CARBON MONOXIDE CONCENTRATION IN ROADWAY(face haulage)

Figure 4. Time Averaged Carbon Monoxide Concentration Profile in Roadway.

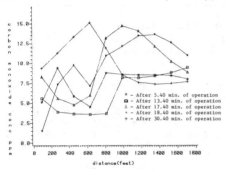

CARBON MONOXIDE CONCENTRATION IN ROADWAY(face haulage)

* – After 5.40 min. of operation
□ – After 13.40 min. of operation
∆ – After 17.40 min. of operation
◦ – After 18.40 min. of operation
+ – After 30.40 min. of operation

Figure 5. Carbon Monoxide Concentration Profiles in Roadway at Various Times.

the production system with these optimal number of teletrams also provided information on the relevant dynamic parameters of the engine duty cycle. Table 1 is an example of the data generated for the teletrams in section SU1. With this and other data, the diesel exhaust model (Equation II) was used to ascertain the cumulative effects of the exhaust contaminations due to multiple diesel unit deployment.

The predicted carbon monoxide concentrations along the roadway at various elapsed times of operation are presented in Table 2. The growth in concentration profiles in the sections at various times is shown in Figures 6, 7, 8 and 9. Peak carbon monoxide concentrations were generally near 25 ppm.

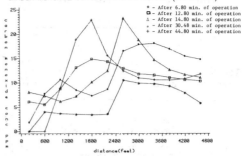

CARBON MONOXIDE CONCENTRATION IN ROADWAY(section SU1)

* – After 6.80 min. of operation
□ – After 12.80 min. of operation
∆ – After 14.80 min. of operation
◦ – After 30.48 min. of operation
+ – After 44.80 min. of operation

Figure 6. Carbon Monoxide Concentration Profiles in Section SU1 at Various Times.

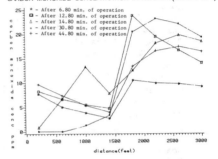

CARBON MONOXIDE CONCENTRATION IN ROADWAY(section SD1)

* – After 6.80 min. of operation
□ – After 12.80 min. of operation
∆ – After 14.80 min. of operation
◦ – After 30.80 min. of operation
+ – After 44.80 min. of operation

Figure 7. Carbon Monoxide Concentration Profiles in Section SD1 at Various Times

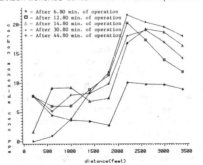

CARBON MONOXIDE CONCENTRATION IN ROADWAY(section NU1)

* – After 6.80 min. of operation
□ – After 12.80 min. of operation
∆ – After 14.80 min. of operation
◦ – After 30.80 min. of operation
+ – After 44.80 min. of operation

Figure 8. Carbon Monoxide Concentration Profiles in Section NU1 at Various Times.

633

Table 1. Duty Cycle Information: Face and Intermediate Haulage--Section SU1.

Teletram number	Engine mode	Cycle time minutes	Average speed ft/min (m/s)	Engine rpm (rev/s)	Maximum tractive effort, lb (kg)
1	Load	1.82	0.0 (0.0)	1100 (18.3)	6199 (2811)
	Travel load face to WP	5.23	478.0 (2.4)	2089 (34.8)	11366 (5154)
	Wait at WP	0.0	0.0 (0.0)	2200 (36.6)	11366 (5154)
	Turn car and travel load WP to dump	0.80	250.0 (1.2)	2089 (34.8)	11366 (5154)
	Dump	0.58	0.0 (0.0)	2200 (36.6)	11366 (5154)
	Travel empty dump to CP	3.23	743.0 (3.7)	2089 (34.8)	6199 (2811)
	Wait at CP	0.73	0.0 (0.0)	2200 (36.6)	6199 (2811)
	Travel empty CP to face	0.40	350.0 (1.7)	2089 (34.8)	6199 (2811)
	Wait on miner	5.57	0.0 (0.0)	525 (8.75)	6199 (2811)
2	Load	1.22	0.0 (0.0)	1100 (18.3)	6199 (2811)
	Travel load face to WP	5.80	431.0 (2.1)	2089 (34.8)	11366 (5154)
	Wait at WP	0.0	0.0 (0.0)	2200 (36.6)	11366 (5154)
	Turn car and travel load WP to dump	0.80	250.0 (1.2)	2089 (34.3)	11366 (5154)
	Dump	0.23	0.0 (0.0)	2200 (36.6)	11366 (5154)
	Travel empty dump to CP	5.0	480.0 (2.4)	2089 (34.3)	6199 (2811)
	Wait at CP	6.45	0.0 (0.0)	2200 (36.6)	6199 (2811)
	Travel empty CP to face	0.40	375.0 (1.9)	2089 (34.3)	6199 (2811)
	Wait on miner	5.45	0.0 (0.0)	525 (8.75)	6199 (2811)
3	Load	3.05	0.0 (0.0)	1100 (18.3)	6199 (2811)
	Travel load face to WP	5.35	467.0 (2.3)	2089 (34.3)	11366 (5154)
	Wait at WP	0.0	0.0 (0.0)	2200 (36.6)	11366 (5154)
	Turn car and travel load WP to dump	0.80	250.0 (1.2)	2089 (34.3)	11366 (5154)
	Dump	0.40	0.0 (0.0)	2200 (36.6)	11366 (5154)
	Travel empty dump to CP	2.75	880.0 (4.4)	2089 (34.3)	6199 (2811)
	Wait at CP	9.35	0.0 (0.0)	2200 (36.6)	6199 (2811)
	Travel empty CP to face	0.40	380.0 (1.9)	2089 (36.6)	6199 (2811)
	Wait on miner	2.0	0.0 (0.0)	525 (8.75)	6199 (2811)

It may be observed from these figures that the trend in concentration growth are the same, since the same air quantity (40,000 cfm, (18.8 cum/s)) is employed in all the sections and the geometric configurations of these sections are similar. The cumulative effect and the location of the highest peaks differ, however, due to the differences in the haul distances and in travelling speeds at various sections. Therefore, the location of the diesel-engine is changing at a different rate in each section. Higher concentrations in the above figures occur when the concentrations due to a vehicle travelling loaded is superimposed by those of another vehicle travelling empty.

Analysis of concentration profiles at the faces indicates that the highest concentrations in section SU1; is 23.23 ppm and occurs after 14.80 minutes of operation; in section NU1, 22.60 ppm after 30.80 minutes; in section NU2 22.09 ppm after 14.80 minutes, and in section SD1 24.34 ppm after 28.80 minutes of operation. The maximum time-averaged concentrations in the sections are 14.85 ppm in section SU1; 14.34 ppm in section NU1; 20.75 ppm in section NU2 and 14.08 ppm in section SD1. Times over which these averages were computed are the

Table 2. Concentration of Carbon Monoxide (ppm) in the Sections.

Distance, ft., from dump	Section				Simulation time (minutes)
	SU1	SD1	NU1	NU2	
400 (121.92m)	0.0	0.09	0.002	0.0	4.80
800 (243.84m)	4.96	12.29	6.79	3.86	8.80
1200 (365.76m)	11.32	5.94	7.45	14.04	12.80
1600 (487.68m)	7.49	13.99	13.35	7.66	16.80
2000 (609.60m)	10.84	12.14	12.97	12.77	20.80
2400 (731.52m)	11.52	10.50	10.83	8.43	24.80
2800 (853.44m)	18.72	15.25	17.73	15.67	28.80
3200 (975.36m)	16.68	21.68	20.71	18.06	32.80
3600 (1097.28m)	20.97	22.11	18.40	21.42	36.80
4000 (1219.2m)	15.90	-	15.77	19.59	40.80

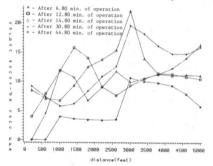

Figure 9. Carbon Monoxide Concentration Profiles in Section NU2 at Various Times.

times that are required to complete one cut in the individual sections. Typically this period is less than an hour.

Multiple Diesel Engines in a Network of Roadways

In a mine, the branches and junctions where the sources of diesel exhaust are located are known a priori. These are the haulage roadways and face areas. In this situation, the contaminant concentration of the air at any points in the mine is typically represented by peaks superimposed on a lower background. Therefore, it is necessary to determine the cumulative effect of all engine exhaust gases on those segments of the ventilation system where miners are working. In any multiple opening situation, it is also necessary to determine the dilution effect of parallel openings. Since multiple openings are usually interconnected, splitting is predominant. This requires the development of an integrated model in which engineering data on equipment, the mining system, and the ventilation network are considered. However, as a first approximation, one can take the most critical condition in each source location and then use a suitable ventilation network model (Didyk, et al., 1977) to analyze the dilution and distribution of the contaminant in branches and junctions.

If the present ventilation system is capable of handling this most critical situation, subcritical conditions will result with the air quantities in parallel entries. Proceeding with the approach outlined here, most of the data were generated from the simulation of the production systems and the diesel exhaust simulation, measured from the projected mine layout, and used for simulation of the entire mine ventilation system.

These kinds of analyses can reveal the adequacy of the projected ventilation plan to handle the production and haulage plan. Mine areas considered critical with regards to air quality, air quantity, and air speed can be highlighted and designed accordingly.

SUMMARY

Underground mining operations involve arduous requirements for motive power units with even higher requirements on safety, reliability and productivity. Various power sources such as electrical, battery, compressed air, steam, gasoline and diesel have been used. Diesel equipment offers some advantages that are not obtainable with the other power sources. In underground mining diesel-powered units were essential factors for the successful introduction of high performance load-haul-dump machines, wheel loaders and trucks, and service vehicles into many mining operations. The most important factors in considering diesel-powered vehicles for underground coal mining are the potential to improve safety, through the elimination of trailing cable in the face haulage systems and the potential to improve production, through the reduction or elimination of haulage related delays and the development of run-around designs for the tram-path.

Mine planning with diesel-powered equipment encompasses many fields such as production, mine ventilation, material transport, exhaust gas and particulate analysis, engine testing and approval, and mine atmospheric monitoring. Ventilation and system structure can only be defined on the basis of a production plan. However, the iterative and feedback nature of this process must be noted since production planning must consider the services required to support the operation.

This paper has provided an unique approach for the analysis of diesel-powered equipment deployment schemes. In this paper, mathematical models of production systems, engine exhaust air quality, and exhaust dilution and transport in mine airways have been integrated to study and plan mine ventilation systems.

REFERENCES

Alcock, K., 1977, Safe use of diesel equipment in coal mines. Mining Congress Journal, vol. 63, no. 3, 53-62.

Bandopadhyay, S., 1982, Planning with diesel-powered equipment in underground mining - a computer analysis of exhaust dispersions. (Unpublished) Ph.D. thesis, The Pennsylvania State University.

Bandopadhyay, S. and R. V. Ramani, 1983, Computer aided analysis of diesel exhaust dispersions in underground airways, The CIM Bulletin, vol. 76, no. 858, 69-74.

Bandopadhyay, S. and R. V. Ramani, 1984, Convection diffusion equations in mine ventilation planning. Proceedings of the 3rd International Mine Ventilation Congress, M. J. Howes and M. J. Jones (eds.), 397-404.

Daniel, J. H., 1984, Diesels in underground mining, A review and an evaluation of an air quality monitoring methodology. U.S. Bureau of Mines Report of Investigation No. 8884, 36 p.

Didyk, M., R. V. Ramani, R. Stefanko and G. W. Luxbacher, 1977, Advancement of mine ventilation network analysis from art to science, vol. III, User's Manual for PSU/MVS, NTIS Publication No. BP 290194/AS, 86 p.

Manula, C. B. and E. K. Albert, 1980, Evaluation of operational constraints in continuous mining systems -- underground material handling simulator, Vol. IV, User's Manual, Final Report to the Department of Energy on Grant No. ET-76-G-01-8982, 217 p.

NIOSH (National Institute of Occupational Safety and Health), 1977, Workshop on the Use of Diesel Equipment in Underground Coal Mines, Morgantown, West Virginia, Sept. 19-23.

Ramani, R. V. and G. W. Kenzy, 1978, Evaluation of diesel equipment in underground coal mines, volume III, analysis of diesel deployment schemes, open-file report, U.S. Bureau of Mines, 127-81.

Control of diesel engine exhaust emissions in underground mining

S.SNIDER & J.J.STEKAR
Engine Control Systems Ltd, Aurora, Ontario, Canada

ABSTRACT: The regulated components of diesel engine exhaust emissions are discussed relative to the emission control devices currently available to underground mining applications. The control devices are compared and contrasted in terms of efficiency, maintenance, cost and practicality. Future product developments involving diesel particulate trapping devices and their practical applications are presented.

1 INTRODUCTION

Diesel exhaust emissions have been a primary concern for mine operators and employees since the advent of dieselized underground operations. This is particularly evident in operations where the available ventilation air may be of an insufficient volume or poorly distributed through the workings. It is quite common to find that the majority of underground mines utilize a variety of emission control devices on their diesel powered equipment. These emission control devices reduce the human exposure to the various toxic and irritating components of diesel exhaust. These regulated toxic and irritating components of diesel exhaust are differentiated on the basis of chemical nature, toxicity, raw emissions volume and dilution ratio requirement.

The various regulated diesel exhaust emission components are:

a) Carbon Dioxide (CO_2). CO_2 is not considered to be dangerously toxic as it is a product of combustion. CO_2 has a TLV (threshold limit value) of 5000 ppm (equivalent to .5%). A typical underground diesel (Deutz F6L714) will emit between 5.6 to 10.2% CO_2 during various load speed operating conditions. These emissions will require a minimum dilution ratio of approximately 20:1 in order to bring CO_2 emissions to a safe level.

b) Carbon Monoxide (CO). CO is considered to be a dangerous diesel exhaust component and it is a product of incomplete combustion. CO has a provincial TLV of 35 ppm and a federal TLV of 50 ppm. A typical underground diesel (Deutz F61714) will emit between 200 and 400 ppm CO at various load speed

operating conditions when properly maintained. A dilution ratio of approximately 11:1 will be required to maintain an adequate exposure level. However, if maintenance procedures are less than stringent, operating problems may result which will increase CO emission to dangerous levels in very short periods of time. (Stawsky, Lawson & Vergeer 1984).

c) Unburned Hydrocarbons (HC). Unburned Hydrocarbons are a potentially dangerous diesel exhaust component best known for their contribution to diesel odour. Unburned Hydrocarbons consist of various irritants including aldehydes, acrolein and formaldehyde. Aldehydes, which are a product of incomplete combustion have a TLV of 3 ppm. Aldehyde emissions are primarily formed in the combustion process of direct injection diesel engines. Typical direct injection diesel engines will emit between 2 and 15 ppm of aldehydes. Other research (Reyl 1977) has shown aldehyde emissions to reach levels as high as 25 ppm. In order to reduce the toxicity level to the 3 ppm TLV a minimum dilution ratio of 8:1 would be required.

There are recommendations under consideration which request that the aldehyde TLV be reduced to .3 ppm (ACGIH 1981). This proposed action would result in a minimum dilution ratio of 80:1 which would be very expensive in terms of ventilation cost.

d) Oxides of Nitrogen (NOx). NOx is made up of the diesel exhaust components NO and NO_2. These exhaust components have TLV's of 25 and 3 ppm respectively. NO is formed during the combustion process as a result of high combustion temperatures and readily available O_2. NO_2 is formed by the post oxidation of NO in the diesel exhaust stream.

Prior Research has indicated that this post oxidation process will stop once NO has been diluted to below 32 ppm and exhaust temperatures cooled to below 160F. NO emissions will be significantly higher during low load/operating conditions where the time for formation of oxides is available (Holtz 1960). Testing has indicated that indirect injection diesels will emit less NOx than comparable direct injection diesels. Typical NO emissions from a well maintained underground diesel (Deutz F6L714) will vary from 530 to 680 ppm during various load/speed conditions. NO2 emissions over identical load/speed conditions will vary between 2 and 41 ppm. The minimum required ventilation ratio for NOx emissions for this particular diesel engine would exceed 27:1.

e) Sulphur Dioxide (SO2). SO2 is a dangerous diesel exhaust component which is formed during the combustion process of a diesel engine. It is a well documented fact that each 0.5% of fuel sulphur content will produce approximately 140 ppm of SO2. Typical diesel #2 will range between 0.2 and 0.5% sulphur. The TLV for SO2 is 3 ppm and a typical underground diesel (F6L714) will produce between 23 and 46 ppm SO2 at various load/speed operating conditions (using a .108% fuel sulphur content).

f) Respirable Combustible Dust (RCD) RCD is composed of respirable diesel particulate (RDP) and various airborne particulates typically found in the mine environment. The formation of respirable diesel particulate has been well documented (Johnson, Lipkea & Vuk 1979). Combined with the diesel particulate matter are a variety of absorbed liquid, sulphate, high molecular weight hydrocarbons and inorganic compounds. Polynuclear Aromatic Hydrocarbons (PAH's), which are high molecular weight hydrocarbons have been found to be potentially carcinogenic in animal studies (EPA 1981). Since 90% of all RDP is considered to be in the Human respirable range, a significant health threat may be posed to underground mine workers. The TLV for Respirable Combustible Dust (RCD) is 2.0 mg/cubic metre and it is assumed that the TLV for respirable diesel particulate (RDP) is 75% of the RDC or 1.5 mg/cubic meter. A typical underground diesel (Deutz F6L714) will emit between 90 and 150 mg/cubic metre of RDP at various load/speed operating conditions.

2 CONTROL DEVICES

The toxic and irritating diesel exhaust components mentioned can be reduced to safe levels by a variety of methods. The earliest form of diesel exhaust emission control was accomplished through the use of water scrubbers.

2.1 Water Scrubbers

Water Scrubbers are utilized in many underground operations around the globe. A well designed, properly maintained water scrubber can remove anywhere from 30% to 60% of diesel particulate matter, 20% of NO2 emissions and 80% of SO2 emissions. (Mogan, Katsuyama & Dainty 1982). An added benefit of water scrubbers is their apparent effectiveness in cooling exhaust gases as well as quenching flames and sparks. The capital cost of a water scrubber is quite high in relation to other available exhaust control devices. A water scrubber does require a considerable amount of maintenance in order to keep it operational, corrosion-free and effective. The high water consumption rates of the majority of water scrubber designs makes it necessary to refill water tanks at regular hourly intervals. Water consumption rates vary dramatically, although new scrubber technologies such as pre-cooling and venturi devices have reduced consumption rates by up to 50%. Water scrubbers are typically large in size and do occupy an inordinate amount of space on typical mining vehicles. Another problem commonly associated with water scrubbers is the increased amount of released water which accumulates in the mine air resulting in atmospheric fogging thus reducing underground visibility considerably. Lastly, there is a fuel consumption penalty to be realized due to the relatively high backpressure associated with present water scrubber designs.

2.2 Catalytic Pellet Scrubbers

Pellet scrubbers utilizing alumina based catalyzed spheres are commonly used for underground diesel engine exhaust control. Under optimum conditions catalytic pellet scrubbers will remove up to 90% of CO in addition to 80% of hydrocarbons and aldehydes. The pellet scrubbers are relatively bulky and have a tendency to plug with diesel particulate matter and engine oil after extended use. Periodic baking of the catalyzed alumina spheres is required in order to burn off accumulated deposits. Additionally, the catalyzed alumina spheres are susceptible to disintegration which will lead to short-circuiting of untreated raw exhaust gas through the scrubber. Of further concern is the interaction of the catalytic alumina spheres with Polynuclear Aromatic Hydrocarbons (PNA's). PNA's are defined as complex high molecular weight hydrocarbons. Certain PAH's are known carcinogens such as Benzo-a Pyrene. PNA's have the capacity to induce a mutagenic change in certain bacteria. This mutagenic capacity has been standardized by the Ames Mutagenic Bio-Assay test. The results of the Ames test provide a fast and relatively effective method of determining the mutagenic nature of various chemical compounds.

Tests conducted in 1983 indicated that catalytic pellet scrubbers produced a mutagenic concentration increase of between 100 - 500 times the level of raw diesel exhaust. (Mogan, Westway, Horton, Dainty, 1983). This may be due in part to the storage release phenomena of diesel particulate matter. Recent EPA studies have indicated that NO_2 may react with the stored PNA's and form very strong mutagens known as dinitropyrenes. (Sato 1983). Further testing is required to determine the health hazard resulting from these PNA concentrations.

2.3 Fume Diluters

Fume Diluters are aerodynamic devices which rely upon the "Coanda" principle to dilute and expel potentially toxic exhaust gases from the vicinity of the underground equipment operator. Exhaust gases are fed into the fume diluter body and released through a pre-set annular gap. Following the annular gap the exhaust gases pass over an aerofoil surface which accelerates the exhaust gases thereby creating an area of low pressure near the diluter inlet. This low pressure induces a secondary airflow which rapidly mixes with the exhaust gases thereby cooling and diluting their toxic concentrations. Typically the induced secondary airflow is 10-15 times the exhaust gas flowrate volume.

The fume diluter is effective in reducing the post-oxidation of NO to NO_2 due to its ability to rapidly dilute raw exhaust gases. Researchers have concluded that the conversion of NO to highly toxic NO_2 is greatly diminished if the NO tailpipe concentrations are less than 32 ppm (Mogan, Dainty 1974). Due to the high volume of induced mine air the fume diluter has the capability of projecting exhaust gases up to 150 ft. from their source. This is advantageous when machinery may be operating in a dead heading situation where ventilation airflow and velocity are low. Fume diluters have a relatively low

capital cost, long service life and require little maintenance.

Conversely, fume diluters do require careful installation to insure that they operate in an efficient manner. With the exception of NO_2, they do not chemically alter or reduce the toxic exhaust gas components into less harmful substances. Fume diluters do impose a backpressure restriction onto the diesel exhaust gas flow thereby increasing fuel consumption.

2.4 Diesel Exhaust Purifiers

Diesel exhaust purifiers typically will oxidize 90 per cent of CO in addition to 80 per cent of unburned hydrocarbons and partially oxidized hydrocarbons. These potentially toxic exhaust gases are converted by the catalyst into relatively harmless components. The NO and NO_2 emissions are left uncharged by the diesel exhaust purifier (Lawson, Vergeer 1977). However the diesel exhaust purifier will convert SO_2 to SO_3 at a conversion rate of approximately 30 per cent. (Mogan, Dainty 1984).

Diesel exhaust purifiers are relatively small in size and require little installation labour. The capital cost is not high in relation to alternate control devices and typically they will function effectively for several thousand operating hours. Diesel exhaust purifiers do require periodic maintenance in order to prevent catalyst plugging and related high exhaust backpressures. Generally high exhaust backpressure is not a problem if exhaust purifiers are correctly sized and regularly maintained.

Two types of diesel exhaust purifiers are in common use underground. The first type is based on a metallic catalyst substrate, the second on a ceramic catalyst substrate. In most circumstances the metallic catalyst substrate is better suited for underground use because of its faster catalyst light-off capabilities and low exhaust backpressure restriction. The metallic substrate is considerably more durable than the ceramic based substrate in terms of thermal and mechanical shock capabilities.

A drawback to the use of diesel exhaust purifiers is the formation of sulphate in the exhaust gas stream due to the oxidation of SO_2. This previously mentioned drawback is in part due to the catalyst formulation. However, new catalyst formulations have considerably lowered the percentage conversion of SO_2 to SO_3.

2.5 Diesel Particulate Filters

Diesel particulate filters are emission control devices which are designed to capture and oxidize trapped particulate matter from diesel exhaust. Although the long term health effects of diesel particulates and associated synergistic pollutants are still under investigation, there is considerable evidence which points to detrimental health effects upon the human respiratory system. The constraints placed upon diesel particulate filters (DPF) are severe. The DPF must be capable of trapping a high percentage of total particulate emissions. It must be able to withstand severe thermal and mechanical stress atypical of underground LHD duty duty cycles. It must not produce a high exhaust backpressure which would increase fuel consumption and alter the baseline raw exhaust emissions. The DPF must be able to regenerate periodically by oxidizing the accumulated particulate matter. Lastly, a DPF system must be safe, reliable and provide a good service life.

At the present time there are three major DPF strategies available to the market or in advanced prototype stages. The first system is described as a Catalytic Trap Oxidizer. (CTO)

2.5.1 Catalytic Trap Oxidizer

The Catalytic Trap Oxidizer is a diesel particulate filter which utilizes a radial flow, stainless steel, wire mesh media to capture particulate. This wire mesh media is precious metal catalyzed and packed under compression into a stainless steel can structure. The CTO is capable of being mounted into an exhaust manifold or various remote mounted configurations. Particulate trapping is accomplished in the CTO filter media by impingement of particulate onto the random wire mesh surfaces. (Enga, Buchman, Lichtenstein 1982). Laboratory investigations and related field experiments have indicated that the CTO is an effective device in reducing the level of diesel exhaust contaminants. Due to the precious metal coating the CTO is very effective in reducing CO and HC emissions. CO reduction over similar hot and cold LHD duty cycles indicated average reductions of 95% over baseline emissions. Total hydrocarbons over cold LHD duty cycles were reduced by 82%. Aldehydes and NO_2 indicated a 74% and 36% reduction respectively. In terms of particulate captive efficiency, soluble particulate matter was reduced by 90 + % and insoluble particulate by

65% (Mogan, Dainty & Vergeer 1985). Ames mutagenicity testing indicated that the CTO produced a negative response and decreased the specific activity considerably. Reductions in Ames mutagenicity ranged from 75 - 90% over continuous regeneration hot cycles and forced regeneration cold cycles. The only diesel pollutant to indicate a marked increase over baseline emissions was SO_4. In the laboratory tests SO_4 emissions increased 1180% over baseline.

The CTO exhibits a very high resistance to thermal stress. This resistance was attained through product developments which control particulate heat release rates and minimize filter media breakdown. Mechanical stress is minimized due to the metallic nature of the filter media. The high mechanical stresses which are typical of underground LHD operations have little effect on the CTO canning. The backpressure restriction imposed by the CTO is primarily a function of available filter area and sizing constraints. The CTO does exhibit a high backpressure rise rate in relation to related competitive products. Under cold LHD duty cycle testing the backpressure rise rate of approximately 5"H_2O/hr. was observed. In order to regenerate continuously the CTO will require a duty cycle with an average exhaust temperature of 420°C.

2.5.2 Catalyzed Ceramic Trap

The catalyzed ceramic trap utilizes a ceramic substrate, manufactured by Corning Glass as a trapping media. The ceramic trap incorporates interception and diffusion of diesel particulate as mechanical trapping mechanisms. The substrate is catalyzed with precious metal catalyst and canned in a stainless steel container. Laboratory tests conducted at the Ontario Research Foundation have concluded that the catalyzed ceramic traps are effective in reducing diesel pollutants. (Vergeer & Lawson 1984) Over cold LHD duty cycle testing CO was reduced by over 95% and total hydrocarbons were reduced by 51%. NO was reduced by 21% while NO_2 showed an increase over baseline emissions of 79%. This may be due in part to the exothermic activity of the catalyzed ceramic trap. The ability of the ceramic trap to capture diesel particulates is very good. In terms of total particulate emissions (SOLUBLE PLUS UNSOLUBLE) the ceramic trap reduced particulates by 92% in comparison to baseline emissions. No published information was available with respect to Ames activity of the ceramic traps. In addition to the increase in toxic NO_2 emissions, the catalyzed ceramic trap increased SO_4 emissions fifteen fold over baseline SO_4 emissions. Similar results were reported with the catalyzed trap oxidizer. The catalyzed Corning has exhibited a good resistance to severe thermal stress. However, when overloaded with particulate matter, there have been some reported failures during regeneration periods. (Ullman, Hare & Baines 1979). Due to the possibility of such failures, the ceramic trap may require frequent regenerations to prevent high particulate loadings. Recent field work has determined that the ceramic trap is durable and resistant to mechanical shocks during underground LHD operations. The Corning ceramic trap has a very low backpressure rise rate of approximately 1" H_2O/hr. This rise rate was observed over a relatively hot duty cycle with an average cycle temperature of 409^0C.

2.5.3 Corning Ceramic Trap with Fuel Additives

This diesel particulate control strategy incorporates a Corning ceramic trap as a particulate trapping media. As mentioned previously the trap utilizes particle interception and diffusion as trapping mechanisms. In order to promote filter regeneration, fuel additives are mixed with the diesel fuel supply. The fuel additive is typically a concentration dosage of 80 mg. manganese and 20 mg. copper per litre of diesel fuel. The use of the fuel additives permits a reduction in particulate ignition temperatures of approximately 150^0C. This enables the Corning/fuel additives system to regenerate over cold LHD duty cycles. (Vergeer & Drummond 1984).

The Corning/fuel additives particulate control strategy is not completely developed yet it shows considerable promise. However, some doubts have been raised with respect to the use of fuel additives in an underground environment. Firstly, the Corning fuel additives system will require an on-board method of metering and injecting the additive into the fuel system of the underground mining vehicle. Such a system will initially be relatively complex and difficult to maintain for underground maintenance personnel. Secondly, there is the risk that in the event of a possible ceramic filter failure, the fuel additives will be released into the underground atmosphere instead of captured by the filter. Stringent controls will be required in order to prevent such occurrences.

The Corning/fuel additives system has little if any effect upon gaseous diesel emissions. No measurable reductions in CO, hydrocarbon and NO_x emissions were noted. However NO_2 emissions did increase, but this may possibly be due to variations in engine test conditions. The Corning/fuel additives system proved to be very effective by trapping 90% of total particulates emissions. In addition to the reduction of particulates, there was no detectable manganese/copper pass through after the Corning filter. This indicates that the filter had captured the total fuel additive emissions. The Ames Mutagenic testing indicated that the Corning/fuel additives system reduced mutagens from a baseline of 14,248 revertants/m^3 to 2,946 revertants/m^3. This would indicate a 79% reduction over the cold LHD duty cycle. With respect to sulphate emissions, the use of manganese/copper fuel additives did not convert any SO_2 to SO_4. This essentially confirms that base metal fuel additives, unlike precious metal catalyst, do not promote SO_2 to SO_4 conversion.

In recent field testing the Corning/fuel additives particulate control system proved to be a reliable method of controlling particulate emissions in underground mining applications. The use of fuel additives is very effective in promoting frequent, low temperature regeneration which prevents the buildup of potentially destructive particulate loadings.

3 SUMMARY

The new generation of diesel emission control devices offer considerable additional benefits over currently available technologies. The Catalytic Trap Oxidizer and Catalyzed Corning Traps are very effective in the capture of particulate matter, reduction of Ames Mutagenic activity and oxidation of gaseous contaminants. However, both products do increase the conversion of SO_2 to SO_4. In the case of the Catalyzed Corning Trap there was an increase in potentially toxic NO_2 emissions. The Corning Trap with fuel additives is very effective in the reduction of particulate matter and Ames Mutagenic activity. This control system is capable of regenerating at very low temperature LHD duty cycles. However, this control system has little effect upon gaseous contaminants and will possibly require increased service and maintenance due to the use of the fuel additives.

In general, each particulate control system has its own positive and negative operating aspects. These aspects, when viewed with respect to the underground vehicular application, will provide the basis upon which mine operators can determine which control system is best suited for their needs.

REFERENCES

Stawsky, A., Lawson, A., Vergeer, H. 1984, Diesel Exhaust Emissions Control Using EGR and Particulate Filters.

Reyl, G. 1977, Deutz Diesel Engines Operating in Underground Mines.

American Conference of Governmental Industrial Hygenists 1981.

Holtz, J.C. 1960, Safety with Mobile Diesel Powered Equipment Underground.

Johnson, J.H., Lipkea, W.H., Vuk, C.T. 1979, The Physical and Chemical Character of Diesel Particulate Emission - Measured Techniques and Fundamental Considerations.

Toxicology Effects of Emission of Diesel Engines 1981.

Mogan, J.P., Katsuyama, K., Dainty, E.D. 1982, The Development of Water Scrubbers for Diesel Exhaust Treatment.

Mogan, J.P., Westway, K.C., Horton, A.J., Dainty, E.D. 1983, Polynuclear Aromatic Hydrocarbons in the Air of Underground Dieselized Mines; Proceedings of the specialized meeting of the Tenth World Congress on the Prevention of Occupational Accidents and Diseases, Ottawa.

Sato, F. 1983, Carcinogenicity of Nitroarenes, U.S.-Japan Cooperative Program, Workshop on Carcinogens and Environmental Factors, Dedham.

Mogan, J.P., Stewart, D.B., D'Aoust, A., Dainty, E.D. 1974, Dilution Efficiencies of Exhaust Gas Dispersion Devices.

Lawson, A., Simmons, E.W., Piett, M. 1979, Emission Control of a Deutz F6L714 Diesel Engine, Derated for Underground use, by Application of Water/oil Fuel Emulsions, FINAL REPORT 2722/02 for Contract 1SQ 78-00022, Canmet, Department of Energy, Mines and Resources.

Stewart, D.B., Ebersole, J.A.D., Mogan, J.P. 1979, Measurement of Exhaust Temperatures on Operating Underground Diesel Equipment, Can. Min. Metall. Bul 70, 801, 70-79.

Enga, B.E., Buchman, M.F. Lichtenstein, I.E. 1982, Catalytic Control of Diesel Particulate. SAE paper.

Mogan, J.P., Dainty, E.D., Vergeer, H.C., Lawson, A., Westaway, K.C., Weglo, J.K., Thomas, L.R. 1985, Investigation of CTO Emission Control System Applied to Heavy-Duty Diesel Engines used in Underground Mining Equipment. SAE paper.

Vergeer, H.C., Lawson, A. 1984, Performance Engelhard Catalyzed Ceramic and Johnson Matthey Catalyzed Mesh Diesel Particulate Traps, ORF report.

Ullman, T.L., Hare, C.T., Baines, T.M. 1984, Preliminary Particulate Trap Tests on a Two Stroke Diesel Bus Engine, SAE.

Vergeer, H.C., Drummond, W. 1984, Assessment of Auto-Regeneration of Diesel Particulate Filters by Fuel Additives, Final Report 4422 ES/CAFU.

16. Ventilation network analysis

Linear analysis for the solution of flow distribution problems in mine ventilation networks

SASTRY S.BHAMIDIPATI & JAMES A.PROCARIONE
University of Utah, Salt Lake City, USA

ABSTRACT: Linear analysis provides a new approach to solve the problem of distributing air in underground mine ventilation systems. The set of nonlinear equations corresponding to the conservation of energy in the network are linearized, without involving Taylor series approximation as in the case of Hardy Cross or Newton-Raphson methods. The approximated linear system is solved iteratively, with the coefficients corrected with each iteration, until some convergence condition is met. Two methods are proposed, the first one based on chord flow rates as primary variables, and the second one based on nodal heads as primary variables. This paper outlines the related mathematics, presents algorithms for computer implementation, and analyzes some cases of flow distribution to illustrate the absence of initialization problems and the simplicity and elegance of computer implementation of the linear analysis theory.

1 THE PROBLEM OF NETWORK FLOW DISTRIBUTION

Physical flow situations as encountered in electric circuits, hydraulic and natural gas pipelines, and mine ventilation systems, are conveniently modeled as networks. The flow distribution in these networks is guided by two laws, the law of conservation of mass (flow), and the law of conservation of energy. Known as Kirchhoff's laws in electric circuits, these laws for a fluid flow network can be described as follows. Based on the assumption of constant fluid density, the conservation of mass requires that at each node of the network, the flow into the node must equal the flow out of the node. The conservation of energy requires that the algebraic sum of pressure losses contained in the branches of each loop or mesh of the network, must equal zero. The pressure loss, or the pressure drop, on a directed branch is given by the difference in nodal pressures between the originating and terminating node of that branch.

Thus, the network flow problem has two fundamental sets of variables, the nodal pressures, and the branch flow rates (node potentials and currents respectively in an electric circuit). Emperical relationships are developed to establish the relation between the two fundamental sets of variables. Ohm's law in DC circuits relates the node potentials of a branch with current. In fluid flow networks an equivalent relationship is provided by many equations such as the Darcy-Weisbach equation, Hazen-Williams equation, and Manning equation. In mine ventilation systems, where turbulent flow invariably prevails in roadways, the Darcy-Weisbach equation came to be known as Atkinson's equation. It essentially states that, for a directed branch, the pressure loss due to friction between the originating and terminating node, is proportional to the square of the branch flow rate. The constant of proportionality is known as branch resistance. Such a nonlinear relationship complicates the analysis of fluid flow in networks, causing the adoption of iterative techniques inevitable.

For a network with NN nodes and NB banches, it is well known that there exist (NN-1) independent linear equations containing flow rates (defining conservation of flow), and (NB-NN+1) nonlinear equations defining conservation of energy. It is a common practice to solve the problem of flow distribution by treating either (NN-1) nodal pressures as the primary variables (one node is at reference or datum pressure), or the (NB-NN+1) chord flows as the primary variables. By knowing one set of values, the other set, or the secondary variables, are obtained from the

emperical relationship between the two
sets as described above.

Any text book containing hydraulic network analysis (Watters 1984) describes the
Hardy Cross method. Having been in use for
nearly 50 years, it is the earliest systematic and algorithmic approach known to
solve the problem of steady state flow
distribution in hydraulic networks. A
popular version of Hardy Cross iterative
technique is based on applying correction
for flow rate to each mesh of the network
individually, until some convergence
requirement is attained. It is an easily
understood technique and suitable for hand
calculations as it does not involve large
sets of simultaneous equations. Over time
this method of individual mesh flow
corrections is proved to be less attractive. With the advent of computer applications and efficient numerical analysis
techniques, new methods came into
existence. The Newton-Raphson method
(Shamir & Howard 1968, Epp & Fowler 1970,
Chandrasekhar & Stewart 1975, Jeppson
1976, Wood & Rayes 1981), and the linear
analysis theory (Wood & Charles 1972, Collins & Johnson 1975, Jeppson 1976, Isaacs
& Mills 1980, Wood & Rayes 1981, Watters
1984), are presently very much established
and widely adopted in the industry. There
is a general agreement in the above references that the two procedures are superior
to Hardy Cross method in several respects.

In addition to the difficulty of finding
a solution method for the nonlinear system
of equations, mine ventilation networks
are complicated by certain unique
features. The problems of leakage, need
for fixed flow branches, government regulations, and thermodynamic influence of
the surroundings on the system, make the
modeling of ventilation networks difficult. Hardy Cross method based on corrections to loop or mesh flows is generally
adopted with appropriate modifications to
reflect the above problems. While the
modeling elegance of the modifications
remains subjective, Hardy Cross method
appears to be the only technique ever used
to analyze mine ventilation systems in
practice. The paper by Wang and Li (1985),
presents the application of Newton-Raphson
method to ventilation networks. Newton-
Raphson technique being a method of quadratic convergence, and simultaneous
corrections for all mesh flows, undoubtedly holds a great potential for for ventilation network analysis. This paper on
the other hand, looks into the applicability of linear analysis to mine ventilation
networks.

2 INTRODUCTION TO LINEAR ANALYSIS

Network flow modeling in mine ventilation
usually follows the closed and connected
network approach. It is common to define
an 'atmospheric node' to which all the
intake and exit portals of the mine are
connected. This approach is some times
rigid because nodes lose their identity.
The modeling procedure considered here may
be known as open network approach. Here
the nodal external flows and potentials
are treated in an explicit manner. This
provides an important means of representing the outside world connections to the
system being modeled. Conditions that can
be represented as a nodal external flow
include a known mine quantity, a fixed
flow branch, emission of gases at any
point in the mine, and leakage into a
neighboring mine through a crack. The unknown for a fixed flow branch is the amount
of external pressure, which is computed
after the flow distribution is obtained
for the rest of the network. By modeling
fixed flow branches as nodal external
flows at the originating and terminating
nodes of that branch, in fact, one branch
is removed from the network, resulting in
simplification. Natural ventilation pressure, or fixed mine head, can be efficiently modeled as a node potential.

For a network that has NN nodes, and NB
branches, any spanning tree contains (NN-
1) branches. Each spanning tree defines a
unique set of meshes or loops in the network, known as the fundamental meshes.
There is a one to one correspondence
between the spanning tree and the fundamental meshes in a network. Each branch
out of the tree, known as a chord, is contained in exatly one fundamental mesh of
the network. Thus the number of chords or
fundamental meshes for a network is NC (NC
= NB-NN+1). For network analysis, choosing
the fundamental meshes corresponding to
the minimal spanning tree (spanning tree
over which sum of the branch resistances
is minimum) is known to improve convergence (Wang 1982, Wang & Li 1985).

Part a of the following figure illustrates a network where the mine quantity
Q_m is known, and the flow on branch 4-5,
Q_b is fixed. The node 8 is the atmospheric
node. It is required to find the flow distribution in the mine for this situation.
The transformed network in part b illustrates the concept of nodal external
flows. The branches marked by the heavy
lines form a spanning tree for this network. The chords of the network are given
as branches 2 and 7. For the network, the
fundamental meshes and their directions
are as shown.

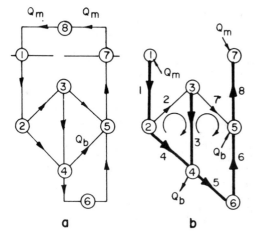

Fig 1. Modeling with nodal external flows
a. original network b. transformed network

For the lack of a better approach, the remainder of the paper resorts to matrix notation to explain the methodology of linear analysis. The reader must note that the parenthesis following a matrix or vector contain its size. The definitions adopted in this paper for basis incidence matrix and fundamental mesh matrix are from Wang (1982). Let $A(NN-1,NB)$ be a basis incidence matrix, and $B(NC,NB)$ be the fundamental mesh matrix corresponding to the spanning tree chosen in the network. Define $Q(NB)$ as the vector of branch flow rates, $H(NB)$ as the vector of branch pressure drops, $N(NN-1)$ as the vector of nodal external flows corresponding to the nodes of A, and $D(NC)$ as the vector of external pressures on the chords. All the branches containing fans and regulators are assumed to be part of the chords. The vector $D(NC)$ contains chord pressure drops or gains, that result from the location of regulators or fans in these chords.

Based on these definitions the following systems of equations are obtained. The conservation of flow in a network is given by

$$A \cdot Q + N = \emptyset \qquad 1$$

where $\emptyset(NN-1)$ is a null vector. External flow out of a node is considered positive in the vector N, to make it compatible with the definition of A. The conservation of energy in a network is given by

$$B \cdot H = D \qquad 2$$

Let us consider for the network in figure 1 b, the conservation of flow at node 7 is discarded in forming the basis incidence

matrix. Then the system described by equation 1, can be written as follows

$$
\begin{bmatrix}
1 & 0 & 0 & 0 & 0 & 0 & 0 & 0 \\
-1 & 1 & 0 & 1 & 0 & 0 & 0 & 0 \\
0 & -1 & 1 & 0 & 0 & 0 & 1 & 0 \\
0 & 0 & -1 & -1 & 1 & 0 & 0 & 0 \\
0 & 0 & 0 & 0 & 0 & -1 & -1 & 1 \\
0 & 0 & 0 & 0 & -1 & 1 & 0 & 0
\end{bmatrix}
\begin{bmatrix}
Q_1 \\ Q_2 \\ Q_3 \\ Q_4 \\ Q_5 \\ Q_6 \\ Q_7 \\ Q_8
\end{bmatrix}
+
\begin{bmatrix}
-Q_m \\ 0 \\ 0 \\ Q_b \\ -Q_b \\ 0
\end{bmatrix}
=
\begin{bmatrix}
0 \\ 0 \\ 0 \\ 0 \\ 0 \\ 0
\end{bmatrix}
$$

The matrix A is also known as unimodular matrix, since each column in A contains no more than two non-zero entries, and all the coefficients are integers; either 0, +1, or -1. For the same network, the loop equations resulting from equation 2, are written as follows

$$
\begin{bmatrix}
0 & 1 & 1 & -1 & 0 & 0 & 0 & 0 \\
0 & 0 & -1 & 0 & -1 & -1 & 1 & 0
\end{bmatrix}
\begin{bmatrix}
H_1 \\ H_2 \\ H_3 \\ H_4 \\ H_5 \\ H_6 \\ H_7 \\ H_8
\end{bmatrix}
=
\begin{bmatrix}
0 \\ 0
\end{bmatrix}
$$

The vector D in the above equations contains zero terms because the chords, branches 2 and 7 do not contain fans or regulators. For a fan in a chord the corresponding term in D takes up a positive value, and vice versa for a regulator.

The pressure loss on branch i in the network, H_i, is related to the flow rate Q_i, according to the Atkinson's equation as follows

$$H_i = R_i \cdot |Q_i| \cdot Q_i$$

the above convention preserves compatibilty of signs between head loss and flow rate. Linear theory fundamentally consists of transforming the above nonlinear equation into a linear equation by approximating the head loss as

$$H_i = R_i \cdot |Q_{io}| \cdot Q_i = R_i' \cdot Q_i \qquad 3$$

in which Q_{io} is the approximate flow rate in branch i. When Q_{io} approaches the actual flow rate Q_i, equation 3 becomes an exact expression for the head loss on branch i. Thus, by using the approximate flow rate Q_{io} to find the modified coefficient R_i' for banch i, the mesh equations can be expressed as linear equations. The flow rate Q_i obtained from one iteration contributes to finding Q_{io} for the next iteration. Thus the R'_i values are modified with each iteration, until the approximate flow rate Q_{io} is close to the actual flow rate Q_i.

Two possibilities exist in applying linear analysis to solve the flow distribution problems in ventilation networks. The first is the method of flow rate equations. In this method, the quantity flow rates on chords $Q_1, Q_2, \ldots, Q_{NC}$ are the primary variables of the model. The other approach is to use the nodal head equations, where the nodal pressures (gage or absolute) $P_1, P_2, \ldots, P_{NN}$ are the primary variables in which the problem is modeled. These two approaches are described in the following sections.

3 THE METHOD OF FLOW RATE EQUATIONS

We have seen that there are $(NN-1)$ linear equations in branch flow rates for continuity (conservation of flow) in the network. Also by substituting the $H_i = R'_i \cdot Q_i$ relationships into the mesh equations given in 2, an additional NC linearized equations are obtained. It is a common practice in linear analysis to solve the system of NB (NB = NN-1+NC) equations in NB unknown branch flow rates at each iteration (Wood & Charles 1972, Jeppson 1976). The coefficient matrix associated with the system of NB linear equations is rather large. In this paper we propose an alternative approach that combines the $(NN-1)$ continuity equations into the NC mesh equations. The resulting linear system of size NC is then solved in terms of the NC chord flow rates.

To begin with, order the chords and tree branches of the network separately. Define the following vectors. QC(NC) is the vector of chord flow rates, QT(NN-1) is the vector of tree branch flow rates, HC(NC) is the vector of pressure losses in chords, and HT(NN-1) is the vector of pressure losses on tree branches. Define RC(NC,NC) as a diagonal matrix corresponding to chords such that

$$RC(i,i) = R'_i \text{ for all } i = 1,2,\ldots,NC$$

Define RT(NN-1,NN-1), as a diagonal matrix corresponding to the tree branches such that

$$RT(j,j) = R'_j \text{ for all } j = 1,2,\ldots,NN-1$$

Based on equation 3, we obtain the following relationships from the above definitions;

$$HC = RC \cdot QC \qquad\qquad 4$$

and
$$HT = RT \cdot QT \qquad\qquad 5$$

Let us partition the basis incidence matrix A(NN-1,NB) as

$$A = \begin{bmatrix} AC & AT \end{bmatrix}$$

so that the submatrix AC(NN-1,NC) consists of columns for chords from A, and the submatrix AT(NN-1,NN-1) consists of columns for tree branches from A. Using this partition procedure, the set of equations given by 1, is written as

$$\begin{bmatrix} AC & AT \end{bmatrix} \begin{bmatrix} QC \\ QT \end{bmatrix} + \begin{bmatrix} N \end{bmatrix} = \emptyset$$

that is, $\qquad AC \cdot QC + AT \cdot QT = -N \qquad 6$

For the network in figure 1 b, the above relationship is expressed as

$$\begin{bmatrix} 0 & 0 \\ 1 & 0 \\ -1 & 1 \\ 0 & 0 \\ 0 & -1 \\ 0 & 0 \end{bmatrix} \begin{bmatrix} Q_2 \\ Q_7 \end{bmatrix} + \begin{bmatrix} 1 & 0 & 0 & 0 & 0 & 0 \\ -1 & 0 & 1 & 0 & 0 & 0 \\ 0 & 1 & 0 & 0 & 0 & 0 \\ 0 & -1 & -1 & 1 & 0 & 0 \\ 0 & 0 & 0 & -1 & 1 & 0 \\ 0 & 0 & 0 & -1 & 1 & 0 \end{bmatrix} \begin{bmatrix} Q_1 \\ Q_3 \\ Q_4 \\ Q_5 \\ Q_6 \\ Q_8 \end{bmatrix} = \begin{bmatrix} Q_m \\ 0 \\ 0 \\ -Q_b \\ Q_b \\ 0 \end{bmatrix}$$

Likewise let us partition the fundamental mesh matrix B as

$$B = \begin{bmatrix} BC & BT \end{bmatrix}$$

so that BC(NC,NC) is the submatrix of B containing columns of chords, and BT(NC,NN-1) is the submatrix of B containing tree branch columns. A partitioning of this nature produces submatrix BC that is always an identity matrix. This is illustrated through rewriting the relationship 2 for the network in figure 1 b, in terms of BC and BT.

$$\begin{bmatrix} BC & BT \end{bmatrix} \begin{bmatrix} HC \\ HT \end{bmatrix} = D$$

that is, $\qquad BC \cdot HC + BT \cdot HT = D \qquad 7$

$$\begin{bmatrix} 1 & 0 \\ 0 & 1 \end{bmatrix} \begin{bmatrix} H_2 \\ H_7 \end{bmatrix} + \begin{bmatrix} 0 & 1 & -1 & 0 & 0 & 0 \\ 0 & -1 & 0 & -1 & -1 & 0 \end{bmatrix} \begin{bmatrix} H_1 \\ H_3 \\ H_4 \\ H_5 \\ H_6 \\ H_8 \end{bmatrix} = \begin{bmatrix} 0 \\ 0 \end{bmatrix}$$

For a network, the flow rates in the tree branches are uniquely defined by the flow rates in chords. We notice this by rewriting the equation 6, as follows

$AT \cdot QT = - AC \cdot QC - N$

that is, $\qquad QT = - AT^{-1} \cdot AC \cdot QC - AT^{-1} \cdot N$

One of the useful relationships between the submatrices A and B (Wang 1982), that is considered here is

$$BT^T = - AT^{-1} \cdot AC \qquad\qquad 8$$

Substituting this into the previous equation provides

$$QT = BT^T \cdot QC - AT^{-1} \cdot N \qquad 9$$

Equation 7 is expanded by combining the relationships expressed as 4 and 5 as follows

$$BC \cdot RC \cdot QC + BT \cdot RT \cdot QT = D$$

BC is an identity matrix, and QT is given by equation 9. Thus

$$RC \cdot QC + BT \cdot RT \cdot BT^T \cdot QC - BT \cdot RT \cdot AT^{-1} \cdot N = D$$

that is,

$$\left[RC + BT \cdot RT \cdot BT^T \right] \cdot QC = D + BT \cdot RT \cdot AT^{-1} \cdot N \qquad 10$$

The above is a linear system of equations of size NC, in terms of chord flow rates. The diagonal matrices, RC and RT, are recomputed at each iteration using the new R_i' values for all the branches from equation 3. The system of equations 10, is thus solved iteratively until two sets of consecutive chord flow rates are close.

The iteration procedure is initiated with each R_i' value equal to R_i, the resistance of the branch i. This is equivalent to considering each flow rate Q_{io} as unity. The solution obtained in the first iteration corresponds to an approximate laminar flow distribution in the network. The result is thus a feasible estimate of the actual turbulent flow distribution in the network (Wood & Charles 1972). This is precisely how the application of linear theory circumvents the need for providing initial estimates for chord flow rates.

Making the approximate flow rate Q_{io} for the present iteration equal to the flow rate Q_{i-1} from the previous iteration, was seen to be an inefficient means of correcting flows. Two successive results are noticed to oscillate around the final solution. To avoid this Wood and Charles (1972) proposed the following for Q_{io} after two iterations

$$Q_{io} = 0.5 (Q_{i-1} + Q_{i-2})$$

which is the arithmetic mean of the flow rates from the previous two iterations. This method of finding Q_{io} is very much in practice in linear analysis applied to hydraulic networks. The convergence is unfortunately slow when the same procedure is tried in mine ventilation networks. However, we noticed a dramatic improvement in convergence by considering geometric mean to evaluate Q_{io} by

$$Q_{io} = (Q_{i-1} Q_{i-2})^{0.5} \qquad 11$$

If there is a change in sign between Q_{i-1} and Q_{i-2}, then Q_{io} is made equal to the square root of the absolute difference between Q_{i-1} and Q_{i-2}.

To recollect from the previous section, the k th element of the vector D(NC), D_k represents the pressure drop or gain on the chord k, due to external pressure sources. Known mine head, or NVP are easily modeled by substituting appropriate constants for the D_k values. But when a fan characteristic is taken into the model as a second degree polynomial it solwed the convergence process. To avoid this, Jeppson (1976) proposed a transformation. If the fan in chord k has the following characteristic

$$H_k = aQ_k^2 + bQ_k + c$$

then define a new variable G_k in place of Q_k such that

$$G_k = Q_k + b/2a \qquad 12$$

the effect of which results in the following modification

$$R_k \leftarrow R_k - a$$

and $\quad D_k = R \cdot b/2a + c - b^2/4a$

At each iteration a value is obtained for G_k. Then the flow rate Q_k is found from the equation 12, and for the next iteration, R_k is replaced by $R_k \cdot G_k$.

The following algorithm is developed for the linear theory method proposed in this section.

Step 1. Set the number of nodes. Read nodal external flows. Read branch info; orig, term, resist. Read chord info; fan char, fixed head.
Step 2. Find min span tree and order chords and tree branches.
Step 3. Form Q,D,N, and AT matrices and invert AT.
Step 4. Compute BT^T (eqn 8), form initial RC,RT.

Step 5. Transform R_k and Q_k for fan branches.

Step 6. Set up the system of linear eqns.

Step 7. Call an equation solver routine.

Step 8. Compute Q_k(new)$\sim Q_{ko}$(old) for all k=1,2,..,NC; if converged goto Step 11.

Step 9. Find approx flow rates Q_{ko} for chords (eqn 11).

Step 10. Recompute QT (eqn 9), RC and RT and goto Step 5.

Step 11. Print solution and stop.

We see that step 3 in the algorithm consists of inverting the basis incidence matrix corresponding to tree branches. Elegant procedures are available to accomplish this from graph theory. We adopted one such techniques described by Swamy and Thulasiraman, 1981 (p.141,142) in the computer program. We are not sure of the contribution of the minimal spanning tree toward convergence in the linear analysis. It was included however, to avoid any possible inefficiencies. The computer implementation requires the storage of at least two matrices. One integer matrix of approximate size (NN,NN), and one real matrix of about the same size, account for the largest portion of the computer storage required.

4 THE METHOD OF NODAL HEAD EQUATIONS

The solution of network flow distribution problems based on nodal heads, either by Hardy Cross method or Newton-Raphson method, is viewed with skepticism by many researchers. It is generally criticized for producing convergence problems. At least in mine ventilation networks there has not been enough experience to subscribe to this outlook. On the other hand, the method of nodal head equations offers some amazing simplifications in model development as well as in the solution of the resultant equations.

This method is based on considering nodal heads (gage or absolute) as the primary variables. Obviously when we solve the network for nodal heads, it is necessary to fix at least one nodal pressure at reference or datum value. The other (NN-1) nodal heads are computed based on this reference. Otherwise the solution becomes indeterminate. In view of this, we need to consider the full incidence matrix of the network in this method. The incidence matrix A(NN,NB). defined for this method, contains the A(NN-1,NB) submatrix considered in the previous section. Correspondingly, the nodal external flow matrix N(NN) defined here contains

external flows for all the NN nodes of the network. Define P(NN) as the vector of nodal pressures.

Consider a branch k in the network with i as its originating node, and j as its terminating node. Then we can write the linearized equation 3, for pressure loss on the branch k in terms of P_i and P_j as follows

$$P_i - P_j = H_k = R_k \cdot |Q_{ko}| \cdot Q_k \qquad 13$$

where it is known that Q_{ko} is the approximate flow rate on the branch k. Let us define a coefficient C_k for the branch k as

$$C_k = 1/(R_k \cdot |Q_{ko}|) \qquad 14$$

Then the equation 13 can be written as

$$Q_k = C_k \cdot (P_i - P_j) \qquad 15$$

This is the most important transformation involved in the method of nodal head equations. Expression 15 implicitly accounts for the conservation of energy in the network. We know that the flow conservation in the network is given by

$$A \cdot Q = -N$$

If we substitute the Q_k values from equation 15 into the Q vector above, some very interesting results are acheived. The above system of equations is transformed into

$$E \cdot P = -N \qquad 16$$

The vector P is known to be the column matrix of nodal heads. The coefficient matrix E becomes a symmetric, positive definite, and usually banded matrix. For the network in figure 1 b, the coefficient matrix E can be easily verified to be the following

├half band width┤

$$
\begin{bmatrix}
C_1 & -C_1 & 0 & 0 & 0 & 0 & 0 \\
-C_1 & C_1{+}C_2{+}C_4 & -C_2 & -C_4 & 0. & 0 & 0 \\
0 & -C_2 & C_2{+}C_3{+}C_7 & -C_3 & -C_7 & 0 & 0 \\
0 & -C_4 & -C_3 & C_3{+}C_4{+}C_5 & 0 & -C_5 & 0 \\
0 & 0 & -C_7 & 0 & C_6{+}C_7{+}C_8 & -C_6 & -C_8 \\
0 & 0 & 0 & -C_5 & -C_6 & C_5{+}C_6 & 0 \\
0 & 0 & 0 & 0 & -C_8 & 0 & C_8
\end{bmatrix}
$$

Due to the symmetry, and banded nature of the matrix, it is only necessary to store the coefficients contained within the half band width as shown above, to define the entire matrix. This results in significant savings in computer storage. However, the half band width is directly affected by node numbering in the network. The size of the half band width, NHBW, is given as

$$NHBW = 1 + \max(|NO(k) - NT(k)| , k=1,2,..,NB)$$

where NO(k) and NT(k) are the origin and terminal node numbers of the branch k respectively. Consequently it is a good practice to keep the origin and terminal node number difference for all branches as small as possible. There are also some algorithms available for node renumbering, to obtain optimal band width for a network. We propose a straight forward means of computing the E matrix. Define a diagonal matrix C(NB,NB) such that

$$C(i,i) = C_i \quad \text{for all } i = 1,2,..,NB$$

$$\text{then} \quad E = A \cdot C \cdot A^T$$

Based on this transformation equation 16 is rewritten as

$$A \cdot C \cdot A^T \cdot P = -N \qquad 17$$

The above is a linear system of NN equations for which, some of the nodal external flows are known, and some of the node potentials are known. At least one node potential needs to be fixed at datum level to obtain a numerical solution for the above system of equations.

The matrix E is analogous to the assembly matrix encountered in two dimensional truss problems. The above equations require to be transformed to account for the known values in the P vector. One procedure for such a transformation is given by Reddy (1984 p.173).

The system of equations in 17, is solved iteratively with the diagonal matrix C recomputed at each iteration. Similar to the procedure adopted in the previous section, the iteration process begins with

$$C_k = 1/R_k$$

for each network branch. The approximate flow rate Q_{ko} is computed for the third iteration onwards based on geometric mean as explained by equation 11. After finding the nodal pressures the branch flow rates are computed as follows

$$Q = C \cdot A^T \cdot P \qquad 18$$

We have not yet decided on a procedure to incorporate fan characteristics into the model. For the present, based on node potentials and nodal external flows as input, the following algorithm is developed for the method.

Step 1. Set number of nodes. Read nodal external flows and potentials. Read branch info; orig, term, resist.
Step 2. Form P,N,A, and C matrices and set reference node potential.
Step 3. Compute the E matrix $(A.C.A^T)$.
Step 4. Transform eqn 17 for known node potentials.
Step 5. Call Choleski's decomposition algorithm.
Step 6. Compute P_i(new) $\sim P_i$(old) for all i = 1,2,..,NN. if converged goto Step 9.
Step 7. Find approx flow rates Q_{io} (eqn 18).
Step 8. Recompute diagonal matrix C and goto Step 3.
Step 9. Print solution and stop.

The Choleski's decompostion algorithm provides an efficient technique to solve the linear system consisting of symmetric and positive definite matrix. The procedure described above is a very simple approach. A little bit of programming effort can eliminate the need to explictly store the matrix A. The E matrix can be stored more efficiently in a size (NN,NHBW), rather than as a (NN,NN) matrix. Band width reduction can be accomplished through careful node numbering, or by using certain algorithms available from finite element methods.

The nodal head equations method does not require the identification of tree branches and chords. We believe it is the simplest technique available for understanding as well as for computer implementation of the network flow distribution problem. The memory requirements can be amazingly low if the nature of the E matrix is exploited and the explicit definition of the incidence matrix is avoided. The network flow distribution problem can be solved with about half the memory as it would take for the method of flow rate equations.

5 COMPUTATIONAL EXPERIENCE

Experience with the application of the linear analysis method has not been extensive. The general impression we formed is, for a well conditioned network consisting of external flows and fixed heads, both the techniques converge in 4 to 5 iterations. The use of geometric mean to compute approximate branch flow rates has greatly contributed to convergence. When the arithmetic mean correction procedure is applied, the following problem required 35 iterations to converge. However, as it can be seen, the geometric mean technique produced convergence in 4 iterations. In the following two simple cases, the solution at each iteration is presented so that one may verify the application of the two linear analysis techniques discussed.

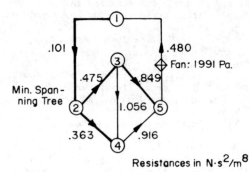

Fig 2. Simple network to illustrate the method of flow rate equations.

The network in figure 2, is adapted from Wang (1982). The problem is to obtain the flow distribution when the fan generates a head of 1991 Pa(8.0 in water gage). The result of solving the problem by the flow rate equations method is listed in the table below.

Table 1. Solution by flow rate equations method.(flows in cu meters per second).

chord	iter1	iter2	iter3	iter4	HC-mtd
3-4	-48.59	-0.07	-1.93	-1.88	-1.88
4-6	807.64	0.66	23.06	23.05	23.05
5-1	1618.58	1.36	46.91	46.91	46.91

We know that the chord flow rates uniquely describe the distribution of flow in the entire network. The chords for the above network are given by, 3-4, 4-6, and 5-1. The Hardy Cross method converged after 4 iterations, based on an initial estimate of flow rates 9.44, 18.89, and 33.04 cubic meters per second for the three chords respectively. The linear analysis method also converged in 4 iterations. Further it did not require initial estimates.

In the figure below the same network is considered again. Here the objective is to determine the flow distribution when the

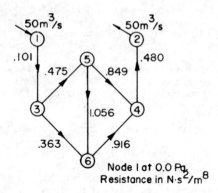

Node 1 at 0.0 Pa, Resistance in $N \cdot s^2/m^8$

Fig 3. Simple network to illustrate the method of nodal head equations.

mine quantity is 50 cubic meters per second.

The problem is solved by the nodal head equations method. To obtain solution, node 1 is chosen to be at a reference pressure of 0.0 Pascals.

Table 2. Solution by nodal head equations method.(nodal pressures are in Pascals)

node	iter1	iter2	iter3
1	0.	0.	0.
2	-62.	-2262.	-2262.
3	-5.	-252.	-252.
4	-38.	-1062.	-1062.
5	-16.	-513.	-513.
6	-14.	-509.	-509.

Since node one is set at zero pressure, the resultant flow distribution corresponds to an exhaust system. The direction of flow is simply from high pressure node to the low pressure node.

In the nodal head equations method, it is essentially a one step procedure to determine the secondary variables that are the branch flow rates. But in the case of flow rate equations method, to determine nodal heads is not all that easy. After one node is assigned a reference head, an algorihmic procedure has to be followed to determine nodal heads for all other nodes.

The following ventilation system has 26 branches and 16 nodes. Node 8 is the intake portal, and node 12 is the return portal for the mine. The branch resistances are shown in figure 4. An exhaust fan is located at node 12 in the branch 10-12. The objective is to obtain the flow distribution in the network corresponding to a fan head of 2500 Pa. In order to solve the problem by the method of flow rate equations, nodes 8 and 12 are assumed to

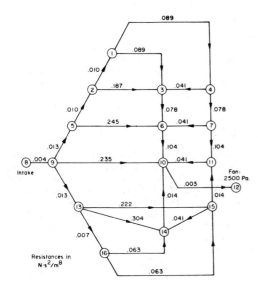

Fig 4. Network schematic to illustrate the application of linear theory methods.

be connected. Thus there are 12 chords in the network. Branch 10-12 is the chord containing an external pressure of 2500 Pa.

Table 3 below illustrates the results of solving the problem by flow rate equations method.

Table 3. Solution by flow rate equations method.(flow in cu meters per second)

chord	iter1	iter2	iter3	iter4	iter5
10-12	59103.4	3.0	405.8	411.6	412.0
13-14	440.8	0.3	39.0	37.7	37.4
5-6	3773.6	0.4	41.0	40.2	40.1
9-10	8877.8	0.6	75.8	74.8	74.7
13-15	4234.2	0.3	40.5	38.8	38.5
2-3	2203.5	0.3	27.8	27.1	27.0
6-10	9447.0	0.7	82.1	81.3	81.3
7-11	4342.8	0.6	54.3	51.9	51.7
1-3	3752.1	0.3	33.3	32.4	32.4
3-6	5258.9	0.4	47.8	46.9	46.8
12-14	16369.3	0.3	72.4	70.9	70.2
11-10	13334.7	0.8	98.4	99.6	99.7

The chords in the table correspond to the minimal spanning tree of the network. The flow rates presented at each iteration are Q_i values. However, for convergence, comparison is made between Q_{io} (old) and Q_i (new), because Q_{io} is the best available approximation for solution before a new iteration begins. Convergence is assumed when the difference in flow rates is less

than 0.1 cubic meters per second, for all the chords. Note the solution closely approximates the final result by the end of third iteration. We obtained this pattern in many cases.

The method of nodal head equations also produced very rapid convergence to the above problem. Here the network is modeled slightly differently. Node 8 is set at a reference pressure of 0.0 Pa, and node 12 is set at -2500. Pa. The table below illustrates the results obtained.

Table 4. Solution by nodal head equations method.(nodal pressures are in Pascals)

node	iter1	iter2	iter3	iter4	iter5
1	-594.0	-1049.1	-1044.1	-1041.6	-1040.3
2	-515.8	-1008.5	-1002.1	-998.9	-997.3
3	-927.9	-1136.9	-1134.7	-1133.5	-1133.0
4	-955.4	-1141.8	-1140.3	-1139.6	-1139.2
5	-415.7	-925.3	-917.4	-913.3	-911.3
6	-1340.2	-1300.3	-1301.8	-1302.6	-1303.0
7	-1324.3	-1302.2	-1303.4	-1304.0	-1304.3
8	0.0	0.0	0.0	0.0	0.0
9	-236.4	-703.6	-691.5	-685.4	-682.3
10	-2322.7	-1972.3	-1981.4	-1986.0	-1988.3
11	-1776.0	-1551.8	-1567.2	-1575.0	-1579.0
12	-2500.0	-2500.0	-2500.0	-2500.0	-2500.0
13	-710.0	-1239.4	-1231.1	-1226.7	-1224.4
14	-1938.5	-1620.8	-1634.3	-1641.2	-1644.7
15	-1650.0	-1524.6	-1537.5	-1544.1	-1547.5
16	-907.2	-1341.4	-1339.7	-1338.6	-1338.1

Convergence is determined by comparing each nodal head of the present iteration with that of the previous iteration. If the absolute difference is within 5 Pa, the iteration procedure is terminated. Again, by the end of the third iteration, the solution very closely approximates the final result.

The introduction of a fixed flow branch is observed to slow convergence in both the methods. The network in figure 4, is solved by treating branch 10-12 as a fixed flow branch with a value of 20.0 cubic meters per second. The result was that the method of flow rate equations required 7 iterations to converge, whereas the method of nodal head equations required 9. However, it must be noted that the convergence criterion is not the same in both the methods. On a Univac-1100 computer, CPU times of 860 and 740 milliseconds were required respectively, to solve the problem. It appears that for one iteration by the method of flow rate equations it takes twice as much time as it would by the method of nodal head equations. It may also be interesting to note that the same problem required 20 iterations by the Hardy Cross method to reach the same level of convergence.

653

6 CONCLUSION

In addition to introducing linear theory to mine ventilation networks, this paper makes some important contributions to the general linear theory technique. We presented a geometric mean correction procedure which contributes greatly to reducing the number of iterations, proposed a method of reducing the system of equations to a size equal to the number of chords (eqn 10), and identified the relationship between the coefficient matrix E and the incidence matrix A of a network.

The advantages of linear theory approach are several. It eliminates the need for initial estimates of flow, it is an easily understood procedure, it is very flexible as it handles nodal external flows and nodal heads directly, and finally the computer implementation of the linear analysis method is elementary. The last factor offers the necessary freedom for a ventilation engineer to write a simple program that is appropriate to his specific needs, rather than depending on a generalized and invariably complex computer package.For example, the code we have written for nodal head equations consists of less than 150 fortran statements.

Our experience indicates that both the linear analysis methods converge very rapidly. By the end of third iteration the solutions appear to be within 10 percent of the final result. Although no particularly ill conditioned cases were analyzed in this paper, the proposed methods required less number of iterations to converge when compared with the Hardy Cross method. There has not been enough experience to compare the two linear analysis methods in terms of convergence characteristics. For most purposes, the method of nodal head equations is a far superior process. It is noticed to take half as much computational time as the method of flow rate equations per iteration, it is the simplest available approach to network analysis, the coding and storage requirements for its computer implementation are minimal, and it uses very elegant numerical analysis techniques.

7 ACKNOWLEDGEMENT

The work has been supported by William C. Browning Scholarship of the Department of Mining Engineering, at the University of Utah.

8 REFERENCES

Chandrasekhar,M. and Stewart,K.H. 1975. Sparsity oriented analysis of large pipe networks. Journal of the Structural Division. Proc ASCE. vol 101,no.HY4, April, p.341-355.

Collins,A.G. and Johnson,R.L. 1975. Finite element method for water distribution networks. AWWA Journal. vol 67, July, p.385-389.

Epp,R. and Fowler,A.G. 1970. Efficient code for steady state flows in networks. Journal of the Hydraulics Division. Proc ASCE. vol 96,no HY1, January, p.43-56.

Isaacs,L.T. and Mills, K.G. 1980. Linear theory methods for pipe network analysis. Journal of the Hydraulics Division. Proc ASCE. vol 106,no HY7, July, p.1191-11201.

Jeppson,R.W. 1976. Analysis of flow in pipe networks. Ann Arbor Science Publication.

Reddy,J.N. 1984. An introduction to the finite element method. McGraw Hill Book Company.

Shamir,U. and Howard,C.D.D. 1968. Water distribution systems analysis. Journal of the Hydraulics Division. Proc ASCE. vol 94,no HY1, January, p.219-234.

Swamy,M.N.S. and Thulasiraman,K. 1981. Graphs, networks, and algorithms. Wiley Interscience Publication.

Wang,Y. and Li,S. 1985. The newton method of calculating the ventilation network. Proc 2nd conf on the use of computers in coal industry. Univ of Alabama, Tuscallosa, April 15-17. p.388-392.

Wang,Y.J. 1982. Ventilation network theory. Mine ventilation and air conditioning.(eds) Hartman,H.L., Mutmansky,J.M., and Wang,Y.J., Wiley Interscience Publication, p.483-517.

Watters,G.Z. 1984. Analysis and control of unsteady flow in pipelines. Ann Arbor Science Book. Butterworths Publcation.

Wood,D.J. and Charles,C.O.A. 1972. Hydraulic network analysis using linear theory. Journal of the Hydraulics Division. Proc ASCE. vol 98,no HY7, July, p.1157-1170.

Wood,D.J. and Rayes,A.G. 1981. Reliability of algorithms for pipe network analysis. Journal of the Hydraulics Division. Proc ASCE. vol 107,no HY10, October, p.1145-1161.

Ventilation models for longwall gob leakage simulation

DANIEL J.BRUNNER
Mine Ventilation Services Inc., Lafayette, CA, USA

ABSTRACT: The application of two detailed network models to an actual mine site and their correlation with survey data are described. The models, designed for use with conventional network simulation programs, were developed to improve current cave leakage modeling techniques for the migration of leakage air through longwall waste areas. The initial model involved construction of a laminar resistance element grid over the gob. Resistance values were assigned in relation to the stress redistribution over the gob. The model was correlated to measured data and the resistance values were adjusted to obtain the required simulation. From the simulated leakage patterns obtained, a simplified model was constructed which only represented the main leakage routes. Both models described accurately the magnitude, distribution and direction of the gob leakage. The simulation algorithms that were developed may be incorporated into current network programs in order to improve their accuracy when applied to longwall mines.

1 INTRODUCTION

In the summer of 1982, the Thompson Creek No. 1 Mine, located in Carbondale Colorado, was made available by the Snowmass Coal Company for detailed experimental ventilation studies. Among the analyses conducted was an effort to develop a detailed network representation of the leakage patterns in the caved waste area behind the longwall face. This was accomplished with data measured in the mine and with the aid of a conventional network simulation program. It was intended that one of the models generated would be generalized in order for it to be applied by ventilation planners to improve the accuracy of current longwall ventilation simulations.

1.1 Site characterization

The Thompson Creek No. 1 mine, at the time of the studies, consisted of a single longwall panel retreating along the strike of a 30 degree dipping seam approximately 2.1m thick. The depth of cover was variable due to mountainous surface terrain; however, at the time of this analysis, the face was approximately 380 meters below the surface. The face was on standbye at the time of the survey and had been idle for two weeks.

Just before shutdown, the face had encompassed a fault running at an angle of about 30 degrees to the face line. It was apparent that the gob caved well up to the back of the shields due to the combined effect of the fault and the standstill time.

The longwall was ventilated by intake air drawn along the second of two entryways and flowed ascensionally along face line toward the tailgate. At the tailgate the air was split before being drawn toward the No. 2 fan as shown on Figure 1.

The caved waste area was ventilated, as opposed to being sealed off, by allowing the intact second entryways to divert air around the entire perimeter of the gob. Stoppings constructed in the cross-cuts along the bleeder intake airway maintained the gob at a reduced pressure, encouraging leakage from the face line into the gob. This maintained the waste area gases present in the gob away from the face. Leakage air was drawn from the bleeder cross-cuts at the top of the waste to the bleeder return.

2 DATA ACQUISITION

The state of the longwall (on standbye) allowed detailed measurements of airflow

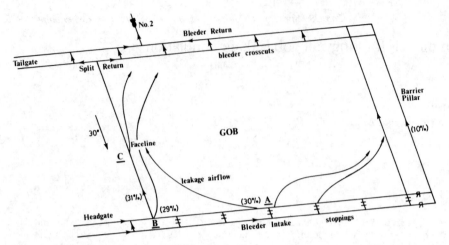

Fig. 1. Isometric showing locations of leakage airflow into the Thompson Creek gob.

to be taken at all the leakage points around the gob area. Measurements of velocity and cross-sectional area were obtained at every stopped cross-cut along the intake bleeder, at increments along the face line and in each bleeder cross-cut along the bleeder return.

Of the air designated for the longwall panel (36.26 m^3/s), 31% entered the face line, 29% leaked directly into the gob at the junction between face and the headgate (B on Figure 1), 30% leaked through the stoppings in the cross-cuts leading from the bleeder intake (A on Figure 1), while 10% continued around the barrier pillar. Additionally, of the air entering the face line, 31% leaked through the shields (C on Figure 1) before reaching the tailgate.

The measured airflows are shown in detail on the network representation of the gob area in Figure 2.

3 PURPOSE OF MODELING LEAKAGE THROUGH CAVED WASTE AREAS

The significant proportions of air found leaking through the gob suggest that the caved region must be considered when simulating longwall panels by network analysis. This substantial amount of leakage affects the economics as well as the efficiency of the entire ventilation system. Additionally, leakage air plays an important role in controlling the environmental conditions in the mine.

High concentrations of methane may be present in the caved region, released by desorption from broken coal or emitted from roof or floor sources following caving and relaxation of stresses in the adjacent strata. In gassy situations, ventilation of the gob may be used in order to prevent gas migration onto the face line.

In longwall mines where seams are liable to spontaneous combustion, the velocity of the air leaking through the gob is important. The conditions that produce spontaneous combustion are easily met in shadow zones of ventilated longwall gobs.

These factors prompted analysis of current leakage simulation practice and resulted in the formulation of more accurate models. The intent was to produce a model which would increase the accuracy of network simulations. This would also provide a means of investigating methane migration in the gob or the effects of leakage control techniques.

3.1 Current methods of modeling waste areas

In current network modeling practice, gob leakage is often represented inadequately. This is especially common when several longwalls are simulated within the general mine infrastructure. Such a simulation may give an acceptable approximation to airflows in the main airways but may be considerably in error on, or close to, each face.

When significant leakage through the gob is anticipated, the flow paths may be represented by a few representative leakage branches positioned in the gob area. Sometimes a more elaborate "sink" arrangement is used where several leakage branches are connected symmetrically to a

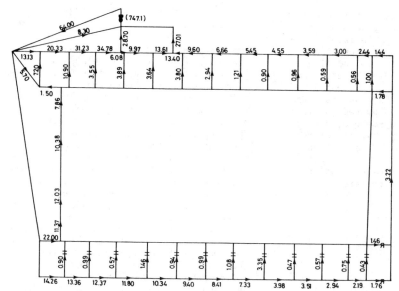

Fig. 2. Airflows (m³/s) measured at the Thompson Creek No. 1 mine
which were used for the model correlation exercises.

center node (or nodes). Resistance values
are ascribed to these branches such that
they represent the total gob resistance.
This method may correctly represent the
magnitude of leakage flowing through the
gob but does not model the path or distribution of the leakage.

3.2 Basis for the improved gob models

The improved models, developed in this
analysis, were based on the presumption
that the caved waste area can be described
as a porous medium of variable grain size,
thickness and consolidation.

The size of the particles, the thickness
of the permeable medium and the consolidation of the medium are all functions
of geology and physical site specific
characteristics of the mine. Specifically,
these three parameters depend upon the
extent and characteristics of the fragmentation of the caved strata as the face
advances. From a generalized perspective
of the redistributed vertical stress over
longwall gobs, the relative conditions in
the gob can be visualized.

The redistributed vertical stress field
over a longwall gob is dome shaped. Vertical stress gradients initiating at the
gob edges proceed to the center where,
depending upon the width/depth ratio of
the panel and the geology, the cover load
pressure may be attained. The consolidation of the permeable medium in the cave is

related to the redistributed vertical
stresses.

Additionally, the extent of the cave in
the vertical direction, increases toward
the center of the gob. The breakage here
is more efficient and thus produces particles of a reduced diameter.

For modeling purposes, the gob can be
envisioned as a medium which increases
in thickness and consolidation, with decreasing particle sizes, towards its
center.

The air flowing through the medium is
assumed to be laminar, that is, the
Reynolds number based on external flow
considerations,

$$Re = q_f D / \nu n$$

where: D = average particle diameter (m)
     ν = kinematic viscosity of fluid (m²)
     n = porosity of the medium
     q_f= Q/A : Darcy velocity (m/s)
     Q = volume flow rate (m³/s)
     A = cross-sectional area of
          medium (m²),

is assumed to be in the laminar range
(less than 1).

An elaborate method of simulating the
distribution of leakage air in the medium
would be to conduct a finite element
analysis. A value of permeability would

657

be assigned to elements distributed in the medium with the element code producing the resulting airflow patterns. However, in order to utilize conventional network simulation programs for the modeling process within acceptable computer run time, the medium can be represented by a finite number of leakage branches. Each branch represents a designated volume of the medium and must be assigned a value which represents the resistance of that volume to airflow. With the assumption that the flow remains in the laminar regime, this resistance can be calculated from Darcy's equation for flow through a permeable bed (neglecting the compressibility of air):

$$Q = k_p Ap / \mu L \qquad 3.1$$

where: Q = volume flow rate (m^3/s)
k_p = liquid permeability of medium (m^2)
μ = dynamic viscoscity (Ns/m^2)
A = cross-sectional area (m^2)
p = pressure drop (Pa)
L = length of bed (m)

The pressure drop through the medium, (p), as a result of the airflow, (Q), by rearrangement of Equation 3.1, is

$$p = (\mu L/k_p A)Q$$

The resistance to laminar flow, R_L is then

$$R_L = \mu L/k_p A \quad Ns/m^5 \qquad 3.2$$

From Equation 3.2, the laminar resistance for flow through a permeable medium, R_L, is a function of the geometry and the permeability of the medium, as well as the fluid properties. For air leaking through a caved zone, the fluid properties (dynamic viscosity) are known, as well as the dimensions of the volumes represented. The permeability is dependent upon the redistributed vertical stress and the efficiency of the breakage (particle size). The permeable cross-sectional area is a function of the extent of the breakage which, as with the permeability, also depends upon location in the medium.

4 NETWORK MODELS APPLIED TO THE THOMPSON CREEK GOB

The Modeling exercises investigated the application of two improved models to the gob area of the Thompson Creek No. 1 mine. Correlation studies were undertaken in order to compare the results from each of these models to the leakage airflows measured around the gob area. (Figure 2). These models were incorporated into the ventilation network analysis program, VNET, which was modified to accommodate airflow in the laminar flow regime for designated branches.

For both models, the region behind the longwall face was isolated from the remainder of the mine and analyzed separately. Appropriate airflows to and from the face were fixed as regulated quantities to simulate the airflow from the rest of the mine. This separation permitted detailed simulations of the gob to be analyzed without running full simulations of the entire mine.

4.1 Resistance element model

This initial modeling effort involved constructing a laminar resistance grid over the caved zone (Figure 3). Each resistance element in the grid was assigned an initial value of resistance based on idealized gob conditions and assumed initial values of permeability and permeable cross-sectional area. These resistances were adjusted until an acceptable correlation with measured data was obtained.

The isolated gob area was divided into a 12 x 6 grid leaving uniformly sized rectangles measuring 18 x 54 meters. Airways representing the airflow through each element were constructed diagonally across each rectangle(Figure 4). Assumed permeability values in the range 10^{-5} to 10^{-7} m^2 were assigned to the elements. The lowest values of permeability were assigned to the element region along the center of the gob, while the higher values were assigned to the perimeter elements. The permeability values of the remaining blocks were chosen such that symmetrical decreasing gradients of permeability existed, extending from the rib sides to the center.

The permeable cross-section was determined for each element by assuming that the medium height was equivalent to the seam thickness for the perimeter elements, increasing to four times the seam thickness for the center elements. From the geometry of the elements,

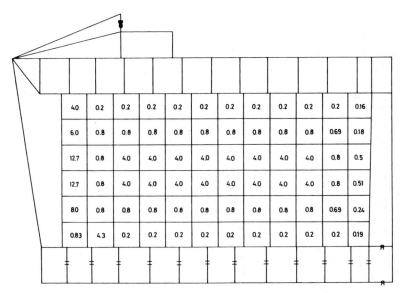

Fig. 3. Laminar resistance grid with resistances (Ns/m^5) required to correlate resistance element model with the measured data.

permeability and permeable cross-sections, laminar resistance values were calculated for each representative airway using equation 3.2.

The airflow in the caved perimeter airways was assumed turbulent. In addition, it was assumed that these airways caved to one half their initial cross-section.

By conducting correlation exercises with the actual data from the Thompson Creek Mine, finalized resistance values were obtained for the branches representing the elements in the gob and the turbulent perimeter airways. The laminar resistances required to simulate the Thompson Creek data are shown on Figure 3 and the resulting airflow patterns on Figure 4.

Larger resistances for the leakage branches leading from the face line were required. These resistances reflect the sealing effect of the longwall shields which limits leakage from the face line to between the shields. Additionally, the resistances of the caved head and tailgate airways required adjustment. A gradual increase in resistance on moving away from the face was required for the simulation. This consequence suggests that the cross-section of these airways converges behind the face. The extent of this zone is a function of the dynamics of the face.

From the laminar resistance distribution on Figure 3, it is evident that most of the leakage air will flow around the consolidated center of the gob. From Figure 4, the airflow is seen to be concentrated in the convergence zones behind the shields and along the bleeder airways, and tends to detour the consolidated area.

Concentrating on the leakage patterns observed from the results of the resistance element model, an endeavor was made to produce a more practical model, one that would represent the main leakage flow paths and magnitudes, but represent this in less detail with fewer branches.

4.2 Simplified model

The simplified model represents the concentrated leakage airflow pattern observed in the gob and additionally accounts for the low leakage flows without the detail of the resistance element model.

Utilizing the resistance grid produced from the resistance element correlation, branches were constructed elliptically around the center gob zone to represent the concentrated perimeter leakage flows. These were, in turn interconnected with branches leading through the gob center to represent the small amounts of leakage through the consolidated center zone. The branch configuration is shown on Figure 5.

The resistance of each branch was calculated by determining the new volumes represented and using equation 3.2 with the values employed in the previous section. These values were then modified

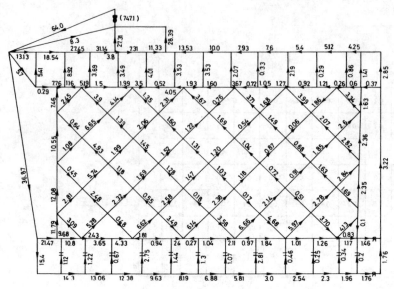

Fig. 4. Simulation of the airflow distribution (m^3/s) in the gob obtained with the resistance element model.

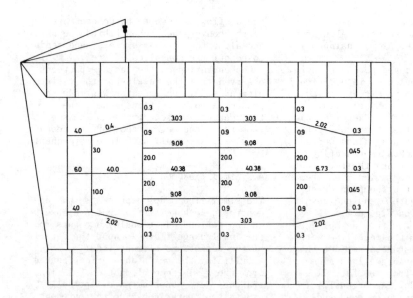

Fig. 5. Laminar resistances (Ns/m^5) and the network configuration used to correlate the simplified model with the measured data.

until a satisfactory correlation with the actual data was obtained. The resistances required for the simulation are shown on Figure 5 and the airflows produced on Figure 6. As with the element model, the final laminar resistances are not quite symmetrical. This effect is believed to be caused by the contribution of the fault, the standstill time and the dipping seam. The resistances required for the caved head and tailgate airways were similar to those required with the resistance element model. The length of the convergence of cross-section was approximately 150 meters beyond the face line.

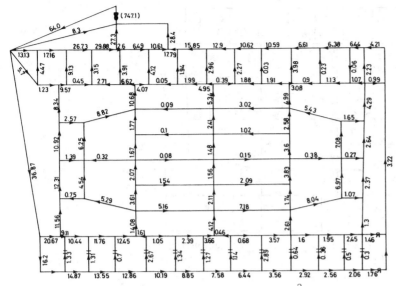

Fig. 6. Simulation of the airflow distribution (m³/s) in the gob obtained with the simplified model.

The airflows produced from this simulation correlate satisfactorily with the output from the resistance element model, confirming that an adequate representation can be attained with the simplified model. This, as well as the practicality of the model, prompted generalizing the modeling procedure for rapid application to other gob areas.

5 GENERALIZATION OF THE SIMPLE MODEL

This concluding section describes the flow net geometry and the resistances required in order to represent a caved longwall waste area with the simple model described in the previous section. The modeling procedure is based on ideal gob conditions, that is, fault free, continuous face advance and symmetrical stress redistribution patterns over the gob.

5.1 Flow net geometry

The flow net for the simple model is shown on Figure 7. The pattern consists of a center grid labeled A and two end grids B and C. Since the resistance values to be given in the following section are based on a known volume, the construction pattern should be well adhered to.

The center grid is constructed by dividing the length of the gob into equal increments of approximately 150 meters. The width is separated into equal increments approximately 17 meters in length.

The two symmetrical end grids are constructed by dividing the face and stationary end into equal increments measuring about 25 meters. The end grids extend one third of the distance to the nearest center grid branch as shown on Figure 7.

5.2 Resistance values

This model assumes that the airflow throughout the entire gob remains in the laminar flow regime. The air flowing in the caved perimeter airways, however, remains turbulent. The caved airways are assumed to consolidate to a height equivalent to half their original dimensions. For the head and tailgate airways, this final height is attained at about 150 meters beyond the face line. The first 150 meters incorporate a convergence zone in which it is approximated that the airways decrease linearly in height. The resistance of any increment within the 150 meter linear convergence zone can be determined by equation 5.1.

661

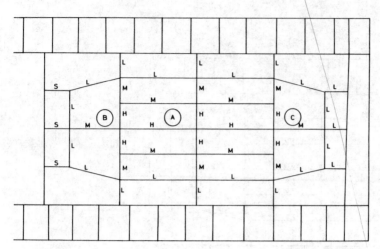

Fig. 7. Network configuration and resistance distribution required to represent longwall gob.

$$R_{x_1}^{x_2} = - \frac{2k}{aw^3} \left[\frac{1}{ax_2+h} \left(\frac{w}{2(ax_2+h)} + 1 \right) - \frac{1}{ax_1+h} \left(\frac{w}{2(ax_1+h)} \right) \right]$$ 5.1

where: x_2, x_1 – distance from face (m)
w – width of airway (m)
k – friction factor (kg/m³)
h – airway height (m)
a – slope of convergence =
(final height – h) /150m =
$(-\frac{1}{2} h/150)$

Owing to the dependence on cross-sectional area, the laminar resistance values for the branches representing the gob, are based on the branch spacing previously suggested.

Four different resistance values are present, each assigned to represent a different degree of consolidation in the gob.

S – Shield Resistance
H – High Resistance
M – Medium Resistance
L – Low Resistance

Shield resistances are assigned to the branches leading into the gob from the face. High resistance values are assigned to the center branches in the gob, while medium values are distributed around this zone. Low resistance values are assigned to branches along the perimeter of the gob. The relative locations of these resistances are shown on Figure 7. Note that these values are distributed in symmetrical fashion.

The values of laminar resistance suggested, are based on the analysis performed with the Thompson Creek data. Assuming that the values of permeability and the distribution of the permeable cross-section are typical of longwall gobs, the values of laminar resistance can be extrapolated. From Equation 3.2, the permeable cross-sectional area is

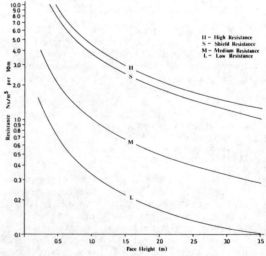

Fig. 8. Nomogram to determine laminar resistances as a function of face height and location in the gob.

662

inversely proportional to the laminar
resistance. Making the assumption that
the permeable cross-section is directly
proportional to the amount of coal removed (the seam height) produces the
laminar resistance curves as functions
of seam height and location in the gob.
(Figure 8).

REFERENCES

Bacharach, J.P.L., Chamberlain, E.A.C.,
  Hall, D.A., Lord, S.B., Steele, D.J.,
  A review of spontaneous combustion
  problems and controls with application
  to U.S. coal mines. D.O.E. Report TID-
  28879, September, 1978.
Bear, J. 1972, Dynamics of fluids in porous
  media. American Elsevier, N.Y., N.Y.
McPherson, M.J., Brunner, D.J., An investigation into the ventilation characteristics of a longwall district in a
  coal mine, D.O.E. Report, Contract No.
  DC AC03-768 F00098, September, 1983.
McPherson, M.J., Wallace Jr., K.G., Mine
  ventilation economics. D.O.E. Report,
  Contract No. DC AC03-768 F00098, September
  1982.
Whittaker, B.N., An appraisal of strata
  control practice, Trans. Inst. Mining
  Engineers, October 1974.

Mine ventilation network design

RANDAL J.BARNES & THYS B.JOHNSON
Colorado School of Mines, Golden, USA

Abstract: The ventilation network modelling tools available today are, for the most part, variations on the ventilation simulator. Using these tools the engineer must manually complete the design of the system, then the numerical simulator computes the resulting air flows. This is analysis and not design.

This paper details research which addresses and solves the general ventilation network design problem. This problem is as follows.

Given: a well defined ventilation network topology, upper and lower bounds on allowed flow in each branch, allowed and forbidden locations for ventilation equipment, specifications and locations of all existing fans.

Find: the optimal (minimum power consumption) locations and sizes of all mine fans, booster fans, and ventilation regulators, and the resulting network air flow.

This approach starts with the basic design criteria and selects the best locations and sizes for all ventilation equipment; thus, it is a design approach and not merely one of analysis.

The computation methodology presented was developed specifically for the ventilation network design problem; this is not a "new application of an old technique". The algorithm is based on a new mathematical decomposition of the describing non-linear network flow problem, resulting in two easily solved linear sub-problems. It is computer time and space efficient, allowing application on the largest of practical mine ventilation problems.

The paper formally states the problem solved, rigorously develops a solution algorithm, and present both in a clear and understandable light. As developed and presented, the algorithm is based on quantity flow modelling; however, it is easily modified to work with mass flow instead.

Recent developments in mine ventilation network theory and analysis

Y.J.WANG
West Virginia University, Morgantown, USA

HOWARD L.HARTMAN
University of Alabama, University, USA

JAN M.MUTMANSKY
The Pennsylvania State University, University Park, USA

ABSTRACT: This paper discusses recent developments in the following areas of mine ventilation network theory and analysis: (a) a concept of subsystems for multiple-fan systems based on a special type of cutset of the network; (b) cutset operations for obtaining alternative network solutions; (c) network properties including the conditions for constant system and subsystem resistances, minimal system resistance, and minimal system air power; (d) diagrams for showing relationships among subsystem parameters for two-fan ventilation systems; and (e) the formulation and solution of mine ventilation network problems using mathematical programming.

1 INTRODUCTION

Mine ventilation networks can be grouped into three categories: (a) networks with natural splitting, (b) networks with natural splitting and fixed-quantity branches, and (c) networks with controlled splitting. Problems associated with each type of network include determination of pressure and quantity, analysis of power consumption, and minimization of network operating costs. There are many published papers dealing with the numerical solutions of these problems using digital computers. However, there is a need to integrate important recent developments and solution methodologies into a comprehensive theory of mine ventilation networks, emphasizing the analysis of generalized network properties for single-fan and multiple-fan systems.

This paper is an attempt to present such a comprehensive discussion on the following areas of development in mine ventilation network theory: (a) a concept of subsystems for multiple-fan systems based on a special type of cutset of the network; (b) cutset operations for obtaining alternative network solutions; (c) network properties including the conditions for constant system and subsystem resistances, minimal system resistance, minimal system air power; (d) diagrams for showing relationships among subsystem parameters for two-fan ventilation systems; and (e) the

formulation and solution of mine ventilation network problems using mathematical programming.

2 SINGLE-FAN SYSTEMS

An example of a mine ventilation network is shown in Fig. 1, with the node numbers (in circles), branch numbers, and branch directions indicated. It represents a ventilation system with two downcast and two upcast shafts. For convenience in modelling, the downcast shafts (branches 5 and 6) are shown as if they are connected to a single inlet node

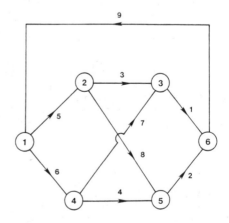

Fig. 1. An example network.

(node 1). Similarly, the upcast shafts (branches 1 and 2) are shown as a common outlet node (node 6). A return dummy branch (branch 9), which is a hypothetical airway with zero resistance, is included to connect the outlet node to the inlet node. Therefore, the air quantity q_9 for the return branch equals the total mine quantity or system quantity Q_T.

The fan total pressure, which is defined by the Air Moving and Conditioning Association (AMCA) as the rise in total pressure from the fan inlet to the fan outlet, will be used in this paper and referred to as the fan pressure. Furthermore, the resistance of the branch corresponding to the upcast shaft will include the equivalent resistance for the outlet velocity. The equivalent resistance r_E for the outlet velocity can be expressed as (Mutmansky, Wang, and Hartman, 1984):

$$r_E = \rho/(2A^2) \tag{1}$$

where ρ is the air density and A is the outlet area.

Proceeding toward the solution of the network in Fig. 1, let the resistance r_j for $j = 1, \ldots , 9$ be given as listed in Table 1. Also, let the total quantity Q_T be 100 m^3/s. If a single fan is assumed to be operating in the return branch, then there exists a unique set of air quantities, pressure losses, and node pressures, called a network solution, satisfying Kirchhoff's current law (KCL) and Kirchhoff's voltage law (KVL). Any solution that can be generated by assuming a fan in the return branch will be referred to as a single-fan solution. The air quantities q_j and pressure losses h_j for the single-fan solution of the example network, with $Q_T = 100$ m^3/s, are listed in Table 1. The corresponding set of node pressures is given in Fig. 2. The node pressures are often expressed in terms of pressure drops with reference to the pressure at the inlet node (node 1). Therefore, the system pressure loss H_T for a single-fan solution equals the node pressure at the outlet node. If desired, the total pressure gradient of a selected path can be readily plotted from the node pressures.

2.1 Cutsets

Increasing interest in the concept of cutsets is one of the most important recent developments in mine ventilation network theory (Wang, 1982a, 1982b). A

Table 1. The single-fan solution for the example network.

j	r_j $N \cdot s^2/m^8$	q_j m^3/s	h_j Pa
1	0.1531	42	270
2	0.1040	58	350
3	0.6447	27	470
4	0.5222	30	470
5	0.1190	55	360
6	0.1383	45	280
7	2.4444	15	550
8	0.4984	28	390
9	0	100	0

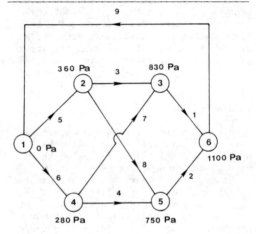

Fig. 2. Nodal pressures for the single-fan solution of the example network.

cutset of a ventilation network is a set of branches whose removal from the network separates the network into two components, provided no proper subset of these branches also separates the network. (A subset V, of a set W, is a proper subset if $V \neq W$.) Here a component may consist of an isolated node. For example, in the network of Fig. 1, the set of branches 5, 6, and 9, is a cutset. The set of branches 3, 6, 7, 8, and 9, on the other hand, is not a cutset because one of its proper subsets, branches 3, 6, 8, and 9, is a cutset. To choose the direction of a cutset, let N_1 and N_2 be the two sets of nodes partitioned by a cutset. We assign the direction of a cutset to be from N_1 to N_2. Thus a branch in a cutset is said to be a positive or negative branch depending on whether it is directed from N_1 to N_2 or from N_2 to N_1, respectively.

For the purposes of analyzing multiple-fan systems, Wang (1983) has introduced the definition of a major cutset. A cut-set that contains the return branch as its only negative branch is called a major cutset. The removal of any major cutset from a ventilation network destroys all paths between the inlet node and the outlet node. All major cutsets of the network of Fig. 1 are identified in Fig. 3. Each set of branches intersected by a broken line is a major cutset. Because the major cutset is dependent on the choice of branch directions, we shall assume that the directions of all branches in the major cutset coincide with the flow directions of the single-fan solution.

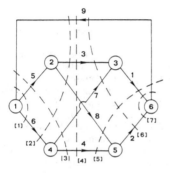

Fig. 3. Major cutsets of the example network.

2.2 Cutset operations

A cutset operation is a technique for manipulating a network in which pressure gains (fans) or pressure losses (regulators) are hypothetically added to a cutset so that Kirchhoff's laws remain satisfied. The mathematical basis for a cutset operation has been outlined previously (Wang, 1983). Cutset operations have been applied both to controlled-flow ventilation networks (Wang, 1982a, 1982b; Wang and Mutmansky, 1982) and to natural-splitting networks (Wang, 1983). The procedure for employing a cutset operation is as follows:

1. Select a cutset compatible with the purpose of the cutset operation.

2. Assume the cutset has K branches and let t_j be the pressure loss of the regulator or fan pressure in branch j of the cutset (for j = 1, ..., K). For

generality, let fan pressure gains be negative.

3. Based upon some network manipulation purpose (generally the specific purpose is to change a particular t_j value to 0), select one specific pressure loss, t_m, from the t_j and let $d_p = t_m$ (or $d_p = -t_m$).

4. Add d_p to the t_j values for all positive branches of the cutset.

5. Subtract d_p from the t_j values for all negative branches of the cutset.

6. Add d_p to the node pressures for all nodes contained in N_2.

An example of a cutset operation is presented in Fig. 4. It is important to note that KCL and KVL are satisfied before a cutset operation and remain satisfied after the operation. The total air power of the network may be increased, decreased, or unchanged by the cutset operation. However, if there is no fan pressure in the cutset before and after a cutset operation (i.e., all t_j's are non-negative and remain non-negative), the air power of the network is unchanged by the cutset operation.

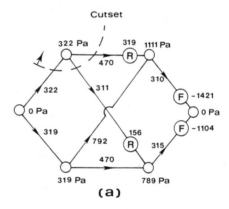

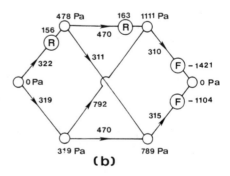

Fig. 4. An example of a cutset operation. (a) Before the cutset operation. (b) After the cutset operation.

3.1 The concept of subsystems

The concept of subsystems has been introduced to facilitate the analysis of multiple-fan systems (Wang, 1984b). Associated with fan i of a multiple-fan system is subsystem i, which is represented by the following subsystem ventilation parameters:

Q_i air quantity

H_i pressure loss

P_i fan pressure

R_i resistance

Z_i air power.

Because they are directly associated with the fan, air quantity Q_i and fan pressure P_i of a subsystem can be measured in practice; at an operating point, the value for the subsystem pressure loss equals that for the subsystem fan pressure (i.e., $H_i = P_i$). The subsystem resistance is defined in a manner similar to that of an airway or single-fan system, that is,

$$R_i = H_i/Q_i^2 = P_i/Q_i^2 \qquad (2)$$

It is important to recognize that the subsystem resistance R_i in Eq. 2 is not constant whereas resistance of a single-fan system is constant.

Application of the concept of subsystems to the total or multiple-fan system, where a fan is located in every positive branch of a major cutset, is a significant development. If there are F+1 branches in the major cutset, then there are F subsystems. Using the subsystem parameters outlined above, we obtain the following corresponding expressions for the total or multiple-fan system parameters (Wang, 1984b):

$$Q_T = \sum_{i=1}^{F} Q_i \qquad (3)$$

$$H_T = \sum_{i=1}^{F} u_i H_i \qquad (4)$$

$$P_T = \sum_{i=1}^{F} u_i P_i \qquad (5)$$

$$R_T = \sum_{i=1}^{F} u_i R_i^3 \qquad (6)$$

$$Z_T = \sum_{i=1}^{F} Z_i \qquad (7)$$

where the u_i values are the following quantity ratios:

$$u_i = Q_i/Q_T \qquad i = 1, \ldots , F \qquad (8)$$

and

$$u_1 + \ldots + u_F = 1 \qquad (9)$$

As implied by Eqs. 4, 8, and 9, the system pressure loss H_T equals the weighted average of pressure losses for the subsystems, using the air quantities for the subsystems as the weights. Similarly, the system fan pressure P_T (Eq. 6) equals the weighted average of subsystem fan pressures.

3.2 Subsystem and system resistances

The existence of multiple values of system and subsystem parameters is one of the most important properties of multiple-fan networks. As an example, the system and subsystem resistances of the example network (Fig. 1) are shown in Fig. 5, assuming fans in branches 1 and 2. Because they are dependent on the quantity ratios and their values are constant for a set of fixed-quantity ratios, the resistances for the example network can also be represented by a family of straight lines on the coordinate system of subsystem air quantities. This is illustrated in Fig. 6. Although it will be impossible to display this concept graphically for any network with four or more fans, this property of resistances is mathematically applicable to a general multiple-fan system.

Another distinctive feature is the existence of a set of quantity ratios $u_1, \ldots , u_F$ under which system resistance R_T is minimized. This set of quantity ratios is the same as that of the single-fan solution. Furthermore, this set of quantity ratios corresponds to the point where system and subsystem pressure losses and fan pressures are equalized, and also to the point where system air power Z_T is minimized for a given quantity Q_T (Wang, 1983, 1984b). These properties are illustrated in Fig. 7, using the example network of Fig. 1 with fans at branches 1 and 2.

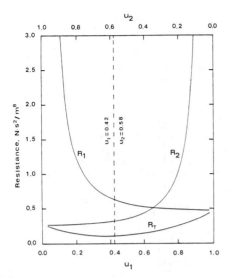

Fig. 5. System and subsystem resistances for the two-fan system. Fans are assumed for branches 1 and 2 in the example network.

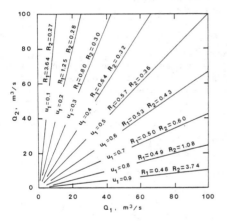

Fig. 6. A diagram showing subsystem resistances as a family of straight lines. Fans are assumed for branches 1 and 2 in the example network.

3.3 Operating points

With the concept of subsystems as an analytical tool, the operating point for a multiple-fan system can be mathematically represented as a point in F-dimensional space with subsystem air quantities defining the bases. However, the graphical display of the operating point is limited to a two- or three-fan system.

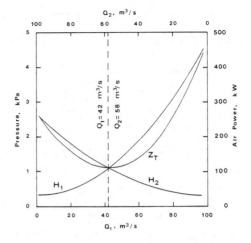

Fig. 7. Curves of pressure losses and air power under a constant total quantity. Fans are assumed for branches 1 and 2 in the example network.

In a recent paper, Wang (1985) has discussed the procedure for graphical representation of the operating point and ventilation parameters of the two-fan system. As shown in Figs. 6 and 8, resistances, pressure losses, fan characteristic curves, and the operating point can be plotted on a plane of two subsystem air quantities. The modified fan curves in Fig. 8 are obtained from the fan characteristic curve of Fig. 9. It should be mentioned, however, that Wang's procedure is not for a graphical solution and that a variety of network solutions are required for plotting the ventilation parameters.

4 MATHAMETICAL PROGRAMMING

4.1 Networks with natural splitting

In the problem of mine ventilation networks with natural splitting, the resistance factors, fan locations, and fan characteristic curves are given. The unknown values to be determined are air quantities for all branches in the network. A common approach to this problem is to formulate it based on KCL and KVL and to obtain the solution by an iterative technique known as the Hardy Cross method (Cross, 1936; Scott and Hinsley, 1951; Wang, 1982b). However, as has been done to a water distribution problem (Collins et al., 1978), the ventilation network problem can be

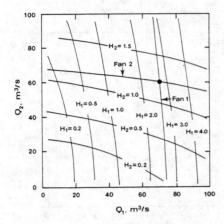

Fig. 8. Representation of subsystem pressure losses, fan curves, and the operating point on a plane of subsystem air quantities. Fans with characteristics shown in Fig. 9 are assumed for branches 1 and 2 of the example network.

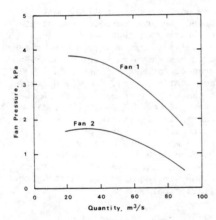

Fig. 9. Fan characteristic curves employed in constructing the modified fan curves in Fig. 8.

formulated as a nonlinear programming (NLP) problem by applying a theorem of "content" for nonlinear systems (Miller, 1951).

Let us consider a ventilation network with B branches, N nodes, and F fans. Let M (M = B − N + 1) be the number of fundamental meshes with respect to a spanning tree, and assume that the fans are located in the branches 1, ... , F (F < M), which are contained in the chord set. The remaining M − F branches that are contained in the chord set are labeled with integers F + 1, ... , M; other branches are labeled with

M + 1, ... , B. If the fan characteristic curves are approximated by second-degree polynomials

$$p_i = \alpha_i + \beta_i q_i + \gamma_i q_i^2,$$

$$i = 1, \ldots, F \qquad (10)$$

then the ventilation network problem can be stated in the following NLP form (Wang, 1984a).

Determine air quantities $q_1, \ldots, q_B$, which minimize the value of

$$U = (1/3) \sum_{i=1}^{B} r_i |q_i^3|$$

$$- \sum_{i=1}^{F} (\alpha_i q_i + \beta_i q_i^2$$

$$+ \gamma_i q_i^3) \qquad (11)$$

subject to

$$\sum_{j=1}^{B} a_{ij} q_j = 0,$$

$$i = 2, \ldots, N \qquad (12)$$

where r_i is the resistance factor and a_{ij} is an element of the incidence matrix.

Expressing the air quantities for tree branches as a linear combination of those for the chords, we obtain

$$q_i = \sum_{k=1}^{M} b_{ki} q_k,$$

$$i = M + 1, \ldots, B \qquad (13)$$

where b_{ki} is an element of the fundamental mesh matrix with respect to the spanning tree (Wang, 1982a). Using Eq. (13), the NLP problem of Eqs. 11 and 12 is simplified to the minimization of the following function of M air quantities:

$$U = (1/3) [\sum_{i=1}^{M} r_i |q_i^3|$$

$$+ \sum_{i=M+1}^{B} r_i |(\sum_{k=1}^{M} b_{ki} q_k)^3 |]$$

$$- \sum_{i=1}^{F} [\alpha_i q_i + (1/2)\beta_i q_i^2$$

$$+ (1/3)\gamma_i q_i^3]. \qquad (14)$$

In this form, the minimization is an unconstrained nonlinear optimization problem. In summary, the solution to the problem of mine ventilation networks with natural splitting can be obtained by first determining the M air quantities such that Eq. 14 is minimized, and then calculating the air quantities of the remaining B − M or N − 1 branches by Eq. 13 and the fan pressures by Eq. 10.

An application of the conjugate gradient algorithm (Fletcher and Reeves, 1964) to minimize Eq. 14 has been reported by Ueng and Wang (1984). Their results clearly demonstrate the applicability of the algorithm to this mine ventilation problem, although the effectiveness of the algorithm compared to other methods has yet to be evaluated.

4.2 Networks with fixed-quantity branches

A fixed-quantity branch is a branch whose air quantity q_i is assigned or given. Because the air quantity is assigned and known, the direction of a fixed-quantity branch can always be chosen so that $q_i > 0$. In the conventional approach, except for the fixed-quantity branches, the unknowns and knowns in the problem of networks with fixed-quantity branches are the same as those in the problem of networks with natural splitting. The unknown to be solved for a fixed-quantity branch is either the regulator pressure loss s_i or the fan pressure p_i, where $s_i \geq 0$, $p_i \geq 0$, and $(s_i)(p_i) = 0$. The fan pressure p_i for the fixed-quantity branch is not treated as a function of q_i in the solution process.

The solution to the problem of networks with fixed-quantity branches is generally obtained by choosing the fixed-quantity branches as chords with respect to a spanning tree and using the Hardy Cross method, but excluding the meshes containing the fixed-quantity branches from the iteration process (Wang, 1982b). The regulator pressure loss or the booster fan pressure is computed for each fixed-quantity branch without considera-tion of the possibility of regulating the airflow in other branches.

A method that minimizes the air power of the network by increasing the resistances of non-fixed-quantity branches has recently been described by Wang and Yao (1984). Given the air quantity for each branch in a major cutset, the problem is to determine the increments of resistance (regulator resistance) and air quantities for branches that are not

in the major cutset so that KCL and KVL are satisfied and the required system air power is minimized. Because the air quantity is a given constant, the minimization of air power is equivalent to the minimization of system pressure loss or fan pressure and no fan curves are involved in the problem. The solution procedure, however, requires repeated application of the Hardy Cross method at every trial value of individual regulator resistance.

Beginning with the conventional solution to the problem with fixed-quantity branches, the direction of each branch is chosen to coincide with that of air quantity. Let us denote B and K + 1 as the numbers of branches in the network and in the major cutset, respectively; label the return dummy branch as branch B and the remaining branches in the major cutset as branches 1, ... , K. Next, choose a set of M paths (directed chains) from the inlet node to the outlet node; form the path matrix $[e_{ij}]$, where e_{ij} is 1 or 0 depending on whether branch j is contained in path i. We now state the problem in the following NLP form:

Minimize

fan pressure p

subject to

$$\sum_{j=1}^{K} e_{ij} r_j q_j^2 + \sum_{j=K+1}^{B-1} e_{ij}(r_j + s_j) q_j^2$$

$$- p = 0 \qquad i = 1, \ldots , M \quad (15)$$

$$\sum_{j=1}^{B} a_{ij} q_j = 0 \qquad i = 2, \ldots , N \quad (16)$$

$$s_j \geq 0 \qquad j = K + 1, \ldots , B - 1$$
$$(17)$$

where s_j is the resistance of the regulator in branch j, and air quantities q_j (j = 1, ... , K) and airway resistances r_j (j = 1, ... , B − 1) are given.

4.3 Networks with controlled splitting

The problem of mine ventilation networks with controlled splitting considered by Hartman (1961: 140-143) is a classic problem that arises in mine ventilation planning. This problem, which is also known as the controlled-flow problem, deals with the use of fans and regulators to control the flow in a network for achieving a preassigned distribution of air. Air quantity and pressure loss for each branch are considered as known values. Therefore, KCL is assumed to be implicitly satisfied and KVL can be expressed as a set of linear constraints in terms of pressures. As a result, the controlled splitting problem does not have a unique solution.

By introducing the condition that the air power provided to (or consumed by) the ventilation system is to be minimized, Wang and Pana (1971) have formulated the problem in a standard linear programming (LP) format and demonstrated the solution by the simplex algorithm. They have generalized the problem to include booster fans. However, if booster fans are not considered in the formulation, the longest-path algorithm of the critical path method (CPM), as discussed in Wang (1982a, 1982b) and Wang and Mutmansky (1982), offers a simpler and more efficient solution. The CPM approach provides a simple arithmetic procedure for analyzing the problem. It is suitable for hand calculations, which can be conveniently performed on a drawing of the ventilation network. Knowing a longest-path solution, other solutions including those with booster fans can be explored by cutset operations.

The application of the out-of-kilter algorithm to the controlled-flow problem have been presented by Johnson et al. (1982) and Barnes and Johnson (1983). The out-of-kilter algorithm is considered to be the most general and widely used algorithm when dealing with capacitated, deterministic network flows. The algorithm is carefully developed using the concepts of LP duality theory and complementary slackness conditions (Phillips and Garcia-Diaz, 1981). The dual formulation of the controlled flow problem is given below, using the node numbers to denote the branches and associated variables and parameters.

Minimize

$$w = - \sum_{ij} h_{ij} x_{ij} \qquad (18)$$

subject to

$$\sum_j x_{ij} - \sum_j x_{ji} = 0$$

$$\text{for all nodes } i \qquad (19)$$

$$L_{ij} \le x_{ij} \le U_{ij}$$

$$\qquad (20)$$

$$\text{for all branches } (i,j)$$

where x_{ij} is the dual variable for branch (i,j); h_{ij} is the pressure loss for branch (i,j), excluding the loss for the regulator; and L_{ij} and U_{ij} are, respectively, upper and lower bounds for x_{ij}. Equation 18 represents the negative of the sum of the products of h_{ij} and x_{ij} for all branches (i,j). Furthermore, the upper and lower bounds in Eq. 20 are defined as follows:

$L_{ij} = 0$ for all branches allowing a regulator

$L_{ij} = - \infty$ for all branches not allowing a regulator

$U_{ij} = q_{ij}$ for all branches allowing a fan

$U_{ij} = + \infty$ for all branches not allowing a fan

where q_{ij} is the given air quantity for branch (i,j).

The application of the out-of-kilter solution technique will yield the optimal values for dual variables x_{ij} and nodal pressures n_i. The optimal fan pressures p_{ij} and regulator pressure losses v_{ij} are then determined as follows:

If $x_{ij} = L_{ij}$, then

$$p_{ij} = 0 \text{ and } v_{ij} = n_i - n_j - h_{ij}$$

If $x_{ij} = U_{ij}$, then

$$p_{ij} = n_j - n_i + h_{ij} \text{ and } v_{ij} = 0$$

If $L_{ij} < x_{ij} < U_{ij}$, then

$$p_{ij} = 0 \text{ and } v_{ij} = 0.$$

5 SUMMARY AND CONCLUSIONS

In recent years, a number of developments in the analysis of mine ventilation network problems has occurred, resulting in an increased number of techniques for mine ventilation network analysis. Some of these procedures have simply been concepts from the general theory of networks adapted for application to ventilation networks while other techniques, particularly the mathematical programming techniques, have been expanded and developed from previous procedures to be applied to ventilation problems.

An important advance that has been made in the past several years is the use of cutsets in the analysis of mine ventilation circuits, particularly in the placement of fans and regulators in a controlled-flow ventilation network. Other advancements have contributed to the fundamental understanding of multiple-fan networks through the development of subsystem concepts and methods for determining the operating points for fans in these systems. In addition, the application of nonlinear programming techniques and the out-of-kilter algorithm offer diversity in the applicable analysis procedures. In conclusion, the methods developed in recent years from more general network and mathematical programming procedures have contributed to a better understanding of mine ventilation networks and a greater number of solution methods for the mine ventilation engineer.

6 REFERENCES

Barnes, R. J. & T.B. Johnson, 1983. The out-of-kilter algorithm and its applications in mining engineering. SME Preprint No. 83-102, SME-AIME Annual Meeting, Atlanta, GA.

Collins, M., L. Cooper, R. Helgason, J. Kennington & L. LeBlanc 1978. Solving the pipe network analysis problem using optimization techniques. Management Science 24:747-760.

Cross, H. 1936. Analysis of flow in networks of conduits or conductors. Bull. 286, Engineering Experiment Station, University of Illinois.

Fletcher, R. & C.M. Reeves 1964. Function minimization by conjugate gradients. Computer Journal 7:149-154.

Hartman, H.L. 1961. Mine ventilation and air conditioning. New York: Ronald Press.

Johnson, T.B., R.J. Barnes & R.J. King 1982. The optimal controlled flow mine ventilation problem: an operations research approach. SME Preprint No. 82-353, First International SME-AIME Fall Meeting, Honolulu, HI.

Millar, W. 1951. Some general theorems for non-linear systems possessing resistance. Philosophical Magazine, 7th Series 42:1150-1160.

Mutmansky, J.M., Y.J. Wang & H.L. Hartman 1984. A comprehensive method for determining the operating point for a mine fan installation. SME Preprint No. 84-407, SME-AIME Fall Meeting and Exhibit, Denver, CO.

Phillips, D.T. & A. Garcia-Diaz 1981. Fundamentals of network analysis. Englewood Cliffs, NJ: Prentice-Hall.

Scott, D.R. & F.B. Hinsley 1951. Ventilation network theory. Colliery Engineering 28:67-71, 159-166, 229-235, 497-500.

Ueng, T.H. & Y.J. Wang 1984. Analysis of mine ventilation networks using non-linear programming techniques. International Journal of Mining Engineering 2:245-252.

Wang, Y.J. 1982a. A critical path approach to mine ventilation networks with controlled flow. Trans. SME-AIME 272:1862-1872.

Wang, Y.J. 1982b. Ventilation network theory. In H.L. Hartman, J.M. Mutmansky & Y.J. Wang (eds.), Mine ventilation and air conditioning, 2nd ed., pp. 483-516. New York: John Wiley.

Wang, Y.J. 1983. Minimizing power consumption in multiple-fan networks by equalizing fan pressure. International Journal of Rock Mechanics and Mining Sciences 20:171-179.

Wang, Y.J. 1984a. A nonlinear programming formulation for mine ventilation networks with natural splitting. International Journal of Rock Mechanics and Mining Sciences 21:43-45.

Wang, Y.J. 1984b. Characteristics of multiple-fan ventilation networks. International Journal of Mining Engineering 2:229-243.

Wang, Y.J. 1985. Graphical representation of the operating points for two-fan ventilation systems. SME Preprint No. 85-80, AIME Annual Meeting, New York.

Wang, Y.J. & J.M. Mutmansky 1982. Application of CPM procedures in mine ventilation. In H.L. Hartman (ed.), Proceedings of the 1st mine ventilation symposium, pp. 159-168. New York: SME-AIME.

Wang, Y.J. & M.T. Pana 1971. Solving mine ventilation network problems by linear programming. SME Preprint No. 71AU132, AIME Annual Meeting, New York.

Wang, Z. & Yao, E.Y. 1984. Optimum method of regulating a ventilation network. In M.J. Howes & M.J. Jones (eds.), Proceedings of third international mine ventilation congress, pp. 53-55. London: Institution of Mining and Metallurgy.

Fixed flow network solutions using linear equation and linear equivalence

W.H.GRIFFIN
University of Alberta, Edmonton, Canada

ABSTRACT: The linear equation method (Wood 1972) for solution of mine ventilation networks is complemented by the addition of linear equivalence theory (Chojcan 1975) for estimation of the first order network sensitivities. The methods are illustrated by the solution of flow distributions in a simple network. Extensions to more complex networks are indicated. A short APL program is presented to illustrate the method.

1. INTRODUCTION

The mathematical basis for solution of ventilation network problems by linear equation iteration is introduced. First order sensitivity, as defined by variance propagation (Hahn and Shapiro, 1967), is obtained using linear equivalence. The simple network used to illustrate these methods, when subjected to Monte Carlo simulation, verifies the linear equivalence method and gives a measure of higher order effects.

2. LINEAR THOERY METHOD

In a network of n branches with m nodes and ℓ fundamental meshes one can write n-1 linear flow continuity equations

$$\sum_{i=1}^{n} Q_i \cdot a_{ji} = E_j \quad (j = 1,2...m-1)$$

$$(1)$$

and ℓ non-linear energy equations

$$\sum_{i=1}^{n} R_i \cdot b_{ki} \cdot Q_i \cdot |Q_i| = H_k$$

$$(k = 1,2...\ell) \quad (2)$$

where, Q_i = flow of the ith branch

E_j = external flow into node j

H_k = internal head applied to mesh k

R_i = resistance of the ith branch

a_{ji} = -1 if branch **originates** at node j

a_{ji} = 1 if branch i terminates at node j

a_{ji} = o otherwise

b_{ki} = 1 if branch i flows clockwise in mesh k

b_{ki} = -1 if branch i flows counterclockwise in mesh k

b_{ki} = o if branch i is not in mesh k

Due to the non-linearity of ℓ of the n equations one must solve the system iteratively. Linear theory (Wood 1972), accomplishes this by the following algorithm:
1. Choose any non-zero initial flow estimate. $Q_{i,o} = 1$ (i = 1,2...n) is acceptable.
2. Rewrite (2) in linear form. That is

$$\sum_{i=1}^{n} R_i' \cdot b_{ki} \cdot Q_{i,1} = 0$$

$$(k = 1,2...\ell) \quad (3)$$

where, $R_i' = R_i \cdot |Q_{i,p-1}|$

3. Combine (1) and (3) and solve the $m - 1 + k = n$ linear equations for $Q_{i,p}$ $(i = 1, 2 \ldots n)$ unknowns.

4. Compare $Q_{i,p}$ with $Q_{i,p-1}$. If within desired accuracy stop, if not replace $Q_{i,p-1}$ with $(Q_{i,p} + Q_{i,p-1})/2$ and recycle to 2.

A compact matrix representation of the above process is

$$Q_p = [M_p]^{-1} \cdot E \qquad (4)$$

where, $Q_p = \begin{bmatrix} Q_{1p} \\ Q_{2p} \\ \vdots \\ Q_{np} \end{bmatrix}$ The vector of n flow distributions at the pth iteration.

(If $p \leq 0$, $Q_i = 1$.)

$E = \begin{bmatrix} E_1 \\ E_2 \\ \vdots \\ E_{m-1} \\ H_1 \\ H_2 \\ \vdots \\ H_\ell \end{bmatrix}$ The vector of external nodal flows and internal mesh heads.

$M_p = \begin{bmatrix} A' \\ B_p \end{bmatrix}$ An n x n matrix whose first m-1 rows are the a_{ji} of (1) and whose remaining ℓ elements are $b_p(ki) = b_{ki} \cdot R_i (Q_{i,p-1} + Q_{i,p-2})/2$.

Lines 81 to 85 of the APL program, Fig. 1, incorporate the above algorithm with the stopping criteria at line 83 set such that exit is made when all flows compare with respect to the comparison tolerance set in the APL interpreter.

The major advantage of linear theory is that convergence to a solution is certain, independent of initial flow estimates and unaffected by the choice of fundamental mesh components. The calculation process, if not optimized to take advantage of the sparse and ordered matrices, could in particular cases be less efficient than current methods. In general, however, as computation costs decrease, linear theory should be the preferred method.

3. SENSITIVITY

In any calculation process one must be prepared to estimate the sensitivity of the result with respect to variations in the data. The evaluation of central moments of the result as a function of the calculation algorithm and the estimated moments of the data is one of the least subjective and generally applicable methods of sensitivity testing. In this method it is assumed that while the calculation algorithm is exact, the data is subject to chance variation or error. This error is quantified by reference to data moments, i.e. expectations of the positive integral power and cross products of all possible data values. The propagation of error in the result is approximated by expressing the calculation function as a Taylor series, taking the requisite expectations and truncating at some derivative and/or expectation order.

The general expressions, where data errors are uncorrelated and derivative terms of third and higher order are neglected, are (Hahn and Shapiro Eq. 7-3, 7-6)

$$E(F(x_1, x_2 \ldots x_n) \approx F(\bar{x}_1, \bar{x}_2 \ldots \bar{x}_n) + \tfrac{1}{2}$$

$$\sum_{i=1}^{n} \frac{\partial^2 F}{\partial x_i^2} \cdot VAR(x_i) \qquad (5)$$

$$VAR(F(x_i, x_2 \ldots x_n) \approx \sum_{i=1}^{n} \left(\frac{\partial F}{\partial x_i}\right)^2 VAR(x_i)$$

$$+ \sum_{i=1}^{n} \left(\frac{\partial F}{\partial x_i}\right)\left(\frac{\partial^2 F}{\partial x_i^2}\right)\mu_3(x_i) \qquad (6)$$

In engineering practice the second order derivatives are usually neglected and error is assumed proportional to the mean. That is,

$$C(x_i) = \frac{\sqrt{VAR(x_i)}}{\bar{x}_i} \qquad (7)$$

where $C(x_i)$ is known as the coefficient of variation. With the additional assumption of a common coefficient of variation

$$C(x_i) = C(x_j) = C(x) \qquad (8)$$

equations (5) and (6) reduce to

$$E(F(x_1,x_2 \ldots x_n)) \approx F(\bar{x}_1, x_2 \ldots \bar{x}_n) \qquad (9)$$

$$VAR(F(x_1,x_2 \ldots x_n)) \approx C^2(x) \sum_{i=1}^{n} (\frac{\partial F}{\partial x_i})^2 (\bar{x}_i)^2 \qquad (10)$$

The combination of (9) and (10) gives the coefficient of variation of the result.

$$C(F) = C(x) \cdot \sqrt{\frac{\sum_{i=1}^{n} (\frac{\partial F}{\partial x_i})^2 \bar{x}_i^2}{F(\bar{x}_i, \bar{x}_2 \ldots \bar{x}_n)}} \qquad (11)$$

An additional assumption of normality gives the two tailed 95% confidence interval as

$$F(\bar{x}_1, \bar{x}_2 \ldots \bar{x}_n) \cdot (1 \pm C(F) \times 1.96) \qquad (12)$$

For example, with a fixed flow Q_T to a parallel network with n equal and independent resistances having a common coefficient of variation $C(R)$, the application of (11) to the flow distribution equation

$$Q_i = \frac{Q_T \, R_i^{-1/2}}{\sum_{i=1}^{n} R_i^{-1/2}} \qquad (13)$$

gives

$$C(Q_i) = \frac{C(R)}{2} \sqrt{\frac{n-1}{n}} \qquad (14)$$

Unfortunately general network problems do not have closed form equations from which partial derivatives can be evaluated directly, however a calculation algorithm is available. Chojcan using the Kirchoff nature of ventilation networks and Tellegen's theorem has developed a method he calls linked network or linear equivalence that will calculate the first order partial derivatives for any of the variables in a general ventilation network.

The method dovetails very nicely with the linear theory method of network solution and is illustrated by application to the fixed flow problem.

For the network in which the external flows are fixed and internal heads are generated solely by branch resistances, the first order partials of flow distri-

bution with respect to resistance are (Chojcan Eq. 6)

$$\frac{\partial Q_i}{\partial R_j} = -Q^*_{j,i} \cdot Q_j \cdot |Q_j|$$

$$(j = 1,2 \ldots n)$$
$$(i = 1,2 \ldots n) \qquad (15)$$

The vector $Q^*_{j,i}$ (j = 1,2...n) is the "flow" in a network whose (a) topology is exactly equal to the original network to which $Q_j, j = 1,2 \ldots n$ is the solution, (b) "head loss" is a linear function of Q^* and resistance, (c) $R^*_j = 2 \cdot R_j \cdot Q_j$, (d) external flow is zero, (e) a source of constant head of unit magnitude exists in branch i.

It is easily seen that the solution of (15) under these conditions is equivalent to the solution of (4) with

(a) $\quad Q^*_i = Q_p$

(b) $\quad A^* = A'$

(c) $\quad b^*_{p(k,i)} = b_{k,i} \cdot R_i \cdot Q_i \cdot 2$

$\qquad (k = 1,2 \ldots \ell)$
$\qquad (i = 1,2 \ldots n)$

(d) $\quad E^{*T}_i = [0, 0 \ldots 0(m-1), b_{1,i},$

$\qquad b_{2,i} \ldots b_{\ell,i}]$

As the matrix M^*_p is unchanged for all i one can form the complete Jacobian of $\frac{\partial Q_i}{\partial R_j}$ by one matrix inversion and multiplication with n right-hand sides. These right-hand sides are the primary mesh incidence matrix B preceded by m-1 rows of zeros.

Line 93 of Fig. 1 gives the complete algebra for the solution of Q^*. Line 94 multiplies out the $Q_j \cdot |Q_j|$ term and inserts the R_i term as required in (10). Note that as the differential component in (10) is squared we need not retain the sign of the differential by multiplication of Q^* with $-Q_j \cdot |Q_j|$.

Line 95 calculates standard deviation and multiplies by 1.96 for 95% standardized confidence interval. Line 96 adds and subtracts this interval from the calculated mean to output the actual 95% limits.

4. TESTING

While the calculation of flow distributions using linear theory are easily verified, the arithmetic verification of the sensitivity using linear equivalence, even for the simplest topology, is very difficult. One may, however, use Monte Carlo simulation to cross validate the results for a given network. While correspondence of results for one topology does not prove that the method is valid for all networks it does lend some credance to the results.

It therefore was decided to simulate a simple fixed flow network, Fig. 2, whose resistances are log normal variates. The reason for the log normal choice are; (a) resistances cannot be negative and (b) chance conditions that tend to increase resistances will out number those the decriment. Simulation was done using the SLAM (Pritsker and Pegen 1979), system driven by single event that chose a random log normal variate for each resistance and then calculated the flow distribution for each branch using linear theory iteration. This event was repeated 500 times with frequency distribution, mean, standard deviation, coefficient of variation, maximum and minimum collected for each internal flow. Simulation runs were done with common resistance coefficients of variation at 10% intervals from 10% to 80% with each run repeated with a second random number seed for a total of 16 runs. The mean statistics with the $\frac{\partial Q_i}{\partial R_j}$ calculated at the mean and, from $\frac{\partial Q_i}{\partial R_j}$ and $\bar{R}_j$, the

calculated 95% limits for the mean at 40% resistance coefficient of variation are given in Table 1. The frequency distributions of the simulated flows when plotted on normal probability paper were invariably linear. The 95% confidence intervals given on the final lines of Table 1 were interpolated from probability plots of a simulation run at 40% resistance coefficient of variation. The correspondence between linear equivalence theory and simulation is clear.

To correlate theory with simulation over a range of resistance variation the simulation sample coefficient of variation statistics for the flows were regressed against the resistance co-

efficients of variation. An orthogonal polynomial regression model with the independent variable weighted inversely proportional to its magnitude and an intercept constrained to zero was run using the BMDP (Dixon 1983), system. It was found that the linear term was highly significant followed by significant quadratic term. The linear term should be equal to

$$\sqrt{\sum_{j=1}^{n} \left(\frac{\partial Q_i}{\partial R_j} \cdot R_j\right)^2} \div Q_i$$

if (10) is a reasonable approximation of the network sensitivity to fixed flow. Comparison of Table 2 columns 2 and 3 shows that, overall, the calculation of sensitivity via linear theory is slightly conservative. The quadratic regression coefficients of columns 4 and 5 arise from the second order differentials that were dropped going from (6) to (10). Note that a similar second order term was discarded going from (5) to (9). With a common resistance coefficient of variation $C(R)$ the difference between (5) and (9) in the fixed flow case is

$$Q_i(5) - Q_i(9) = \frac{1}{2} C^2(R) \sum_{j=1}^{n} \frac{\partial^2 Q}{\partial R_j^2} \cdot \bar{R}_j^2$$

$$(16)$$

An estimate of $\sum_{j=1}^{n} \frac{\partial^2 Q_i}{\partial R_j^2} \cdot \bar{R}_j^2$ can be

made from the simulation results by regressing the differences of the mean flows at various levels of $C(R)$ from those at $C(R) = 0$ against $\frac{1}{2} C^2(R)$. This was done for the 16 simulation runs using the same polynomial models employed previously. The results were generally only linearly significant. The linear coefficient estimates given in column 6 of Table 2, can be thought of as mean flow adjustments to be made with respect to a resistance coefficient of variation at 100%. Remembering that the factor of $\frac{1}{2}$ must also be applied, it is seen that at $C(R) = 100\%$ the departure of mean flows from that at $C(R) = 0$ in most branches will be less than 10%. In branch 10 however, the effects of resistance variation will give on average a mean flow 25% below that of exact calculations. When we note that branch 10 has the highest $C(Q_i)$ coefficients, it is

evident that this branch is the most
sensitive part of the network. Control
of this flow will be difficult with
imprecise control of network resistances.

5. CONCLUSIONS

Mine ventilation network solution by
linear linear theory is nicely complemented by the linear equivalence solution
for first order sensitivity analysis.
While this calculated first order sensitivity will identify the major components affecting the flow distribution
in a fixed flow problem, higher order
terms are operative even in simple
networks. Direct Monte Carlo simulation
is, at present, a possible, but expensive
method for quantification of such
effects. Further mathematical research
in the style of the linear equivalence
method could be rewarding.

REFERENCES

Chojcan, J. "A method of calculating
the effect of changes in the value
of ventilation network elements
upon air flow intensity in mine
workings applying a so-called
'linked' network". Prezglad Gorniczy;
31, (1); 1975; pp. 17-24. In Polish.
(English translation, Central Index
of Translations No. T 3875/BGC/1977.)

Dixon, W.J. (ed). "BMPD Statistical
Software". University of California
Press, (1983).

Hahn, Gerald J., Shapiro, Samuel S,
"Statistical Models in Engineering".
Wiley, N. York, (1967).

Wood, Don J., Charles, Carl O.A.
"Hydraulic Network Analysis Using
Linear Theory". Journal of the
Hydraulics Division, Proc. ASCE,
Vol. 98, No. Hy 7, Proc. paper 9031,
July 1972, p. 1157-1170.

```
Z+RECVQ QNWFLW AB;N;W;L;R;E;CV;QO;Q2;A;B;LHSBR
∆ TO CALCULATE THE AIRFLOW IN THE BRANCHES OF A NETWORK WHOSE EXTERNAL
∆ FLOW IS FIXED AND WHOSE RESISTANCES ARE KNOWN TO A COEFFICIENT OF
∆ VARIATION.
∆**********************************************************************
∆ INPUT
∆ AB IS A N+1 ROWS BY N COLUMNS MATRIX OF +1,‾1 OR O.N IS THE NUMBER OF
∆ BRANCHES OF THE NETWORK.THE FIRST W ROWS ARE THE BRANCH INCIDENCE
∆ MATRIX WITH W THE NUMBER OF NODES IN THE NETWORK.LET A+THIS MATRIX.
∆ IF A[I;J]=1 THEN BRANCH J TERMINATES AT NODE I.IF A[I;J]=‾1 THEN
∆ BRANCH J EMANATES FROM NODE I.OTHERWISE A[I;J]=0.
∆ THE LAST N-W+1=L ROWS OF AB ARE THE MESH-BRANCH DIRECTIONALITY
∆ MATRIX B.IF B[I;J]=+1 BRANCH J FLOWS CLOCKWISE IN MESH I.IF B[I;J]=‾1
∆ BRANCH J FLOWS COUNTERCLOCKWISE IN MESH I.OTHERWISE B[I;J]=0.
∆ TESTS FOR THE VALIDITY OF A AND B ARE AS FOLLOWS:
∆ 1. THE COLUMN SUMS OF A=0. SINGLE BRANCHES ONLY BETWEEN NODES.
∆ 2. THE COLUMN SUMS OF |A|=2. THERE ARE NO DANGLING NODES.
∆ 3. EVERY BRANCH OCCURS AT LEAST ONCE IN B.THE MESHES ARE COMPLETE.
∆ 4. THE INNER PRODUCT A+.×B IS NULL.WITH 1 2 AND 3 THIS TEST SATISFIES
∆    THE KIRCHOFF TOPOLOGY REQUIREMENTS FOR A DIRECTED FLOW NETWORK.
∆
∆ RECVQ IS CONCATENATED VECTOR OF 2N+W+1 LENGTH OF RESISTANCE,EXTERNAL
∆ FLOW,COEFFICIENT OF VARIATION FOR RESISTANCES AND INTIAL FLOW
∆ ESTIMATE.LET THE FIRST W MEMBERS OF RECVQ BE R THE RESISTANCE FACTOR
∆ OF THE W BRANCHES. ALL R[I]'S MUST BE GREATER THAN ZERO AND OF SUCH
∆ UNITS THE TURBULENT FLOW EQUATION H=R×Q×|Q| IS SATISFIED.THE NEXT W
∆ MEMBERS OF RECVQ ARE E THE EXTERNAL FLOW INTO OR OUT OF THE W NODES
∆ OF THE NETWORK.THE SUM OF THE COMPONENTS OF E MUST BE ZERO (BALANCED
∆ FLOW) AND THE VECTOR E MUST NOT BE NULL (NON ZERO FLOW).
∆ IF THE EXTERNAL FLOW IS INTO NODE I THEN E[I]<0.
∆ IF THE EXTERNAL FLOW IS OUT OF NODE I THEN E[I]>0.
∆ OTHERWISE E[I]=0.
∆ RECVQ[N+W+1] IS THE COEFFICIENT OF VARIATION FOR ALL RESISTANCES.
∆ THIS VALUE MUST ≥0. ZERO INDICATES EXACT VALUES ARE KNOWN FOR THE
∆ RESISTANCES. THE LAST W MEMBERS OF RECVQ ARE THE INITIAL ESTIMATE
∆ OF THE FLOW DISTRIBUTION IN THE W BRANCHES.THE CLOSER THESE VALUES
∆ TO THE SOLUTION THE FEWER ITERATIONS REQUIRED TO SOLUTION.HOWEVER
∆ ANY NON NULL VECTOR IS PERMISSABLE.
∆
∆**********************************************************************
∆ OUTPUT
∆ AN EXPLICIT 2 ROWS BY W COLUMNS ARRAY OF THE INDIVIDUAL 95 PCT
∆ CONFIDENCE INTERVALS OF THE BRANCH FLOW DISTRIBUTION IS GIVEN.UNITS
∆ ARE THOSE OF THE EXTERNAL FLOW VECTOR.
∆
∆**********************************************************************
∆
∆ UNRAVEL THE INPUT.
∆
N+ρAB
+((W[1]-1)≠W[2]j)/ER1
N+W[2]
W+(⍴RECVQ)-1+2×N
L+N+1-W
R+N+RECVQ
E+N+N+RECVQ
CV+RECVQ[N+W+1]
QO+(-W)+RECVQ
A+(W,N)+AB
B+(-L),W)+AB
∆ TEST INPUT FOR VALIDITY.
∆
+((|/R)≤0)/ER2
+((∨/1E)=0)/ER3
+((+/E)≠0)/ER3
+(CV<0)/ER4
+((+/|QO)=0)/ER5
+((+/+/|↓A)≠0)/ER6
+((+/((+/|↓A)=2))≠N)/ER6
+((1+/1B)≠0)/ER7
+((+/+/A+.×B)≠0)/ER8
∆
∆ SET UP THE RHS AND THE CONSTANT PART OF THE LEFT HAND SIDE.
E+(‾1+E),L↓0
LHSBR+B×(L,W)⍴R
∆ DELETE THE LAST ROW OF A.
A+(‾1,0)↓A
∆ ITERATE LINEAR SOLUTION TO THE QUADRATIC EQUATION SET.THE SOLUTION
∆ IS COMPLETE WHEN SEQUENTIAL TRIAL FLOWS VARY BY LESS THAN FUZZ.
∆ TO REDUCE SOLUTION TIME ONE CAN RESET FUZZ EXTERNALLY USING ⎕CT←.
∆
ST1:Q2+E⌹A,[1] LHSBR×(L,W)⍴|QO
Z+(Q2+QO)÷2
+((+/QO=Q2)=W)/ST2
QO+Z
+ST1
ST2:W+W-1
∆ CALCULATE PARTIAL DIFFERENTIALS OF Q WITH RESPECT TO R BY USE OF
∆ LINKED NETWORK THEORY.USE THESE PARTIALS IN FIRST ORDER VARIANCE
∆ COMPONENT ESTIMATION TO ESTIMATE STANDARD DEVIATIONS OF THE FLOWS.
∆ NORMAL THEORY VIA AN APPEAL TO THE CENTRAL LIMIT THEOREM CONCLUDES
∆ SETTING OF 95 PCT CONFIDENCE LIMITS AND PROGRAM EXIT.
∆
QO+(((W,W)⍴0),[1] B)⌹A,[1] B×(L,W)⍴R×Z×2
QO+(½QO)×(W,W)⍴R×Z×Z
QO+((+/QO×QO)×0.5)×CV×1.96
Z+(2,W)⍴(Z+QO),Z-QO
+0
∆ ALL THE FOLLOWING STATEMENTS ARE INPUT ERROR EXITS WHICH FILL THE
∆ VECTOR WITH CHARACTERS AS WELL AS WRITING TO THE SCREEN.
∆
ER1:⎕+Z+'DIMENSIONS OF AB FAULTY'
+0
ER2:⎕+Z+'NEGATIVE OR ZERO RESISTANCES'
+0
ER3:⎕+Z+'INVALID EXTERNAL FLOW'
+0
ER4:⎕+Z+'COV NEGATIVE'
+0
ER5:⎕+Z+'FLOW ESTIMATE IS NULL'
+0
ER6:⎕+Z+'INVALID BRANCH INCIDENCE MATRIX'
+0
ER7:⎕+Z+'INVALID MESH MATRIX'
+0
ER8:⎕+Z+'MATRIX PRODUCT A+.×B IS NOT NULL'
```

Fig. 1:   APL Program for a Fixed Flow
          Ventilation Network

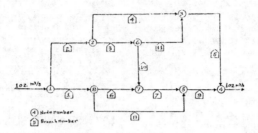

Fig. 2: Simple Fixed Flow Network

					Branch No.						
	1	2	3	4	5	6	7	8	9	10	11
$R_i\ (\frac{P \cdot s^2}{m^5}) \times 10^2$	1.06	9.12	22.3	363.	33.2	31.7	12.2	34.9	2.62	149.	91.2
$\bar{Q}_i$ (m³/s)	67.8	34.2	24.3	9.93	25.7	30.6	39.1	37.2	76.3	8.49	15.8
$\Delta'_i\ (P_u)$	48.7	107.	132.	358.	219.	297.	187.	483.	153.	109.	228.
$\frac{\partial Q_i}{\partial R_j}$ i = 1	-108.9	27.71	11.10	0.4770	6.515	-12.74	0.3430	-13.96	-57.38	0.9977	1.250
i = 2	108.9	-27.71	-11.10	-0.4770	-6.515	12.74	-0.3430	13.96	57.38	-0.9977	-1.250
i = 3	86.68	-22.06	-14.63	0.5902	-2.935	10.69	-4.572	10.29	25.85	-1.040	-2.593
i = 4	22.22	-5.654	3.522	-1.067	-3.580	2.045	4.229	3.667	31.53	0.042	1.343
i = 5	45.31	-11.53	-2.615	-0.5346	-11.45	3.150	16.55	8.987	100.8	0.5386	-2.963
i = 6	-62.56	15.92	6.727	0.2156	2.224	-24.74	-24.73	17.75	-19.59	0.7386	0.2037
i = 7	1.032	-0.2626	-1.762	0.2732	7.160	-15.15	-41.62	22.73	-63.96	-0.7975	2.007
i = 8	-46.35	11.79	4.377	0.2614	4.291	12.00	25.07	-31.70	-37.79	0.2589	0.9563
i = 9	-45.31	11.53	2.615	0.5346	11.45	-3.150	-16.55	-8.967	-100.8	-0.5386	2.963
i = 10	63.59	-16.18	-8.489	0.0576	4.936	9.588	-16.89	4.969	-43.47	-1.530	1.713
C(R) = 40% i = 11	23.09	-5.876	-6.137	.5327	-7.871	1.105	12.37	5.320	69.32	0.4965	-4.306
Calc. = 95% upper m³/s	74.2	40.6	30.0	13.6	31.2	39.1	47.9	46.9	81.8	13.1	20.6
Conf. Int. lower m³/s	61.4	27.8	18.5	6.3	20.2	22.1	30.3	27.5	70.8	3.9	10.9
Sim. 95% upper m³/s	72.9	39.6	29.4	13.7	31.6	38.2	46.0	45.6	80.4	13.0	21.0
Conf. Int. lower m³/s	60.7	27.9	18.5	6.8	20.6	22.1	30.8	29.7	70.2	3.1	11.0

Table 1: Flow Distribution of Network

Branch No.	$\frac{\sum_{j=1}^{n} (\frac{\partial Q_i}{\partial R_j} \cdot R_j)^2}{Q_i}$	Regression Results $C(Q_i) = a \cdot C(R) + b \cdot C(R)^2$		Reg. Est. of $\sum_{j=1}^{n} \frac{\partial^2 Q_i}{\partial R_j^2} \cdot R_j^2$ m³/s
		Linear only a =	Linear + Quadratic a = b =	
1	0.121	0.120	0.139 -4.4×10^{-4}	1.55
2	0.240	0.232	0.252 -4.6×10^{-4}	-1.54
3	0.303	0.298	0.318 -4.4×10^{-4}	3.12
4	0.471	0.447	0.485 -8.5×10^{-4}	1.65
5	0.274	0.259	0.281 -4.9×10^{-4}	2.81
6	0.352	0.323	0.357 -7.4×10^{-4}	-0.712
7	0.287	0.263	0.285 -5.0×10^{-4}	-5.02
8	0.333	0.296	0.333 -8.0×10^{-4}	2.21
9	0.092	0.094	0.111 -3.7×10^{-4}	-2.85
10	0.691	0.723	0.662 -1.4×10^{-3}	-4.46
11	0.390	0.376	0.398 -5.0×10^{-4}	1.16

Table 2: Test Results from SLAM Simulation
of Network Figure 2

17. Miscellaneous

Tunnel boring machines in Norway
Ventilation and dust problems

TOM MYRAN
University of Trondheim, Norway

ABSTRACT: This paper describes the ventilation and dust problems, and requirements to maintain adequate environmental conditions at full-face, hard rock tunnel boring machines (TBM) in Norway. Special references are made to a 6,3 m diameter TBM operating in rocks with a high content of free silica.

1 INTRODUCTION

Several types of full-face, hard rock boring machines have been introduced in Norway, namely tunnel and blindhole borers. The first full-face tunnel boring machine (TBM) was put into operation in Norway in 1972, and until to-day approximately 140 kilometres are excavated. In 1984 15,4 kilometres were excavated. This makes a total of 0,32 million cubic metres of solid rocks, and by volume about 10% of the total excavated volume, a trend expected to continue. The rocks bored are often hard, with a high content of free silica. This creates considerable dust problems due to the health risk.

Five full-face tunneling machines are presently in operation in Norway. Many problems had to be solved before an adequate environmental standard of full-face drivages can be achieved. One of greatest problems connected with TBM drivages in Norway is overcoming the dust problems. Other environmental problems are noise, vibration and heat dissipation.

Tunnelling work is characterized by more frequent occurence of accidents and problems of occupational diseases caused by dust and vibration compared to other industries. The contractors are compulsorily required to provide special training for those workers who work in dust tunnelling and lining. Each worker must be qualified for the work he is involved in.

2 THRESHOLD LIMIT VALUES

The threshold limit values (TLV) for air pollutants in Norway are with few exceptions, based on the list from ACGIH (American Conference of Governmental Industrial Hygienists). For quartz, however, the TLV in Norway are (1985):

Total dust   0,6 mg/m^3
Respirable dust : 0,2 mg/m^3

This is twice the american value. The Norwegian State Inspectorate has proposed to reduce these values down to 0,3 mg/m^3 and 0,1 mg/m^3 respectively, as from 1986, i.e. the same values as in the U.S.

3 TUNNEL BORER VENTILATION AND DUST EXTRACTION

3.1 Dust extraction

The whole cutter head of a TBM is closed off with a steel shield, rubber lined along the edges, to form a barrier preventing excess dust from being liberated into the air. Rock chips produced by the cutter head are loaded automatically through the dust shield onto a conveyor belt. At this point air is exhausted via a duct through dust extraction unit by means of fans.

Two different types of dust extraction units have been in use:

- Rotovent wet scrubber. The air laden with dust flows directly from the cutter head through one ore more parallell agglomeration zones, into which the washing medium is infused. Through contact between dust and washing liquid the wettened particles agglomerate. For separation of the muddy water from the air the mixture is put into a rotating flow in a radial diffusor.

- Turbofilter Jet-Bagfilter dry type self-cleaning fabricfilter. The dust laden air is exhausted from the cutter head into a chamber with many small filter bags. The filter bags are cleaned by an air pulse, and the dust is removed by a screw fed to chamber where it is mixed with water. This slurry is pumped out of the tunnel.

Some years ago the wet scrubber was dominating in Norway, but now it is the dry filter. The decision to use the self-cleaning dry filter instead of the wet scrubber was mainly made because of the superior properties of such a filter over the wet scrubber in the fields of efficiency, presure loss and maintenance.

3.2 Dust suppression

A number of different head spray design has been tried. At present sprays are for dust suppression and cutter cooling. Nozzles mounted on the cutter head and on the top of the steel shield spray chilled water onto the face being advanced and on the side walls. An automatic moved nozzles also sprays water over the TBM nearest the face. On some TBM's water is also supplied to the intake ventilation air and conveyor belts at transfer points.

Unlimited amounts of water cannot be used at the cutter head, because of the danger of excessive sludge on the machine conveyors and the problem of removing a large quantity of water from the face.

3.3 Ventilation

The schematic layout is shown in Fig. 1. The main force ventilation consists of an axial fan handling air from the start of the tunnel to the machine area.

The machine exhaust system draw air from the head of the machine into twin ducts mounted on each side of the conveyor. The ducts are joined into a single column,

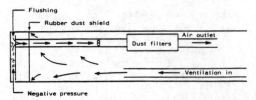

Fig. 1. General layout of ventilation and dust extraction unit.

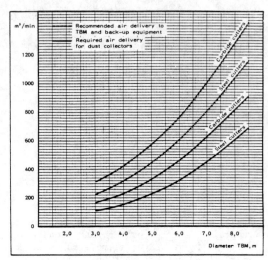

Fig. 2. Ventilation requirements for TBM and back-up equipment.

which carries dust laden air to the dust extraction unit.

Fig. 2 shows recommended air delivery to TBM and back-up equipment, and air exhaust capacity for dust collectors.

4 OBSERVATIONS

4.1 Dust

Dust and ventilation measurements have been taken since 1972 when the first TBM came into operation in Norway. The experiences from TBM with operating diameters from 2,3 metres up to 7,8 metres show that the dust situation generally is kept under control when operating in rocks with low content of free α-quarts (< 5%). With increasing quarts-contents the dust problems became more and more unacceptable.

In special cases the dust concentrations can be 1,5-5 times higher than the 8-hours TLV.

A direct comparison of dust measurements was made on two identical 3,0 m dia. TBM operating on opposite faces in the same tunnel, with closely related conditions concerning rock type, main ventilation capacity and drilling rate. A wet scrubber dust extraction unit was mounted on one TBM, and a dry filter on the other. The dust concentration recorded in the tunnel during operation have been satisfactory because of the very low quarts content in the dust (< 0,5%). Table 1 gives typical values of dust concentration.

Table 1. Dust concentration in 3,0 m dia. TBM tunnel.

Position	Dust concentration mg/m^3	
	Total dust	Respirable dust
WET SCRUBBER		
Operator's place	12,4	4,0
Behind back-up equipment	4,1	2,3
Locomotive, driver's place	6,1	2,7
DRY FILTER		
Operator's place	6,3	2,3
Behind back-up equipment	2,7	1,2
Locomotive, driver's place	2,0	0,9

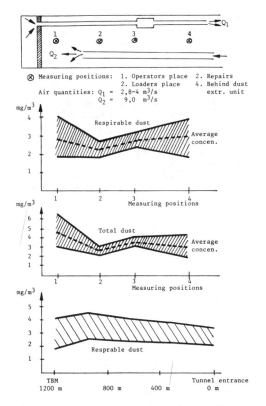

Fig. 3. Dust concentration, 6,3 m dia. TBM tunnel

4.2 Experience with a 6,3 m dia. TBM

The results of dust measurements on a 6,3 m dia. TBM are shown in Fig. 3.

Force ventilation is supplied by a 2 x 75 KW axial fan via a 1200 mm dia. PVC pipe. The fan delivery volume in the mea-suring period was 15 m^3/s and the face de-livery volume 9 m^3/s. This corresponds to an average leakage in the ventilation duct of 2,7% per 100 m.

From the cutterhead 2,8-4 m^3/s of air is exhausted via two 40 mm dia ducts through a reverse-pulse dry dust bag filter by means of two 2 x 15 KW axial fans. The in-crease in the exhaust volume from 2,8 to 4 m^3/s is due to cleaning up the dust fil-ter and the ducts. The filter chamber con-tains 648 bags, giving a filtration area of 50 square metres and a filtration speed 0,08 m/s. The presure drop across the fil-ter including ducts with an 8 second pulse setting, is approximately 6 kPa during boring.

The dust concentrations shown in Fig.3 are unacceptable, mainly due to the high content of free silica in the dust (> 20%). Efforts to reduce the concentration will be discussed in the following.

1. The rubber lined steel shield prevent-ing excess dust from being liberated from the cutter head of the TBM to the working zone is not closing off good enough. A minimum leakage area of 0,6 square metres is anticipated. The same order of leakage area is also anticipated for the inspection flap and the man hole in the cutter head. The exhausting velocity through these leak-age areas will be about 3 m/s. This rela-tively low velocity can allow dust to re-turn from cutter head, where the dust con-centration can be in order of 3 g/m^3 total, into the working zone in front of the dust shield. This leakage should be reduced.

2. Too little water was used to the dust suppression. More water should be used onto the face. Better watercleaning of TBM and tunnel walls is necessary.

3. The bag filter used has a theoretical dust collection efficiency of 99,99%, docu-mented from continuous acceptance test and

687

trials. On this 6,3 m dia. TBM the efficiency was lower. The reason was some small hole observed through the filter material in the intake top of the filter bags, probably due to mechanical damage.

4. Due to space considerations the exhausting ducts from the cutter head to the dust extraction unit have bend and diameter changes which cause too low conveying velocities in parts of the duct, and dust settling in other parts. The duct system has to be kept as straight as possible, and the conveying velocity above 20 m/s.

5. The flow of the main and secondary ventilation arrangement should be more optimized from a dust point of view. Neither the TBM operator nor the loading operator are sitting in a cab. Work is now going on to improve the air distribution to these two stationary working places.

New dust measurements in the same positions three months later indicated that the dust concentrations were reduced. The average reduction was approx. 30% (18-50%). These results indicate that efforts putting in have been successful so far. However, still more work to reduce the dusting will be done in the near future.

4.3 Methane

Problems with methane gas occuring at the cutter head during TBM-operations in the hard Nowegian non-sedimentary rocks were considered not to happen. However, one year ago methane ignited in front of the cutter head on a 3,5 m dia. TBM, and a local explosion happened. No persons were hurted. The methane is supposed to come from sedimentary zones in granitegneiss. The ignition probably occured because of sparks caused by boring the hardrock.

The capacity of the dust suppression unit, which was initially only supposed to prevent dust being emitted from the cutter head, was too low to dilute the CH_4 concentration under the explosive limit. This shows that conditions can occur in which a TBM has to be stopped or can only be continued when elaborate additional measures have been taken, even in such types of non-sedimentary rocks.

Measurements show concentrations of methane along roofline on the face over the explosive limit, and levels of oxygen below the recommended working level. The precautionary measures taken include systematic control of the methane liberation, and use of extraordinarily ventilation on the face in front of the cutter head.

5. SUMMARY AND CONCLUSION

Examples of a number of typical environmental problems concerning dust and ventilation by operating full face, hard rock tunnel boring machines in Norway have been described. They demonstrate how obstacles can be removed by modifications, safety related measures and organisational procedures.

The specific performances of TBM's have reached a high standard. But it is still possible to make TBM more suitable for special environmental demands and generally improve their design and construction.

REFERENCES

Wet, de B.G.J. 1984. Ventilation requirements for developing with a tunnel borer. Journ. of the Mine Soc. of S.A. April 1984.

Myran, T. 1983. A comparison of the environmental conditions at tunnelling boring machines and conventional blasting. Proceedings Fjellsprengningskonferansen, pp. 7.1-7.11. Tapir forlag, Oslo, Norway.

Myran, T. 1984. Methane - only a coal mine problem? Proceedings Bergmekanikkdagen, pp. 36.1-36.8. Tapir forlag, Oslo, Norway.

Myran, T. 1985. Kursdagene NTH. Proceedings NIF-kurs: "Mine and Tunnel Ventilation".

The analysis of post detonation fumes in an underground operating environment – A case study

M.HINE
Kalgoorlie Mining Associates, Australia

I.O.JONES
Western Australian School of Mines, Kalgoorlie

ABSTRACT: This project materialised as a final year undergraduate project for a mature age student at the Western Australian School of Mines and was sponsored by ICI Australia Limited. It was of direct interest to local mining companies, notably Kalgoorlie Mining Associates who provided the test site and the in-mine support facilities. The objectives of the project were to assess the production of CO_2, CO, NO and NO_2 by four commercially available explosives. Whilst only three of the four explosives were tested, and the test work was not as extensive as had been intended, the results are valuable in highlighting the problems of assessing the toxic gas production of explosives.

1 INTRODUCTION

Noxious fumes generated by detonator sensitive explosives (usually factory manufactured and packaged) have been measured in specially designed laboratory scale equipment and in underground test sites. The correlation between laboratory and full scale tests has been acceptable and consequently some Statutory Authorities have classified detonator sensitive explosives according to their fume generating potential as assessed in the laboratory.

The laboratory equipment used in North America is the Bichel Bomb and the Crawshaw-Jones apparatus. However such equipment is limited in its capacity and cannot cope with the large quantities of ANFO mixtures and watergels which would be required to assess their fume generating potential.

Full scale 'on site' testing is the only means of obtaining valid results on the fumes generated by the non detonator sensitive bulk explosives. It is also agreed that on site testing will also provide valuable additional information on the fume characterisitcs of the detonator sensitive paint fine aluminium sensitised watergels (e.g. Molanite) and increase our knowledge of the fume characteristics of NG, PETN and TNT based explosives.

The main factors influencing the fumes generated by an explosive are (ICI Australia Ltd 1979):-

1   Chemical composition i.e. oxygen balance, presence of cooling salts and inert materials
2   Physical nature - size of ingredients, stability, age and deterioration
3   Type and composition of wrapper
4   Nature, size and position of primer
5   Degree of confinement
6   Ambient temperature
7   Presence of or absence of water
8   Nature of confinement

Whilst items 1 to 4 can be readily controlled items 6, 7 and 8 are site dependent. Item 5 (degree of confinement) on the other hand depends on the competence of the driller, amount of stemming and sequence of firing to give progressive relief of burden.

It has been suggested that the 'on site' testing facilities be arranged as follows:

Sealed Heading Test Site: for the testing of detonator sensitive explosives including NG, TNT, PETN sensitised grades, ANFO and ALANFO pneumatically loaded using the venturi principle and a number of cartridged or bulk IREMITES and MOLANITES.

Open Stope Test Site: for testing multihole multirow blasts using detonator insensitive grades of explosives. In this case the fumes produced would be monitored in the return drift using continuous gas sampling techniques.

This paper describes some recent work carried out in a sealed heading located at the Mt Charlotte Gold Mine owned by Kalgo-

orlie Mining Associates. The purpose of
the work was to ascertain the extent to
which CO, CO_2, NO and NO_2 is produced by
four commercial grades of explosives.

2 BACKGROUND INFORMATION

The detonation of explosive charges can be
treated as a fast combustion process.

Oxygen is taken from the oxygen rich compounds in the explosive charge to produce
a variety of reaction products. In the
ideal case these products would consist of
such substances as water, carbon dioxide,
nitrogen and metal oxides. However, because of incomplete reactions due to rapid
changes in temperature and pressure and
reactions between detonation products and
the nitrogen and oxygen in the air many
other products are formed. The most important of these are hydrogen, methane, amm-
onia, carbon, carbon monoxide, nitric oxide
and nitrogen dioxide. Many of these gases
are extremely toxic and care has to be
taken to minimise their inhalation by mine
workers.

Studies to examine the reaction products
produced by explosive charges have been
carried out since the early part of this
century when the Bichel gauge and the Crawshaw-Jones apparatus were developed. These
tests involved detonating the explosive
charges in a closed chamber and analysing
the reaction gases by standard laboratory
techniques. However, it was realised fairly early on that the gaseous products produced in such conditions did not accurately
represent the detonation products released
in the full-scale mining situation (Engsbraten 1980).

Consequently many attempts have been made
to study the production of reactant products under mining conditions. Indeed the
work of Tiffany, Murphy and Harma in the
US Bureau of Mines Laboratory concluded
that the Bichel Gauge could not be used to
predict with certainty the quantity of
post-detonation gases released in the mine.

More recently two studies of note have
been carried out. In the late 70's Nitro
Nobel AB carried out test work in the full
scale situation at the Vastra Sund test
site (Engsbraten 1980). The work was carried out in a closed tunnel where single
charges were fired in 38mm x 1.5m boreholes.
Immediately after firing the explosive charge the tunnel door was closed and a fan
was started up to mix the gases. After ten
minutes or so gas samples were drawn through
sampling tubes to the instrument station.
Nitric oxide and nitrogen dioxide were monitored directly and batch samples of the

gas mixture in small cylinders were taken
for laboratory analysis of hydrogen, carbon monoxide, carbon dioxide and methane.

The work was directed towards studying
the influence of confinement and package
materials.

The results indicated that the oxidation
of NO to NO_2 depends on the initial NO concentration and the time delay between blasting and sampling. The NO_x concentration
appeared to be rather insensitive to variations in the wrapping material and oxygen
balance. Watergel explosives of the Reomex
type produced relatively small amounts of
NO_x under confinement (e.g. less than 1 L/
kg) whereas ANFO produced up to 5-12 L/kg
of NO_x. The work demonstrated the great
importance of confinement on the NO_x production. Decreasing confinement led in all
cases to an increase in NO_x production.

It can also be assumed that the grains of
nitrate present in the charge have an influence on the NO_x production since the reaction in the grains takes place well behind
the detonation front. ANFO contains fairly
large grains of ammonium nitrate and consequently has a tendency to produce large
volumes of NO_x. A watergel explosive on
the other hand has most of the nitrate in
the form of a solution and is thus in intimate contact with the fuel. It can therefore be assumed that the reaction takes place near to the detonation front and can be
completed without the production of large
quantities of NO and hence NO_x.

Good confinement causes the reaction products to be kept at a higher temperature
and pressure for a longer period thereby
leading to a reduction in the production
of NO. On the other hand poor confinement
allows the hot reaction gases to come into
early contact with the ambient air which
may lead to the oxidation of atomic nitrogen according to:

$$N + \tfrac{1}{2}O_2 \longrightarrow NO$$

As far as the production of CO and H_2 was
concerned the influence of package material
such as PVC was noticeable in lowering the
oxygen balance and thereby caused a rise in
CO and H_2 concentration.

This work also indicated that the carbon
in the explosive forms CO_2, CO and CH_4 on
detonation. In this regard it is also noteworthy to record the fact that the wrapping
material could of itself produce 20-50 L/kg
of carbonaceous gases. Once again good confinement increases the time available for
converting the wrapping into carbonaceous
gases.

The other study was carried out by the
College of Mines, University of Arizona

the period 1979-82 under contract from the US Bureau of Mines (BuMines OFR 72-83). This work involved measuring the volumes of six gases (NO, NO_2, CO, CO_2, SO_2 and NH_3) produced by the detonation of known amounts of explosives in a mining tunnel. The work was carried out at the University's San Xavier Mine where the gases produced by the detonation were sealed in the tunnel by the closing of a bulkhead door at the mouth of the tunnel. Once again the air and gases sealed in the tunnel were mixed by a recirculation fan before withdrawing samples for analysis. In this case the samples were drawn at a rate of about $2ft^3$ per minute by a vacuum pump through 225ft of 0.5in diameter Teflon tubing to the surface where the analytical instruments were located.

The results obtained in these tests indicated that:

* nitroglycerine based explosives (dynamites) produced greater volumes of each gas than did non-nitroglycerine explosives

* concentration of gases within the test tunnel decreased with time, the decay being more rapid for NO and NO_2. This could be due to the adsorption and absorption of the gaseous products on the solid surfaces and solutions in the tunnel and the reaction of one gas with another.

3 OBJECTIVES OF PROJECT AND EXPLOSIVE PRODUCTS TO BE TESTED

The main objective of the project was to examine the production of NO, NO_2, CO and CO_2 created by the detonation of the following explosive products (ICI Australia Ltd 1984) in a mining tunnel.

AN60
AMEX (94/6 ANFO)
MOLANITE 115
POWERGEL

AN60 is a gelatinous explosive composed primarily of nitroglycerine or nitroglycol, nitrocotton, sodium and ammonium nitrates and cellulosic materials. It is a general purpose explosive of great utility in metalliferous mining where it is necessary to excavate in hard rock. It is also an excellent base charge and primer for ANFO.

AMEX is a premixed ammonium nitrate-fuel oil mixture. It is by far the most popular blasting agent used throughout Australia for blasting in most types of dry rock. This explosive contains the stoichiometric mixture of 94% ammonium nitrate and 6% distillate by weight.

MOLANITE 115 is a high strength detonator sensitive cartridged watergel explosive and is composed entirely of non-explosive ingredients which make it an attractive alternative to nitroglycerine based explosives such as AN60. Some of the advantages of

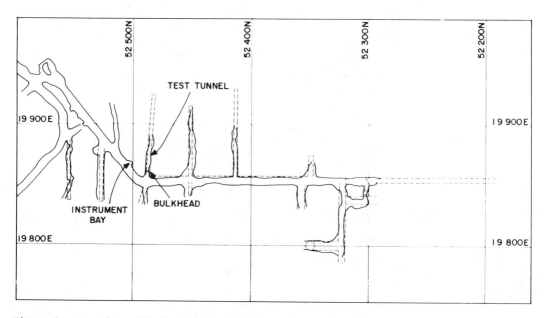

Figure 1: Location of test tunnel and instrument bay

watergel explosives over their nitroglycerine based counterparts are:

* the risk of premature detonation from
  burning, friction or impact is greatly
  reduced
* less risk of hole to hole propagation
* less smoke and toxic fumes produced

POWERGEL 2131 is a new type of explosive.
It is a detonator sensitive emulsion high
explosive which contains no explosive ingredients and subsequently has the following
advantages:

* high efficiency of detonation
* very low impact sensitivity
* excellent loading characteristics due
  to its rheology
* excellent water resistance properties and
  is superior in this regard to watergels
* very high detonation velocity and this
  makes it ideal for the initiation of ANFO
* better post-detonation fume characteristics than most other explosives

4 TEST FACILITIES AND EXPERIMENTAL PROCEDURES

4.1 Test site

The work was carried out in a 3m x 3m development heading located at the 20 level of
the Mount Charlotte Gold Mine (see Figure 1)
A leak-proof bulkhead was constructed at the
entrance of the heading which consisted of
a 2m x 1.8m steel door located in a double
brick wall. Some 10m on the intake side of
the heading an instrument station was constructed from which two 75mm diameter boreholes were drilled to intersect the heading
or test chamber as shown in Figure 2.
   A 10 HP fan and associated ducting was
located in the test chamber to recirculate
the ambient air and post-detonation fumes
after blasting.

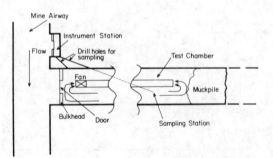

Figure 2: Layout of test site

4.2 General procedure

The drill pattern used to advance the heading was a standard 45 hole drill round uti-
lising a one reamer burn cut as shown in
Figure 3. Holes were drilled to a depth
of 1.8m using a series 11 integrated tung-
sten-carbide bit drill steel. A full face
blast was initiated by using 0.5 sec. electric delays ranging from 0-12.

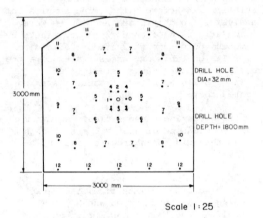

Scale 1:25

Figure 3: Drilling and blasting pattern

   Whilst a number of plugs inserted in the
holes differed for each explosive, attempts
were made to use the same explosive charge
density for each blast. In this regard it
should be noted that two series of tests
were carried out; those involving the firing of a combination of less than 10 holes
and those involving a full face blast.
This was done in an attempt to compare the
production of post-detonation fumes in both
situations.
   The holes were loaded and tamped according to standard mine blasting procedures with a
200mm collar being left on each hole. When
loading cartridged explosives every second
cartridge was tamped in order to compact
the charge.
   The volume of the test chamber was ascertained from survey data. The method involved running a line down the centre of the
drive and measuring distances to the roof,
floor and sides of the heading every metre.

5 EXPERIMENTAL PROCEDURE

Before embarking on the test work an attempt was made to analyse the nature of the
problem and to thereafter design an exper-

imental procedure which would lead to conclusive results. Looking at the problem from the point of view of the normal routine in blasting the face of a development heading the blasting fumes could be found in any one of five locations:

1 Mixed with the ambient air
2 Entrapped and adsorbed within the muckpile
3 In solution with natural mine water or the water used to spray the sides of the heading and muckpile after the blast
4 Adsorbed on the solid surfaces and dust particles in the heading
5 Trapped and/or adsorbed within the micro-fractures and fissures in the strata

Library research revealed that other workers had looked at the problem in a similar way. Rossi (Rossi 1971) pointed out that the blast gases released into the mine air behave differently from those trapped in the muckpile or within the fissures in the rock. He states that the carbon monoxide is retained in the ambient air and has to be removed by ventilation whilst the NO in the atmosphere is converted slowly to NO_2. He further states that the nitrogen oxides in the muckpile may remain as NO for some time which on liberation into the atmosphere may become oxidised to NO_2. Balkovoi and Ostronshko (Balkovoi and Ostronshko 1971) deduced from their laboratory and field experiments that in medium strength rock about 40% of the gaseous reaction products are released into the working atmosphere whilst 20% is trapped in the muckpile and 40% or so remains in the micro-cracks and fissures within the rock mass. Reference is also made in another recent text (Bossard, Le Fever, Le Fever and Stout 1983) to the fact that as much as 60% of the gases or fumes produced remain in the adjacent rock or muckpile.

Other USSR investigators (Dudyrev, Oborin, Ilina and Reshetova 1971) examined the adsorption and desportion of toxic fumes by dust particles in the mine atmosphere. This work led to the conclusion that all the toxic gases produced on detonation and water vapour were readily adsorbed on the surface of the quartz dusts. The results also indicated a strong bond between the adsorbed gas and the dust particles with only 27% of the NO_x and 54% of the CO being desorbed.

It was therefore decided to adopt the following experimental procedures:

* to analyse the post-detonation atmosphere for NO, NO_x, CO and CO_2 concentrations

* to measure as far as possible the gases contained within the muckpile
* to monitor the volume of water used in spraying the muckpile and rock surfaces and to analyse samples of the water for dissolved oxides of nitrogen

Unfortunately no worthwhile samples of spray water could be collected due primarily to the fact that the spray water became mixed with water left from drilling and that produced by the wet scrubber on the diesel loader. However, some success was achieved in the muckpile by drawing samples through a pipe left on the floor which was subsequently buried beneath the muckpile (see Figure 4).

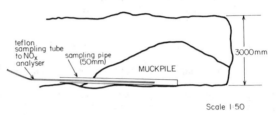

Scale 1:50

Figure 4: Muckpile analysis layout

The following test procedure was designed and adhered to throughout the test work:

1 Drilling the face of the heading according to standard mining practice
2 Surveying the heading in order to calculate the volume of air in the test chamber
3 At least one hour before firing the on-site analytical instruments were turned on and the necessary calibrations carried out
4 Charging the face with explosive to be tested and firing the round
5 Closing the bulkhead door as soon as possible after the last hole had detonated
6 Starting up the recirculation fan within the test chamber
7 Recording NO, NO_x and where possible the CO concentrations using continuous monitoring equipment
8 Collecting bag samples for laboratory analysis of CO_2 and CO
9 Opening the bulkhead door and ventilate the heading using auxiliary ventilation
10 Inserting sampling tube into the pipe partly buried beneath the muckpile to sample residual gases
11 Inspect heading, scale and rock bolt as necessary

12 Watering down the muckpile and walls and loading out muckpile using an EIMCO 912 loader.

6 ANALYTICAL INSTRUMENTS AND ANALYTICAL PROCEDURES

6.1 On-site equipment

It was thought important to have the NO/NO_x analyser located as near to the test chamber as possible and consequently a Beckman NO/NO_x analyser was obtained and located at the instrument station some 10m on the intake side of the test chamber.

A Riken (Model EC-231) CO analyser was also located on-site with the intention of using it to monitor low concentrations of CO up to a maximum of 300 ppm.

6.2 The Beckman NO/NO_x analyser

This analyser employs the chemi-luminescent method of detecting NO. This technique is based on the reaction of NO with ozone, O_3, to produce NO_2 and O_2 i.e.

$$NO + O_3 \longrightarrow NO_2^* + O_2$$

$$NO_2^* \longrightarrow NO_2 + hf$$

where h = Planck's constant
and   f = frequency

As the NO and O_3 mix in the reaction chamber the chemi-luminescent reaction produces an emission of light which is directly proportional to the concentration of NO. This light emission is measured by a photomultiplier tube and a direct readout is obtained in ppm. The NO_x determination is carried out in the same way except that prior to entry into the reaction chamber the sample is routed through a converter where NO_2 is converted to NO i.e.

$$2NO_2 \longrightarrow 2NO + O_2$$

Needless to say a dust and moisture filter was needed in the sample line to safeguard the on-site analytical equipment.

Whilst a Riken CO analyser was available on-site its use was not required due to the fact that in all cases the CO concentration in the test chamber was well above the operating range of this instrument (i.e. 0-300 ppm).

A schematic view of the sampling arrangements at the on-site instrument station is given in Figure 5.

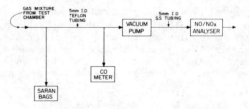

Figure 5: Sampling system layout

7 LABORATORY EQUIPMENT

The analysis of the CO and CO_2 content of the post-detonation atmosphere was carried out on batch samples using a Pye Unicam GCV Gas Chromatograph. A schematic view of the gas chromatograph is given in Figure 6. It consisted of a temperature controlled oven and a thermal conductivity detector.

The helium carrier gas transports a 0.5ml sample from the sample loop on to an adsorbent column where the gas mixture is fractionated into its component parts. The quantity of the pollutant gas is thereafter assessed by the output of the detector which produces distinctive peaks on a potentiometric recorder.

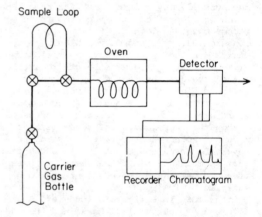

Figure 6: Schematic of gas chromatographic unit

Unfortunately two solid adsorbent columns were required to analyse the CO and CO_2 content.

Molecular Sieve 5A in a 1.5m x 4mm column was required for the analysis of CO. It fractionated the sample to give clean pulses of oxygen, nitrogen and carbon monoxide, whilst the CO_2 is irreversibly adsorbed by the column packing.

Poropak Q in a 2m x 4mm glass tube was used to fractionate the sample into clean

pulses of air (oxygen and nitrogen), methane (if present) and CO_2 and was therefore
used in the analysis of CO_2.

Whilst the CO_2 analysis was carried out
expeditiously within 3 minutes or so the
analysis of CO using molecular sieve was a
much slower process taking at least 15 mins
per sample.

The operating conditions for the analysis
of both CO and CO_2 were identical and are
listed as follows:

Carrier Gas (Helium) Flowrate	25ml/minute
Sample Loop Size	5ml
Detector Temperature	90°C
Column Temperature	70°C
Filament Temperature	240°C

Of course prior to use, the gas chromatograph was calibrated using standard mixtures of CO and CO_2 in air. Using the
calibration curves so produced the concentration of these gases in the test chamber
atmosphere could be determined.

Typical chromatograms for the analysis
of CO and CO_2 are given in Figures 7 and 8.

8 SAMPLING TUBES AND BAGS

Considerable thought was given to the sampling tubes and the technique to be adopted

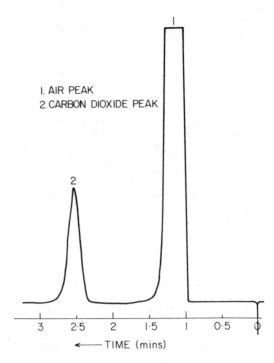

1. AIR PEAK
2. CARBON DIOXIDE PEAK

← TIME (mins)

Figure 8: CO_2 chromatogram

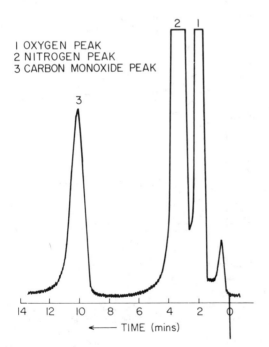

1 OXYGEN PEAK
2 NITROGEN PEAK
3 CARBON MONOXIDE PEAK

← TIME (mins)

Figure 7: CO chromatogram

in taking batch samples of the test-chamber
atmosphere for subsequent laboratory analysis. It was thought unwise to use steel
sampling cylinders such as Gresham or GFG
tubes because of the possibility that the
gases would either condense out on the cold
metallic surfaces or simply be adsorbed on
the surface. Consequently either SARAN
bags or wine-cask liners were used. Both
had proved effective as a medium for storing such toxic gas as the oxides of nitrogen. Once again because of the corrosive
nature of these gases the sampling tubes
were made of TEFLON a polymer-plastic material.

9 EXPERIMENTAL RESULTS

Following the blast, alternate NO and NO_x
readings were visually recorded from the
Beckman Analyser at one minute intervals.
When the readings became reasonably stable
eight four litre bags were filled with
samples of the test-chamber atmosphere for
laboratory analysis of CO and CO_2.

NO and NO_x were taken continuously over
a 30 minute period after which time the
bulkhead door was opened and the heading
scoured by auxiliary ventilation.

695

Some typical records of the NO/NO_x concentrations recorded are shown in Figures 9, 10 and 11.

In the laboratory sixteen analyses were carried out for each gas using four of the eight sample bags. The finally accepted values were the arithmetic mean of the 16 values.

The computed volumes of toxic gas produced by each explosive were reduced to STP conditions before determining the volume of each gas produced per unit mass of explosive (litres/kilogram).

The NO/NO_x readings obtained from within the muckpile provided stable readings for each of the four tests carried out and the values obtained are given in Table 1.

Table 1. NO and NO_x concentrations in muckpile atmosphere

No of Holes Fired	Explosive Type	Concentrations in Muckpile (ppm)	
		NO	NO_x
44	AN60	355	395
44	POWERGEL	125	155
44	MOLANITE	115 345	405

Fairly extensive tests were carried out on AN60 in order to examine the relationship which may exist between the total mass of explosive used and the production of toxic gases. Hence a number of tests were carried out using less than 10 holes at a time as well as tests which involved a full face blast. The results of these tests can be seen in Tables 2 and 3.

With the other explosives only full scale blasts were carried out and the mean result obtained for each of the three explosives tested are given in Table 4.

Finally a comparison of the calculated volumes of NO_x and CO produced with those values obtained by ICI Australia Ltd in their testing facility is given in Tables 5 and 6.

10 ANALYSIS OF RESULTS

10.1 Influence of changing mass of explosive

The scaled tests using less than 10 charged holes of AN60 resulted in higher values of NO_x, CO and CO_2 measured in litres per kilogram of explosive than was produced by a

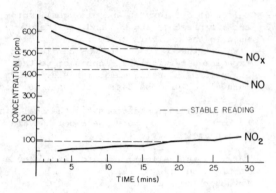

Figure 9: NO_x plot (test number 6) "AN60"

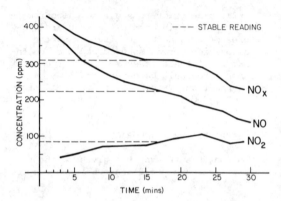

Figure 10: NO_x plot (test number 9) "Powergel"

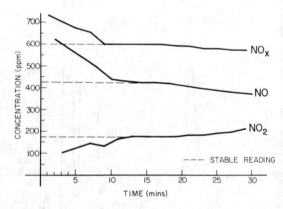

Figure 11: NO_x plot (test number 10) "Molanite"

full face blast. This could be due in part to the lack of confinement in the smaller scale blast and to the higher shock and bubble energy produced by a full face blast causing more pronounced fracturing of the

Table 2. Computed toxic gas values for tests no 1 to 5 using AN60

TEST NO	EXPLOSIVE AMOUNT (kg)	CONTAINMENT VOLUME (m³)	COMPUTED TOXIC GAS VALUE (L/Kg)				
			NO_x	NO	NO_2	CO	CO_2
1	6.84	455	9.31	7.64	1.66	66.52	152.86
2	6.08	456	9.00	5.63	3.38	79.65	249.90
3	6.84	459	7.05	4.50	2.55	92.94	227.42
4	6.08	464	8.77	5.72	3.05	109.67	223.61
5	5.89	470	9.58	6.70	2.87	100.94	238.35
	MEAN VALUES		8.74	6.04	2.70	89.94	218.43

Table 3. Computed toxic gas values for full face blasts

EXPLOSIVE TYPE	TEST NO	AMOUNT OF EXPLOSIVE USED (kg)	CONTAINMENT VOLUME (m³)	COMPUTED TOXIC GAS VALUE (L/Kg)				
				NO_x	NO	NO_2	CO	CO_2
AN60	6	33.44	474	8.50	5.95	2.55	65.31	198.59
AN60	7	33.44	493	8.55	5.90	2.65	66.59	192.16
POWERGEL	8	36.96	510	2.48	1.86	0.62	25.97	99.85
POWERGEL	9	48.62	526	3.35	2.43	0.92	17.51	105.02
MOLANITE 115	10	49.80	546	5.70	4.66	1.04	27.84	70.15

Table 4. Summary of toxic gas values (at 20° and 1 atm) for full face blasts

EXPLOSIVE TYPE	COMPUTED TOXIC GAS VALUE (L/Kg)				
	NO_x	NO	NO_2	CO	CO_2
AN60	8.59	5.98	2.62	66.46	196.90
POWERGEL	3.38	2.45	0.93	17.65	105.84
MOLANITE 115	5.74	4.70	1.04	28.06	70.69

Table 5. Computed toxic gas values (at 20° and 1 atm) for different explosives as supplied from ICI

EXPLOSIVE TYPE	COMPUTED TOXIC GAS VALUE (L/Kg)	
	NO_x	CO
AN60	37.0	51.2
POWERGEL	12.5	13.1
MOLANITE 115	14.4	9.7

Table 6. Modified computed toxic gas values (at 20° and 1 atm) for full face blasts

EXPLOSIVE TYPE	NO_x (L/Kg)	CO (L/Kg)
AN60	21.48	166.15
POWERGEL	8.45	44.13
MOLANITE 115	14.35	70.15

rock mass where more of the toxic gases could be stored. The findings of Balkovoi et al are supportive of this assertion since he found that up to 40% of the post detonation fumes are forced into the surrounding strata.

10.2 Comparison of explosives

Of the three explosives tested the worst results were obtained from AN60 and the best from POWERGEL with MOLANITE 115 producing intermediate results except in the case of CO_2 where it was the best performer. This may be due to a difference in the oxygen balance characteristics of the MOLANITE 115.

10.3 Comparison of predicted and measured values of toxic gas production

On the face of it considerably less toxic gas is produced in the full scale situation than is predicted by the manufacturer. Of course the manufacturer's figures are based on model tests and do not include the effects of the rock mass and muckpile. Accepting the results of Balkovoi et al, that 60% of the gaseous product is contained in the muckpile and surrounding strata and using this figure the measured values can be amended to give those in Table 6. Whilst the modified NO_x values agree fairly well with those predicted by the manufacturer those for CO are very much larger.

Of course it may be that less of the CO is held in the muckpile and surrounding strata than is the case with NO and NO_2.

10.4 Gases contained in the muckpile

The readings obtained from the muckpile confirmed the fact that a considerable volume of NO_x is stored within it. That apart the readings also indicated that most of the NO_x was in the form of NO which is not readily soluble in water and would not therefore be removed to any great extent by watering.

11 CONCLUSIONS

The results obtained lead the authors to the following conclusions.

1 Of the three explosives tested (ie AN60, Molanite 115 and Powergel) the best and worst results (ie minimum and maximum production of toxic fumes) were obtained from the newly available Powergel and the popular AN60 respectively.

2 On site analysis of fumes generated by explosives must include an assessment of the fumes contained within the muckpile. The results quoted in this paper support the assertion of Balkovoi et al, that 60% of the fumes produced from a blast are held within the muckpile and surrounding strata and that nitrogen oxides in the muckpile may remain as NO for some time. However, more work needs to be carried out on this problem since it is a serious hidden source of toxic fumes.

3 In accepting the validity of the findings of Balkovoi et al, it is assumed that 40% of the fumes produced are contained within the fissures and fractures in the surrounding rock mass or adsorbed on dust particles. Once again more work needs to be done to investigate these possibilities.

4 It appears that in keeping with the findings of previous workers that the degree of confinement and the prescence or absence of water have a significant effect on the quantity and nature of the toxic fumes remaining after a blast.

REFERENCES

ICI Australia Ltd 1979. Standardisation of methods and equipment used in measuring post detonation fumes in mines - internal report.

Engsbraten, B. 1980. Fumes from detonation of commercial explosives in boreholes and steel tubes. Proceedings of the sixth conference on explosives and blasting techniques. Society of explosives engineers, annual meeting.

Garcia, M.M., Jucevic, E.D. and Kittrell, W.C. 1983. Monitoring of gases from explosives detonated in an underground mine. US Bureau of Mines Report, OFR 72-83.

Personal communication with ICI Australia Limited.

Rossi, B.D. 1971. A hundred years use of ammonium nitrate explosives - control of noxious gases in blasting work and new methods of testing industrial explosives. Translated from Russian under auspices of the Israel Program for Scientific Translations.

Balkovoi, P.I. and Ostronshko, I.A. 1971. The entrapment of noxious gases in rocks. Translated from Russian under the auspices of the Israel Program for Scientific Translations.

Bossard, F.C., LeFever, J.J., LeFever, J.B. and Stout, K.S. 1983. A manual of mine ventilation design practices, 2nd edition. Floyd C. Bossard and Assoc., Inc.

Dudyrev, A.M., Oborin, V.V., Ilina, I.M. and Reshetova, V.A. 1971. Desorption of noxious gases from mining dust. Translated from Russian under auspices of the Israel Program for Scientific Translations.

Application of biotechnology for methane control in coal mines

R.N.CHAKRAVORTY
Energy, Mines & Resources, Calgary, Canada

P.I.FORRESTER
Gemini Biochemical Research Ltd, Calgary, Canada

ABSTRACT: Effective control of methane during all phases of mining is essential to ensure safer mining environment. This paper reports on work done to date on an unconventional method of methane control based on biotechnology i.e. the use of microorganisms to oxidize methane into carbon dioxide. Methane oxidizing bacteria which can be grown easily under controlled laboratory conditions have been successfully isolated from some of the Canadian coal mines. Laboratory tests have shown that these organisms can remain active over an extended period with low sulphur coals and that the rate of methane oxidation is quite appreciable. The use of microorganisms is considered to have potential for large scale application in methane degasification.

1 INTRODUCTION

Most coal mine explosions have been caused by the emission and accumulation of methane. Inspite of major advances in mining technology it has not been possible to eliminate this hazard completely. As production increases and the workings become more extensive the problem can become more serious. Effective control of methane is therefore essential to improve coal mine safety.

A review of techniques developed for dealing with methane in coal mines indicates that control by ventilation alone is not adequate, particularly during mining of highly gassy seams. Methane drainage during mining operations and in advance of mining has proved successful in some cases (Suarez and Chakravorty 1976).

Attempts have been made to use unconventional methods for methane control in coal mines. One such method based on biotechnology has been tried in Eastern Europe and more recently in India to supplement methane control measures. In general, this method would involve bringing methane oxidizing microorganisms in contact with coal and allow them to oxidize the methane present to non-inflammable products in a controlled manner. CANMET has recently initiated research work on this particular approach to evaluate its potential application in Canadian coal mines.

2 REVIEW OF PAST WORK

2.1 Occurrence of methane oxidizing bacteria

Methane oxidizing microorganisms have been isolated from a variety of locations, including coal mines. The very wide range of habitats from which these organisms have been isolated suggests that they are ubiquitous.

Reports on the isolation of methane oxidizing bacteria from coal mines originate mostly from Eastern Europe. Malashenko et al. (1978) examined 500 samples of water collected from coal mines in the USSR and found that the majority of the samples contained methane oxidizing bacteria. Ivanov (1978) reported that out of 76 water samples taken from 14 coal mines in the USSR, 57 contained methane oxidizing bacteria. Studies in Poland (Godlewska-Lipowa 1979) showed that all of the 200 water samples taken from Polish coal mines were found to contain methane oxidizing bacteria. These studies strongly indicate that conditions suitable for the growth of these organisms exist in most coal mines. Canadian studies on the isolation of methane oxidizing organisms are rather limited. Mueller (1969) has found them in sewage and Roth-Wood (unpublished results obtained by Gemini Biochemical Research Ltd.) has found them in the soils of Southern Alberta.

2.2 Metabolic characteristics of the organisms

The metabolic and physical characteristics of methane oxidizing microorganisms will effect their potential use for biological control of methane. In general, the organisms must be able to grow and oxidize methane found in coal mining environments. Specifically they must be active at the ambient temperatures, pH, pressure and ionic conditions found in the mines or silos. They must also be active in the presence of any organic constituents of the coal and must not adversely affect the coal or its structural integrity. A suitable electron acceptor for the oxidation of methane must also be present.

Some of the pertinent factors affecting the activity of the methane oxidizing bacteria are summarized below.

2.2.1 Temperature and pH effects

Methane oxidizing bacteria grow over a wide range of temperatures (15°-60°C) but most of the organisms studied to date will grow within a range of 25°C (Leadbetter 1974). Methane oxidizing bacteria will grow best in a pH range of 6.0-8.5 (Leadbetter 1974). Methane oxidizing yeasts are active between pH 4.0 and 6.0 (Wolf and Hanson 1979).

2.2.2 Ion effects and requirements

Methane oxidizing bacteria require a number of inorganic ions for growth, particularly nitrogen and phosphorous. These elements can be supplied in a variety of forms. Some of the organisms have the ability to fix atmospheric nitrogen.The majority of the methane oxidizing bacteria will grow on a simple inorganic medium without supplementation with complex organic compounds such as vitamins and aminoacids (Foster and Davis 1966). Some metal ions such as copper have been shown to be inhibitory.

The exact requirements for each organism will vary so that particular strains can be selected for each mine.

2.2.3 Effect of organic components of the coal

Methane oxidizing bacteria normally do not lose any activity when they are in contact with coal. In fact, the coal was shown to stimulate the activity of two species of methane oxidizing bacteria. Methylosinus trichosporium and Methylococcus ucrainicus were found to oxidize 2.1 and 1.5 times, respectively, the methane of an equal amount of non-adsorbed bacteria (Nesterov and Nazerenko 1975). Methylomonas methanica was also shown to have increased activity when adsorbed onto coal. This species was 1.14 times as active in the adsorbed state as when unadsorbed (Abramov et al. 1977).

2.2.4 Oxygen requirements

The complete bio-oxidation of methane to carbon dioxide follows the pathway shown below (Patel et al. 1972):

$$CH_4 \rightarrow CH_3OH \rightarrow CH_2O \rightarrow HCOOH \rightarrow CO_2$$

The overall equation can be summarized as:

$$CH_4 + 2O_2 = CO_2 + 2H_2O$$

All methane oxidizing bacteria isolated to date are aerobic, that is, they require molecular oxygen as their terminal electron acceptor. Thus the availability of oxygen becomes a key factor in determining where these organisms can be used.

2.2.5 Heat effects

Large amounts of heat are released during the microbial oxidation of methane (Cooney et al. 1968). If the oxidation is carried out within a coal seam, heat generated will raise the temperature of the organisms to such a level that they may die. However, if the organisms are applied to the surface of the coal the heat can be dissipated without raising the temperature to detrimental levels.

2.2.6 Pressure effects

A number of reports show that micro-organisms can be subjected to high pressures without adversely effecting their activity. Methylomonas methanica, which had been isolated under normal conditions, was shown to retain the ability to grow after being injected into a coal seam at pressures up to 150 atmospheres (Abramov et al. 1977).

2.3 Field test results

Field tests carried out in Russia indicate that biologically enhanced methane control has possible uses in coal mines. Ivanov et al. (1981) treated a 30 m x 20 m surface section of coal adjacent to a longwall face with a fermentor grown culture of the methane oxidizing bacterium Methylococcus capsulatus. In this particular mine 70% of the methane in the ventilation air came from worked out areas. Over a 15 day period the bacteria oxidized 6.3% of the methane emitted from the treated area on day one, to a high of 40% by day 9. By day fifteen, the amount of methane oxidized had fallen to 13%.

Thakur et al. (1983), in India, also met with considerable success during field testing in three different mines. In one mine there was an approximately 50% decrease in emission of methane three days after a nutrient solution and methane oxidizing microorganisms were injected into a coal seam. In a second test there was complete removal of methane two months after injection of the organisms and in a third test the methane emission rate was reduced by approximately 66%, 20 days after injection of the organism. However, no details on the apparent mechanism of the reaction were reported.

3 EXPERIMENTAL METHODS

A series of laboratory experiments were initiated to study the possible application of methane oxidizing micro-organisms in controlling methane levels in a coal mining environment.

3.1 Isolation and culturing of the methane oxidizing microorganisms

Methane oxidizing bacteria were isolated from mine waters by spreading 0.1 ml aliquots onto Bushnell Haas (1941) agar plates. After incubation in 15% methane atmosphere at 25°C for a period of 14-28 days, colonies of methane oxidizing bacteria were visible on the plates. Single colony isolates were transferred to fresh plates for the isolation of pure strains of the organisms. Innocula from these pure strains were added to liquid Bushnell Haas medium and larger quantities of cells were harvested after growth in a 15% methane atmosphere for 7 days.

3.2 pH profile

Ten ml aliquots of a culture of methane oxidizing bacteria were placed in 50 ml serum vials. The pH of the cultures were adjusted to various values between 5.0 and 9.0 and the rate of methane oxidation determined after incubation at 25°C for 24 hours.

3.3 Methane oxidation in the presence of partially degassed coal

A 250 ml Erlenmeyer flask, fitted with a sidearm, was filled with 150 grams of partially degassed medium volatile bituminous coal (crushed to 1 cm pieces). Five mls of a concentrated cell slurry and 20 mls of Bushnell Haas medium were added to the coal. The tops of the flasks were sealed with a rubber membrane and the sidearms fitted with a syringe cap to permit addition and removal of gases. Methane gas was added to the flasks to give a final concentration of 12-15%. The flasks were incubated at 30°C for 24 hours when a gas sample was removed and analyzed for methane and carbon dioxide concentrations. All gases were analyzed using gas chromatography. Flasks where no microorganisms were added, served as controls.

3.4 Methane oxidation in the presence of fresh coal

500 ml Erlenmeyer flasks, equipped with sidearms, were filled with 200 g of freshly crushed medium volatile bituminous coal. After addition of 5 mls of cell suspension and 20 mls of Bushnell Haas medium the flasks were sealed as described above. Samples of gas were analyzed for methane and carbon dioxide at 24 hour intervals.

3.5 Longevity of the microbial activity

The methane concentration in flasks, prepared as in 3.2 above, was adjusted to 12%. The flasks were incubated at 30°C for 24 hours and the methane concentrations analyzed. The seals were removed and the flasks were filled once more with 12% methane-air mixture. After a further 24 hours, air samples were analyzed for methane and carbon dioxide. This cycle of events was repeated for a period of 100 days.

4 RESULTS AND DISCUSSIONS

The chemical composition of water samples obtained from western Canadian coal mines appears to be compatible with the requirements for maintaining the activity of methane oxidizing microorganisms. In particular, the pH of the water was within the range (7.0-8.5) required for maximum growth and no potentially toxic ions were present. This was further corroborated by the fact that large numbers of methane oxidizing bacteria of different types could be isolated and grown from all the water samples obtained from the western Canadian mines.

A number of different colony types were present on the agar plates used for the isolation procedure. However, only one colony type was selected for further study. This strain was the fastest growing and produced the largest colonies on the agar plates. The pigmentation of the strain changed from pink to salmon after growth in submerged culture. Microscopic examination showed that this culture appeared to be a mixed culture containing two cell types resembling Methylomonas methanooxidans and M. methanica.

The methane oxidizing bacteria were more active in the presence of coal than when suspended in a liquid culture. This is in keeping with the observation that most microorganisms prefer to grow on a solid support rather than in a free flowing suspension (Geesey and Costerton 1979). These authors showed that 95% of the microorganisms present in a natural river environment were attached to suspended solids and only 5% were unsupported. Another reason for the increased activity is that the organisms on the coal were growing in a thin film of water surrounding the coal which allowed better access of oxygen and methane to the organisms.

The pH profile of a strain of methane oxidizing bacteria isolated from a western Canadian coal mine is illustrated in Figure 1. This strain shows good activity between pH 5.5 and 8.0.

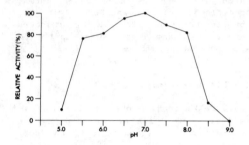

Fig. 1. The pH profile of a strain of methane oxidizing bacteria isolated from a Western Canadian coal mine.

The rate of methane oxidation obtained from static experiments is illustrated in Figure 2. The rates are significant since levels of 12% methane can be reduced to zero within a period of 24 hours.

Figure 3 shows that all of the methane which is emitted from fresh coal in the test model can be oxidized before it could accumulate to measureable levels.

For successful application of this technique under actual mining conditions it is essential that the bacterial activity have a long half-life. The results of experiments to determine the longevity of the microbial activity are shown in Figure 4. The results clearly show that the bacteria remained active for extended periods. Even after a period of 70 days they had lost only 40% of their initial activity. The longevity of the organisms was reduced when the Bushnell Haas medium used to wet the coal was replaced with distilled water. This is illustrated in Figure 5.

The only parameter which appeared to have a negative effect on the microbial activity was the presence of high levels of sulphur (3-5%) in the coal. The activity remained high with all Western Canadian coals (0.5% sulphur) tested but was reduced very significantly in the presence of Cape Breton coals (5.2% sulphur). It is believed that Thiobacilli, present in the coal, oxidized the pyrites in the coal producing sulphuric acid (Colmer et al. 1950) which lowered the pH of the liquid to levels where the particular strain used

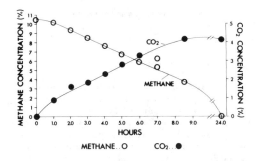

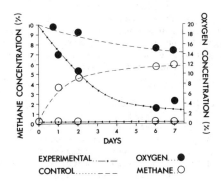

Fig. 2. Rate of bacterial oxidation of
methane in the presence of coal obtained
from a western Canadian coal mine

Fig. 3. Oxidation of methane emitted from
freshly ground coal with and without addition of the methane oxidizing bacteria

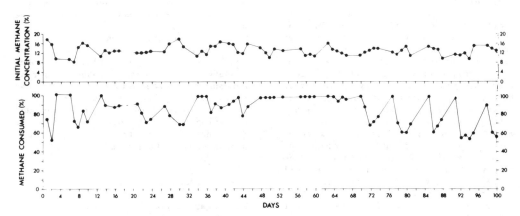

Fig. 4. Longevity of the methane oxidizing bacteria in the presence of coal

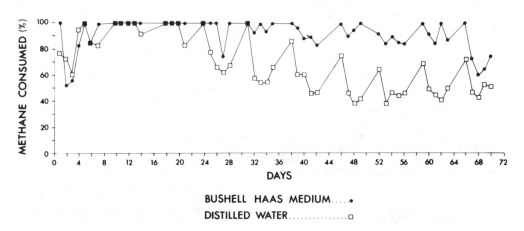

Fig. 5. Comparison of the bacterial activity when Bushnell Haas medium used to wet coal
is replaced with distilled water

in this study was not active. This hypothesis was confirmed by the finding that the pH of the liquids in the flasks containing Cape Breton coals was found to be in the range 2.6-4.0.

5 MINING APPLICATIONS

The work described shows that the methane oxidizing bacteria isolated from coal mines and grown under laboratory conditions can oxidize methane when in contact with coal. If scaled up tests prove successful the bacteria could be used in a number of mining situations, e.g.:
  a. underground mines
  b. coal storage silos
  c. ship holds.
The rate at which microorganisms will demethanate coal will depend on the oxygen availability, amount of culture, the rate of methane emission and the bacterial activity. The temperature of coal mines, coal storage silos and coal transportation systems may also affect the activity of methane oxidizing microorganism. Most organisms used in laboratory experiments so far have been mesophilic having an optimum temperature range from 15°-30°C. Because of auto-oxidation and low thermal conductivity it is expected that the coal storage facilities will maintain this temperature during winter months. However, should the need arise another group of organisms, the psychrophilic group, which are active at a temperature range of 0°-15°C could be developed and tested for low temperature applications. Organisms applied to coal under certain conditions would have to be pressure tolerant. If organisms were applied to coal during hydraulic mining, they should be able to tolerate, but not grow, at pressures up to 100 atmospheres and be able to survive rapid decompression back to atmospheric pressure. ZoBell and Johnson (1949) determined that many bacteria can survive cyclical changes of pressures up to 600 atmospheres. Organisms introduced in the fluid during water infusion should also be pressure resistant.

5.1 Application methods

Essentially two different methods can be used in coal mine applications:
  a. Direct infusion into the coal seam

  b. Spraying organisms onto the surface of the coal.

5.1.1 Direct infusion

Water infusion is a procedure used in coal mines to aid in the control of dust. The microorganism could be added to the infusion water and pumped into the coal seam. This method will have two serious drawbacks. Firstly, the availability of an adequate supply of oxygen would be a major problem. Feeding the oxygen in the gaseous form may not be desirable as this will increase the risk of explosion. The cost of supplying the oxygen in the form of sulphate or nitrate may be prohibitive. In addition, some of the products of metabolism may prove to be a health hazard.

The second limitation arises due to the possible temperature rise in the coal bed. The coal being a poor conductor the heat produced during methane oxidation will dissipate rather slowly and as a result the organism may die. Thus, pumping the methane oxidizing bacteria into the coal seam does not appear to be a practical proposition at this time.

5.1.2 Surface spraying

The organism could be sprayed easily on all strategic locations. In a coal mine they could be sprayed on the rib, roof and other accessible areas. They could also be sprayed on the coal prior to being loaded into a silo or a ship hold.

Heating effects and oxygen availability would not be a problem with surface spraying. This method appears to be the simplest and would certainly be the method of choice for initial field tests.

6 CONCLUSIONS

A review of the scientific literature shows that microbiological control of methane in coal mines is feasible. The physiological properties of methane oxidizing microorganisms appear to be consistent with the conditions existent in most coal mines in western Canada. As reported in this paper, the methane oxidizing bacteria have been successfully isolated from a number of water and coal samples collected from Canadian mines. Laboratory tests have shown that these organisms can remain active over an extended period with low sulphur coals and that the rate of methane oxidation is quite appreciable. However their activity was considerably reduced in presence of high sulphur coals.

Particular application of this technique is foreseen in underground mining operations, gob areas, mine roadways and in coal pillars. The method may also prove useful in controlling methane in coal storage silos and in ship holds provided an adequate supply of oxygen can be maintained.

No single method for methane control would be able to provide complete protection during mining of gassy seams. Method based on biotechnology appear to have potential applications in coal mining operations. However, these methods should not be used as a substitute for methane drainage or good ventilation practice but as a supplement to provide additional measures of safety.

7 ACKNOWLEDGEMENT

Work on this project was carried out by Gemini Biochemical Research Ltd., Calgary, under the auspices of the Canada Centre for Mineral and Energy Technology, Energy, Mines and Resources Canada.

The authors would like to express their thanks to Miss H. MacKeigan for her assistance in the laboratory work. The cooperation of the coal mine operators is gratefully acknowledged.

8 REFERENCES

Abramov F.A., Yu.R. Malashenko, V.I. Myaken'kii, I.K. Kurdish, G.A. Shevelov, P.S. Litvinov & N.V. Voloshia 1977. Vital activity of methane oxidizing bacteria after filtration through coal porous structures. Mikrobiologichnyi Zhurnal (Kiev) 39(3):290-293.

Bushnell, L.D. & H.F. Haas 1941. The utilization of certain hydrocarbons by microorganisms. J. Bacteriol. 41:653-673.

Colmer, A.R., K.L. Temple & M.E. Hinkle 1950. An iron-oxidizing bacterium from the acid drainage of some bituminous coal mines. J. Bacteriol, 59:317-328.

Cooney, C.L., D.I.C. Wang & R.I. Mateles 1968. Measurement of heat evolution and correlation with oxygen consumption during microbial growth. Biotechnol. Bioeng. 11:269-281.

Foster, J.W. & R.H. Davis 1966. A methane dependent coccus, with notes on classification, and nomenclature of obligate, methane utilizing bacteria. J. Bacteriol. 91:1924-1931.

Geesey, G.G. & J.W. Costerton 1979. Microbiology of a northern river: bacterial distribution and relationship to suspended sediment and organic carbon. Can. J. Microbiol. 25:1058-1062.

Godlewska-Lipowa, W.A. 1979. Control of gas hazards in coal mines with bacteria. Przeglad Gornicz. 35(6):-257-261.

Ivanov, M.V., A.I. Nesterov, B.B. Namsaraev, V.F. Gelchenko & A.V. Nazarenko 1978. Distribution and geochemical activity of methanotrophic bacteria in coal mine waters. Mikrobiologiya. 47(3):489-494.

Ivanov, M.V., A.M. Zyakum, V.A. Bondar, P.S. Litvinov, A.P. Petukh, V.I. Myaken'kii & B.I. Myagkii 1981. Use of the results of analysis of the isotopic composition of methane and carbon dioxide as an indicator of the microbiological oxidation of methane in coal mines. Doklady Akademii Nauk SSSR. 257(6): 1470-1473.

Leadbetter, E.R. 1974. Family IV. Methylomonadaceae. Bergey's Manual of Determinate Bacteriology. Editors Buchanan R.E. and Gibbons N.E. The Williams and Wilkins Co., Baltimore, USA. 267-269.

Malashenko, Yu R. & V.A. Romanovskaya 1978. Theoretical and applied aspects of the study of methane oxidizing bacteria. Mikrobiologichnyi Zhurnal (Kiev). 40:275-292.

Mueller, J.A. 1969. Fermentation of natural gas with a cyclone column fermenter. Can. J. Microbiol. 15:1047-1050.

Nesterov, A.I. & A.B. Nazerenko 1975. Activity of methane oxidizing bacteria in the absorbed state. Mikrobiologiya. 44:851-854.

Patel, R.N., H.R. Bose, W.J. Mandy & D.S. Hoare 1972. Physiological studies of methane and methanol oxidizing bacteria: comparison of a primary alcohol dehydrogenase from Methylococcus capsulatus (Texas strain) and Pseudomonas species M27. J. Bacteriol. 110:570-577.

Suarez, J. & R.N. Chakravorty 1976. Methane control in coal mines - A review. CANMET Report ERP/MRL 76-108(TR) EMR Canada. 30 pp.

Thakur, D.N., K.K. Saroj, A. Gupta & S.K. Srivastava 1983. Laboratory and in-situ investigation into microbial degasification of coal seams. Trans. Min. Geol. Met. Inst. of India, Calcutta. 79(2):67-92.

Wolf, H.J. & R.S. Hanson 1979. Isolation and characterization of methane utilizing yeasts. J. Gen. Microbiol. 114:187-194.

ZoBell, C.E. & F.H. Johnson 1949. The influence of hydrostatic pressure on the growth and viability of terrestrial and marine bacteria. J. Bacteriol. 57:179-189.

Analysis of friction factors in mine ventilation systems

M.K.GANGAL
CANMET, Energy, Mines & Resources Canada, Ottawa, Ontario

K.R.NOTLEY & J.F.ARCHIBALD
Queen's University, Kingston, Ontario, Canada

ABSTRACT: A program of field measurement work to determine airway friction factors was carried out in underground Canadian Mines. The work was confined to straight, unobstructed lengths and included steel arched gangways, and gunited surfaces as well as raw rock airways. Standard techniques were employed for airflow and pressure drop measurements and a photoprofiling method was used to characterize the airways. The results obtained in the igneous rock sites were generally lower than those in the well-known standard table of airway friction factors, reflecting the larger size and smoother conditions in the modern airways. The hydraulic surface roughness values for airways were calculated using the basic Darcy equation for duct flows. The friction factor of airways with different dimensions can be predicted if the roughness value is known.

1 INTRODUCTION

In the design of an underground mine ventilation system, the first step is to determine the volume of airflow required in all parts of the mine. A number of factors are involved in this determination such as quantities of dust and noxious gases to be diluted to safe levels, provision of adequate oxygen levels and maintenance of a comfortable temperature range. Once the flow distribution has been determined, it is necessary to estimate the pressure required to force these air quantities through the respective mine openings for the selection and placement of fans and other ventilation controls.

The principles underlying the accepted theory of mine ventilation were put forward by J.J. Atkinson in 1854. In its simplest form, Atkinson's equation states that the pressure drop in a mine airway is proportional to the square of the volume of air flowing, the constant of proportionality being defined as the resistance of the airway, or: -

$$P = RQ^2 \qquad (1)$$

where P is the pressure drop, R is the resistance, and Q the airflow rate.

The airway resistance, R, depends on the dimensions of the airway and the nature of the rubbing surface, such that:

$$R = KCL/A^3 \qquad (2)$$

where K is regarded as a friction factor characterizing the roughness of the rock surface, L is the length of the airway, C is the perimeter of the airway cross section, and A is the cross-sectional area. The dimensional parameters are readily determined but some judgement is required in the selection of K.

Many mining companies measure representative K-factors for their own internal use but, for planning purposes, the most widely used reference source dates back to a series of carefully conducted measurements carried out in the mid 1920's (McElroy & Richardson 1927). Measurements of K-factors carried out in modern mine openings have generally indicated values lower than those of the 1920's, particularly for the case of airways in igneous rock. This is understandable considering the mine openings in which the original work was done. Photographs of these airways show that they were mostly between 2 and 2.5 m in diameter, with extremely rough rock surfaces. Modern mine openings, particularly in trackless mining zones, are generally much larger, and modern drilling and blasting techniques generally produce smoother walls. The relative roughness is thus reduced and, since modern hydraulic theory shows that for fully turbulent flow the friction factor is purely a function of relative roughness of the conduit, we should expect lower values of the K-factor.

There is, therefore, a need to produce a table of airway friction factors more applicable to modern mining conditions. This need is augmented by the increasing trend to the use of computer network programs to solve mine ventilation problems. Accurate resistance values for each branch of the network are required for the basic data set for such programs.

Examination of equations (1) and (2) shows that the value of the K-factor, as used in mine ventilation work, depends on the units chosen for length, pressure, and airflow rate. In the SI system, with lengths in m, pressure in Pa, and flow rates in m^3/s, the K-factor has units of Ns^2/m^4, or kg/m^3 (i.e., units of density).

In other unit systems, K contains various unit conversion factors depending on the pressure and flow units used. Thus, K takes different values in different systems.

It has become customary, in mine ventilation work, to include standard air density in the K-factor and apply corrections to calculated pressure drops according to the actual air density in the mine. When using SI units, however, a case can be made for separating the density term from the K-factor, making the latter dimensionless.

The Darcy equation for head loss caused by the friction in long, straight, uniform ducts is given by

$$H_f = fLV^2/2gD \qquad (3)$$

where f is the dimensionless hydraulic friction factor, L is the length of the duct, D is the hydraulic diameter (4 A/C), V is the velocity, and g is the acceleration due to gravity. This equation is applicable for all units, as long as the quantities are consistently expressed.

The equation (3) is valid for a duct of any shape of cross section and applies for either laminar or turbulent flows. In general, the value of the hydraulic friction factor, f, is a function of Reynolds number and the relative roughness, e/D. The roughness, e, is a length characterizing the hydraulically effective roughness of any duct surface and can be derived experimentally. The results of tests (Nikuradse 1933) on turbulent flow in rough ducts has shown that at high Reynolds numbers, the value of f of rough ducts becomes constant, depending wholly upon the roughness of the duct surface and is thus independent of the

Reynolds number. The relationship between f and e/D is expressed by

$$f = 1/[2 \log (3.7\ D/e)]^2, \quad D = 4A/C. \qquad (4)$$

The dimensional comparison of the Atkinson and Darcy equations shows that

$$K = 0.125\ f. \qquad (5)$$

Since airflow in mines is usually well into the turbulent range, this is a means of calculating K-factors from airway dimensions and an assessment of surface roughness.

The roughness value, e, for an airway can be determined from the following equation.

$$e = 14.8\ A/C\ Alog\ (1/\sqrt{32K}) \qquad (6)$$

The advantage of this equation is that once the roughness value is known for an airway, the K-factor for similar types of airways, but with different cross sections, can be determined using equations (4) and (5).

In this study, all measurements and calculations were carried out in SI units. These units are very convenient for mine ventilation work and are becoming well established in the mining industry. Friction factor and roughness values were determined for straight unobstructed airways only. No attempt was made to assess shock factors due to bends, contractions or other obstructions.

2 TEST METHODS

In principle, the determination of mine airway K-factors is quite simple. Using equation (1), the resistance of an airway can be calculated from a measurement of the pressure drop associated with a measured flow rate. From this resistance value and the relevant airway dimensions, the K-factor can be calculated using equation (2).

In practice, of course, the process is much more difficult due to the irregularity of the airways, fluctuations in airflow rate and the practical difficulties associated with measuring flow rates and pressure drops to the required degree of accuracy.

2.1 Selection of representative airway

The first pre-requisite is the selection of a representative length of airway.

Since the objective is to characterize the frictional nature of the airway surface, the airway chosen must be as uniform in nature as possible throughout its length. To avoid the masking influence of shock factors, it should be reasonably straight and unobstructed. Ideally, the longer the airway the more accurately the pressure drop can be measured.

The accuracy of both pressure drop and flow rate measurements is improved at higher velocities of airflow. A minimum velocity of 3 m/s should be sought. In some cases it may be possible to arrange to have more airflow temporarily diverted to the survey section for the purposes of the measurements. Measuring pressure drops at two or more different flow rates gives a useful check on the results.

2.2 Preparation of test section

The length over which the pressure drop was to be determined was accurately measured with a tape, and the ends marked on the wall with a paint spray can. The perimeter and area were recorded at regular intervals along the section using a photo-profile technique described in the next section. Representative photographs were also taken to illustrate the general nature and condition of the airway.

Two suitable locations for airflow measurement were selected, marked, and photo-profiled for each airway. Where possible, the flow measuring stations were selected outside the test section to cause minimum interference with the pressure drop measurement.

2.3 Cross section measurement

In airways of irregular area the airflow does not occupy the total area. The total area measurement may lead to some errors in the calculations. In theory, the effective area through which air actually flows should be used rather than the total area of the sections. However, in practice it is difficult to measure the effective areas of the sections. In this work the airway dimensions were measured at various locations in test section using the photoprofile method to reduce the error due to inaccurate estimations of effective areas. The average of airway dimensions measured at various locations in the test section was used in all the calculations. The most important effect on the K-factor is that of area determination, and areas and perimeters are

calculated from the same data. The errors in such data will involve an error in the friction factors proportional to A^3/C. For example, a 10% increase in area and a 5% increase in perimeter will increase the value of K by about 27%.

The measurement of areas and parameters in this work were conducted by a photographic method devised previously by the Mining Department of Queen's University. In principle this involves the projection of a narrow band of light to illuminate the perimeter of an airway on a plane perpendicular to the airway axis. A polaroid camera is then used to photograph the illuminated band. A scale of known length is also illuminated in the plane of the profile to relate the dimensions of the photograph to field dimensions. Perimeters and areas were determined from the photographs by a simple computer program after tracing the outline and scale by means of a digitizer.

To illustrate the variability of the airways, composite plots were made of the digitized profiles for each test section. A small computer program was written to find the coordinate centre of each section, adjust all sections to the same scale, and plot the super-imposed profiles at a convenient size together with an appropriate scale bar. Typical cross sections of airways are shown in Fig. 1.

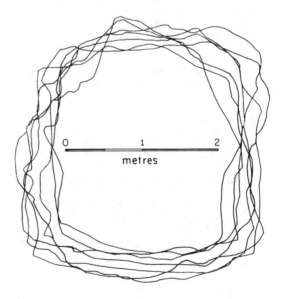

Fig. 1 - Typical cross sections of airways.

709

2.4 Airflow measurement

Air flow rate is not measured directly but is determined as the product of velocity and cross-sectional area, each of which is measured separately. The area measurement has already been discussed.

There are several methods of measuring air velocity and, since the velocity is not constant across the full cross-section, a technique must be used which gives a satisfactory average. The standard rotating vane anemometer is commonly used for routine mine ventilation work, and, although it is subject to large errors if carelessly used, by observing proper procedures a good average velocity can be obtained.

Since the velocity tends to be lower near the walls than in the centre of the airway, it is important to attempt to sample the airflow over the full cross-section so that the measurement represents a weighted average velocity. This is sometimes achieved by traversing the anemometer slowly in a zigzag fashion over the area. The problem with this technique is that the operator has to move around to cover the full area and this affects the air currents. If extension rods are used to enable the operator to reach the far corners of the airway from one location, then they get in the way when trying to cover the closer parts.

The preferred technique is to hold the anemometer for equal lengths of time at a number of positions, each of which represents an equal area of the airway, using extension rods where necessary, and keeping the body out of the airstream if possible.

The latter technique is more conveniently used with the second type of anemometer, namely the Electronic Direct Reading anemometer. This instrument also has a rotating vane, and is equipped with a power supply and associated circuitry which enables it to respond to the rate of rotation of the vanes so that it gives a continuous reading of velocity. Besides being more suitable for measuring velocity at selected points, this instrument is useful for observing the actual variations of velocity over the cross-section and monitoring the stability of the airflow over time. Both of these types of anemometer were used at independent stations for each test airway.

2.5 Pressure drop measurement

There are two methods of measuring pressure drop over a length of mine airway; the barometric method and the gauge and tube method. The barometric method is favoured for complete mine resistance surveys where points may be widely separated. The gauge and tube method, which measures pressure drop directly, is inherently more accurate than the barometric method and was used in this study. In operation, a length of rubber tubing is stretched out along the axis of the airway and connected at one end to one limb of a sensitive pressure difference measuring device. A length of tubing is connected to the other limb of the pressure gauge.

An electronic micromanometer was used for the measurement of differential pressures, except in the coal mines, where a magnehelic gauge was used.

The air flowing through the airway between the ends of the tubing suffers a loss of total pressure due to friction while the air in the tube is stationary. If the open ends of the tubing are pointed upstream at positions of mean velocity in the cross-section of the airway, the total pressure difference between the ends of the tube will be registered by the gauge. In practice, it is often difficult to select a point of mean velocity and it is more convenient to use a static tip at the ends of the tubes so that the gauge registers the difference in static pressure between the ends of the tube. For this survey, a pitot tube was mounted on a tripod and pointed upstream at the centre of the airway at each end of the test section. The static connections of the pitot tubes were connected to the manometer tubing. Since, for this work with uniform cross-sections of airway, the air velocity was approximately the same at both ends, the difference in static pressure was assumed to be the same as the difference in total pressure.

2.6 Air density

In order to adjust the measured K-factors to common terms, either in dimensionless form or at standard density, it is necessary to know the air density at which the pressure drop measurements are made.

K-factor values calculated from equations (1) and (2) using measured P and Q values must be adjusted by a factor of $1.2/d$ for SI units, or by a factor of $1/d$ for dimensionless units, where d is the measured density in kg/m^3.

Air density was determined from barometric pressure, and wet and dry bulb temperature readings, using standard psychrometric formulae. For this purpose, the barometric pressure need not be as accurately determined as the pressure difference measurement.

2.7 Procedure

Before making any measurements, the proposed test location by the mine official was visited to assess the suitability of the site. It was desired to have a reasonable straight, uniform airway, at least 150 m in length, and with a minimum air velocity of 3 m/s. In many cases, it was found impossible to meet all of the desired conditions at any one site. A shorter section could be tolerated if the velocity was high, and conversely, a low velocity could be offset by a longer length of airway. At many locations the air velocity was too low to give a reasonable pressure drop over the lengths of airway available. During the reconnaissance trips the direct reading anemometer was used to check the speed and uniformity of the air current.

Selected test airways were measured, marked up, and a series of photoprofile pictures taken at suitable intervals. A description of the airway was written and, in some cases, illustrative photographs were also taken of the general appearance.

The tubing was stretched out along the airway between the marked end points, taking care to avoid kinks, and the pitot-static tubes were mounted on tripods and pointed upstream at corresponding positions in the airway cross-section. The micromanometer was connected to the tubing and the barometer and psychrometer were set up nearby.

One team member read the pressure drop at five-minute intervals, while velocity measurements were made independently by the other two members. Generally, each observer made two complete sets of velocity measurements at each of the flow measuring stations. Barometric pressure and wet and dry bulb temperatures were read at the beginning and end of the work at each test section and a mean air density was calculated from the two sets of readings.

3 DISCUSSION AND RESULTS

Even though the average velocity may be determined accurately by taking a large number of readings across the cross-section of the airway, the quantity of air flow depends on the flow area used. The effective area of flow is not necessarily the accurately measured area at the plane where the velocity is measured. The effective flow area seems to be dependent on conditions upstream of the measuring station.

It is well established that the air flow is disturbed for some distance downstream of a bend or an obvious obstruction, and, in this work, velocity measuring stations were carefully selected to ensure that they were well away from the influence of any such disturbances. Nevertheless, at many of the test sites, there was a significant difference in the mean flows measured at the two stations, although consistent measurements were obtained at each individual flow station.

Probably the only way to measure airflow accurately is to build a smooth-lined duct of smaller area than the airway and use a converging inlet to minimize turbulence (McElroy & Richardson 1927). However, such an installation is expensive and impractical for most work of this type.

Mine openings are often extremely variable in shape, size, surface texture and sinuosity. The purpose of the K-factor is to characterize these variables by a single parameter, such that airway resistance can be calculated by the formula:

$$R = KCL/A^3$$

Multiplying the resistance value by the square of the airflow rate will then give the pressure drop through the airway. The resultant value of pressure drop through a given circuit then defines the fan duty required to move the specified quantity of air.

The airway length, L, can generally be determined quite accurately, but C and A, the perimeter and area, are more difficult to define. Besides contributing to R, A is also significant in the definition of flow rate, Q, since Q = Area x Velocity, and velocity is normally what is measured. Area appears in the resistance equation to the third power, so any error in the value used for A will have an exaggerated effect on the resistance. Since it is evident that, in airways of irregular shape, the airstream does not occupy the total area, accurate determination of effective area is not possible by direct measurement. The photoprofile method, for instance, although giving an accurate measurement of actual area at a particular plane, gives an area larger than that which is effectively available in an irregular airway.

Table 1. Correction factor for K-values of different cross sections at various roughness values.

Roughness, e m	$\dfrac{K_{0.75}}{K_{0.50}}$	$\dfrac{K_{1.0}}{K_{0.50}}$
0.01	0.89	0.82
0.02	0.88	0.80
0.50	0.85	0.77
0.10	0.84	0.74
0.15	0.82	0.72
0.20	0.81	0.70
0.30	0.79	0.68

It was hoped that the photoprofile pictures, when digitized and super-imposed, could be used as a measure of surface roughness. Although they give an idea of the variability of the airways, they do not really provide a measure of quantifying the roughness. Further, the K-factor of airways varies widely, depending on the roughness of walls; the roughness in turn depends on the properties of the rock, on the location of shotholes, on the obstructions in the airways, etc. Therefore, it is difficult to estimate the value of roughness in very irregular airways. However, once the K-factor is measured, the

Table 2. K-Factor and roughness values for straight airways.

Rock/surface type Airway description	A/C m	K-Factor Dimensionless $(10^{-10}$ lb min^2/ft^4)	Roughness Value, e m
Gunited Rectangular, slight obstruction	0.68	0.00491 (32)	0.03
Gunited Rectangular, slight obstruction	0.64	0.00596 (39)	0.05
Gunited Rectangular, slight obstruction	0.64	0.00597 (39)	0.05
Footwall gneiss Rectangular, ramp, clean	1.36	0.00969 (63)	0.32
Gunited Rectangular, blocky	0.76	0.00903 (59)	0.15
Quartz pebble conglomerate Rectangular, clean	1.0	0.00531 (34)	0.06
Quartz pebble conglomerate Rectangular, slight obstruction	0.98	0.00767 (50)	0.14
Sediments & conglomerate Rectangular, obstruction, slightly curved	0.67	0.00888 (58)	0.13
Basaltic andesite Rectangular, slight obstruction	1.08	0.00710 (46)	0.13
Dacite, mafic flow Rectangular, medium obstruction	0.88	0.00717 (46)	0.11
Andesite, dacite Rectangular, medium obstruction	0.82	0.01008 (65)	0.21
Flow breccia Rectangular, medium obstruction	0.86	0.00925 (60)	0.19
Steel arches with corrugated sheeting, slight obstruction, uneven floor	0.91	0.01017 (66)	0.24

roughness value can be estimated from equation (6). This value of e can now be used to calculate the K-factor for similar airways with different cross sections. The K-factor table (McElroy & Richardson 1927) was produced mostly for smaller airways with an average A/C value of 0.5 m. The values of A/C for the airways under this study varied from 0.6 m to 1.4 m. The increase in A/C will increase the K-factor, if other conditions remain the same. Table 1 shows the correction factors for K-values for A/C values of 0.75 m and 1.0 m compared to A/C value of 0.50 m at various roughness of the airways. For example, the K-factor for an airway with A/C = 1.0 m will be 0.74 times the K-factor of similar airway with A/C = 0.5 m at the roughness value of 0.1 m.

The K-factors measured are given in Table 2. The K-factors are given in dimensionless form and in old units (10^{-10} lb min^2/ft^4) along with the values of area/perimeter and surface of roughness. The value of roughness is a useful parameter to calculate the K-factor for similar airways with different dimensions. This can be done using equations (4) and (5).

4 CONCLUSIONS

Systematic evaluation of mine airway friction factor parameters has been performed across a broad array of mining environments. Measurements were carried out using standard methods and equipment in reasonably straight and unobstructed airways.

The photoprofile technique for characterizing airway cross-section is simple and fast in execution but tedious in the data processing stage. Superimposing the pictures gives a good idea of the variability of an airway but does not characterize the surface roughness unless the pictures are taken closely together.

The roughness values for irregular airways can be adequately determined from the field experiment. This is a valuable parameter in predicting the K-factor of airways with different cross sections.

The range of data presented in this study is not intended to be suitable for application by all mining operations. Due to limitations of time, and the number of sites available, the data gathered offer only general applicability to many mining operations. The methodology used, however, may be readily adopted by individual operators and used to perform site specific measurements. The experiment was carefully planned, but the friction factors are best estimates only.

5 ACKNOWLEDGEMENTS

The authors would like to thank H.J. Hackwood, P. Lausch, D. Turbitt, G. Tucker, P. Satchwell and R. McPherson for their assistance in data collection and processing. The assistance and co-operation of the management and staffs of the various mines visited throughout the survey are greatly appreciated.

6 REFERENCES

McElroy, G.E. & A.S. Richardson 1927. Resistance of metal mine airways. Bureau of Mines. Bulletin 261.
Nikuradse, J. 1933. Stromungsgesetze in rauhen Rohren. V.D.I. Forschungsheft.

A survey of English language mine ventilation literature
with emphasis on literature sources and databases

MARY B.ANSARI & ANNE AMARAL
University of Nevada-Reno, USA

ABSTRACT: Emphasis of this survey of mine ventilation literature is on the field's major English language abstracts, indexes, and bibliographic databases. Also discussed are ventilation's primary publication formats, including journals, government documents, conference proceedings, and books. Addresses for selected ventilation publishing, database and information sources are given. It is concluded that the literature is so interdisciplinary and dispersed that the only way to search it thoroughly is by computer.

1 INTRODUCTION

Mine ventilation literature is diverse in format, widely interdisciplinary, and extremely international in scope, with many important foreign language writings emanating from the Soviet Union, Eastern Europe, West Germany, China, and Japan. Because of the worldwide provenance of the subject and the attendant publication deluge, this survey is limited to coverage of current English language literature sources for metal and non-metal mine ventilation with emphases on the engineering aspects of ventilation and on United States and Canadian publication sources.

This paper updates a survey of English language ventilation literature written in 1979 (Ansari 1980). Emphasis of this update is on the field's major abstracts, indexes and bibliographic databases. In addition, the authors have examined the field's English language periodicals, government reports, books, and conference proceedings. Addresses for many of the publication and information sources and database vendors examined in the paper are listed at the end.

2 INDEXES, ABSTRACTS, AND ACCOMPANYING DATABASES

Abstracts and indexes are the key to finding literature on mine ventilation. They list journal articles, books, reports, dissertations, and many other types of literature. The basic listing is usually by subject, but most also allow searching by author, institution, or various other points of access. Anyone looking for information on an aspect of mining ventilation should start with a check of these important sources.

Researchers may do hand searches through the printed versions of indexes and abstracts, or they may turn to computerized literature searching. Many of the bibliographies published today are produced by typing indexing data onto computer tapes. Publishers then produce printed editions from the tapes, and database vendors produce online versions from them. The vendors, such as DIALOG or QL, have each developed unique software that enables their users to access the data on the tapes via computer terminal. Searchers simply dialup via telephone the computers of their choice and search the indexes and abstracts directly from the tapes. Computerized searching is faster and usually more thorough than manual searching of printed indexes. For this reason, mine ventilation literature searchers should certainly consider computer searching.

Most large academic and public libraries offer computer searching, as do many private information brokers, such as The Information Store or Information on Demand. Some researchers are also doing their own computer searching, which requires special training to learn the commands and intricacies of the various vendors' systems. Menu driven software has been developed for individuals who search only occasionally and do not want to devote a lot of time to learning how to do computer searching. In-Search developed by the Menlo Corporation and Sci-Mate from the Institute for Scientific Information are examples of such menu driven systems, and other programs are being developed every day in this fast moving field.

Computer searching costs money to do. Prices vary from a few dollars to several hundred, depending on the database being searched, the complexity of the question being researched, and the number of citations retrieved. Hand searching, on the other hand, is free if a library that subscribes to the indexes is nearby, but it takes a lot of time to do hand searching. Researchers must decide whether it is more cost effective to do manual or computer searching when they approach indexes.

The following listings are intended to give an idea of the variety of abstracts and indexes available. A perusal of all of them might supply more than you ever wanted to know about mine ventilation, but a perusal of several is necessary if you want to get the basic information. Mine ventilation information is spread through several disciplines and a single source will not include everything.

Printed versions of indexes are discussed first, followed by information on online versions. Vendors of online versions are indicated, but this is a rapidly changing field and the information given may be incorrect by the time this paper is published. Contact the index publisher for the most up-to-date news on online access.

MINTEC: Mining Technology Abstracts is one of the best indexes available for mine ventilation information. It is produced by CANMET, the Canadian government's Centre for Mineral and Energy Technology, a part of the Department of Energy, Mines and Resources in Ottawa. It contains references to articles, reports, papers, government publications, and books on mining technology and includes a vast number of citations on mine ventilation. Abstracts are given for most materials indexed. CANMET will supply photocopies for a fee of items that cannot be obtained at local libraries. The online version of MINTEC is available from QL/Search, a Canadian vendor. It includes citations to items published from 1973 to the present, with some dating back to 1968, and it is updated every two weeks. Best of all, it is one of the least expensive mining databases available. This is probably the best index to start with when doing a search.

The Engineering Index and the Engineering Conference Index cover all fields of engineering. The Engineering Index began publication in 1906, and is now issued monthly with annual cumulations. This vast work attempts to index everything published worldwide on engineering, and includes a section on mining engineering which in turn includes a section on mine ventilation. It gives bibliographic citations to journal articles, books, reports, standards and conferences. Individual conference papers were included in Engineering Index until 1982 when it was decided to place them in a separate index that is just beginning publication and is called the Engineering Conference Index. Together these two works give in-depth coverage of mine ventilation literature. They are produced by Engineering Information, Inc., which also provides copies for a fee of the articles and other materials indexed. The online version of the Engineering Index is called COMPENDEX and includes materials published from 1970 to current. The online version of the Engineering Conference Index is called EI Engineering Meetings and includes materials from 1982 to the present. They are both available from several vendors, such as DIALOG, SDC, and Pergamon.

IMM Abstracts is an English index that covers the entire field of mining, metallurgy and geological exploration. Produced by the Institution of Mining and Metallurgy in London, it is a good source of information on mine ventilation. It began publication in 1950 and is currently issued every two months. It indexes and includes abstracts of journal articles, books, government publications, reports and papers. The printed version has no specific approach to mine ventilation but merely includes articles under the general subject of mine services or health and safety. Additionally, there are no annual cumulations, so searchers must look through each bimonthly issue to locate materials. Both of these problems are solved in the online version, which is called IMMAGE. It supplies quick and precise access to materials on mine ventilation from about 1982 to the present and is available directly from the Institution of Mining and Metallurgy.

Coal Abstracts includes a large number of references on mine ventilation as part of its coverage on coal mine environment, equipment, and health and safety. It is published monthly by IEA Coal Research, which was established by the International Energy Agency in 1975 to cover projects on coal information and assessment. It provides world-wide coverage of books, journal articles, reports, dissertations and conference proceedings, and gives abstracts of materials indexed. It will provide copies of items indexed if other available sources cannot do so. Online versions are available from QL and from DIALOG where it is issued as part of DOE Energy.

Great Britain's National Coal Board also publishes abstracts on coal mine ventilation in National Coal Board Abstracts A: Technical Coal Press. Listings on ventilation are classified under the letter F, which covers ventilation, methane, dust, fires, and illumination. There is no online version of this.

GEOARCHIVE is an online index only. It has no printed equivalent, although several secondary publications are printed from the information contained in it. It emphasizes geology but also includes mining engineering in its subject coverage and contains a quantity of references on mine ventilation, most of them dating from 1981 to the present. It indexes journal articles, books, government publications, and dissertations. GEOARCHIVE is produced by Geosystems, an English company, and is available through DIALOG.

The Applied Science and Technology Index (ASTI) is a good index to use for quick searches for information. It is published by the H.W.Wilson Company which also produces the Readers' Guide to Periodical Literature, that bulwark of the public libraries that indexes popular magazines. ASTI uses the same format as the Readers' Guide but indexes key journals in science and technology. It lists articles under mine ventilation or related headings, such as mine air or mine gas. It is easy and quick to use. Best of all, if you can find articles on your subject, you will probably have instant access to the articles because ASTI indexes the journals most commonly owned by libraries. It is published monthly and is cumulated into an annual volume. It began publication in 1913, so it is possible to do quite a lengthy search in it. Most people use it, though, to check only the last 5 or 6 years in order to see what noteworthy things are happening in their field. ASTI is available for computerized searching from November 1983 to current through WILSONLINE directly from the H.W.Wilson Company.

Science Citation Index (SCI) is a good index to use if you want to know the most recent work being done in an area that you have not researched in several years. SCI is an index to approximately 2600 of the most used journals in science and technol-

ogy. It provides a subject approach to items based on keywords from the titles of the articles, and it also provides a unique citation approach. If you know a key article that was published several years ago about an aspect of mine ventilation, you can find if anyone has cited it in the current literature indexed in SCI by using the citation index. The new article that cites your original article will frequently reveal what research has occurred on the subject since the original was published. SCI has been published by the Institute for Scientific Information since 1961. The online version is called SCISEARCH and covers the years 1974 to current. It is available from DIALOG.

Dissertation Abstracts International (DAI) and Comprehensive Dissertation Index (CDI) index doctoral dissertations, which contain original research and should be included in comprehensive searches on mine ventilation. DAI is the original index for checking dissertations. It is published monthly and includes abstracts of each item listed. It has gone through many changes since it began publication in 1938, but it now includes all dissertations produced at accredited institutions in the United States and Canada and also some from Europe and other parts of the world. CDI lists dissertations published in the United States from 1861 to the present, and currently also includes all dissertations produced in Canada and many from Europe and other parts of the world. It is produced annually and acts as an index to DAI. Items are listed under broad subjects such as engineering, and then arranged by keywords taken from the titles of the dissertations. It does not have abstracts, but it gives references to the abstracts published in DAI. Both DAI and CDI are published by University Microfilms International, which sells microfilm copies of most of the dissertations indexed. The online version is called Dissertation Abstracts Online, and is available from several vendors, such as DIALOG and BRS.

Conference Papers Index (CPI) and the Index to Scientific and Technical Proceedings (ISTP) both index papers presented at scientific meetings, symposia, etc. They cover a broad range of disciplines and include a limited amount of information on mine ventilation. CPI includes both published and unpublished papers; ISTP indexes published papers only. CPI is published by Cambridge Scientific Abstracts and is available online from DIALOG. ISTP is published by the Institute For Scientific Information and is available online from a

German vendor, DIMDI.

Safety Science Abstracts Journal covers the emerging discipline of identifying, evaluating, and eliminating or controlling hazards in the environment, in transportation, aviation, medicine, health, and in industry. It contains a limited amount of information on mine ventilation as it relates to safety in mining operations. For instance, it indexes an article that reports on the quantity of quartz dust at various mines. It is published quarterly by Cambridge Scientific Abstracts, Inc. for the University of Southern California Institute of Safety Systems Management, and it indexes and gives abstracts for journal articles, governments reports, books, conference papers, etc. It currently has no online version.

Occupational Safety and Health (NIOSH) is another source of information on mine ventilation as it relates to safety. Developed by the U.S. National Institute for Occupational Safety and Health as an in-house file, it covers all occupations including mining. Articles on such subjects as dust control in mines or the design of safe mine ventilating systems can be found here. Journal articles, books, reports and government documents are indexed. NIOSH makes its file available for general use through online searching from vendors such as DIALOG or Pergamon. There is no printed version.

Government publications are an excellent source of information on mine ventilation. The Monthly Catalog of United States Government Publications lists materials published by all U.S. agencies and includes items produced by the U.S. Bureau of Mines. It lists materials on mine ventilation under such headings as mine accidents, mine dusts, mine explosions, mine fires, or mine safety. It is published by the Superintendent of Documents, and has an online version called GPO Monthly Catalog that's available through DIALOG or BRS.

The Government Reports Announcements and Index covers government research reports, an important source for the latest research in any field and an essential source to check for comprehensive coverage in mine ventilation. It is published biweekly by the National Technical Information Service (NTIS) of the United States Department of Commerce. NTIS serves as the central source for public sale of U.S. and foreign government sponsored research, development, and engineering reports and other studies prepared by federal and local government agencies or their grantees and contractors. Citations include abstracts along with price codes for microfiche or hard copy of the reports. The online version is called NTIS and is available from DIALOG, BRS, and SDC.

Federal Research In Progress provides online access to information about ongoing research projects funded by the United States government. It has no printed version. The U.S. Bureau of Mines unfortunately does not contribute to the database, but such agencies as the National Institute for Health and Safety do, and some information is given on research in progress on mine ventilation. Federal Research in Progress is produced by NTIS and is available online from DIALOG.

National, state and provincial government agencies dealing with mining are all likely to publish reports on ventilation and to issue publication lists of their available publications. In addition to the Monthly Catalog and the NTIS indexes, the United States government publishes the List of Bureau of Mines Publications and Articles, which is available from the Superintendent of Documents. The list is in several volumes covering the years 1910-1960, 1960-64, 1965-69, 1970-74, 1975-79, and annual volumes cover the years after 1979. Current publications are listed monthly in New Publications – Bureau of Mines, which is available directly from the U.S. Bureau of Mines. The monthly lists are cumulated into the annual lists, although with considerable delay. There is no online version of these lists.

There is no good index to U.S. state publications on mining, although they are sometimes listed in the Monthly Checklist of State Publications produced by the U.S. Library of Congress and available from the Superintendent of Documents. The best way to keep on top of these publications is to write the various state agencies that deal with mining and ask for copies of their publication lists.

British government publications are listed in the Annual Catalogue: Government Publications available from HMSO. There is no online version of this.

CANMET, which produces MINTEC, issues its own publications lists, and they are also listed in the Catalogue of Government of Canada Publications published by the Canadian Government Publishing Centre. MINTEC, either online or in the printed version, is a good source for listings of Canadian government publications on mine ventilation.

3 JOURNALS

The field's periodical literature consists of one specialized journal and a great deal of material scattered throughout many mining, health and safety journals.

The Journal of the Mine Ventilation Society of South Africa continues to be the field's only periodical title devoted completely to the subject of mine ventilation. Now in its thirty-eighth year, it is published monthly by the Mine Ventilation Society of South Africa. Each issue contains an average of three or four referenced articles on ventilation as well as announcements of forthcoming meetings and news of the Society and its members. The preponderance of contributors of articles are from South Africa, but the subjects covered are of general interest.

A number of periodicals carry occasional articles on ventilation, e.g. American Industrial Hygiene Association Journal, British Journal of Industrial Medicine, Canadian Mining Journal, Canadian Mining & Metallurgical (CIM) Bulletin, Coal Age, Coal Mining & Processing, Colliery Guardian, Engineering & Mining Journal, Glueckauf & Translation, International Journal of Rock Mechanics, Institution of Mining & Metallurgy Transactions, Journal of the South African Institute of Mining & Metallurgy, Mining Congress Journal, Mining Engineer (London), Mining Engineering, Mining Magazine, Mining Technology, Soviet Mining Science (in English), and Tunnels & Tunneling, to mention some of the more common ones. Since general mining, health and safety journals publish on a wide range of topics in addition to ventilation, it is recommended that you use the indexes/abstracts/databases discussed in the preceding section to find references to ventilation articles.

4 GOVERNMENT REPORTS

Much of the field's publishing appears in the form of reports generated by national and state/provincial government agencies. Some of the more common of these will be discussed below.

The U.S. Bureau of Mines (USBM) is the source of a large number of publications dealing with health, safety and engineering aspects of ventilation. Several of the computer databases discussed earlier, such as COMPENDEX and MINTEC, may be used to identify specific USBM publications. For those doing manual searching, the USBM issues its own indexes and publication lists, which also have been described earlier.

Material on or relating to ventilation appears in most series issued by the USBM, e.g. Reports of Investigations (RI's), Information Circulars (IC's), Open File Report (OFR's), and Technical Progress Reports (TPR's), so no particular series can be easily identified as being the best source. This is particularly significant when looking for their newest publication releases, as items not yet appearing in an annual cumulated publication list, complete with detailed index, must be identified through their monthly listing, New Publications - Bureau of Mines. This listing has no index and is arranged by series and publication number. USBM indexes also list patents which have been granted to the USBM. Additionally, USBM issues an information sheet called Technology News, which sometimes devotes its coverage to various aspects of ventilation technology. Many USBM publications, such as RI's and IC's, are available for free from the USBM office in Pittsburgh, PA. Much information on ventilation appears in the form of OFR's, most of which may be purchased from the National Technical Information Service (NTIS).

The U.S. Mine Safety and Health Administration (MSHA), formerly Mining Enforcement and Safety Administration, has recently issued several ventilation/ventilation-related documents in its Informational Reports series. Its Safety Manual series includes manuals on mine gases, heat stress, radiation hazards, accident prevention, dust, and ventilation. MSHA's Mine Safety & Health magazine often carries short articles on the health and safety aspects of ventilation; it also lists new MSHA publications in the Focus Section under "MSHA Library". MSHA publications are released through the Superintendent of Documents or NTIS.

The U.S. National Institute on Occupational Safety and Health (NIOSH) publishes reports and sponsors research useful in describing and identifying hazards resulting from ventilation-related problems. Federal Research In Progress, NTIS and NIOSH databases, discussed earlier, contain references to a significant amount of NIOSH-sponsored research relating to ventilation. Most NIOSH publications are available from the Superintendent of Documents or NTIS.

Many reports issued by the above U.S. government agencies are likely to become available from NTIS. NTIS materials are accessible through a printed index, the Government Reports Announcements & Index, which is fairly widely available, and a computer database known as NTIS, both of which were detailed in the section on indexes and abstracts. Items released through the Superintendent of Documents will be listed in the Monthly Catalog of United State Government Publications, which is quite widely available. Back issues and backsets of USBM and MSHA publications are available on microfiche from UPDATA in Los Angeles, CA.; its "Catalog of U.S. Government and R & D Documents on Microfiche" is available upon request.

The Canadian government issues reports on ventilation and other aspects of mining technology through CANMET's Coal Research Laboratory Reports, Contract Reports, and Mining Research Laboratory Reports. These reports and other publications are accessible through CANMET's own publication lists and announcements, MINTEC: Mining Technology Abstracts, and the MINTEC database as well as the Catalogue of Canadian Government Publications.

Many other government sources are worthy of note. Great Britain's Safety in Mines Research Establishment (SMRE) carries out research into various safety problems arising within the mining industry. Many of their research projects result from investigations and recommendations following mining accidents. SMRE publishes several series, such as its Research Reports, Reports, and Bibliographies, which include ventilation-related material. Its index, Health and Safety Laboratories Abstracts, ceased publication in 1983, but a publications catalog is available; also, SMRE's major reports are published by Her Majesty's Stationery Office and are listed in HMSO's Annual Catalogue: Government Publications. Britain's National Coal Board also has several useful series, including basic training aids. A publications list is available.

Publications emanating from Australian government agencies on mining research are listed in Australian Government Publications, published by the National Library of Australia. The Chamber of Mines of South Africa publishes reports on ventilation, refrigeration and the physiological effects of heat on miners. Publication lists are available.

Agencies of various states/provinces and local governments may occasionally issue useful materials. These are more difficult to identify and will usually require identifying relevant agencies and writing to ask for a list of publications. Unfortunately, many of these agencies do not maintain continuing mailing lists.

Because NTIS serves as the single largest source for public sale of U.S. and foreign government-sponsored research, development, and engineering reports and other studies prepared by federal and local government agencies or their grantees and contractors, it is the best single source for obtaining technical reports that are based on domestic and foreign government-sponsored research. Additionally, NTIS has available two important bibliographies of citations on ventilation from its NTIS database: (1) Mine Ventilation 1964 - February 1982, Report # PB83-807909, and (2) Mine Ventilation March 1982 - June 1983, Report # PB83-807917.

5 CONFERENCE PROCEEDINGS

Proceedings of conferences and symposia are a very fruitful source of information. Proceedings solely on the subject of ventilation are published by the International Mine Ventilation Congress (IMVC), the Underground Ventilation Committee of the Coal/Mining and Exploration divisions of the Society of Mining Engineers of the American Institute of Mining, Metallurgical, and Petroleum Engineers (AIME), and the Mine Ventilation Society of South Africa (MVSSA). The proceedings of the the First IMVC held in Johannesburg in 1975 were published by the MVSSA in 1976. Papers of the Second IMVC held in Reno, Nevada USA were published in New York by AIME in 1980. In 1984 the Institute of Mining & Metallurgy (IMM) in London published the proceedings of the Third IMVC held in Harrogate in 1984.

Proceedings of the First U.S. Mine Ventilation Symposium (USMVS) held in 1982 in Tuscaloosa, Alabama USA were published by AIME in 1982. Papers from the Second USMVS to be held in Reno, Nevada in 1985 will be published for AIME by A.A. Balkema Publishers in Rotterdam, Netherlands.

The Mine Ventilation Society of South Africa holds an annual symposium on mine ventilation. The papers are published in looseleaf format by the Society.

Papers on ventilation frequently are included in the published proceedings of conferences, congresses, institutes, and symposia that are not devoted entirely to the subject of ventilation. A few of these include papers or proceedings from the American Mining Congresses; Australasian Institute of Mining & Metallurgy symposia and conferences; Canadian Institute of Mining & Metallurgy annual meetings; Commonwealth Mining & Metallurgical Congresses; Institution of Mining Engineers (London) colloquia and symposia; International Symposia on Crisis Management of Mine Fires, Explosions, and Other Emergencies; International Symposia on the Application of Computers in the Mineral Industry (APCOM); Mining & Metallurgical Institute of Japan joint meetings and symposia; Rapid Excavation & Tunneling conferences; Safety in Mines Research Institute International Conferences; Society of Mining Engineers (SME)-AIME annual and fall meetings; and World Mining Congresses.

The published proceedings of Virginia Polytechnic Institute & State University's annual coal mine health and safety institutes normally contain several papers on ventilation. The Association of Mine Managers of South Africa Papers and Discussions, published biennially by the Chamber of Mines of South Africa, usually contain a section on ventilation and dust control. Because of the extremely scattered nature of the material, conference, congress, symposia, colloquia and institute papers are most readily accessed through the Engineering Index, Engineering Conference Index, IMM Abstracts, Conference Papers Index, and Index to Scientific & Technical Proceedings or their attendant databases.

6 BOOKS

Surprisingly little of the field's recent publication appears in traditional book format. A consultation of the 1984/85 Subject Guide to Books in Print published by R.R. Bowker & Co., reveals only one listing under the subject of mine ventilation, that of the Proceedings of the First (U.S.) Mine Ventilation Symposium. Also, because of the relatively small size of the discipline there is almost no publication of traditional type reference books, such as handbooks, dictionaries, encyclopedias, directories, or guides to the literature. Frequently, material on ventilation is included in chapters or sections of books covering more general mining engineering topics. The American Institute of Mining Engineers, Mine Ventilation Society of South Africa and Britain's National Coal Board are good sources of books/booklets on ventilation. Some recent noteworthy books are listed below:

Bossard, F.C. 1983. A manual of mine ventilation design practices. 2d ed., 552 p. Butte Montana: Floyd C. Bossard & Associates, Inc.
Burrows, J. (ed.) 1982. Environmental engineering in South African mines. 987p. Marshalltown, South Africa: Mine Ventilation Society of South Africa.

Hall, C.J. 1981. Mine ventilation engineering. 344p. New York: AIME.
Hartman, H.L. 1982. Mine ventilation and air conditioning. 2d ed., 791p. New York: Wiley.

7 CLOSING OBSERVATIONS

Since ventilation research is reported primarily in general mining journals, government reports and conference proceedings, it is necessary to consult a large number of indexes and abstracts in order to conduct a thorough search of the literature. Prior to the advent of computerized literature searching, a thorough manual search of the literature was a major undertaking. Increasing availability of computerized literature searching has made what was once a formidable task for the researcher or librarian a very fast and efficient method of searching the literature.

8 LISTING OF INFORMATION, PUBLICATION, AND DATABASE SOURCES

Listed below are addresses and phone numbers for mine ventilation information sources, publishers and database vendors.

AIME
345 E. 47th St., 14th Floor
New York, NY 10017, USA
Ph. 212/705-7695
or
Society of Mining Engineers of AIME
Caller No. D
Littleton, CO 80127, USA
Ph. 303/973-9550

Australasian Institute of Mining & Metallurgy
Clunies Ross House, 191 Royal Parade, Parkville, Victoria, Australia 3052
Ph. 03/347-3166

BRS
1200 Route 7
Latham, New York 12110, USA
Ph. 800/833-4707; 800/553-5566 (in NY); 518/783-1161 (outside USA)

Cambridge Scientific Abstracts
5161 River Road
Bethesda, MD 20816, USA
Ph. 800/638-8076;
301/951-1400 (in MD and outside USA)

CANMET (Canada Centre for Mineral & Energy
   Technology)
555 Booth St.
Ottawa, Ontario, Canada K1A 0G1
Ph. 613/995-4029

Canadian Government Publishing Centre
Supply and Services Canada
Ottawa, Ontario, Canada K1A 0S9

Chamber of Mines of South Africa
5 Hollard St.
Johannesburg 2001
Republic of South Africa
Ph. 838-8211

Colorado School of Mines Library
Golden, CO 80401, USA
Ph. 303/273-3665

DIALOG Information Services, Inc.
3460 Hillview Avenue
Palo Alto, CA 94304, USA
Ph. 800/227-1927; 800/982-5838 (in CA);
   415/858-3785 (outside USA)

DIMDI
Weisshausstrasse 27
P.O. Box 420580
5000 Cologne 41
Federal Republic of Germany
Ph. 221/4724-1

Engineering Information, Inc.
345 East 47th Street
New York, NY 10017, USA
Ph. 800/221-1044;
   212/705-7600 (in NY and outside USA)

Engineering Societies Library
345 E. 47th St.
New York, NY 10017, USA
Ph. 212/705-7611

Geosystems
P.O. Box 573
Cambridge, MA 02139, USA
   or
Geosystems
P.O. Box 1024
Westminster, London SW1, UK
Ph. 01/222-7305

HMSO Books (Her Majesty's Stationery Of-
   fice)
P.O. Box 276
London SW8 5DR, UK
Ph. 01/211-5656

IEA Coal Research
14/15 Lower Grosvenor Place
London SW1W 0EX, UK
Ph. 01/828-4661

Information on Demand, Inc.
Box 9550
Berkeley, CA 94709, USA
Ph. 415/644-4500

Information Store
140 Second Street
San Francisco, CA 94105, USA
Ph. 415/543-4636

Institute for Scientific Information
3501 Market Street
Philadelphia, PA 19104, USA
Ph. 800/523-1850;
   215/386-0100 (in PA and outside USA)

Institution of Mining & Metallurgy
44 Portland Place
London, W1N 4BR, UK
Ph. 01/580-3802

Institution of Mining Engineers
Hobart House
Grosvenor Place
London SW1X 7AE, UK

Mackay School of Mines Library
University of Nevada
Reno, NV 89557-0044, USA
Ph. 702/784-6596

Menlo Corporation
4633 Old Ironsides, Suite 400
Santa Clara, CA 95050, USA
Ph. 408/986-1200

Mine Ventilation Society of South Africa
Kelvin House, 2 Hollard St.
Johannesburg 2001
Republic of South Africa
Ph. 832-2177

National Coal Board
Hobart House, Grosvenor Place,
London, SW1X 7AE, UK
Ph. 01/235-2020 ext. 34683

National Library of Australia
Parkes Place
Canberra, ACT 2600, Australia
Ph. 62-1111

National Technical Information Service
   (NTIS)
5285 Port Royal Road
Springfield, VA 22161, USA
Ph. 703/487-4600

Pergamon Infoline, Inc.
1340 Old Chain Bridge Road
McLean, VA 22101, USA
Ph. 800/336-7575; 703/442-0900 (in VA)
   or

Pergamon Press Ltd.
150 Consumers Road, Suite 104
Willowdale, Ontario, Canada M2J 1P9
Ph. 416/497-8337
  or
Pergamon Infoline Ltd.
12 Vandy St.
London EC2A 2DE, UK
Ph. 01/377-4650

QL Systems Limited
205 Tower B
112 Kent Street
Ottawa, Ontario, Canada K1P 5P2
Ph. 613/238-3499

Safety in Mines Research Establishment
Health and Safety Executive,
Red Hill off Broad Lane
Sheffield S3 7HQ, UK
Ph. Sheffield 78141 ext. 3115

SDC Information Services
2500 Colorado Avenue
Santa Monica, CA 90406, USA
Ph. 800/421-7229; 800/352-6689 (in CA);
    213/453-6194 (outside USA)

Superintendent of Documents
U.S. Government Printing Office
Washington, DC 20401, USA
Ph. 202/275-3204

U.S. Bureau of Mines
4800 Forbes Ave.
Pittsburgh, PA 15213, USA
Ph. 412/621-4500

U.S. MSHA Information Services Library
P.O. Box 25367
Denver, CO 80225, USA
Ph. 303/236-2729

U.S. National Institute for Occupational
  Safety and Health
4676 Columbia Parkway
Cincinnati, OH 45226, USA
Ph. 513/684-8326

U.S. National Mines Health and Safety
  Academy
Learning Resource Center
P.O. Box 1166
Beckley, WV 25801, USA
Ph. 304/255-0451 ext. 266

University Microfilms International
300 North Zeeb Road
Ann Arbor, MI 48106, USA
Ph. 800/521-3042 (in USA);
    800/268-6090 (in Canada);
    313/761-4700(in MI and outside USA &
    Canada)

UPDATA
1746 Westwood Blvd.
Los Angeles, CA 90024, USA
Ph. 213/474-5900

H.W. Wilson Company
950 University Avenue
Bronx, New York, 10452, USA
Ph. 800/622-4002;
    212/588-8400 (in NY and outside USA)

9 REFERENCE

Ansari, M.B., S.J. Hoyle, & A.H. Reynolds. 1980. Recent English language literature of mine ventilation. In Mousset-Jones, P. (ed.) Proc. 2nd Int. Mine Ventilation Cong., Reno, Nev., 1979, p.63-69, New York: American Institute of Mining, Metallurgical & Petroleum Engineers)

18. Face ventilation II

Laboratory investigation on the effectiveness of an air spray system for dust control on longwall faces

SANDIP K.MUKHERJEE, ANTHONY W.LAURITO & MADAN M.SINGH
Engineers International Inc., Westmont, IL, USA

NATESA I.JAYARAMAN
US Bureau of Mines, Pittsburgh, PA, USA

ABSTRACT: Engineers International, Inc. (EI), under an U.S. Bureau of Mines sponsored study, conducted a laboratory investigation to evaluate the effectiveness of an air spray system for dust control on longwall faces. The investigation consisted of constructing a full scale 30.48 m (100 ft) long, 1.83 m (6 ft) high longwall mock-up gallery with a 4.57 m (15 ft) wide headgate, Dowty 2-leg shield support units, and a simulated Eickhoff double-drum shearer. An exhausting fan was located at the "tailgate" end of the model to induce realistic quantities of air across the face. Spray nozzles, delivering air provided by an air compressor, were mounted on the face side of the supports and oriented in the direction of the airflow.

This paper describes in detail all aspects of the laboratory investigation conducted. Construction of the longwall mock-up gallery, the pneumatic circuitry utilized, and the smoke and tracer gas (sulfur hexafluoride) tests conducted to determine system effectiveness are described. The results obtained for varying face air velocities are presented.

1 INTRODUCTION

One of the primary constraints toward achieving higher production from longwall faces is compliance with the federal dust standard of 2 mg/m^3. In recent years, the Bureau of Mines and private industry have spent considerable effort to reduce dust concentrations in longwall faces. Some of the noteworthy developments are:

- improved ventilation schemes including homotropal ventilation and the use of gob and cut-out curtains
- development of a proportioning valve to regulate the flow of water to the drums
- an external spray system (sprays mounted on the body of the machine) known as the Shearer Clearer which utilizes the air moving characteristics of ordinary water sprays to confine the dust cloud to the face and away from the machine operators
- wet-head drums in which water is supplied to nozzles located ahead or behind each bit or along the scrolls. Encouraging results have been recently obtained with a system flushing water through the bits
- improved drum and bit design incorporating deeper cutting and reduced drum rotational speed

- machine mounted scrubbers
- modified cutting sequences aimed at keeping the machine operators upstream of the cutter drums.

Despite these developments, longwall operators face a difficult task in consistently complying with the federal dust standard while concurrently exploiting the full production potential of the shearer. Clearly, a need exists to evaluate the effectiveness of available and new and novel dust control techniques which would guide the longwall operators toward the best available technology to control respirable dust without adversely affecting production. One proposed new and novel dust control concept investigated by EI for the Bureau of Mines is the "air spray" system which is the subject of this paper.

2 CONCEPT DESCRIPTION

The "air spray" concept is predicated on the principle that air circulation not only causes dilution of contaminants but can also keep the contaminants away from the face personnel.

Spray nozzles are available which deliver only air. These spray nozzles properly mounted on the face side of the supports and oriented in the direction of the airflow could help to confine the dust along the face and keep the shearer operator and downstream face personnel in compliance by providing these locations with additional local dilution capability. The air to the nozzles can be supplied from an air compressor located in the headgate area. It was expected that this approach will provide a simple and rugged system comfortable and easy to maintain, and which can be applicable under a wide range of geological conditions. Alternatively, air spray nozzles could also be mounted directly on the shearer to reduce dust concentrations at the shearer-operator position. The air spray system was not intended to replace wet cutting and the use of water sprays mounted on the machine. Rather it was perceived to work in conjunction with the existing dust control system with perhaps a reduction in the number of external sprays. The concept is depicted in Figure 1.

3   CONCEPT INVESTIGATION

The effectiveness of the air spray system was investigated in the laboratory using tracer gas (sulfur hexafluoride) to simulate respirable dust. The laboratory investigation consisted of the following:
- construction of a longwall mock-up
- selection of an adequate capacity fan to provide sufficient airflow through the mock-up
- compressor selection based on anticipated air requirements for the air spray nozzles
- completing the pneumatic circuitry from the compressor to the model and pneumatic layout within the model, and
- smoke and tracer gas tests to determine the effectiveness of the air spray system.

3.1   Construction of the longwall mock-up

The longwall mock-up constructed in the laboratory was patterned after a standard face utilizing an Eickhoff double-drum shearer and Dowty 2-leg shield support units. It was constructed of timber (2x4's) and polyurethene sheeting. Furring strips were utilized to support the sheeting and cross members were used for additional support and reinforcement. The model length was approximately 38.1 m (125 ft) with a 4.57 m (15-ft) headgate, 3 m (10-ft) fan transition and 30.48 m (100-ft) "actual" face. Face height was 1.83 m (6-ft). The simulated model of the Eickhoff shearer was also constructed from 2x4 timber and polyurethane sheeting. Dimensions of the shearer are approximately 7.6 m (25-ft) long by 1.07 m (3½-ft) wide and 1.22 m (4-ft) high with lead and tail drums each 76.2 mm (30-in.) wide mounted at each end of the shearer body. Figure 2 illustrates the details of the mock-up assembly. The shearer was positioned to cut against the airflow.

Airflow through the mock-up model was provided by an exhausting fan located at the "tailgate" end of the model. A Powerline tube-axial fan (Model BT-48) was selected which was more than adequate to satisfy the necessary air quantity and pressure requirements. Fan specifications are as follows:

Type of fan: Tube axial.
Propellers: Cast aluminum, non-overloading airfoil section blades.
Motors and drives: Standard NEMA frame motors, equipped with ball bearings. Motors mounted outside the airstream on top of the housing. Belt drive mechanism

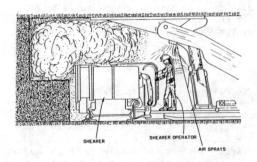

Fig. 1.   The air spray longwall dust control concept.

Fig. 2.   Laboratory longwall mock-up facility with shearer and fan in the background.

also located out of the air stream, enclosed by steel tube open at both ends for cooling the shielded bearing and for easy belt maintenance. Belt tensioning facilities provided.

Frame: The fan drum is constructed of heavy rolled steel with welded seam and wide mounting flanges.

Discharge: 1,076 m^3/min (38,000 cfm) at 6.35 mm (0.25-in.) static pressure.

RPM: 1,130

Motor power: 7.46 kw (10 hp).

Electrical: 460 v, 60 Hz, 3 phase.

During actual testing, the fan was able to induce about 1,133 m^3/min (40,000 cfm) of air into the model. Air velocity, measured at the headgate was about 2.79 m/s (550 fpm). Mounting of the fan was accomplished with channel steel and angle iron. The transition between the longwall mock-up and fan was a fire resistant canvas tarpaulin. The tarpaulin could withstand pressure differences in excess of 25.4 mm (1-in.) water gauge.

The basic pneumatic circuit consisted of two trunk lines, four face lines and 20 support lines to provide maximum flexibility and coverage (Figure 3). The trunk line from each compressor consisted of 30.48 m (100 ft) of 76.2 mm (3-in.) line, 15.2 m (50 ft) of 50.8 mm (2-in.) line and 22.86 m (75-ft) of 38.1 mm (1½-in/2-in.) line. The 38.1 mm (1½-in.) line was split to feed two face lines of 12.7 mm (½-in.) diameter. Support lines were connected to the face lines in banks of five. In addition to the two banks of two sprays on each support unit (one on the canopy and the other near the walkway), four sprays were mounted in the headgate area. A total of eighty nozzles were installed in the longwall model. Spray nozzles were of the air diffusion type. The nozzles were of brass

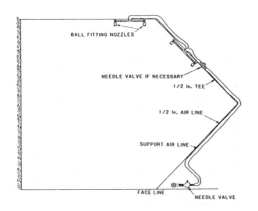

Fig. 3. Pneumatic circuitry along face.

material and capable of delivering up to 0.76 m^3/min. (27 scfm) at 414 kPa (60 psi). All support and face hoses were made of Uniroyal P-980, which is resistant to oil, weather, and abrasion. Maximum working pressure was 1724 kPa (250 psi) for the double braid hoses.

Two Joy diesel powered compressors, models PTS-900 and PTS-1200 were utilized to provide the required 59.5 m^3/min. (2100 scfm) of compressed air to the longwall model. These compressors delivered oil free air for human use and were silenced for quiet operation. Working pressure for the two units was 703 kPa (102 psi) obtained through two compression stages. The actual free air deliveries for the PTS-900 and PTS-1200 were 25.5 m^3/min. (900 scfm) and 34 m^3/min. (1200 scfm), respectively. These units were operated in parallel with each unit feeding a trunk line. The PTS-1200 was always used with the trunk line responsible for the air sprays near the shearer operator positions, regardless of the shearer's position in the model.

3.2 Laboratory testing of air sprays system

The effectiveness of the air sprays system was tested using the longwall mock-up facility. Tracer gas (SF$_6$) samples were collected with and without the air sprays system in operation in order to determine the laboratory effectiveness of the air sprays system.

Prior to sampling with the air sprays system in operation, smoke tests were conducted for orienting the air spray nozzles to minimize dust dispersion at the shearer operator positions and maximize confinement of the dust cloud toward and along the face. Although some rollback of the smoke cloud toward the operator positions was apparent, it was found in almost all instances that orienting the nozzles in the direction of airflow at about 45 degrees from the perpendicular to the face yielded the best results.

The laboratory investigation was conducted for two shearer positions (shearer near the headgate and at midface) and varying air velocities from 1 m/s (200 fpm) to 2.8 m/s (550 fpm). For a particular shearer position and face velocity, tracer gas samples were collected with and without the air sprays in operation.

Prior to base system and air sprays system sampling, grab samples were collected in the laboratory and from within the mock-up facility to determine any ambient concentration of SF$_6$. If any presence of SF$_6$ was detected, the mock-up was flushed

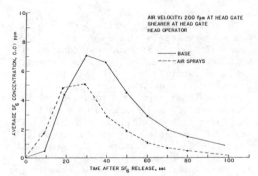

Fig. 4. Time versus concentration curve for head operator.

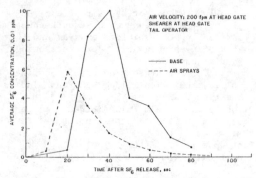

Fig. 5. Time versus concentration curve for tail operator.

by keeping the fan running until there was no ambient SF_6 concentration. It was possible to run two tests per day but on occasions (at lower velocities) only one test could be conducted.

The comparatively high face velocities on longwall faces necessitates quick collection of samples after release of the tracer at the lead drum. Initially, samples were collected every 30 seconds but dilution was so rapid that peak concentrations were not detected. It was decided therefore, that in order to make a realistic assessment of the air spray system, samples need to be collected as quickly as possible. The minimum time to collect a single sample was about 10 seconds and all the tests conducted in the laboratory had this sampling frequency for the lead and tail shearer operator positions.

The SF_6 samples collected prior to, and during testing were analyzed on a gas chromatograph to determine their concentrations. These concentrations were plotted as a function of time from the moment SF_6 was released.

4 LABORATORY TEST RESULTS

As mentioned earlier, after release of the specific amount of tracer gas at the head drum of the shearer, samples were collected at the head and tail operator locations. A typical time versus average concentration plot is shown in Figures 4 and 5 for the head and tail operators position, respectively, when the shearer was located near the headgate and face velocity was 1 m/s (200 fpm). The average effectiveness or efficiency of the air spray system for a particular face velocity and shearer location was determined using the following simple arithmetic equation:

$$E = \frac{B_{t_1} - A_{t_1}}{B_{t_1}} + \frac{B_{t_2} - A_{t_2}}{B_{t_2}} \cdots \frac{B_{t_n} - A_{t_n}}{B_{t_n}} \div n$$

where,

E = average effectiveness of the air sprays system, decimal fraction

$B_{t_1} \cdots B_{t_n}$ = average base concentrations at times $t_1, \cdots t_n$

$A_{t_1} \cdots A_{t_n}$ = average air spray concentrations at times $t_1 \cdots t_n$

n = number of concentration readings

In general it was found that the effectiveness of air sprays decreases as air velocity increases (Figure 6). A maximum of 75 percent effectiveness was obtained for the air spray system with an air velocity of 1 m/s (200 fpm).

5 DISCUSSION OF RESULTS

The air spray effectiveness at 1 m/s (200 fpm) ranged from 66 to 75 percent. It appears that air spray effectiveness is slightly better for the tail operator position. This is probably due to dilution of concentration between the sampling positions by the main airflow across the face and by the airflow from the air spray nozzles.

The SF_6 concentration versus time graphs at 1.52 m/s (300 fpm) air velocity for the base and air sprays system suggest that the initial dust concentration with the air sprays increases very slightly compared to the base system. The effectiveness of air

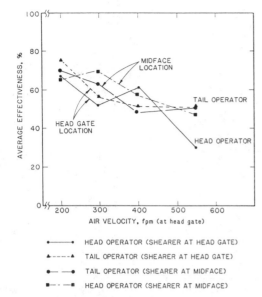

Fig. 6. Effectiveness of air spray system
for varying air velocities and shearer
positions.

sprays at this air velocity ranged from 53
to 70 percent. For both operator positions,
air spray performance appears to be better
when the shearer is in the middle of the
longwall. Effectiveness at the headgate
position is lower because of definite
rollback and turbulence around the face
corner and shearer.

When the air velocity is 2 m/s (400 fpm)
the initial concentrations were higher for
the air sprays system. This is possibly
due to turbulence, but thereafter the dilution rate is faster. There does not
appear to be any significant difference in
the effectiveness values for variation in
shearer position. This could be because
the airflow patterns around the shearer
have not been significantly altered due to
the change in shearer position.

The SF_6 concentration versus time for 2.5
m/s (550 fpm) showed that although the
concentration with air sprays increases
momentarily as compared to the base system,
it ultimately decreases and the rate of
dilution is much faster. In general the
air spray effectiveness at 2.5 m/s (550
fpm) varied from 30 to 52 percent. At
this high velocity, when the shearer is
cutting at the headgate, significant
rollback has caused the low effectiveness
value at the head operator's position.
It appears, however, that this rollback
subsides quickly. This is possibly because
the air sprays are beginning to influence

the dust cloud, and consequently produce a
higher effectiveness value for the tail
operator. With the shearer at midface,
effectiveness value at both operator
positions are nearly the same with a
slightly higher value for the tail operator, possibly due to the dilution effect
from the main airflow and nozzle airflow.

6 CONCLUSIONS

The results of this investigation reveal
that the air sprays system is quite effective at lower face velocities (less than
1.52 m/s [300 fpm]). At these low velocities, dilution due to the air sprays is
more pronounced than the dilution caused
by the primary airflow across the face.
At higher face velocities, however, the
effectiveness of the air sprays system
diminishes due to high turbulence and
insufficient momentum transfer from the
air sprays to the dust cloud for a fixed
capacity air compressor.

A maximum of 75 percent reduction in
concentration was achieved during these
tests at the lowest air velocity of 1 m/s
(200 fpm). At the highest air velocity of
2.5 m/s (550 fpm), the reduction obtained
was the lowest (about 30 percent).

These tests were conducted for two
shearer positions. Perhaps the worst case
with regard to dust control arises when
the shearer is cutting out at the headgate. In this situation, there is significant rollback of dust over the shearer
operator positions. The laboratory testing
of the air sprays system under these
conditions produced reductions between 30
to 75 percent, which is quite significant.

The air sprays system for dust control
possesses several unique advantages compared to other methods of dust control.
These are:

- the system uses compressed air for dilution of dust concentration at shearer
  operator position(s). This augments
  the primary airflow across the face
  resulting in improved ventilation
  conditions at the face
- compared to water sprays and other
  "wet" methods of dust control, the air
  sprays system does not add any water
  to the coal.
- the system can be easily installed on
  any existing longwall face. The pneumatic circuitry is totally independent and
  its interference with face operations
  is minimal. The air line does not have
  to pass through the cable-handler nor
  are there any plumbing complications
  associated with the system

731

- the system requires little maintenance once installed.

   The performance of the air sprays system can be readily improved if consideration is given to:

- mounting the air spray nozzles strategically on the shearer body. This arrangement would also reduce compressor size and total air quantity requirements
- synchronization of the air delivery from sprays mounted on the roof supports with the position of the shearer along the face. This would also reduce compressed air requirements and cause more dilution at the shearer operator positions.

7 ACKNOWLEDGEMENTS

The work reported here was conducted under U.S. Bureau of Mines Contract No. J0318095, and this support is gratefully acknowledged.

The use of water-powered scrubbers on NMS Marietta^R drum miners

JOHN JEFFREY SARTAINE
National Mine Service Company, Ashland, KY, USA

ABSTRACT: This paper will cover the technical and operational aspects of the adaptation of a water-powered scrubber to a continuous miner. The scrubber contains no fan or fibrous screen and is totally contained within the cutter boom frame of the continuous miner with no telescoping joint or long sections of ductwork on the machine. It is totally self-cleaning except for the water sprays and is virtually maintenance free. The paper will include the initial testing of the system which was conducted at a mine in Southern Illinois during 1982, as well as subsequent installations at other locations throughout the United States. Test results showing operational efficiency will be discussed as well as a comparison of the water-powered or JSAM^{TM} (Jet Spray Air Mover) scrubber to the conventional flooded bed type scrubber.

1 DEVELOPMENT OF MACHINE MOUNTED DUST SCRUBBERS

The installation of dust scrubbing devices on continuous miners began in Southern Illinois in January, 1971. The first system used a centripetal hydraulic fan mounted on the cutter boom and water sprays to collect and capture the dust. (Hill, 1974) Later attempts at a different Illinois coal mine in the mid 1970's resulted in a rear fender mounted electric fan to pull the airflow from the face into ductwork on the cutterboom. (Campbell, 1979) The ductwork extended from the cutterboom to the rear of the tractor frame and contained a wire mesh screen wetted by water sprays. As the air flow and dust passed through the wetted or "flooded bed" of the screen, it became entrapped by the water and removed from the airflow. This system resulted in the "flooded bed" type scrubber.

A different type of dust scrubber was developed for coal mines in the early 1970's but was applied at belt transfer points and other locations instead of at the face. This system became known as the JSAM^{TM} (Jet Spray Air Mover) scrubber because it was totally water powered with no

electric fan to pull the air through the scrubbing process.

The JSAM^{TM} scrubber was first used at the face in 1982 on a National Mine Service Marietta^R Drum Miner at a third Illinois coal mine. The testing of this scrubber system indicated that not only did it efficiently clean the dust from the face area, but that it also had several operational benefits over the flooded bed scrubber because of its simplicity and low maintenance requirements. (Page and Hughes 1982)

This system used a series of high pressure water sprays to both induce the airflow and also to perform the scrubbing action. The JSAM^{TM} used no fibrous screen and was self-cleaning. Because no fan was required, the entire scrubber could be built into and protected by the cutter boom without significantly increasing the overall machine height. During 35 production shifts of testing for respirable dust, the following data was obtained in Illinois:

Position Sampled *Avg. Exposure (Mg/M^3)

Continuous Miner   1.2

Off-side Shuttlecar   1.5

Subsequent testing of the JSAM^{TM} at a Virginia coal mine indicated that reductions in methane peaks could be obtained

* Averaged approximately 450-500 TPUS

with the scrubber thus allowing greater curtain holdback distances, deeper cuts and increased productivity. Other tests involving the JSAM™ scrubber occurred in Alabama in 1984 and resulted in approval of the JSAM™ scrubber for 40 foot (12.19 meters) deep cuts with blowing line curtain. (Richardson, 1985)

2 SCRUBBER FACE VENTILATION SYSTEMS

When a scrubber is added to the face ventilation system of a coal mine, the existing style of face ventilation must frequently be altered to accommodate the scrubber. There are several important topics to consider when making these alterations. These include:

    1. A decision on whether to use blowing or exhausting line curtain to ventilate the working face.

    2. The effects of recirculation on methane levels in the face area.

    3. The desired curtain holdback distance.

    4. The desired depth of cut for the mining system.

Each of these topics will be discussed only briefly because of time limitations.

Prior to the introduction of the JSAM™ scrubber, blowing line curtain was a requirement for a scrubber section due to the rear bumper discharge of the flooded bed scrubber. Because the discharge from the JSAM™ scrubber is inby the cutterboom pivot, it can be directed into the exhaust curtain (i.e. exhaust curtain can be used in conjunction with the JSAM). Overall, however, a scrubber is best suited for blowing line curtain for several reasons. The most significant of these reasons involves the methane dilution efficiency of blowing line curtain as compared to exhausting line curtain. A 1969 Bureau of Mines study reported that blowing line curtain with a mean ventilation efficiency of 54 percent was consistently more effective than comparable exhaust line curtain with a mean ventilation efficiency of only 11 percent. (Luxner, 1969) Studies by MSHA Technical Support group have also acknowledged the superiority of blowing line curtain for methane control. (Haney, Smith, and Denk, 1983)

The added efficiency of blowing curtain is attributed to the air momentum advantage it holds over exhaust curtain because of the smaller cross-sectional area relative to the incoming air flow. The increased velocity carries the fresh air to the face instead of allowing it to short circuit into the return side of the curtain. When compared to blowing line curtain, exhaust-ing line curtain is more of a dust control measure and is a compromise for methane control. Thus, to dilute methane most efficiently, blowing line curtain is the best choice. The big problem when using blowing line curtain is the exposure of the workers to dust being blown back out of the face. This is the obvious application for the machine mounted dust scrubber which prevents the dust from leaving the working face.

A big concern for most persons first being introduced to a scrubber/blowing line curtain face ventilation system is the recirculation of methane gas due to the presence of the scrubber. Recirculation is a quite natural occurrence in face ventilation systems and occurs in both exhausting and blowing systems, even without a scrubber because of turbulence. (Luxner, 1969) When the scrubber is added to the system, care must be taken to ensure that the capacity of the scrubber and the volume of air behind the curtain are properly matched. If the capacity of the scrubber is significantly greater than the curtain airflow, true recirculation can and will occur. If the curtain airflow is significantly greater than the scrubber capacity, the airflow will overpower the scrubber and blow the dust past the scrubber before it has time to capture the dust.

The real question is whether or not a recirculation problem exists when the capacity of the scrubber is properly matched to the curtain airflow quantity. According to a 1975 Bureau of Mines Report, "It appears that recirculation becomes dangerous when recirculated air is substituted for fresh air. In situations where an adequate flow of fresh air is maintained, and continues to be maintained, the addition of some recirculated air does not increase the average methane levels in the face area." (Kissell and Bielicki, 1975) Thus, a scrubber will not manufacture fresh air at the face to help dilute the methane but does not cause a dangerous situation if an adequate quantity of air is delivered to the face. Even if the scrubber is not present, an inadequate airflow will create a methane buildup in the face. The addition of the scrubber does not change this fact.

The next important consideration for a scrubber/blowing line curtain system is the appropriate curtain holdback distance (i.e. the distance from the end of the curtain to the point of deepest penetration). In the past the magic number has always been 10 feet (3.05 meters) for exhausting line curtain. In an MSHA Technical Support study, however, it was found

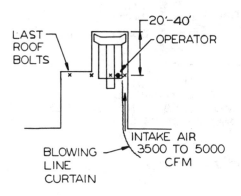

TYPICAL JSAM
SCRUBBER SECTION

20'-40'
OPERATOR
LAST
ROOF
BOLTS
INTAKE AIR
3500 TO 5000
CFM
BLOWING
LINE
CURTAIN

SECTION PUMP TO PROVIDE 40
GPM @ 450 PSI

FIGURE-I.

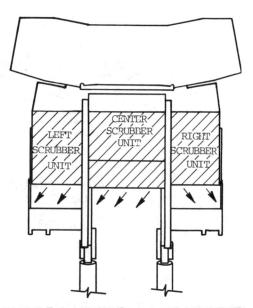

FIGURE 2.-CUTTER BOOM-MARIETTA
DRUM MINER

that "For a constant air quantity, the methane dilution capacity of a blowing face ventilation system fluctuates but does not show any great deterioration as the brattice to face distance increases from 10 to 50 feet (3.05 to 15.24 meters)." (Gigliotti, 1981)

Because the desired end result of the addition of a dust scrubber to a continuous miner is increased productivity, the desired depth of cut must also be considered in the specification of the curtain holdback distance. For a center operated shuttlecar, the shuttlecar operator's compartment is normally about 35 feet (10.67 meters) from the face. For an end operated shuttlecar, pushout car, or mobile bridge, this distance is normally about 40 feet (12.19 meters). Both of these distances have been found to be acceptable in terms of curtain holdback distances during actual field testing. Thus, with remote control mining, the added safety feature of not being required to extend ventilation by means of installing temporary roof supports is gained.

To summarize, a scrubber system is best suited for blowing line curtain because of the increased methane dilution capability of the blowing system as compared to the exhausting system. Recirculation only becomes a problem when an inadequate supply of fresh air is supplied to the working face. The capacity of the scrubber and the quantity of air behind the curtain

should be roughly equal to produce the optimum results for both methane dilution and dust scrubbing. And finally, curtain holdback distances of up to 40 feet (12.19 meters) do not significantly decrease the methane dilution capability of the system and provide for deeper cuts (added productivity) without the possible dangers of extending ventilation past permanently supported roof.

Overall, the scrubber/blowing line curtain face ventilation system has been widely accepted by mine operators, mine workers, and regulatory personnel as a significant improvement over previous methods of ventilating the working face. The dust is no longer pushed behind the exhaust curtain to drift and settle in the return entries creating an explosive hazard for the workers and a rerockdusting task for the mine operator. Rather it is captured at the face and disposed of harmlessly providing not only a safer environment for the workers, but also a more productive mining system (See Figure 1).

3 DESCRIPTION OF THE JSAM[TM] SCRUBBER AND
  A COMPARISON TO THE FLOODED BED SCRUBBER

The JSAM[TM] scrubber was adapted to Marietta drum miners by National Mine Service Company because of several distinct disad-

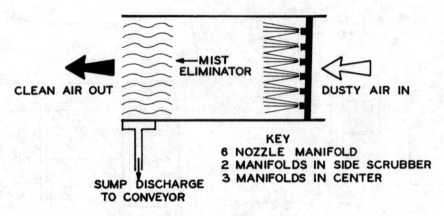

KEY
6 NOZZLE MANIFOLD
2 MANIFOLDS IN SIDE SCRUBBER
3 MANIFOLDS IN CENTER

CLEAN AIR OUT

MIST ELIMINATOR

DUSTY AIR IN

SUMP DISCHARGE TO CONVEYOR

FIGURE 3-SCRUBBER ACTION

vantages of the flooded bed scrubber--
mainly its cleaning requirements, noise
level, and location on the machine. The
JSAM™ scrubber provides the same dust
scrubbing capabilities as the flooded-bed
scrubber without the shortcomings. More
than 95 percent of the airborne respirable
dust and 99 percent of the float dust which
passes through the JSAM™ is removed.

The JSAM™ consists of a center throat
unit which is mounted between the boom
legs and either a left side or right side
unit or both depending upon the application.
At a given time only the center unit and
one side unit (the side opposite the intake air) will be used (See Figure 2).

The JSAM™ is totally water powered with
no electric fan. It uses high-pressure
water sprays to induce a high velocity
airflow through a short section of ductwork. Once induced into the ductwork, the
dust-ladened air mixes with water droplets
which capture the dust. The water/dust
droplets then pass into a low restriction
mist eliminator which removes the droplets
from the air stream. The discharge out
the rear of the mist eliminator is clean,
dry air. The slurry from the mist eliminator flows onto the conveyor where it
mixes with the coal and is loaded out
(See Figure 3). Thus, no jet pump is required to pump the slurry.

The JSAM™ is therefore the ultimate in
simplicity consisting of a box, a series
of water sprays, and a mist eliminator.
The flooded bed scrubber, although effective in capturing dust, is considerably
more bulky and complicated. Its components include:
A. Two long avenues of ductwork which
join at the cutter boom pivot to extend
the length of the machine.

B. A slide joint at the cutter boom
pivot.
C. A series of water sprays.
D. A wire mesh screen to entrap particles
of coal dust.
E. A jet pump to move the water/coal dust
slurry back to the cutterhead.
F. A mist eliminator to remove water
vapor from the air flow.
G. An electric motor and fan to create
the airflow through the system.
These components are shown in Figure 4.

Because no fan is required, the JSAM™
is totally contained within the cutter
boom frame and is well protected. No
bulky slide joint or exposed ductwork is
added to the tractor frame. This enables
the JSAM™ to be used in low-seam applications down to 35" (89 CM) on particular
models of continuous miners, and no ductwork has to be removed to work on other
machine components. Because the system has
no fan, the noise level of 75 dba for the
JSAM™ is lost in the sound of the continuous miner. The JSAM™ is the ultimate in
simplicity with no moving parts!

The JSAM™ is virtually maintenance free,
requiring only that the water sprays be
kept clean. To facilitate this, a spin-off
disposable filter is placed on the machine
with an indicator to show when it needs to
be replaced. Because no fibrous screen is
used (a routine maintenance item on a
flooded-bed scrubber), the JSAM™ is self-
cleaning. This allows the intakes of the
JSAM™ to be located closer to the face,
capturing the dust as soon as it is generated. The larger particles are simply
washed through the JSAM™ and deposited
onto the conveyor. Because the JSAM™ is
located closer to the face, adequate dust
collection is performed with lower scrubber

FLOODED BED DUST SCRUBBER

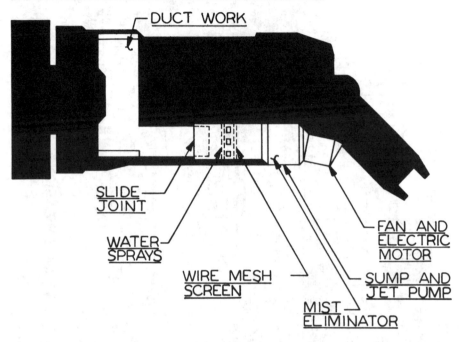

DUCT WORK

SLIDE
JOINT

WATER
SPRAYS

WIRE MESH
SCREEN

MIST
ELIMINATOR

FAN AND
ELECTRIC
MOTOR

SUMP AND
JET PUMP

FIGURE-4.

air volumes. Therefore, the mine operator can supply less intake air to the face than would be required with a flooded-bed scrubber.

The JSAM[TM] scrubber can readily accommodate right or left hand face ventilation systems as often found on "super" sections or in room-and-pillar applications. This can be achieved by operating either the right or left side collector along with the center unit, depending upon the direction of face ventilation. This side unit opposite the intake air would be used and a selector valve is located in the operator's pit. The JSAM[TM] is flexible and is tuned to individual air volume requirements by adjusting the pressure and flow rate of the water supplied to the water sprays. The JSAM[TM] can be used with either blowing or exhausting curtain at the face. Because the air is discharged at the rear of the cutter boom, the discharge can still be directed into the exhaust curtain.

To prevent wetting of the entry, the operation of the scrubber has been made manual, so that it can be turned off when dust is not being generated, such as during cleanup.

In summary, some of the advantages of the JSAM[TM] scrubber are:

A. No fan or fan motor to maintain - Totally water powered with no moving parts.

B. Low 75 dba noise level.

C. No bulky slide joint or exposed duct-work on the tractor frame - Can be used in seams as low as 35" (89 CM) on particular models of continuous miners.

D. Virtually maintenance free - Self-cleaning, except for the water filter - Uses no fibrous screen.

E. Requires less intake air to be supplied to the face.

F. Can accommodate right or left hand face ventilation.

G. Can be tuned to individual air volume requirements.

In general, the JSAM[TM] scrubber is a highly flexible and efficient dust collector which has been neatly packaged into the limited space on a continuous miner. It effectively "scrubs" the dust without requiring a significant amount of maintenance. The Marietta[R] Drum Miner equipped with the optional Jet Spray Air Mover (JSAM)[TM] dust scrubber system provides to the mining industry state-of-the-art technology for improved productivity (See Figures 5 and 6).

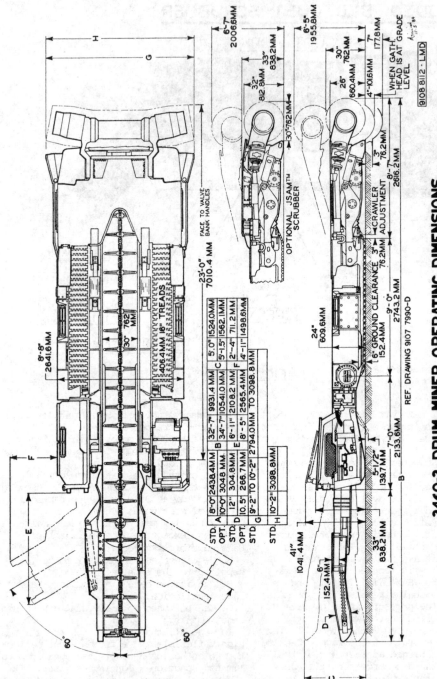

2460-2 DRUM MINER OPERATING DIMENSIONS

FIGURE-5.

STD.	A	8'-0"	2438.4MM	32'-7"	993I.4 MM	5'-0"	I524.0MM
OPT.		I0'-0"	3048 MM	B	34'-7"	I0541.0 MM	C	5'-I.5"	I562.IMM
STD.	D	I2"	304.8MM	E	6'-II"	2I08.2 MM	F	2'-4"	7II.2 MM
OPT.		I0.5"	266.7MM		8'-5"	2565.4MM		4'-II"	I498.6MM
STD.	G	9'-2" TO I0'-2"	2794.0MM TO 3098.8 MM
STD.	H	I0'-2"	3098.8MM

FACE TO VALVE
BANK HANDLES

23'-0" 7010.4 MM

8'-8"
264I.6MM

30" 762 MM

4064 MM I6" TREADS

F

E

60° 60°

6'-7" 2006.6MM

32" 8I2.8MM

33" 838.2MM

30" 762 MM

OPTIONAL JS AM™
SCRUBBER

6'-5" I955.8MM

26" 660.4MM

30" 762 MM

4"-I0.6MM

7" I77.8MM

WHEN GATH
HEAD IS AT GRADE
LEVEL

9108 8II2 · LMD

3"
76.2MM

CRAWLER
ADJUSTMENT

8'-
266I.2MM

6" GROUND CLEARANCE
I52.4MM

3"
76.2MM

24"
609.6MM

9'-0"
2743.2MM

7'-0"
2133.6MM

B

5-I/2"
I39.7 MM

REF. DRAWING 9107 7990-D

4I"
I04I.4 MM

33"
838.2MM

A

6"
I52.4MM

D

C

738

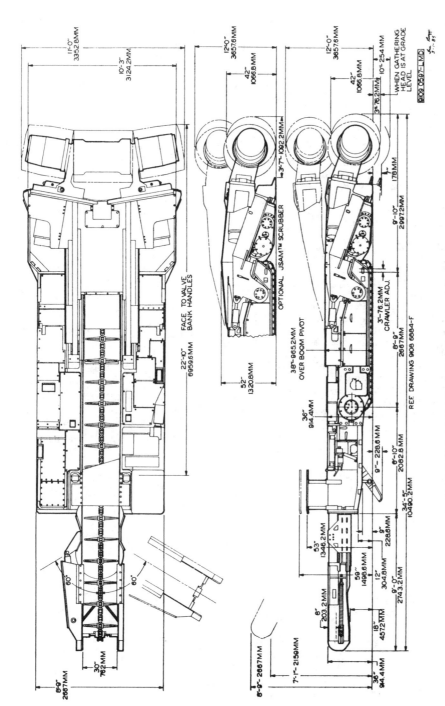

3612 DRUM MINER OPERATING DIMENSIONS

FIGURE-6.

739

REFERENCES

Hill, R. 1974. Dust control with collectors
on continuous miners. Mining Congress
Journal.

Campbell, J.A.L. 1979. Peabody resolves
respirable dust emissions at Camp 11.
Mining Congress Journal.

Page, S.J. and Hughes, B. 1982. Water-
powered scrubber controls dust on continuous miner.

Richardson, J. 1985. Use of remote control
continuous miners at Alabama By-Products
Corporation. 7th annual Alabama Mining
Institute Meeting.

Luxner, J.V. 1969. Face ventilation in
underground bituminous coal mines, R.I.
7223, USBM.

Haney, R.A., Smith, G.E. & Denk, J.M. 1983
Evaluation of a face ventilation system
using a machine mounted dust collector,
Pittsburgh Health Technology Center
Ventilation Division, MSHA.

Kissell, F.N. and Bielicki, R.J. 1975.
Methane buildup hazards caused by dust
scrubber recirculation at coal mine
working faces, a preliminary estimate,
R.I. 8015, USBM.

Gigliotti, S.J. 1981. Laboratory evaluation
of an unassisted blowing face ventilation
system in a low height coal seam, Investi-
gative report P203-V109, Ventilation
Division of Pittsburgh Health Technology
Center.

Extended advance of continuous miner successfully ventilated with a scrubber in a blowing section

JON C.VOLKWEIN & EDWARD D.THIMONS
Pittsburgh Research Center, Bureau of Mines, PA, USA

GEORGE HALFINGER, Jr.
Ingersoll-Rand Research Inc., Princeton, NJ, USA

ABSTRACT: Underground testing was carried out by Ingersoll-Rand, Inc., under contract to the Bureau of Mines to determine the effectiveness of a machine-mounted scrubber system for ventilating the face during an extended advance. A continuous miner equipped with an integral flooded-bed dust scrubber system was instrumented with methanometers and Real-time Aerosol Monitor (RAM) dust monitors. Methane and respirable dust data were collected at brattice setbacks of 7.5 m (current operating distance), 10.5 m, and 15 m (blowing ventilation) during production shifts. Results showed that a suitable machine-mounted scrubber system can adequately ventilate the face at brattice setbacks up to 15 m. No deterioration in ventilation performance was observed as brattice setbacks were increased from 7.5 m to 15 m. The scrubber system effectively controlled face methane levels at large setbacks, though respirable dust levels increased as much as 33 pct at the operator's cab at setbacks greater than 7.5 m.

1 INTRODUCTION

Conventional methods of face ventilation in underground coal mines use either line brattice or auxiliary fans and tubing to direct air up to and away from the working face. This air, moving across the face, dilutes the methane liberated during the working cycle to safe concentrations, and diffuses and carries respirable dust away from the face workers. In exhausting type systems, air enters through the main entry, sweeps over the equipment and workers, and exits behind the line brattice or through tubing. In contrast, with blowing ventilation, air enters behind the line brattice, sweeps over the face, and exits via the main entry. The exhausting system has obvious advantages in allowing only fresh air to course over workers in the entry, thereby greatly reducing exposure to harmful dust levels. The blowing system, though carrying dust over face workers, has the advantage of greatly improving air delivered to the face when methane dilution is a major concern.

Regardless of the system employed, stringent regulations governing face ventilation design were called for following passage of the 1969 and 1977 Federal Coal Mine Health and Safety Acts. Federal Mine Safety and Health Administration (MSHA) regulations currently limit worker exposure to respirable coal dust to under 2 mg/m^3 measured over a working shift, and limit methane concentrations to less than 1 pct by volume at any point more than 30 cm from the working face, roof or rib. Additional regulations define ventilation system design more specifically and limit brattice position to within 6 m of the face for blowing ventilation and to within 3 m for exhaust ventilation; variances to these distances are occasionally granted based on various factors.

The amount of advance that a continuous mining machine can make is limited by the need to maintain adequate face ventilation, and the requirement that the machine operator must stay under permanently supported roof. Consequently, cutting is limited to 3 to 6 m before roof bolting and curtain advance are required, and the miner is moved to a new face. Studies have shown that this movement between entries is a significant portion of the miner working cycle. Productivity could be greatly enhanced if the miner spent more time in the entry cutting coal and less time tramming between entries.

The response to this problem has been the deep cutting (Wellman 1981) or extended advance system. Significant increases in safety and production have been estimated for mining systems utilizing radio-remote controls and auxiliary ventilation systems with cut depths up to 20 m. Extensive research by Peabody Coal and the Bureau of Mines has shown the capability of machine-mounted scrubber systems to remove dust at the face, while utilizing a blowing curtain system to control methane (Chironis 1977, Campbell 1979). Studies, using aboveground modeling, have shown the potential of an extended advance system that utilizes a scrubber to control dust and provide adequate face ventilation for methane control (Gillies 1982).

Because of insufficient information on methane ventilation in extended advance, Ingersoll-Rand Research, under contract to the Bureau of Mines, initiated a program to investigate the use of a suitable machine-mounted scrubber system to ventilate the face during an extended advance. This involved utilizing brattice setbacks in excess of the current MSHA limits. Since extensive research had been conducted with machine-mounted scrubbers with regard to dust control, the present work centered on methane control and the ability of the machine-mounted scrubber to control methane at the large brattice setbacks necessary in an extended advance.

2 UNDERGROUND TESTING

Laboratory scale model studies indicated the feasibility of testing the ventilation effectiveness of a machine-mounted flooded bed scrubber system in a production section of an underground coal mine. The in-mine testing involved monitoring methane and respirable dust levels over approximately 30 shifts of continuous miner operation. The ability of the scrubber system to adequately ventilate the face during simulated extended advance with brattice setbacks up to 15 m was investigated.

Extended advance was simulated by setting the line curtain back from the face in a bolted entry to a predetermined distance. The miner was then allowed to take a full 6-7 m cut to reach the maximum setback

distance to be tested. A total of 41 successful test cuts were carried out at 3 setback distances:

Average Setbacks	Number Of Tests
7.5 m	15
10.5 m	13
15.0 m	13

An additional five tests were performed during left and right crosscut advances.

All testing was done utilizing line brattice and nominally 1.89 m^3/s of blowing type ventilation with the scrubber system operating. The test mine used a room-and-pillar system, with an 8-entry advance. Crosscuts were turned on 22.5-m centers, with 6-m entries and a 2.00-cm entry height. The maximum permissible blowing curtain setback of 7.5 m from the face was temporarily waived by MSHA to allow simulated extended advance testing. The mine operated an Ingersoll-Rand LN-800 continuous miner with an integral flooded bed dust scrubber. The scrubber system had 4 inlets located on ducting mounted on the cutterhead boom (fig. 1). During operation, dust-laden air was drawn away from the face, through the flooded bed scrubber (fig. 2) and exited to the left rear of the machine (fig. 3). An axial vane fan provided up to 3.3 m^3/s air moving capacity for the system. Scrubber ducting and the flooded bed screen were water flushed at the start of every shift.

Methane was monitored during a test cut by drawing air samples directly from the face and from the entry. This system utilized two permissible vacuum pumps and tubing to draw an air sample into a 4-L container. At a pump flowrate of 2 L/min, the air volume in the can was exchanged every 2 min. Evacuated bottle samples were taken at each air exchange and analyzed using gas chromatography. This method reduced the number of samples taken, yet accounted for all methane present.

Methane samples were drawn from the right and left sides of the miner boom (at the location of machine methane monitor) and combined in the 4-L container to give the average face methane level. The evacuated bottle sample was taken as shown in figure 4. The methane in the return air was drawn from 3 equal length tubes suspended 30 cm from the roof bolts at 1.2-m intervals across the return. Bottle

FIGURE 1. Boom duct work.

FIGURE 3. Scrubber exhaust.

FIGURE 2. Flooded bed scrubber panel.

FIGURE 4. Sampling at cab.

samples were taken opposite the brattice in the intake (fig. 5).

The results of samples taken from the face and return were used to determine an average face ventilation effectiveness (FVE) value for the setbacks employed. Face ventilation effectiveness at the extended advance setbacks (10.5 m, 15 m) were then compared directly to similar results obtained from the baseline, or allowable setback, of 7.5 m. The face methane and dust data obtained at all three setbacks were compared in the same way.

The testing sequence was identical for all setbacks tested. Before the miner began a cut, the curtain was positioned a set distance from the original face (for example, 6 m for a final setback of 12 m at the end of the cut). At the start of the first miner sump into the face, methane and dust sampling was begun. Face methane samples at the miner were taken simultaneously with return samples taken in the entry. Dust level readings

FIGURE 5. Sampling in return.

were taken with a real time aerosol monitor at the operator's cab at 1-min intervals, in conjunction with the methane sampling. During the course of a test, additional readings were taken of the intake air velocity at the curtain mouth and miner downtime, scrubber operation, and cut duration were noted.

743

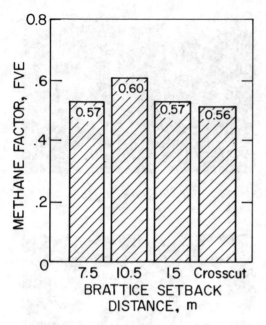

FIGURE 6. Face ventilation effectiveness versus brattice setback.

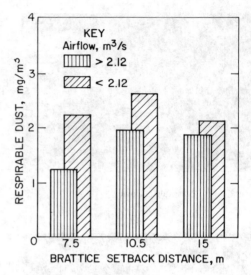

FIGURE 7. Dust levels.

3 RESULTS

A machine-mounted scrubber system can adequately ventilate the face at brattice setbacks up to 15 m. Figure 6 shows no decrease in face ventilation effectiveness as setbacks were increased from 7.5 m to 15 m. Data for all setbacks had standard deviations ranging from 0.11 to 0.13.

Although the number of crosscut tests was low for an accurate FVE determination, the data suggest that ventilation was comparable to straight entry ventilation.

Face ventilation effectiveness was independent of intake airflows. All data with airflows greater than 2.12 m^3/s were compared to data with airflows less than 2.12 m^3/s. No difference in methane FVE factors were evident for the two data sets. Scrubber airflow for these tests was 2.36-3.30 m^3/s.

Primary intake airflow was an important variable in controlling respirable dust levels at the operator's cab. As brattice setback distance increased, respirable dust levels also increased by as much as 33 pct. Figure 7 shows that if intake airflows are maintained above 2.12 m^3/s then short-term respirable dust levels at

the operators can be maintained at less than 2 mg/m^3.

Since this system is ultimately meant for radio-remote operation, the miner has the option of occupying an even less dusty location.

The effectiveness of any dust collection system is the product of its capture efficiency times its scrubbing efficiency. The dust data from this study show that when the scrubber intake duct is inside the confined area of the box cut, cab dust levels are low as expected, however, when the intake dust is less confined while cutting the slab, dust levels rise. Figure 8 shows that as brattice setback distance increases, slab dust levels, which are normally higher than box levels, begin to approach box cut levels.

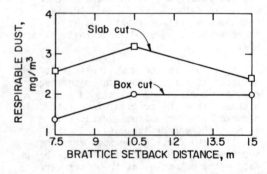

FIGURE 8. Slab versus box cut dust levels.

Decreased turbulence from the blowing
curtain at increased brattice setback
probably increases the dust capture
efficiency and reduces entry dust levels.

A radio-remote scrubber equipped
continuous mining machine can take an
extended mining advance, and maintain safe
methane and respirable dust levels with
brattice setbacks up to 15 m. Placing
the operator outby the mining machine and
eliminating the need for extension of the
brattice curtain under temporarily
supported roof increases worker safety.
Productivity is also improved by reducing
the number of moves required to mine an
equivalent quantity of coal.

4 REFERENCES

Wellman, T. S. Deep Cut Remote Control
    Mining. Mining Congress Journal,
    Oct. 1981, pp. 27-30.

Chironis, N. P. Peabody Mines Find Better
    Way to Reduce Dust. Coal Age Operating
    Handbook of Underground Mining, 1977,
    pp. 220-224.

Campbell, J. A. L. Peabody Resolves
    Respirable Dust Emissions at Camp 11.
    Mining Congress Journal, Mar. 1979,
    pp. 23-26.

Gillies, A. D. S. Studies in Improvements
    to Coal Face Ventilation with Mining
    Machine Mount Dust Scrubber Systems.
    SME/AIME Ann. Meeting, Dallas, TX,
    Feb. 1982, 82-24, 13 pp.

Pressure diffuser device for use with auxiliary face fans

DAVID L.BARTSCH
Quarto Mining Company, Powhatan Point, OH, USA

ABSTRACT: The increased use of auxiliary fans for ventilating working places in underground coal mines has increased the occurrences of recirculating air. The standard solution to recirculation in the past has been to increase the amount of available section air at the last open crosscut to 1.5 to 2.5 times the fan quantity. Federal mean entry air standards require a minimum velocity of 60 fpm (0.30 mps) at the place where coal is being mined, and in order to achieve this velocity, the fans are commonly operated at high blade settings. This research has found that although adequate air quantities are delivered to the fan through the main ventilating current, recirculation may still occur due to the pressures created by a fan that operates at such high settings.

The author has developed a diffuser device (Figure 1), much like an evasé found on the end of a main mine fan, that reduces or eliminates the pressure-caused recirculation. The device is low in cost, light-weight, portable, and can be mounted on the end of the fan or attached to the end of the exhaust tubing. Additional benefits to the use of the device are increased fan quantities and decreased respirable dust as measured at the miner operator's station.

Data is presented showing the effect of the fan on the section ventilation with and without the diffuser. The design of the device is also discussed with regard to size and efficiency.

1 INTRODUCTION

The use of auxiliary face fans has again increased in order to meet the requirements of U.S. Federal mining regulations with regard to mean entry air velocity where coal is being mined. The use of these fans diminished at one point in time due to improvements in brattice and line curtain materials, but to a larger degree, to inherent disadvantages with auxiliary fans. Among the drawbacks is recirculation of air, prohibited under Title 30 of the Code of Federal Regulations Part 75.302-4(a). Some states outlaw the use of auxiliary fans or make the regulations to use these devices so stringent that, from an operation's standpoint, it is impractical. Past experience shows that maintaining mean entry air velocities in excess of 60 fpm (0.3 mps) with line brattice is difficult at best. Constant interruptions to the section ventilation system caused by ancillary equipment is increased with the use of brattice curtains. In addition, less clearance can be maintained with the use of curtains, diminishing safety, and discouraging the use of larger and more productive equipment.

In order to make the use of auxiliary fans and tubing safer and more acceptable, ways to eliminate recirculation must be discovered. This paper investigates one successful method that involves the use of a pressure diffuser device. The design and field trials of the device will be discussed.

2 CAUSES OF RECIRCULATION

Before the solution to the recirculation problem can be addressed, a total understanding of recirculation is required. Typically, recirculation has been treated almost entirely as a "quantity" problem. In other words, the auxiliary fan requires more air than the return can handle. The severity of this

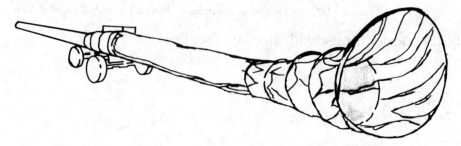

Figure 1. Conceptual view of tubing, fan, and diffuser

problem varies according to the condition and position of the temporary and permanent ventilation controls (curtains and stoppings), as well as the difference between the section return quantity and the auxiliary fan quantity. The causes of "quantity recirculation" include an undersized main ventilating fan, excessive leakage that consumes the return capacity outby the active area of mining, short circuits or improperly placed controls in the active mining area, or over-regulation of the return.

Past research work on "quantity recirculation" recommended that the section return capacity at the last open crosscut be 1.5 to 2.5 times the rated fan capacity. Given the marginal air quantities that existed in many mines in the early 1970's and the above-mentioned criteria, it is easy to see why the use of auxiliary fans diminished. Fans were designed to meet the mean entry air standards, but the return capacity was deficient and recirculation occurred.

It is safe to say that most coal mining sections are better ventilated today, and thus auxiliary fans are being used in more instances. However, even with section quantities that meet the recommended criteria to stop "quantity recirculation," we still find air recirculating. An examination of the auxiliary fan system in a pressure-distance diagram is needed to determine the cause of the problem.

The pressure-distance diagram for an exhaust fan is shown in Figure 2. Air enters the intake tubing at the mine atmospheric datum pressure and is discharged to air that is at approximately the same pressure (neglecting losses for a short distance of airway). The velocity pressure at the discharge of the fan (or exhaust tubing) is appreciable and tends to increase the absolute pressure on the downstream (return) side

of the fan. If the pressure differential between the next adjacent entry and the return when the fan is not operating is minimal, the pressure created by the fan will cause the air to flow from the return into the adjacent entry. Depending upon the direction of the airflow in the adjacent entry, this air may find its way back to the face. This type of recirculation is termed "pressure recirculation."

It must be noted that "pressure recirculation" is greatly influenced by several factors, including section configuration, stopping line location, number of open crosscuts at the face, number of intake and return airways, and proximity of other mining units using auxiliary fans. Our studies show that the section configuration is important because it is easier for recirculation to occur from return to neutral than from return to intake, since the pressure differential is generally smaller between return and neutral entries.

The extent of the recirculation is influenced by the number of airways in common with both the intake (or neutral) and return. The pressure in the intake air is inversely proportional to the square of the airway resistance. Therefore, the pressure differential between a multiple entry intake and the return is smaller than the differential between a single entry intake and the return, and "pressure recirculation" can occur more easily. Stopping lines from intake to neutral must be extended regularly and open crosscuts must be kept to a minimum in order to limit this effect.

It is common to mine rooms on the return side of a section on advance, and this practice also affects recirculation. A single entry return tends to dissipate the velocity pressure produced by the fan more quickly than multiple airways (including rooms). Thus, when rooms are

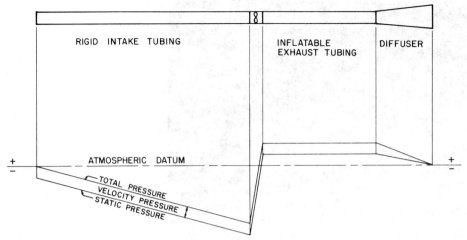

Figure 2. Pressure-distance diagram

driven on the return side, our observations show that pressure recirculation can extend over 1500 ft (457.2 m) outby the last open crosscut.

The pressure caused by an auxiliary fan can disrupt the normal flow of air in adjacent mining sections. In one of our observations, an auxiliary fan in section A actually caused air to flow toward the face in the return entry of section B.

Other problems associated with "pressure recirculation" include increased respirable dust and, in gassy mines, increased methane in the work area. Also, the velocity pressure consumes fan power that could be used to provide more air quantity which would increase the mean entry air velocity at the working face. Another problem arises when a battery charging station that must be vented to return is within the range of the recirculation. Hydrogen gas could accumulate if this situation exists over long periods of time.

3 COMMON SOLUTIONS

Several solutions to this problem have been tried. Many auxiliary fans are equipped with adjustable blades that are commonly adjusted to reduce the discharge pressure. Unfortunately, decreasing the blade setting also decreases the air quantity that the fan produces, thus making the mean entry air

velocity standard of 60 fpm (0.3 mps) more difficult to achieve.

Another solution that has been tried is to direct the fan discharge at a rib or at a brattice curtain to dissipate the velocity pressure by causing a "shock loss." This solution, although effective, does not convert the velocity pressure into usable fan power that could increase mean entry air velocities at the working face. Also, a brattice curtain subjected to the force of the exhaust air is difficult to maintain in good condition. A curtain may also act like a regulator, decreasing the section return quantity. Furthermore, it may not always be possible to direct the discharge of the fan at a rib depending upon the configuration of the entry and the method used to attach the intake tubing to the fan.

4 DESIGN OF PRESSURE DIFFUSER DEVICE

Since solutions that have been previously tried have serious disadvantages, we began to investigate other ways to solve the pressure problem. Our work led us to the development of a diffuser device (or more properly - an evasé). In theory, a diffuser converts the velocity pressure into usable static pressure, thus reducing the energy lost and restoring the discharge air to the pressure of the surrounding atmosphere. The diffuser (Figure 3) is made of light-weight, fiber reinforced brattice

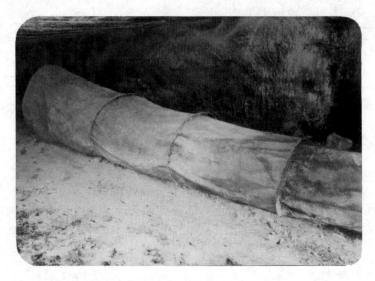

Figure 3. Diffuser installation in coal mine

material with stiffener rings to maintain its shape. The diffuser can be attached directly to the end of the fan or to exhaust tubing (extending from the discharge end of the fan) by means of a cinch strap. The discharge end of the diffuser must be attached to the mine roof in order to prevent its collapse. Figure 4 shows a diagram of the diffuser in place.

The diffuser was designed on the basis of achieving a 4:1 ratio of discharge area to inlet area and a side divergence angle of 3 degrees for maximum efficiency (Hartman 1961). Since our inlet diameter was 2.5 ft (0.76 m), the discharge diameter required was 5 ft (1.52 m). To achieve this transition with a 3 degree divergence angle, a length of 25 ft (7.62 m) was needed. Although this length may seem excessive, we commonly use up to 100 ft (30.48 m) of exhaust tubing on the discharge end of our auxiliary fans.

5 FIELD TRIALS

Although now a standard production item from Peabody ABC Corporation, we fabricated our own prototype by cutting out two wedge shaped pieces of brattice cloth and stapling the edges together. The initial test of this device proved that 1) the pressure created by the fan was causing the recirculation, 2) the diffuser device eliminated the problem, and 3) stiffener rings were needed to

maintain the intended shape of the device.

Once we obtained diffusers with stiffener rings, we transported them to the mining sections and collected the data presented in Table 1. The mining sections were 3 or 4 entry longwall gate development units with a belt entry, a combined intake and track entry, and a return entry (3 entry configuration) or with a belt entry and track entry (in common), an intake entry, and a return entry with rooms driven on the return side (4 entry configuration).

Figure 5 shows the following: A) "typical" section ventilation without the fan in operation, B) section ventilation disrupted by the use of the auxiliary fan, and C) section ventilation with the use of the diffuser. Since the "typical" section pressure differential between intake and return is low, recirculation can occur easily.

Perhaps the biggest problem with the "pressure recirculation" is the fact that dust-laden air is coursed back to the continuous miner, posing a serious health threat. Measurements show that respirable dust at the continuous miner operator's station can be reduced by as much as 45 percent with the use of the diffuser (Wirch 1985). This reduction in respirable dust will undoubtedly help keep the operator in compliance with the 2 mg/m^3 standard.

Because our mines use the inflatable exhaust tubing to carry the dusty air to the return side of the section, the

Table 1. Summary of diffuser field tests

Cut	Section description	Section intake (cfm)	Section return (cfm)	Fan quantity (cfm)	Blower tubing discharge pressure Static (in. wg)	Velocity (in. wg)	Diffuser Discharge Pressure Static (in. wg)	Velocity (in. wg)	Recirculation Static pressure (in. wg)	Direction	Extent
1-D[1]	Entry configuration[4]	22,496	14,759	8,493	.27	.28	0	0	0	I-R[2]	7 Crosscuts
1	Entry configuration[4]	22,496	14,759	8,493	.32	.25	N/A	N/A	.008	R-I[3]	7 Crosscuts
2-D[1]	Entry configuration[4]	22,496	14,759	7,920	.35	.28	-0.1	0	.001	I-R[2]	7 Crosscuts
2	Entry configuration[4]	22,496	14,759	7,920	.39	.39	N/A	N/A	.008	R-I[3]	7 Crosscuts
3-D[1]	Entry configuration[4]	22,496	14,759	9,371	.35	.25	N/A	0	.001	I-R[2]	7 Crosscuts
3	Entry configuration[4]	22,496	14,759	9,536	.22	.35	0	0	.008	R-I[3]	7 Crosscuts
4-D[1]	Entry configuration[4]	15,538	17,699	12,269	.60	.35	N/A	N/A	.005	I-R[2]	7 Crosscuts
4	Entry configuration[4]	15,538	17,699	12,396	.60	.37	N/A	0	0	R-I[3]	7 Crosscuts
5-D[1]	Entry configuration[5]	17,040	12,112	12,011	.30	.33	-0.1	0	0	NM[6]	2 Crosscuts
5	Entry configuration[5]	17,040	12,112	12,522	.60	.27	N/A	N/A	.01	R-I[3]	8 Crosscuts
6-D[1]	Entry configuration[5]	17,040	12,112	10,735	1.00	.30	0	0	0	NM[6]	2 Crosscuts
6	Entry configuration[5]	17,040	12,112	10,625	1.20	.27	N/A	N/A	.01	R-I[3]	8 Crosscuts
7-D[1]	Entry configuration[5]	17,040	12,112	11,477	1.20	.30	-0.1	0	0	NM[6]	2 Crosscuts
7	Entry configuration[5]	17,040	12,112	10,916	.80	.30	N/A	N/A	.01	R-I[3]	8 Crosscuts
8-D[1]	Entry configuration[5]	16,200	14,170	9,699	1.00	.20	0	0	0	NM[6]	2 Crosscuts
8	Entry configuration[5]	16,200	14,170	9,699	1.00	.19	N/A	N/A	.01	R-I[3]	8 Crosscuts
9-D[1]	Entry configuration[5]	16,200	14,170	10,625	1.60	.22	0	0	0	NM[6]	2 Crosscuts
9	Entry configuration[5]	16,200	14,170	10,477	1.60	.22	N/A	N/A	.01	R-I[3]	8 Crosscuts
10-D[1]	Entry configuration[5]	16,200	14,170	9,030	2.20	.21	0	0	0	NM[6]	2 Crosscuts
10	Entry configuration[5]	16,200	14,170	8,854	2.00	.21	N/A	N/A	.01	R-I[3]	8 Crosscuts

Average with diffuser   10,163
Average without diffuser 10,143

1 D = With diffuser
2 I-R = Intake to Return (Normal flow)
3 R-I = Return to Intake (Recirculation)
4 4 Entry configuration = return, intake, track, belt
5 3 Entry configuration = return, intake along track, belt
6 NM = No movement of air for 2 crosscuts; outby that point movement is from intake to return

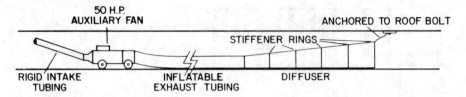

Figure 4. Diagram of diffuser in place

diffuser is not as effective as expected in changing the velocity pressure into usable static pressure. During the field trials, an average of only 20 cfm (.009 m^3/s) of additional air (0.2 percent) was observed through the fan when the diffuser was used. However, in Cut 3, the diffuser was placed directly on the discharge side of the fan. The fan quantity increased to 9860 cfm (4.65 m^3/s) or an improvement of 5.2 percent over the diffuser mounted on the exhaust tubing (9371 cfm). In addition, the pressure moving from intake to return with the diffuser on the fan itself increased to 0.003 in. wg (0.75 Pa) an improvement of 0.002 in. wg (.50 Pa). This increase, coupled with the ability to operate the auxiliary fan at higher blade settings, will help achieve minimum mean entry air velocity at the working place without the associated recirculation problems.

There are disadvantages and short-comings to almost anything, and the diffuser is no exception. The last thing that is needed on a mining section is another piece of equipment. However, since a number of mines already use exhaust tubing, the diffuser fits in quite well. It is light-weight and portable and only requires about one minute to attach. Improved designs make it as durable as the exhaust tubing already in use.

6 CONCLUSIONS

6.1 Recirculation is related to the capacity of the section return and the fan, the resistance of the intake and return airways in the vicinity of the active working face, the pressure differential between the return and the next adjacent entry, and the pressure created by the auxiliary fan.

6.2 If little can be changed with regard to quantity, resistance, and pressure differentials on the section, the use of a diffuser on the dis-

charge of an auxiliary fan can solve the problem of recirculation in many cases.

6.3 The elimination of recirculation with the use of a diffuser can also reduce concentrations of respirable dust in the working place.

6.4 When the diffuser is used, the quantity of air produced by the auxiliary fan may be increased as a result of changing velocity pressure into static pressure.

6.5 When the diffuser is used, the blade settings on the auxiliary fan may be increased to maintain mean entry air velocity at the working face without increasing the occurrences of recirculation provided there is a sufficient quantity of air on the section.

7 ACKNOWLEDGMENTS

The author wishes to acknowledge the valuable assistance of Mr. Wilber Heslep, Ventilation Technician and co-developer of the diffuser, who spent many hours testing the devices in the coal mine. In addition, the author wishes to acknowledge Mr. William Huffer, Regional Sales Manager of Peabody ABC Corporation, for his cooperation and guidance in the fabrication of the diffuser.

REFERENCES

Hartman, Howard L., 1961 Mine Ventilation and Air Conditioning, p. 83, 206-210, New York, The Ronald Press Company.
Coal Mine Ventilation Awareness Program, September, 1984, U.S. Department of Labor, Mine Safety and Health Administration, National Mine Health and Safety Academy, Beckley, West Virginia, Section 5, p. 16-31.
Notes from short course entitled "Computer Analysis of Mine Ventilation Systems" given at The Pennsylvania State University, May 21-23, 1984,

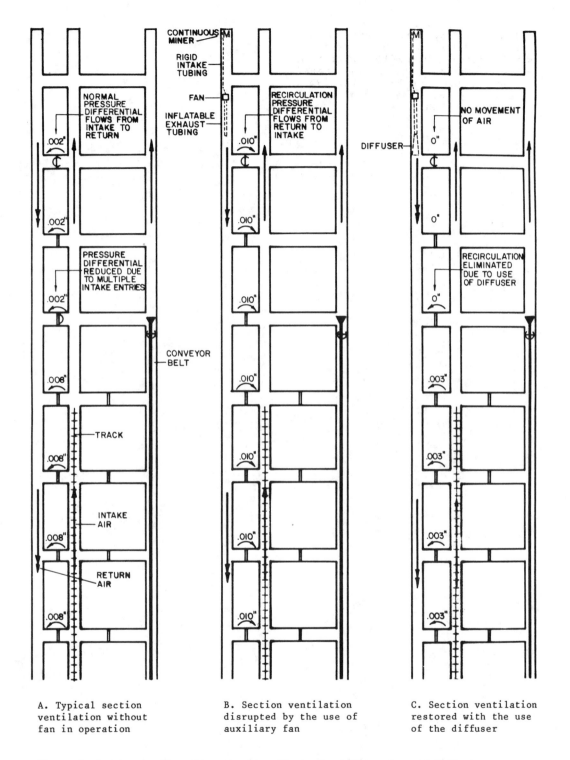

A. Typical section
ventilation without
fan in operation

B. Section ventilation
disrupted by the use of
auxiliary fan

C. Section ventilation
restored with the use
of the diffuser

Figure 5. 3 entry configuration showing effect of auxiliary fan and diffuser

University Park, Pennsylvania, Section
10, p. 212-213.
Wirch, Steven, Foster Miller Associates,
Boston, Massachusetts, personal
conversation on March 12, 1985,
regarding dust samples taken in
Powhatan No. 4 Mine.

19. Monitoring

Ventilation experience using computer monitoring and control systems at major US coal mines

ALBERT E.KETLER
Rel-Teck Corp., Pittsburgh, PA, USA

Abstract: The introduction of monitoring and control systems in U.S. mines provides a new tool to enhance safety and productivity of the mining operations.
Reasons for these installations varied, but justification ususally had a strong tie
to improving ventilation and fire detection. Particular problems surfaced in
regard to:

    1) maintaining the computers and sensors in the dusty, moist
          environment of a mine,

    2) complying with new MSHA rules and the terms of the $101°C$
          petitions.

The experience in solving these problems and achieving better ventilation is
discussed from the viewpoint of the mine operator. Cost and payback benefits are
evaluated.

Justification of a chromatograph for mine air analyses

T.E.McNIDER
Jim Walter Resources Inc., Brookwood, AL, USA

ABSTRACT: Jim Walter Resources realized the need for complete mine air analyses in 1982 when low concentrations in parts per million of carbon monoxide were detected in the airstream from an active gob area. Carbon monoxide alone does not mean an incipient underground mine fire. The Mining Division of Jim Walter Resources, Inc., is interested in carbon monoxide together with other gob gases; and a chromatograph was purchased to identify and establish a historical base at parts per million levels of gob gas con- taminents. This paper discusses the type of chromatograph purchased, the air analyses being done, historical data collected and other uses for the chromatograph.

1 INTRODUCTION

Jim Walter Resources started its first retreat longwall in February, 1979. Seven longwall panels are currently being mined in Blue Creek No. 3, 4 and 7 Mines with eleven faces projected. Overburden depths range from 1300-2100 feet (396.24-640.08m). Panels have been mined 600 feet (182.88m) by 6000 feet (1828.8m) long with both dimen- sions expected to increase in future mining.

The No. 1 Longwall at No. 3 Mine is mining in the seventh consecutive long- wall panel without sealing. Active long- wall gob areas are open and ventilated. In October, 1982, low concentrations in parts per million of carbon monoxide were detected by Jim Walter Resources operating personnel in the airstream ventilating this gob area. As a precautionary measure, the mine was closed for eight days to de- termine if a fire existed in the gob. There was no fire, but natural actions occurring in the gob resulted in chemical reactions that produced relatively large quantities of carbon dioxide together with small quantities of carbon monoxide that were detected in the airstream ventilating the gob. Although many coal beds are known to generate carbon monoxide, this was the first occurrence of this phenomenon doc- umented in the Blue Creek coal bed of the Mary Lee series. This prompted Jim Walter Resources to purchase a chromatograph to identify and establish a historical base

at parts per million levels of gob gas contaminents.

2 THE CHROMATOGRAPH PURCHASED

In February, 1984, Jim Walter Resources purchased a chromatograph and shortly thereafter hired a technician with sole responsibility of operating the chro- matograph.

The chromatograph selected was the Perkin-Elmer Sigma 300 FID/HWD. It is configured with dual heated packed in- jectors. The chromatograph is equipped with a heated automatic gas sampling valve, 10 port, with a 0.5 CC sample loop. The valve, which is air operated, injects a sample and switches columns automatically to enable the separation of the component gases. The chromatograph operates on a power source of 115 volts AC.

The Sigma 300 is tied into a model LCI- 100 Laboratory Computing Integrator. It is a single channel integrator and includes 128 Kilobytes RAM, a full alphanumeric keyboard, liquid crystal display (LCD), and an 11-1/4 inch (.2858m) printer/plotter with multiple method/multiple peak file storage with the ability to replot and re- integrate the last chromatogram. It is also equipped with external start, external device ready logic, and an integrator ready- out signal. In addition, a series kit is provided to operate the hot wire detector and flame ionization detector simultaneously. The integrator automatically switches from the hot wire detector to the flame ionization detector for an incoming signal.

3 GASES ANALYZED

The Sigma 300 and LCI-100 are capable of separating and integrating the following gases to the limits specified: 1ppm-100% methane, propylene, ethane, ethylene, and carbon monoxide; 10ppm-100% carbon dioxide, nitrogen, oxygen, and hydrogen. Carbon monoxide can be converted to methane with a methanizer, which is the catalytic accessory for the flame ionization detector. Gas samples will be collected by Jim Walter Resources personnel from the various mines to be analyzed by the chromatograph. Sample analyses of certain gases will be plotted in percentages, parts per million, or cubic feet per minute versus time to establish bases and trends. They can also be used to calculate ratios and indices which will be discussed in greater detail in a subsequent section of this paper.

4 DETECTOR GAS FOR HISTORICAL BASE

According to E.A.C. Chamberlain, a noted fire expert, the presence of carbon monoxide provides the earliest indication of incipient spontaneous combustion or heating. Figure 1 shows that the rate of evolution of carbon monoxide from coal begins to increase rapidly at about 50 degrees C. The fire gases, hydrogen, ethylene and propylene, do not show a similar rapid rate of increase until temperatures of over 100 degrees C are reached. Although these gases can be used to trace the progress of a heating, an increase in carbon monoxide is the first indicator of a heating (Chamberlain, 1970).

Since carbon monoxide is the most sensitive detector gas of a heating, it is being monitored and plotted in cubic feet per minute. One would think that monitoring carbon monoxide concentrations alone would provide sufficient information to analyze the progress of a heating, however, carbon monoxide samples can show changes in concentration with a change in the mine ventilation circuit and a different differential pressure across the gob. Therefore, Jim Walter has decided to also monitor the Index for Carbon Monoxide or Graham Ratio vs. time. This index depends upon the constant ratio of oxygen to nitrogen of .265 in fresh air and is the ratio of the production of carbon monoxide to the oxygen consumed. It is calculated

by the following formula:

$$Ico = \frac{CO}{.265N_2 - O_2} \times 100$$

As heat is generated in an underground environment, the Index for Carbon Monoxide will change. For normal and safe conditions, the change will be slow, small, and distinquishable. The normal index fluctuates slightly around its average value, which will generally be less than 0.3, although it could be as high as 1.0 in some areas of the mine. It responds primarily to the rising temperature associated with the heating or burning materials. Thus, small sustained increases in the index warn of a developing fire. A developing or active fire will add at least 0.5 to the normal index (Mitchell, 1973).

Figure 2 shows how the Ico, or Graham Ratio, varies with the oxidation rate and how they are both related to the temperature at the heating during the development

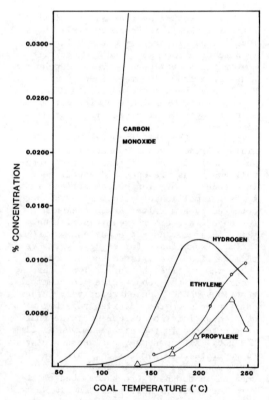

Figure 1. Production of "Detector Gases" with increase of temperature

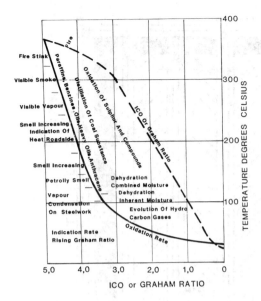

Figure 2. Distillation products of oxidation

of spontaneous combustion. The slope of the Graham Ratio line is greatest where that of the oxidation line is least, so it is in this region where the Graham Ratio is most sensitive to an increase in oxidation. As the oxidation rate increases and an open fire situation is approached, the slope of the Graham Ratio line decreases because the production of carbon dioxide increases relative to carbon monoxide (Holding, 1982).

In addition, the CO/CO_2 ratio vs. time can be plotted, and an increase in this ratio would indicate heating due to oxidation and a possible fire. Using samples collected and analyzed by Jim Walter Resources personnel, it has been determined that for the Blue Creek Seam, as CO/CO_2 ratios climb to exceed 0.02, for a sustained rise in CO, or for a sustained decrease in O_2, investigations into possible causes will be initiated.

The time required for a significant fire to develop due to spontaneous combustion in a mine might be hours or even days. Therefore, an analysis of the cubic feet of carbon monoxide in conjunction with the Index of Carbon Monoxide and the ratio of CO/CO_2 might provide an early warning of possible fire conditions in sufficient time to locate and isolate the developing fire.

5 PROGRESSIVE SEALING

Jim Walter Resources, working closely with MSHA, has proposed a planned program, Phase I, to erect seals and partial seals or regulators to ventilate gob areas, and in the event of a fire on the active face or in the gob, the affected area could be speedily isolated and quickly sealed from the rest of the mine. All of Jim Walter Resources longwall faces are mined on the retreat, and in Phase II the proposed plan is to only ventilate the gob area behind the active face and the adjacent gob created by the previously mine panel. Bleeder connectors in panels adjacent to the vented gobs will be sealed (See Fig. 3). The chromatograph will be a valuable tool for monitoring this type of gob, particularly the sealed portion of the gob, to determine the progression in which this area becomes inert. Hand held detectors lose thier reliability in an inert atmosphere; therefore, as a sealed or partially sealed gob becomes inert, the chromatograph becomes a necessity.

6 SAMPLE RELIABILITY AND INTERPRETATION

Trickett's Ratio (TR) can be used to determine the reliability of gas analysis data. Trickett's Ratio is based on the principle that the volumes of gas produced in a fire are porportional to the molecular weight of each material involved in the combustion process, as calculated by the following formula:

$$TR = \frac{CO_2 + 0.75CO - 0.25H_2}{0.265N_2 - O_2}$$

Where:

CO_2, CO, H_2, N_2 and O_2 represent the gas sample percentages by volume of carbon dioxide, carbon monoxide, hydrogen, nitrogen, and oxygen

Fire related TR's should range from 0.4 to 1.6. A TR value less than 0.4 generally indicates no fire, the sample gases being the result of residual rather than active fire gases or of oxidation without significant heating, 0.4 - 0.5 indicates the fuel is methane, 0.5 - 1.0 indicates the fuel is coal, oil, conveyor belting, insulation, or polyurethane foam, and 0.9 - 1.6 the fuel is wood. Low values of fire related TR's are strong evidence of little to no carbon evolution. A TR value higher than 1.6 indicates the sample is suspect. Where no possibility exists of wood burning, then a TR value greater than 1.0 should be suspect.

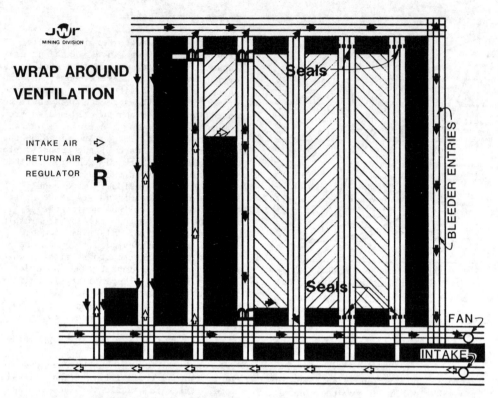

Figure 3. Progressive ventilation Phase II

Fire-related TR values are customarily considered valid when the CO/CO_2 and afterdamp/blackdamp ratios are high when compared to the normal liberation rate prior to the fire. These high ratios would indicate a thermal reaction rather than CO^2 or blackdamp being generated from a normal gob (Mitchell, 1984).

7 EMERGENCY SITUATIONS

Although carbon monoxide has been detected in the airstream from Jim Walter Resources longwall gob areas, the Blue Creek coal bed of the Mary Lee series is not subject to spontaneous combustion. However, materials used in longwall areas such as wood cribs and timber, hydraulic oil, etc. are subject to spontaneous combustion. Therefore Jim Walter Resources wants to be prepared in the event that an emergency situation should arise. In emergency situations where a fire exists, the faster that fire fighting techniques can be started or the decision to seal an area can be made, the greater the probability that the fire can be brought under control and put out. Several explosions and fires have occurred in underground workings

throughout the country in recent years requiring that a chromatograph be brought in from an outside source in order to assess the situation and plan how to combat the fire. Waiting on equipment of this nature to be transported to the fire site would obviously delay recovery operations. Since Jim Walter Resources already has its own chromatograph, gas samples can immediately be analyzed. Immediate analyses of samples can provide much needed answers on the size of the fire, how to go about safely combating the fire, or when to seal.

Another formula which can be used in fire situations to determine the intensity of the fire and placement of boreholes to insert inerting gases is "Relative Intensity" (RI), which can be computed from the following relationship:

$$RI = (1-3.83 \; O_2/N_2) \; (I_{co})$$

The "Relative Intensity" is a measure of the available air to burn a unit weight of fuel, the percentage of air consumed in the process, and the effect of that consumption on temperature. In practice, the RI value should increase as the distance to the thermal reaction (the fire) decreases (Mitchell, 1984).

The Explosibility Index (E) of the fire gases should be closely monitored and plotted to determine if fire fighting personnel can be sent underground to combat the fire. Once fire fighting personnel have been sent underground, the Explosibility Index should be continually monitored to determine trends and how long the men can be safely allowed to combat the fire. The Explosibility Index can be calculated by the following formula:

$$E = EC \left(\frac{0_2}{MAO} \right)$$

Where:

EC(Effective Combustible) = $\%CH_4 + 1.25(\%H_2) + 0.4(\%CO)$

MAO (Maximum Allowable Oxygen) =

$$5 + 7 \left(\frac{CH_4}{CH_4 + H_2 + CO} \right)$$

Generally, the people involved in fighting the fire should be evacuated from the mine when the Explosibility Index reaches 3.5 (Mitchell and Burns, 1979).

Accuracy of the sampling and gas analysis is of the utmost importance. If inaccurate gas analyses are obtained, resulting in false indices and ratios, or gases are mistaken for fire gases when they are not, an improper evaluation of the situation and possible sealing of the mine could result when it would not be necessary. A stationary chromatograph with a trained operator will give more accurate results and would be invaluable in an emergency situation.

8 DEGASIFICATION

Shortly after mining was started in the deeper portion of the Blue Creek seam, large volumes of methane gas were encountered. In order to more safely mine this seam and maximize production, it was determined that a degasification program would be desirable. As a result, Jim Walter Resources is now producing 18,000,000 FT^3/day (509,703.24m^3/day) and selling approximately 15,000,000 FT^3/day (424,752.7m^3/day) of methane gas from its underground mines. The gas is produced from three different sources: horizontal and vertical holes in advance of mining, and gob wells producing gas from above the longwall gob area. The methane gas is collected and sold to Southern Natural Gas. Pipeline quality of the gas must be at least 950 BTU/FT^3 (1,002,307J/.0283m^3) saturated and can have no more than 4% non-hydrocarbon gases. The heat value of methane gas is 1012 BTU/FT^3 (1,067,720.7J/.0283m^3). The gob wells are drilled to within approximately 30 feet (9.144m) of the mined

(Blue Creek) seam. These wells are drilled in front of the longwall face, and, as the face advances beyond the well, the top rock falls, opening fissures and passages for the gas to collect and to migrate to the well. Normally the wells are operated under negative pressure to produce the maximum amount of gas. The wells are producing from the top of the gob area; but, depending on the degree of communication with the ventilation circuit and the degree of negative pressure placed on the well, the methane can be diluted with oxygen and nitrogen (fresh air). Depending on the age of the gob, carbon monoxide and carbon dioxide can also be generated.

The gas generated from the Gob Degasification Program is mixed with the gas produced from the horizontal and vertical wells that generally is 98 - 100% methane. As much gas as possible is produced from the gob wells to prevent this gas from migrating into the ventilation circuit. Therefore, the use of the chromatograph to assure pipeline quality gas is essential. At the present time, this is the primary use of the chromatograph, making it cost effective.

9 ENGINEERING

The chromatograph, in addition to the uses outlined previously, can also serve as a valuable tool for conducting a number of engineering studies. These studies include analyzing emissions from diesel engines and LP gas heaters, determining liberation patterns while cutting coal, and monitoring tracer gas concentrations to evaluate leakage preformance of ventlation controls. In any engineering study which there is a need to analyze the component gases of the mine atmosphere, a chromatograph would be useful.

10 SUMMARY

Jim Walter Resources is committed to longwall mining in order to maximize reserves and to minimize mining cost. Therfore, it is believed that the gases being generated from longwall gob areas should be closely monitored. Even though the interior of the longwall gob area cannot be inspected, gob gases can be evaluated to determine change and thereby maintain a high degree of safety.

Jim Walter Resources has justified the use of a chromatograph, not only through the monitoring of gob areas for potential heating, but also through its Degasification Program. To maintain pipeline

quality gas and produce the maximum amount
of gas without a chromatograph would be
difficult.

The use of a chromatograph as previously
discussed, coupled with routine mine air
analyses for engineering studies, have
more than justified the purchase of a
chromatograph at Jim Walter Resources.

11 REFERENCES

Chamberlain, E.A.C., Hall, D.A.,
Thirlaway, J.T. 1970,
    The ambient temperature oxidation of
    coal in relation to the early detection
    of spontaneous heating. The Mining
    Engineer, October, 1970.
Holding, W.,
    Fires in coal mines, in the environ-
    mental engineering in South African
    Mines, ed. J. Burrows, Cape & Transvaal
    Printers, Cape Town, South Africa.
Mitchell, D.W., 1973, The index for car-
    bon Monoxide-What it can mean to you.
    Proceedings American Mining Congress,
    Pittsburgh, Pennsylvania.
Mitchell, D.W., and Burns, F.A., 1979,
    Interpreting the state of a mine fire
    MSHA IR1103
Mitchell, D.W., 1984, Understanding a
    fire - case studies. Proceedings
    Combatting Underground Mine Fires -
    Two Day Workshop, Section 6, Denver,
    Colorado
Mitchell, D.W., 1985, Personal Communi-
    cation
Schlick, D.P., and Stevenson, J.W., 1983
    Longwall ventilation at Jim Walter
    Resources. Proceedings 14th Annual
    Institute on Coal Mining Health, Safety
    and Research, pp. 149-155, Virginia
    Polytechnic Institute & State University,
    Blacksburg, Virginia

An analysis of decision-making aspects in mine ventilation systems

R.V.RAMANI, R.BHASKAR & R.L.FRANTZ
The Pennsylvania State University, University Park, USA

ABSTRACT: The purpose of this paper is to examine the various facets of information flow in a mine environmental monitoring system linked to a computerized management information system. The discussion centers on the role of such a system in aiding managerial decision-making in the planning and operation of mine ventilation systems. The paper is in three parts. The first deals with ventilation decisions, outlining the situational aspects impacting ventilation decisions. The second part discusses aspects of ventilation management, including the role and levels of management, the various activities performed, the organization of the ventilation functions, and information needs for decision-making. The last section describes a ventilation management information system, stressing data collection, processing, information use and management reporting procedures.

Ventilation decisions are associated with a high degree of health, safety, productivity, and legal implications. The development of an integrated mine atmospheric monitoring and management information system must be sensitive to these issues and commensurate with the responsibility, authority, and accountability of the decision-makers.

1 INTRODUCTION

With the increasing popularity of management information systems (MIS) in business and of continuous monitoring systems in the mining industry, considerable thought has to be given for integrating the two systems for their use in management of mines. Extensive planning is necessary to use the data collected by the monitoring systems as direct input to computerized management information systems both for timely and effective control of operations and for increased effectiveness and efficiency of planning and engineering functions. The results of such an integration will indeed be synergistic, leading to increased decision quality. It will also be a step in the direction of total systems approach to mine planning and mine management. This is particularly true for mine ventilation systems which play an all pervasive role in worker health, safety and productivity.

2 VENTILATION DECISIONS

The objective of ventilation decisions is to improve safety and productivity through planning, engineering, and operating the ventilation system according to the recommended methods. The type of operating situations that warrant ventilation decisions vary extensively with the nature of mining operations such as metal, non-metal or coal mines, gassy or non-gassy mines and deep or shallow mines. Other factors influencing the complexity of the decisions to be made are the available time, the personnel involvement, the scale of operations, the mining method, the type of equipment, and spatial extent and age of the mine. The decisions themselves can range from those that have high local significance such as airflow changes at a single mining face to those having mine-wide ramifications such as the establishment of corporate policies, the sinking of a new ventilation shaft or the installation of a new fan.

2.1 Situational aspects of decisions

The situations warranting decisions may be of two types: repetitive and highly predictable, and infrequently occurring and less predictable. An example of the first type is the need to maintain the quality and quantity of air in the face areas of a coal mine through frequent adjustments to the face ventilation system as the coal cutting machine moves from one cut to

another in a working section. Another
example is the panel regulator adjustment
as the panel is advanced and retreated
during development and pillaring, respectively. In these situations, the parameters that describe the condition are well
defined, and show little variability. The
consequences of the decisions are generally predictable. The decisions to be
made and the procedures to be followed can
be described by a set of rules and imparted to the foreman and other personnel
through training procedures. The key to
this category of decision environment is a
high degree of predictability.

On the other hand, situations occur
where the variability in the parameters of
the ventilation system is higher than
anticipated. Methane emission from the
strata is a case in point. Despite the
predictability of methane emission rates
from seams, local variations occur.
Release rates, far in excess of those
anticipated, can cause a severe upset to
the production operations. It may not be
possible to deal with these widely varying
emissions by simple modifications to the
face ventilation system. The cause and
effect relationships may not be obvious or
easily defined. The greater the variability in the parameter, the less predictable
is the likely value of the parameter in
practice. A set of predefined rules may
not be adequate to bring the system under
control. As a result, decisions to be
made under these circumstances require a
certain degree of judgment, intuition, and
creativity.

Decisions relating to mine ventilation
can also be classified as those made under
normal mine operating conditions and those
made during mine emergencies. Decisions
made under normal conditions can be identified as those made by line or staff personnel. Decisions relating to design are
usually made by staff personnel under the
direction of top and middle management
personnel. By nature, these decisions
often involve exploring new concepts or
ideas and have a certain time frame to
them. It could involve policy decisions.
Planning and engineering decisions have
medium- to long-term implications. Decisions will typically relate to spatial
layout of the mine, number and locations
of shafts, selection of fans, and struc-
turing out emergency escapeways. Thus
considerable interaction with several
other aspects of mine design as well as
top management commitment are required.

After a system is engineered and implemented, the line personnel have the
responsibility to operate the system and
preserve the integrity of the system by
exercising operating controls. Line oper-

ating decisions relate to matters that
concern the operating and resource utilization aspects of the system. The implications of the decisions are in the
immediate or short term. Decisions are
generally made in a more structured
environment.

Emergency situations can be classified
into two categories -- emergencies or
abnormal situations where emergency operating plans have been drawn up and emergencies where the plans are not appropriate for tackling the situation. A mine
emergency plan in its simplest form is a
set of rules to be followed by the various
personnel in the event of an emergency to
ensure the health and safety of workers
and include training in emergency procedures, and evacuation and escape. The
federal regulations require the anticipation of emergencies and development of
mine specific plans to minimize the threat
to the health and safety of workers.
Emergencies falling in the second category
present extreme difficulties in decision-
making. Perception of a hazard or an
emergency is a necessary condition for the
development of evasive measures. Since
the emergency falling in the second category may not have been perceived, the
existing emergency plans may be of limited
use. The situation may call for the creation of a new set of rules or for improvi-
sation on the existing ones as more data
becomes available during the course of the
emergency situation.

2.2 General guidelines and limitations

Regardless of the type of decision environment, all decision makers have to operate within a broad set of guidelines. The
guidelines are of three types: legal
requirements, company procedures and policy, not necessarily minimum legal limits,
and research and development findings.
Higher levels of management, when making
decisions related to ventilation, will
evaluate the decision options in the
context of company policy, any recent
research findings, and capital requirements. On the other hand, operating
management will be bound by legal requirements and acceptable practices. The large
number of constraints at the operating
level, including time, legal requirements,
and resource utilization responsibilities
reduces the number of options available to
operating management for analyses, evaluation and resolution.

3 VENTILATION MANAGEMENT

Decision making authority and the information needs are closely interrelated and

can be defined only with respect to the role and levels of management (Frantz and Ramani, 1982; Ramani and Frantz, 1984). Management is the component of an organization that brings together the capital, labor, and raw materials necessary to accomplish the purpose of the business. Management hierarchy can be divided into three broad levels -- top, middle, and front-line. Top management is heavily focused on planning with a broad scope and a large time frame. Generally, its activities are complex and unstructured (Donaldson and Lorsch, 1983). It seeks and deals with information about events and activities external to the organization. By comparison, middle management's activities are functional, shorter range, better structured, and more often, implementational by nature. The information sought is mostly internal to the organization. In essence, the plans, policies and strategies of top management are interpreted by middle management to produce plans, schedules, and procedures. The activities of the operating or front-line management, in this framework, are highly structured and of very short duration so as to require minimum planning. The information sought is internal and generally historical. The major focus of management at the middle and operating levels is on direction and control as their activities are performed in an environment which is relatively more restricted and better defined than that of top management. It is not implied that there are no planning activities at these levels; only that planning is not the major activity.

Mine ventilation, by its very nature being primarily an engineering and service function, may not prominently figure in the top management's strategic planning process. This is not to minimize its critical importance. Ventilation planning and design is a means of achieving the corporate policies and objectives in production and worker health and safety, areas where top management is vitally concerned. Therefore, even if ventilation is not necessarily a principal focus of the strategic planning process, ventilation planning acquires considerable importance in the implementation of production, health and safety policies and of long and medium range plans. Top management is interested in evaluating the performance of the company in terms of health and safety with respect to those of other companies operating in the same environment. Investment outlays in capital assets such as ventilation shafts receive the attention of top management and is a significant part of top management resource allocation responsibilities.

3.1 Line and staff activities

The organizational activities can also be viewed as a line-staff function to further spell out the role of personnel involved with decision-making in ventilation. The responsibilities can be considered within the functions of planning, organizing and staffing, and directing and controlling. Planning, organizing and staffing functions are supported by staff departments in an organization. The technical staff departments range from planning and engineering to safety and training. The engineering departments are staffed with project engineers some of whom may have special assignments to deal with ventilation matters. Major ventilation planning and design analyses for a mine will generally be done in consultation with corporate engineering staff. For an operating mine, modifications to the ventilation system may not be a frequent exercise. Once the alternatives to the ventilation system are analyzed by a project engineer, much of the decision and implementational aspects are left to personnel at the mine, e.g., mine engineer and line management.

To a degree, operating decision areas -- directing and controlling functions -- are better delineated. Directing and controlling responsibilities may be considered under both normal and emergency conditions. The operating management is responsible for keeping the working places safe for people and equipment to operate utilizing the available resources. With respect to ventilation, they have to ensure that all safety regulations -- state, federal and company -- are met and that all regulatory requirements such as quantity and quality measurements are satisfied. This also includes maintaining pre- and on-shift examination reports, filed by examiners and certified personnel. Mine managers and mine superintendents, working with mine foremen and mine engineers, ensure smooth operations. Given the general nature of their duties and the emphasis on production, the line managers delegate staff engineering responsibilities to the mine engineer.

During emergencies, the mine official who is in charge or present at the time of occurrence of the emergent situation is required to adopt the procedures outlined in the emergency plans. When emergencies occur, experienced management personnel work with specialists and line personnel as a crisis management team to ensure safety of workers. Under such circumstances, the decisions made and the approaches taken are largely dependent on the experience of the team members, the speed of appropriate decision making, the availability of resources, and the execution of the emergency plans and decisions by the miners and management alike (National Academy of Sciences, 1981).

3.2 Organization of ventilation functions

The importance of ventilation is well recognized in the mining industry. The need for good planning, proper engineering and adequate control is also well recognized. However, the organizational structure and lines of authority for design and control of ventilation systems are very diverse. Structured ventilation departments staffed with certified personnel can be found in some countries (Burrows and Roberts, 1980; Rose, 1980). United States mining laws and regulations provide guidelines as to the ventilation standards to be maintained, the frequency and kinds of measurements, and maintenance of records (Code of Federal Regulations, 1981). Instructions on organizational structure and lines of authority are not very specific although statutory responsibilities are defined in terms of certified and authorized persons. In some cases and for some engineering and design work, an engineer must have been certified by a State or National Board of Professional Registration.

Mines facing severe gas and ventilation problems or having extensive methane drainage programs often have engineering and management positions staffed by personnel with extensive knowledge and experience in these areas (Stevenson, 1980). This type of staffing can also be found in coal mining companies having a large number of mines. The specialized services provided by these staff personnel are often requested and utilized by the line management in the mine. On the other hand, these specialized personnel routinely perform special studies to provide technical support to top management in setting policies and procedures on matters dealing with mine control.

From a management standpoint, ventilation planning and design are staff functions. Ventilation is also an auxiliary service function to the production function. However, the actual management of the ventilation system in a mine is a line function, intertwined with the production operations. Thus, whereas ventilation planning and engineering can be and often is located within the corporate engineering departments, at the mine the mine engineer is generally responsible for carrying out the needed changes to the system.

Since ventilation is an integral part of the mining activity, the responsibility for maintaining the integrity of the system lies with the mine operating personnel, viz. the mine and section foreman as well as the mine superintendent, assisted by the mine engineer. In the same vein, ventilation planning and design is a part of mine planning and is closely related with other planning activity. Therefore, the planning engineer responsible for the design of the mine should also be responsible for ventilation planning. Whatever the manner in which ventilation functions are organized, there must be a clear cut definition of the responsibility, authority and accountability for the functions.

3.3 Information needs

All management personnel from section foremen to higher officials have to rely on a good flow of information for decision-making. While information and data are frequently used interchangeably, from the management point of view, it is important to draw a distinction. Data is the raw material which is converted into information through processing. As data flows upwards in an organization, the amount of raw data flow is reduced but summarized information content increases. The greater the interaction between management and information, the better are the decisions made (Alter, 1980). Since mine ventilation is primarily an engineering problem, in many cases, the decisions to be made deal with technical problems. However, with the increasing emphasis on health, safety and productivity aspects and on total system planning of mines, management decision making has become complex. Decisions have to be made in the larger context of the company's objectives such as health and safety, quality of life and profits. This requires that mineral managers view any engineering design problem in its context to the overall mine design. Considerations of other mine-related data such as mining costs, resources, hydrology, geologic factors, social aspects, and governmental requirements become important. Fortunately, the developing management information technology blending monitoring, computing, and communicating functions provides an opportunity to alter fundamentally the way in which data and information are acquired, processed and distributed for improved effectiveness of mine management.

4 VENTILATION MANAGEMENT INFORMATION SYSTEM

The functions of an information system are to record, store, recall and process data

768

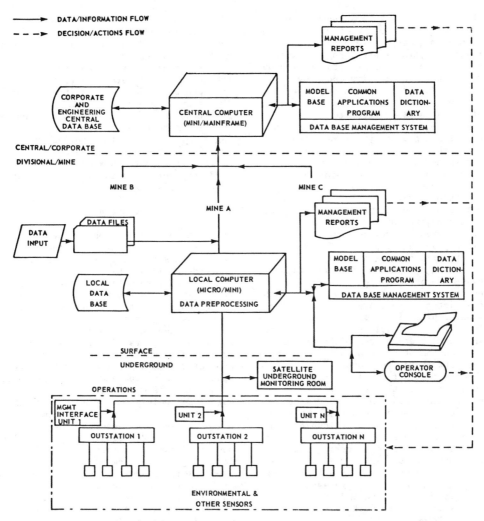

Figure 1. Integrated mine monitoring and management information system.

and information. Management information system is an information system which supports the decision-making functions of the management in an organization (Zmud, 1983). The ventilation management information system is therefore a system which has data, information and models to support ventilation decision-making. Therefore, the role of data acquisition, analysis, synthesis and presentation are all important interrelated components of management information systems. The various aspects of data collection and processing, information use and management interaction with information are discussed with reference to Figure 1 which shows the physical structure of the various elements of an integrated mine monitoring and management information system.

4.1 Data collection and processing

The data collection and processing system has three subsystems, viz, the underground subsystem, the surface divisional or mine subsystem, and the central or corporate subsystem. The data collected in a system is dependent on a number of factors, the important ones relating to health and safety and legal requirements. In an industry such as mining where the degree of regulatory control is high, legal requirements often transcend other factors in importance for data collection.

The outstations underground receive data from the various sensors. The first level of response to changing parameter values can be located here. The outstation may also incorporate a management interface which in its simplest form can be a video display unit with function keys to enable a foreman to query and respond to the mine ventilation system status. A satellite underground monitoring station can improve coordination of activities and provides an additional interface for underground management personnel. The data from the outstations is transmitted to a surface intermediate computer which has its own data base management system, application software and model base. Data will be processed and stored in the local data base and also used to produce management reports. Much of this may be accomplished with little operator involvement. Data from intermediate computers at the various mines can be transmitted to a central computer for creating engineering data bases for mine planning purposes. The model base here will be more extensive with models being designed with mine-wide or company-wide optimization in view.

Reports are generated at various points in the system. The degree of processing of a particular set of data, however, varies. The data will be subject to minimal processing at the console operator's level. While not being a part of the management structure, the console operator may be instructed to respond in a certain defined manner to the system status. Often, transmission links to the mine foreman and higher officials in the mine are provided for immediate communication of exception levels in the monitored data. Information needs of engineering and research staff, however, differ. The data procured over a period of time from a monitoring system have to be processed for generating the causative and trend relationships for use with planning and design decisions.

4.2 Information use in operations, planning and design

The data collected by the front end of an information system, such as a mine monitoring system, is used by the operating, engineering and planning departments. At the operating level, the monitoring and MIS system ensures that information on all mine parameters is readily available. The use of such information may be purely of the default type i.e. no decision needs to be taken if the system is operating within specifications.

The availability of extensive data on various mine ventilation parameter enables comprehensive planning exercises to be carried out. Most mine environment related computer models currently available (Ramani, 1982) that would be a part

of the model base require extensive input data considered expensive to collect manually. The data from a mine monitoring system can be expected to increase not only the quality of the analysis but also the number of different types of analyses that can be performed. An important contribution of a mine monitoring system is the reduction in time between the appearance and detection of problems. Also, regulatory reports are more easily filed as all required data can be automatically dumped into a data base or printed out in hard copy form. Overall, better management control is achieved.

4.3 Management reporting

The availability of sensors for most mine environmental parameters and the development of sensors for still others, indicate that one would be able to obtain almost all desired information through a monitoring system in the near future. Since an enormous amount of data can be collected by the monitoring system, summarization is needed to produce meaningful management reports. The degree of summarization is a function of several factors. Some of the more important ones are the amount of data received for a particular parameter in a given period of time, the use of an individual piece of data, the type of decision that has to be made using the data, and the manager's method of using data for decision-making. The last mentioned aspect is particularly important for designing management information systems. Some managers tend to be very data-oriented in their decisions. Others tend to rely more on intuition and feelings. But for a technical system such as a mine ventilation monitoring system, the management information system has to be designed to generate quantitative information.

At the operating level, data is not subject to a high degree of summarization because individual pieces of data may be as important as the overall system status. Also, during an abnormal state, the individual pieces of data are most important for any level of management. Research and engineering personnel, however, may require access to both raw and summarized data. This demands that during the design of the MIS, system analysts should assess the various uses for the data. The outputs of the MIS should be developed in conjunction with management, engineering, operating and research personnel.

Time intervals between reports and the information content and manner of reports play a significant role in management's attitude to and interaction with information. Time considerations for data presentation can be one of three types: periodic, exception, and on-demand. Periodic presentation of reports is the normal reporting procedure to management on the status of an operation. Exception reporting is the way to effectively deal with the large volumes of data, as data that meet specified conditions such as concentration levels and trends do not demand action. Exceptions to the norm have to be brought to management's attention for action. The basis for exception reporting may be legal requirements or company policies and procedures. Often an exception report may trigger a reaction from management in the form of reevaluation of the system. For example, frequent gassing out of a face may prompt management to design alternate mining cut plans or face ventilation techniques. Management may demand a report containing specific data to aid in the evaluation of the various systems. Such reports are called on-demand reports and the frequency of these reports is dependent on the need for the information.

The information in a report can be of three types: operating status data, summary data, and planning and engineering data. Operating status data are displayed at the operator's console in the monitoring room. The data to be presented and the best format for data presentation are:

1. Mine layout in skeletal form with sensor locations, transmission paths, and outstation locations for all sensor types. Separate frames should be provided for data such as methane and CO, again superimposed on schematics. In case the output of the sensor at any point exceeds specified levels then the value should be flashed on the screen in a different color. The lower few lines should indicate trends in the readings.

2. Summary of all sensor outputs in a tabulated form, if requested. An excursion should be flashed in a different color.

3. Quantity and velocity values superimposed on a mine schematic. Variations from a set minimum should be flashed on the screen.

4. All ventilation devices should be displayed on a mine schematic in symbolic language.

5. Fan performance. Head generated, quantity of flow, pitch setting, horsepower, speed, temperature and vibration.

6. Emergency facilities. A display of emergency evacuation routes including location of all ventilation devices, maintenance shops, refuge chambers, tracks, etc.

7. Drainage map. A graphical display of pump locations, maximum pumping capacity, pump on/off status and water levels in sumps.

8. Status of major facilities. Operating status of surface refrigeration plants, spot coolers, methane drainage system, etc.

9. Sensors operating status. Details of sensors and transmission lines.

All exceptions to preset standards should be printed on a hard copy terminal and also dumped onto a data base with system time, location details, and details on action taken.

Presentation of summary data can be in the form of tables, charts, graphs, or maps depending on the data presented. If accurate readings of data are important, e.g., methane concentration levels above legal limits, a table or a multi-colored chart would be desirable. If the data is to indicate trends, charts or graphs are better. Again, variance and direct comparisons are better presented by graphs, bar charts, pie charts and area graphs. Summary reports are usually on hard copy and have formal user lists. Examples of summary reports are:

1. all required regulatory reports,

2. ventilation system power costs, itemwise for fans, refrigeration units and methane drainage data,

3. ventilation system maintenance costs -- supplies and personnel,

4. log of deviations of ventilation parameters from preset levels -- dynamic data base dump.

While operating status summary reports outlined above are routine reports, i.e., the frequency of report generation is predetermined, presentation of planning and engineering data in report form is a more infrequent exercise dependent upon the activity. However, ventilation system plans have to be updated with every major development in the mine. Data for updates can be collected and collated in input form by the management information system as and when desired by the engineers for planning and engineering.

No amount of discussion can completely cover the scope, content and manner of management reporting procedures. However, the reports must be closely tied to an organizational framework and developed with an intent to fulfill the management's planning and control needs. They should contain information commensurate with responsibility and authority.

5 SUMMARY

The development in sensors, telemetry

systems and microcomputers has provided a great impetus for major applications of monitoring and control systems in mining. While there are under twenty major mine atmospheric monitoring applications in the U.S., the trend is clearly towards increased automatic monitoring of such mine environmental parameters as air velocity, methane concentration, temperature, humidity and fan pressure and quantity. Some of the impediments to rapid growth are the lack of suitable data to clearly document the benefits of Electronic Data Processing applications in this area. The longer term uses of the acquired information from a mine atmospheric environmental monitoring system are also not very clear. In addition, the data in the atmospheric environmental monitoring system is updated at very short intervals leading to voluminous data handling needs. The paper has discussed, in some detail, these aspects. Despite these limitations, the importance of the role that automatic monitoring can play in mine atmospheric environmental planning and control is unquestioned. Several recent publications have presented the results of cost-benefit studies of environmental monitoring systems (Welsh, 1985; Brown, 1982; Anonymous, 1985). The wider use of these systems in other industries and for several functions in mining is an indication that these systems are gaining greater acceptance (Weiss, 1982). However, the environmental data for the data base in mines and its applications to short-term and long-term mine planning require greater consideration and justification as compared to those for production, delay and cost reporting and control. An integrated mine monitoring and management information system can play an important role in the development of timely and accurate data for operations control, data base and models for ventilation systems design. These in turn offer great potential for increasing the quality of decisions at all levels of management.

6 REFERENCES

Alter, S.L. 1980. Decision Support Systems, Current Practices and Continuing Challenges. Addison Wesley, 316 pp.
Anonymous 1985. Automatic Monitoring of Underground Environment. British Coal International Newsletter, n. 5, p. 5.
Brown, W.A. 1982. Real-time Monitoring in an Underground Coal Mine -- Production and Safety. Technology Exchange Seminar, Mining Information Systems, Final Report to US Bureau of Mines on Contract J0113093, pp. 201-220.
Burrows, J. & B.G. Roberts 1980. Managing and Manning Ventilation Departments of South African Gold Mines. Second International Mine Ventilation Congress, Reno, NV, Society of Mining Engineers, New York, pp. 42-50.
Code of Federal Regulations 1981. Federal Coal Mine Safety Standards. CFR, Title 30:75.
Donaldson, G. & J.W. Lorsch 1983. Decision Making at the Top, The Shaping of Strategic Direction. Basic Books, Inc., New York, 208 pp.
Frantz, R.L. & R.V. Ramani 1982. Mineral Management -- Its Unique Aspects. Earth and Mineral Sciences, v. 52, n 1, pp. 1-4.
National Academy of Sciences 1981. Underground Mine Disaster Survival and Rescue: An Evaluation of Research Accomplishments. Report of the Committee on Underground Mine Disaster Survival and Rescue, National Academy Press, Washington, DC, 122 pp.
Ramani, R.V. 1982. Application of Computers to Mine Ventilation. Mine Ventilation and Air-Conditioning, Ed. H.L. Hartman, John Wiley and Sons, pp. 517-545.
Ramani, R.V. & R.L. Frantz 1984. Information Systems and Management Information Systems for Mineral Management. Proceedings, 18th International Symposium on the Application of Computers and Mathematics in the Mineral Industries, Institute of Mining and Metallurgy (U.K.), pp. 723-734.
Rose, H.J.M. 1980. Management of Ventilation Departments. Second International Mine Ventilation Congress, Reno, NV, Society of Mining Engineers, New York, pp. 17-22.
Stevenson, J.W. 1980. Establishing a Mine Ventilation Department in an Operating Mine. Second International Mine Ventilation Congress, Reno, NV, Society of Mining Engineers, New York, pp. 23-26.
Weiss, A. 1982. Technology Exchange Seminar: Mining Information Systems. Final Report to US Bureau of Mines on Contract J0113093, 335 pp.
Welsh, J.H. 1985. Costs and Benefits of Mine Monitoring. Proc. of the 2nd Conference on the Use of Computers in the Coal Industry, AL, Society of Mining Engineers, New York, pp. 107-114.
Zmud, R.W. 1983. Information Systems in Organizations. Scott, Foresman and Company, 445 pp.

List of authors

M.E. Adams
J.F.T. Agapito & Assoc., Inc.
715 Horizon Dr., Suite 340
Grand Junction, CO 81501

J.F.T. Agapito
J.F.T. Agapito & Assoc., Inc.
715 Horizon Dr., Suite 340
Grand Junction, CO 81501

Zacharias G. Agioutantis
Polytechnic Inst. & St. Univ.
Dept. of Min. Minerals Eng.
Blacksburg, VA 24061

Anne Amaral
Mines Library
University of Nevada-Reno
Reno, NV 89557

Mary Ansari
Mines Library
University of Nevada-Reno
Reno, NV 89557

James F. Archibald
Queen's University
Dept. Mining Engineering
Kingston, Ontario K7L 3N6

G. P. Badenhorst
Dept. of Mineral & Energy Affairs
P.O. Box 1132
Johannesburg, 2000, Rep. S.A.

A.A. Baklanov
Mining Institute, Kola Branch
184200 Apatity, 24 Fersman St.
Murmansk Region, USSR

Sukumar Bandopadhyay
University of Alaska, Fairbanks
Dept. of Mining & Engineering
Fairbanks, AK 99701

Randal J. Barnes
Colorado School of Mines
Dept. of Mining Engineering
Golden, CO 80401

P. Barton
Central Mining Institute
pl. Gwarkow 1
40-951 Katowice, Poland

David Bartsch
Quarto Mining Co

Powhatan Point, OH 43942

S. Phillip Battino
BHP Collieries Steel Div.
P.O. Box 1239
Wollongong, N.S.W., Australia 2500

Sastay Bhamidipati
Univ. of Utah
318 William Browning Bldg.
Salt Lake City, UT 84112

R. Bhaskar
Dept. of Mining Engineering
The Pennsylvania State University
University Park, PA 16802

J. Bigu
Energy, Mines & Resources
P.O. Box 100
Elliot Lake, Ontario P5A 2J6

773

Nuh Bilgin
Dept. of Min. Eng.
Maden Fakultesi I.T.U.
Tesvikiye-Istanbul, Turkey

S.J. Bluhm
Chamber of Mines
P.O. Box 91230
Auckland Park, Rep. S. A. 2006

Floyd C. Bossard
F.C. Bossard & Associates, Inc.
P.O. Box 3837
Butte, MT 59702

P. Bottomley
Chamber of Mines
P.O. Box 91230
Auckland Pk, Rep. S.A. 2006

Carl E. Brechtel
J.F.T. Agapito & Assoc., Inc.
715 Horizon Drive, Suite 340
Grand Junction, CO 81501

William E. Bruce
Mine Safety & Health Adm.
P.O. Box 25367, DFC
Denver, CO 80225

Daniel J. Brunner
Mine Ventilation Ser., Inc.
3717 Mt. Diablo Blvd.
Lafayette, CA 94549

Paul Burgwinkel
Ins. fur Bergwerks
Und Huttenmaschinenkunde
25100, Aachen, Fed. Rep. W.Germany

Felipe Calizaya
Dept. of Mining Engineering
Colorado School of Mines
Golden, CO 80401

A.A. Campoli
U.S. Bureau of Mines
P.O. Box 18070
Pittsburgh, PA 15236

P. Cassini
CERCHAR
PB 2-06550
Verneuil-en-Halatte, France

Andrew B. Cecala
U.S. Bureau of Mines
P.O. Box 18070
Pittsburgh, PA 15236

Joseph Cervik
U.S. Bureau of Mines
P.O. Box 18070
Pittsburgh, PA 15236

R.N. Chakravorty
Energy, Mines & Resources
4500 16th. N.W.
Calgary, Alberta T3B OM6

Xintan Chang
Michigan Technological Univ.
Dept. of Mining Engineering
Houghton, MI 49931

G. Chorosz
Ecole Des Mines
35 Rue Saint Honore
77305 Fontainebleau, France

R.N. Christensen
Nuclear Eng. Prog.
206 W. 18th Ave.
Columbus, OH 43210

C.J. Csatary
Mining Engineering
5 Jacqueline Ave., North Cliff
Johannesburg, Rep. S.A. 2195

N. D'Albrand
CERCHAR
BP 2-60550
Verneuil-en-Halatte, France

Eric P. Deliac
Ecole des Mines
35 Rue Saint Honore
Fontainebleau 77305,   France

F. Djahanguiri
Battelle Memorial Inst.
505 King Ave., Bldg. 13
Columbus, OH 43201

William Donley
Federal Bldg.
515 9th St.
Rapid City, SD 57701

R. Dworok
Central Mining Institute
pl. Gwarkow 1
40-951, Katowice, Poland

Wu Gang
Design & Res. Inst.
P.O. Box 189
Shijiazhuang, Hebei Prov. P.R.O.C.

John Edwards
U.S. Bureau of Mines
P.O. Box 18070
Pittsburgh, PA 15236

M. Gangal, Ph.D.
Energy, Mines & Resources
555 Booth St.
Ottawa, Ontario K1A 0G1

Jindrich Fiala
Ceskoslovenska Akad. Ved
Hornicky Ustav 71000
Ostrava 2, Hladnovska 7 Czech.

F. Garcia
U.S. Bureau of Mines
P.O. Box 18070
Pittsburgh, PA 15236

P.I. Forrester
Gemini Biochemical Research
#12, 3610 29th St. N.E.
Calgary, Alberta T1Y 5Z7

T.W. Goodman
U.S. Bureau of Mines
P.O. Box 18070
Pittsburg, PA 15236

Ralph K. Foster
Mine Safety & Health Adm.
P.O. Box 25367, DFC
Denver, CO 80225

M.G. Grenier
Energy, Mines & Resources
P.O. Box 100
Elliot Lake , Ontario P5A 2J6

William Francart
Mine Safety & Health Adm.
4800 Forbes Ave.
Pittsburgh, PA 15213

Rudolf E. Greuer
Michigan Technological Univ.
Dept. of Mining Engineering
Houghton, MI 49931

John C. Franklin
Dept. of Mining Engineering
Colorado School of Mines
Golden, CO 80401

W.H. Griffin
University of Alberta
Dept. of Mineral Engineering
Edmonton, Alberta T6G 2G6

R.L. Frantz
Dept. of Mining Engineering
The Pennsylvania State University
University Park, PA 16802

H. James Hackwood
Queen's University
Mining Engineering
Kingston, Ontario K7L 3N6

C. Froger
CERCHAR
BP #2 - 60550
Verneuil-en-Hallatte, France

Danny W. Hagood
Jim Walter Resources, Inc.
Rt. 1 Box 321
Brookwood, AL 35444

Jerry L. Fuller
Mine Safety & Health Adm.
P.O. Box 25367
Denver, CO 80225

George Halfinger, Jr.
Ingersoll-Rand Research, Inc.
RD1 Montgomery Rd.
Skillman, NJ 08558

Raymond R. Gadomski
Mine Safety & Health Adm.
4800 Forbes Ave.
Pittsburgh, PA 15213

A.E. Hall
University of British Columbia
Dept. of Mining
Vancouver, B. C. V6T 1W5

Christopher J. Hall
University of Idaho
Metallurgical & Mining Eng.
Moscow, ID 83843

Douglas F. Hambley
Argonne National Lab.
9700 South Cass Ave.
Argonne, IL 60439

S.G. Hardcastle
Elliot Lake Laboratory
P.O. Box 100
Elliot Lake, Ontario P5A 2J6

Satya Harpalani
Mining & Geological Eng.
University of Arizona
Tucson, AZ 85712

Stephen P. Harrison
Consolidation Coal Co.
Consol Plaza
Pittsburgh, PA 15241

Howard Hartman
Dept. of Mining Engineering
University of Alabama
University, AL 35486

Kiyoshi Higuchi
Hokkaido Univ., Engin.
Kita-13, Nishi-8
Sapporo, 060 Japan

M. Hine
WA School of Mines
P.O. Box 597
Kalgoorlie 6430 Australia

Sheng-Shyong Huang
EMRO, ITRI
#1, Tun Hwa South Road
Taipei, Taiwan R.O.C.

Anthony T. Iannacchione
U.S. Bureau of Mines
P.O. Box 18070
Pittsburgh, PA 15236

Robert A. Jankowski
U.S. Bureau of Mines
P.O. Box 18070
Pittsburgh, PA 15236

Natesa I. Jayaraman
U.S. Bureau of Mines
P.O. Box 18070
Pittsburgh, PA 15236

Colin Johnson
Engineering Control Systems
P.O. Box 458
Aurora, Ontario L4G 3L5

Thys B. Johnson
Dept. of Mining Engineering
Colorado School of Mines
Golden, CO 80401

D.L. Johnston
Joy Manufacturing Co.

New Philadelphia, Ohio 44663

I.O. Jones
WA School of Mines
P.O. Box 597
Kalgoorlie, Australia 6430

J.E. Jones
Jim Walter Resources, Inc.
Rt. 1, Box 321
Brookwood, AL 35444

G.V. Kalabin
Mining Institute, Kola Branch
184200 Apatity, 24 Fersman St.
Murmansk Region, USSR

David W. Kennedy
Buffalo Forge Co.
P.O. Box 985
Buffalo, New York 14204

Albert E. Ketler
Rel-Tek Corporation
150 Plum Industrial Court
Pittsburgh, PA 15239

Kuldip Khunkhun
Fundt Corp.
4101 E. Irvington Rd.
Tucson, AZ 85726

Bong J. Kim
Mineral & Resources Eng.
Chosum University
Kwang-Ju, South Korea

Robert H. King
Dept. of Mining Engineering
Colorado School of Mines
Golden, CO 80401

Fred N. Kissell
U.S. Bureau of Mines
P.O. Box 18070
Pittsburgh, PA 15236

Geoffrey Knight
Elliot Lake Laboratory
P.O. Box 100
Elliot Lake, Ontario P5A 2J6

J. Kout
Ceskoslovenska Akademia Ved
Hornicky Ustav 71000
Ostrava 2, Hladnovska 7, Czech.

P. Krzystolik
Central Mining Institute
pl. Gwarkow 1,
40-951, Katowice Poland

Abdula J. Kudiya
Hercules, Inc.
10166 S., 2130 E.
Sandy, UT 84092

Vic Kutay
Consolidation Coal
Moundsville Oper. Eastern Reg.
Moundsville, W. VA 26041

Steven Lautenschlaeger
Hecla Mining Co.
Box 320
Wallace, ID 83873

K. Lebecki
Central Mining Institute
pl. Gwarkow 1
40-951, Katowice, Poland

R.D. Lee
Dept. of Mining Engineering
University of Nottingham
Nottingham NG7 2RD England

I. Longson
Dept. of Mining Engineering
University of Nottingham
Nottingham England NG7 2RD

I.S. Lowndes
Dept. of Mining Engineering
University of Nottingham
Nottingham NG7 2RD England

M.G. Mack
Civil & Mineral Engineering
Univ. of Minnesota
Minneapolis, MN 55455

John Marks
Homestake Mining Co.

Lead, SD 57754

Mervin D. Marschall
Mine Safety Appliance
Research Corp.
Evans City, PA 16033

Michael J. Martinson
Univ. of the Witwatersrand
1 Jan Smuts Ave.
Johannesburg, Rep. S.A. 2001

Hiroaki Matsukura
Hokkaido Univ., Engin.
Kita-13, Nishi-8
Sapporo, 060 Japan

John F. McCoy
Foster-Miller, Inc.
350 Second Ave.
Waltham, MA 02254

Charles McGlothlin
Beaver Creek Coal Co.
P.O. Box 1368
Price, UT 84501

Chuck McLendon
Allied Corp.
P.O. Box 551
Green River, WY 82935

Thomas E. McNider
Jim. Walter Resources, Inc.
P.O. Box 133
Brookwood, AL 35444

Malcolm J. McPherson
Univ. of California, Berkeley
Materials Sci. & Mineral Eng.
Berkeley, CA 94720

777

Paul C. Miclea
Raymond Kaiser Engineers
P.O. Box 23210
Oakland, CA 94623

Byron Miller
Bell Laboratories

Holmdel, NJ 07733

P.B. Mitchell
BHP Collieries Steel Division
P.O. Box 1239
Wollongong, N.S.W. Australia 2500

Gregory M. Molinda
U.S. Bureau of Mines
P.O. Box 18070
Pittsburgh, PA 15236

D.T. Moore
Stauffer Chemical Co.
P.O. Box 513
Green River, WY 82935

R. Morris
Consol. Invest. Co., LTD
P.O. Box 590
Johannesburg 2000, Rep. S.A.

Sandip K. Mukherjee
Engineers International Inc.
98 East Naperville Rd.
Westmont, IL 60559-1595

Jan Mutmansky
Mining Engineering Dept.
The Pennsylvania State Univ.
University Park, PA 16802

Tom Myran
SINTEF
Norwegian Insti. of Tech.
Trondheim-NTH   Norway 7034

L. Nel
Sherritt Gordon Mines Ltd.
P.O. Box 1000
Leaf Rapids, Manitoba ROB 1WO

Robert Nesbit
Mine Safety & Health Adm.
4015 Wilson Blvd.
Arlington, VA 22203

Keith Notley
Mining Engineering Dept.
Queen's University
Kingston, Ontario K7L 3N6

Carl J. Opatrny
VSM Corporation
7515 Northfield Road
Cleveland, OH 44146

R.L. Osborne
Nuclear Eng. Program
206 W. 18th Ave.
Columbus, OH 43210

Steven J. Ouderkirk
U.S. Bureau of Mines
5629 Minnehaha Ave. South
Minneapolis, MN 55417

A. Ouederni
Ecole Des Mines
35 Rue Saint Honore
77305 Fontainebleau, France

Carl Peterson
Massachusetts Insti. of Tech.
Rm 3-455A
Cambridge, MA 02139

William Pomroy
U.S. Dept. of the Interior
5629 Minnehaha Ave. South
Minneapolis, MN 55417

K.R. Price
Jim Walter Resources, Inc.
Rt.1, Box 321
Brookwood, AL 35444

James Procarione
Univ. of Utah
318 William Browning Bldg.
Salt Lake City, UT 84112

R.V. Ramani
The Pennsylvania State Univ.
Mining Engineering
University Park, PA 16802

R. Ramsden
Chamber of Mines
P.O. Box 91230
Auckland Park, Rep. of S.A. 2006

778

Walter Richards
Methane Drainage Ventures
187 W. Orangethorpe
Placentia, CA 92670

Steven K. Ruggieri
Foster-Miller, Inc.
350 Second Ave.
Waltham, MA 02254

John Jeffrey Sartaine
National Mine Service Co.
P.O. Box 1447
Ashland, KY 41105

Steven J. Schatzel
U.S. Bureau of Mines
P.O. Box 18070
Pittsburgh, PA 15236

Bryan K. Schroeder
Rockwell Hanford Operations
P.O. Box 800
Richland, WA 99352

J. Sliz
Central Mining Institute
pl. Gwarkow 1
40-951, Katowice, Poland

A.M. Starfield
Civil & Mineral Engineering
Univ. of Minnesota
Minneapolis, MN 55455

J.W. Stevenson
Jim Walter Resources, Inc.
P.O. Box C-79
Birmingham, AL 35283

Andrew W. Stokes
CANMMET, EMR
210 George St.
Sydney, Nova Scotia B1P 1J3

Richard T. Stoltz
Mine Safety & Health Adm.
4800 Forbes Ave.
Pittsburgh, PA 15213

A. Taufer
Ceskoslovenska Akademia Ved
Hornicky Ustav 71000
Ostrava 2, Hladnovska 7 , Czech.

Edward D. Thimons
U.S. Bureau of Mines
P.O. Box 18070
Pittsburgh, PA 15236

R.W. Thompkins
RRt. 1
641 Upper Horning Rd.
Ancaster, Ontario L9G 3K9

Robert J. Timko
U.S. Bureau of Mines
P.O. Box 18070
Pittsburgh, PA 15236

Yuusaku Tominaga
Hokkaido Univ., Engin.
Kita-13 Nishi-8
Sapporo, Japan 060

Ertugrul Topuz
Virginia Insti. & State Univ.
Mining, Minerals Engineering
Blackburg, VA 24061

Robert N. Torbin
Foster-Miller
350 Second Ave.
Waltham, MA 02154

Michael A. Trevits
U.S. Bureau of Mines
P.O. Box 18070
Pittsburgh, PA 15236

M.A. Tuck
Dept. of Mining Engineering
University of Nottingham
Nottingham, NG7 2RD, England

Jon C. Volkwein
U.S. Bureau of Mines
P.O. Box 18070
Pittsburgh, PA 15236

Andrew M. Wala
Mining Engineering Dept.
University of Kentucky
Lexington, KY 40506

Y.J. Wang
College of Mineral Resources
West Virginia University
Morgantown, WV 26506

779

Robert Waytulonis
U.S. Bureau of Mines
5629 Minnehaha Ave. S.
Minneapolis, MN 55417

Zhao Zi-Cheng
Kunming Institute of Tech.
Dept. of Mine Ventilation
Kunming, Yunnan, P.R.O.C.